TK 6553

Green, C

Troublesh
and theory of AM, FM, and

TROUBLESHOOTING, SERVICING, AND THEORY OF AM, FM, AND FM STEREO RECEIVERS

TROUBLESHOOTING, SERVICING, AND THEORY OF AM, FM, AND FM STEREO RECEIVERS
Second Edition

CLARENCE R. GREEN
Davidson County Community College

ROBERT M. BOURQUE
J. Sargent Reynolds Community College

PRENTICE HALL, Englewood Cliffs, New Jersey 07632

Library of Congress Catologing-in-Publication Data

GREEN, CLARENCE R.
　　Troubleshooting, servicing, and theory of AM, FM, and FM stereo receivers.
　　　Rev. ed. of: The theory and servicing of AM, FM, and FM stereo receivers. © 1980.
　　　Bibliography
　　　Includes index.
　　　1. Radio—Repairing.　2. Stereophonic receivers—Repairing.　I. Bourque, Robert M.　II. Green, Clarence R. Theory and servicing of AM, FM, and FM stereo receivers. III. Title.
TK6553.G72　1987　　　621.3841'87　　　86-12274
ISBN　0-13-931114-9

Editorial/production supervision
and interior design: *Theresa A. Soler*
Cover design: *20/20 Services, Inc.*
Manufacturing buyer: *Carol Bystrom*

©1987, 1980 by Prentice-Hall, Inc.
A Simon & Schuster Company
Englewood Cliffs, New Jersey 07632

All rights reserved. No part of this book may be reproduced, in any form or by any means, without permission in writing from the publisher.

Printed in the United States of America
10　9　8

ISBN　0-13-931114-9　025

PRENTICE-HALL INTERNATIONAL (UK) LIMITED, *London*
PRENTICE-HALL OF AUSTRALIA PTY. LIMITED, *Sydney*
PRENTICE-HALL CANADA INC., *Toronto*
PRENTICE-HALL HISPANOAMERICANA, S.A., *Mexico*
PRENTICE-HALL OF INDIA PRIVATE LIMITED, *New Delhi*
PRENTICE-HALL OF JAPAN, INC., *Tokyo*
PRENTICE-HALL OF SOUTHEAST ASIA PTE. LTD., *Singapore*
EDITORA PRENTICE-HALL DO BRASIL, LTDA., *Rio de Janeiro*

With warm affection and appreciation,
this book is dedicated to

My parents, C.A. and Opha Green of Butler, Tennessee.

Clarence R. Green

My grandparents, John and Lizzie Martin of Kenbridge, Virginia.

Robert M. Bourque

CONTENTS

PREFACE xv

1 INTRODUCTION 1

1-1 Fundamental Principles 1
1-2 AM and FM (Amplitude and Frequency Modulation) 3
1-3 A Radio Receiver 4
1-4 Superheterodyne Receivers 6
1-5 FM Stereo 7
1-6 Organization of This Text 8
1-7 Looking Ahead 8

2 FUNDAMENTALS OF ACTIVE DEVICES 11

2-1 Intrinsic Semiconductors 11
2-2 Doping—Adding Impurities to Semiconductors 13
2-3 The PN Junction 15
2-4 Solid-State Diodes 19
2-5 Bipolar Junction Transistors 21
2-6 Junction Field-Effect Transistors 26
2-7 Insulated Gate FETs—MOSFETs 32
2-8 Operational Amplifiers 36
2-9 The Vacuum Diode 40

Summary 42

Questions and Problems 44

vii

3 POWER SUPPLIES   47

- 3-1 Transformer Fundamentals 47
- 3-2 Half-Wave Rectifiers 50
- 3-3 Full-Wave Rectifier Using Center-Tapped Transformer 54
- 3-4 Bridge Rectifier 57
- 3-5 Voltage Doublers 57
- 3-6 Regulated Power Supplies 60
- 3-7 Three-Terminal Regulators 66
- 3-8 Active Ripple Filters 68
- 3-9 Vacuum-Tube Rectifiers 69

Summary 71

Questions and Problems 72

4 BIASING ARRANGEMENTS   76

- 4-1 General Considerations 76
- 4-2 Biasing Requirements of Bipolar Junction Transistors 78
- 4-3 Base Bias 79
- 4-4 Base Bias with Emitter Feedback 86
- 4-5 Base Bias with Collector Feedback 88
- 4-6 Base Bias with Collector and Emitter Feedback 90
- 4-7 Base Bias with Separate Supply 91
- 4-8 Universal Bias 92
- 4-9 Universal Bias with Collector Feedback 94
- 4-10 Emitter Bias with Two Supplies 97
- 4-11 Miscellaneous Topics 97
- 4-12 JFET Biasing Arrangements 102
- 4-13 MOSFET Biasing Arrangements 112
- 4-14 Vacuum-Tube Biasing 115

Summary 117

Questions and Problems 119

5 AUDIO AND POWER AUDIO AMPLIFIERS   123

- 5-1 Cascaded Stages 123
- 5-2 Direct-Coupled Amplifiers 125
- 5-3 Amplifier Configurations 129
- 5-4 Tube-Type Single-Ended Power Audio Amplifiers 132
- 5-5 BJT Single-Ended Power Audio Amplifiers 135
- 5-6 Vacuum-Tube Push-Pull Audio Amplifiers 136

- 5-7 Harmonic Distortion 139
- 5-8 Solid-State Transformer-Coupled Push-Pull Amplifier 140
- 5-9 Output Transformer-Less Audio-Output Stage 145
- 5-10 Complementary-Symmetry Amplifier 146
- 5-11 Quasi-complementary Amplifier 149
- 5-12 Commercial Amplifiers 150

  Summary 154

  Questions and Problems 156

6 RADIO-FREQUENCY AMPLIFIERS 158

- 6-1 Introduction 158
- 6-2 Resonant Circuits 159
- 6-3 Triodes as RF Amplifiers 163
- 6-4 Transistor (BJT) RF Amplifiers 166
- 6-5 Pentode RF Amplifiers 169
- 6-6 Gain of RF Amplifiers 170
- 6-7 Automatic Volume Control 171
- 6-8 JFET RF Amplifiers 173
- 6-9 MOSFET RF Amplifiers 174
- 6-10 Dual-Gate MOSFET Amplifiers 175
- 6-11 Noise 177

  Summary 183

  Questions and Problems 184

7 FREQUENCY CHANGERS 187

- 7-1 Introduction 187
- 7-2 Principle of Frequency Changers—Heterodyning 187
- 7-3 Basic Oscillator Circuits 189
- 7-4 Separate Mixer–Oscillator Frequency Changer 194
- 7-5 Pentagrid Converters 195
- 7-6 Bipolar Transistor Converters 198
- 7-7 Frequency-Changer Considerations 199
- 7-8 Frequency Changers for FM 201
- 7-9 Automatic Frequency Control 202
- 7-10 FM Frequency Changer with AFC 205
- 7-11 Voltage-Variable Capacitors 206

  Summary 207

  Questions and Problems 208

8 PRINCIPLES OF AMPLITUDE MODULATION 210

8-1 Aspects of Radio Waves 210
8-2 Simple Transmission System 216
8-3 Voice Transmission System 217
8-4 Practical AM Transmitter 220
8-5 Tuned Radio-Frequency Receivers 225
8-6 Superheterodyne Receivers 226

Summary 231

Questions and Problems 232

9 AM RECEIVER ANALYSIS 234

9-1 Preliminary Considerations 234
9-2 AM Receiver; Power Supply 236
9-3 Direct-Coupled, Single-Ended Audio Amplifier 238
9-4 IF Amplifier, Detector, AVC 243
9-5 Variation: Bridge Rectifier; Push-Pull Audio Amplifier 245
9-6 Variation: Transformer Power Supply; OTL Push-Pull Audio Amplifier 247
9-7 Variation: A Complicated Power Supply 249
9-8 Variation: Quasi-complementary AF Amplifier with Darlington Pairs 252
9-9 A Vacuum-Tube Receiver 258
9-10 VT Receiver with RF Amplifier and PP Audio Amplifier 262

Summary 267

Questions and Problems 269

10 INTRODUCTION TO TROUBLESHOOTING 272

10-1 Tools for Electronic Servicing 272
10-2 Equipment Considerations 277
10-3 Failure Mechanisms in Tubes 285
10-4 Failure Mechanisms in Transistors 287
10-5 Ohmmeter Identification and Testing of BJTs 288
10-6 Choosing Replacements for Defective Parts 289
10-7 Tips on Replacing Defective Parts 293

Summary 295

Questions and Problems 297

11 BASIC TROUBLESHOOTING PROCEDURE 300

11-1 Basic Troubleshooting Procedure 300
11-2 Signal Injection 303
11-3 Signal Tracing 308
11-4 Voltage-Resistance Analysis 310
11-5 Voltage-Resistance Analysis of Tube Amplifiers 313
11-6 Voltage-Resistance Analysis of Transistor Circuits 317
11-7 Analysis of an IC Stage 322
11-8 Generator Coupling to Receiver Test Points 324

Summary 325

Questions and Problems 327

12 TROUBLESHOOTING AM RECEIVERS 330

12-1 Introduction 330
12-2 Symptom-Function Diagnosis 331
12-3 Filaments Do Not Light 331
12-4 Dead Receiver 334
12-5 Turn-On Plop Is Present; Little Static 337
12-6 Turn-On Plop with Lots of Static 339
12-7 Loud Ripple Hum with Volume Control Counterclockwise 342
12-8 Distorted Audio 344
12-9 Loss of Sensitivity 346
12-10 Insufficient Volume, Sensitivity Normal 353
12-11 Receiver Oscillations: Motorboating, Squeals, Howls 354
12-12 Noisy Reception 356
12-13 Intermittent Problems 357
12-14 AM Receiver Alignment 359

Summary 364

Questions and Problems 367

13 PRINCIPLES OF FM RECEIVERS 371

13-1 Introduction to FM 371
13-2 FM Sidebands 375
13-3 Superheterodyne Receivers for FM 378
13-4 Limiters 378
13-5 Phasors—Vector Representations of Sine Waves 381

Contents

13-6	Foster-Seeley Discriminator 383	
13-7	Ratio Detector 390	
13-8	Quadrature Detectors 395	
13-9	Phase-Locked-Loop FM Detector 397	
13-10	Preemphasis and Deemphasis 398	

Summary 399

Questions and Problems 400

14 AM-FM RECEIVER ANALYSIS 403

14-1	Typical AM-FM Front End 403
14-2	Typical AM-FM IF Section 406
14-3	Common AM Converter—FM First IF Amplifier 410
14-4	Complex Coupling in IF Sections 413
14-5	Tuning Meter 413
14-6	Two Power Supplies for AM-FM Receivers 417
14-7	Tube-Type FM Front End 419
14-8	AM Front-End and AM-FM IF Sections 422
14-9	AM-FM Front End; Common AM Converter—First FM IF Amplifier 425
14-10	Ratio Detector with Tuning Indicator 431

Summary 433

Questions and Problems 434

15 TROUBLESHOOTING AM-FM RECEIVERS 437

15-1	Introduction 437
15-2	AM Normal; No FM 439
15-3	No AM or FM; Turn-On Plop (or Hum) Present 445
15-4	AM Normal; Insufficient FM Sensitivity 446
15-5	Oscillations on FM: AM Normal 448
15-6	Distorted Audio on FM; AM Normal 448
15-7	Insufficient AM and FM Sensitivity 449
15-8	Noisy FM; AM Normal 449
15-9	FM Alignment—An Introduction 450
15-10	FM IF Alignment; AM-VTVM Method 450
15-11	Sweep/Marker Generator Operation 453
15-12	Sweep/Marker FM IF Alignment 456
15-13	FM Front-End Alignment 460

Summary 460

Questions and Problems 461

16 PRINCIPLES OF FM STEREO — 464

- 16-1 Compatibility Requirements 464
- 16-2 $L + R, L - R$ Scheme of Multiplexing 466
- 16-3 FM-Stereo System 469
- 16-4 Simple Balanced Modulator 471
- 16-5 Frequency Doublers 473
- 16-6 Composite FM-Stereo Signal 474
- 15-7 Matrix Decoder Principles 479
- 16-8 Time-Division Decoding 481
- 16-9 Qualitative Description of TD Decoding 487
- 16-10 Block Diagram of a Typical FM-Stereo Receiver 491

Summary 491

Questions and Problems 493

17 FM-STEREO DECODER CIRCUIT ANALYSIS — 495

- 17-1 Typical Matrix Decoder 495
- 17-2 Matrix-Decoder Variation 498
- 17-3 Standard Time-Division Multiplex Decoder 500
- 17-4 Bridge-Type Time-Division Decoder 503
- 17-5 Biplex Detector 508
- 17-6 Integrated-Circuit Time-Division Decoder 511
- 17-7 Phase-Locked-Loop Decoder Principles 513
- 17-8 AM, FM, FM-Stereo Receiver 515

Summary 518

Questions and Problems 519

18 TROUBLESHOOTING FM-STEREO RECEIVERS — 522

- 18-1 Introduction 522
- 18-2 One Channel Inoperative on All Functions 523
- 18-3 AM Normal; FM and FM Stereo Inoperative 525
- 18-4 One Channel Inoperative in FM-Stereo Mode 527
- 18-5 Monaural FM Stereo 529
- 18-6 Nonfuntioning Stereo Indicator 530
- 18-7 Stereo-Decoder Alignment—An Introduction 532
- 18-8 Matrix-Decoder Alignment 532
- 18-9 Time-Division-Decoder Alignment 536
- 18-10 PLL-Decoder Alignment 538

Summary 538

Questions and Problems 539

SUGGESTED BOOKS FOR FURTHER READING   541

SOLUTIONS TO SELECTED QUESTIONS
AND PROBLEMS   542

INDEX   549

PREFACE

This book is to help a person who knows a little bit about electricity and electronics to learn about AM, FM, and FM-stereo radio receivers. As the title implies, we also explain how to go about servicing a receiver that fails to work properly.

Our emphasis is on the principles involved and how the circuits work. We believe that a good understanding of the circuitry is the best tool a service technician can have. This book should be helpful even if your objective is not to become a service technician because of the detailed circuit descriptions that are included. We claim, for example, that Chap. 16 provides the best and most detailed description of the FM-stereo system yet published at the introductory level, not only in texts on electronics servicing but in texts on electronic communications as well.

It is assumed that the reader has a knowledge of the fundamentals of DC and AC electricity: Ohm's law, series and parallel circuits, magnetic fields, alternating current, capacitors, inductors, and transformers. However, many of these topics are briefly described in this text in order to refresh the memory of the person who might be a little rusty with some of the ideas.

We start from scratch with semiconductors so that no prior knowledge of solid-state devices is required. No prior knowledge of troubleshooting is assumed, and no mathematics beyond simple algebra is required since circuit operation is described primarily in nonmathematical terms. Although formulas are given for calculating many things, no attempt has been made to provide derivations because they are available in standard engineering texts for those who are interested.

Vacuum tube theory is included, but not emphasized, in order to cover all the active devices. Many old tube sets are still around, and we believe a person should know at least a little about them—for the sake of history if for no other reason.

xv

All important principles are presented relative to solid-state devices so that the tube circuits may be skipped if time or interest make it desirable to do so.

Our feeling is that three quarters or two semesters are required to cover the entire book and do it well, with time provided for laboratory work. Moving somewhat faster, it should be possible to do a reasonable job in two quarters, especially if one has some background in semiconductors, amplifiers, and so forth. In only one quarter, the principles of AM, FM, and FM stereo could be covered, but time would not permit extensive work in receiver analysis or troubleshooting.

Due to the length of the text and our treatment of fundamental principles, the scope does not include advanced, high-power amplifiers nor advanced receivers that make extensive use of digital electronics. Those devices are worthy of an entire book in their own right.

While addressed to AM, FM, and FM stereo receivers, this text presents many techniques and procedures that are directly applicable to other systems, such as CB transceivers, audio systems, commercial communications receivers, many sections of a TV receiver, and so forth. Indeed, a major portion of the circuits (or variations) described in this text are widely used in other electronics systems.

We are indebted to Howard W. Sams & Co., Inc. for granting permission to include in this text the many schematic diagrams that are taken from actual service data published by Howard W. Sams & Co., Inc. Also, to Zenth Radio Corporation and to the many organizations which supplied invaluable technical information, we express our sincere gratitude.

Clarence R. Green
Davidson County Community College

Robert M. Bourque
J. Sargent Reynolds Community College

TROUBLESHOOTING, SERVICING, AND THEORY OF AM, FM, AND FM STEREO RECEIVERS

1
INTRODUCTION

The purpose of this introductory chapter is to provide an overview of AM, FM, and FM-stereo receivers and to acquaint the reader with the organization of this text.

1-1 FUNDAMENTAL PRINCIPLES

Radio Waves

We are all vaguely aware of radio waves. After all, that is how the music comes to our radio receiver and how the picture and music together are delivered to the TV set. Amateur radio operators use radio waves to talk over great distances, and a far greater number of CB radio operators use radio waves to talk over short distances. A few CB operators know how to use radio waves to talk over great distances. Radio waves are everywhere, and therefore we do not think about them much.

Strictly speaking, radio waves, television waves, microwaves, and radar waves are all forms of *electromagnetic radiation*. What makes one type different from another is the frequency— the number of waves emitted per second. Radio waves are relatively low-frequency waves; television waves are sort of medium-frequency waves; microwave and radar waves are high-frequency waves.

If we continue upward to even higher frequencies in the electromagnetic spectrum, to the region beyond the realm of electronics, we encounter *infrared* radiation. This consists of radio waves of frequency so high that we perceive it as heat. The warm glow on one's face while standing near a fireplace is due to infrared radiation, a form of electromagnetic radiation.

Going further, until we reach frequencies about 100,000 times as high as that of radar waves, we come to the visible spectrum—light. What our eyes perceive as light is electromagnetic radiation whose frequency is on the order of 10^{14} Hz. Within the visible spectrum, our eyes sense waves of different frequency as different colors. Thus, color is due to frequency. Blue is of higher frequency than red, and green is in the middle.

At only slightly higher frequencies, our eyes cannot respond to the waves. This is the *ultraviolet* region of the spectrum, the so-called black light that makes fluorescent posters glow with a peculiar iridescence. Beyond the ultraviolet region lies the realm of gamma radiation, where the energies are so high and the wavelengths so short that the waves become more like particles than waves. Only people who study nuclear physics need to be concerned about gamma radiation, from the academic point of view, at least.

From all this we conclude that radio waves and light waves are basically the same physical manifestation. All electromagnetic waves travel at the same speed—the speed of light, naturally. They consist of both an electric and a magnetic field and are able to travel through the absolute vacuum of intergalactic space. For our purposes, we need only be concerned with the low-frequency waves that are sent out from radio transmitters.

Transmission and Reception of Radio Waves

At a radio station, high-frequency currents are forced to flow in the transmitting antenna by the power amplifier of the transmitter, which operates at the assigned frequency of the station. Electrons surge back and forth in the antenna, and the accelerations experienced by the electrons give rise to the production of radio waves. When an electric charge is accelerated, an electromagnetic wave is produced. The waves travel outward from the antenna at the speed of light—186,000 miles per second, or one mile in 5.37 microseconds (μs). Thus, high-frequency currents in the transmitting antenna produce radio waves.

When the waves encounter a receiving antenna, they induce currents in the antenna as they pass by. Furthermore, the induced currents are of the same frequency as those in the transmitting antenna, and the strength of the induced currents is directly proportional to the strength of the currents in the transmitting antenna.

Here we have a link established between the transmitting antenna and the receiving antenna, via the radio waves. This link forms the basis for the transfer of information from the transmitter to the receiver, and the information may take the form of voice, music, or in the case of television, a picture.

Tuning

When an inductor is connected to a capacitor, a resonant circuit is formed that produces its maximum response at a particular frequency called the resonant frequency. These circuits form the basis of tuning, the ability to select one transmitter and reject all the others. Without the phenomenon of tuning, there could be only one transmitter on the air at a time, and wars would probably be fought for the privilege of using the transmitter.

Voice Waves

First of all, sound, any sound, is a mechanical vibration of the molecules of the air (or other medium, such as water or wood) through which the sound propagates. The vibrations propagate outward from the disturbance that acts as the source in the form of waves—sound waves. A sound wave entails a variation in pressure, to which our ears will respond.

Sound waves are not electromagnetic; they are mechanical. The highest frequency that can be heard by a human ear is about 20 kHz, and the waves travel at about 1100 f/s in air at room temperature.

A microphone is a device—a *transducer*—that converts variations in pressure (sound waves) to variations in voltage—an electrical signal, called an audio signal. The output voltage variation may be observed on an oscilloscope, and the waveform is called a *voice wave*. Here we use the term "wave" to relate to the shape of the trace on the oscilloscope; no new type of wave phenomenon is implied.

Due to the rapid variations of voltage in an audio signal, it assumes the characteristics of an alternating current. However, the waveforms are highly irregular and seldom resemble the smooth sine wave typically associated with an alternating current. The audio spectrum spans the range from 20 Hz to 20 kHz, but most ears will not respond to frequencies much over 15 kHz. The upper-frequency limit of hearing decreases as the age of the listener advances, a normal progression that is not always considered a disadvantage.

The Dynamic Loudspeaker

A truly marvelous and noteworthy device is a common, ordinary speaker. The poorest and best alike consist of a paper cone attached to a small coil of wire (the voice coil) that is suspended in a magnetic field. A current sent through the coil causes a deflection of the paper cone. A voice wave sent through the coil causes the voice to be reproduced. This simple device can reproduce with amazing fidelity the sound of the ocean, a violin, a tuba, an oboe, the human voice, and so forth, ad infinitum.

Respectively, the sounds of a violin, tuba, and oboe are produced by a vibrating string, a brass tube in conjunction with the vibrating lip of the tuba player, and a vibrating reed. The sound of the ocean is produced by splashing water. Is it not just a little bit amazing that a vibrating paper cone can mimic the sounds of these different types of sound sources? Does it not seem that a gadget that can reproduce the sound of an orchestra, for example, should be enormously complex, consisting of strings, reeds, brass tubes, wooden pipes, and perhaps even a few bells?

1-2 AM AND FM (AMPLITUDE AND FREQUENCY MODULATION)

We know that a transmitter can produce radio waves. But radio waves by themselves carry no information. We need a method for transporting a voice wave (or the equivalent) via a radio wave, called the *carrier*. Commercial radio uses two methods for doing this—AM and FM.

In the AM system, the voice wave is used to control, or *modulate*, the amplitude (strength) of the radio waves being transmitted at that instant. The transmitter circuitry that does this is the *modulator*. As the voice wave varies, so does the amplitude of the transmitted waves. As the voice wave becomes more positive, the carrier amplitude increases, and as it becomes more negative, the carrier amplitude decreases. Thus the instantaneous amplitude of the carrier is proportional to the instantaneous voltage of the voice wave—the signal from the microphone, perhaps. In the AM system, the frequency of the carrier does not change.

In the FM system, the amplitude of the carrier always remains the same, but the frequency transmitted at any instant depends on the level of the voice wave at that instant. When the voice wave goes more positive, the carrier frequency goes up. When the voice wave goes more negative, the carrier frequency shifts downward a proportional amount. Thus, the carrier frequency swings back and forth (deviates) about the center frequency as the voice wave alternately goes positive and negative.

Obviously, to receive an AM station, one needs an AM receiver; similarly for FM. Therefore, we find two different types of receivers on the market, AM and FM. In most cases, however, a receiver will have a switch to select either the AM mode or the FM mode. Such a receiver is called an AM-FM receiver.

The broadcast bands for AM and FM are located far apart in frequency. FM frequencies are about 100 times higher than AM, as set up by the Federal Communications Commission (FCC). The AM band ranges from 540 to 1600 kHz with a channel width of 10 kHz. The FM band ranges from 88 to 108 MHz with a channel width of 200 kHz. Thus, an FM station is alloted 20 times as much spectrum space as an AM station. For purposes of comparison, a TV channel is 6 MHz wide—the equivalent of 30 FM channels or 600 AM channels.

Fidelity

Everyone knows that FM sounds better than AM, and there are two main reasons why this is so. First, because an FM channel is 20 times as wide as an AM channel, an FM station can transmit more of the audio spectrum—up to 15 kHz. The legal limit for AM is 5 kHz. A wider-frequency response translates directly to higher-fidelity transmission.

Second, electrical noise (static) occurs in the form of amplitude variations, and an AM receiver responds directly to amplitude variations. Hence, we hear static in the speaker of an AM radio, especially when the station is weak. An FM receiver, however, responds only to frequency variations and is therefore immune to static. Indeed, FM receivers use limiters which remove any amplitude variation that might appear at the antenna.

A third reason most FM stations sound better than an AM station is that the majority of FM stations transmit in stereo. FM stereo represents an add-on to the FM system for which there is no counterpart in the present AM system. More is said in regard to FM stereo in the following sections.

1-3 A RADIO RECEIVER

At this point we can identify some of the things a receiver must accomplish without worrying about whether the receiver as AM or FM. We shall see that the overall block diagrams of the two types are quite similar.

First, there must be an antenna to receive the waves from the transmitter. Years ago this was a wire strung as high as possible to a tree in the backyard. Now the antenna is built into the cabinet of almost all AM receivers, and FM receivers use a few tricks to avoid an external antenna. For best FM reception, however, a TV-type FM antenna is still preferred.

Next, we must provide our receiver with tuning capability. *LC* resonant circuits do this to perfection, almost. Tuning is no problem.

The big problem, however, is one of amplification. The signal produced by the antenna is very weak, voltages being measured in the range of microvolts and power levels in the nanowatt range at best. So let us provide about three stages of amplification immediately following the antenna. Because these amplifiers operate at the antenna frequency, we call them *RF amplifiers*—*RF* stands for *radio frequency*. Being clever, we will include tuning circuits in these amplifiers so that we tune and amplify together. We then have the antenna followed by three tuned RF amplifiers.

The output signal of the RF section (called the *front end*) is then delivered to the demodulator, where the voice wave (the audio signal) is recovered from the modulated carrier. The demodulator is sometimes called the *detector*. If our receiver is AM, our demodulator must be an AM demodulator. For FM we must have an FM demodulator. In either case, the output of the demodulator is the much-desired audio signal, the music.

The audio signal produced by the demodulator is usually rather small. In no case could it be applied to a speaker to render an audible volume level. It must be amplified first. To do this, we use an audio amplifier, of course. The audio section may utilize one or two stages of voltage amplification to build up the signal for delivery to the power audio amplifier. The power amplifier is designed to deliver the power required by the speaker. We would incorporate a volume control in one of the voltage amplifiers following the demodulator, and perhaps we would even include a tone control to provide for bass-treble adjustments.

Finally, a power supply must be provided to supply DC voltages to each stage, and the receiver is complete. The block diagram is shown in Figure 1-1. Such a receiver is called a *tuned radio frequency* (TRF) receiver, owing to the three stages of RF amplification following the antenna. This type of receiver will work, and

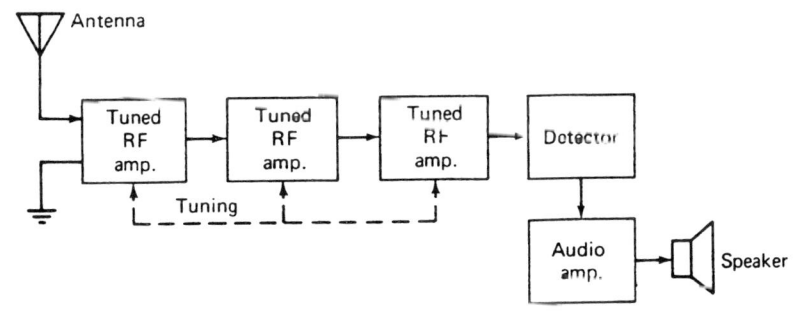

FIGURE 1-1 Block diagram of a tuned radio-frequency (TRF) receiver. (The power supply is omitted.)

Sec. 1-3   *A Radio Receiver*   5

millions have been sold. However, it suffers from several disadvantages when compared to the receiver described in the next section.

1-4 SUPERHETERODYNE RECEIVERS

With the possible exception of simple demonstration receivers or perhaps very specialized receivers that are dedicated to a specific function, all modern AM or FM receivers utilize the superheterodyne principle and are therefore called *superheterodyne receivers*. Whether a receiver is AM or FM, tube-type or solid state, portable or console, rudimentary or elaborate in design, it is almost a certainty that the superheterodyne principle is employed.

Superheterodyne receivers convert the frequency of the signal received at the antenna to a lower, intermediate-frequency signal that is then amplified by a narrow-band amplifier called the *intermediate frequency* (IF) *amplifier*. The IF amplifier provides a major portion of the amplification that is required to increase the level of the antenna signal to the level required by the demodulator. A generalized block diagram of a superheterodyne receiver, without regard to whether it is AM or FM, is shown in Figure 1-2.

The process of converting the antenna signal to a signal of the IF frequency is called either *heterodyning* or *mixing*, and the stage that performs the frequency conversion may be called either a *converter* or a *mixer*.

Frequency conversion is achieved by mixing two signals of different frequencies to obtain a third signal whose frequency equals the difference between the two original frequencies. Thus, two signals must be applied to the mixer. One comes from the antenna via the RF amplifier, and the other is provided by a *local oscillator* (signal generator) that constitutes a portion of the frequency converter stage. The output of the local oscillator heterodynes with the antenna signal to produce the IF signal, which is then amplified by the IF amplifier.

Because the IF amplifier operates at only one frequency, it can be economically designed to produce a high gain with narrow bandwidth. Further, since the IF frequency does not change as the receiver is tuned across the band, the bandwidth of the IF amplifier remains constant as the receiver is tuned across the band. The

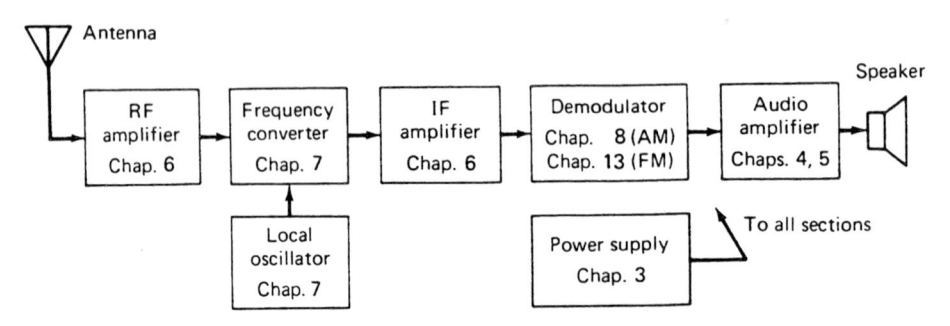

FIGURE 1-2 Generalized block diagram of a superheterodyne receiver. The chapter designations refer to the chapters of the text in which each block is described.

high gain of the IF amplifier contributes to the overall sensitivity of the receiver, and the narrow bandwidth of the IF amplifier contributes to the overall selectivity of the receiver. Since the IF amplifier bandwidth is independent of the receiver tuning, the receiver will exhibit essentially the same selectivity at all portions of the broadcast band.

The superheterodyne principle and frequency conversion are described in detail in later chapters; indeed, the text is devoted almost entirely to circuits encountered in superheterodyne receivers. More is said in this regard in the following sections of this chapter.

1-5 FM STEREO

The FM-stereo system is an enhancement of the original FM monaural system and is capable of transmitting two independent audio channels via one frequency-modulated RF carrier. FM-stereo receivers receive stereo broadcasts in stereo, and monaural receivers reproduce the same broadcast monaurally with negligible degradation of fidelity in comparison to the reproduction of monophonic broadcasts. In other words, the FM-stereo system is *compatible* with monaural FM receivers.

In regular monophonic FM transmission, the highest audio frequency transmitted is 15,000 Hz. An examination of the system capabilities, however, reveals that it is possible to transmit modulating frequencies up to 75 kHz. Obviously, the human ear will not respond to frequencies much beyond 15,000 Hz, so modulating frequency spectrum space is available between 15 and 75 kHz that may be (and is) used to transmit the second channel of audio information required for FM stereo transmissions.

The second channel of encoded audio information is placed above the audio spectrum of the main channel by amplitude-modulating (with the carrier suppressed) a subcarrier whose frequency is 38 kHz. The sidebands of the modulated subcarrier are then added to the modulating signal of the main audio channel to form a composite FM signal that contains audio information that the stereo receiver may process to form independent left and right audio channels for stereo reproduction.

An FM-stereo receiver is not radically different from a monaural receiver. The significant difference is that an FM-stereo receiver includes added circuitry for deriving the two independent channels (the left and the right) from the transmitted FM-stereo signal. The RF amplifier, frequency converter, IF amplifiers, and demodulator sections may be almost identical to those of a monaural receiver. The stereo *decoder* comes after the demodulator, and it is the function of the decoder to form the left and right channel signals from the FM stereo composite signal that comes from the demodulator. For stereo reproduction, of course, two independent audio amplifiers and speaker systems are required.

A generalized block diagram of an FM-stereo receiver is shown in Figure 1-3. Further discussion of the FM-stereo system is postponed because of the rather technical and subtle nature of the system. The development of the composite signal and the operation of stereo receivers are described in detail in Chaps. 16 and 17.

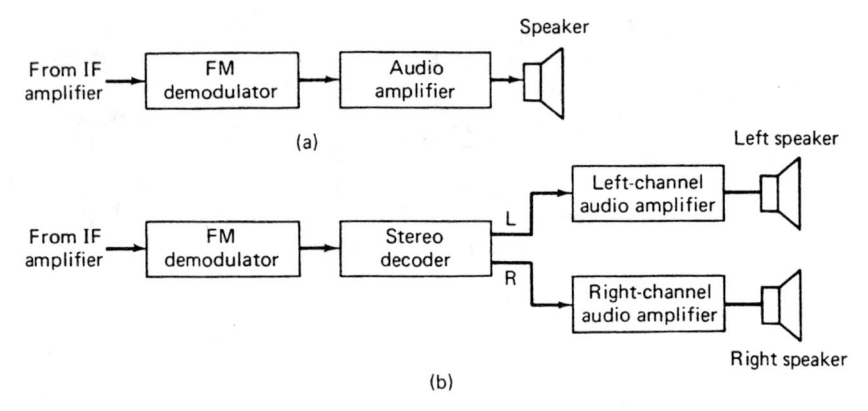

FIGURE 1-3 Block diagrams of the FM demodulator and audio section (a) of a monaural FM receiver, and (b) of a stereo FM receiver.

1-6 ORGANIZATION OF THIS TEXT

This text is divided into four main sections, as shown in the block diagram of Figure 1-4. The first section provides a description of active devices—tubes, transistors, field-effect transistors, and operational amplifiers—and a description of the circuits commonly encountered in AM, FM, and FM-stereo receivers. This includes audio amplifiers of various types, RF amplifiers, frequency changers, and power supplies. The first section forms a base of information that is used extensively in later sections.

The second section is devoted to AM receivers, to the fundamentals of troubleshooting, and to the troubleshooting of AM receivers. The principles of AM systems are given in Chap. 8, and actual receivers are analyzed in Chap. 9. Note that the fundamentals of troubleshooting and the basic troubleshooting procedure are introduced immediately after a complete system (an AM receiver) is analyzed.

The third section parallels the second, being devoted to FM receivers. The principles of FM transmission are given, followed by an analysis of receivers and troubleshooting.

The fourth section is devoted to FM stereo. A complete, detailed explanation of the FM-stereo system is given in Chap. 16. This is followed by the analysis of FM-stereo decoders; the final chapter deals with the troubleshooting of FM-stereo receivers.

1-7 LOOKING AHEAD

A person who is just beginning to study electronic servicing may tend to be impressed by the amount of material that he is required to master, and, indeed, the volume of material is considerable. But, as the longest journey begins with the first step, the devoted student should make rapid progress—even though the amount of progress may not at first be clearly evident. Electronic servicing is a big field,

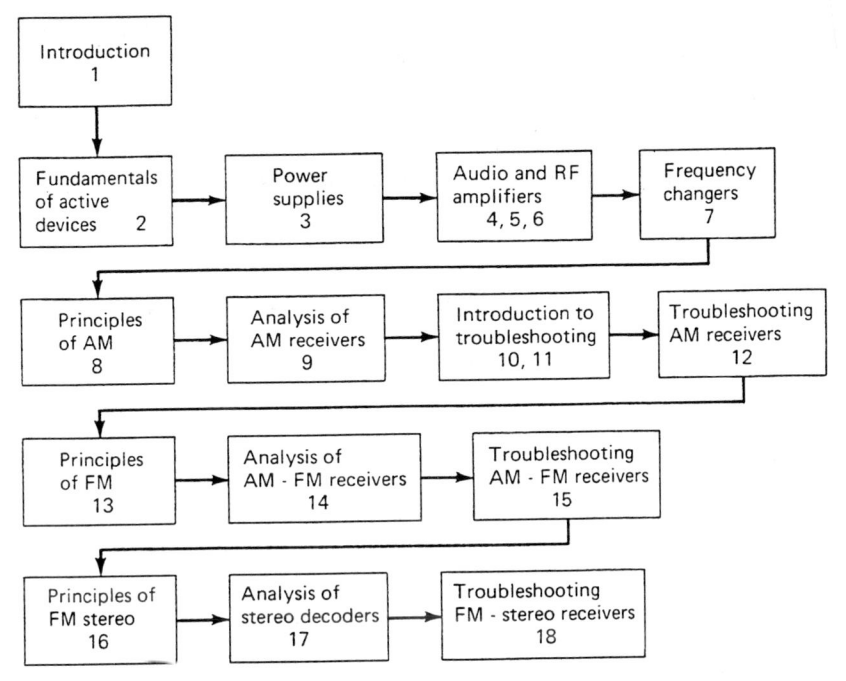

FIGURE 1-4 Block diagram illustrating the organization of this text.

and it is one that can never be mastered in its entirety—partly because the technology is advancing so rapidly and partly because of the tremendous number of devices on the market. Consequently, electronic servicing requires continued study in order to keep up with the technology and to constantly improve and increase one's abilities.

An objective of this text is to emphasize troubleshooting by procedure and to emphasize a systematic approach to troubleshooting rather than to present a collection of symptoms and the probable causes of each. As a technician gains experience, however, he will be increasingly able to recognize a symptom and to pinpoint the defect. However, many receivers exhibit unusual symptoms that can be misleading or deceptive and rather difficult to assess; such receivers are known affectionately as "dogs." With such receivers, only a carefully executed systematic approach is likely to be productive.

In real life, troubleshooting is never as straightforward and forthright as a text on the subject makes it appear that it should be. It is likely that this text is no exception. The defects assumed for purposes of illustration have been carefully selected to illustrate applicable procedures, and, indeed, many, if not most, stem from specific troubleshooting situations experienced by the authors. No defects have been included that have not been encountered in practice.

It is usually very beneficial for the beginning technician to acquire an assortment of junk receivers in order to become familiar with the physical aspects of the circuitry. Such a practice is even more beneficial if service information can be obtained for the receivers. As the technician gains experience, attempts can be made to repair the receivers—there is no better way to learn than by doing. However, it is generally not a good idea for a novice technician to accept service jobs for hire until he has acquired considerable experience on "noncritical" receivers.

In the face of a rapidly advancing technology, a concern naturally arises that any text on troubleshooting is obsolete or soon destined to become so. To some extent this must be true, but many elements of troubleshooting have remained essentially constant for a long time. The basic block diagram of superheterodyne receivers has remained unchanged for decades, as have the basic techniques of signal tracing, signal injection, and voltage-resistance analysis. Although the details of implementation may change or require modification, procedures remain essentially unchanged even as the technology undergoes transitions from vacuum tubes to transistors and onward to an increasing use of monolithic integrated circuits. Another consideration is that, to some extent, an electronic service technician must look not only to the future, but must also be concerned with what has gone before. Millions of older sets utilizing outdated technology by present standards are still in use and will require servicing for years to come.

It is probably true that in most specialized service organizations, the major effort is toward TV servicing. Although this text is not directed toward TV servicing, many of the techniques and procedures elaborated upon here are applicable not only to many sections of TV receivers, but also to citizen's band tranceivers and commercial two-way communications equipment. This text provides a good background for almost any area of electronic servicing.

Electronic servicing is a challenging endeavor that can be very satisfying and rewarding. Frustrations often arise, but the satisfaction derived from conquering a particularly stubborn problem more than compensates for the attendant frustrations.

2
FUNDAMENTALS OF ACTIVE DEVICES*

An active device is one that is capable of amplifying a signal. Accordingly, vacuum tubes, transistors, field-effect transistors, and linear integrated circuits are active devices, whereas resistors, capacitors, inductors, and transformers are passive devices.

Speaking more precisely, an active device must be able to increase the power level of an input signal. A step-up transformer will output a greater voltage than that applied to its input, but a transformer cannot have an output power greater than the input power. Hence, a transformer is not an active device.

In this chapter we present the fundamental principles of four types of active devices: bipolar junction transistors (BJTs), junction field-effect transistors (JFETs), insulated-gate field-effect transistors (MOSFETs), and vacuum triodes. Additionally, we briefly describe a general-purpose operational amplifier (op amp). We begin with a discussion of the physical principles of solid-state devices.

2-1 INTRINSIC SEMICONDUCTORS

The two most important semiconductor materials are silicon and germanium. The properties of the two are similar because atoms of each have four electrons in the outer shell. The outermost electrons are called *valence* electrons.

Considering that the nucleus and the inner electrons play only passive roles in determining the electronic properties of these materials, it is convenient to speak of a silicon *core* or a germanium *core* when referring to the nucleus and inner electrons. Hence, we may visualize a silicon or germanium atom as a core surrounded by four electrons. This picture emphasizes the importance of the four outer electrons.

The crystal structure of silicon and germanium easily accommodates four electrons in the vicinity of each atomic core. However, the electrons that occupy these positions belong more to the electronic structure of the crystal than to individual atoms, even though the electrons were carried into the structure by individual atoms. Each core, taken by itself, contains four positive charges (protons) that are not balanced with negative electrons. Hence, four electrons per core are required to establish electrical neutrality, and these electrons are provided by the electronic

---

* This material is from C. Green, *Technical Physics* (Englewood Cliffs, N.J.: Prentice-Hall, 1984).

structure of the crystal. Figure 2-1 is a two-dimensional representation of a silicon or germanium crystal.

Because silicon has far better properties than germanium for use in electronics, it is now, by far, the most popular semiconductor material. In the following we refer only to silicon, but the principles involved apply also to germanium.

Electron-Hole Pairs

The valence electrons gain energy as the material becomes warmer, and when an electron accumulates about 1.1 eV of thermal energy (for silicon), it is able to break away from its core and become free. We say the electron is *excited* from the *valence band* (of energy) into the *conduction band*. This produces a free electron (the one that was excited) and also a vacancy in the electronic structure where the electron originally resided. The vacancy is called a *hole*. Obviously, a hole is produced each time a free electron is produced, and we therefore speak of *thermally induced electron-hole pairs*. The word *pair*, however, does not imply that the free electron and hole stay together or even near each other after they are produced.

For temperatures above absolute zero, there will exist an equilibrium number of electron-hole pairs in an intrinsic semiconductor. At higher temperatures, the number increases. The presence of the free electrons and holes enables the semiconductor to conduct an electric current because mobile, charged particles are available to form the current. Furthermore, because more electron-hole pairs are present at higher temperatures, the electrical conductivity increases with temperature; warm semiconductors conduct better than cool semiconductors.

⊙ = silicon or germanium core

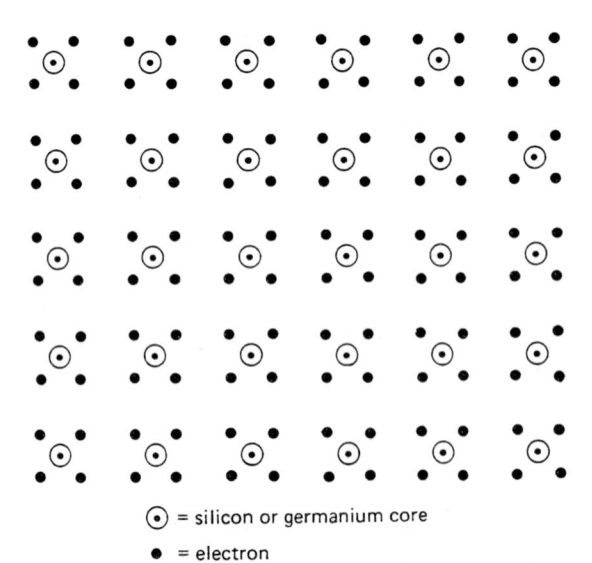

● = electron

FIGURE 2-1 Two-dimensional representation of a silicon or germanium crystal. The electronic structure easily accommodates four electrons near each core.

Hole Conduction

It is easy to understand that the presence of free electrons in a silicon crystal will increase its conductivity, but it is somewhat surprising to find that the holes also increase the conductivity. That is, a hole can act as a charge carrier.

Recall that a hole is a vacancy, a place in the electronic structure where an electron could be but is not. But a hole in conjunction with a silicon core represents an imbalance of electrical charge. The hole-core combination appears positive because of the electron deficiency represented by the hole. The positive charge actually resides in the nucleus of the silicon atom, but it is convenient (and only slightly misleading) to speak of a hole as being positively charged an amount equal to the charge of one proton.

The mechanism of hole conduction is illustrated in Figure 2-2. A hole moves from one core to an adjacent core via an electron jump from one core to the other in the direction opposite to the motion of the hole. Motion of the hole to the left, from core 10 to core 1, is accomplished by successive electron jumps from 9 to 10, then from 8 to 9, and so on. The net result is a transfer of one negative charge from core 1 to core 10. Far less energy is required for an electron to move from a given core to a hole on an adjacent core than is required to excite the same electron to the realm of free electrons (into the conduction band).

2-2 DOPING—ADDING IMPURITIES TO SEMICONDUCTORS

Pure (intrinsic) semiconductors as such are not very useful because they are rather good insulators. The electrical conductivity of semiconductors can be increased dramatically, however, by the addition of minute quantities of selected impurities

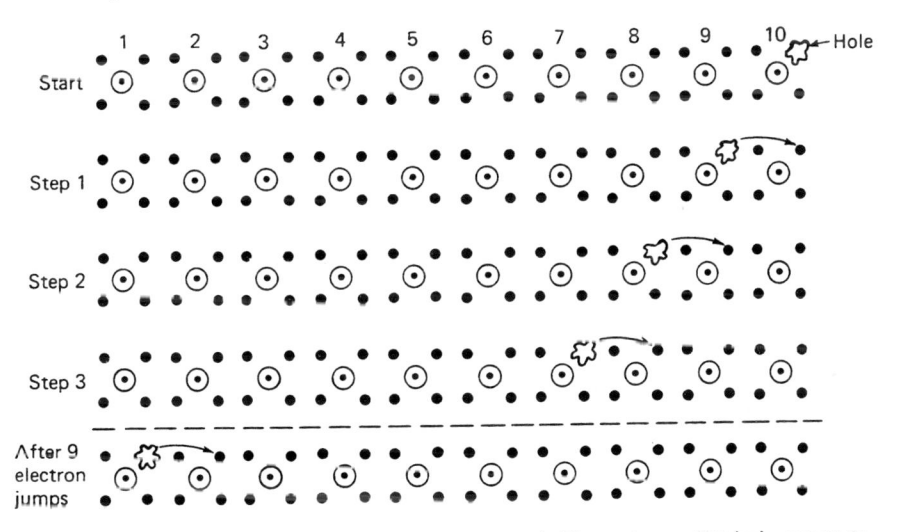

FIGURE 2-2 Illustration of hole motion along a line of silicon atoms. The hole moves to the left as electrons jump to the right.

to the semiconductor crystal. This process is called *doping*, and it serves to introduce either free electrons or free holes into the crystal, depending upon which dopant is used. An impurity concentration on the order of one part per million can result in a millionfold increase in the electrical conductivity of the material.

N-type Material

The elements phosphorus, arsenic, antimony, and bismuth consist of atoms that have five electrons in the outer shell. If one of these elements is used to dope a silicon crystal, a small number of the silicon atoms will be replaced by the impurity atoms. Since the impurity atoms have five electrons in the outer shell and since the electronic structure of the crystal easily accommodates only four electrons per core, the extra electron does not fit conveniently into the electronic structure of the crystal. But it must accompany the impurity atom into the crystal in order for the impurity atom to remain electrically neutral. The result is that the fifth electron is not as tightly bound to the atomic cores as are the four electrons that are easily accommodated. The fifth electron becomes a free electron (at ordinary temperatures) that can move under the influence of an electric field and form an electric current.

Semiconductors doped with impurities that contribute extra electrons are called *N-type semiconductors* because the polarity of the mobile charge carriers (the free electrons) is negative. Similarly, the impurities that produce *N*-type semiconductors are known as *N-type impurities*.

P-type Material

The elements aluminum, boron, gallium, and indium have only three electrons in the outer shell. When these elements are added as impurities to a semiconductor, each impurity atom contributes only three electrons instead of four to the electronic structure of the crystal. Thus, a hole exists at the site of each impurity atom. At ordinary temperatures, the holes become free holes and serve to increase the conductivity of the crystal.

Semiconductors doped with impurities that produce holes are called *P-type semiconductors*, and the impurities are called *P-type impurities*. The polarity of charge carriers in *P*-type materials (holes) is positive. A symbolic representation of *N*-type and *P*-type semiconductors is shown in Figure 2-3.

Current Conduction in Semiconductors

We have now described three types of semiconductors, intrinsic, *N*-type, and *P*-type. In intrinsic (undoped) semiconductors, current flow results from the motion of both electrons and holes that stem from thermally induced electron-hole pairs. Holes move in one direction and electrons move in the other. In comparison with doped semiconductors, the conductivity of intrinsic material is very small.

In *N*-type material, the impurity atoms contribute large numbers of free electrons in addition to the thermally induced electron-hole pairs. Consequently, cur-

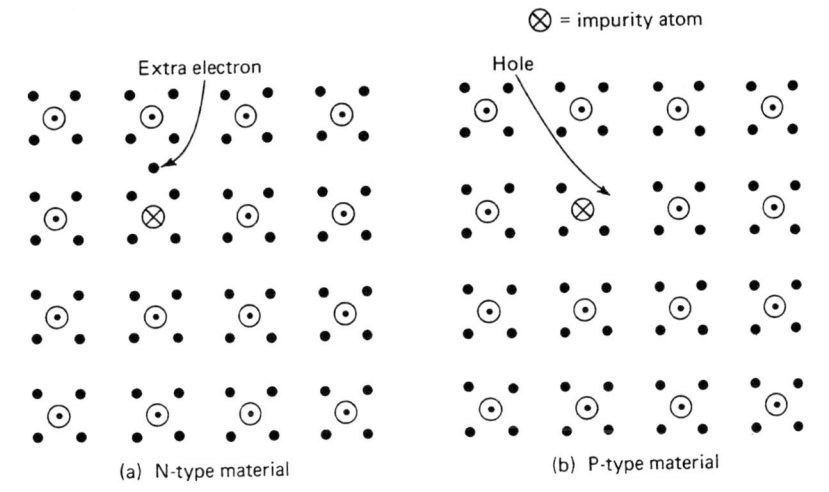

FIGURE 2-3 Representation of (a) N- and (b) P-type materials. The N-type impurity atom contributes an extra electron in N-type material, while the P-type impurity atom contributes a hole in the P-type material.

rent conduction is due primarily to the movement of free electrons. The thermally generated holes also contribute to the total current, but their effect is very small in comparison to that of the free electrons. In N-type material, the free electrons are the *majority carriers* while the holes are the *minority carriers*.

In P-type materials, free holes are the majority carriers. This is to say that current flow occurs primarily via the process of hole conduction. The thermally generated free electrons constitute the minority carriers.

We should always be mindful of the "nothingness" of holes. Hole conduction is actually caused by the movement of electrons in the opposite direction.

2-3 THE PN JUNCTION

A *PN junction* is formed where a region of N-type material is in intimate contact with a region of P-type material with the continuity of the crystal structure maintained at the interface between the two regions. Mechanically speaking, a dramatic change does not occur at the junction. The thing that changes, in crossing the junction, is the type of impurity atoms that are substituted for an occasional atom of the host material. Since the impurity concentrations are small, even this change is hardly noticeable, but the effect upon the electronic properties of the crystal is dramatic.

Suppose that a crystal of N-type material is suddenly joined to a crystal of P-type material, with the contact being such that the geometrical structures of the two crystals match up at the point of contact. The N-type material has an abundance of free electrons whereas the P-type material has an abundance of holes. When the two crystals are joined, what happens at the interface between the two types of material?

The obvious occurs. Free electrons near the interface on the *N*-side cross the interface and fill holes located near the interface on the *P*-side. The result is that free electrons of the *N*-type material disappear from the vicinity of the interface, and the holes near the interface on the *P*-side are filled by electrons coming over from the *N*-side. Thus, all charge carriers near the interface are *depleted*, and the region surrounding the interface, where there are neither electrons or holes, is called a *depletion layer*. This is illustrated in Figure 2-4.

The migration of free electrons across the interface does not continue until all free electrons in the *N*-material have gone over to fill the holes in the *P*-material. When an electron crosses from the *N*-side to the *P*-side, it leaves an unbalanced positive charge behind in the *N*-material and it contributes an extra negative charge to the *P*-material. A layer of positive charge develops near the interface in the *N*-material, and a layer of negative charge develops near the interface in the *P*-material. These layers establish an electric field across the interface that tends to oppose further migration of electrons. This limits the depletion layer thickness to values that are extremely thin in comparison to the overall dimensions of the *N*- and *P*-regions.

The depletion layer has no free electrons or holes, which makes it a good insulator. Thus, the depletion layer represents a layer of insulation that naturally develops between the *N*- and *P*-regions. In the following, however, we shall see that the thickness of this layer may be altered by applying an external voltage to the *N*- and *P*-regions. We shall see that electrons will flow across the junction in one direction but not the other. That is, a *PN* junction exhibits the properties of a *rectifier*.

Forward-Biased PN Junction

In Figure 2-5(a), a voltage source is connected to a *PN* junction with the positive voltage applied to the *P*-type material. Holes are repelled from the positive terminal through the *P*-region toward the interface. On the other side, electrons are repelled from the negative terminal through the *N*-region toward the interface. The result is that the depletion layer becomes narrower, and when the applied voltage reaches about 0.6 V, the depletion layer practically disappears and a comparatively large current will flow across the junction and through the external circuit. Under this condition, the junction is said to be *forward-biased*.

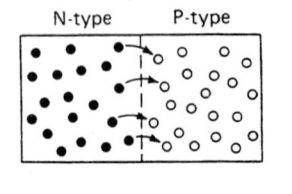

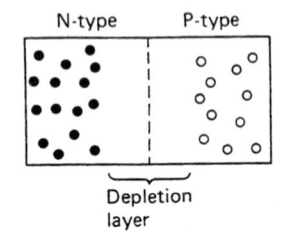

FIGURE 2-4 Formation of the depletion layer at the interface of a *PN* junction. Electrons cross the interface from the *N*-material to fill holes near the interface in the *P*-material.

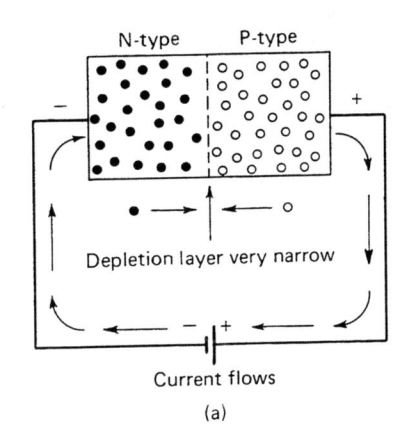

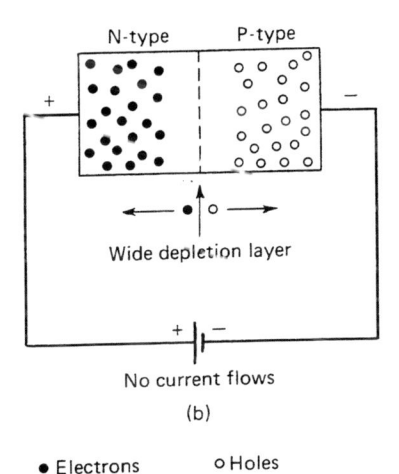

FIGURE 2-5 (a) Forward-biased and (b) reverse-biased PN junction.

Electron flow through the N-material of a forward-biased *PN* junction is by the movement of free electrons through the crystal toward the interface. At the interface, the electrons combine with holes arriving at the interface from the other direction. In the *P*-material, conduction is by the movement of holes in the manner depicted in Figure 2-2.

Reverse-Biased PN Junction

In Figure 2-5(b), the external voltage is connected with the positive voltage applied to the *N*-type material. Free electrons are attracted to the positive terminal and are drawn through the *N*-material away from the interface. On the other side, holes are attracted to the negative terminal and are also drawn away from the interface. The result is that the depletion layer becomes wider; no current flows across the junction due to the absence of mobile charge carriers in the depletion layer. Under this condition, the junction is said to be *reverse-biased*.

Conduction Characteristics of PN Junction

The forward-and-reverse conduction characteristics of a *PN* junction are shown in Figure 2-6. Current flow is plotted along the vertical axis and the applied voltage is plotted along the horizontal axis. The right-hand side of the graph is the region of forward bias; the reverse-bias region is on the left.

As the applied voltage is increased from zero in the direction of foward bias, the current increases very slowly (actually exponentially) until a forward bias of about 0.5 V is reached. The current then increases more rapidly, and the plot turns upward sharply at about 0.65 V. The region where the plot turns upward is called the *knee* of the curve. Beyond about 0.7 V, the current increases dramatically as the applied voltage is increased.

The reverse characteristic is obtained by reversing the polarity of the voltage applied to the junction. As the reverse voltage is increased, a very small *saturation current*, I_s, begins to flow that remains essentially constant until the reverse voltage becomes quite large, perhaps several hundred volts. This saturation current is on the order of 10^{-12} A and is therefore negligible in most practical applications, because the forward current is typically on the order of 10^{-3} A or larger. As the reverse voltage is further increased, the *breakdown region* of the device is ap-

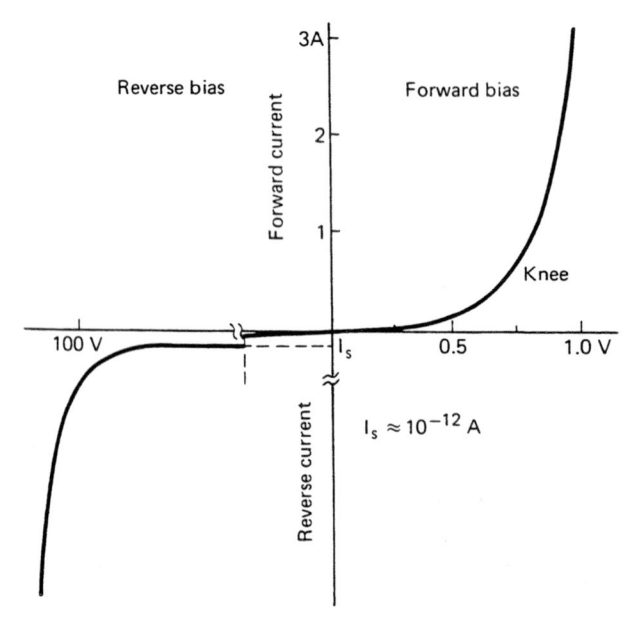

FIGURE 2-6 Forward- and reverse-conduction characteristic of a *PN* junction. Note that the scales are different in the forward and reverse regions.

proached, and it begins to conduct in the reverse direction. In practical applications, such as a rectifier, a *PN* junction (a *diode*) must not be allowed to enter the reverse breakdown region.

For voltages greater than about 100 mV applied in the forward direction, the current that will flow through a diode is given to a good approximation by

$$I = I_s e^{V_d/26} \qquad (2\text{-}1)$$

for a silicon diode at room temperature. In this equation, V_d is the voltage applied to the diode (expressed in millivolts) and I_s is the saturation current. You may verify that when V_d is 700 mV and I_s is 10^{-12} A, the current I is 0.49 A.

2-4 SOLID-STATE DIODES

PN junctions are widely used in electronics and are known as *solid-state diodes* (to distinguish them from the now obsolete vacuum diodes). They serve many functions, one of which is as rectifying elements in power supplies that convert alternating current to direct current for use in TVs, radios, phonographs, and so forth.

Ordinary diodes intended for use in power supplies are characterized by two important parameters. One is the *maximum forward current*, which is the maximum current that the diode can safely conduct in the forward direction. The other is the *peak inverse voltage* (PIV), which is closely related to the reverse breakdown voltage of the junction. Inexpensive diodes may safely handle a current of 10 A while larger, more expensive units may handle currents in the hundreds of amps. PIVs may range from 50 V to 1000 V or more for diodes commonly encountered.

The schematic symbol for a solid-state diode and the direction of electron flow is shown in Figure 2-7.

Zener Diodes

A special class of solid-state diodes called *zener diodes* is fabricated so that the reverse-breakdown characteristic is very sharp and occurs at low voltages. These diodes are used as voltage regulators because the voltage developed across the diode in the reverse-breakdown region is essentially independent of the current

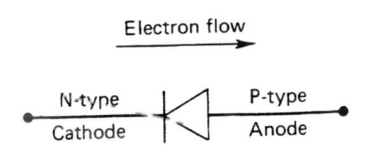

FIGURE 2-7 Symbol for a solid-state diode. Electrons flow from the cathode to the anode (terminology held over from the era of vacuum tubes).

flowing through the diode. The reverse conduction characteristic of a zener diode is shown in Figure 2-8(a) and a simple regulator circuit is shown in Figure 2-8(b). As voltage regulators, zeners are always operated in the reverse breakdown region. Zeners are characterized by their zener voltage V_z, and values of V_z range from slightly less than 3 V to 100 V or more.

As the source voltage in Figure 2-8(b) is increased from zero, the voltage across the zener increases linearly until V_z is approached. No current (practically speaking) flows through the zener until the vicinity of V_z is reached, and then the zener current increases sharply. At this point, voltage $V_s - V_z$ appears across the series resistor R_s. A further increase in V_s causes the voltage across the zener to increase only slightly, but the additional current increases the IR drop across R_s so that the voltage across the zener remains nearly equal to V_z.

Light-Emitting Diodes

These devices, known commonly as LEDs, are widely used as indicator and pilot lamps in a wide variety of electronic devices. The junction is designed so that light (usually red, green, or yellow) is produced when electrons combine with holes as

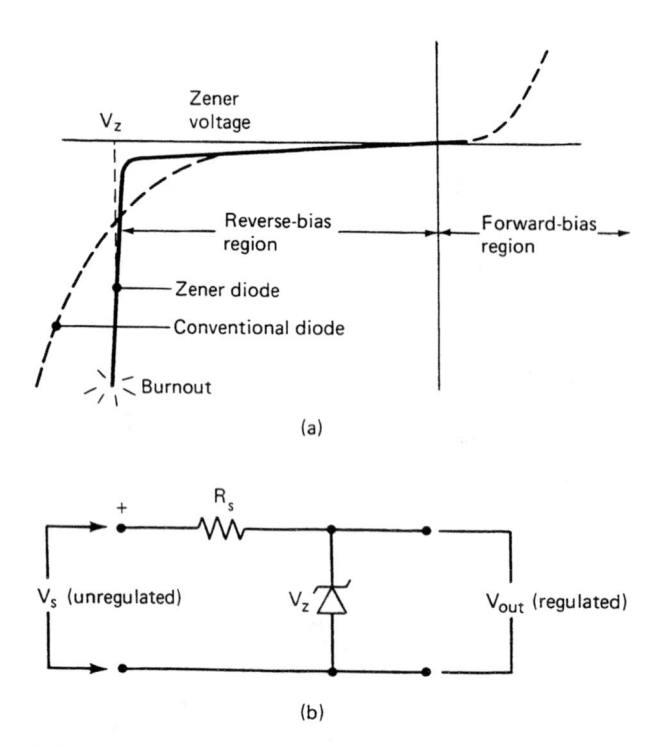

FIGURE 2-8 (a) The reverse breakdown characteristic of a zener diode is much sharper than that of a conventional diode. (b) Simple regulator circuit.

Fundamentals of Active Devices   Chap. 2

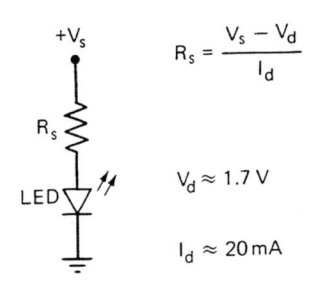

FIGURE 2-9 LED indicator circuit.

the junction conducts in the forward direction.* Typical junction currents range from 15 to 25 mA, and the voltage developed across the junction ranges from about 1.5 to 2.0 V. The junction voltage is higher than the familiar 0.7 V because a different semiconductor material (gallium arsenide, GAs) is used instead of silicon. The basic LED indicator circuit is shown in Figure 2-9. As for zener diodes, a series resistor must be included in order to limit the flow of current through the diode.

2-5 BIPOLAR JUNCTION TRANSISTORS

We now consider a solid-state device that is capable of amplification, the *biopolar junction transistor* (BJT). It is *bipolar* because both N- and P-type materials are used, and it is a *junction* transistor because two PN junctions form the heart of the device.

A BJT is a semiconductor sandwich consisting of a thin layer of N- or P-material sandwiched between two regions of material of the opposite type, as shown in Figure 2-10. Two possible configurations give rise to NPN and PNP transistors as shown. The terminals are identified as the emitter E, base B, and collector C, and we identify the junctions as the base-emitter (B-E) junction the base-collector (B-C) junction. We shall describe BJT operation in terms of an NPN silicon device.

Examine the circuit of Figure 2-11. Note that battery A forward biases the BE junction. Observe that switch S is open so that no external voltage is applied to the collector. When the voltage of A reaches the knee voltage of the junction, electrons will flow from the N-type emitter region into the P-type base region and out into the external circuit. The magnitude of the current depends upon the voltage applied to the junction by battery A, and a graph of the current vs. the applied voltage would resemble the forward conduction characteristic that is typical of PN junctions. When the voltage of A drops below the knee voltage (about 0.6 V for

---

* Because LEDs light up when they conduct, a vivid demonstration of the action of a bridge rectifier, for example, can be performed by using LEDs as rectifier diodes and then using a source of "slow AC" (about 1 Hz) to drive the bridge. The slow AC signal can be obtained from a function generator.

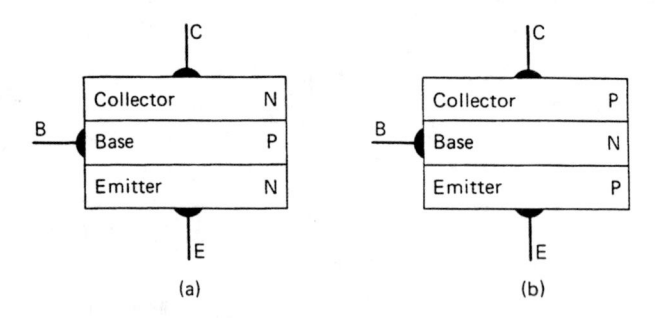

FIGURE 2-10 (a) *NPN* and (b) *PNP* bipolar junction transistors. Actual device geometries differ greatly from this simple pictorial representation.

silicon), electron flow into the base drops to very low levels. If the voltage of *A* were increased to about 0.9 V, a very large current would flow that in many transistors might destroy the B-E junction. Electron flow from the emitter to the base is controlled by the voltage applied to the B-E junction.

With switch *S* open, no voltage is applied to the collector, and the collector does not affect the operation of the B-E junction. When *S* is closed, however, a large positive voltage (9 V in this case) is applied to the collector making the collector region of the transistor much more positive than the base region. This places a reverse bias on the B-C junction, and the B-C depletion layer becomes wide. We do not expect a current to flow across the B-C junction because of the scarcity of charge carriers in the B-C depletion layer. We find, however, that a large current does flow across this junction. Why?

The base region is physically very thin and is lightly doped with impurities in comparison with the emitter and collector. These factors reduce the total number of holes in the base region. When the B-E junction is forward biased, more electrons cross the B-E interface than there are holes in the *P*-type base region for them to

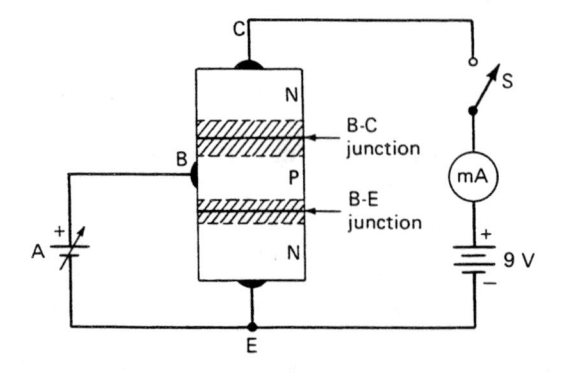

FIGURE 2-11 Battery *A* forward-biases the B-E junction. When switch *S* is closed, a collector current flows even though the B-C junction is reverse-biased.

Fundamentals of Active Devices   Chap. 2

drop into. Electrons "pile up" at the base edge of the B-E depletion layer, and this pile of electrons extends across the thin base region to the base edge of the B-C depletion layer. But the pile of electrons is not of equal depth across the base region.

In other words, the positive voltage applied to the base pulls electrons from the emitter into the base region. A small part of these electrons combine with holes in the *P*-type base material, but most remain in the base region as free electrons. The electron concentration is greatest near the B-E junction; it becomes smaller with increasing distance from the B-E junction and becomes nearly zero at the B-C depletion layer. This is illustrated in Figure 2-12.

The effect of this electron concentration *gradient* is to produce a *diffusion current* of electrons from the region of greatest electron concentration to the region of lower concentration. This produces a current flow from the emitter side to the collector side of the base region.

Recall that a wide depletion layer appears between the base and the collector regions that acts as an insulator due to the scarcity of charge carriers within the layer. The diffusion current serves to inject free electrons into the B-C depletion layer, and the injected electrons are readily drawn across the region to the collector. These electrons ultimately form the collector current in the external circuit.

Thus, a large current flows from the emitter to the collector of the transistor, and this current is under the control of the voltage applied to the base, or equivalently, the collector current is determined by the forward bias applied to the B-E junction. When the forward bias is increased, more electrons are drawn into the base region, the electron concentration gradient is increased, the diffusion current increases, and the collector current increases. The opposite effects occur when the forward bias of the B-E junction is decreased.

If the forward bias is removed from the base, no electrons will be drawn into the base region and no diffusion current will occur. Consequently, the collector current will be zero. Even if the collector is operated at a high voltage (less than the breakdown voltage of the transistor), no appreciable collector current will flow because the collector is not able to extract electrons from the emitter without assistance from the base.

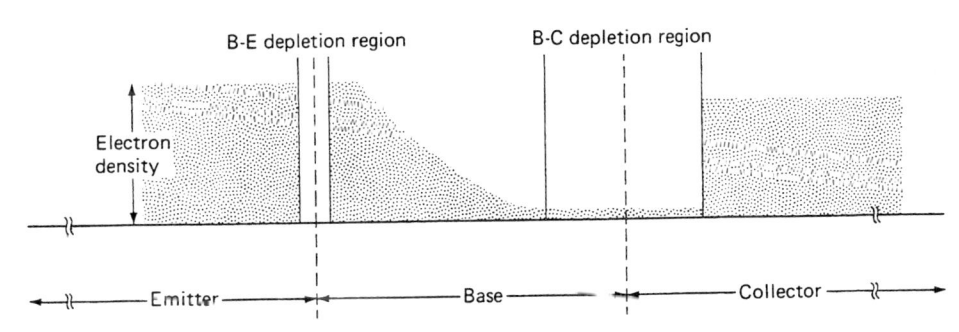

FIGURE 2-12 Electron densities in an *NPN* transistor. Electrons are most dense in the emitter, and somewhat less dense in the collector. The uneven density in the base region produces a diffusion current that introduces electrons into the depletion layer between the base and collector.

Sec. 2-5 *Bipolar Junction Transistors*

Current Gain from Base to Collector

Of the total number of electrons entering the base from the emitter, a small fraction (about 1%) will combine with holes in the base and be removed from the diffusion current. It is these electrons that combine with holes that ultimately constitute the current that flows from the base terminal of the transistor. In practice, it is desirable for the base current to be minimized.

Here is the principle that makes the transistor useful as an amplifier. A small base current, resulting from a voltage applied to the base, can effectively control a much larger current that flows from the emitter to the collector. In terms of energy, the base uses a small amount of energy to control a much greater expenditure of energy in the collector circuit. This is the principle on which amplification is based.

For a given transistor, there is an almost constant relationship between the base current I_b and collector current I_c. To a good approximation, the collector current is a constant multiple of the base current:

$$I_c = \beta I_b \qquad (2\text{-}2)$$

The factor β is called the *beta* of the transistor, defined as the ratio of collector current to base current:

$$\beta = \frac{I_c}{I_b} \qquad (2\text{-}3)$$

Typical values of β range from about 20 to 200 for transistors commonly encountered. The division of emitter current between the base and collector is illustrated in Figure 2-13.

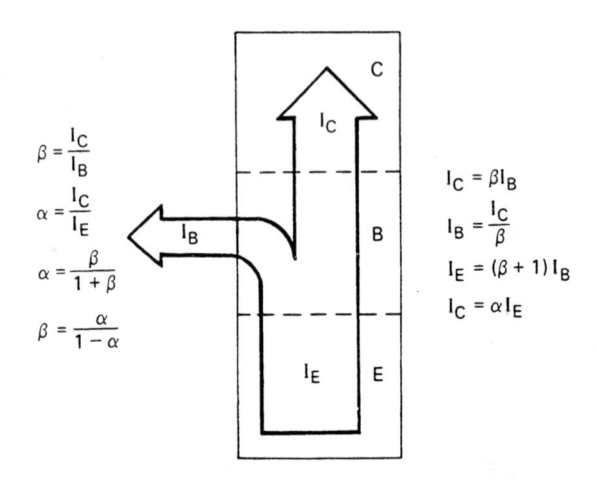

FIGURE 2-13 Illustration of the meaning of alpha (α) and beta (β), and applicable mathematical relationships.

Another parameter, besides beta, that relates to the division of emitter current between the base and collector is *alpha*, α. Alpha is the fraction of the emitter current that flows on to the collector. Its value is related to beta, as shown in Figure 2-13. Typical values of α range from about 0.95, a low value, to very nearly 1.0. Because alpha and beta are related so that one may be obtained from the other, alpha is seldom specified for a transistor; beta is the parameter most commonly encountered.

The operating characteristics of a typical transistor are shown in the graph of Figure 2-14(a), and the circuit used to obtain the curves is shown in part (b) of the figure. Voltage source V_{bb} is adjusted to give the desired base current I_b. Then source V_{cc} is varied to give the desired value of V_{ce} while the resulting value of I_c is recorded. (Curve tracers are available that display these curves automatically on the screen of a CRT.)

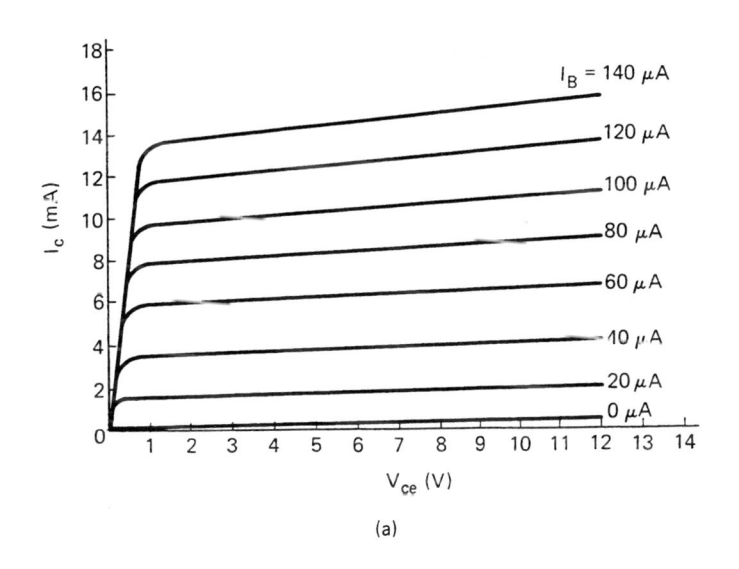

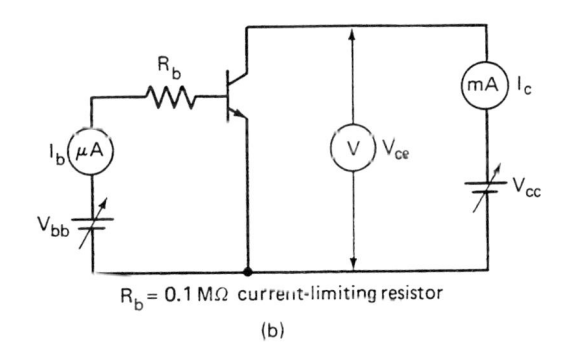

FIGURE 2-14 (a) Typical transistor characteristic curves. (b) Circuit used to obtain the curves.

Sec. 2-5   *Bipolar Junction Transistors*

Note that I_c increases sharply as V_{ce} is increased from zero. But when V_{ce} reaches about 0.2 V, the I_c curve flattens out and increases only slightly as V_{ce} becomes larger. The two regions are called the *saturation region* and the *linear region*, as indicated. In the linear region, V_{ce} is large enough so that the entirety (almost) of the base diffusion current is "collected" by the collector, and because the magnitude of this current is controlled by I_b, I_c increases only slightly with V_{ce}. In the saturation region, however, V_{ce} is not large enough to pull all the diffusion current into the collector.

2-6 JUNCTION FIELD-EFFECT TRANSISTORS

Another type of transistor is the *field-effect transistor* (FET) which works on an entirely different principle from that of the bipolar junction transistor (BJT) described in the previous sections. The family of field-effect transistors is shown in Figure 2-15. The type of FET that we shall describe first is the junction field-effect transistor, the JFET.

Junction FET (JFET)

Figure 2-16 shows a model of an N-channel JFET that illustrates the principle of operation. Actual device geometries differ greatly from this simple model, but the operating principles are the same. A bar of *N*-type material has *P*-type regions embedded on each side. *PN* junctions form where the two types of material meet, and the depletion layers extend into the interior of the bar as indicated. Contact is made to the side regions to form the third contact of the device, called the *gate*. The contacts to the ends of the bar are the *source* and the *drain*, as noted in the figure.

A current flows through the bar when a voltage is applied as shown in Figure 2-17. The resistance of the bar limits the current. The distribution of current is not uniform over the cross section of the bar, however, because the depletion regions formed around the gate junctions are depleted of charge carriers. Current flow can occur only through the interior portion of the bar, called the *channel*, between the

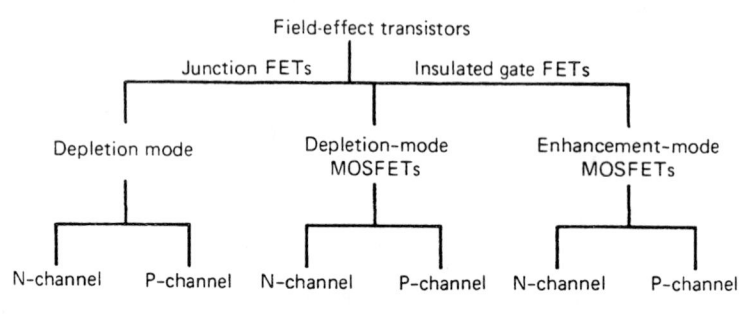

FIGURE 2-15 Family of field-effect transistors.

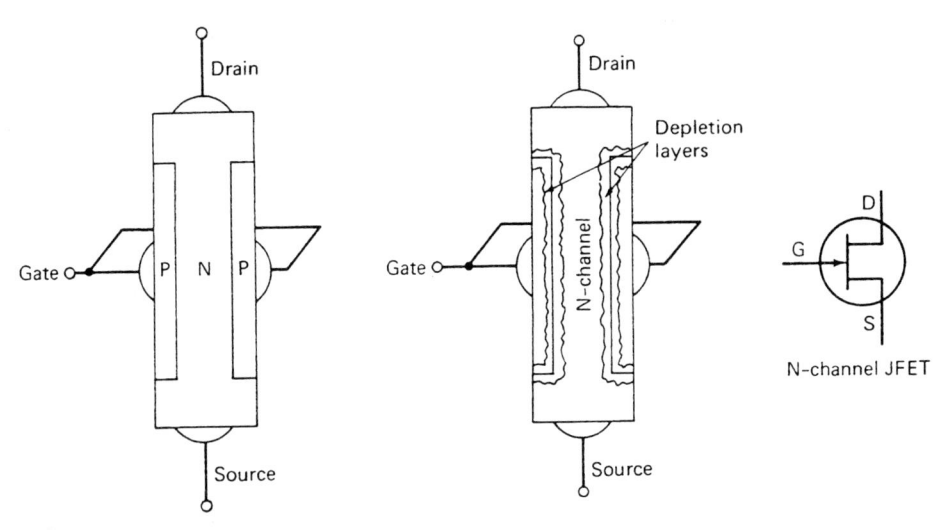

FIGURE 2-16 Model of a junction field-effect transistor (JFET). Depletion layers form at the junctions between the gate and the channel materials.

opposing depletion layers. The current that flows from the source through the channel to the drain is called the *drain current*.

We now consider the effect of a voltage applied to the gate upon the width of the depletion layers extending into the channel. By applying a voltage to the gate (relative to the source) that reverse biases the *PN* junctions, the width of the depletion layers can be made to vary as in an ordinary *PN* junction under the influence of a varying reverse bias. As the reverse bias is increased, the depletion layers extend farther into the channel, reducing its cross-sectional area. Since the

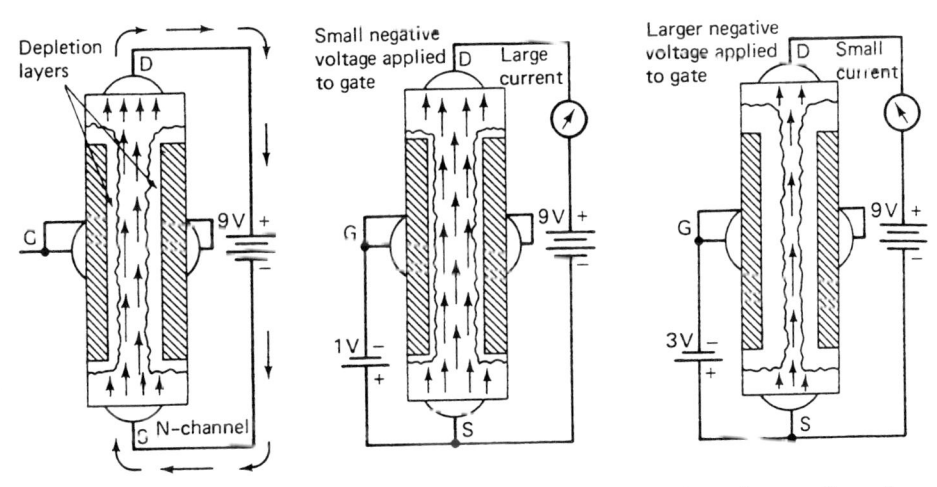

FIGURE 2-17 Dimension of the conductive portion of the channel depends upon the voltage applied to the gate. The gate voltage controls the width of the depletion layers.

Sec. 2-6 Junction Field-Effect Transistors

resistance of the channel is determined, in part, by the cross-sectional area of the channel, the current flowing from the source to the drain due to the application of the drain voltage can be made to vary.

As the reverse bias at the gate is increased, the widening depletion layers cause the channel to shrink in cross-sectional area producing a decrease in the drain current. As the reverse bias is diminished, the cross-sectional area increases and so does the drain current. Thus, the voltage applied to the gate controls the drain current. This is the operating principle of JFETs.

Many details remain to be investigated, however. Since no current flows through the gate, all parts of the gate region are at the same voltage. This is not the case for the channel, because the drain current flows through the channel, and the drain-source voltage is distributed along the length of the channel. Consequently, the reverse bias of the gate-channel junction also varies along the length of the channel. As the reverse bias varies, so does the width of the depletion layers. The net effect is that the width of the depletion layers increases, proceeding from the source to the drain, causing the cross-sectional area of the channel to decrease. This condition is shown in Figure 2-18.

By applying sufficient reverse bias to the gate, it is possible to cause the depletion layers to come together to "pinch off" the conducting channel, and the voltage at which this occurs is called the *pinch-off voltage*. One might expect the drain current to vanish entirely at this point, but this does not happen, even though the drain current is significantly reduced. Although the depletion layers appear to touch at the center of the channel, a small region remains between the layers through which a concentrated stream of electrons flow to maintain the drain current. If the reverse bias is increased, more of the channel is pinched off and the current is further

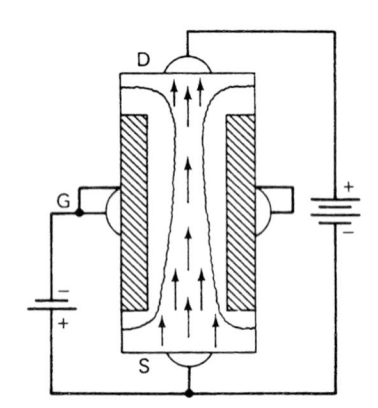

FIGURE 2-18 Width of the depletion layers is not uniform in a JFET because of the voltage variation along the length of the channel. The channel voltage is greater near the drain than at the source.

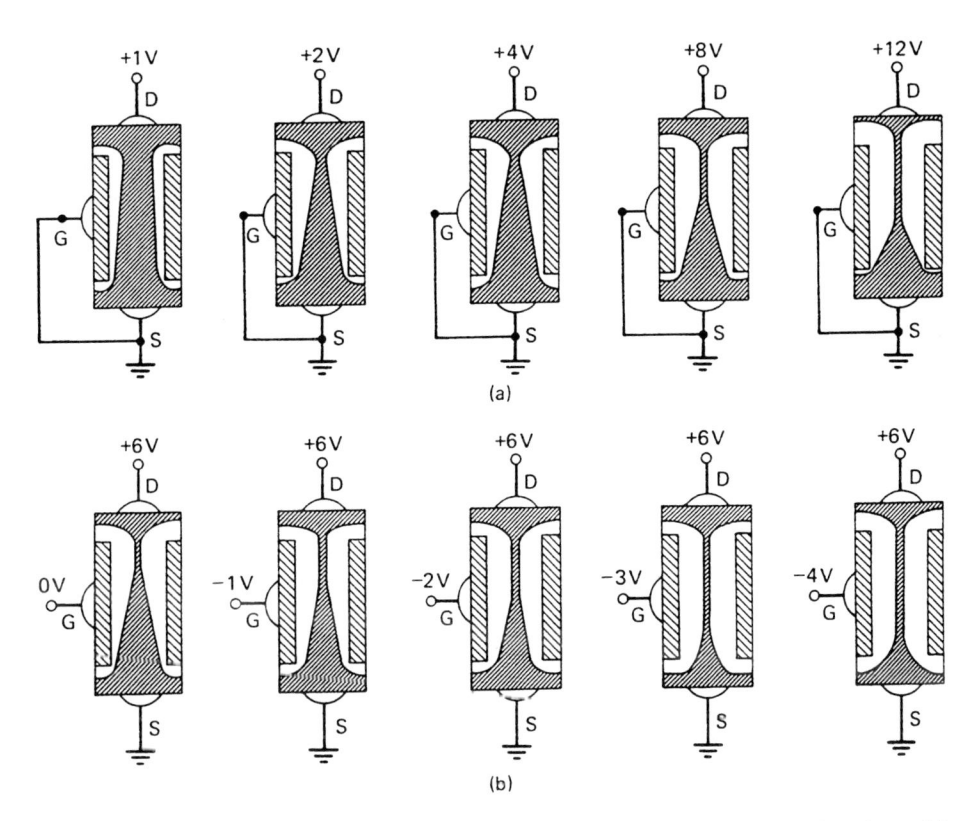

FIGURE 2-19 (a) Pinching off of the channel that occurs due to an increase in drain voltage. (b) Pinching off of the channel due to the application of a gate voltage.

decreased, but a current flow will still exist. By increasing the reverse bias still more, it is possible to reduce to negligible proportions the current flow from source to drain. Figure 2-19 illustrates the pinching-off action as the reverse bias is increased.

Drain Characteristics

Figure 2-20 shows the drain characteristics of a junction FET. At a given bias voltage, the graph of drain current versus drain voltage rises linearly at low values of drain voltage, where the channel acts almost as a resistor of constant value. At higher drain voltages, the graph begins to flatten out as the depletion layers decrease the width of the channel. When the drain voltage increases to the point where the reverse bias near the drain is just sufficient to pinch off the channel, the curve becomes essentially flat and remains so. The region of linear increase is called the *ohmic region* of the curve, and the flat portion is called the *saturation region*. In light of the saturation region of the drain characteristics, we say that a FET exhibits a constant-drain-current characteristic.

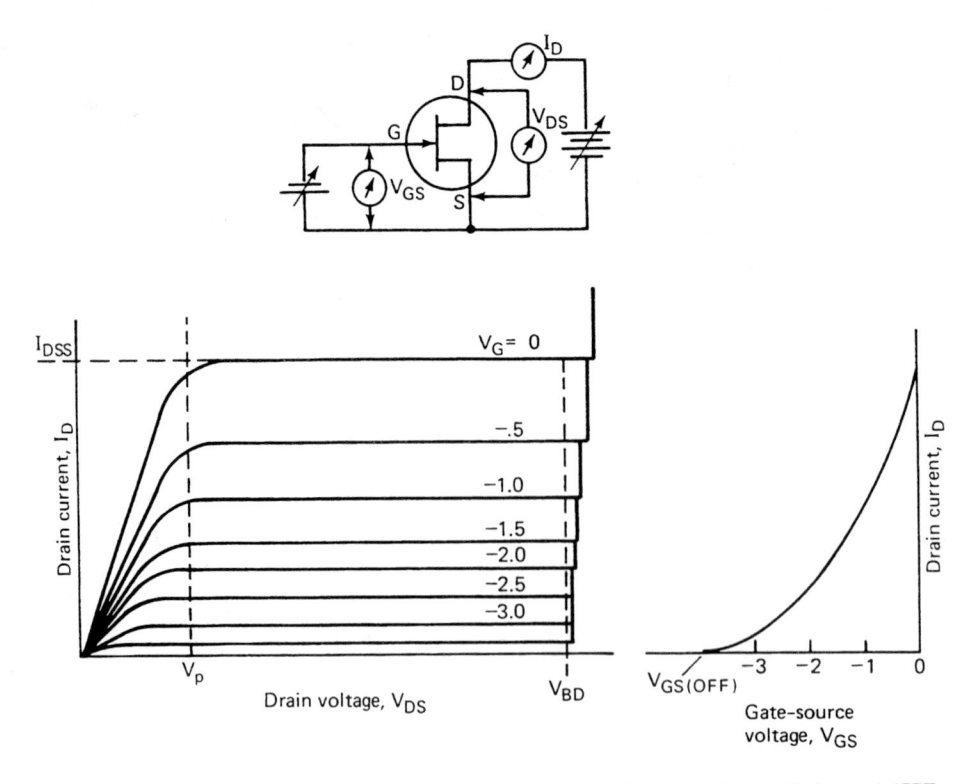

FIGURE 2-20 Typical drain characteristics and a typical transfer curve for an N-channel JFET.

To understand why the drain current does not continually increase with drain voltage, consideration must be given to two factors: (1) as the drain voltage is increased, a greater force is exerted on the electrons (in n-channel devices) to pull them through the channel from source to drain; (2) as the drain voltage is increased, the channel voltage near the drain is increased, and this produces a greater reverse bias on the PN junctions near the drain. This, in turn, reduces the conductivity of the channel as a result of the increased length of the pinched-off region.

These two effects are opposing forces that tend, for the most part, to cancel. As the drain voltage increases, the resistance of the channel increases almost proportionally, and the current flowing through the channel remains almost constant. This is illustrated in Figure 2-21.

Notable features of the drain characteristics include the pinch-off voltage V_p, the drain-source current with the gate shorted to the source I_{DSS}, and the gate-source cutoff voltage $V_{GS(off)}$. Also important is the breakdown voltage V_{BD} which, if exceeded, may lead to destruction of the device. JFETs are normally operated in the region of the characteristics between V_p and V_{BD}, where the characteristics are essentially flat. Note that the transfer characteristic resembles one-half of a parabola; this is called a *square-law response*, and the transfer curve of FETs follows the square law almost perfectly.

30 Fundamentals of Active Devices Chap. 2

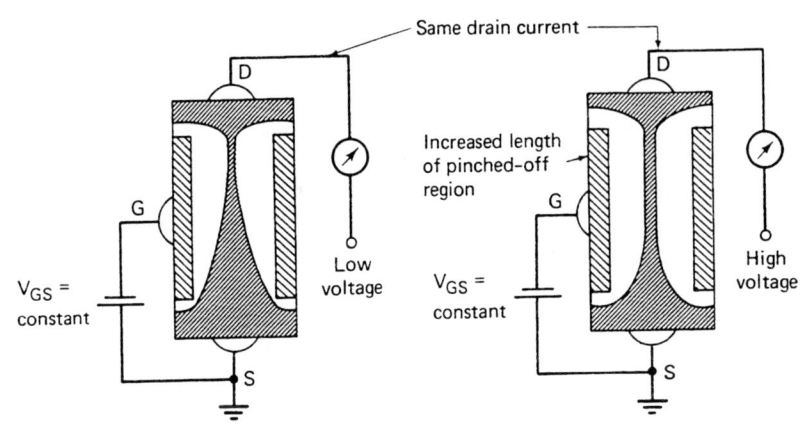

FIGURE 2-21 Increasing the drain voltage does not significantly increase the drain current because more of the channel is pinched off at high drain voltages.

Channel Material

Either N- or P-type material may be used as the channel of an FET, with the gate material always being of the opposite type. N-type materials have greater conductivity than P-type materials for the same impurity concentration, but it is sometimes advantageous to fabricate or use a P-channel device. Thus, both types are commonly encountered. The symbol for a FET indicates the type material used for the channel, and the symbols for both types are indicated in Figure 2-22.

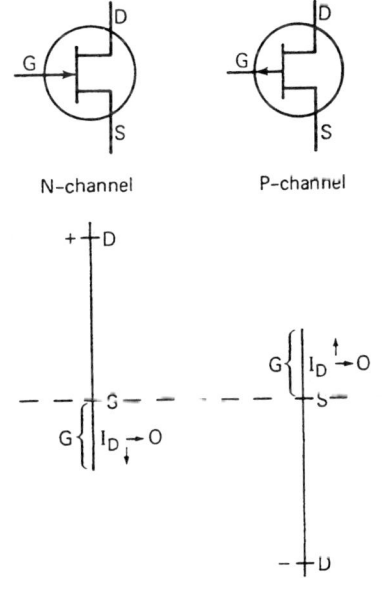

FIGURE 2-22 Symbols and biasing requirements for N- and P-channel junction field-effect transistors.

Sec. 2-6 Junction Field-Effect Transistors 31

DC Voltage Polarities

In operation, the gate of JFETs is always reverse-biased. Hence, if the channel is N-type, the gate must be P-type, and the gate must be biased negative relative to the source. For an N-channel JFET, the drain is held positive relative to the source, and the gate is made more negative in order to reduce the drain current to zero. For a P-channel JFET, the drain is held negative relative to the source, and the gate is made more positive in order to reduce the drain current to zero. The biasing polarities are illustrated in Figure 2-22.

2-7 INSULATED GATE FETs—MOSFETs

Insulated gate FETs make use of the fact that a physical *PN* junction is not required to alter the conduction characteristics of the channel. The basic structure of this type of FET is shown in Figure 2-23, where it may be noted that the metal gate electrode is separated from the channel by a thin layer of insulating oxide. A voltage applied to the gate sets up an electric field that alters the availability of charge carriers in the channel. The acronym MOSFET identifies this type of FET, the metal-oxide-semiconductor field-effect transistor. MOSFETs may be either N-channel or P-channel, and we shall see that for each type of channel there are two modes of operation, the depletion mode and the enhancement mode.

Referring to Figure 2-24, note that a channel of N-type material is formed just underneath the insulating oxide opposite the gate electrode. When a negative voltage is applied to the gate, the holes of the P-type substrate are drawn toward the gate and combine with some of the electrons of the N-channel. This decreases the channel thickness, and current conduction from source to drain is diminished. If however, the gate is made less negative or even positive, the holes are repelled by the gate, and the N-channel becomes thicker, increasing conduction between source and drain. In this manner the voltage applied to the gate controls the current flowing in the channel.

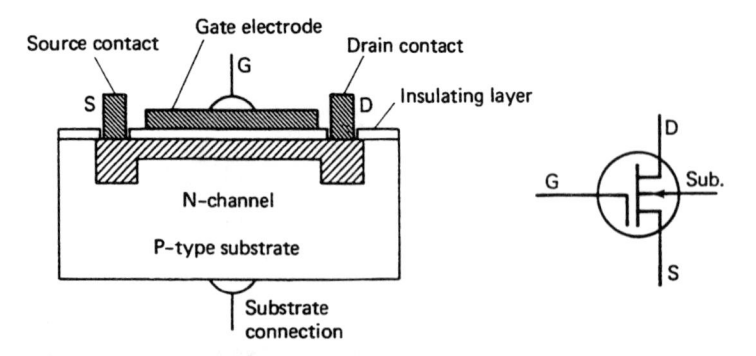

FIGURE 2-23 Structure and symbol for an N-channel depletion-mode MOSFET. Note that the gate is insulated from the channel.

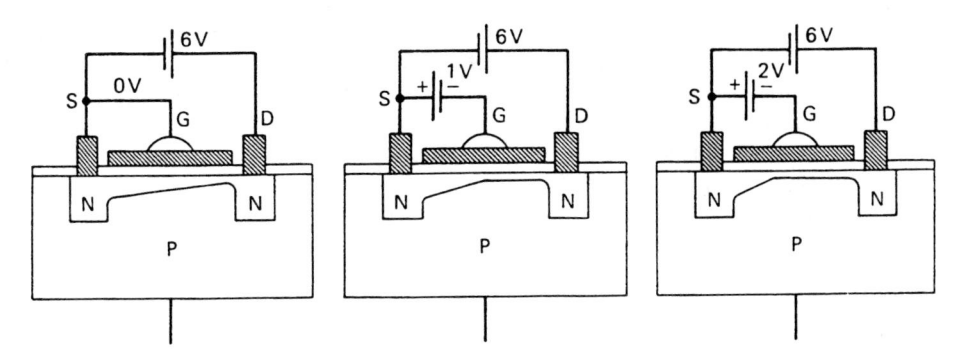

FIGURE 2-24 As the gate is made more negative, holes from the P-region are attracted into the N-channel, reducing its dimension. The pinching-off action is illustrated.

An important difference exists between this device and the junction FET described earlier. In the junction device the gate must always be reverse-biased relative to the channel to avoid a flow of current from the channel to the gate. In the MOSFET, however, the oxide layer provides the separation between the gate and the channel and no such restriction of the gate voltage exists.

Depletion and Enhancement Modes

The FETs discussed thus far exhibit near-maximum conduction with zero bias voltage applied to the gate. Application of a bias voltage causes a narrowing of the channel as the depletion layers become wider and conduction is diminished. Such devices are said to operate in the depletion mode because the bias voltage depletes the channel. We now come to a class of MOSFETs that exhibit minimum or near zero conduction at zero bias. These devices require a bias voltage to produce source to drain conduction and are said to operate in the *enhancement mode*.

Figure 2-25 shows the structure of an N-channel enhancement-mode MOSFET; note the absence of a channel from source to drain. In order for the device to conduct, a channel must be *induced* by a positive voltage applied to the gate

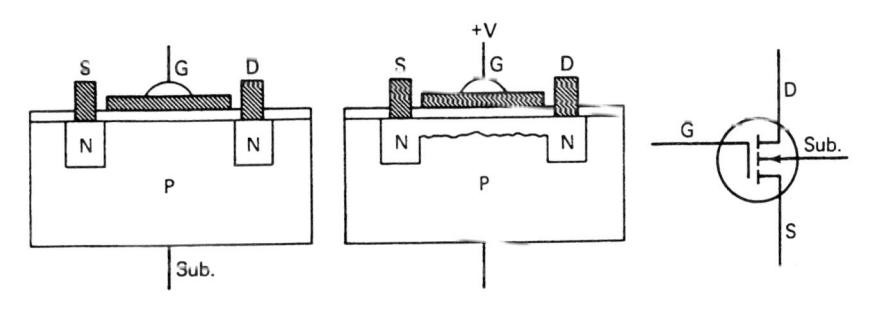

FIGURE 2-25 Structure and action of an N-channel enhancement-mode MOSFET. The gate must be made positive to induce a channel between the source and the drain.

Sec. 2-7   *Insulated Gate FET's—MOSFETs*   33

electrode. A positive voltage on the gate attracts electrons to the vicinity of the gate and forms an effective *N*-channel. When the positive voltage is removed, the channel disappears and conduction through the device will cease. *P*-channel enhancement MOSFETs operate in a similar manner, where a negative voltage must be applied to the gate to attract holes in order to form a channel. The symbol for enhancement MOSFETs denotes the normally open channel by using a broken line to indicate the channel. This and a summary of the voltage polarities required for the different types of MOSFET's are given in Figure 2-26.

Substrate

The bottom supporting layer of planar-type semiconductor devices is called the *substrate*, and in many device geometries, an undesirable *PN* junction is formed between the substrate and the active portion of the device, as shown in Figure 2-27. This undesired junction must be kept reverse-biased in order to be electrically inactive. If the substrate becomes forward-biased relative to the channel, catastrophic failure of the device may result as large currents flow in the forward-biased *PN* junction. Many packages provide an external lead to the substrate which must

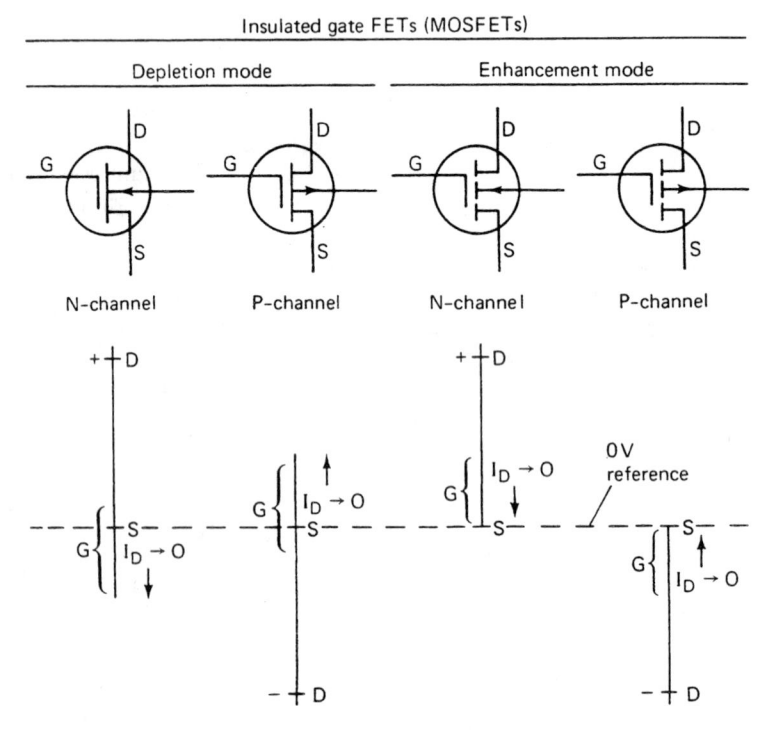

FIGURE 2-26 Symbols and biasing requirements of insulated gate FETs (MOSFETs). The region of allowed gate voltages is indicated, and the polarity of gate voltage required to reduce the drain current I_D to zero is shown.

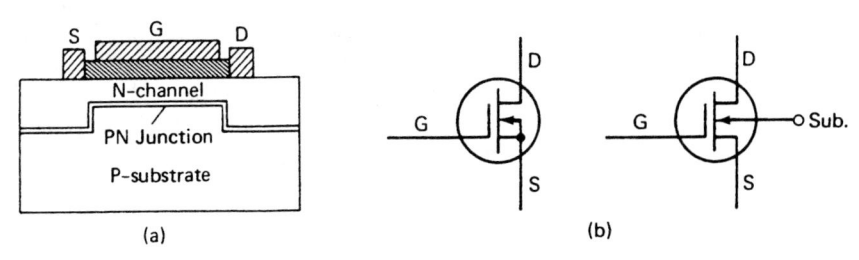

FIGURE 2-27 (a) A *PN* junction is formed between the channel and substrate of a MOSFET. (b) The substrate may be connected internally to the source or an external connection may be provided for the substrate.

be properly connected. For *P*-type substrates, the substrate is commonly connected to ground or to the most negative part of the circuit. In other devices the substrate is internally connected, with no external lead provided. For MOSFETs the substrate connection is shown on the symbol, as may be noted in Figure 2-27.

Dual-Gate MOSFETs

Figure 2-28 shows the structure of a dual-gate, depletion-mode, *N*-channel MOSFET. Note the source contact and the drain contact, and note that an extra region of *N*-type material divides the channel into two sections in series. Each section of the channel has a gate electrode, where the gates are identified as gate 1 and gate 2. Gate 1 is closest to the source. Each gate is capable of cutting off the current in its section, and therefore, either gate can cut off the flow of current through the entire device. Considering the device as two single gate devices in series, the device source serves as the source for section 1 while the central region serves as the drain. The central region, in addition to acting as a drain for section 1, serves as the source for section 2. The device drain serves as the drain for section 2.

The independent pair of gates makes the device applicable to RF amplifiers, gain-controlled amplifiers, mixers, demodulators, and so forth. As a controlled-

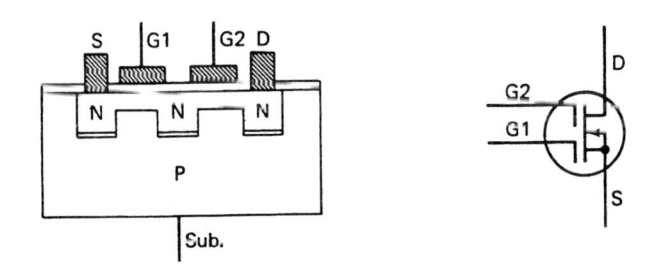

FIGURE 2-28 Structure and symbol of a dual-gate, *N*-channel, depletion-mode MOSFET. In the symbol, the substrate is shown connected internally to the source.

Sec. 2-7 Insulated Gate FET's—MOSFETs 35

gain amplifier, the signal to be amplified is applied to gate 1 while a DC control voltage derived from an automatic-gain-control circuit is applied to gate 2. Also, the presence of the second gate section tends to isolate the device drain from the input signal applied to gate 1, and this allows high-frequency operation well into the UHF range without need of neutralization.

Handling MOSFETs

The gate electrode of a MOSFET is insulated from the channel region by an insulating layer of extremely high resistance. The insulating layer is so thin, however, that a voltage difference on the order of 100 V between the gate and the channel will cause the insulation to break down, giving rise to "punch-through" in which the gate is shorted to the channel, destroying the device.

Small charges of static electricity can easily develop sufficient gate-channel voltage to puncture the insulation. Such charges may come from fingertips, tools, soldering devices, plastic materials, Styrofoam, and so forth. It is possible to destroy a MOSFET simply by removing it from its package. Hence, special precautions must be taken when handling MOSFETs.

MOSFETs are shipped with all leads shorted, either by a metal spring, metal foil, or by a special conductive foam. If possible, metal springs or clips should not be removed until after the device is soldered into the circuit. If the conducting foam is used, it should not be removed until just before the device is inserted into the circuit, and then only after precautions against static electricity have been taken.

Precautionary measures include grounding the soldering device, the receiver chassis, and the technician. Insulating materials—plastic, rubber, nylon—are the most dangerous and should not be allowed to contact the device. A small piece of lightly moistened tissue may be wrapped around the device to short the leads during handling. Equal care is needed when removing a MOSFET from a circuit for testing.

Some MOSFETs are provided with internal protection diodes that limit the gate-channel voltage, but it is good practice to handle all MOSFET devices with special care.

2-8 OPERATIONAL AMPLIFIERS

Integrated circuit (IC) technology has provided a multitude of specialized devices in the form of "chips" so that it now seems that there is a chip for every chore. However, one category of linear IC that is definitely general purpose is that of the *operational amplifier*, the op amp. With only minor variations in the external circuitry, a general-purpose op amp can be made to perform a wide variety of functions. In this section we present the highlights of these devices. The symbol for a general-purpose op amp and the typical power supply connection are shown in Figure 2-29.

Typical features of a general-purpose op amp include: (1) two inputs, one inverting and one noninverting; (2) an input impedance of 1 MΩ or more; (3) an

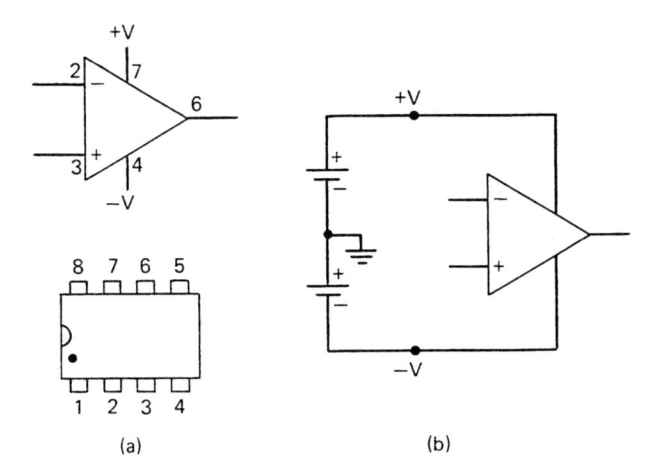

FIGURE 2-29 (a) Symbol and top view of a general-purpose op amp. The pin numbers are given for the familiar 741. (b) Connections for operation with a balanced power supply.

output impedance that typically is less than 100 Ω; and (4) a very large low-frequency gain, typically more than 100,000. Also, most op amps are intended to operate from a balanced power supply so that the output voltage may swing symmetrically above and below zero volts. Some op amps include an offset-null adjustment that permits the output voltage to be adjusted to exactly zero volts when the input voltage is zero.

Strictly speaking, the op amp is a difference amplifier; the output is a greatly amplified version of the *difference* in the voltages applied to the two inputs. A positive voltage applied to the noninverting (+) input drives the output to a more positive value, but a positive voltage applied to the inverting (−) input drives the output to a more negative value. The net output voltage is the difference of these two effects.

In practical applications, the very large gain (about 100,000) is reduced dramatically to usable levels by the use of *negative feedback*. A portion of the output voltage is brought back to the input side, where it is applied to the inverting input terminal. Then, when the output goes more positive, for example, the voltage applied to the inverting input will also go more positive. This will tend to drive the output toward the negative. Any movement of the output voltage will be counteracted (somewhat) by the voltage fed back to the inverting input terminal. The net effect is a reduction in gain of the overall amplifier.

Voltage Follower

An extreme case in which all the output voltage is fed back to the inverting input is represented by the *voltage follower*, shown in Figure 2-30. Suppose that a positive voltage of 1 V is applied to the (+) input, which represents the signal input for the circuit. This will drive the output voltage positive due to the large gain of the

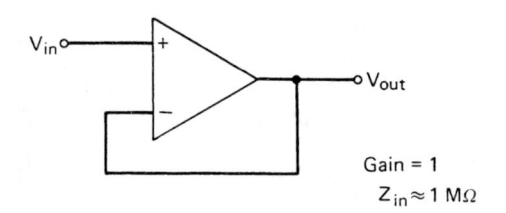

FIGURE 2-30 Voltage follower. The high input impedance and low output impedance makes this circuit useful as an impedance-matching device.

op amp itself. But when the output voltage reaches 1 V, the same voltage will be applied to the inverting input (via the feedback connection) and the output voltage will stabilize at 1 V. At this point, there will be (for practical purposes) zero differences in voltage between the two inputs. This condition of zero difference in voltage between the two inputs is a common feature of all op amp amplifier circuits.

Let us now examine the action of this circuit more closely, paying attention to some tiny voltages that were neglected in the above. Again, suppose that $+1$ V is applied to the $(+)$ input. The output voltage will rise, but it will not rise all the way to 1 V, as we implied above. Instead, it will fall short of 1 V by about 1/100,000 V, which is 10 μV. (We are assuming the gain of the op amp to be 100,000.) Consequently, the voltage fed back to the $(-)$ input will be 10 μV less than 1 V. This means that the difference in voltage between the $(+)$ and $(-)$ inputs will be 10 μV. Then, because an op amp is a difference amplifier, the 10-μV difference will be amplified by 100,000. This gives rise to the (very nearly) 1 V at the output.

From this we conclude that the gain of the voltage follower is 1; the output voltage is the same as the input voltage; or, the output voltage follows the input voltage. The usefulness of the circuit stems from its high input impedance (about 1 MΩ) and very low output impedance (< 100 Ω); it is used as a buffer amplifier or as an impedance-matching device.

Open-Loop and Closed-Loop Gain

The voltage gain exhibited by an op amp without feedback is called the *open-loop gain*, and it is very large, as we have seen. But as amplifiers, op amps are always operated with negative feedback in order to reduce the gain to useful levels. The negative feedback path closes the loop around the op amp itself, and we refer to the gain of the entire circuit as the *closed-loop gain*. For the voltage follower, the closed-loop gain is 1 while the open-loop gain is on the order of 100,000.

Noninverting Amplifier

When 100% of the output voltage V_{out} is fed back to the $(-)$ input, we obtain a closed-loop gain of 1, that is, a voltage follower. If, however, only a fraction of V_{out} is fed back, we achieve a closed-loop gain greater than 1. A convenient way

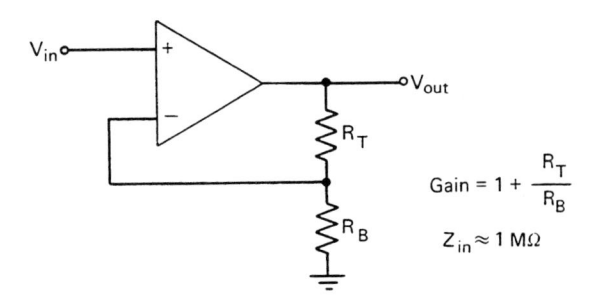

FIGURE 2-31 Noninverting amplifier. When R_T is 10 kΩ and R_B is 1 kΩ, the gain is 11.

to obtain the fraction of V_{out} is to use a voltage divider on the output, as shown in the schematic of Figure 2-31. The resulting amplifier is known as a *noninverting amplifier*.

Inverting Amplifier

Another amplifier, shown in Figure 2-32, has the input signal applied through R_s to the inverting input so that the output signal is inverted relative to the input signal. The net signal appearing at the ($-$) input is due to the combination of V_{in} applied through R_s, and V_{out} applied through R_f. Because V_{in} and V_{out} are inverted relative to each other, the two component voltages at the ($-$) input tend to cancel. Therefore, the voltage at the ($-$) input is always within a few tens of microvolts of zero, and the difference voltage is, accordingly, rather small. Because the ($-$) input is always at very nearly ground potential (the same as the ($+$) input), it is sometimes called a *virtual ground*. The closed-loop voltage gain is determined by the ratio R_f/R_s, and because of the virtual ground at the ($-$) input, the input impedance to the circuit is the same as R_s.

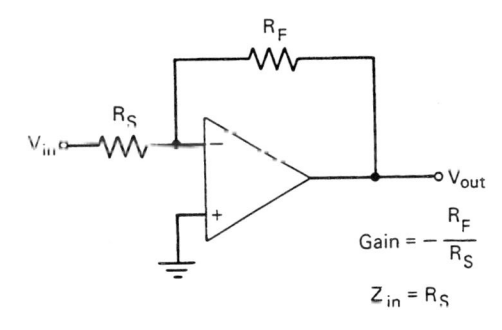

FIGURE 2-32 Inverting amplifier. The minus sign associated with the gain indicates that the output signal is inverted relative to the input signal.

Sec. 2-8   **Operational Amplifiers**

Op Amp Limitations

Even though op amps are suitable for a wide range of applications, they suffer from at least three significant limitations. One is that their high-frequency response is rather limited. The open-loop gain drops off dramatically with increasing frequency. For example, an op amp having an open-loop gain of 100,000 at very low frequencies might have have a gain of only 100 at a frequency of 10 kHz, of 10 at 100 kHz, and of only 1 at 1 MHz. The *gain-bandwidth product* (GBW) expresses this property. If the GBW of an op amp is known, the expected open-loop gain A_{ol} at a frequency f is given by

$$A_{ol} = \frac{\text{GBW}}{f} \tag{2-4}$$

The GBW of a 741 general-purpose op amp, for example, is 1 MHz.

Another limitation arises from the fact that the output voltage cannot change instantaneously from one value to another. The rate at which the voltage changes is called the *slew rate*, and the maximum slew rate is an important op amp specification. For a 741, the maximum slew rate is 500,000 V/s, or 0.5 V/μs.

The maximum slew rate of a sine-wave voltage is given by

$$\text{SR}_{max} = 2\pi f V_p \tag{2-5}$$

where f is the frequency in hertz and V_p is the peak voltage of the sine wave. Note that SR_{max} depends on both the frequency and amplitude of the sine wave. You may verify that a 1-kHz sine wave whose peak amplitude is 10 V has a maximum slew rate of 62,832 V/s. This is well under the 500,000-V/s capability of a 741 op amp, which can reproduce the waveform with no difficulty.

The third limitation arises from the electrical noise produced within the device. In applications where very small signals are to be amplified, better results (signal/noise ratio) are usually obtained with discrete devices. However, special low-noise op amps can be obtained at a somewhat higher price that offer excellent noise characteristics.

2-9 THE VACUUM DIODE

The central structure of a vacuum tube is the *cathode*. It consists of a hollow, oxide-coated metal cylinder that surrounds an electric *heater* which heats it to red-hot temperatures. When heated, the cathode emits electrons from its outer surface, and a cloud of electrons forms around the cathode. The phenomenon of electron emission due to heat is called *thermionic emission*, and the electron cloud that forms around the cathode is called the *space charge*.

A *vacuum diode* is made by surrounding the cathode with a metal cylinder called the *anode*, or more commonly, the *plate*. When the plate is made more positive than the cathode, electrons flow from the cathode via the space charge to

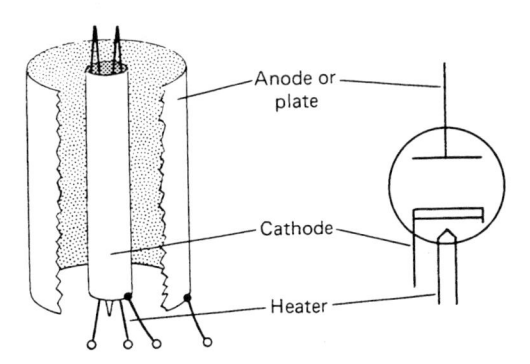

FIGURE 2-33 Structure and symbol for a vacuum diode.

the plate. That is, an electron current flows from the cathode to the plate. But when the cathode is made more positive than the plate, no current flows from the plate to the cathode because the plate is not heated to temperatures sufficiently high for it to emit electrons. Thus, electrons will flow from the cathode to the plate, but they will not flow from the plate to the cathode. The device exhibits the property of a rectifier; its structure and schematic symbol are shown in Figure 2-33.

The Triode

Control over the electron current flowing from the cathode to the plate can be achieved by surrounding the cathode with a loosely wound spiral of thin wire, called the *grid*. When the grid is made negative relative to the cathode, the plate current becomes smaller. This happens because the electrons of the space charge are repelled by the negative charge on the grid. Conversely, the plate current increases when the grid becomes less negative. This ability of the grid voltage to control the plate current makes the *triode* (as the device is called) useful as an amplifier. The structure and schematic symbol of a triode are shown in Figure 2-34.

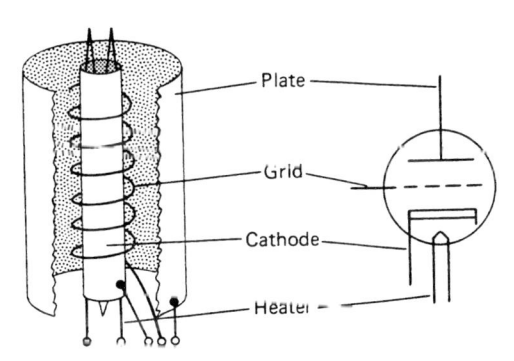

FIGURE 2-34 Structure and symbol for a vacuum triode.

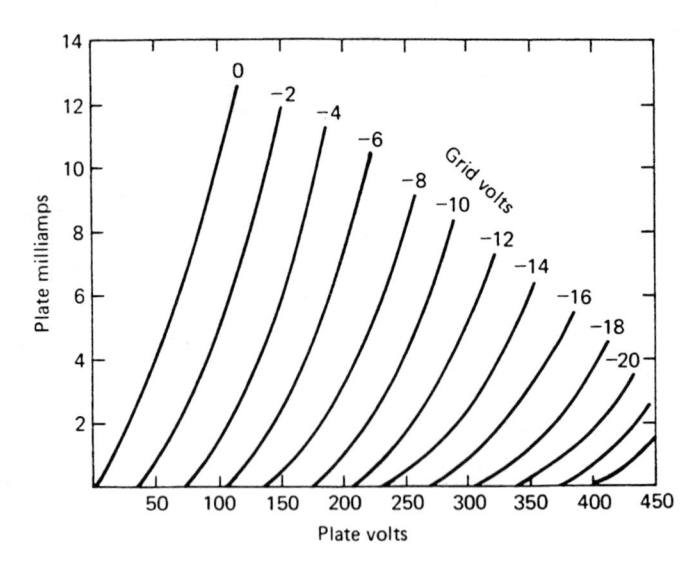

FIGURE 2-35 Plate characteristics of a triode.

Plate Characteristics of a Triode

The performance characteristics of a triode are given by its plate characteristics. The plate characteristics of a typical triode are shown in Figure 2-35. The grid curves illustrate the effect of the grid voltage in controlling the plate current. Note that for a given plate voltage, a larger and more negative grid voltage reduces the plate current. In the figure, for a plate voltage of 150 V, about 6 mA of plate current flows when the grid is −4 V with respect to the cathode, and only about 2 mA flows when the grid is held at −6 V.

SUMMARY

1. The minute conductivity of intrinsic (pure) silicon or germanium is due to thermally induced electron-hole pairs that give rise to free electrons and free holes. The conductivity increases with temperature as more electron-hole pairs are produced.

2. A movement of holes to the left is equivalent to, and in fact is caused by, a movement of electrons to the right. A hole moves by virtue of a succession of electron jumps in the opposite direction.

3. The N-type impurities, phosphorus, arsenic, antimony, and bismuth, have five electrons in their outer shell and contribute free electrons to the semiconductor. The P-type impurities, aluminum, boron, gallium, and indium, have only three electrons in their outer shell and contribute free holes to the semiconductor.

4. A depletion layer forms at the interface of a *PN* junction due to the passage of free electrons from the *N*-side to the *P*-side, where they combine with free holes. The depletion layer may be regarded as a layer of insulation because the free electrons and holes in the layer have been depleted.

5. When a *PN* junction is forward biased (by making the *P*-material positive relative to the *N*-material), the depletion layer becomes thinner and a large current will flow. Under reverse bias, the depletion layer becomes thicker, and only a very small leakage current will flow.

6. *PN* junctions (solid-state diodes) serve as rectifiers in power supplies for converting AC to DC. Zener diodes have a very sharp reverse breakdown characteristic and are used as voltage regulators. Light-emitting diodes (LEDs) are widely used as indicator lamps.

7. About 0.7 V is required to forward bias a silicon diode, but only about 0.25 V is required to forward bias a germanium diode. An LED requires about 1.7 V to forward bias it to the conduction region.

8. The B-E junction of a bipolar junction transistor is forward-biased in normal operation. Current flows across the base region by a process of diffusion.

9. Fundamentally, a BJT is a current multiplier: the base current is multiplied by the beta of the transistor to yield the collector current. The base current is controlled by the voltage applied to the B-E junction.

10. In the linear region, the collector current increases only slightly as the collector voltage is increased.

11. The terminals of a BJT are the emitter, base, and collector. These correspond to the source, gate, and drain of a FET, and to the cathode, grid, and plate of a vacuum triode.

12. Whereas the B-E junction of a BJT is forward-biased, the gate-channel junction of a JFET is always reverse-biased.

13. Current flow from the source to the drain of a JFET is controlled by the dimensions of the channel, which, in turn, are controlled by the voltage appearing between the gate and source.

14. The biasing requirements of an *NPN* transistor, *N*-channel JFET, and vacuum triode are similar in that the collector, drain, or plate is held positive relative to the emitter, source, or cathode.

15. In a MOSFET, the gate electrode is completely insulated from the channel by a layer of oxide. Consequently, the gate voltage may be of either polarity relative to the source, and this makes it possible to fabricate both depletion-mode and enhancement-mode MOSFETs.

16. Conduction from source to drain occurs normally and naturally in a depletion-mode FET; a voltage must be applied to the gate to reduce the level of conduction. In enhancement-mode FETs, however, no conduction will occur until a voltage applied to the gate causes it to occur.

17. The gate-channel junction of a JFET must never become forward-biased. Therefore, the gate voltage of a JFET is restricted to one polarity. In MOSFET devices, however,

Summary 43

no such restriction exists. Consequently, MOSFETs occur of both depletion-mode and enhancement-mode types.

18. Dual-gate MOSFETs have two gates, which makes the devices useful as controlled-gain RF amplifiers, mixers, demodulators, and so on.

19. Special care is needed in handling MOSFET devices because static discharges can easily destroy the device by puncturing the insulation between the gate and the channel.

20. An operational amplifier is a very high gain general-purpose amplifier that may be tailored to a wide variety of application by adding a few external components. It is a difference amplifier. Negative feedback is used to reduce the gain to practical levels.

21. A voltage follower, which uses 100% negative feedback to reduce the circuit gain to unity, is often used as an impedance-matching device.

22. A noninverting op amp amplifier may utilize a voltage divider to return only a fraction of the output signal voltage back to the input as negative feedback.

23. An inverting op amp amplifier incorporates resistances in series with the input signal path and in series with the feedback path from output to input.

24. Op amp limitations include a limited high-frequency response, a limited slew-rate capability, and a tendency to produce greater amounts of electrical noise than amplifiers using discrete components.

25. A vacuum diode consists of a cathode, a plate, and a heater inside the cathode to heat the cathode to temperatures at which thermionic emission will occur.

26. A triode includes a grid situated between the cathode and plate. Making the grid more negative reduces electron flow from the cathode to the plate.

QUESTIONS AND PROBLEMS

2-1. What is an electron-hole pair? Why do free electrons and holes occur in pairs in intrinsic semiconductors? What is responsible for the generation of electron-hole pairs?

2-2. Describe the mechanism by which the presence of electron-hole pairs increases the electrical conductivity of a semiconductor.

2-3. Why does a hole appear to have a positive charge even though a hole is really nothing?

2-4. Explain why the presence of holes in a *P*-type semiconductor increases the electrical conductivity of the material.

2-5. What are the majority and the minority carriers in *N*-type semiconductor material?

2-6. What causes a depletion layer to form at the interface between the *N* and *P* regions of a *PN* junction?

2-7. What happens to the depletion layer when a *PN* junction is forward biased?

2-8. When a *PN* junction made of silicon is forward-biased, what approximate voltage appears across the junction under conditions of moderate conduction?

2-9. Describe the reverse-breakdown characteristic of a zener diode. What is the primary application of zener diodes?

2-10. In normal operation of an *NPN* transistor, what voltage polarities are applied to the base and collector regions? Which is more positive?

2-11. Why is it necessary to forward-bias the B-E junction?

2-12. In normal operation the B-C junction of a BJT is reverse-biased, yet a large current easily flows across the junction. What is responsible for this effect?

2-13. Why do most of the electrons entering the base region go on to the collector?

2-14. A transistor is often called a current multiplier. What current is multiplied? What is the beta of a transistor?

2-15. How do the emitter and collector currents compare? How does the base current compare with the collector current?

2-16. In normal operation of a BJT, the collector current is controlled by the voltage existing between which two elements of the transistor?

2-17. Explain how the structure of a JFET differs from that of a BJT.

2-18. Why must the gate-channel junction of a JFET always be reverse-biased?

2-19. Why are there no enhancement-mode JFETs?

2-20. To reduce the drain current of an *N*-channel JFET, should the gate voltage be made more positive or more negative?

2-21. In normal operation, should the gate of a *P*-channel JFET be positive or negative relative to the source?

2-22. How does the structure of a MOSFET differ from that of a JFET?

2-23. With the gate terminal shorted to the source of an *N*-channel device, will any drain current flow when a moderate voltage (say, 6 V) is applied between the drain and source:

(a) of a JFET?

(b) Of a depletion-mode MOSFET?

(c) Of an enhancement-mode MOSFET?

2-24. The channel-substrate junction of MOSFETs must always be reverse-biased. Why?

2-25. To stimulate an enhancement-mode, *N*-channel MOSFET into conduction, must the gate be made positive or negative relative to the source?

2-26. Why must MOSFETs be handled carefully in regard to static electricity?

2-27. What does it mean when we say that op amp is a difference amplifier?

2-28. What technique or principle is used to reduce the large, open-loop gain of an op amp to practical levels?

2-29. Why are most general-purpose op amps operated from a balanced power supply?

2-30. Describe the three major limitations of op amps.

2-31. In a vacuum diode, why will electrons not flow from the plate to the cathode even when the cathode is made positive relative to the plate?

2-32. Does the plate current of a vacuum triode increase or decrease as the grid is made more negative?

2-33. In normal operation, is the grid voltage of a triode held negative or positive relative to the cathode?

3
POWER SUPPLIES

The power supply section of a receiver supplies the proper AC or DC voltages to the various stages of the device. Line-operated receivers require a rectifier to convert the alternating voltage of the AC line to a DC voltage for use by the plate or collector circuits of the tubes or transistors. Filament power must be provided in tube-type receivers. Small, solid-state, portable receivers are almost always battery-operated, and the battery constitutes the major portion of the power supply circuitry.

In this chapter we describe basic half-wave and full-wave rectifier circuits, filters for smoothing the pulsating DC produced by the rectifier, voltage doublers, regulated power supplies, active ripple filters, and a half-wave vacuum tube power supply found in many older receivers.

3-1 TRANSFORMER FUNDAMENTALS

Power supplies commonly involve transformers. In this section we briefly review the relationships between the primary and secondary voltages and currents of a transformer. Also, the isolation properties of a transformer are discussed.

Secondary Voltage and Current

The secondary voltage V_s depends upon the primary voltage V_p and upon the number of turns T_s in the secondary winding relative to the turns T_p in the primary winding. If the secondary contains more (or fewer) turns than the primary, the transformer is a step-up (or step-down) transformer, and the secondary voltage will be greater (or smaller) than the primary voltage. Mathematically,

$$V_s = \frac{T_s}{T_p} V_p \qquad (3\text{-}1)$$

The ratio T_s/T_p is called the *turns ratio*, denoted n. Thus, $V_s = nV_p$.

The current that flows in the secondary is determined by the secondary voltage and the resistance (or circuitry) connected to the secondary. Ohm's law is used with root-mean-square (rms) values of current and voltage.

Multiple Secondaries

Transformers frequently have more than one secondary. In such cases the voltage of each secondary is calculated by determining a turns ratio for each secondary and then using the turns-ratio formula, given above, for each secondary. The same transformer may have both step-up and step-down windings. Power transformers for tube-type receivers, for example, frequently have a high-voltage secondary to supply power for the plate circuitry of the tubes, and a low-voltage secondary to provide power for the tube filaments.

Power Relationships

In an ideal transformer the power delivered to the primary equals the power that leaves the transformer via the secondaries. The applicable unit of power is the *volt-amp*, VA, obtained by multiplying the voltage of a given secondary by the current flowing through it. Hence, the volt-amps delivered to the primary equal the volt-amps flowing out of the secondary:

$$V_p I_p = V_s I_s \qquad (3\text{-}2)$$

This relation gives us a formula for calculating the primary current, I_p:

$$I_p = \frac{V_s}{V_p} I_s \qquad (3\text{-}3)$$

Note that the primary current depends upon the secondary current. If no current flows through the secondary, no current flows through the primary (in an ideal transformer). In practice, owing to losses in the iron core, a small current flows through the primary even when the secondary is open.

For multiple secondary transformers, the preceding formula for the primary current is

$$I_p = \frac{V_{s1} I_{s1} + V_{s2} I_{s2} + \cdots}{V_p} \qquad (3\text{-}4)$$

where as many terms appear in the numerator of the right-hand side as there are secondaries. A sample calculation is given in Figure 3-1.

Isolation Properties of a Transformer

Since there is no direct electrical connection between the primary and secondary windings of a transformer, the secondary is isolated from the primary. Any DC voltage that might be superimposed upon the AC component of the primary voltage will not be passed on to the secondary winding.

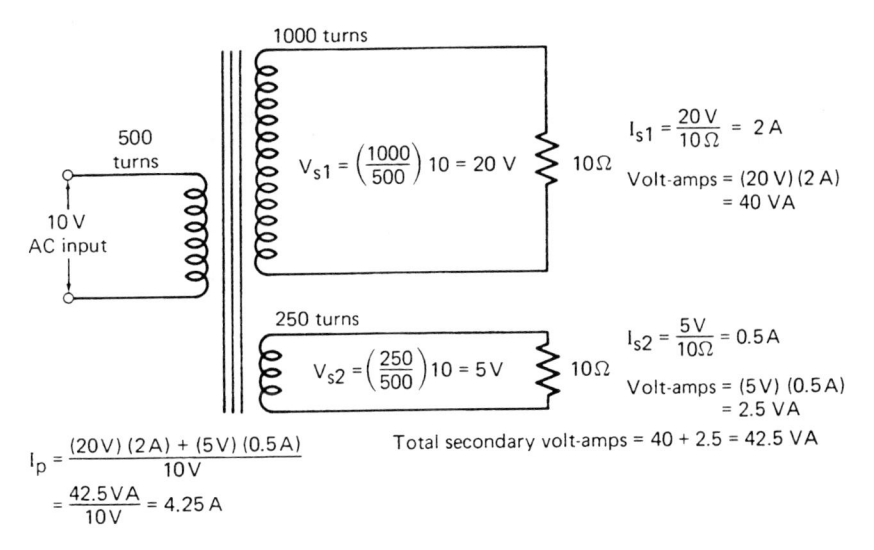

FIGURE 3-1 Example showing the calculation of the secondary voltages, secondary currents, and primary current.

The isolating property of a transformer is useful in preventing disastrous ground loops on the service bench. Many receivers have one side of the AC line connected directly to the chassis, which plays the role of common ground for the receiver circuitry. Typical VTVMs* have the common or ground lead connected to the AC ground associated with the AC line. The VTVM common lead must be connected to the chassis of the receiver under test. Thus, there is a 50-50 chance that the receiver chassis will be charged to the opposite polarity of the VTVM ground. When this extremely dangerous situation exists, a ground loop will be developed when the ground lead of the VTVM is connected to the receiver chassis, and the result is always a rather impressive and damaging short circuit directly across the AC line that tends to melt the clip off the VTVM ground lead.

From the standpoint of personal safety, a serious shock can result if a technician should touch the ground clip of the VTVM while holding the receiver chassis with his other hand. This is dangerous because current flow through the body is from arm to arm, through the chest, fearfully close to the heart. These hazards can be avoided by faithfully using an *isolation transformer*.

An isolation transformer is a transformer with a 1:1 turns ratio for use as shown in Figure 3-2. The isolation transformer isolates the line circuitry of the receiver from the raw AC line voltage at the wall socket. Such transformers are essential for a service shop, no matter how small. All receivers to be serviced should be routinely powered via an isolation transformer.

---

* A VTVM is a vacuum-tube voltmeter. It uses vacuum tubes in a circuit that provides power to drive the swinging-needle meter movement. Transistorized versions of the same instrument may be called TVMs.

Sec. 3-1   Transformer Fundamentals

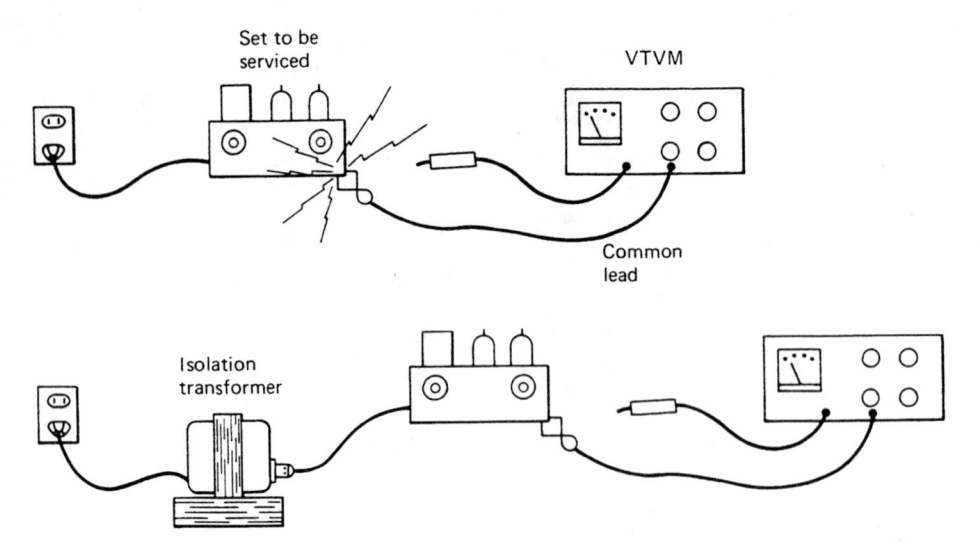

FIGURE 3-2 Isolation transformer is inserted between the wall socket and the set to be serviced.

3-2 HALF-WAVE RECTIFIERS

Basic Circuit

Figure 3-3 is a basic half-wave rectifier that utilizes a transformer as a source of AC voltage and a solid-state diode as the rectifier. Resistor R serves as the load. Point B is grounded, and all voltages are measured relative to this point. During the positive alternation of the AC input voltage, point A goes positive, and the diode is forward-biased into conduction. Electrons flow from ground through the resistor to point C and then through the diode to the secondary winding. Electron flow through the secondary winding is from point A toward ground, point B. The ground connection returns the electrons to point C, completing the circuit.

During the negative alternation, point A becomes negative, and the diode is reverse-biased. No electrons flow through the diode, and consequently no electrons flow in the series circuit, consisting of the diode, the resistor, and the transformer secondary. The diode acts as an open switch, and the entire negative voltage of the transformer secondary is developed across the diode rather than across the load resistor R.

An AC waveform appears at point A, but the waveform at C is that of a pulsating direct current. It is a direct current because the voltage pulses are only of one polarity, but the voltage is not a pure DC voltage because it is not constant. It is a series of positive half-cycles or half-waves. Hence, this is a half-wave rectifier.

The output of this rectifier is not smooth enough for use as the plate voltage in a radio receiver. Something more is required to "fill in the gaps" between the half-wave pulses.

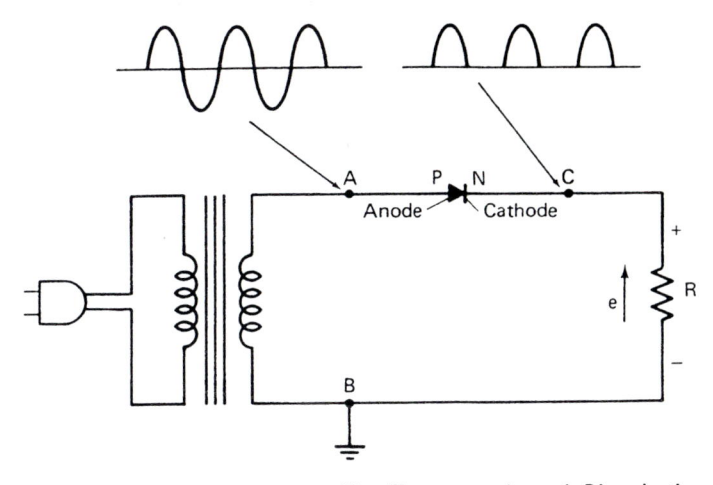

FIGURE 3-3 Basic half-wave rectifier. The current through R is pulsating direct current.

Filter Capacitors

Examine the circuit of Figure 3-4(a), in which a switch S is pulsed on for ½s out of every 3s. The voltage appearing across the 10-kΩ load resistor is shown in part (b) of the figure. The series of pulses may be described as pulsating DC, since the polarity is always the same, but the waveform is anything but that of a pure DC voltage.

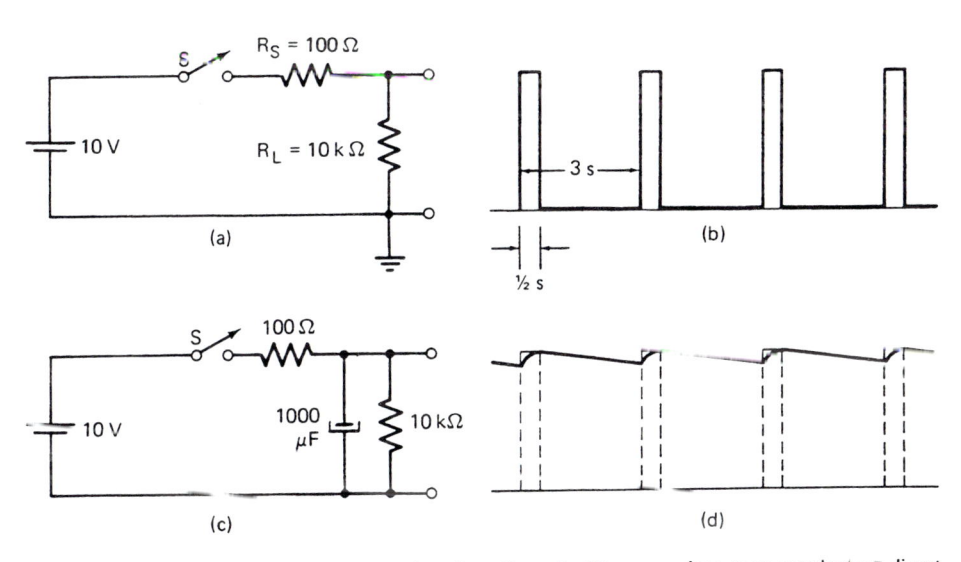

FIGURE 3-4 Demonstration circuit showing the effect of a filter capacitor upon a pulsating direct current.

Sec. 3-2   Half-Wave Rectifiers   51

Figure 3-4(c) is the same circuit except that a *filter capacitor* has been connected across the load resistor. The switch action is the same as before, but note that in Figure 2-4(d), the spaces between the pulses have been filled in. A bit of "roughness" called *ripple* remains in the waveform, but it is a much closer approximation to a pure DC voltage. The filter capacitor is responsible for "smoothing out" the pulses.

The operation of the circuit with the capacitor included is as follows (refer to Figure 3-5). When the switch is closed, the battery attracts electrons through resistor R_S. Electrons are drawn off the positive plate of the capacitor, and the capacitor is quickly charged to maximum voltage. As the voltage is established across the capacitor, current begins to flow through the load resistor. When the switch is opened, the battery is disconnected, but the voltage across the capacitor remains because it is charged. The positive plate has a deficiency of electrons, while the negative plate has an excess of electrons. The positive plate attracts electrons through the load resistor from the negative plate of the capacitor, and current flow through R_L is maintained. As electrons flow from the negative plate through R_L to the positive plate of the capacitor, the charge on the capacitor is slowly depleted. The voltage across the capacitor would eventually go to zero, but before the voltage can decrease significantly, the switch is closed again, recharging the capacitor.

The pulses are smoothed out because the voltage across a capacitor cannot change quickly. A movement of electrons from the one plate of the capacitor to the other is required for the capacitor voltage to change. How quickly this occurs depends upon the resistance between the plates through which the electrons must flow. The battery and R_S constitute the "charging circuit" of the capacitor, and note that the series resistance is only 100 Ω. The charging process therefore proceeds quickly. The capacitor must discharge through the 10-kΩ load resistor, and the discharging occurs much more slowly. In the present circuit the capacitor charges almost completely in about 0.4 s, while about 40 s is required for the capacitor to discharge.

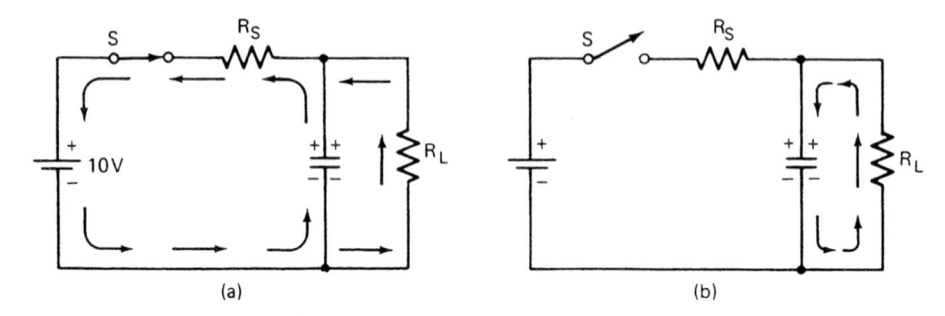

FIGURE 3-5 When the switch is closed, the capacitor is charged quickly to maximum voltage. When the switch is opened, the voltage established across the capacitor maintains the current through R_L.

Half-Wave Rectifier with Filter Capacitors

The circuit of Figure 3-6(a) shows the addition of a filter capacitor to the circuit of Figure 3-3. The action of this circuit is directly analogous to the preceding demonstration circuit with the switch. The diode plays the role of the switch, in that it conducts only during the positive alternation of the secondary voltage at point A. The voltage waveform at point C is shown in the figure. Note the vast improvement. Some ripple voltage remains, but in many applications this simple filter provides enough filtering of the pulsating DC to be acceptable. Generally speaking, however, additional filtering is required.

Figure 3-6(b) illustrates a half-wave rectifier with a two-section filter. Another capacitor has been added, and a filter resistor separates the positive plates of the filter capacitors. The second filter capacitor provides additional reduction of the ripple voltage.

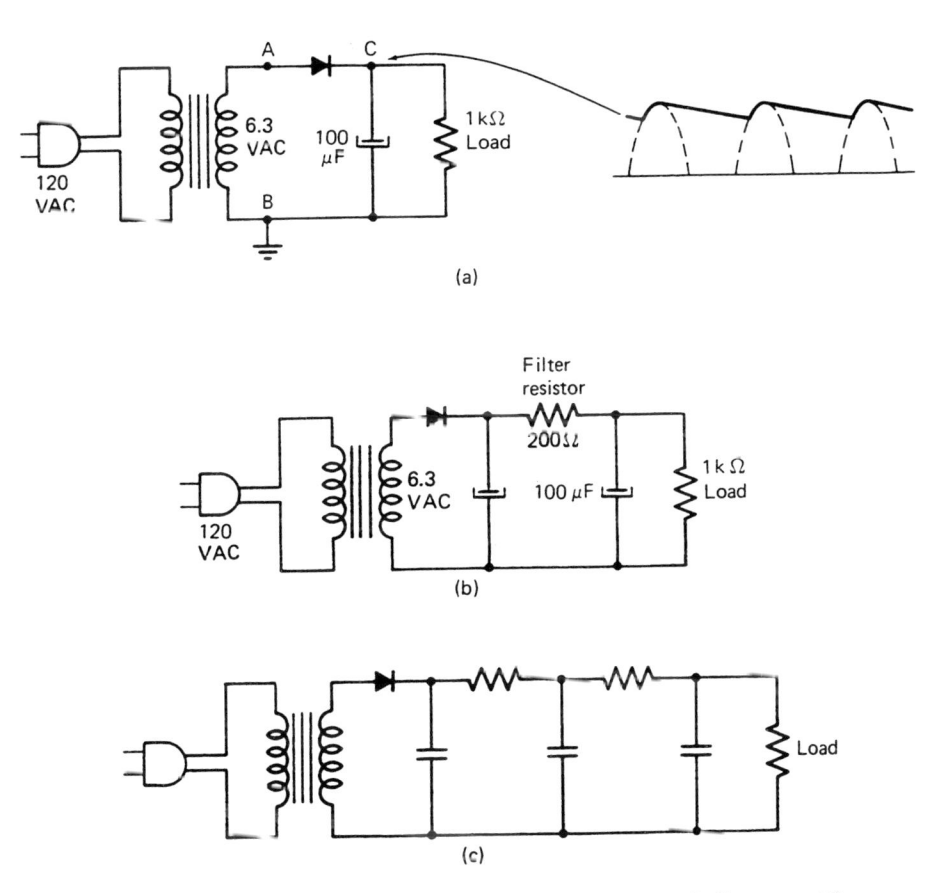

FIGURE 3-6 Simple filter and a two-section filter connected to a half-wave rectifier.

Sec. 3-2   Half-Wave Rectifiers

Surge Current

When power is first applied to a rectifier circuit, all the filter capacitors are uncharged. During the first positive alternation of the input voltage, a large surge current flows through the rectifier to charge the input filter capacitor. The other filter capacitors are separated from the rectifier by the filter resistors, which limit the initial charging current of these capacitors. If the input capacitor is large, and if high voltages are involved, the surge current places considerable stress on the rectifier. Problems may be avoided by choosing a rectifier that can withstand the initial surge of current or by including a resistor in series with the charging path of the input filter capacitor to limit the surge current to safe values. We shall encounter "surge-limiting resistors" in the power supplies of receivers that follow.

Effect of Ripple

The effect of the ripple voltage remaining at the filter output is to introduce a 60-Hz hum into the signal path. This produces a noticeable hum in the speaker. If an unfiltered, pulsating DC voltage were applied to the circuitry of a receiver, the hum level would be so great that the set would be completely inoperative. An experienced technician can listen to the hum of a receiver and determine fairly accurately whether the power supply and audio circuitry are functioning normally.

3-3 FULL-WAVE RECTIFIER USING CENTER-TAPPED TRANSFORMER

The basic full-wave rectifier utilizing a center-tapped transformer is shown in Figure 3-7(a). The center tap of the transformer is grounded, which places the voltage at the tap at 0 V. Voltages at points A and B are measured relative to the center tap, or ground, and are 180° out of phase. When point A is positive, point B is negative. This is a characteristic of center-tapped transformers.

Figure 3-7(b) shows the electron flow at the instant when point A becomes positive. Diode $D1$ is forward-biased and conducts due to the positive voltage at point A, while diode $D2$ is reverse-biased, because of the negative voltage at point B. Note the direction of electron flow through the load resistor, and observe that no current flows through the lower half of the transformer.

Figure 3-7(c) shows the electron flow when the polarity reverses and point B becomes positive while point A becomes negative. $D2$ is now forward-biased while $D1$ is reverse-biased. Current now flows through the lower half of the transformer, but the direction of electron flow through the load resistor is the same as before.

The voltage applied to the load resistor is a series of pulses, as indicated in Figure 3-7(d). Compare this waveform with that of the half-wave rectifier shown in Figure 3-3. The full-wave circuit delivers two pulses per cycle of input voltage, whereas the half-wave system delivers only one. If the line frequency is 60 Hz, the *ripple frequency* will be 120 Hz for the full-wave circuit but only 60 Hz for the half-

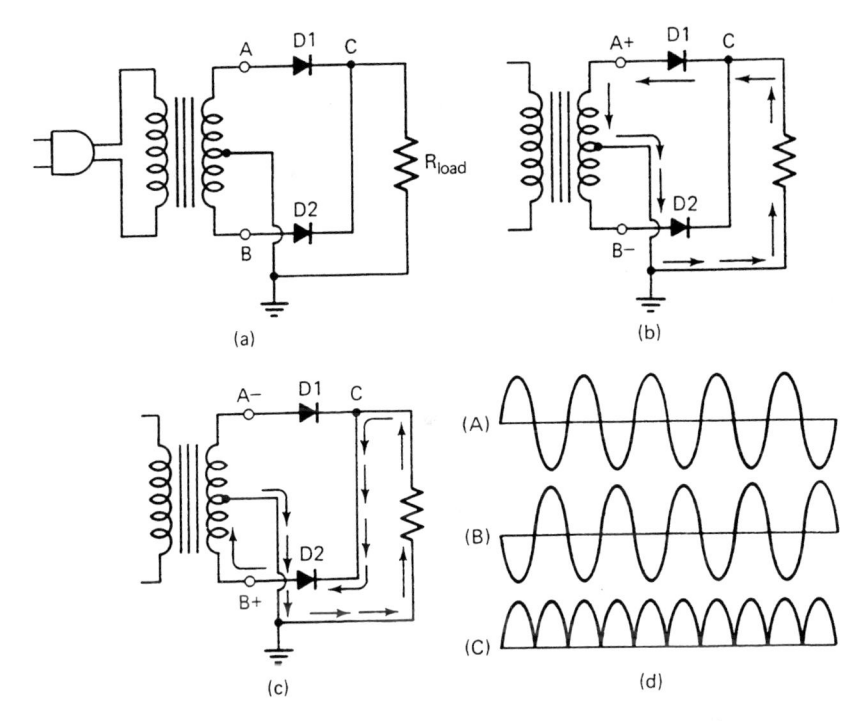

FIGURE 3-7 Basic full-wave rectifier using a center-tapped transformer.

wave circuit. The higher frequency is easier to filter than the lower, and this is one advantage of the full-wave system.

As with the half-wave rectifier, the full-wave rectifier requires a filter to smooth out the pulsating DC to (almost) pure direct current. The same type of filter may be used, and a complete full-wave power supply is shown in Figure 3-8. Component values may be slightly different in light of the higher ripple frequency of the full-wave system, but the operation of the filter circuit is exactly the same as before.

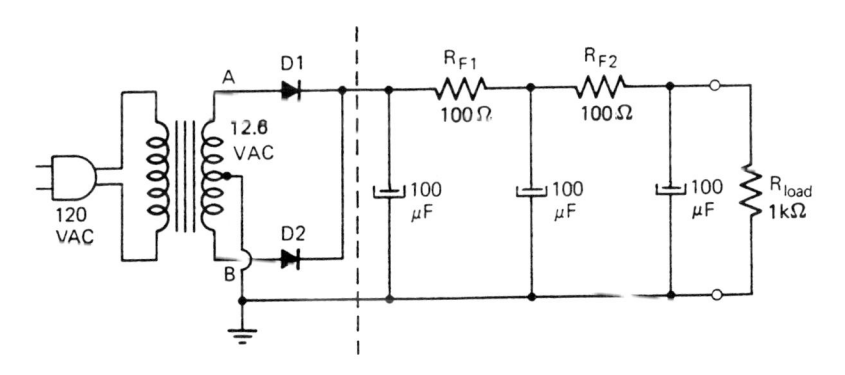

FIGURE 3-8 Complete full-wave power supply.

Sec. 3-3 *Full-Wave Rectifier Using Center-Tapped Transformer* 55

Use of an Inductor as a Filter Element

Filter resistors are commonly used to separate capacitors in a filter circuit, but an inductor is actually better suited to the application than a resistor. An inductor tends to oppose any change in the level of current flowing through it. Consequently, an inductor actively participates in the smoothing of the ripple pulses, whereas a resistor plays a passive role.

A disadvantage of using a resistor as a filter element is that a voltage drop occurs across the resistor because of the current flowing through it, and considerable power is dissipated in the resistor. On the other hand, the DC resistance of an inductor is small in comparison with the filter resistor, so that both the voltage drop and the power dissipation are minimized. The one disadvantage of an inductor for this application is the increased size, weight, and expense.

When used in a power supply, an inductor is called a *choke*, or *filter choke*, because it tends to choke any ripple that tries to sneak through. Typical values of inductance used as power supply filters are 8 or 10 H.

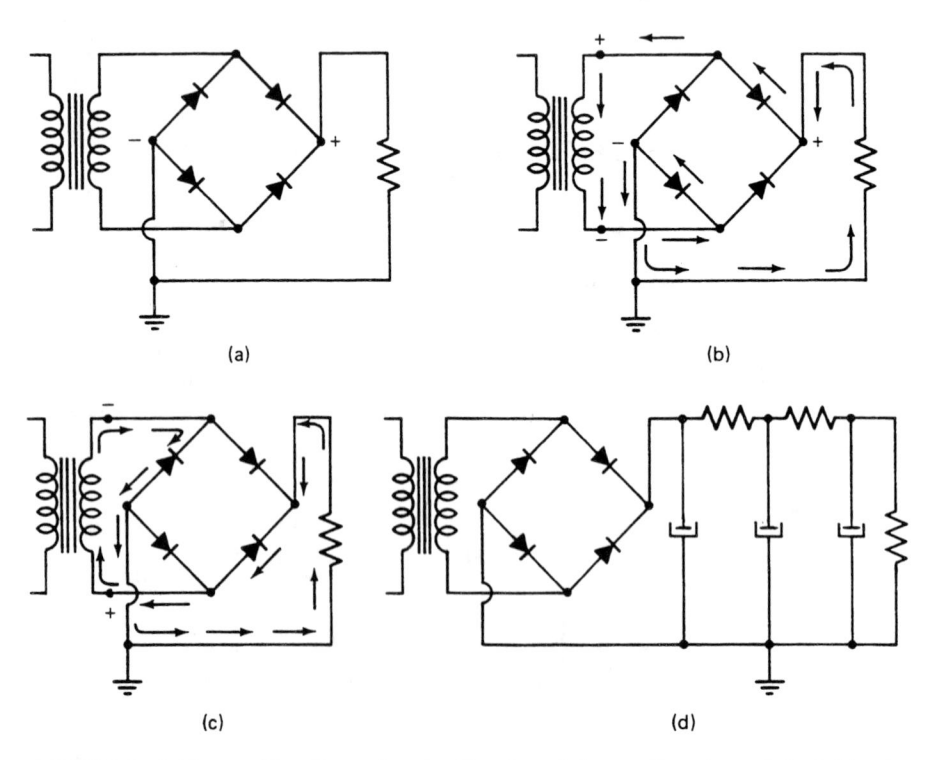

FIGURE 3-9 Bridge rectifier, illustrating the direction of electron flow through the bridge for the positive and negative alternation of the transformer voltage. A complete bridge rectifier power supply is shown in part (d).

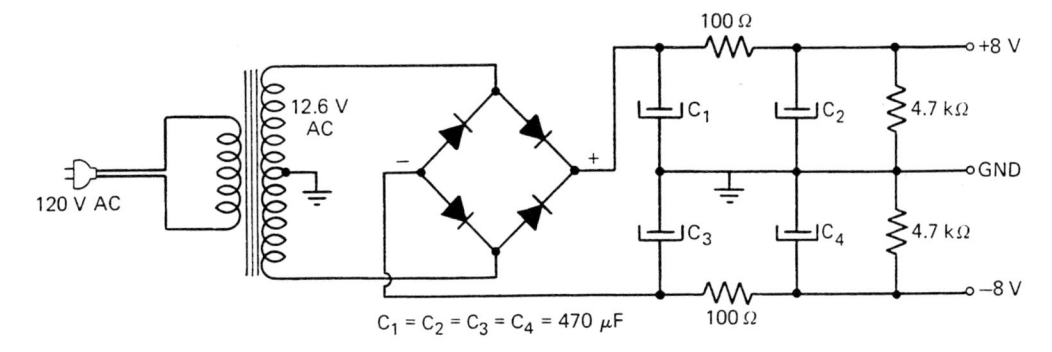

FIGURE 3-10 Variation of the bridge rectifier circuit that provides both positive and negative output voltages relative to ground.

3-4 BRIDGE RECTIFIER

Figure 3-9(a) shows a full-wave bridge rectifier circuit that utilizes four solid-state diodes to achieve full-wave rectification without requiring a center-tapped transformer. Bridge rectifiers are commonly encountered in transistorized receivers. Although it is possible to construct a bridge rectifier using vacuum tubes, such a circuit is never used in consumer devices because of the expense involved in providing suitable filament power. Note that no center tap is present on the transformer secondary, and neither side of the secondary is grounded.

When terminal A of the secondary goes positive, electrons flow through the bridge as indicated in Figure 3-9(b). When the polarity reverses, the electron flow is as indicted in Figure 3-9(c). Note that in both cases the direction of the electron flow through the load is the same. The output waveform is pulsating DC at a ripple frequency of 120 Hz.

A variation of a bridge rectifier circuit is shown in Figure 3-10 that provides both a positive and a negative output voltage. This circuit uses a center-tapped transformer and two sets of filter components. Such circuits are commonly encountered in conjunction with high-power solid-state audio amplifiers that require a balanced power supply. The power supply shown in the figure will serve very nicely as a power supply for an operational amplifier circuit that requires a balanced supply.

3-5 VOLTAGE DOUBLERS

A modification can be made to an ordinary rectifer circuit to produce a *voltage doubler*, which has an output voltage roughly twice that of an ordinary rectifier. Voltage doublers depend upon the ability of a capacitor to store a charge and maintain a voltage across the capacitor until the charge is depleted. There are two types of doubler circuits, the half-wave and the full-wave.

A *half-wave voltage doubler* is shown in Figure 3-11. The transformer secondary provides an AC voltage to the doubler circuit consisting of $D1$, $D2$, and C_S. Resistor R_L constitutes the load, and capacitor C_F smooths the voltage pulses coming from the doubler. The operation of the circuit can be understood by noting what happens during positive and negative alternations of the input voltage (refer also to Figure 3-12).

When point A goes negative, electrons flow to the negative plate of C_S (C_S-) and force electrons to flow off the positive plate of C_S (C_S+). The electrons from C_S+ cannot pass through $D1$, but they can and do pass through $D2$ to ground. The result is that C_S is charged to the peak of the AC voltage with the polarity indicated in the figure. As the negative voltage at A subsides, electrons cannot flow back to C_S+ because of $D2$, and the voltage remains across C_S. As point A becomes less negative, point C gets pushed up in voltage because the voltage across the capacitor remains constant while point A is proceeding upward from the negative peak, toward 0 V.

As point A goes positive, the voltage across the capacitor C_S is in series, adding with the voltage at A. When point A reaches the peak voltage, the voltage at point C will be roughly twice the peak voltage of the AC input. Electrons will be drawn through $D1$ and the load resistor will see the doubled voltage. The electrons that flow through $D1$ pass momentarily onto the positive plate of C_S, and as electrons arrive at C_S+ the voltage of C_S+ decreases somewhat.

When point A goes negative again, the electrons that entered C_S+ via $D1$ are repelled from the capacitor and flow to ground through $D2$. $D1$ prevents electron flow through the load in the reverse direction.

A *full-wave voltage doubler* is shown in Figure 3-13, and this circuit operates by charging capacitors $C1$ and $C2$ alternately to the peak AC voltage. The output voltage is taken from across the series combination of the capacitors and is therefore roughly twice the peak AC voltage. Less filtration is required for the output of a full-wave doubler because the doubler capacitors, $C1$ and $C2$, also serve the function of filtering.

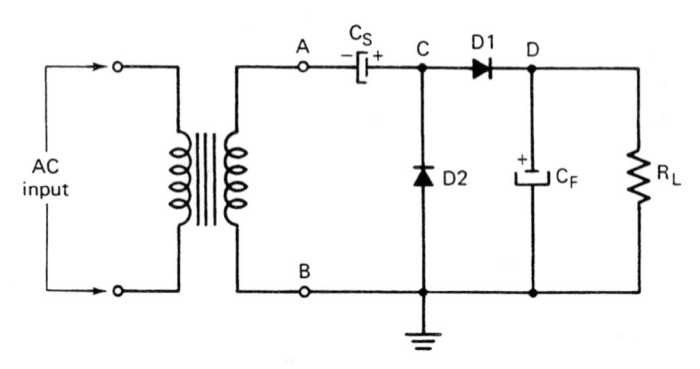

FIGURE 3-11 Basic half-wave voltage doubler.

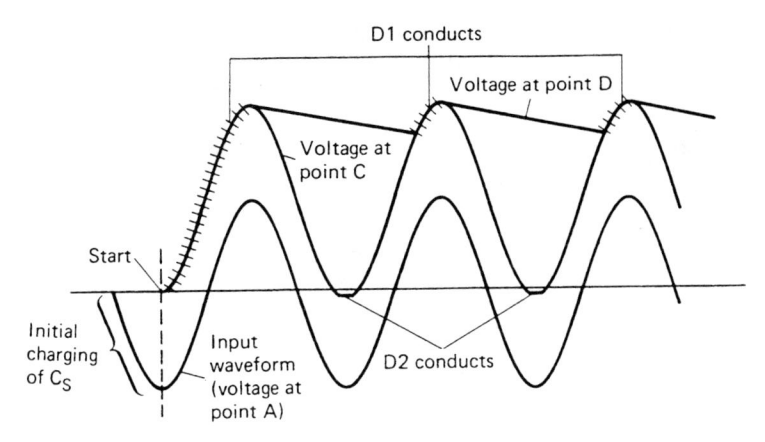

FIGURE 3-12 Waveforms associated with the half-wave voltage doubler.

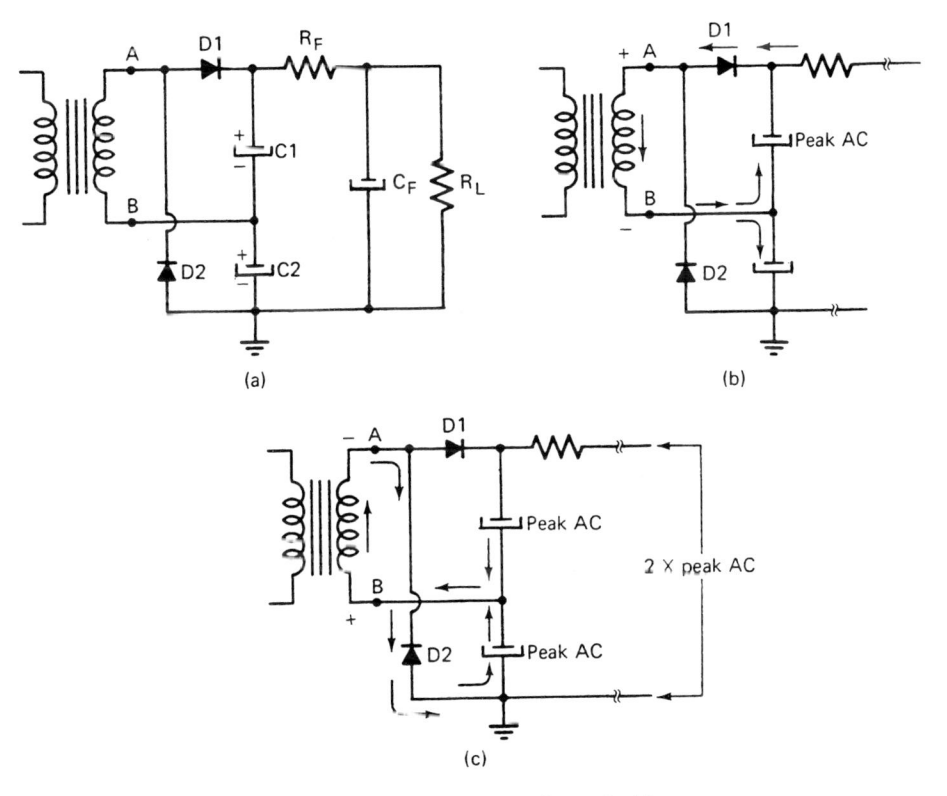

FIGURE 3-13 Full-wave voltage doubler.

Sec. 3-5 　　　　　　　　　　　　　*Voltage Doublers* 　　　　　　　　　　　　　59

When point A goes positive, D1 conducts and charges C1 to the peak AC value. D2 is reverse-biased and does not conduct. Electrons are drawn from C1+ and are forced onto C1−, which connects to the other end of the transformer winding. When the polarity reverses and point B becomes positive, D2 conducts and capacitor C2 is charged to the peak AC value. Electrons that were forced onto C1− are now drawn toward point B and flow through the secondary to point A and then to ground via D2. Thus, C1 is charged on the positive alternation while C2 is charged on the negative alternation.

The output voltage doubler depends rather strongly upon the load current. If a large current flows through the load, the output voltage will be considerably less than twice the peak AC value.

3-6 REGULATED POWER SUPPLIES

The output voltage of a regulated power supply remains essentially constant against variations in either the input voltage or the load current. Voltage-regulator (VR) circuits range from the very simple to the very complex, depending upon the requirements of a given situation. Gaseous VR tubes, such as the OB2 and OC3, constitute the regulating elements of vacuum-tube technology, whereas zener diodes perform the same function in the realm of solid state.

Zener Regulators

Zener diodes are characterized by the sharpness of the reverse breakdown characteristic as compared with that of a conventional diode. This is illustrated in Figure 3-14(a). In the breakdown region, the voltage developed across the zener is essentially independent of the current flowing through it. This property makes zeners useful as voltage regulators, and as such, they are always operated in the reverse breakdown region.

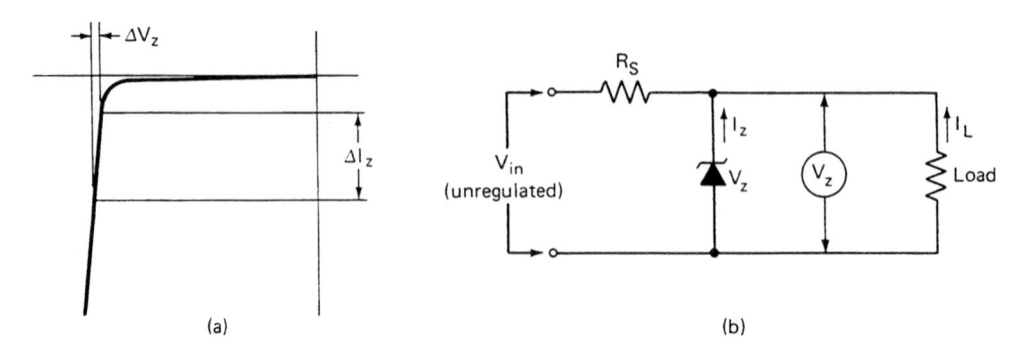

FIGURE 3-14 (a) In the reverse breakdown region of a zener diode, a small change in voltage ΔV_z produces a large change in zener current ΔI_z. (b) Basic zener regulator circuit. The voltage applied to the load is regulated at the zener voltage V_z.

Figure 3-14(b) is the basic zener regulator circuit. An unregulated DC input voltage V_{in} is applied to a series resistor R_S and the zener diode. Note the symbol for the zener and that it is reverse-biased. The output voltage is the zener breakdown voltage V_Z. A current I_L is drawn by the load, and the sum of the load current I_L and the current flowing through the zener flows through the series resistor R_S.

The output voltage is regulated against variations in load current I_L and input voltage V_{in}. First, consider the effect of an increase in V_{in}, assuming that I_L remains constant. As V_{in} increases, the voltage across the zener increases "microscopically," but this small increase produces a significant increase in zener current. This gives a larger current through R_S, and as a result of the larger current, the IR drop across R_S increases. The larger voltage drop across R_S almost exactly offsets the increase in input voltage, so that a constant voltage is maintained at the output (neglecting the "microscopic" increase in the voltage across the zener).

Conversely, if the input voltage decreases, the zener current decreases, owing to a microscopic decrease in the voltage across the zener. This produces a small IR drop across R_S, and once again the output voltage is held very close to the original value.

An upper limit exists for the input voltage because, as V_{in} increases, so does the power dissipation in the zener. The power dissipated by the zener equals the product of the zener current and zener voltage. If the maximum power dissipation capability (in watts) of the zener is W_{max}, the maximum zener current is W_{max}/V_Z. The following formula gives the upper limit for the input voltage:

$$V_{in(max)} = \left(I_L + \frac{W_{max}}{V_Z}\right) R_S + V_Z \qquad (3\text{-}5)$$

If the input voltage exceeds this value, the zener is apt to be overheated and destroyed.

The zener current decreases as V_{in} decreases, and the minimum allowed V_{in} occurs when the zener current becomes zero. Further reduction of V_{in} brings the zener out of the breakdown region, and regulation is lost as the output voltage falls below V_Z, the zener voltage. The following formula gives the minimum allowed value of the input voltage:

$$V_{in(min)} = I_L R_S + V_Z \qquad (3\text{-}6)$$

Variation in Load Current

We now turn our attention to the effect of a variation in I_L, assuming that the input voltage V_{in} is held constant. As I_L increases, a "microscopic" drop in voltage across the zener occurs as a result of increased current through R_S. The small drop in voltage across the zener produces a reduction in zener current almost exactly equal to the increase in I_L. Thus, the current through R_S increases only microscopically, and the output voltage remains essentially constant at V_Z. Conversely, when I_L decreases, the zener current increases in an effort to hold the same current flow through R_S.

Limits also exist for allowed variations in the load current I_L. As I_L increases, I_Z decreases, and the maximum permitted load current occurs when I_Z becomes zero. A further increase in load current causes a loss of regulation, because the increased IR drop across the series resistor reduces the voltage applied to the zener to a value below the regulating voltage V_Z. The following formula gives the maximum allowed value of the load current:

$$I_{L(max)} = \frac{V_{in} - V_Z}{R_S} \qquad (3\text{-}7)$$

Note that an increasing load current produces a smaller power dissipation in the zener. Conversely, a decrease in load current imposes a greater power dissipation upon the zener. This is because I_Z must increase to compensate for the decreasing load current. The following formula gives the minimum load current allowed for a given circuit:

$$I_{L(min)} = \frac{V_{in} - V_Z}{R_S} - \frac{W_{max}}{V_Z} \qquad (3\text{-}8)$$

If, in a given situation, the minimum load current calculates to be a negative number, the interpretation is that the load current could be reduced to zero with a margin of safety.

Pass Transistor Regulator

From the foregoing discussion, it is apparent that the simple zener circuit is somewhat limited in regulating ability, especially in those situations where the load current is large in comparison with the zener current. A major improvement is represented by the circuit of Figure 3-15, which utilizes a pass transistor as a control element for the current flowing through the load. The zener circuit consisting of R_S and V_Z holds the base voltage of the pass transistor constant at the zener voltage

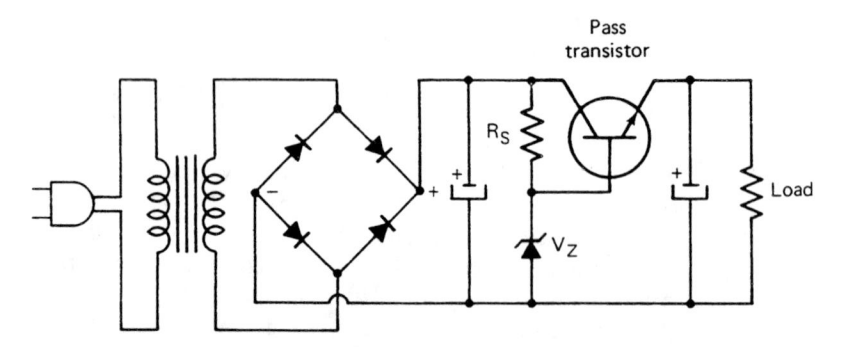

FIGURE 3-15 Regulated power supply that uses a pass transistor to increase the current capability of the zener circuit.

V_Z. The current load placed on the zener circuit by the pass transistor is only the base current of the transistor; most of the load current flows to the collector. Thus, the zener circuit current is much smaller than the load current, and regulation over a much wider range of load currents can be achieved.

In operation, the action of the transistor causes the emitter voltage to "pull up" to within a fraction of a volt of the base voltage. For a silicon transistor, the emitter voltage will be less than the base voltage by about 0.7 V. If a variation in load resistance occurs so that the load current increases, the emitter voltage will drop slightly. The increase in base-emitter voltage then causes the transistor to conduct a greater current, preventing a further drop in emitter voltage.

The maximum current capability of this circuit depends primarily upon the current-carrying capability of the pass transistor. Considerable power may be dissipated in the transistor if the input voltage is high and if the current is large. The power dissipation in watts of the pass transistor is the product of the voltage across the transistor (from collector to emitter) and the current flowing through it.

Op Amp Regulated Power Supply

The pass-transistor regulator of the preceding section allows the output voltage to decrease slightly as the current load increases. The circuit of Figure 3-16 uses an op amp to overcome this difficulty. Also, it presents a constant load to the zener circuit so that zener voltage variations are almost eliminated.

The input transformer and bridge rectifier are standard. Note the symbol for the bridge rectifier. Only one filter capacitor C_F is needed because the regulator reduces the ripple to an insignificant level. Observe how supply power for the op amp is obtained.

The zener circuit supplies a constant +6 V to the (+) input of the op amp. The op amp output is driven positive, and the pass transistor conducts due to the positive voltage applied to its base by the op amp. As the transistor conducts, the emitter terminal is pulled up to a positive voltage, and this constitutes the output of the power supply.

Just how positive will the output become? It will go more and more positive until the voltage applied to the (−) input of the op amp equals the voltage applied to its (+) input, namely 6 V, from the zener circuit. This implies that the output of the voltage divider (R_T and R_B) must equal V_Z. Because the voltage at B is only two-thirds that at A, the output voltage must therefore be 9 V (at point A) in order to have 6 V appear at point B.

Note that the base-emitter voltage drop (about 0.6 V) of the pass transistor does not enter into the determination of the output voltage. The op amp output rises above the voltage at point A just enough to exactly cancel the B-E voltage drop. Recall that the op amp will do whatever is necessary to achieve the equality of voltages at the (+) and (−) inputs. By obtaining the voltage for negative feedback from the emitter of the pass transistor rather than from the op amp output directly, the B-E voltage drop is avoided.

To see how voltage regulation occurs under a current load, suppose that a small resistance (say about 100 Ω) is connected to the output terminals of the supply.

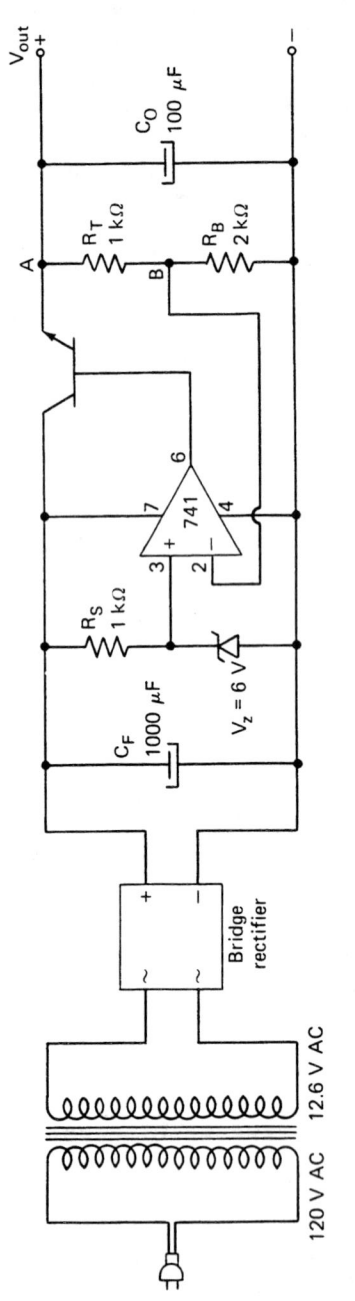

FIGURE 3-16 Regulated power supply that uses an op amp to drive a pass transistor.

Then, about 90 mA worth of electrons will be dumped onto the output terminal and onto point A, and the voltage at point A will decrease a small amount. This decrease will be felt at point B, and the voltage at the $(-)$ input of the op amp will fall under the 6 V at the $(+)$ input. The op amp, a difference amplifier, will respond to this difference in voltage at the $(-)$ and $(+)$ inputs, the result being that the op amp output voltage will rise. As it rises, the pass transistor conducts more, pulls more electrons from its emitter, and the output voltage rises. This corrects the decrease in voltage that occurred when the additional load was connected.

If the load current suddenly becomes less, the output voltage will tend to rise. However, the rise will quickly be counteracted by a sequence of events just the opposite of those described above.

The output voltage is set by the voltage-divider resistors R_T and R_B in conjunction with the zener voltage V_Z. It may be calculated via

$$V_{out} = \left(1 + \frac{R_T}{R_B}\right)V_Z \qquad (3\text{-}9)$$

which is the same formula as for the gain of a noninverting op amp amplifier. Indeed, this circuit may be regarded as a noninverting amplifier that amplifies the voltage from the zener circuit.

Two-Transistor Voltage Regulator

In Figure 3-17 is shown a zener-stabilized voltage regulator that uses a zener diode to stabilize the voltage at the emitter of a driver transistor $Q2$. The voltage at the base of $Q2$ is determined by a voltage divider consisting of R_T and R_B that is driven by the output voltage V_{out}.

The purpose of the zener diode is to hold the emitter of $Q2$ at the zener voltage V_Z. For this to happen, however, the zener must be broken down so that it conducts.

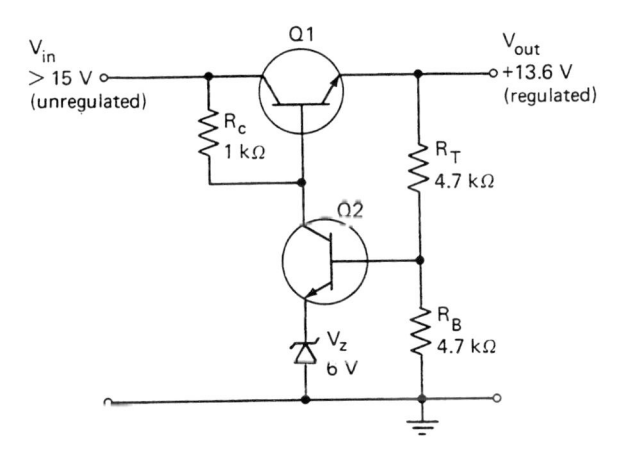

FIGURE 3-17 Two-transistor regulated power supply.

This means that $Q2$, as well, must conduct due to the series connection of the zener diode and $Q2$.

For $Q2$ to be turned on and conduct, the voltage at the base of $Q2$ must be V_{BE} (about 0.7 V) higher than the voltage at the emitter of $Q2$. Because the base voltage is provided by the R_T-R_B voltage divider, the output of the divider must be higher than V_Z by about 0.7 V. This implies that the output voltage V_{out} must be even higher in order to turn $Q2$ on.

At this point, let us see how the conduction of $Q2$ affects the level of conduction in the pass transistor $Q1$. Ignoring $Q2$ for the moment, we note that resistor R_c is connected from the high-voltage input to the base of $Q1$. This, by itself, turns $Q1$ on, and conduction through $Q1$ will make the output voltage rise. But when $Q2$ conducts, electrons are dumped onto the base of $Q1$ so that the voltage at the base of $Q1$ is lowered. In other words, the conduction of $Q2$ opposes the action of resistor R_c in turning $Q1$ on and in making the output voltage go more positive.

We can now see how the circuit acts as a voltage regulator, and a good one. If V_{out} is too low, $Q2$ will not be conducting, and the resistor R_c will turn $Q1$ on, making the output voltage V_{out} rise. But when V_{out} rises to its regulating value, $Q2$ is turned on just enough to stabilize the voltage at the base of $Q2$, preventing further increase in the output voltage.

If, for some reason, the output voltage becomes too high, then $Q2$ will conduct more and bring the voltage at the base of $Q1$ down. This, in turn, lowers the output voltage.

A formula for calculating the output voltage is

$$V_{out} = \left(1 + \frac{R_T}{R_B}\right)(V_Z + V_{BE}) \qquad (3\text{-}10)$$

Note the similarity of this formula to that given in the preceding section for the op amp regulated supply.

3-7 THREE-TERMINAL REGULATORS

Complete voltage regulators are now available in IC form that include all the regulator circuitry in a single package. These regulators have only three terminals, an input for unregulated DC, the output, and ground. The voltage at which they regulate is set at the time of manufacture. For example, an LM309 is a 5-V, 1-A regulator in a TO-3 case intended for digital systems requiring a regulated 5-V supply. The 7805, 7808, and 7812 regulate at 5 V, 8 V, and 12 V, respectively. Because packages are used of the same style as for transistors, it is easy, at first glance, to mistake a three-terminal regulator for a power transistor. But, obviously, they are not the same; one is a transistor and the other is an IC.

Three-terminal regulators are easy to use, almost foolproof, but at least three precautions need to be observed. First, the input unregulated DC must always be

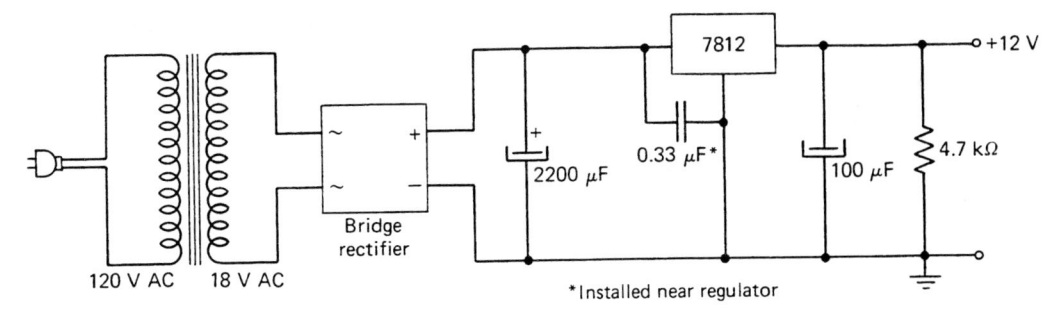

FIGURE 3-18 A 12-V regulated power supply that utilizes a 7812 three-terminal regulator.

about 2 V higher than the regulated DC output voltage. This includes the bottom of the ripple valley, the point of greatest concern. Failure to maintain the 2 V or so of "headroom" will allow the regulator output voltage to decrease.

The second precaution stems from the high-gain amplifier inside the regulator, the error amplifier. It can break into oscillation at very high frequencies if long wires (more than just an inch or so) separate the regulator terminals from the power supply filter capacitors. The oscillation can lead to destruction of the regulator, or it can contaminate the output voltage with a large high-frequency signal that can bring chaos to the circuits the regulator is driving. The oscillation can be avoided by connecting small tantalum or disk ceramic capacitors (0.1 µF to 1 µF) from the regulator terminals to ground, installing the capacitors as close to the regulator terminals as possible.

The third consideration is true for any power-handling semiconductor: you must keep the device cool. Heat sinks and adequate ventilation are essential even though semiconductors can run impressively hot (to the touch) and still survive. Most regulators have internal circuitry that will shut the device down if the internal temperature exceeds the rated limit. A good rule is that if holding your thumb against the device becomes unpleasant after a few seconds, it is running too hot and should be provided with a better heat sink.

A typical circuit is shown in Figure 3-18, in which a 7812 provides a regulated 12 V at the output. With a good heat sink, the supply will handle current loads up to 1 A.

Negative Regulators

A negative regulator must be used if a power supply is to provide a negative output voltage. Typical of these are the 7905, 7908, and 7912, which are the complementary regulators to the 78XX series. The balanced power supply of Figure 3-19 utilizes an LM320, the negative regulator, in conjunction with an LM340, the positive regulator.

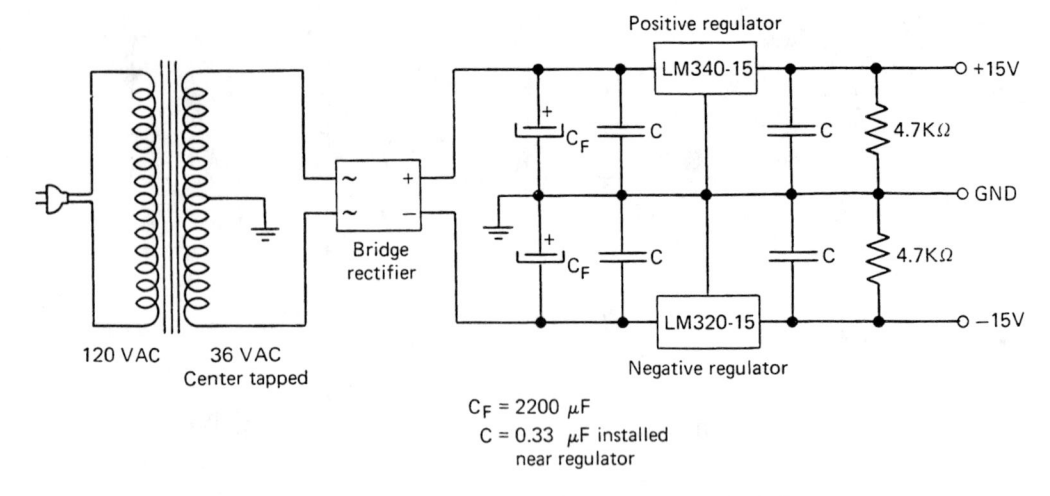

FIGURE 3-19 Balanced power supply that utilizes both a positive and a negative three-terminal regulator.

3-8 ACTIVE RIPPLE FILTERS

A capacitor provides the basic physical mechanism for removing ripple from the output of a rectifier: the voltage across the capacitor cannot change quickly. An inductor may also be used because the current through an inductor cannot change quickly. As we have seen, very effective filters can be designed using these components.

In practice, the current rating of a power supply has a direct bearing on the size of the filter capacitors required to reduce the ripple voltage to a satisfactory level. Large current levels require large capacitances; capacitances of thousands of microfarads are not unheard of. Large capacitors are heavy and expensive.

The idea behind an active ripple filter is this: use an active device such as a BJT to assist the capacitor in smoothing the ripple. That is, use a transistor to make a small capacitor to appear larger. We might even refer to this as an amplification of capacitance.

A circuit in which this effect occurs is shown in Figure 3-20. It is a zener-regulated pass transistor, but note the capacitor connected from the base to ground. If the capacitor can help to hold the base current constant, the emitter current will also be held constant. Because the base current is only about 1% of the emitter current, the capacitor connected to the base can be comparatively small. But it produces the same effect on the ripple as a much larger capacitor connected between the emitter and ground. In fact, the capacitance appears to be multiplied by a factor equal to the beta of the transistor.

If the beta of the pass transistor is 100, then a 10-μF capacitor connected to the base acts like a 1000-μF capacitor connected to the emitter. Circuits such as this are often encountered in modern receivers.

Actually, the three-terminal regulators described earlier use active filtering. This is why the input circuits typically use only a single filter capacitor. A typical regulator reduces the ripple by a factor of 1000 or more.

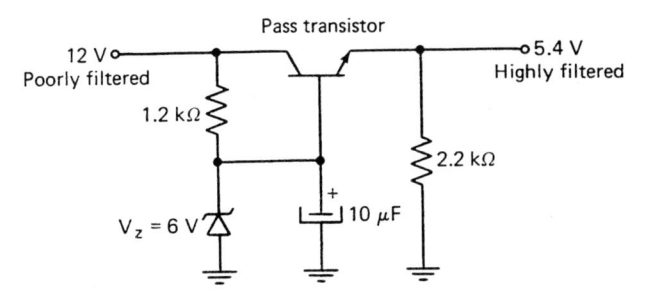

FIGURE 3-20 Active ripple filter.

3-9 VACUUM-TUBE RECTIFIERS

Basic vacuum-tube (VT) and solid-state power supplies are very similar. The major differences are that a vacuum diode requires a source of power for heating the cathode, and VT circuits typically operate at much higher voltages than solid-state circuits. VT power supplies are found only in older receivers in which power supplies tend to be rather unsophisticated in comparison to modern solid-state supplies. Half-wave rectifiers are commonly encountered. Full-wave rectifiers almost always are those using a center-tapped transformer in conjunction with a dual-diode rectifier tube. Bridge rectifiers are almost never found because of the expense involved in providing heater power for the four rectifier cathodes.

In this section we describe only one type of VT power supply, a transformerless half-wave circuit. The other circuits are straightforward in their operation so that it is easy to see how they work when they are encountered. Therefore, we will not dwell on them here.

Vacuum-Tube Half-Wave Rectifier

Half-wave rectifiers are seldom found in transformer-type power supplies but are quite common in small, five-tube AC/DC receivers. The power supply section of one of these receivers is shown in Figure 3-21.

Near the line plug is an interlock, a plug-and-socket combination built into the rear panel of the cabinet. When the panel is removed, power is automatically disconnected, for reasons of safety. The on-off switch is mounted on the volume control and is in series with the low side of the AC line which connects to the chassis. As is typical of these receivers, one side of the AC line connects directly to the chassis, and therefore the chassis is hot. Connected to the other side of the line is a 0.05-μF 600-V line filter capacitor that attenuates electrical noise that may be present on the line.

Filament String

The filaments of all tubes in the receiver are connected in series so that the series combination can be supplied directly by the AC line. This is possible because the sum of the filament voltages of all the tubes is equal to the line voltage; the filament

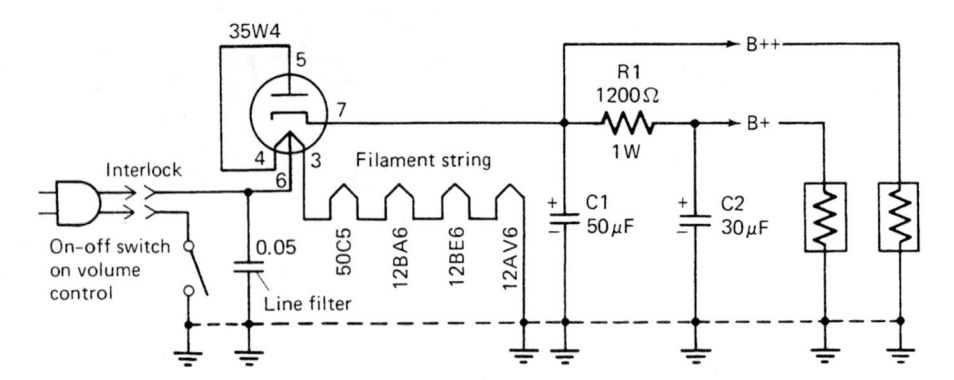

FIGURE 3-21 Half-wave power supply of a small five-tube AC/DC receiver.

of each tube requires the same filament current (0.15 A) as all others in the string; and the warm-up times of all the tubes are very nearly the same. It is standard practice to draw the filament string in the power supply section of the schematic rather than to draw the filaments inside the tube symbols. This avoids a cluttered schematic.

The rectifier tube is the familiar 35W4, which has a tapped filament. The tap is not in the electrical center of the filament, however, but is located about 6 V away from pin 4 when 35 V is applied across pins 4 and 3. In this circuit the high side of the AC line connects to the tap (pin 6), and, therefore only the portion of the filament between pins 6 and 3 is part of the filament string. Hence, only about 29 V is dropped across the rectifier filament.

Plate Circuit

The portion of the filament between pins 4 and 6 acts as a resistor between the high side of the AC line and the rectifier plate. It functions as a surge-limiting resistor to limit the initial charging current that flows through the rectifier tube to input filter capacitor $C1$. It also serves as a fuse in the event that a serious short circuit develops in the rectifier circuitry. A voltage drop of only a few volts occurs across the resistance, and therefore an AC voltage nearly equal to the line voltage is applied to the rectifier plate.

On positive alternations of the line voltage, the rectifier plate attracts electrons from the cathode, making it rise to a positive voltage that nearly equals the voltage applied to the plate. On negative alternations, no conduction through the rectifier occurs, since the plate is negative relative to the cathode. Thus, a pulsating DC voltage appears at the cathode, and this voltage is applied to the filter network, consisting of $C1$, $C2$, and filter $R1$.

Filter Circuit

The filter circuit is a two-section filter with two voltage sources, the B++ and the B+. The B++ voltage is somewhat higher than the B+ and is applied to the plate circuit of the audio-output stage. The B++ has a higher ripple content than the B+, but the audio-output stage is less sensitive to ripple voltage than the other stages, which are supplied by the more highly filtered B+.

In Figure 3-21 the receiver circuitry is denoted by resistances enclosed in boxes. Dashed lines connect the ground symbols to emphasize that they are connected together. It is instructive to trace the electron flow around the circuit at the instant the rectifier plate is positive. This is left as an exercise.

The B++ voltage may vary between 110 and 130 V, depending upon the current drawn by the load in receivers of various designs. The fact that the DC output voltage of the rectifier (130 V) may be greater than the AC input voltage (120 V) might be surprising, but recall that the AC voltage given is the rms value, not the peak value. If the rms voltage is 120 V, the peak voltage is 1.414 times this, or nearly 170 V. A voltage drop occurs across the rectifier tube from plate to cathode, but the input filter capacitor easily charges to levels above the rms voltage.

SUMMARY

1. The secondary voltage of a transformer is calculated using the turns-ratio formula. If more turns appear on the secondary than on the primary, the secondary voltage will be greater than the primary voltage and the transformer is a step-up transformer.

2. A transformer may have more than one secondary winding. In such a case, the turns-ratio formula is applied to each secondary in turn.

3. In step-down transformers, the secondary current is larger than the primary current by a factor equal to the turns ratio. The reverse is true for a step-up transformer.

4. The volt-amps delivered by a secondary of a transformer is calculated by multiplying the voltage produced by that secondary by the current flowing through that secondary.

5. The primary current is computed by dividing the sum of the secondary volt-amps by the primary voltage.

6. An isolation transformer has a turns ratio of 1 and is an important item on any service bench.

7. A half-wave rectifier involves only one diode and is active only on the positive alternations of the input AC waveform. The ripple frequency of a half-wave rectifier is 60 Hz.

8. A full-wave rectifier may involve a center-tapped transformer or a bridge circuit. The rectifier circuit delivers a pulse to the filter on both the positive and negative alternation of the input waveform. The ripple frequency is 120 Hz.

9. A power supply having both a positive and a negative output voltage can be made by using a bridge rectifier in conjunction with a center-tapped transformer.

10. In applications where current demands are small, voltage doublers may be used to provide DC voltages on the order of twice the rms voltage of the input waveform. The actual voltage output from a voltage doubler depends strongly upon the current loading of the device.

11. A voltage doubler functions through the use of capacitors in conjunction with the rectifier diodes. In the half-wave doubler, the capacitor is charged and is then placed in series with the input waveform (effectively). In the full-wave doubler, the two capacitors in series are alternately charged to near the peak AC input voltage; the output voltage is taken from across the series combination.

12. A zener diode may be used to form a simple voltage regulator. As such, the zener is operated in the reverse-bias region, and a series resistance is required. The regulating ability of the zener is limited in the simplest circuit.

13. Voltage regulation over a wider range of currents can be achieved with a regulator circuit that utilizes a pass transistor.

14. An op amp can be used to drive the base of a pass transistor in a regulated power supply. The zener circuit drives the (+) input of the op amp while a fraction of the regulated output voltage is fed back to the (−) input.

15. Three-terminal regulators are complete regulators in IC form that regulate to a fixed voltage set by the internal circuitry. A negative regulator is required if a negative output voltage is desired.

16. An active ripple filter uses an active device to multiply the effect of a comparatively small filter capacitor.

17. In vacuum-tube sets, filament power supplies are usually AC. Typically, they are simply series or parallel connections of tube filaments to a source of AC voltage.

18. With the exception of the filament (heater) circuitry, vacuum-tube power supplies are similar to their solid-state counterparts, but they typically operate at much higher voltages.

QUESTIONS AND PROBLEMS

3-1. A step-down transformer whose turns ratio is 6:1 has 120 V AC applied to the primary.
   (a) What voltage appears on the secondary?
   (b) If the transformer is an ideal transformer, what current will flow in the primary when the secondary circuit is open?
   (c) If a 10-Ω resistor is connected to the secondary, what current will flow in the secondary?
   (d) What current will then flow in the primary?

3-2 The primary winding of a transformer consists of 2000 turns. The secondary has only 500 turns. An AC voltage of 100 V is applied to the primary winding. The secondary is connected to a 5-Ω resistor.
   (a) What voltage appears on the secondary?
   (b) What current flows in the secondary?
   (c) Calculate the volt-amps delivered by the secondary.

(d) Compute the volt-amps delivered to the primary by the 100-V source.

(e) What current flows in the primary winding?

3-3. The primary of a transformer is connected to a 120-V AC source. The transformer has two identical secondaries, and each outputs a voltage of 12 V AC. A 6-Ω resistor is connected to one secondary; a 4-Ω resistor is connected to the other.

(a) Draw the schematic diagram of the circuit.

(b) What is the turns ratio for each secondary?

(c) Calculate the current flowing in each secondary.

(d) Compute the volt-amps delivered by each secondary.

(e) What is the total volt-amps delivered to the primary winding?

(f) Find the primary current.

3-4. A 3-A fuse is located in the primary circuit of a transformer whose step-down ratio is 10:1. If there is only one secondary on the transformer, what current will have to flow in the secondary in order to blow the fuse in the primary circuit?

3-5. A step-down transformer steps 120-V line current down to 12 V for application to the rectifier of a battery charger. It is desired that the charger be fused on the primary side of the transformer, and it is desired that the circuit should be interrupted when the secondary current reaches 10 A. What size of fuse should be incorporated into the primary circuit?

3-6. (a) What is the turns ratio of an isolation transformer that is used on the bench of a good service technician?

(b) What is the purpose of an isolation transformer?

3-7. The filter capacitor of a certain half-wave rectifier power supply charges to 90% of the peak AC voltage applied to the rectifier. If the rectifier is powered by the 12-V secondary of a transformer, what voltage will appear across the capacitor?

3-8. How many pulses per second come from:

(a) A half-wave rectifier?

(b) A full-wave rectifier?

3-9. (a) When power is first applied to a power supply, why does a surge of current occur through the rectifier?

(b) What may be done to limit this surge current to a safe level?

3-10. Suppose that a 12-V center-tapped transformer is used to build the following types of rectifiers. Estimate the output voltage of each. (Assume moderate current loading.)

(a) A half-wave rectifier as in Figure 3-6 (center tap open).

(b) A full-wave rectifier that uses two diodes, as in Figure 3-8.

(c) A full-wave rectifier that uses a bridge, as in Figure 3-9.

(d) A balanced power supply as in Figure 3-10.

3-11. What is the ripple frequency for each circuit in Prob. 3-10?

3-12. If one diode in a bridge rectifier were to become "open" so that it does not conduct, what would be the effect on the operation of the circuit? Would the bridge continue to operate at all?

3-13. Draw diagrams of a bridge rectifier and show that when even one diode of the four becomes shorted, the secondary of the transformer is shorted one-half the time. What would be the effect upon the transformer? Upon the diodes of the bridge?

3-14. What characteristic of a zener diode makes it useful as a voltage regulator?

3-15. In normal operation as a voltage regulator, is a zener diode operated in the reverse-bias region or in the forward-bias region?

3-16. A zener diode whose zener voltage is 12 V conducts a current of 24 mA. What is the power dissipation in the zener?

3-17. If the reverse current through a zener diode is doubled, by what percentage is the power dissipation in the zener increased?

3-18. When a silicon rectifier diode conducts moderately in the forward direction, about 1 V (or slightly less) is developed across the diode. If such a diode conducts a current of 500 mA, what will be the power dissipation in the diode?

3-19. Why is it that a simple zener regulator, as in Figure 3-14(b), drops out of regulation after the load current increases beyond a certain level?

3-20. Explain the advantage of using a pass transistor in conjunction with a zener regulator.

3-21. From what source is power supply voltage obtained for the op amp in the op amp-regulated power supply of Figure 3-16?

3-22. Can the regulated power supply of Figure 3-16 output a greater voltage than the voltage V_z at which the zener breaks down?

3-23. Why is it good practice to solder small tantalum or disk ceramic capacitors close to the terminals of a three-terminal regulator?

3-24. What determines the output voltage of a three-terminal regulator?

3-25. Explain the difference in application between a positive regulator and a negative regulator.

3-26. How is it that the active ripple filter of Figure 3-19 appears to amplify the capacitance connected to the base terminal of the transistor?

3-27. What will be the effect in the circuit of Figure 3-8 if one diode becomes open? If one diode shorts completely?

3-28. What would be the effect in the circuit of Figure 3-20 if a short circuit develops between the filament and cathode of the rectifier tube?

3-29. In the circuit of Figure 3-20, what would be the likely effect of a shorted filter capacitor? What would be the effect of an open filter resistor?

3-30. In the circuit of Figure 3-21, what would be the effect of a shorted line filter capacitor?

3-31. What is the likely effect of application of an alternating current to an electrolytic capacitor? Of connecting an electrolytic capacitor into a DC circuit with the polarity reversed?

3-32. What is the advantage of using an inductor as a filter element in a power supply? What are the disadvantages?

4
BIASING ARRANGEMENTS

In order for a tube or transistor to function properly as an amplifier, the device must be properly biased. That is, the proper DC voltages must be applied to the terminals of the device to establish the desired DC operation under no-signal conditions. The circuit elements primarily concerned with the DC parameters constitute the biasing arrangement. Typically, circuit elements used as part of the biasing arrangement also affect the AC operation of a given amplifier stage so that the biasing and signal-handling aspects are not completely separable. Nevertheless, the first step in understanding a given amplifier is to recognize and understand the type of biasing arrangement used.

This chapter is devoted to biasing arrangements for bipolar junction transistors (BJTs), junction field-effect transistors (JFETs), metal-oxide-silicon field-effect transistors (MOSFETs), and vacuum triodes. The material is very important because almost all discrete amplifier stages in receivers use one of the biasing arrangements (or a variation) described in this chapter. If biasing arrangements are well understood, the underlying principles of the receiver are greatly simplified. That is worth working for.

4-1 GENERAL CONSIDERATIONS

Amplifiers and Signals

In Chap. 2 we saw that a BJT is capable of acting as a current multiplier: the base current is multiplied by the beta β of the transistor to give the collector current. Here is an extension of this idea: if the base current is caused to vary, the resulting variation of the collector current will be beta times as large as the variation in base current. This provides the basis for designing an amplifier using a BJT as the active device.

Everyone knows what an amplifier does. If a small signal is applied to the input of an amplifier, a larger version will appear at the output. We say the signal has been *amplified* (made larger in amplitude) by the amplifier. This is the purpose of amplifiers—to make signals larger.

But what, exactly, is a *signal*? We often hear of signals. Signal generators, microphones, and phonograph cartridges produce signals, and a signal applied to an earphone or to a speaker will produce sound. We can even see a signal displayed on an oscilloscope. So what is a signal? A signal is a variation in voltage. More precisely, a signal is a voltage variation that is used to convey some type of information.

There are a few situations where a variation in current may constitute a signal, but in almost all cases in dealing with radio receivers, a signal is a variation in voltage.

We can now restate the function of an amplifier. If a small variation in voltage is applied to its input, a larger variation in voltage will be produced at its output.

The Requirements of a Biasing Arrangement

Before a BJT will act as a current multiplier, a positive (for *NPN* devices) voltage of at least several volts must be applied to the collector, and the base-emitter junction must be forward-biased so that a base current will be established. When these two conditions are met, the collector current will be beta times as large as the base current.

The circuit that provides the proper voltage for the collector and for the base-emitter junction is called a biasing arrangement, and it is an important part of any amplifier circuit. Vacuum-tube and field-effect transistor (FET) amplifiers also require proper biasing arrangements. These may differ somewhat in detail, but many similarities may be noted among biasing arrangements for different types of active devices.

The Q-point

The "*Q*" in *Q*-point stands for quiescent—calm, peaceful—and in electronics, "quiescent conditions" refers to the situation where no signal is applied to the input of an amplifier. Accordingly, the *Q*-point collector voltage, for example, is the steady voltage that appears at the collector when no signal (variation in voltage) is present. The design of a particular biasing arrangement will set the *Q*-point values of collector voltage, collector current, base voltage, base current, and so forth.

The collector represents the point at which the output signal is developed for many BJT amplifier circuits. That is, the collector voltage rises and falls as directed by the input signal to form the output signal. The *Q*-point collector voltage represents the center point of the variation. Therefore, it is necessary for the *Q*-point collector voltage to be centrally located so that the collector voltage can rise and fall equal amounts in reproducing a symmetrical signal. Similar considerations apply to the *Q*-point plate voltage for tubes, and to the *Q*-point drain voltage for FETs.

Dynamic Range

This expression refers to the range over which the input and output voltages can vary in the course of amplifying a signal. On the output side, the collector voltage cannot rise to a level greater than the power supply voltage, and it cannot fall to a level lower than the voltage at the base. Hence, these two voltage levels represent the ends of the dynamic range for the output. On the input side, the dynamic range is determined, in part, by the forward conduction characteristic of the base-emitter junction, and it may be as small as 100 mV for some circuits.

4-2 BIASING REQUIREMENTS OF BIPOLAR JUNCTION TRANSISTORS

The biasing arrangement of a BJT must provide the proper voltages to the base, emitter, and collector in order to establish the proper Q-point collector current and to provide an adequate dynamic range for the amplifier stage. Further, because of the variability of transistor parameters, the biasing arrangement should be such that the operating characteristics of the stage should not be strongly dependent upon the transistor parameters. The gain and input impedance, for example, should not change significantly when the beta changes from, say, 80 to 125.

Most single-transistor biasing arrangements can be identified as one of about seven or eight basic types, the exact number depending upon what one considers basic. In more complicated direct-coupled amplifiers in which two or more transistors are DC-connected, the biasing schemes are not so easily categorized, and the amplifiers do not lend themselves to simple analysis.

The relationships between the emitter, base, and collector voltages required for *NPN* and *PNP* transistors are illustrated in Figure 4-1 using voltage plots. Observe that the voltage relationships required for *PNP* transistors are identical to those required for *NPN* transistors, with the exception that the polarity is inverted. Also, note the base-emitter voltage, denoted V_{BE}.

Biasing the Base-Emitter Junction

A primary function of a biasing arrangement is to provide forward bias for the base-emitter junction. An *NPN* silicon transistor requires that the base be held about 0.7 V more positive than the emitter (0.25 for germanium). The constancy

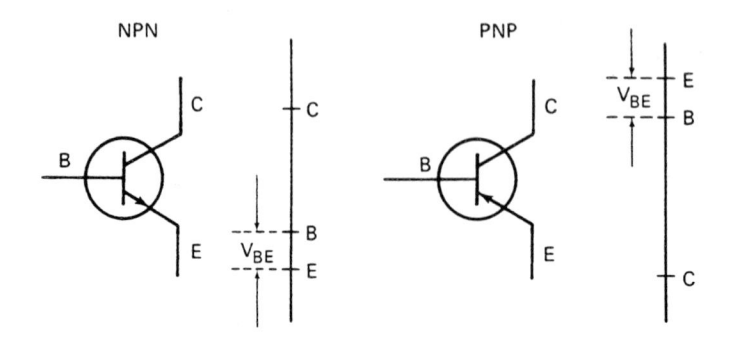

FIGURE 4-1 Biasing requirements of bipolar junction transistors (BJTs).

of the base-emitter voltage V_{BE} greatly simplifies the analysis of a biasing arrangement. When properly biased, a silicon transistor will always exhibit a V_{BE} of about 0.7 V, and we may assume this at the beginning of an analysis of a particular biasing arrangement.

V_{BE} varies slightly as current conduction through a transistor changes, and power transistors exhibit larger values of V_{BE} than do small, low-power voltage amplifiers. For silicon, a V_{BE} of 0.60 V implies a small conduction, while a V_{BE} of 0.80 (or slightly greater) implies heavy conduction. Corresponding voltages for germanium are on the order of 0.15 and 0.40 V. Maximum normal V_{BE} values are seldom greater than about 1.1 V for silicon and about 0.5 V for germanium. If larger values are encountered, the device exhibiting the large V_{BE} should be considered to be suspect. This provides the service technician with a quick check of a transistor that may be made without removing the transistor from the circuit.

4-3 BASE BIAS

Figure 4-2 shows the simplest biasing arrangement applicable to a bipolar transistor (BJT). The name *base bias* derives from the fact that a single resistor connects to the base, providing the positive (for *NPN*) voltage required by the base in order to establish the *Q*-point collector current. The collector load resistor R_L serves as a current-to-voltage converter; it causes the variations in collector current to produce a varying voltage—a signal—at the collector. Hence, the amplified signal is developed at the collector. Capacitors *C1* and *C2* are coupling capacitors. They block any DC component that is present in the signal while letting the AC component, the variation in voltage, pass through. *C2* blocks the DC component of the collector voltage, preventing it from appearing at the output terminal of the amplifier.

$$I_B = \frac{V_{CC} - V_{BE}}{R_B}$$

$$I_C = \beta I_B$$

$$V_B = V_{BE}$$

$$V_C = V_{CC} - I_C R_L$$

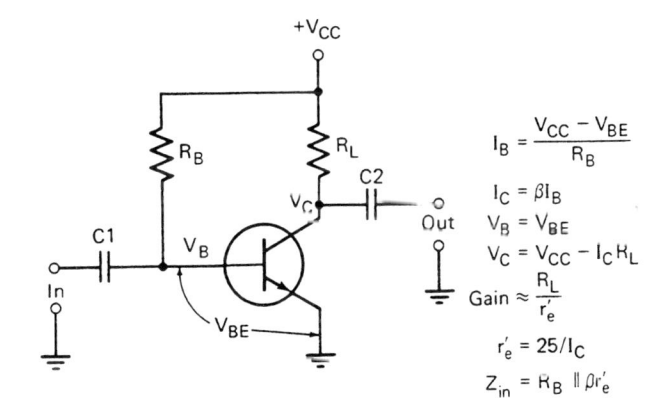

$$\text{Gain} \approx \frac{R_L}{r'_e}$$

$$r'_e = 25/I_C$$

$$Z_{in} = R_B \parallel \beta r'_e$$

FIGURE 4-2 Base-bias arrangement. This is the simplest biasing arrangement for a bipolar transistor, but it has no natural stability against variations in transistor parameters.

The voltage at the base must be above ground an amount V_{BE} since the emitter is grounded. This implies that for a silicon transistor the voltage at the base will be 0.7 V. Since the voltage across the base resistor is $(V_{CC} - V_{BE})$, and since the current flowing from the base terminal is the same as the current flowing through the base resistor, the base current of the transistor is given by $I_B = (V_{CC} - V_{BE})/R_B$. From this we conclude that the base current is determined by the power supply voltage and the ohmic value of the base resistor. Since the collector current of a transistor is always beta times the base current, the collector current for the base-bias circuit is $I_C = \beta I_B$.

The voltage at the collector is the power supply voltage minus the voltage drop that occurs across the collector load resistor. Since the collector current flows through the collector load resistor, the voltage developed across that resistor is $I_C R_L$, and therefore the collector voltage V_C is given by $V_C = V_{CC} - I_C R_L$.

Signal Amplification

The signal voltage arriving at the base via the coupling capacitor is a small AC voltage that is superimposed upon the DC bias voltage present at the base. The relatively small voltage variation on the base produces a corresponding variation of the base current. The current-amplifying action of the transistor converts the small variations of base current to much larger variations of the collector current. The collector current rises and falls in phase with the signal voltage applied to the base.

The collector-current variations produce a varying voltage drop across the collector load resistor R_L. The voltage at the top of the R_L is fixed at the power supply voltage, and since the voltage drop across R_L varies, the voltage at the collector varies also. The collector-voltage variation constitutes the output signal of the stage.

The signal voltage at the collector is inverted in phase relative to the signal applied to the base. An increase in base voltage produces an increase in base current. An increase in base current produces an increase in collector current, which causes the collector voltage to decrease. Thus, as the base voltage increases, the collector voltage decreases. Waveforms illustrating the basic process of amplification are shown in Figure 4-3.

AC Junction Resistance

The signal component of the base current results from the signal component of the voltage applied to the base-emitter junction. An AC junction resistance for the base-emitter junction may be defined as the ratio of the applied signal voltage to the resulting signal current. A given signal voltage applied to a large AC junction resistance produces a small signal current, while the same signal voltage applied to a small AC junction produces a larger signal current.

The ohmic value of the AC junction resistance depends upon the DC voltage applied to the junction. Refer to the forward conduction characteristics of a *PN* junction shown in Figure 4-4, and note that the slope of the curve is not constant.

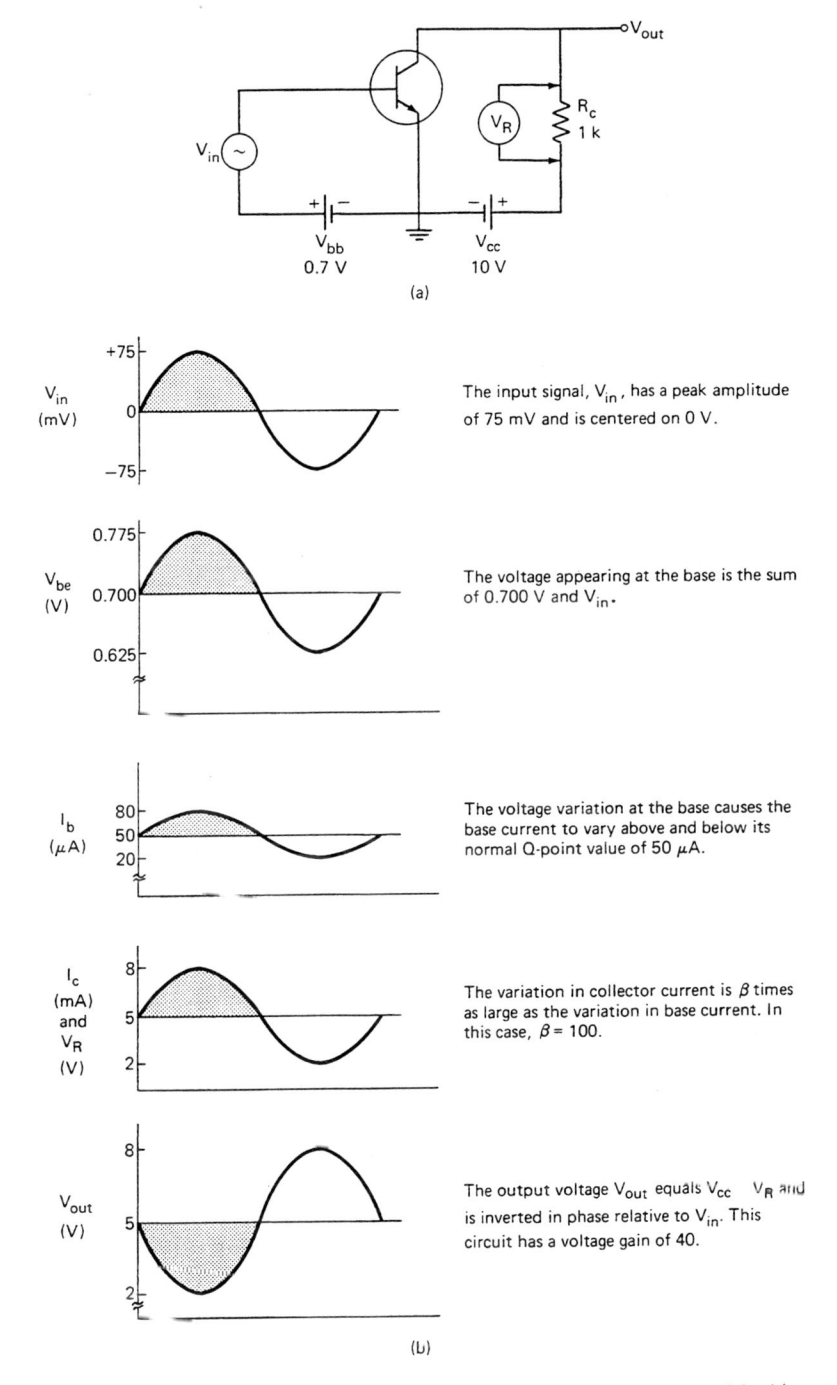

FIGURE 4-3 (a) Demonstration circuit using two batteries (or power supplies) for biasing; (b) waveforms associated with a simple amplifier.

The input signal, V_{in}, has a peak amplitude of 75 mV and is centered on 0 V.

The voltage appearing at the base is the sum of 0.700 V and V_{in}.

The voltage variation at the base causes the base current to vary above and below its normal Q-point value of 50 µA.

The variation in collector current is β times as large as the variation in base current. In this case, $\beta = 100$.

The output voltage V_{out} equals $V_{cc} - V_R$ and is inverted in phase relative to V_{in}. This circuit has a voltage gain of 40.

Sec. 4-3 Base Bias

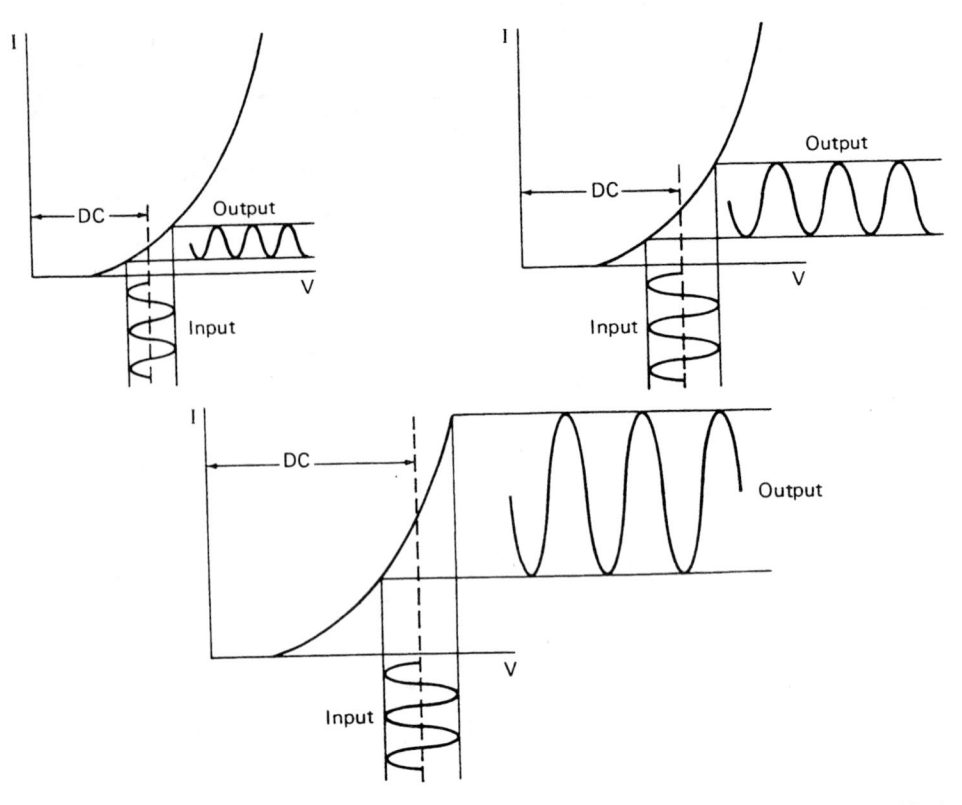

FIGURE 4-4 AC current resulting from an AC voltage applied to a *PN* junction varies with the average DC level of the applied signal. Consequently, the AC junction resistance varies with the DC level.

When a small AC signal is superimposed upon a DC voltage and the combination is applied to the junction, the resulting AC signal current depends upon the DC level at which the signal is applied. This illustrated in Figure 4-4, in which the AC signal is applied to different portions of the curve.

As long as the signal component is small, the portion of the curve traversed by the signal voltage appears nearly linear and the junction behaves like a resistance. Changing the DC voltage component moves the signal component to a different portion of the curve, where the slope is different. Consequently, the resistance appears to change as the DC level is shifted.

The AC junction resistance of an ideal junction at room temperature is given approximately by $r'_e = 25/I_{mA}$, where r'_e is the AC junction resistance in ohms and I_{mA} is the DC current flowing through the junction, measured in milliamps. When 1 mA flows through an ideal junction, the AC junction resistance is 25 Ω. In practice, the junction resistance is a variable and unpredictable parameter and varies over a range of about 2 to 1. Consequently, the numerator in the preceding formula varies from about 25 to 50.

Demonstration Circuit

A circuit that demonstrates the effects of the AC junction resistance of a diode is shown in Figure 4-5, in which the variable AC resistance of the diode is used to vary the output voltage. A small AC signal is coupled to a diode that has a DC voltage applied by variable, low-voltage power supply V. Resistor $R1$ keeps the incoming AC signal from being shunted to ground through the power supply. $R2$ plays the role of a load resistor. The direct current through the diode is given by $V/(R_1 + R_2)$. If an AC signal in the millivolt range is applied to the input, the amplitude of the signal appearing across $R2$ can be varied by changing the DC voltage of the power supply.

Certain precautions are in order in operating the circuit. If the AC signal is too large, severe distortion will occur as the diode begins to act as a half-wave rectifier. If the applied DC voltage is too large, the operating point will be in the linear region of the characteristic, and no significant change in signal level will occur as the DC level is varied.

Input Resistance to Base

We can visualize the AC junction resistance as shown in Figure 4-6, where a fictitious resistance has been added between the base and the emitter. The AC resistance of the base-emitter junction of an operating transistor is much greater than r'_e, the AC junction resistance of an isolated PN junction. To an AC signal applied to the base, the junction appears as a resistance of $\beta r'_e$. The amplifying action of the transistor causes r'_e to be multiplied by the beta of the transistor.

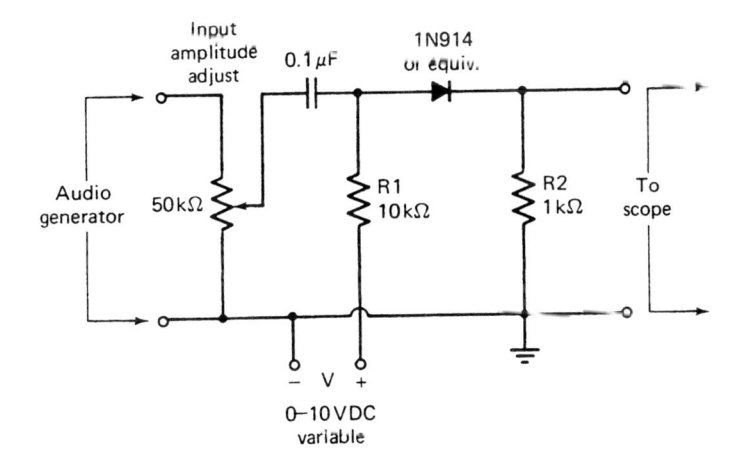

FIGURE 4-5 Circuit that demonstrates the variation of AC junction resistance with the DC level. With the input amplitude set at a constant level, the amplitude of the waveform displayed on the scope will vary with the DC voltage applied to the diode.

Sec. 4-3　　　　　　　　　　Base Bias

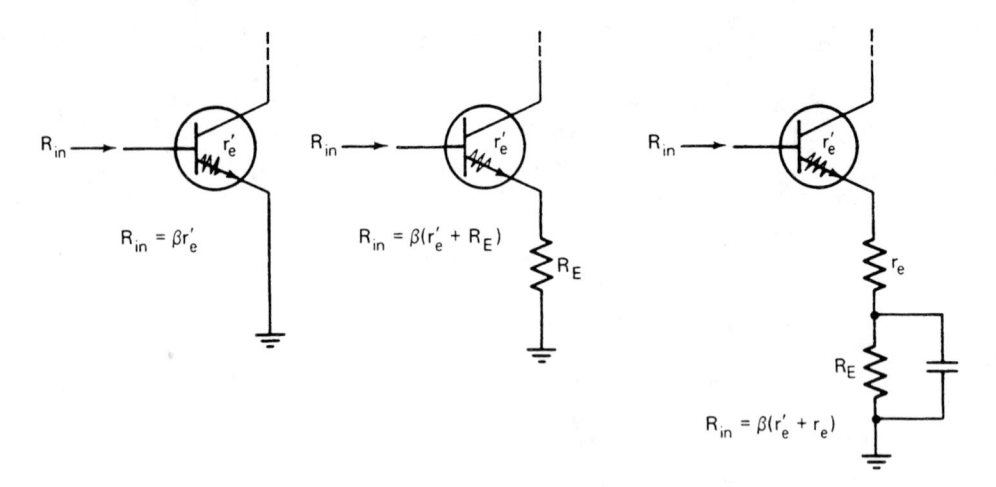

FIGURE 4-6 Input resistance to the base R_{in} depends upon the beta of the transistor, the AC junction resistance, and any unbypassed resistance in the emitter lead of the transistor.

Therefore, for the simple base-bias circuit, the input resistance to the base is $\beta r'_e$. For circuits in which the emitter is not grounded, a slightly more complicated expression results, given in Figure 4-6. In short, the input resistance to the base equals β times the sum of r'_e and any unbypassed resistance that appears in the emitter circuit. This is illustrated in the following sections.

Input Impedance Z_{in} for Base Bias

A signal appearing at the input of the circuit of Figure 4-7 can flow to ground through two paths as indicated in the figure. These paths are in parallel, and therefore the input impedance of the amplifier is the parallel equivalent of the two paths. The formula for the input impedance is given in the figure. The parallel bars $\|$ indicate that Z_{in} is the parallel equivalent of R_B and $\beta r'_e$.

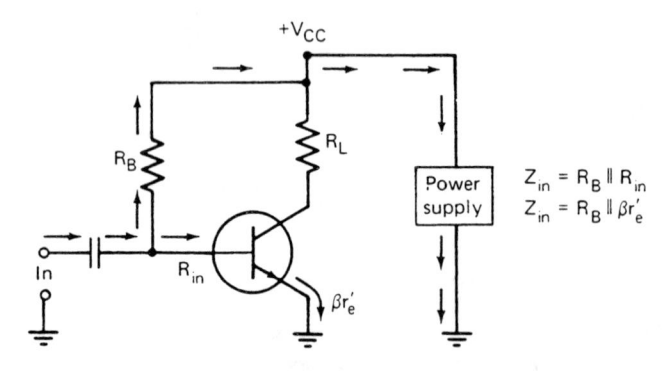

FIGURE 4-7 AC signal applied to the input can flow to ground through two parallel paths, as shown.

The input impedance is important because, for a given input voltage, it determines the power that must be supplied to the amplifer to drive it properly. A low input impedance requires more power than a high input impedance. This fact becomes important for certain sources, such as crystal microphones, for example, which are capable of delivering very little power. A low-input-impedance amplifier can load the source to the point where the output voltage of the source is seriously diminished.

Gain Formula for Base Bias

The AC junction resistance r'_e plays an important role in determining the voltage gain of the base-biased amplifier stage. Specifically, the gain is given by the ratio of the collector load resistor R_L to the AC junction resistance r'_e: gain (base bias) = R_L/r'_e. This is an approximation, given without proof.

EXAMPLE 4-1: Figure 4-8 is an analysis of a base-bias circuit. Note the *voltage plot*. Such plots are helpful in visualizing the voltage relationships of a circuit.

Disadvantages of the Base-Bias Circuit

The base-biasing arrangement suffers from disadvantages of such significance that it is seldom found in commercial circuits. The collector current depends directly upon the beta of the transistor. Variations in beta cause the Q-point collector current and voltage to be rather unpredictable, since the base-bias circuit has no inherent stability.

The beta of a transistor increases with temperature. If a transistor begins to overheat, the beta increases and produces a corresponding increase in the collector

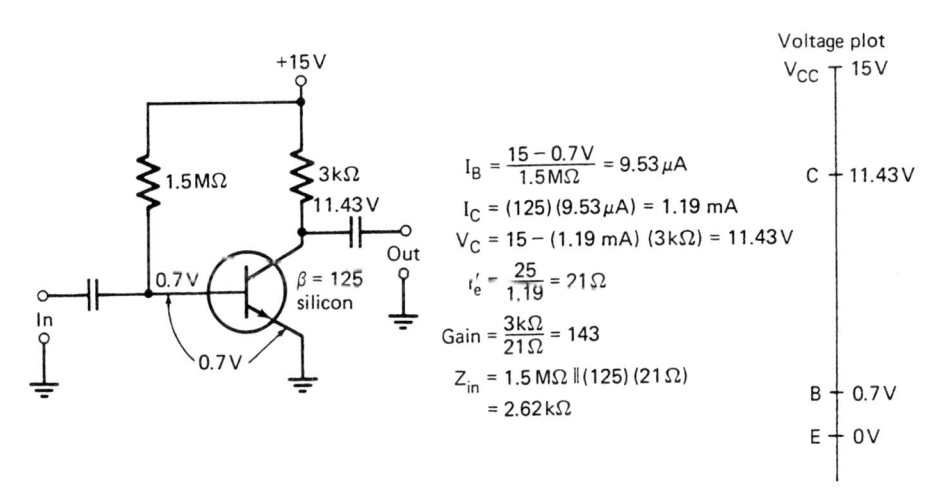

$$I_B = \frac{15 - 0.7V}{1.5M\Omega} = 9.53\,\mu A$$

$$I_C = (125)(9.53\,\mu A) = 1.19\text{ mA}$$

$$V_C = 15 - (1.19\text{ mA})(3k\Omega) = 11.43V$$

$$r'_e = \frac{25}{1.19} = 21\,\Omega$$

$$\text{Gain} = \frac{3k\Omega}{21\Omega} = 143$$

$$Z_{in} = 1.5\,M\Omega \parallel (125)(21\,\Omega)$$
$$= 2.62\,k\Omega$$

FIGURE 4-8 Sample calculation and voltage plot for a base-bias arrangement. The gain actually obtained in practice will depend somewhat upon the output impedance of the signal source supplying the input signal to the amplifier.

current. This, in turn, increases the power dissipation in the transistor, which causes it to heat up even more. This process, called *thermal runaway*, can continue until the transistor is destroyed.

The gain of the base-bias circuit depends strongly upon the AC junction resistance, which is capable of wide variation. Thus, the gain tends to be unpredictable. This is an undesirable trait from the point of view of a manufacturer who desires identical performance of all circuits produced.

4-4 BASE BIAS WITH EMITTER FEEDBACK (BBEF)

The frequently encountered circuit of Figure 4-9 is similar to the simple base-bias circuit, but emitter resistor R_E gives the circuit a stability against variations in beta and the AC junction resistance the previous circuit does not have.

A voltage given by $I_C R_E$ is developed across the emitter resistor so that the emitter voltage V_E is given by $I_C R_E$. Here we assume the emitter and collector currents are equal, a good assumption since the base current is small in comparison.

By writing, mathematically, the fact that V_{BE} plus the voltages developed across R_B and R_E must equal the power supply voltage, a formula for calculating the collector current I_C can be obtained. This formula is given in the figure. The collector current depends upon the power supply voltage, the emitter resistor R_E, and the term R_B/β. The collector current is much less sensitive to changes in beta than in the simple base-bias circuit.

The collector voltage V_C is calculated with the same formula as before, but the base voltage is the sum of V_E and V_{BE}. The base current may be calculated by dividing the collector current by the β of the transistor.

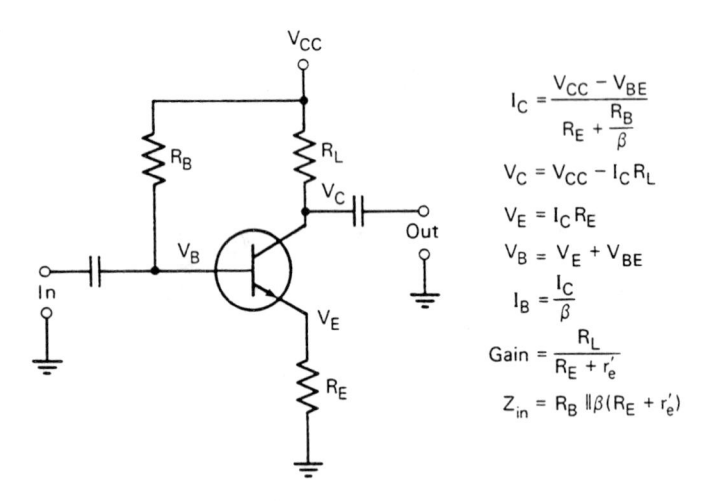

$$I_C = \frac{V_{CC} - V_{BE}}{R_E + \frac{R_B}{\beta}}$$

$$V_C = V_{CC} - I_C R_L$$

$$V_E = I_C R_E$$

$$V_B = V_E + V_{BE}$$

$$I_B = \frac{I_C}{\beta}$$

$$\text{Gain} = \frac{R_L}{R_E + r'_e}$$

$$Z_{in} = R_B \| \beta(R_E + r'_e)$$

FIGURE 4-9 Base bias with emitter feedback arrangement (BBEF). The emitter resistor stabilizes both the collector current and the gain against variations in the transistor parameters.

Collector-Current Stability

Emitter resistor R_E provides negative feedback that tends to stabilize the collector current against changes in beta. Assuming that a constant voltage is applied to the base, let us suppose that the beta increases so that a larger collector current flows. The current through R_E increases and produces a larger voltage across R_E. This decreases the base-emitter voltage and tends to counteract the increase in collector current.

AC Characteristics

The AC junction resistance r'_e is calculated by the same formula as before, namely, $r'_e = 25/I_C$. The external emitter resistor R_E appears in series with r'_e, and the two are added when calculating the gain and input impedance. Hence, terms like $(R_E + r'_e)$ appear in the formulas of Figure 4-9.

The gain of the present circuit is much less than the previous circuit, because R_E is added to r'_e in the gain formula. R_E produces degeneration and negative feedback in the same manner as the cathode resistor of vacuum-tube circuits.

R_E is typically large in comparison with r'_e and tends to "swamp" variations in r'_e. This accounts for the much more stable and predictable gain of the present circuit. Note, however, that stability is obtained at the expense of a reduction in gain.

The input impedance is obtained as before except that R_E is added to r'_e. Consequently, R_E causes the input impedance to be much higher than in the previous circuit.

EXAMPLE 4-2: An example analysis of a base bias with emitter feedback circuit is presented in Figure 4-10.

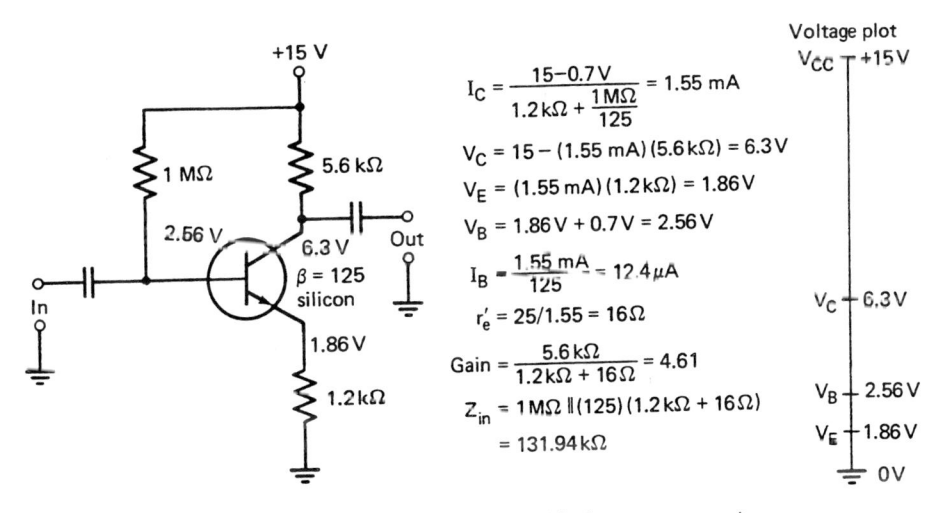

$$I_C = \frac{15 - 0.7\,V}{1.2\,k\Omega + \frac{1\,M\Omega}{125}} = 1.55\,mA$$

$$V_C = 15 - (1.55\,mA)(5.6\,k\Omega) = 6.3\,V$$

$$V_E = (1.55\,mA)(1.2\,k\Omega) = 1.86\,V$$

$$V_B = 1.86\,V + 0.7\,V = 2.56\,V$$

$$I_B = \frac{1.55\,mA}{125} = 12.4\,\mu A$$

$$r'_e = 25/1.55 = 16\,\Omega$$

$$\text{Gain} = \frac{5.6\,k\Omega}{1.2\,k\Omega + 16\,\Omega} = 4.61$$

$$Z_{in} = 1\,M\Omega \,\|\,(125)(1.2\,k\Omega + 16\,\Omega) = 131.94\,k\Omega$$

FIGURE 4-10 Analysis of a BBEF-biasing arrangement.

4-5 BASE BIAS WITH COLLECTOR FEEDBACK (BBCF)

By connecting the base resistor to the collector rather than to the positive power supply, we obtain the BBCF arrangement shown in Figure 4-11. This is possible because the collector is at a positive potential, and a positive voltage for the base can be obtained from the collector. The collector voltage is not a fixed or constant voltage, however. It varies with the input signal, and it will vary as the collector current varies due to changes in beta. Hence, the voltage applied to the base is not constant.

This connection produces negative feedback from the output (the collector) back to the input (the base) of the amplifier and stabilizes the collector current.

Formulas are given in Figure 4-11 for calculating the important circuit parameters. Note that the collector current depends upon the ohmic value of the collector load resistor R_L. This was not the case in preceding circuits. This dependence of I_C upon R_L occurs because the collector voltage, which is the source of base voltage, depends upon R_L. Large values of R_L produce low collector voltages, and low collector voltages imply low baising voltages applied to the base. Incidentally, I_C depends appreciably upon R_L only in biasing arrangements that utilize collector feedback.

Input Impedance, Z_{in}

We run into a small complication as we start to determine the input impedance of this amplifier. The high-voltage end of R_B does not connect to a point at signal ground. The signal appearing on the collector affects the flow of the input signal through R_B. As the input signal goes positive, the collector signal goes negative. This increases the net signal voltage across R_B, and this effect results in increased signal current flow through R_B. Thus, the ohmic value of R_B appears to be reduced

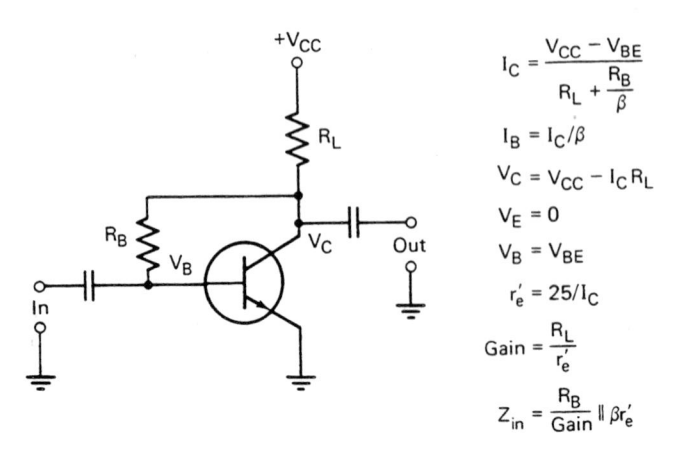

$$I_C = \frac{V_{CC} - V_{BE}}{R_L + \frac{R_B}{\beta}}$$

$$I_B = I_C/\beta$$

$$V_C = V_{CC} - I_C R_L$$

$$V_E = 0$$

$$V_B = V_{BE}$$

$$r_e' = 25/I_C$$

$$\text{Gain} = \frac{R_L}{r_e'}$$

$$Z_{in} = \frac{R_B}{\text{Gain}} \| \beta r_e'$$

FIGURE 4-11 Base bias with collector feedback arrangement (BBCF).

by the action of the transistor. Specifically, the value of R_B appears to be reduced by a factor equal to the gain of the amplifier. The effective resistance of R_B to the input signal is R_B/A, where A is the gain of the amplifier. Hence, the input impedance of this amplifier is the parallel equivalent of R_B/A and the input resistance to the base, $\beta r'_e$. The input impedance is reduced slightly by the collector feedback.

Collector-Current Stability

Collector feedback causes the circuit to oppose changes in collector current for the following reason. If the collector current increases as a result of an increase in beta, for example, the collector voltage decreases due to the increased voltage drop across the load resistor. As the collector voltage decreases, the driving voltage to the base is diminished, which opposes the increase in collector current.

Gain Dependence upon AC Junction Resistance

Note that the formula for the gain of this circuit has the AC junction resistance r'_e appearing alone in the denominator. This causes the gain to be very sensitive to variations of r'_e, making the gain of a particular amplifier rather unpredictable, as in the base-bias circuit. This can be avoided by connecting an external *swamping resistor* into the emitter circuit to achieve gain stability at the expense of a reduction in gain. The swamping resistor masks variations in r'_e, and the resulting amplifier has both collector and emitter feedback.

EXAMPLE 4-3: A sample circuit is analyzed in Figure 4-12.

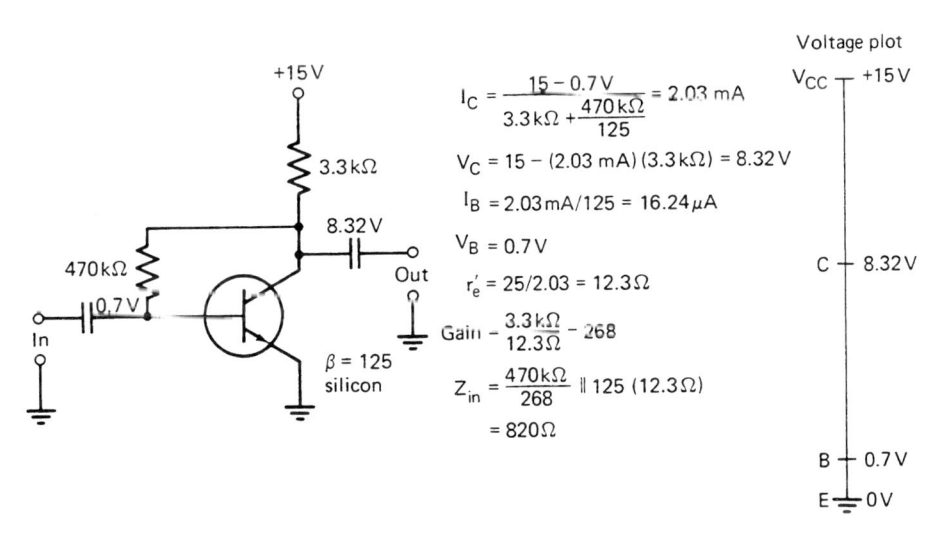

$$I_C = \frac{15 - 0.7\,V}{3.3\,k\Omega + \frac{470\,k\Omega}{125}} = 2.03\,mA$$

$$V_C = 15 - (2.03\,mA)(3.3\,k\Omega) = 8.32\,V$$

$$I_B = 2.03\,mA/125 = 16.24\,\mu A$$

$$V_B = 0.7\,V$$

$$r'_e = 25/2.03 = 12.3\,\Omega$$

$$\text{Gain} = \frac{3.3\,k\Omega}{12.3\,\Omega} = 268$$

$$Z_{in} = \frac{470\,k\Omega}{268} \parallel 125\,(12.3\,\Omega)$$

$$= 820\,\Omega$$

Voltage plot: $V_{CC} = +15\,V$; $C = 8.32\,V$; $B = 0.7\,V$; $E = 0\,V$.

FIGURE 4-12 Analysis of a BBCF-biasing arrangement. Note that the large gain is obtained at the expense of a very low input impedance.

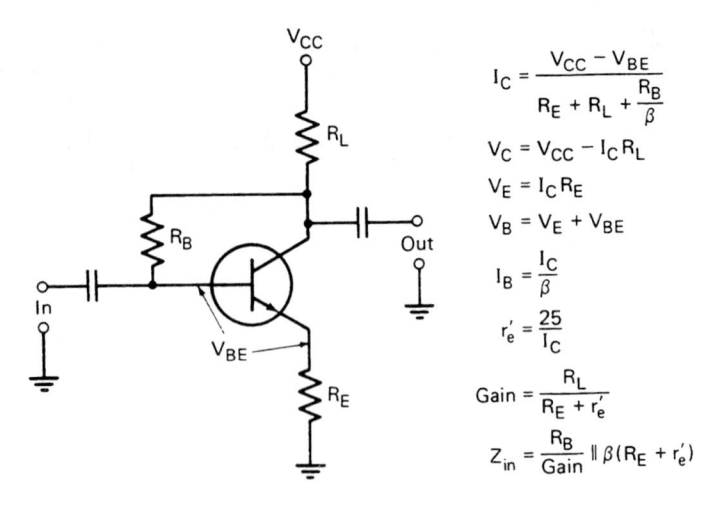

FIGURE 4-13 Base bias with collector and emitter feedback arrangement (BBCEF).

$$I_C = \frac{V_{CC} - V_{BE}}{R_E + R_L + \frac{R_B}{\beta}}$$

$$V_C = V_{CC} - I_C R_L$$

$$V_E = I_C R_E$$

$$V_B = V_E + V_{BE}$$

$$I_B = \frac{I_C}{\beta}$$

$$r'_e = \frac{25}{I_C}$$

$$\text{Gain} = \frac{R_L}{R_E + r'_e}$$

$$Z_{in} = \frac{R_B}{\text{Gain}} \parallel \beta(R_E + r'_e)$$

4-6 BASE BIAS WITH COLLECTOR AND EMITTER FEEDBACK (BBCEF)

This biasing arrangement, shown in Figure 4-13, is a combination of the two preceding circuits. The emitter resistor provides stability against gain variations, and the emitter resistor and the collector feedback connection stabilize the collector current. Appropriate formulas are given in Figure 4-13, and a sample calculation is given in Figure 4-14.

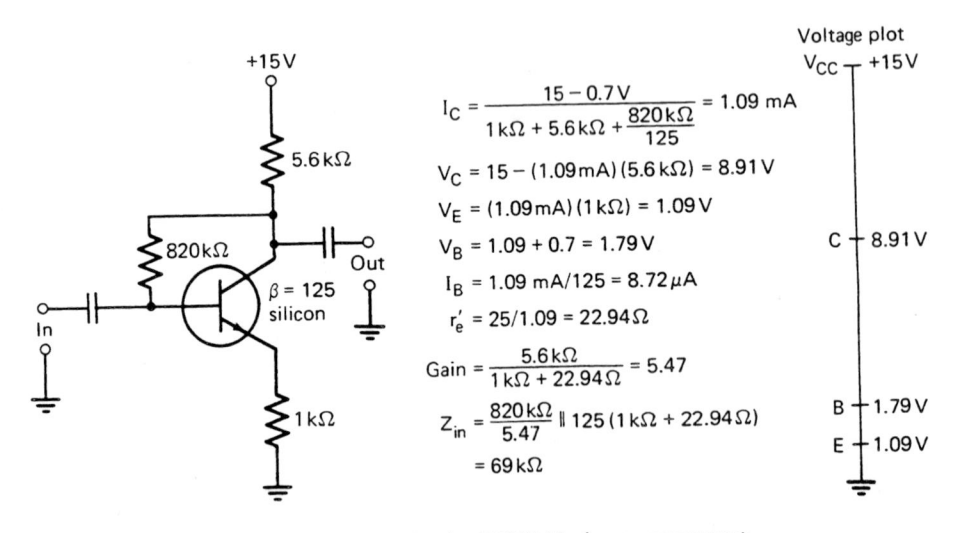

$$I_C = \frac{15 - 0.7\,V}{1\,k\Omega + 5.6\,k\Omega + \frac{820\,k\Omega}{125}} = 1.09\text{ mA}$$

$$V_C = 15 - (1.09\,\text{mA})(5.6\,k\Omega) = 8.91\,V$$

$$V_E = (1.09\,\text{mA})(1\,k\Omega) = 1.09\,V$$

$$V_B = 1.09 + 0.7 = 1.79\,V$$

$$I_B = 1.09\text{ mA}/125 = 8.72\,\mu A$$

$$r'_e = 25/1.09 = 22.94\,\Omega$$

$$\text{Gain} = \frac{5.6\,k\Omega}{1\,k\Omega + 22.94\,\Omega} = 5.47$$

$$Z_{in} = \frac{820\,k\Omega}{5.47} \parallel 125(1\,k\Omega + 22.94\,\Omega)$$
$$= 69\,k\Omega$$

Voltage plot: V_{CC} +15 V; C 8.91 V; B 1.79 V; E 1.09 V.

FIGURE 4-14 Analysis of a BBCEF-biasing arrangement.

[Figure 4-15 circuit diagram with formulas:]

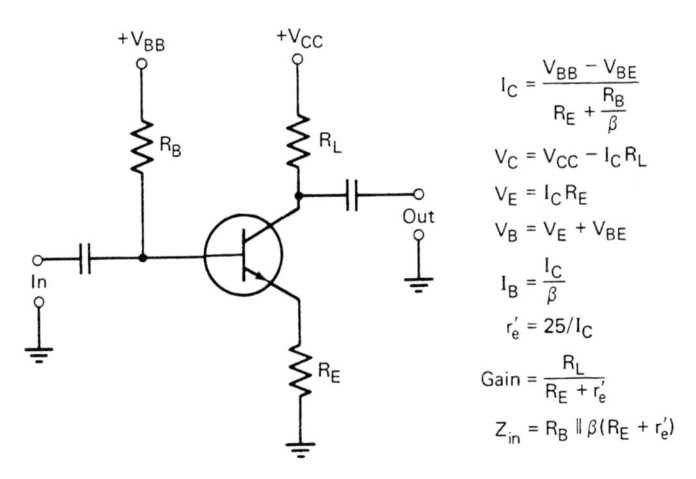

$$I_C = \frac{V_{BB} - V_{BE}}{R_E + \frac{R_B}{\beta}}$$

$$V_C = V_{CC} - I_C R_L$$

$$V_E = I_C R_E$$

$$V_B = V_E + V_{BE}$$

$$I_B = \frac{I_C}{\beta}$$

$$r'_e = 25/I_C$$

$$\text{Gain} = \frac{R_L}{R_E + r'_e}$$

$$Z_{in} = R_B \parallel \beta(R_E + r'_e)$$

FIGURE 4-15 Base bias with separate supply. This arrangement is very similar to the BBEF arrangement.

4-7 BASE BIAS WITH SEPARATE SUPPLY (BBSS)

The base resistor in base-biased circuits is frequently returned to a source of positive voltage other than the main power supply. Such a circuit is illustrated in Figure 4-15, where the base voltage source is designated V_{BB}. This circuit is a variation of the BBEF circuit, and formulas for calculating the important parameters are given in the figure. This circuit can be converted to a modified simple base-bias circuit merely by setting R_E equal to zero, both physically and mathematically. A sample calculation is shown in Figure 4-16.

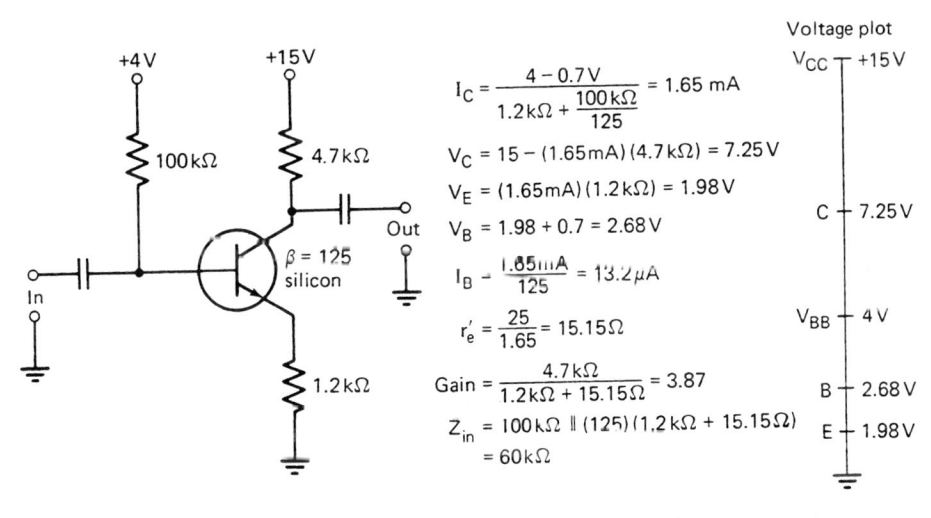

[Figure 4-16 circuit: +4V, +15V supplies, 100 kΩ, 4.7 kΩ, β = 125 silicon, 1.2 kΩ]

$$I_C = \frac{4 - 0.7\,V}{1.2\,k\Omega + \frac{100\,k\Omega}{125}} = 1.65\,mA$$

$$V_C = 15 - (1.65\,mA)(4.7\,k\Omega) = 7.25\,V$$

$$V_E = (1.65\,mA)(1.2\,k\Omega) = 1.98\,V$$

$$V_B = 1.98 + 0.7 = 2.68\,V$$

$$I_B = \frac{1.65\,mA}{125} = 13.2\,\mu A$$

$$r'_e = \frac{25}{1.65} = 15.15\,\Omega$$

$$\text{Gain} = \frac{4.7\,k\Omega}{1.2\,k\Omega + 15.15\,\Omega} = 3.87$$

$$Z_{in} = 100\,k\Omega \parallel (125)(1.2\,k\Omega + 15.15\,\Omega)$$
$$= 60\,k\Omega$$

Voltage plot:
- V_{CC} +15 V
- C 7.25 V
- V_{BB} 4 V
- B 2.68 V
- E 1.98 V

FIGURE 4-16 Analysis of a base bias with separate supply arrangement.

4-8 UNIVERSAL BIAS

We now consider a biasing arrangement that is widely used because of its stability and because its operating characteristics are essentially independent of the transistor parameters. The circuit, shown in Figure 4-17, is sometimes referred to as *voltage-divider bias, emtter bias with one supply*, or, as we shall call it, the *universal bias*. In the following discussion we neglect the base current in order to simplify the analysis. This is permissible since the base current is small in comparison with the collector current and since the base current is usually small in comparison with the current flowing in the voltage-divider portion of the biasing arrangement.

Two resistors connect to the base, one of which connects to the positive supply. We call this resistor the *high-base resistor* R_{HB} because it usually is of higher ohmic value than the other base resistor and because it connects to the "high voltage" of the power supply. The other resistor is the *low-base resistor* R_{LB} because it is lower in ohmic value and connects to ground, a low voltage. These resistors form a voltage divider, and therefore the voltage on the base is given by

$$V_B = \left(\frac{R_{LB}}{R_{LB} + R_{HB}}\right) V_{CC} \qquad (4-1)$$

Once the voltage on the base is known, the voltage on the emitter can be determined by subtracting V_{BE} from the voltage on the base. Hence, $V_E = V_B - V_{BE}$.

The voltage on the emitter is developed by the current flow through the emitter resistance. Further, this current very nearly equals the collector current, being in

$$V_B = \left(\frac{R_{LB}}{R_{LB} + R_{HB}}\right) V_{CC}$$

$$V_E = V_B - V_{BE}$$

$$I_C = \frac{V_E}{R_E}$$

$$V_C = V_{CC} - I_C R_L$$

$$I_B = \frac{I_C}{\beta}$$

$$r'_e = \frac{25}{I_C}$$

$$\text{Gain} = \frac{R_L}{R_E + r'_e}$$

$$Z_{in} = R_{HB} \parallel R_{LB} \parallel \beta(R_E + r'_e)$$

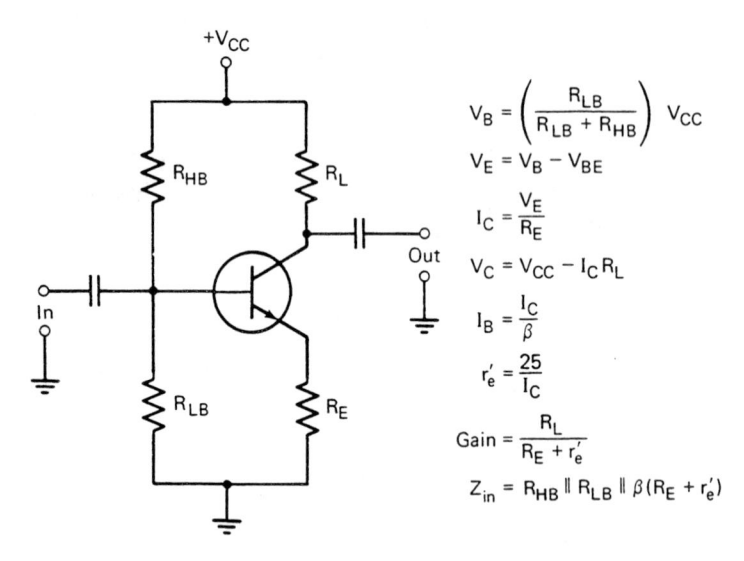

FIGURE 4-17 Universal-bias arrangement. This biasing arrangement is widely used and is one of the most stable against variations of the transistor parameters.

Biasing Arrangements — Chap. 4

error only by the base current. Therefore, we write the following expression for the collector current: $I_C = V_E/R_E$.

The voltage on the collector is then determined by subtracting the voltage drop across R_L from the power supply voltage: $V_C = V_{CC} - I_C R_L$.

A formula is given in Figure 4-17 for calculating the gain of the stage.

Input Impedance

The input impedance is obtained by considering that the signal may pass from the input to ground by three paths instead of two as in previous circuits. Resistive paths consist of the base resistors, which, in this capacity, appear in parallel. The other path is through the base-emitter junction of the transistor, as in previous circuits. The input impedance is easily calculated by first calculating the parallel equivalent of R_{HB} and R_{LB} and then placing this equivalent resistance in parallel with the input resistance to the base. The expression for the input impedance is

$$Z_{in} = (R_{HB} \| R_{LB}) \| \beta(R_E + r'_e) \qquad (4\text{-}2)$$

A typical circuit and sample calculation is shown in Figure 4-18.

Stability of the Universal-Bias Arrangement

The base voltage of the universal circuit is determined by the base resistors rather than by the parameters of the transistor. It is not prone to vary with temperature or transistor parameters as long as the base current remains small in comparison

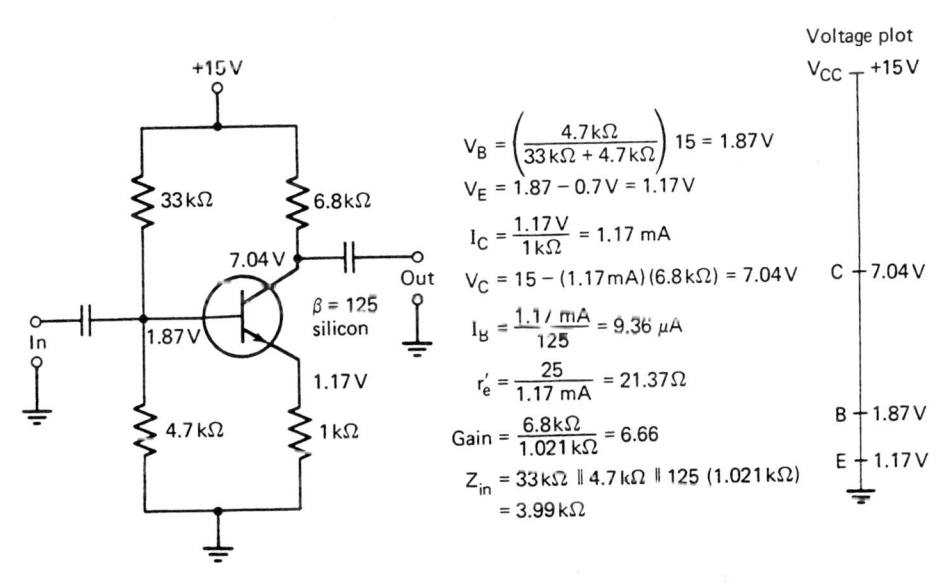

$V_B = \left(\dfrac{4.7\,k\Omega}{33\,k\Omega + 4.7\,k\Omega}\right) 15 = 1.87\,V$

$V_E = 1.87 - 0.7\,V = 1.17\,V$

$I_C = \dfrac{1.17\,V}{1\,k\Omega} = 1.17\,mA$

$V_C = 15 - (1.17\,mA)(6.8\,k\Omega) = 7.04\,V$

$I_B = \dfrac{1.17\,mA}{125} = 9.36\,\mu A$

$r'_e = \dfrac{25}{1.17\,mA} = 21.37\,\Omega$

Gain $= \dfrac{6.8\,k\Omega}{1.021\,k\Omega} = 6.66$

$Z_{in} = 33\,k\Omega \| 4.7\,k\Omega \| 125\,(1.021\,k\Omega)$
$= 3.99\,k\Omega$

FIGURE 4-18 Analysis of a universal-biasing arrangement.

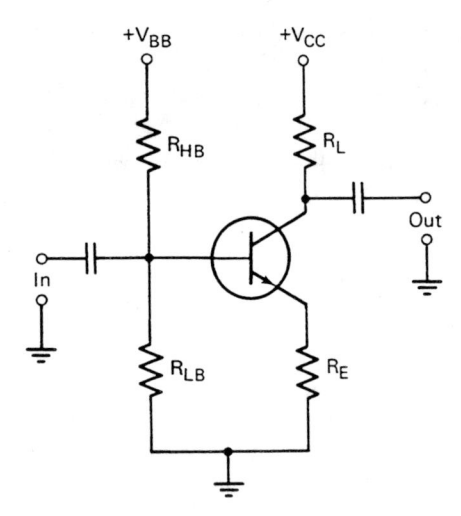

FIGURE 4-19 Modified universal-bias arrangement. The high-base resistor is supplied by a separate supply.

to the current flowing through R_{LB}. Thus, the base voltage is stable. The resistance R_E in the emitter lead provides negative feedback that opposes any tendency for the collector current to change, and R_E also stabilizes the gain if R_E is large in comparison with r_e'. The fixed base voltage and the negative feedback account for the increased stability of this circuit.

Modified Universal Bias

Sometimes a circuit is encountered that appears to be a universal bias with the exception that the high-base resistor connects to a source of positive voltage other than the power supply. We call this a *modified universal bias*. Such a circuit is shown in Figure 4-19. The expressions for the various circuit parameters are identical to those for the unmodified universal bias with the exception that V_{BB} is substituted for V_{CC} in the expression for the voltage on the base.

4-9 UNIVERSAL BIAS WITH COLLECTOR FEEDBACK (UBCF)

Figure 4-20 shows the universal-bias circuit modified to provide collector feedback by connecting the high-base resistor to the collector rather than to the positive power supply. The collector feedback lowers the input impedance slightly and provides additional collector-current stability.

Analysis of the circuit is complicated by the feedback from the collector to the base. The collector voltage provides the bias voltage for the base; conversely, the bias voltage on the base indirectly determines the collector voltage. Thus, there is no obvious place to begin unless certain simplifying assumptions are made.

$$y = \frac{R_{LB}}{R_{LB} + R_{HB} + R_L}$$

$$I_C = \frac{yV_{CC} - V_{BE}}{R_E + yR_L}$$

$$x = \frac{R_{LB}}{R_{LB} + R_{HB}}$$

$$V_C = (V_{CC} - I_C R_L)\left(\frac{y}{x}\right)$$

$$V_B = xV_C$$

$$V_E = I_C R_E$$

Note: x and y are dummy variables to simplify calculations. Gain and impedance formulas are given in the text.

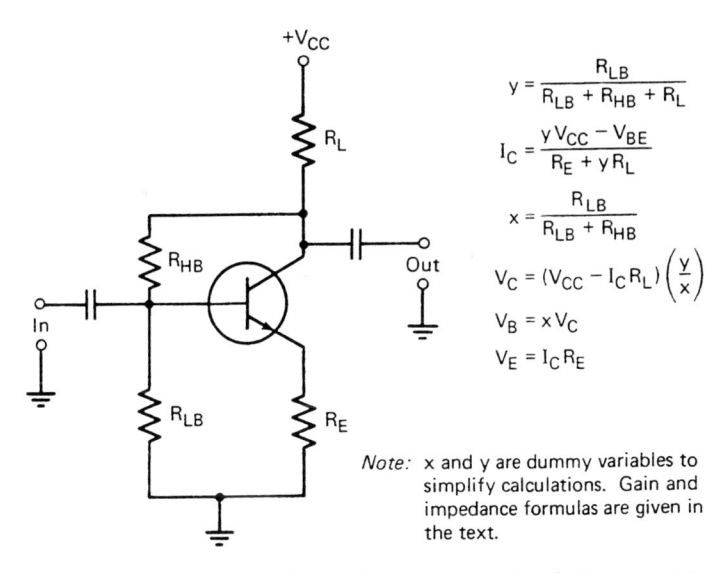

FIGURE 4-20 Universal bias with collector feedback arrangement.

We assume that the base current is small in comparison with the current flowing through the voltage divider consisting of R_{HB} and R_{LB}. This will be a good assumption if the transistor beta is high and if the ohmic values of the voltage-divider resistors are not extremely large. This allows us to ignore the base current that flows through R_{HB}, and the resulting mathematical expressions for the DC parameters are fairly simple and give reasonable accuracy.

The formulas are given in Figure 4-20 in the order in which they are used, and an example is shown in Figure 4-21.

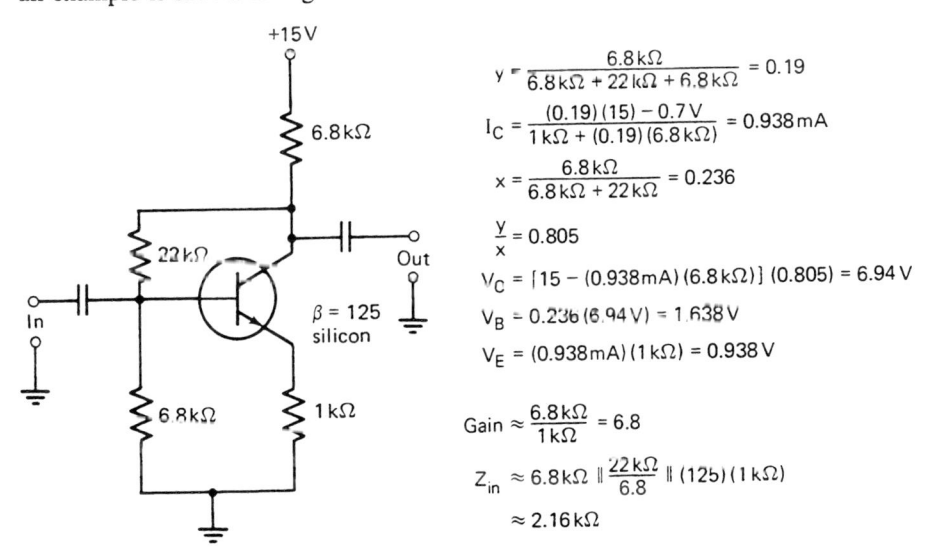

$$y = \frac{6.8\,k\Omega}{6.8\,k\Omega + 22\,k\Omega + 6.8\,k\Omega} = 0.19$$

$$I_C = \frac{(0.19)(15) - 0.7\,V}{1\,k\Omega + (0.19)(6.8\,k\Omega)} = 0.938\,mA$$

$$x = \frac{6.8\,k\Omega}{6.8\,k\Omega + 22\,k\Omega} = 0.236$$

$$\frac{y}{x} = 0.805$$

$$V_C = [15 - (0.938\,mA)(6.8\,k\Omega)](0.805) = 6.94\,V$$

$$V_B = 0.236\,(6.94\,V) = 1.638\,V$$

$$V_E = (0.938\,mA)(1\,k\Omega) = 0.938\,V$$

$$\text{Gain} \approx \frac{6.8\,k\Omega}{1\,k\Omega} = 6.8$$

$$Z_{in} \approx 6.8\,k\Omega \parallel \frac{22\,k\Omega}{6.8} \parallel (125)(1\,k\Omega)$$

$$\approx 2.16\,k\Omega$$

FIGURE 4-21 Analysis of a universal bias with collector feedback circuit.

Sec. 4-9 *Universal Bias with Collector Feedback (UBCF)*

A meaningful AC analysis is rather involved, and goes beyond our needs. Therefore, we give the following simplified formulas for the input impedance and gain:

$$\text{Gain} \equiv A \approx \frac{R_L}{R_E + r'_e} \qquad Z_{in} \approx R_{LB} \left\| \frac{R_{HB}}{A} \right\| \beta(R_E + r'_e) \qquad (4\text{-}3)$$

These formulas provide only "ballpark" accuracy, but they are sufficient for our purposes.

$$I_C = \frac{V_{EE} - V_{BE}}{R_E + \frac{R_B}{\beta}}$$

$$V_C = V_{CC} - I_C R_L$$

$$V_E = -V_{EE} + I_C R_E$$

$$V_B = V_E + V_{BE}$$

$$r'_e = \frac{25}{I_C} \qquad I_B = I_C/\beta$$

$$\text{Gain} = \frac{R_L}{R_E + r'_e}$$

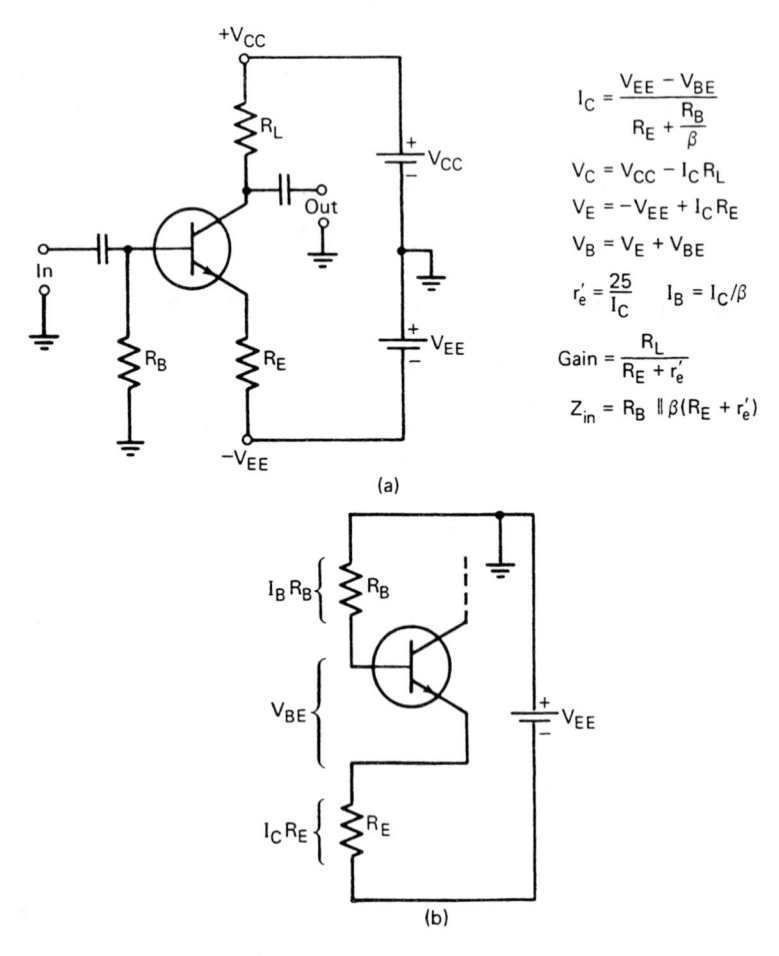

$$Z_{in} = R_B \,\|\, \beta(R_E + r'_e)$$

FIGURE 4-22 Emitter bias with two supplies (EBTS). In part (a) the power supply connections are shown, and in part (b) the base-emitter circuit is drawn to show how the forward bias on the base is controlled by V_{EE}.

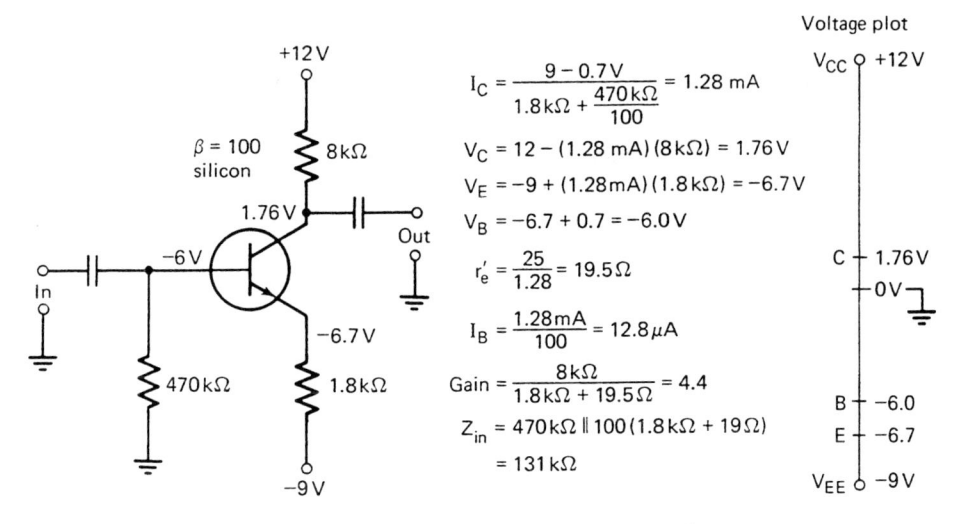

FIGURE 4-23 Analysis of an emitter bias with a two-supplies arrangement.

4-10 EMITTER BIAS WITH TWO SUPPLIES (EBTS)

As the last circuit of this series, we describe the EBTS circuit shown in Figure 4-22(a). We have included the two power supplies to illustrate the connection. Note that the base reactor R_B connects to ground.

Figure 4-22(b) shows the base and emitter circuit redrawn to show that it resembles the BBEF circuit, even though the ground point has been relocated. Appropriate formulas are given in the figure. It is not uncommon for the base voltage to be negative or positive by several volts even though R_B is grounded.

It is noteworthy that neither the collector load resistor nor the collector power supply V_{CC} affects the value of collector current I_C. These components determine the collector voltage V_C, but as long as the collector is a volt or so more positive than the base, the same collector current will flow as determined by the negative supply, V_{EE}, R_E, R_B, and beta. A typical circuit and sample calculation is presented in Figure 4-23.

4-11 MISCELLANEOUS TOPICS

Biasing PNP Transistors

The biasing arrangements of the preceding section were described in terms of *NPN* transistors because they are much more prevalent in commercial circuits. The biasing of *PNP* transistors is identical to that of *NPN* transistors, with the important exception that voltage polarities are reversed. Whereas the collector of an *NPN* transistor is positive relative to the base and the emitter, the collector of a *PNP* transistor must be negative. In short, to use *PNP* transistors, simply reverse the polarity of the power supply. As examples, two *PNP* circuits are shown in Figure 4-24.

PNP transistor operation is identical to that of an *NPN* transistor, except that holes play the role of charge carriers. Electrons enter via the emitter of an *NPN*

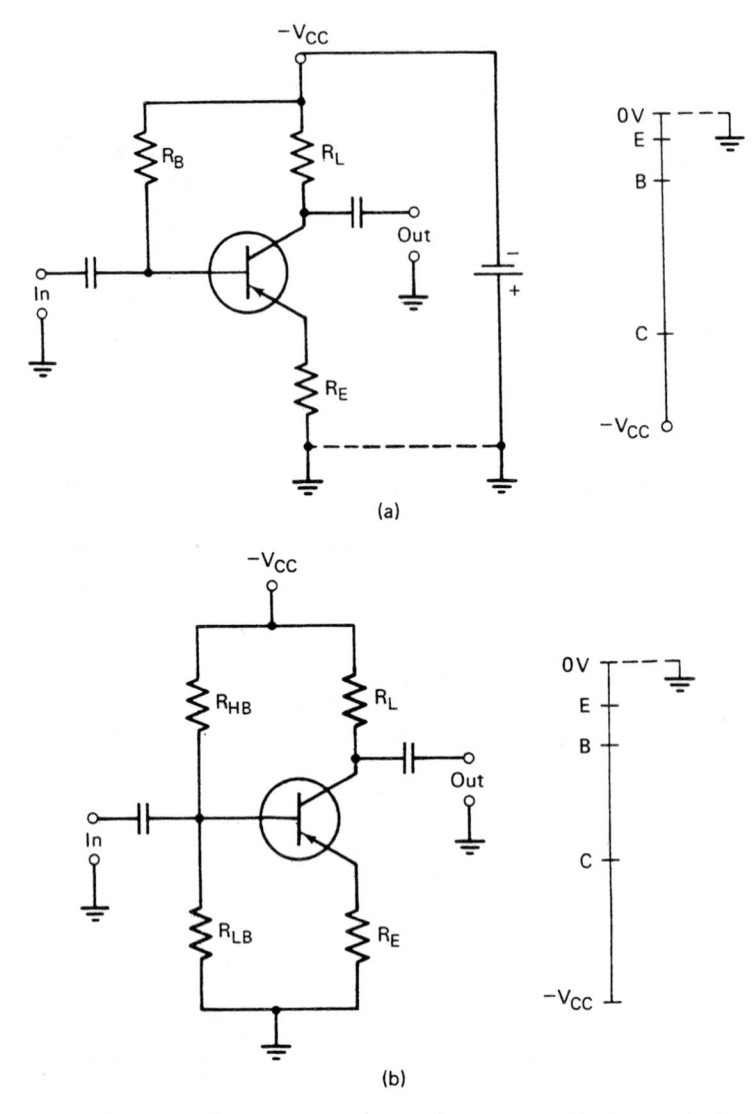

FIGURE 4-24 Only the power supply voltages need be inverted when *PNP* transistors are used in the preceding biasing arrangements.

98   *Biasing Arrangements*   Chap. 4

device and leave via the base and collector. In a *PNP* device, holes enter via the emitter and leave by the base and collector. To avoid confusion, remember that a flow of holes is equivalent to a flow of electrons in the opposite direction. Consequently, electrons enter via the base and collector of a *PNP* transistor and leave via the emitter.

Relocation of Ground

Figure 4-25 illustrates an *NPN* circuit in which the ground point has been moved to the collector side of the transistor, grounding the point where V_{CC} is usually applied. The emitter resistor, usually grounded, is connected to a power supply that provides a negative voltage. The emitter is negative relative to the collector, as must be the case, and all voltage relationships in the circuit are properly maintained. The voltages have been shifted downward an amount equal to V_{CC}. The shift requires slight modification of the formulas given earlier, and the modified formulas are given in the figure.

The circuit of Figure 4-26 utilizes a *PNP* transistor in the BBEF arrangement. Study the voltage relationships very carefully. The input signal appears to be applied between the base and the collector, but this is not the case. Recall that the voltage source $+V$ is connected to circuit ground through a large capacitor so that the voltage source point is at signal ground potential. Hence, connecting the low side of the input signal to ground, as shown, effectively applies the signal between the base and the emitter.

The ground point of other biasing arrangements may be shifted as in the circuit presented here. We shall see many examples of "shifted ground points" in the latter chapters in which receiver circuits are described in detail. Incidentally, the usefulness of the voltage plot as an aid in visualizing the voltage relationships should now be evident.

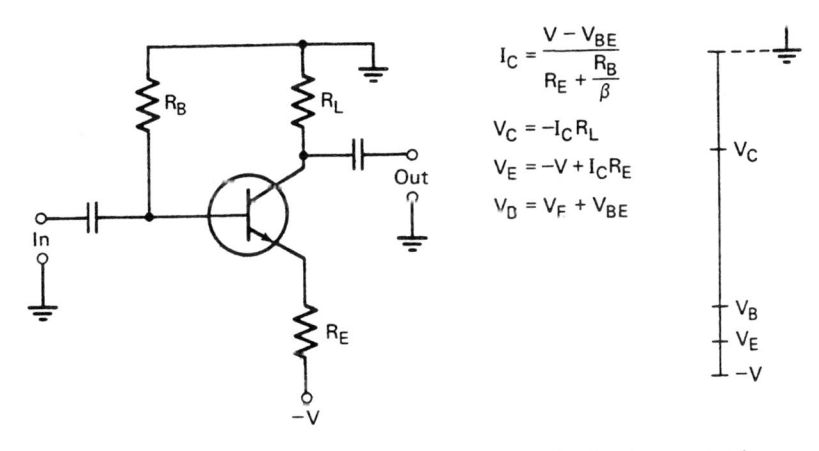

$$I_C = \frac{V - V_{BE}}{R_E + \frac{R_B}{\beta}}$$

$$V_C = -I_C R_L$$

$$V_E = -V + I_C R_E$$

$$V_D = V_E + V_{BE}$$

FIGURE 4-25 *NPN* circuit in which the ground point has been moved.

Sec. 4-11   *Miscellaneous Topics*

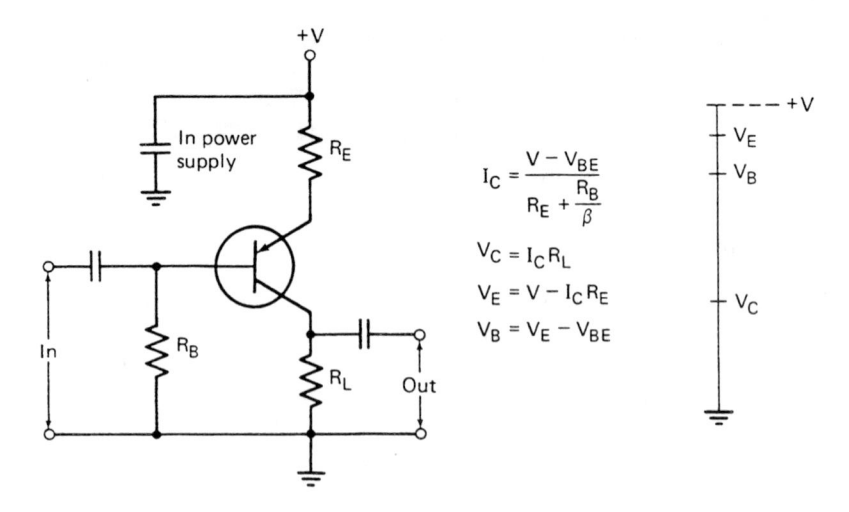

FIGURE 4-26 *PNP* transistor used in a base-bias-with-emitter-feedback arrangement.

Emitter Bypass Capacitor

Signal degeneration (gain reduction) produced by the emitter resistor can be eliminated by connecting an emitter bypass capacitor in parallel with the emitter resistor. The principle is the same as that of the cathode bypass capacitor in vacuum-tube circuits. The emitter bypass capacitor does not affect the DC parameters; the formulas given earlier may still be used. However, the bypassed resistance is not included in formulas for calculating gain and input impedance to the stage.

Sometimes only a portion of the emitter resistance is bypassed, as in Figure 4-27. The total resistance, $R_E + r_e$, produces DC feedback and stabilizes the DC parameters, but only the unbypassed resistance produces AC feedback. If r_e is fairly large in comparison with the AC junction resistance r'_e, variations in r'_e will not affect the gain as much as if r_e were omitted from the circuit. The relatively large value of r_e tends to "mask" or "swamp out" variations in r'_e. We shall call the unbypassed resistance r_e the *swamping resistor*.

To clarify the action of r_e in stabilizing the gain, consider a circuit in which the collector load resistor is 300 Ω and assume that r'_e is 25 Ω. The gain, given by R_L/r'_e, is 300/25 equals 12. If r'_e doubles and becomes 50 Ω, the new gain will be 300/50 equals 6. Doubling r'_e reduces the gain by 50%.

If a swamping resistor of 50 Ω is included in the circuit, the gain is given by $R_L/(r_e + r'_e)$. When r'_e is 25 Ω, the gain is 300/75 equals 4. If r'_e should increase to 50 Ω, the new gain will be 300/(50 + 50) equals 3. In this case, doubling r'_e reduces the gain by only 25%. Note that the gain is reduced from 12 to 4 by the swamping resistor. Gain stability is obtained at the expense of a reduction in gain.

The capacitance required to bypass the emitter resistance depends upon the lowest operating frequency and upon the ohmic value of the resistance being by-

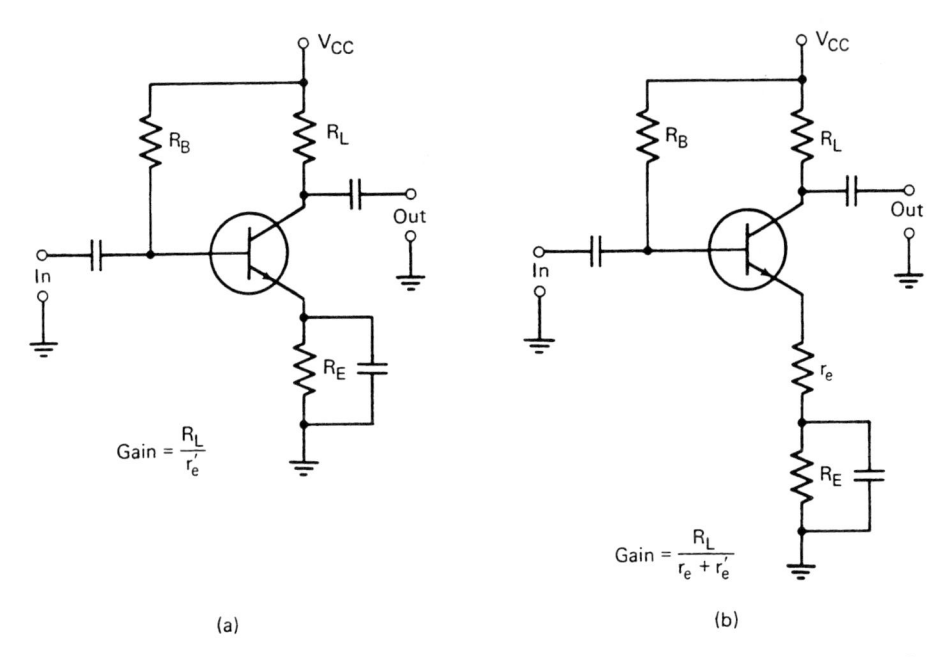

FIGURE 4-27 Degenerative (gain-reducing) effects of the emitter resistor can be avoided by bypassing the emitter resistor with an emitter bypass capacitor. In part (b), only a part of the emitter resistance is bypassed.

passed. The capacitance is typically chosen so that the capacitive reactance at the lowest frequency of operation is not more than 10% of the resistance being bypassed. Thus, the capacitance C required to bypass resistance R is given by $C = 10/(2\pi f R)$, where f is the lowest frequency of operation in kHz, C is in μF, and R is in kΩ.

Emitter Bypass Capacitance versus Input Impedance

Bypassing the emitter resistor increases the gain of a stage at the expense of reducing the input impedance. The impedance seen looking into the base is $\beta(R_E + r'_e)$ as shown in Figure 4-28. The bypass capacitor effectively removes R from the signal path, and the impedance drops to $\beta(r_e + r'_e)$, when r_e is any emitter resistance that is left unbypassed. The input impedance to the stage is formed by placing the biasing resistors that connect to the base in parallel with the input resistance to the base.

We may state a general rule for the amplifiers considered here in regard to the relationship between input impedance and gain. The rule is simply that as the gain goes up, the input impedance goes down, and as the input impedance goes up, the gain goes down. For example, we have seen that bypassing the emitter resistance reduces the input impedance by removing the emitter resistance from the signal

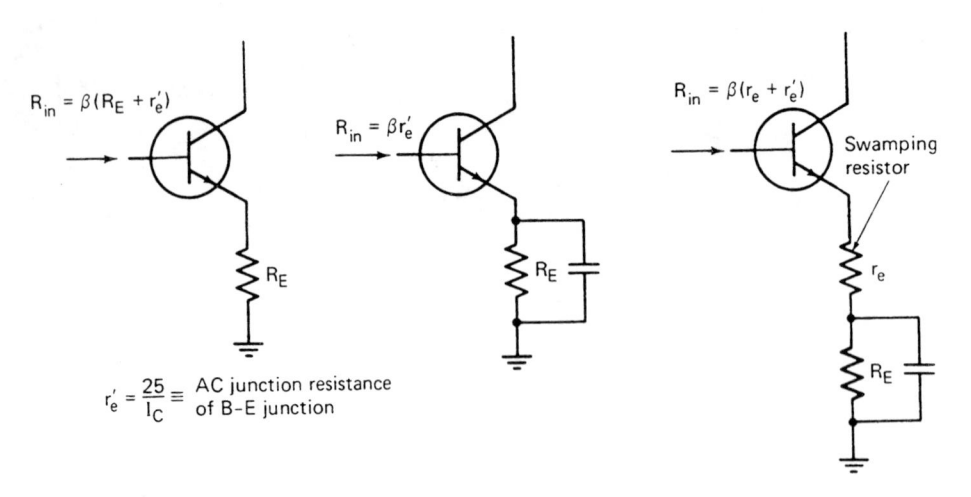

FIGURE 4-28 Bypassing the emitter resistance lowers the AC input resistance to the base. This, in turn, lowers the input impedance of the stage.

path. At the same time, the gain is increased from $A = R_L/(R_E + r'_e)$ to $A' = R_L/(r_e + r'_e)$ as a result of changing the resistance in the signal path from R_E to r_e.

4-12 JFET BIASING ARRANGEMENTS

JFET biasing requirements differ from those of BJTs in that the gate of a JFET must be held negative (for N-channel devices) relative to the source, whereas the base of a BJT must be held about 0.7 V more positive (for *NPN* devices) than the emitter. In this section we investigate three JFET biasing arrangements: self-bias, source bias, and voltage-divider bias.

In the analysis of BJT circuits, a considerable simplification results from the fact that V_{BE}, the voltage between base and emitter, varies only slightly from one transistor to another. Unfortunately, no such simple relationship holds for FETs. The gate-source voltage varies over a wide range from one device to another. This means that we must have certain data for the FET we are using before we can embark upon a meaningful analysis or design of a biasing arrangement for the device.

The data we need are easily obtained from the drain characteristic curves described in Chap. 2. The curves may be obtained from manufacturer's data, from a curve tracer if one is available, or by constructing the curves using the circuit of Figure 2-20. In the following, we assume that the curves are available. The discussion is based on an N-channel JFET, but it is equally applicable to P-channel devices if complementary voltages are used.

Self-Bias

The self-bias circuit is shown in Figure 4-29. The drain resistor R_D serves the same function as the collector resistor in BJT circuits; it causes variations in drain current to produce variations in voltage (the signal) at the drain. The gate resistor R_G establishes a conducting path between the gate and ground so that the gate voltage is not allowed to float. The source resistor R_S causes the voltage at the source to rise to a small positive value due to the flow of drain current through it.

The voltage at the gate under no-signal conditions (the Q-point gate voltage) will be zero because only a negligibly small leakage current flows from the gate terminal through R_G to ground. Hence, no voltage is developed across R_G, and the gate voltage is therefore zero.

Because R_S causes a small positive voltage to develop at the source, the source is positive relative to the gate. Viewed from the other direction, the gate is negative relative to the source. This is the condition required for biasing a JFET.

Self-Biased JFET Circuit Design

To further illustrate the operation of the self-bias circuit, we now give a procedure for designing a self-biased JFET amplifier. That is, we shall use the drain curves to help us determine the ohmic values of R_S, R_D, and R_G required to build an amplifier. We first give the procedure (in steps) and then explain the reasoning behind each step. We use the curves of Figure 4-30.

Procedure

1. Pick a power supply voltage and plot the *power supply point* (PSP) on the drain curves. (In this example, PSP = 20 V.)

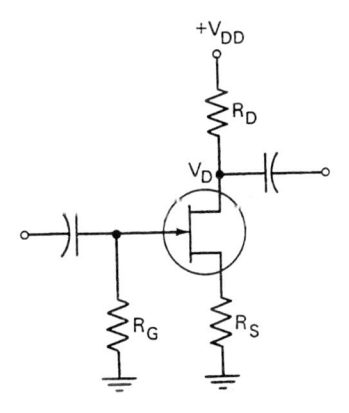

FIGURE 4-29 Self-biased JFET amplifier.

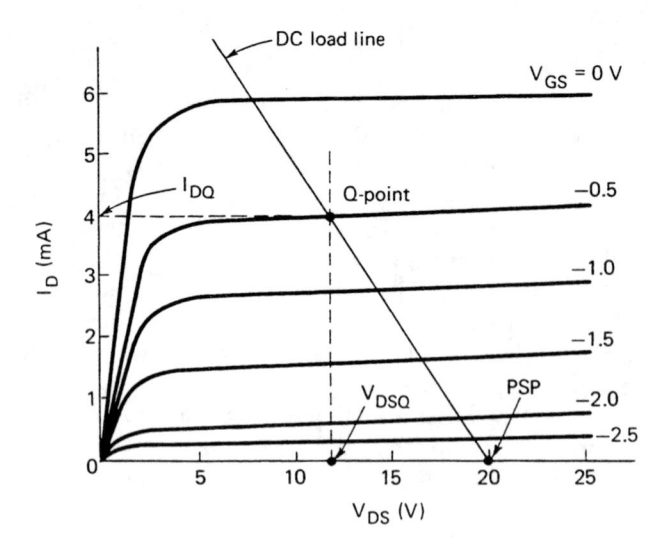

FIGURE 4-30 JFET drain characteristic curves used to design a self-biased JFET amplifier.

2. Pick a Q-point in the upper-left-center region of the curves, on a V_{GS} curve for convenience. Then determine I_{DQ}, V_{DSQ}, and V_{GSQ} as indicated. (In this example, $I_{DQ} = 4$ mA, $V_{DSQ} = 12$ V, and $V_{GSQ} = 0.5$ V.)

3. Draw the DC load line by constructing a straight line from the PSP through the Q-point.

4. Calculate R_S using $R_S = V_{GSQ}/I_{DQ}$.
   (In this example, $R_S = 0.5$ V/4 mA $= 125\ \Omega$.)

5. Calculate $(R_D + R_S)$ using $(R_D + R_S) = (V_{DD} - V_{DSQ})/I_{DQ}$. (In this example, $(R_D + R_S) = (20 - 12$ V$)/4$ mA $= 8$ V/4 mA $= 2000\ \Omega$.)

6. Calculate R_D via $R_D = (R_D + R_S) - R_S$.
   (In this example, $R_D = 2000\ \Omega - 125\ \Omega = 1875\ \Omega$.)

7. Pick R_G to give the desired input impedance. Values up to 10 MΩ or higher can be used. (In this example, R_G was chosen as 1 MΩ.)

Reasoning. For experiment and demonstration, the choice of power supply voltage in step 1 is rather arbitrary. In the design of an actual receiver, the power supply voltage may be set by factors other than the design of the JFET stage.

The choice of Q-point affects the gain of the amplifier and the size of the largest signal the amplifier can handle. In the following, we shall see that Q-points near the top of the curves produce maximum gain, but the Q-point should be centered to produce the maximum dynamic range.

The DC load line (step 3) lets us see at a glance what the effect will be of establishing various values of voltage between the gate and source. For example, if the gate is 1.5 V more negative than the source, I_D will be about 1.4 mA and V_{DS} will be about 17 V. The slope of the DC load line is inversely proportional to $(R_D + R_S)$. Greater slopes give rise to amplifiers of lower gain.

In step 4 we use Ohm's law to determine the value of R_S that will produce the Q-point we selected in step 2. Here is the question: What resistance is required in order to develop a voltage equal to V_{GSQ} across it when a current of I_{DQ} flows through it?

Steps 5 and 6 give us the ohmic value of R_D that is required in order to produce the DC load line drawn in step 3. Our choice of Q-point and power supply voltage V_{DD} fixes the slope of the load line, which is related to $(R_D + R_S)$. The formula in step 5 comes from the fact that the sum of the voltages developed across R_S, R_D, and the JFET must equal the power supply voltage:

$$I_{DQ}R_S + I_{DQ}R_D + V_{GSQ} = V_{DD} \qquad (4\text{-}4)$$

The choice of R_G in step 7 has little effect upon the operation of the JFET as long as it is not so large that the minute leakage current across the gate-channel junction develops a significant voltage across R_G.

AC Load Line

Now that the DC analysis is complete, we turn our attention to the AC parameters to see how the circuit will behave as an amplifier. Refer to the more-general circuit of Figure 4-31(a). A load resistance R_L is included, and the source resistance has been split into two components, one of which is bypassed by capacitor C_S.

The net resistance r_D seen by an AC signal appearing at the drain is the parallel equivalent of R_D and R_L:

$$r_D = R_D \| R_L \qquad (4\text{-}5)$$

The AC resistance in the source lead is simply r_S because R_S is bypassed by C_S. We use r_D and r_S in constructing the AC load line.

The AC load line shows us how the drain-source voltage varies under the influence of an input signal. It is a graph of V_{DS} vs. I_D, considered for an AC signal. If r_S is fairly small, which is usually the case, the variation of V_{DS} will be very nearly equal to the variation of the drain voltage alone, which is the output signal of the amplifier.

Because the AC drain resistance r_D is smaller than R_D ($r_D = R_D \| R_L$, and is therefore smaller than R_D alone) and because r_S is smaller than the DC resistance in the source lead ($r_S < R_S + r_S$), the slope of the AC load line is greater than that of the DC load line. However, the two always intersect at the Q-point.

The equation for the AC load line can be written as

$$V_{DS} = V_{DSQ} - (I_D - I_{DQ})(r_D + r_S) \qquad (4\text{-}6)$$

where V_{DS} and I_D are instantaneous values of the drain-source voltage and drain

(a)

$V_{DD} = 20$ V
$R_D = 1875\ \Omega$
$R_L = 4.7$ kΩ
R_G 1 MΩ
$r_s = 25\ \Omega$
R_s 100
$C_s = 1\ \mu$F
$C_1 = C_2 = 0.22\ \mu$F

$r_D = R_D \| R_L$
$= (1875\ \Omega) \| (4700\ \Omega)$
$= 1340\ \Omega$

From Fig 4-30:

$V_{DSQ} = 12$ V
$I_{DQ} = 4$ mA
$V_{DS(max)} = V_{DSQ} + I_{DQ}(r_D + r_S)$
$= 12\text{ V} + (4\text{ mA})(1340 + 25\ \Omega)$
$= 17.46$ V

(b)

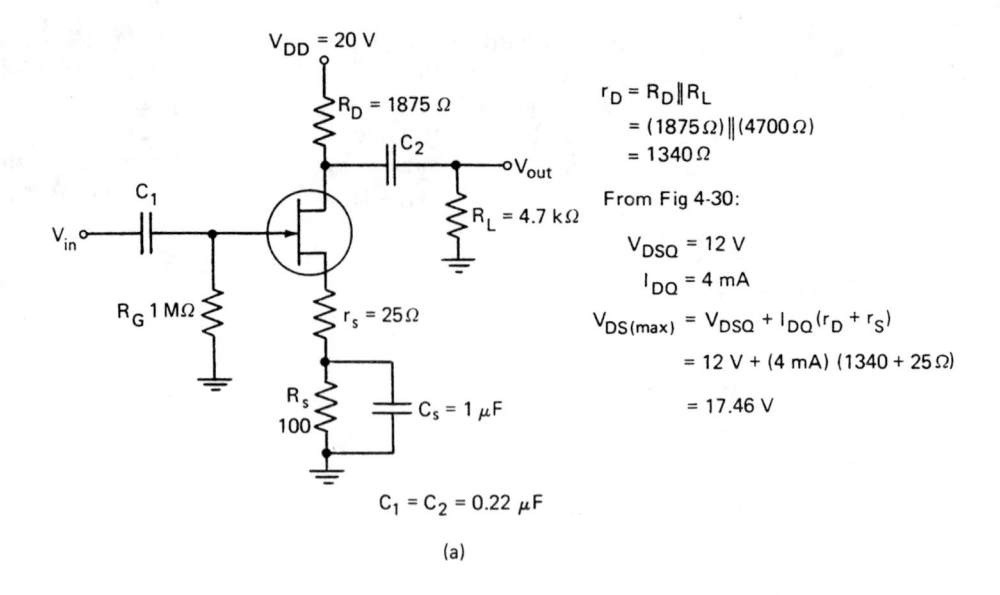

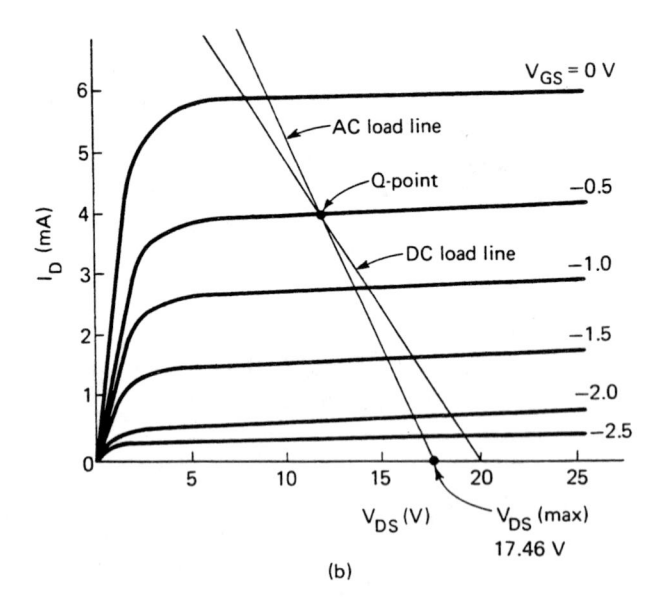

$V_{DS(max)}$ 17.46 V

FIGURE 4-31 (a) A more general circuit for a self-biased JFET. (b) Construction of the AC load line using the results of the calculations in part (a).

106   Biasing Arrangements   Chap. 4

current, respectively. V_{DS} will reach its maximum when I_D is instantaneously zero. By setting $I_D = 0$ in equation (4-6), we obtain

$$V_{DS(max)} = V_{DSQ} + I_{DQ}(r_D + r_S) \qquad (4\text{-}7)$$

This equation is useful in constructing the AC load line; it gives a point on the horizontal axis, as shown in Figure 4-31(b). A second point is the Q-point itself, and these two points are all we need to draw the line.

The Dynamic Transfer Curve (DTC)

Further insight into the AC operation can be obtained by constructing the DTC, as shown in Figure 4-32. Points where the AC load line cross the various V_{GS} curves are noted and transferred to the plot of I_D vs. V_{GS} to form the DTC. An obvious feature of the DTC is that it is indeed curved. As a matter of fact, it will closely approximate a parabola. The implications of this curvature are discussed below.

Once the DTC is available, a graphic view of the circuit operation can be obtained by constructing an input signal on a vertical axis that passes through the Q-point on the DTC. Each point of the input signal is reflected upward to the DTC, then to the right to the AC load line, and then downward where the output waveform is constructed on a vertical axis. Note that the output signal is larger in amplitude than the input signal, a sign of amplification.

The curvature of the DTC causes large signals to be distorted. The slope of the DTC is much less on the negative alternation of the input signal so that the effect on the drain current is less, and the output waveform is not an exact (although

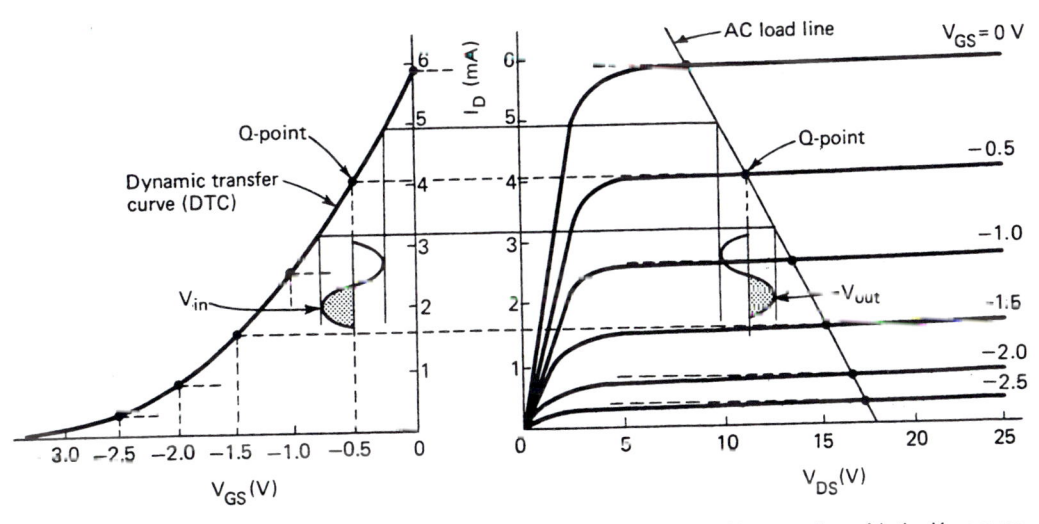

FIGURE 4-32 The DTC is constructed from the AC load line and the points of intersection with the V_{GS} curves of the drain characteristics.

inverted) duplicate of the input. This is not a desirable situation for an amplifier, so JFETs are used primarily as small-signal amplifiers.

If the input signal is small, only a small portion of the DTC will be used. In this small portion, the DTC is approximately straight and the distortion is minimal.

Transconductance, g_m

This parameter tells how effective the input signal is in controlling the output current, the drain current in this case. By definition, it is

$$g_m = \frac{\Delta I_D}{\Delta V_{GS}} \qquad (4\text{-}8)$$

which is the ratio of the change in drain current ΔI_D to the change in gate-source voltage ΔV_{GS} that caused the change in drain current. We can turn this around to obtain

$$\Delta I_D = g_m(\Delta V_{GS}) \qquad (4\text{-}9)$$

which says that the transconductance times the change in gate-source voltage gives the change in drain current. The unit of transconductance is the siemens (S). Smaller units are the millisiemens (mS) and microsiemens (μS). For example, if a change in V_{GS} of 1 V produces a change in I_D of 1 mA, the transconductance is

$$g_m = \frac{\Delta I_D}{\Delta V_{GS}} = \frac{1 \text{ mA}}{1 \text{ V}} = 1 \text{ mS} \qquad (4\text{-}10)$$

Grapically, the transconductance g_m is the slope, at a given point, of the DTC. Since the DTC is not straight (not linear), the transconductance varies from one point to another on the DTC. It will be greater for larger values of I_D. By shifting the Q-point to a greater value for I_{DQ}, a larger g_m can be utilized.

Determination of g_m

The transconductance can be determined from the drain characteristic curves as shown in Figure 4-33. However, because the V_{GS} curves are not uniformly spaced (a general property of FETs), the value obtained will vary significantly from one region of the curves to another. For the curves used in our design example, g_m is about 4 mS for $I_D = 5$ mA, but it is only about 2 mS for $I_D = 1$ mA. For a small-signal amplifier, g_m should be determined in the general vicinity of the Q-point.

Gain

The gain can be determined graphically by taking the ratio of V_{out} to V_{in}, but an easier way is to make use of g_m. Suppose that an input signal V_{in} is applied to a JFET circuit in which $r_S = 0$ (all the source resistance is bypassed). Then from

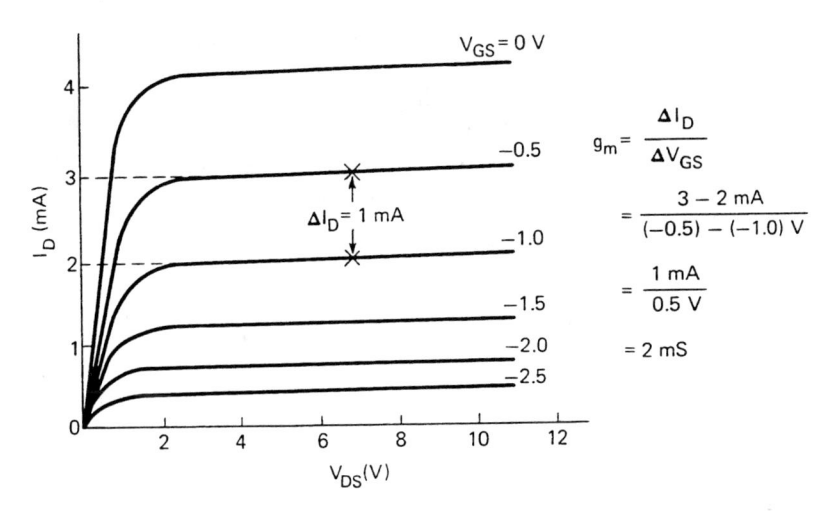

FIGURE 4-33 Determination of the transconductance g_m from the drain characteristics of a JFET.

$$g_m = \frac{\Delta I_D}{\Delta V_{GS}} = \frac{3-2 \text{ mA}}{(-0.5)-(-1.0) \text{ V}} = \frac{1 \text{ mA}}{0.5 \text{ V}} = 2 \text{ mS}$$

equation (4-9), $\Delta I_D = g_m V_{in}$. The corresponding variation at the output (the drain terminal) is

$$V_{out} = (\Delta I_D)r_D = (g_m V_{in})r_D \qquad (4\text{-}11)$$

which yields, because gain = V_{out}/V_{in};

$$r_S = 0: \qquad \text{Gain} = g_m r_D \qquad (4\text{-}12)$$

If r_S is not zero, a slightly more complicated analysis gives

$$r_s \neq 0: \qquad \text{Gain} = g_m r_D/(1 + g_m r_S) \qquad (4\text{-}13)$$

In our example, we determined g_m to be about 4 mS; r_D is 1340 Ω and r_S is 25 Ω. Hence,

$$\text{Gain} = \frac{(4 \text{ mS})(1340 \text{ }\Omega)}{1 + (4 \text{ mS})(25 \text{ }\Omega)} = 4.87$$

If r_S had been zero, the gain would have been

$$\text{Gain} = (4 \text{ mS})(1340 \text{ }\Omega) = 5.36$$

Controlled-Gain Amplifiers

At this point we have seen that shifting the Q-point will change the value of g_m exhibited by a JFET. Further, the gain of the stage depends directly upon g_m. Thus, by shifting the Q-point, we can vary the gain of the stage.

JFETs are often encountered as RF amplifiers in radio receivers. In this application, the variable gain feature is used to advantage as part of the automatic gain control circuit. This circuit, as the name implies, automatically controls the gain of several cascaded stages so that a signal of more-or-less constant amplitude is delivered to the demodulator, even when the signal at the antenna varies widely. These circuits are described in more detailed in later chapters.

Dynamic Range

The dynamic range, the range of permitted input voltage variations, can be determined from the DTC. Obviously, V_{GS} should not be driven more negative than the cutoff point, $V_{GS(off)}$, and V_{GS} cannot be allowed to become positive. Hence, the dynamic range must span a voltage range less than $V_{GS(off)}$, and in practice, the curvature of the DTC limits it to an even smaller value. Clipping or severe distortion occurs if the dynamic range is exceeded.

Input Impedance

Because the gate-channel junction is always reverse-biased, only a very small leakage current flows from the gate terminal. Therefore, the resistance seen looking into the gate can usually be considered infinite, and, in any event, it is well into the megohm region. Therefore, the input impedance is determined primarily by the gate resistor R_G. At high frequencies, however, the small interelectrode capacitance between the gate and channel comes into play, so that the input impedance to the gate is much smaller than its low-frequency value. But in the audio range, the capacitive effects are negligible.

Source Biasing

Source biasing, shown in Figure 4-34, provides for I_D to be less sensitive to V_{GSQ} than in the self-bias circuit. If the source supply voltage V_{SS} is considerably larger than V_{GSQ} (V_{GSQ} is typically 1 or 2 V), I_{DQ} will depend more upon V_{SS} and R_S than upon V_{GSQ}. The result is that I_{DQ} is much more stable than in the self-bias circuit because it depends less upon the parameters of the FET.

If we can estimate the V_{GSQ} a particular FET requires, we can calculate with reasonable accuracy the Q-point drain current I_{DQ} and the Q-point drain and source voltages, V_D and V_S. If $V_{GS(off)}$ is known, we may take V_{GSQ} to be one-half its value. In most cases, V_{GSQ} will be 1 or 2 V. Once an estimate is made of V_{GSQ}, we can calculate the other DC parameters by the formulas given in Figure 4-34.

Voltage-Divider Bias

The voltage-divider bias circuit, shown in Figure 4-35, resembles the universal-bias arrangement of BJTs. It offers improved stability over that of the self-bias circuit. The voltage at the gate V_G is the output of the voltage divider consisting of R_1 and

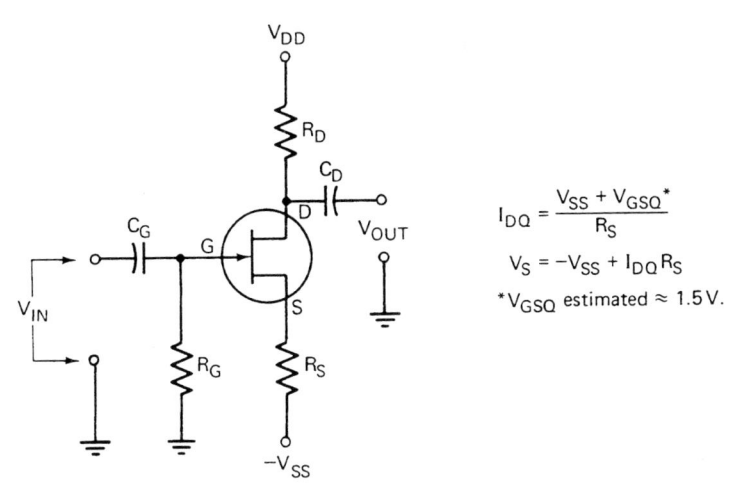

FIGURE 4-34 Source biasing of an N-channel JFET.

$$I_{DQ} = \frac{V_{SS} + V_{GSQ}^*}{R_S}$$

$$V_S = -V_{SS} + I_{DQ} R_S$$

*V_{GSQ} estimated ≈ 1.5 V.

R_2. By pulling the gate to a high positive voltage, the circuit achieves the same effect the source biasing arrangement achieves by using a separate supply to put the low end of the source resistor at a negative voltage. Appropriate formulas for an approximate analysis are given in the figure. Once again, an estimate of V_{GSQ} is required.

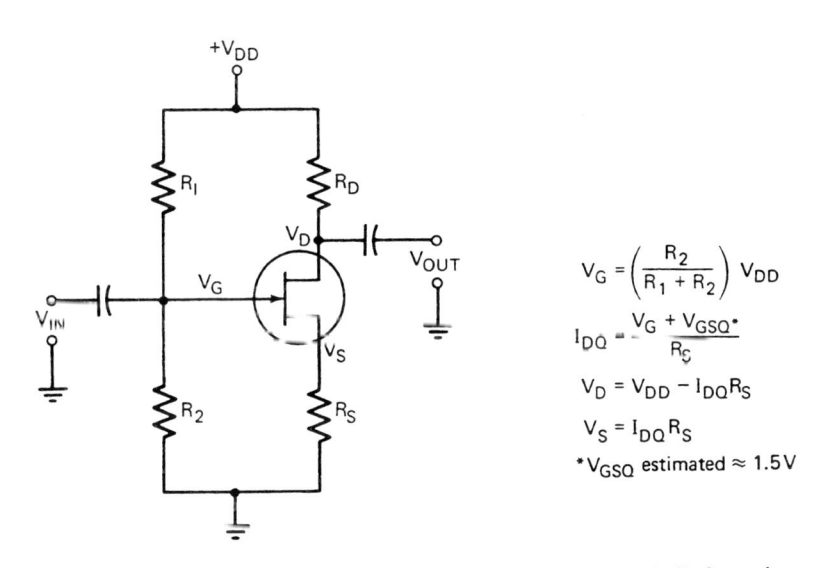

$$V_G = \left(\frac{R_2}{R_1 + R_2}\right) V_{DD}$$

$$I_{DQ} = -\frac{V_G + V_{GSQ}^*}{R_S}$$

$$V_D = V_{DD} - I_{DQ} R_S$$

$$V_S = I_{DQ} R_S$$

*V_{GSQ} estimated ≈ 1.5 V

FIGURE 4-35 Voltage-divider biasing of an N-channel JFET. The similarity to the universal-bias arrangement is evident.

Modification

A modification of the preceding arrangements is frequently encountered in which the low end of the gate resistor is connected to a source of DC voltage rather than to ground. This is done so that the operating point of the JFET can be shifted to control the gain of the stage. Almost always, the DC source supplying the gate is a part of the AVC circuit.

4-13 MOSFET BIASING ARRANGEMENTS

Depletion-mode MOSFETs may be used with the same biasing arrangements used for JFETS—self-bias, source bias, or voltage-divider bias—as shown in Figure 4-36. In these devices, the gate must be maintained negative relative to the source.

Enhancement-mode (E-mode) devices require that the gate be held positive relative to the source (*N*-channel devices), and because of the difference in required

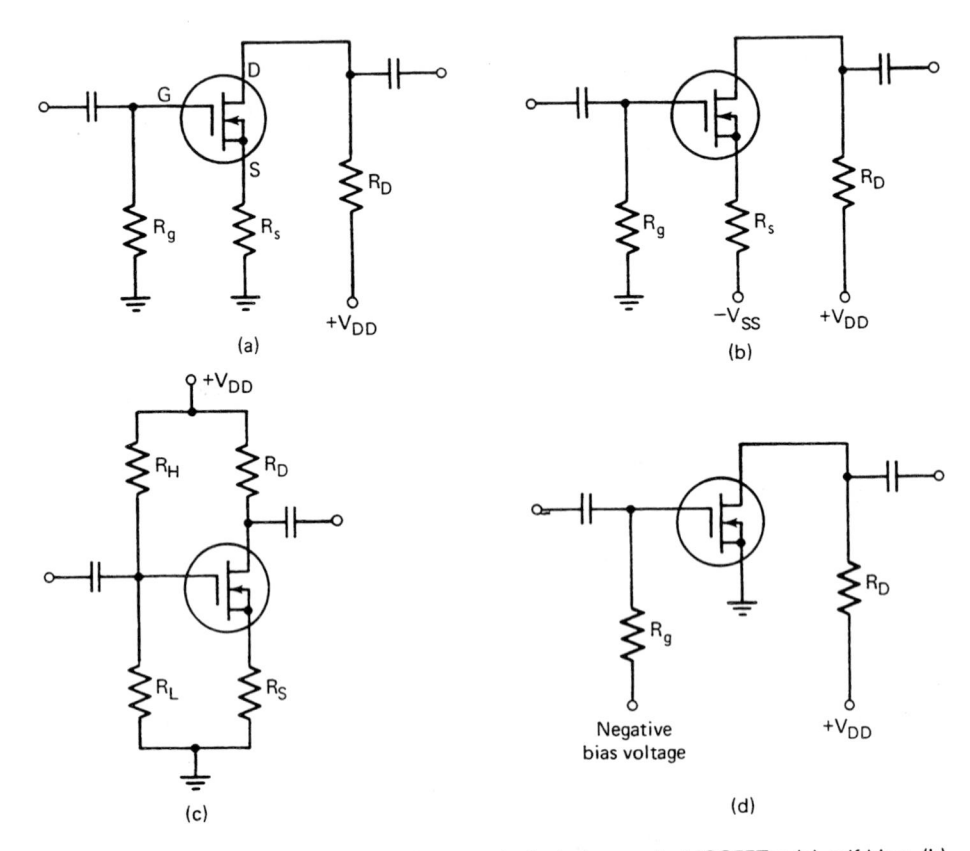

FIGURE 4-36 Bias arrangements for *N*-channel, depletion-mode MOSFETs: (a) self-bias; (b) source bias; (c) voltage-divider bias; (d) bias derived from an external supply.

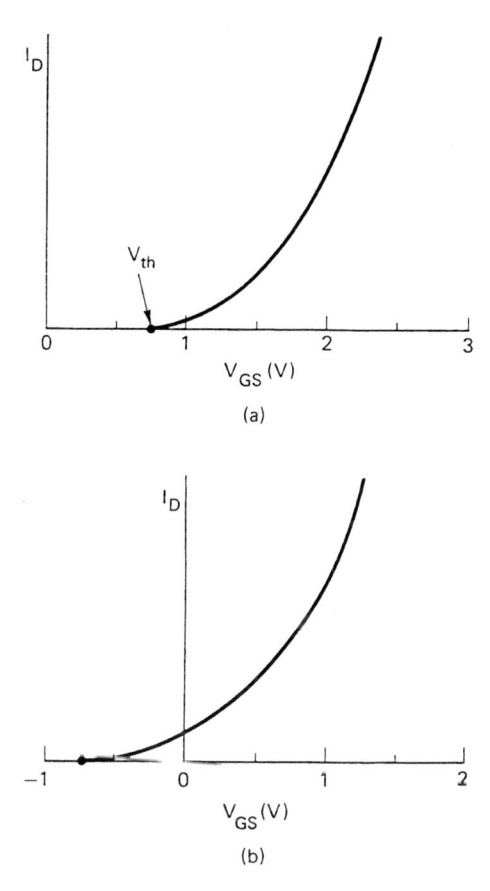

FIGURE 4-37 A transfer curve is a plot of I_D as it depends upon the gate-source voltage V_{GS}. (a) Transfer curve for an enhancement-mode MOSFET. (b) Transfer curve for a D-E mode MOSFET.

bias voltage polarity, the self-bias scheme cannot be used for E-mode MOSFETs. A typical transfer curve for an E-mode MOSFET is shown in Figure 4-37(a). Note that the threshold voltage V_{th} must be exceeded before current flow from source to drain is established. Because of their similar bias voltage requirements, E-mode MOSFETs and BJTs use bias arrangements that are similar. This may be noted in the typical bias arrangements for E-mode MOSFETs shown in Figure 4-38.

MOSFETs are available that are neither purely depletion- or enhancement-mode devices. They may be operated in either mode because their transfer curve occupies the region between the pure depletion and pure enhancement types. The transfer curve of a DE-mode MOSFET is shown in Figure 4-37(b). Note that the gate may be either positive or negative relative to the source. Such devices are frequently operated with essentially zero gate-source voltage, but they are also used with the arrangements used for pure D- or E-mode devices.

Sec. 4-13   *MOSFET Biasing Arrangements*   113

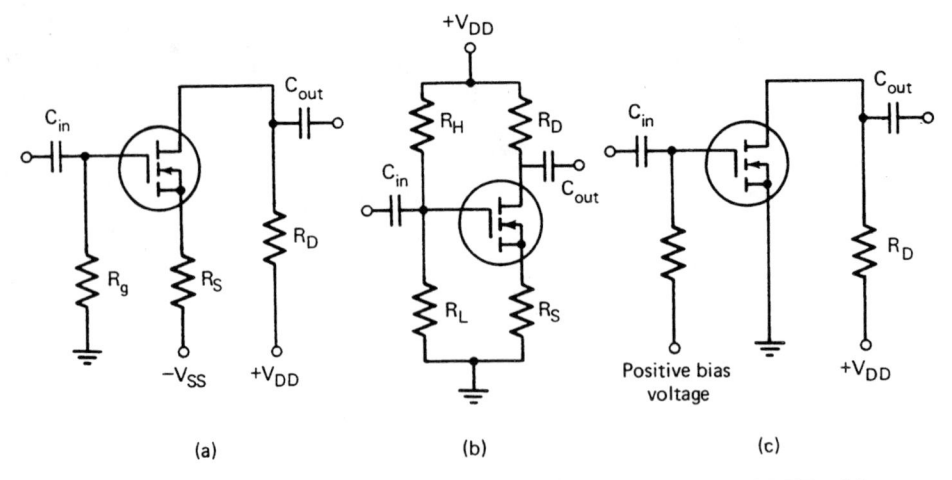

FIGURE 4-38 Biasing arrangements for N-channel, enhancement-mode MOSFETs: (a) source bias; (b) voltage-divider bias; (c) bias voltage derived from an external supply.

Biasing Dual-Gate MOSFETs

It is required that both gates of a dual-gate MOSFET be biased, and almost any combination of the arrangements described above may utilized. Almost always, the input signal is applied to gate $G1$ while a DC gain-control voltage or simply a fixed DC voltage is applied to gate $G2$. In mixer applications, in the frequency converter section of a superheterodyne receiver, both gates may have an RF signal applied. In RF applications in which a signal is applied only to gate $G1$, gate $G2$

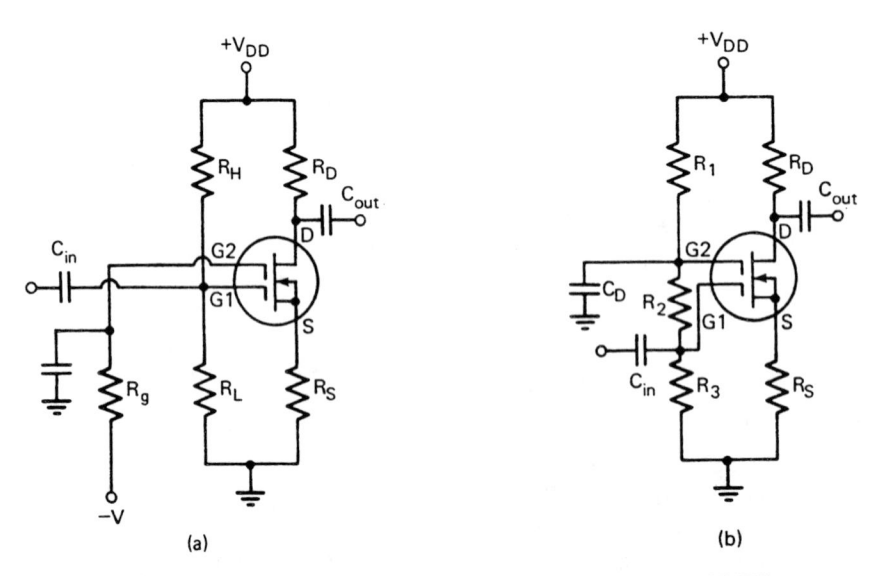

FIGURE 4-39 Typical biasing arrangements for dual-gate MOSFETs.

Biasing Arrangements   Chap. 4

is held at RF signal ground by a capacitor connected from *G2* to ground. Two typical arrangements are shown in Figure 4-39.

Almost all MOSFETs encountered in receivers will be used as RF amplifiers and therefore will usually be used in conjunction with tuned circuits associated with both the input and output of the device. Thus, the drain load will typically be an inductance instead of the drain load resistor R_D shown, for simplicity, in the figures cited above. Further, all the circuits in the figures utilize *N*-channel devices. If *P*-channel devices are used, voltages of opposite polarities are applicable.

4-14 VACUUM-TUBE BIASING

By far, the most common biasing arrangement for vacuum tubes is the self-bias arrangement. Moreover, a vacuum triode in operation acts almost the same as an *N*-channel JFET, although the physical principles are different. A typical self-biased triode amplifier is shown in Figure 4-40. In Figure 4-41 is shown an AC load line and a dynamic transfer curve for a triode amplifier having a power supply voltage of 350 V, R_L of 35 kΩ, R_{LOAD} of 30 kΩ, and R_K of 1.2 kΩ. The analysis parallels that of a JFET almost exactly.

Contact (Grid Leak)-Biased Triode Amplifier

A second method of arranging for the grid to be maintained at a negative potential relative to the cathode utilizes a large resistance in the grid circuit that opposes the leaking off of electrons that tend to accumulate on the grid. The circuit is shown in Figure 4-42(a). No cathode resistor is used, but the large grid resistance traps electrons on the grid, giving it a net negative charge. This method of biasing is called *contact biasing* or *grid leak biasing*.

The grid picks up electrons from the cathode-to-plate current for two reasons. First, the grid wires are physically in the path of the electron flow, and a few

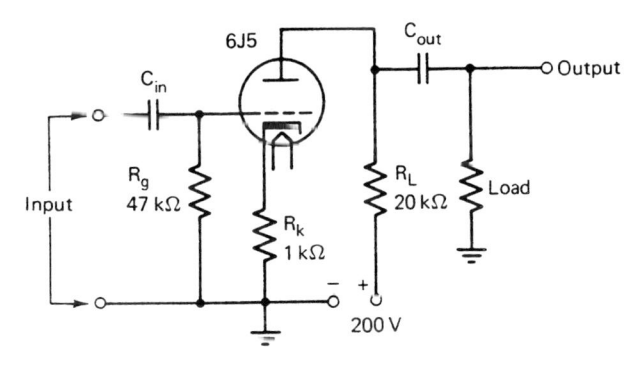

FIGURE 4-40 *R-C*-coupled triode amplifier that uses cathode biasing.

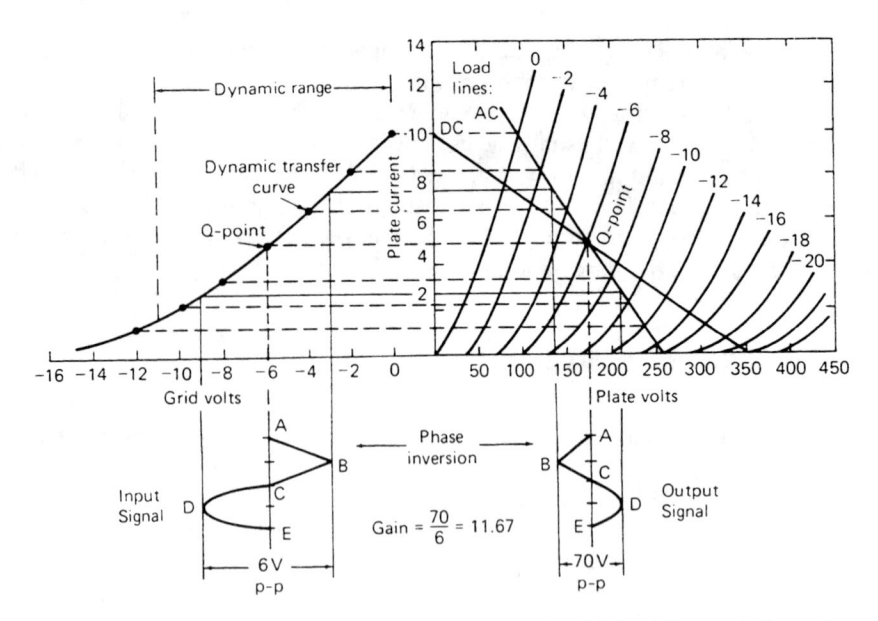

FIGURE 4-41 Dynamic transfer curve is obtained from the AC load line and shows how the plate current varies with the AC component of the grid voltage. The curvature of the dynamic transfer curve introduces distortion into the waveform being amplified.

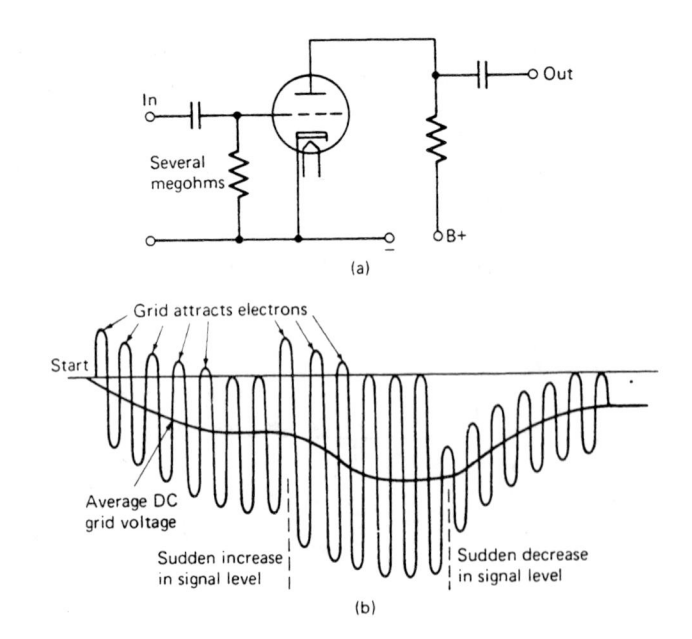

Gain = $\frac{70}{6}$ = 11.67

FIGURE 4-42 Contact bias or grid-leak bias. (a) Basic circuit; note that the cathode resistor is missing, and note that a large resistance is connected into the grid circuit. (b) Variation of the average value of the grid voltage as the input signal amplitude varies.

Biasing Arrangements   Chap. 4

electrons strike the grid and become bound to it. Second, if a signal is applied to the grid, the grid will be driven positive on the first positive half-cycle of the input signal and will attract electrons. The presence of these electrons on the grid causes it to assume a net negative charge which increases with subsequent positive half-cycles until an equilibrium value is obtained, as shown in Figure 4-42(b). At the equilibrium value, the positive peaks of the grid voltage lie very near the zero voltage line. The grid bias varies with the amplitude of the input signal as shown, but the bias voltage so derived prevents the grid from becoming sufficiently positive to introduce serious distortion.

SUMMARY

1. A BJT is basically a current multiplier: the base current is multiplied by the beta to give the collector current.

2. A signal is a variation in voltage that conveys some type of information.

3. The Q-point refers to the particular combination of voltages and currents that exist in an amplifier circuit when no signal is applied to the input of the amplifier.

4. An *NPN* BJT requires that the base be held about 0.7 V more positive than the emitter, and that the collector be maintained at least a volt or so more positive than the base. Hence, the base-emitter junction is forward-biased while the collector-base junction is reverse-biased.

5. The relative voltages for biasing a *PNP* BJT are the same as for an *NPN* BJT, but the polarity is different. The collector of a *PNP* BJT is negative relative to the base and emitter.

6. The simplest biasing arrangement for a BJT is simple base bias in which a resistor connects from the base to the power supply, V_{cc}. This circuit, however, has no natural stability against variations in beta, gain, and so forth, and is not often encountered in commercial circuits.

7. The simple base-bias circuit can be improved by adding a resistor in the emitter circuit to obtain a biasing arrangement known as base bias with emitter feedback.

8. By connecting the base resistor to the collector rather than to the power supply, the simple base-bias circuit is converted to the base bias with collector feedback arrangement.

9. When the base resistor is connected to the collector, and when an emitter resistor is included, the biasing arrangement is that of base bias with collector and emitter feedback.

10. The universal-bias arrangement uses a voltage divider to obtain the voltage applied to the base. It is widely used because of its stability.

11. By connecting the high-base resistor of the universal bias arrangement to the collector rather than to the power supply, collector feedback is added to the circuit and the stability is increased even further.

12. Both a positive and a negative power supply are used in the biasing arrangement known as emitter bias with two supplies.

13. The resistance exhibited by a *PN* junction (i.e., the base-emitter junction of a bipolar transistor) to an applied AC signal voltage is given, for an ideal junction, by 25 divided by the current through the junction expressed in milliamps. Practical junctions give a somewhat larger resistance.

14. The same biasing arrangements may be used for *PNP* transistors by simply reversing the polarity of the power supply. *NPN* transistors were utilized in the circuits of this chapter because of their wider use in commercial equipment.

15. The basic biasing arrangements are sometimes caused to appear more complex by relocating the ground point of the circuit by inverting the power supply. A power supply inversion for an *NPN* stage might involve grounding the collector circuit and supplying the emitter with a negative voltage.

16. The degenerative effects of the emitter resistor may be avoided by bypassing the emitter resistor with an emitter bypass capacitor. This practice increases the gain of the stage at the expense of a lowered input impedance.

17. Field-effect transistors exhibit much higher input impedances than BJTs because FETs are voltage-operated devices, whereas a BJT may be considered to be current driven.

18. Control over the drain current in a FET is achieved by varying the gate voltage relative to the source. The gate voltage controls the thickness of the depletion layers associated with the gate, and the depletion layers, in turn, control the dimensions of the conductive portion of the channel.

19. There are three commonly encountered biasing arrangements for JFETs: self-bias, source bias, and voltage-divider bias.

20. The gate-channel junction of a JFET must never become forward-biased. Therefore, the gate voltage of a JFET is restricted to one polarity. In MOSFET devices, however, no such restriction exists. Consequently, MOSFETs occur of both depletion-mode and enhancement-mode types.

21. Dual-gate MOSFETs have two gates, which make the devices useful as controlled-gain RF amplifiers, mixers, demodulators, and so on.

22. Special care is needed in handling MOSFET devices because static discharges can easily destroy the device by puncturing the insulation between the gate and the channel.

23. Three biasing arrangements are commonly encountered for JFETs: self-biasing, source biasing, and voltage-divider biasing. Biasing arrangements for JFETs are a bit more difficult to analyze than BJT arrangements because of the variability of the Q-point gate-source voltage.

24. Depletion-mode MOSFETs may utilize the same type of biasing arrangements as JFETs.

25. Enhancement-mode MOSFETs require that the gate be held at a potential (relative to the source) of the same polarity as that of the drain. Consequently, the self-biasing arrangement cannot be used for enhancement-mode MOSFETs.

26. The vacuum triode is typically found to use either cathode biasing or grid-leak biasing. Cathode biasing is similar to the source biasing of a JFET. Grid-leak biasing uses a

large grid resistor to trap electrons on the grid, thereby producing a negative voltage on the grid, as required.

27. The dynamic transfer curve is constructed from the AC load line, and it shows how the output current varies with a changing voltage at the input.

28. The degenerative effects of the emitter, source, or cathode resistor can be avoided by connecting a bypass capacitor in parallel with the resistor. Sometimes, however, a small amount of degeneration is desirable because degeneration (reduction in gain) tends to reduce distortion. Degeneration is also called negative feedback.

29. Higher input impedances can be achieved in BJT circuits by including an unbypassed resistance in the emitter circuit.

QUESTIONS AND PROBLEMS

4-1. Give the polarity of the voltage applied to the collector or drain of the following:

   (a) *NPN* transistor.

   (b) *PNP* transistor.

   (c) *N*-channel JFET.

   (d) *P*-channel JFET.

   (e) *N*-channel MOSFET.

   (f) *P*-channel MOSFET.

4-2. A silicon transistor is conducting a moderate current from the emitter to the collector. What approximate voltage appears between the base and the emitter?

4-3. The beta of a silicon transistor is 125. The base current is 5 μA.

   (a) What is the collector current?

   (b) What is the emitter current?

   (c) What is the alpha (α) of the device?

4-4. A power transistor whose beta (β) is 40 conducts a collector current of 1 A. What is the base current?

4-5. A transistor whose β is 80 has a base current that varies from 6 to 11 μA. Over what range does the collector current vary?

4-6. How do biasing arrangements for *PNP* transistors differ from biasing arrangements for *NPN* transistors?

4-7. Why is the base-emitter voltage V_{BE} very nearly the same for all *NPN* silicon transistors in all biasing arrangements? What is the typical value for V_{BE} for silicon transistors? For germanium transistors?

4-8. What is the AC junction resistance of a transistor? Does this resistance depend upon the collector current? Does this resistance affect the input impedance of a transistor stage? Does this resistance affect the gain of a stage?

4-9. Why does the emitter resistor stabilize the gain of a stage against variations in the AC junction resistance of the transistor?

4-10. If the ohmic value of the emitter resistor is increased, does the collector current increase or decrease? Does the collector voltage increase or decrease?

4-11. If the base resistor of a base-biased arrangement is increased in value, do the following increase, decrease, or stay the same?

(a) Collector current.

(b) Collector voltage.

(c) Input impedance.

4-12. Explain how a swamping resistor in the emitter lead of a transistor can stabilize the gain of an amplifier stage.

4-13. In what biasing arrangement does the ohmic value of the collector load resistor affect the collector current?

4-14. Does the collector voltage depend upon the collector load resistor in all biasing arrangements? Why or why not?

4-15. What is the effect on the following parameters of bypassing the emitter resistor with an emitter bypass capacitor?

(a) The collector current.

(b) The input impedance.

(c) The gain.

(d) The collector voltage.

4-16. Why is only a portion of the emitter resistance bypassed in many circuits?

4-17. A simple base-bias circuit (Figure 4-2) using a silicon transistor whose beta is 100 has the following component values: $R_L = 3.3$ kΩ, $R_B = 820$ kΩ; $V_{CC} = 9$ V. Calculate the collector current, the collector voltage, the AC junction resistance, the gain, and the input impedance for the stage.

4-18. In Prob. 17, change the beta of the transistor to 175 and repeat the calculations. Compare the new values of collector current, collector voltage, and gain with the previous results. Why does the input impedance not change with beta?

4-19. A base bias with emitter feedback arrangement is as follows: $V_{CC} = 9$ V, beta = 100, silicon, $R_B = 820$ kΩ, $R_L = 1$ kΩ, $R_L = 3.3$ kΩ. Calculate the collector current, the collector voltage, the gain, and the input impedance.

4-20. In Prob. 19, change the beta of the transistor to 175 and repeat the calculations. What parameters are not effected by the change in beta?

4-21. In Figure 4-12, suppose that the supply voltage changes from 15 V to 12 V. Repeat the calculations of the figure for $V_{CC} = 12$ V and note the effect on the important parameters of the circuit. What happens to the input impedance? Why?

4-22. In Figure 4-14, change R_E from 1 kΩ to 2 kΩ and repeat the calculations. What is the effect on the collector current and voltage? Upon the gain and input impedance?

4-23. A silicon transistor of beta 125 is operated in a universal bias arrangement (Figure 4-17) for which the power supply voltage is 12 V. The component values are $R_{HB} = 22$ kΩ, $R_{LB} = 2.2$ kΩ, $R_E = 390$ Ω, and $R_L = 5.5$ kΩ. Perform an analysis of the circuit as in Figure 4-18.

4-24. In Figure 4-21, change R_E from 1 kΩ to 500 Ω and repeat the calculation.

4-25. A silicon transistor of beta 125 is operated in an emitter-bias-with-two-supplies arrangement whose power supplies are equal in voltage, 9 V. The component values are $R_B = 330$ kΩ, $R_E = 270$ Ω, $R_L = 2.7$ kΩ. Perform an analysis of the circuit as in Figure 4-23.

4-26. Suppose that the transistor in a universal-bias arrangement develops an open base-emitter junction. What will be the effect of this upon:

(a) The collector current?

(b) The collector voltage?

(c) The emitter voltage?

(d) The base voltage?

4-27. When a transistor becomes leaky, it conducts far more than it should. What effect will this have upon the collector voltage?

4-28. Suppose that the collector resistor in a universal-bias arrangement is short-circuited by a tiny strand of wire that becomes lodged across the terminals of the resistor. What will be the effect of this upon:

(a) The collector current?

(b) The collector voltage?

(c) The gain of the stage?

(d) The power dissipation in the transistor?

4-29. An amplifier stage uses an N-channel JFET as the active device. The voltage at the gate is 2.0 V. Would you expect the voltage at the source to be somewhat greater or somewhat less than 2.0 V?

4-30. What accounts for the fact that the gate current of a JFET is extremely small?

4-31. In a self-biased JFET amplifier as in Figure 4-29, what DC voltage would you expect to measure at the gate?

4-32. As the source resistor in a self-biased JFET stage is made smaller, will the drain voltage increase or decrease? Why?

4-33. As the drain resistor R_D is made larger in ohmic value, does the DC load line for a self-biased JFET develop a greater of smaller slope? What effect does this have upon the Q-point drain voltage?

4-34. As the source resistor R_S is made larger in ohmic value, does this shift the Q-point toward or away from the power supply point on the DC load line?

4-35. What effect, if any, does lowering the ohmic value of the gate resistor R_G of a self-biased JFET have upon the location of the Q-point?

4-36. Suppose that a JFET is encountered for which the gate voltage has little effect in controlling the drain current. Would this device have a very large or a very small transconductance?

4-37. Calculate the transconductance for a JFET at an operating point where a change in gate-source voltage of 0.25 V produces a change in drain current of 0.75 mA.

4-38. Does a JFET exhibit a greater transconductance at larger or smaller values of drain current?

4-39. Calculate the gain of a JFET amplifier if the transconductance is 4 mS and the AC drain resistance is 2.2 kΩ.

4-40. In the source-biased amplifier of Figure 4-33, would you expect the source voltage to be slightly greater than or less than 0 V?

4-41. In the universal-bias arrangement for a BJT, the emitter voltage is always slightly less than the base voltage. In the voltage-divider bias circuit of Figure 4-34, will the source voltage be greater than or less than the voltage at the gate?

4-42. An *N*-channel, enhancement-mode MOSFET is utilized in a voltage-divider circuit as in Figure 4-37(b). Will the source voltage be more or less positive than the voltage at the gate?

4-43. In normal operation, is the grid voltage of a vacuum triode maintained more positive or more negative than the voltage at the cathode?

4-44. What is the approximate voltage on the grid relative to ground in an *RC*-coupled triode amplifier?

4-45. Suppose that the cathode resistor of an *RC*-coupled triode amplifier is not bypassed with a capacitor. Will a signal voltage be found at the cathode? Explain.

4-46. Explain how a grid bias voltage is obtained using contact (grid-leak) bias.

4-47. Why does the average level of grid bias voltage obtained by grid leak bias vary with the amplitude of the input signal?

5
AUDIO AND POWER AUDIO AMPLIFIERS

Broadly speaking, amplifiers may be classified as either voltage amplifiers or power amplifiers. The purpose of a voltage amplifier is simply to increase the amplitude of the input signal with little regard or concern for the absolute power levels involved. Typically, voltage amplifiers operate into fairly high impedances. On the other hand, power amplifiers are designed to deliver significant power to a fairly low impedance, such as an earphone or speaker, and they may contribute only small amounts of voltage gain. Of course, most, if not all, amplifiers encountered will increase to some extent the power level of the input signal.

The amplifiers considered in Chap. 4 were voltage amplifiers not primarily intended to be power amplifiers. In this chapter we consider the modifications that are necessary to construct power amplifiers that are capable of driving the speaker of a receiver. Both vacuum-tube and solid-state circuits are considered, and the emphasis is placed on push-pull (*P-P*) amplifiers.

The first three sections of this chapter are devoted to cascaded stages, direct-coupled amplifiers, and amplifier configurations such as the common-base and common-collector circuits. The body of the chapter treats power amplifiers, and a description is given of three audio amplifier sections taken from actual receivers. The circuits described in this chapter are commonly encountered in radio receivers and other audio amplifiers.

5-1 CASCADED STAGES

When two stages are connected so that the signal amplified by stage 1 passes on to stage 2 for further amplification, the stages are said to be *cascaded*. When the stages are transformer- or *RC*-coupled, the DC voltages of the stages do not interact, and the biasing arrangements may be considered separately. Such is not the case for direct-coupled amplifiers, however, because the direct connection between the stages allows DC levels to interact. Direct-coupled amplifiers are described in the following section, and several examples of transformer-coupled stages appear in later sections of this chapter. We now investigate the factors that affect the total gain of two *RC*-coupled stages connected in cascade as shown in Figure 5-1. This discussion is directed toward BJT amplifier stages, but the same principles are applicable to cascaded vacuum-tube and FET stages as well.

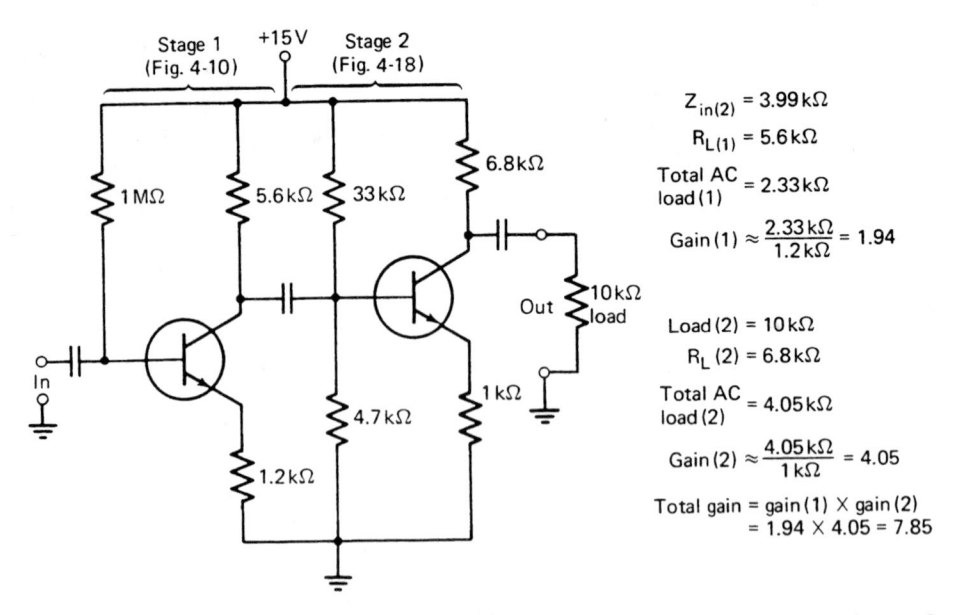

FIGURE 5-1 Cascaded stages to form a two-stage amplifier. The input impedance of stage 2 forms part of the load of stage 1 and reduces the gain of stage 1. The total gain is the product of the separate stage gains.

$Z_{in(2)} = 3.99\,k\Omega$
$R_{L(1)} = 5.6\,k\Omega$
Total AC load (1) $= 2.33\,k\Omega$
Gain (1) $\approx \dfrac{2.33\,k\Omega}{1.2\,k\Omega} = 1.94$

Load (2) $= 10\,k\Omega$
$R_L(2) = 6.8\,k\Omega$
Total AC load (2) $= 4.05\,k\Omega$
Gain (2) $\approx \dfrac{4.05\,k\Omega}{1\,k\Omega} = 4.05$

Total gain = gain (1) × gain (2)
= 1.94 × 4.05 = 7.85

The voltage gain provided by the various biasing arrangements described in Chap. 4 was determined under the assumption that no load was placed on the output of the stage. When a load (a resistance, for example) is connected to the output, the total AC load resistance in the collector circuit becomes the parallel equivalent of the collector load resistance R_L and the load. We denote this parallel equivalent resistance as r_L. Then, to calculate the gain of the loaded stage, r_L is used in the gain formula in the place of R_L. The effect of connecting a load is to reduce the gain of the stage.

When stages are cascaded, the input impedance of stage 2 becomes the load for stage 1, and the gain of stage 1 will be reduced. The total AC load of stage 1 is the parallel equivalent of R_{L1} and Z_{in2}. If a load is placed on stage 2, the effect of the load on the gain of stage 2 must be calculated. The overall gain of the cascaded amplifier is the product of the gains of the stages, with stage loads considered.

To illustrate, refer to the calculations shown in Figure 5-1. The DC analysis of the two stages were given in Figures 4-10 and 4-18. The analysis of the cascaded amplifier begins with the calculation of the total AC load of stage 1, which is the parallel equivalent of 5.6 kΩ and 3.99 kΩ, namely 2.33 kΩ. Hence, the gain of stage 1 is obtained by dividing 2.33 kΩ by the emitter resistance of 1.2 kΩ, and the gain is found to be 1.94. This is to be compared with the no-load gain of stage 1 (Figure 4-10) of 4.61. Connecting stage 2 to the output of stage 1 lowered the gain of stage 1.

In a similar manner, the AC load of stage 2 is calculated to be 4.05 kΩ, and the loaded gain of stage 2 is 4.05. Consequently, the gain of the entire amplifier composed of both stages is the product of the loaded gains of each stage, namely 7.85, as indicated.

Anything that may happen to stage 2 that serves to reduce the input impedance to stage 2 will, in turn, give rise to a smaller AC load impedance for stage 1 and will reduce the gain of stage 1. This will reduce the gain of the entire amplifier. For example, if the beta of the second-stage transistor decreases, so will the input impedance of the second stage, since the input impedance depends upon the beta of the transistor. A technician should be alert to the fact that reduced gain of a given stage might be due to a defect in the following stage that tends to "load down" the stage being investigated.

5-2 DIRECT-COUPLED AMPLIFIERS

Figure 5-2 illustrates a two-stage amplifier that utilizes *direct coupling* between the stages. The coupling capacitor has been omitted so that the collector of $Q1$ connects directly to the base of $Q2$. Bias voltage for the base of $Q2$ is provided by the DC voltage at the collector of $Q1$, and no biasing resistors are required for the $Q2$ base.

A disadvantage of direct coupling is that the DC voltage levels of the two stages are not independent. Any variation in voltage at the collector of $Q1$ will be passed on to and amplified by $Q2$. This makes careful design essential, and particular attention must be given to the thermal stability of the amplifier.

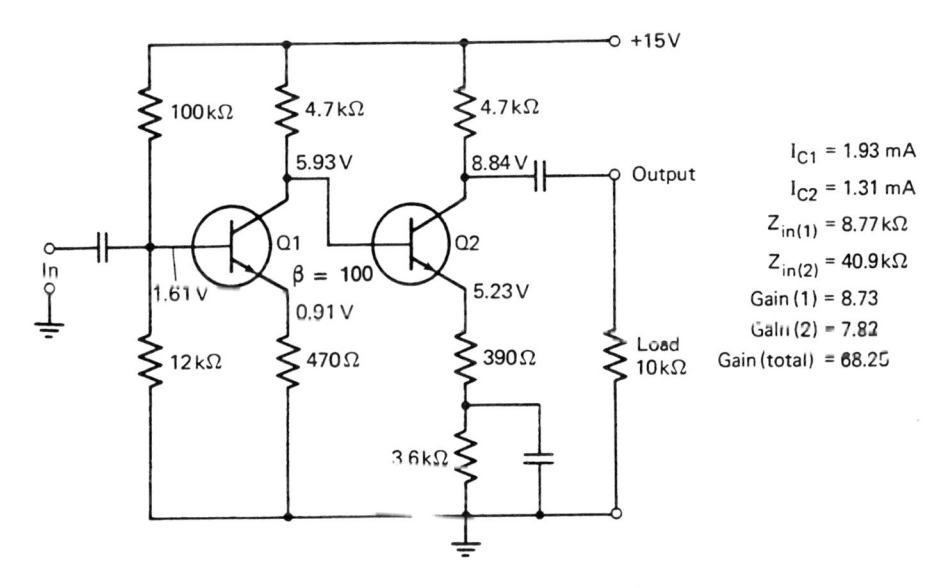

FIGURE 5-2 Direct-coupled amplifier.

The first stage is a universal-bias arrangement that may be analyzed with formulas given earlier provided that the base current of $Q2$ is small in comparison with the collector current of $Q1$. After proceeding with the analysis until the $Q1$ collector voltage is known, the voltage at the emitter of $Q2$ is found by subtracting V_{BE} (assumed to be 0.7 V) from the $Q2$ base voltage, which is the same as the $Q1$ collector voltage. The collector current of $Q2$ is then obtained by applying Ohm's law to the emitter resistance of $Q2$. Finally, the $Q2$ collector voltage is calculated by subtracting the voltage drop across the $Q2$ collector load resistor from the power supply voltage.

The AC analysis is essentially the same as for capacitively coupled stages. The input impedance to stage 1 is calculated as before. The input impedance to the second stage is the impedance "seen" looking into the base. The gain of the first stage is calculated by considering that the input impedance to the second stage is in parallel with the load resistor of stage 1. The gain of the second stage is obtained by calculating the parallel equivalent of the collector load resistor and the 10-kΩ amplifier load. The results of the AC analysis are given in the figure.

When *RC* coupling is used in cascaded stages, the reactance of the coupling capacitors reduces the low-frequency gain and places a lower limit on the operating frequency of the amplifier. By eliminating the coupling capacitors in direct-coupled amplifiers, the low-frequency limit is removed, and the amplifiers will operate at frequencies as low as pure DC.

In the amplifier of Figure 5-2, note that the collector voltage of $Q2$ is higher than that of $Q1$. This must be the case if the proper voltage difference is to be maintained between the base and collector of $Q2$. We say the collector voltages are "stacked" because they must become higher and higher as we proceed from the input to the output stage of a multistage amplifier. This can limit the dynamic range of such an amplifier and may be considered a disadvantage of this circuit. We now consider a circuit that avoids the stacking of the collector voltages.

Complementary Amplifier

The complementary amplifier of Figure 5-3 utilizes an *NPN* and a *PNP* transistor to avoid stacking of the collector voltages. The *PNP* transistor operates "upside down" in the circuit because of the voltage polarities required by the *PNP* transistor. Note that the collector voltage of $Q2$ is lower than that of $Q1$.

Analysis of the first stage is straightforward, but the second stage requires a bit of care. The emitter voltage of the second-stage transistor $Q2$ is higher than the base voltage by an amount V_{BE}. The voltage across the $Q2$ emitter resistor is found by subtracting the emitter voltage from the power supply voltage, and then the $Q2$ collector current is found by applying Ohm's law to the emitter resistor. The $Q2$ collector voltage is simply the voltage developed across the collector load resistor by the flow of the collector current, since the low end of the load resistor is grounded. The gain and input impedance calculations proceed in the same manner as for the amplifier of Figure 5-1. The results are given in the figure.

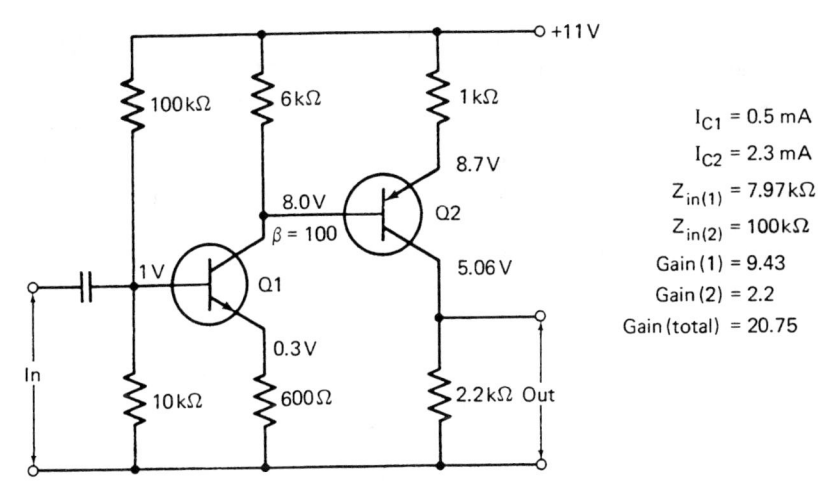

FIGURE 5-3 Direct-coupled amplifier that uses an *NPN* and *PNP* transistor combination to avoid stacking of the collector voltages.

DC-Stabilized Direct-Coupled Amplifier

The configuration of a direct-coupled amplifier frequently encountered is shown in Figure 5-4. The bias voltage for $Q1$ is derived from the emitter of $Q2$ in a modified universal-bias arrangement in which the high-base resistor R_{HB} connects to the positive voltage at the $Q2$ emitter rather than to the positive power supply. This connection is made in order to enhance the DC stability of the amplifier.

Suppose that the DC collector voltage of $Q1$ rises undesirably for some reason. This increases the forward bias of $Q2$ and will thereby cause the $Q2$ collector current to increase. The increased collector current of $Q2$ could result in either nonlinear operation of $Q2$, or, if the current increase is sufficiently large, it could result in the overheating and eventual destruction of $Q2$.

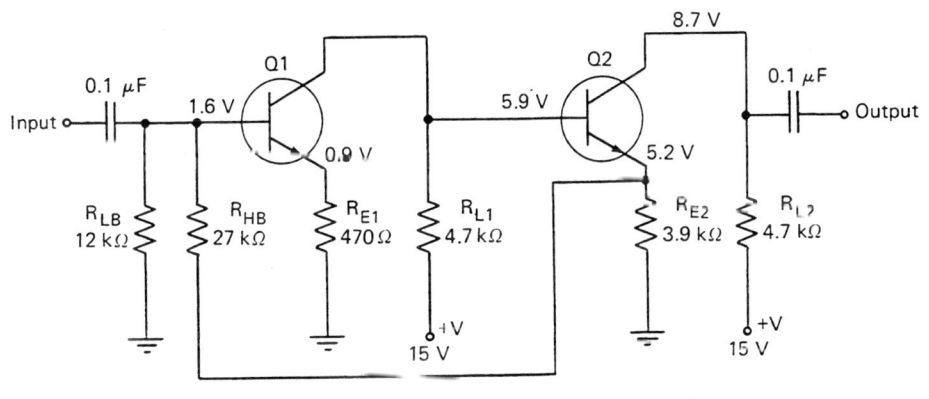

FIGURE 5-4 DC-stabilized direct-coupled amplifier.

Sec. 5-2 Direct-Coupled Amplifiers 127

As the Q2 collector current increases, the positive voltage at the emitter of Q2 rises also in proportion to the current as a result of the increased voltage drop across R_{E2}. Consequently, the voltage applied to the base of Q1 via R_{HB} increases and, in turn, casues the Q1 collector current to increase. The increased collector current of Q1 then causes the Q1 collector voltage to decrease, offsetting the initial, undesired increase in voltage at the collector of Q1. Thus, by connecting R_{HB} to the emitter of Q2, a DC feedback path is created that tends to stabilize the DC voltage levels in the amplifier.

Darlington Pair

A useful combination of transistors known as a *Darlington pair* acts as a single transistor whose beta equals the product of the betas of the two transistors that form the pair. Consequently, betas in the thousands are easily achieved, and the high beta translates directly into a high input impedance for the device. Darlington pairs are frequently encountered, and two possible combinations are shown in Figure 5-5.

Note that the equivalent transistor of the pair shown in Figure 5-5(b) is of the same type as the input transistor of the pair. We refer to Darlington pairs that use transistors of opposite types as being *complementary* Darlington pairs.

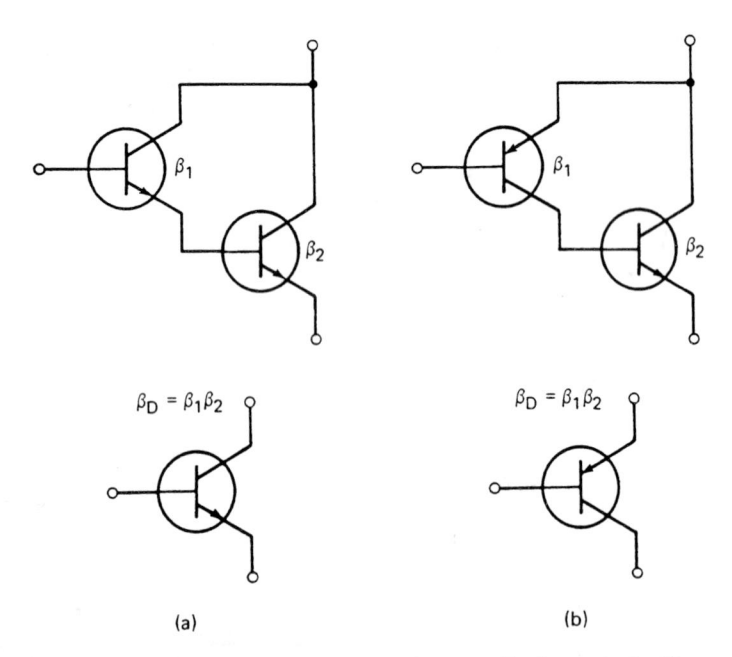

FIGURE 5-5 (a) Standard and (b) complementary Darlington pairs. The effective beta of the Darlington pair is the product of the betas of the individual transistors.

5-3 AMPLIFIER CONFIGURATIONS

The amplifiers considered thus far have been common cathode, common emitter, or common source amplifiers for vacuum-tube, BJT, and FET technologies, respectively. Momentarily directing our discussion to BJTs, the input signal, in the common-emitter configuration, is applied to the base while the output signal appears on the collector. It is possible to operate BJTs as common base (CB) and common-collector (CC) amplifiers using many of the same biasing arrangements that are used for the common-emitter (CE) circuits. Likewise, vacuum tubes can be operated as common-grid and as common-plate (cathode follower) amplifiers, and FETs can be operated as common-gate and as common-drain amplifiers. We now consider the common-base configuration for BJTs.

Common-Base Configuration

A BBEF arrangement operated as a CB amplifier is shown in Figure 5-6. The input signal is applied to the emitter while the output signal appears at the collector. Note the capacitor that holds the base at signal ground.

In all BJT configurations, the controlling parameter is the forward bias applied to the base-emitter junction. In the common-base configuration, the signal applied to the emitter causes the emitter voltage to vary. Since the base voltage is held constant, the voltage variation at the emitter gives rise to a variation in voltage across the base-emitter junction. This has the same effect upon the collector current as if the emitter voltage were held constant while the base voltage varies, as in the CE configuration. The net forward bias applied to the base-emitter junction varies

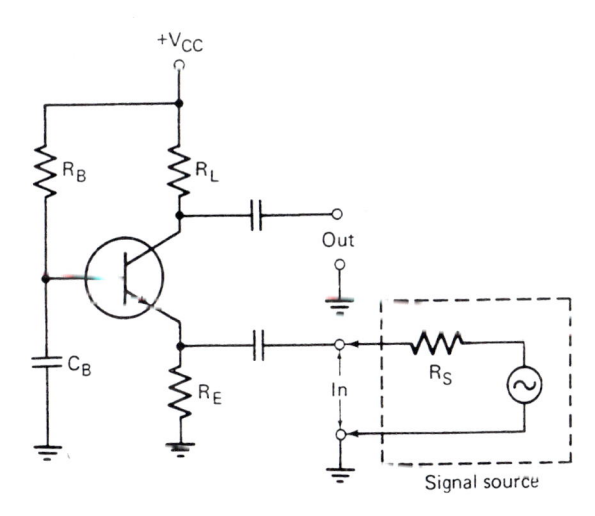

FIGURE 5-6 BBEF-biasing arrangement configured as a common-base amplifier. The gain depends strongly upon the source resistance R_S.

with the input signal, and this produces a variation in the collector current. The variation of collector current produces the output signal in the same manner as in the CE configuration.

Since the input signal is applied to the emitter, only those biasing arrangements that have an emitter resistor may be used in the CB configuration. When no emitter resistor is present, the emitter is at signal ground, and application of the input signal to this point would merely ground the input signal. Further, the emitter resistor may not be bypassed with an emitter bypass capacitor, for similar reasons.

No phase inversion occurs between the input and output (from emitter to collector) of a common-base amplifier. An increase in emitter voltage tends to reduce the forward bias of the BE junction, which, in turn, reduces the collector current. As the collector current decreases, the collector voltage increases. Hence, the input voltage to the emitter and the output voltage developed at the collector are in phase.

Common-base amplifiers typically have high gains and low input impedances. They are used primarily as RF amplifiers and are seldom found in audio applications.

The same principles may be applied to vacuum-tube and FET technology to produce common-grid (grounded-grid) amplifiers and common-gate amplifiers. A common-grid amplifier is shown in Figure 6-10. Common-gate FET amplifiers are not frequently encountered.

Cathode Follower

The common-plate configuration of vacuum-tube technology is called a cathode follower, shown in Figure 5-7. The input signal is applied to the grid and the output signal is taken from the cathode. The plate is held at signal ground, typically by a

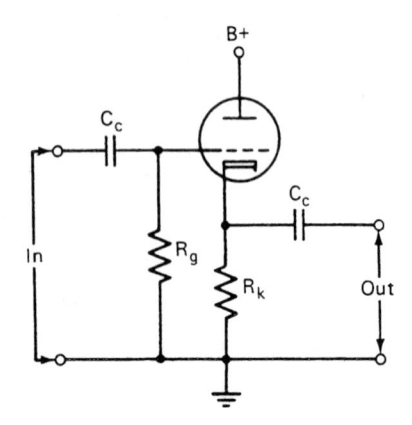

FIGURE 5-7 Cathode follower. The primary use of this circuit is that of an impedance-matching device. It has a high input impedance and a low output impedance. Its gain is 1, at most.

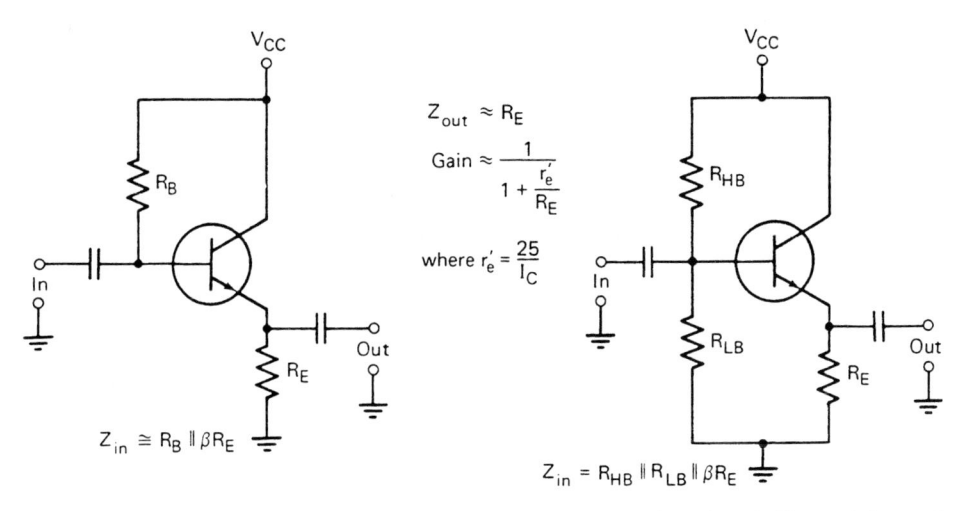

FIGURE 5-8 Typical emitter-follower circuits. The gain is always less than 1. The usefulness of the circuit lies in its impedance-matching properties; it has a high input impedance and a low output impedance.

filter capacitor associated with the power supply. The circuit can, at most, have a gain of 1 (unity), and its primary use is that of an impedance matching device. The input impedance is high while the output impedance is low. In the realm of AM, FM, and FM stereo receivers, cathode followers are frequently encountered only in the older tube-type stereo (matrix) decoders.

No phase inversion occurs from the input to the output. As the grid goes positive, plate current increases so that the signal voltage developed across R_K goes more positive also. Obviously, a cathode resistor must be present, and no cathode bypass capacitor can be used in an effort to increase the gain of the stage.

The term "cathode follower" arises from the fact that the cathode signal voltage tends to follow, almost exactly, the signal voltage applied to the grid. Formulas for the input impedance, the output impedance, and the gain are given in the figure.

Emitter Follower (Common Collector)

This circuit is the BJT counterpart of the cathode follower discussed above. The input signal is applied to the base, and the output signal is taken from the emitter. The collector load resistor is missing entirely, and therefore the collector is held at signal ground by the capacitors associated with the power supply. The circuit is characterized by a voltage gain of less than 1, a high input impedance, and a low output impedance. The circuit is valuable as an impedance matching circuit.

DC voltage and current calculations follow the same pattern as for the common-emitter configuration. Typical circuits and formulas for calculating the gain, input impedance, and output impedance are given in Figure 5-8.

5-4 TUBE-TYPE SINGLE-ENDED POWER AUDIO AMPLIFIERS

An audio amplifier designed to drive a speaker is a power amplifier as opposed to a voltage amplifier, whose function is to increase the amplitude of a signal with little concern for increasing its power level. In radio receivers, the power amplifier that drives the speaker is called the *audio-output stage*, and the tube involved is called the *audio-output tube*.

Tubes designed to deliver appreciable power are more sturdily constructed than voltage amplifiers. The cathode and plate are larger, and the plate may be designed to dissipate considerable heat. This, of course, depends upon the amount of power actually involved, which may range from a fraction of a watt to 25 W or more. Refer to a tube manual and compare a 3V4 and a 6L6. Triodes are almost never used as output tubes because of the higher efficiency and power sensitivity of beam power pentodes.

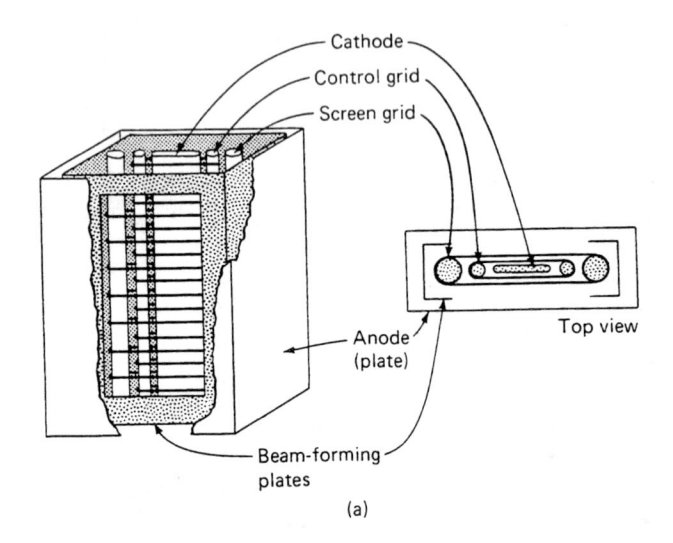

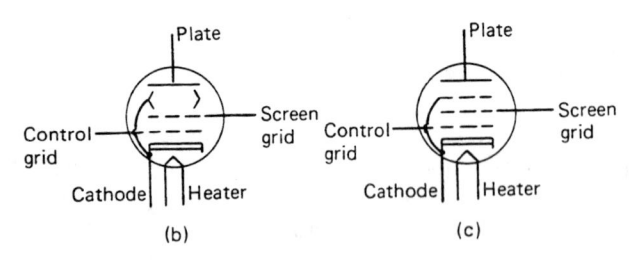

FIGURE 5-9 (a) Structure of a beam power pentode in which the electrons traveling from cathode to plate are channeled through an opening in the beam-forming plates. (b) Symbol showing the beam-forming plates. (c) Symbol in which the beam-forming plates are represented as a suppressor grid.

Audio and Power Audio Amplifiers   Chap. 5

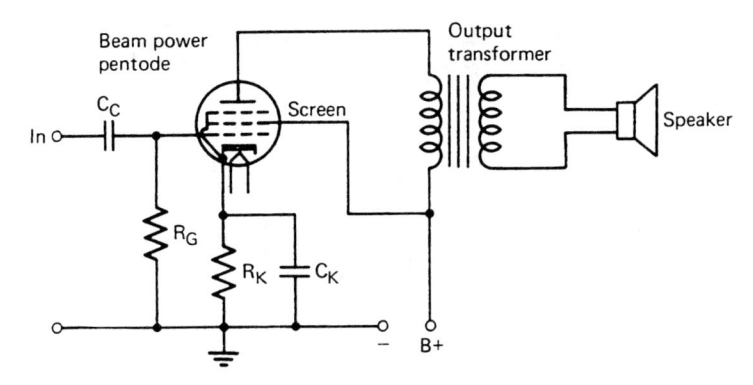

FIGURE 5-10 Basic circuit for a single-ended power amplifier. The output transformer is an essential element since it matches the impedance of the tube to the impedance of the speaker.

Compared to a triode, a beam power pentode has two additional elements. As shown in Figure 5-9, one is a screen grid, and the other is a set of beam-forming plates. In actuality, the beam-forming plates consist of a rectangular, box-like structure with openings on opposite side through which electrons pass in concentrated beams as they travel from cathode to plate. The screen grid is located between the signal grid and beam-forming plates, and is similar to the signal grid in structure. A high positive voltage applied to the screen helps to accelerate electrons from the cathode to the plate, and it also causes the plate current to be less dependent upon the plate voltage.

The beam-forming plates surround the cathode and the two grids and are internally connected to the cathode. The plates concentrate the electrons into beams, thereby giving the signal grid more control over the electrons as they flow from the cathode to the plate.

The basic single-ended output stage is shown in Figure 5-10. The conspicuous item is the audio output transformer in the plate circuit. The tube is a beam power pentode. Note that the screen grid is connected directly to the B+ power supply. The input circuit is the same as the *RC*-coupled circuit of the triode voltage amplifier, and self-biasing is utilized.

Output Transformer

The output transformer matches the high impedance of the plate circuit to the low impedance of the speaker, a requirement that must be met if maximum power is to be transferred from the plate circuit to the speaker. Relatively small currents flow in the plate circuit, but the signal voltage changes at the plate are quite large. Thus, the ratio of voltage change to current change is large at the plate, and we say that the plate circuit is a high-impedance circuit. A speaker, on the other hand, requires a relatively large current and a correspondingly small voltage. Thus, the impedance of a speaker is low. Unless an impedance matching transformer is used between the plate circuit and the speaker, efficient power transfer cannot be obtained.

Let us now investigate the impedance matching function of a transformer. Study the circuit and calculations of Figure 5-11. The transformer is a step-down transformer of turns ratio 10:1. Therefore, the secondary voltage is 1/10 the primary voltage of 100 V, or 10 V. A 5-Ω resistor is connected to the secondary, and a current of 2 A flows through the resistor. The power dissipation in the resistor is 20 W, and 20 VA of power are delivered by the secondary to the resistor. Since the volt-amps remain constant from the input of a transformer to the output, 20 VA of power must be delivered to the primary by the 100-V source. Dividing this by 100 V gives the primary current of 0.2 A.

Here is a question. What resistance is required so that when 100 V is applied to it, 0.2 A of current results? The answer is 500 Ω. The point is that the 5-Ω resistor and transformer combination acts as a 500-Ω resistance as far as the 100-V source is concerned. We say the source "sees" 500 Ω. The 5-Ω resistor appears to have been multiplied by a factor of 100 by the action of the transformer. The 500-Ω equivalent resistance seen by the source is the "reflection" of the 5-Ω resistor from the secondary to the primary, and it is called the *reflected impedance*.

An impedance in the secondary circuit is multiplied by the square of the turns ratio as it is reflected into the primary circuit. In the example above, the turns ratio was 10, and the square of 10 is 100. Therefore, the 5-Ω resistor was reflected into the primary as 500 Ω.

If the transformer is a step-down transformer, the reflected impedance will be higher than the actual impedance in the secondary circuit. If the transformer is a step-up transformer, the converse is true. The following formulas relate the primary impedance Z_p to the secondary impedance Z_s in terms of the turns on the primary T_p and the turns on the secondary T_s:

$$Z_p = \left(\frac{T_p}{T_s}\right)^2 Z_s \qquad Z_s = \left(\frac{T_p}{T_s}\right)^2 Z_p$$

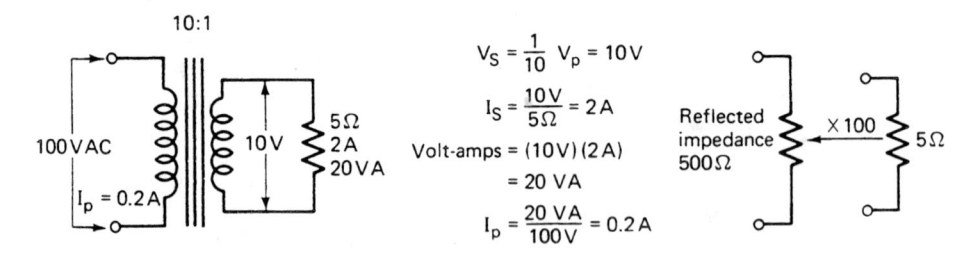

FIGURE 5-11 Illustration of reflected impedance. Current flows in the primary circuit as if the primary exhibited a resistance of 500 Ω. The impedance reflected into the primary equals the secondary resistance times the square of the turns ratio.

Plate-to-Speaker Impedance Matching

Typical power-output tubes are generally provided with an impedance of about 3000 Ω, but a common speaker impedance is 8 Ω. The ratio of the impedance is 375, obtained by dividing 3000 by 8. Taking the square root of this gives the required turns ratio, 19.36, of the necessary impedance matching transformer. The winding of fewer turns must be connected to the speaker.

Output Circuit Operation

Since the DC resistance of the primary winding of the output transformer is fairly small (on the order of 100 Ω), the DC voltage appearing at the plate is practically the same as the B+ voltage of the power supply. Only a small DC voltage drop occurs across the output transformer primary.

The input signal, RC coupled to the grid, causes the plate current to vary in accordance with the grid voltage. The varying plate current flows through the output transformer primary, producing a varying magnetic field in the transformer core. The changing magnetic field induces the output voltage in the secondary.

The resistance presented by the transformer primary to the AC signal currents is much greater than the DC resistance because of the reflected impedance of the speaker. Thus, a large signal voltage appears at the plate. It is important to remember that the output transformer is a step-down transformer; the signal voltage applied to the speaker is much less than the signal voltage appearing at the plate. However, a larger current flows in the speaker circuit so that the power levels in the primary and secondary circuits are (ideally) the same.

5-5 BJT SINGLE-ENDED POWER AUDIO AMPLIFIERS

The BJT amplifiers considered thus far have been RC-coupled voltage amplifiers. These amplifiers may be used to drive a speaker by replacing the collector load resistor with an output transformer that couples the amplified signal to the speaker. The transformer matches the impedance of the collector circuit to the lower impedance of the speaker, and also provides DC isolation. The transistor must be compatible with the power level required by the speaker, and power transistors mounted on *heat sinks* are frequently encountered in such applications. Two circuit possibilities are shown in Figure 5-12.

The impedance matching properties of the transformer are determined by the turns ratio as described in the preceding section. The impedance reflected into the primary circuit forms the AC collector load impedance and determines the gain of the amplifier, from the base to the collector. Recall that a step-down of voltage occurs in passing from the primary to the secondary side of the output transformer.

The primary winding offers little resistance to the DC component of the collector current, and the voltage at the collector is, therefore, nearly equal to the power

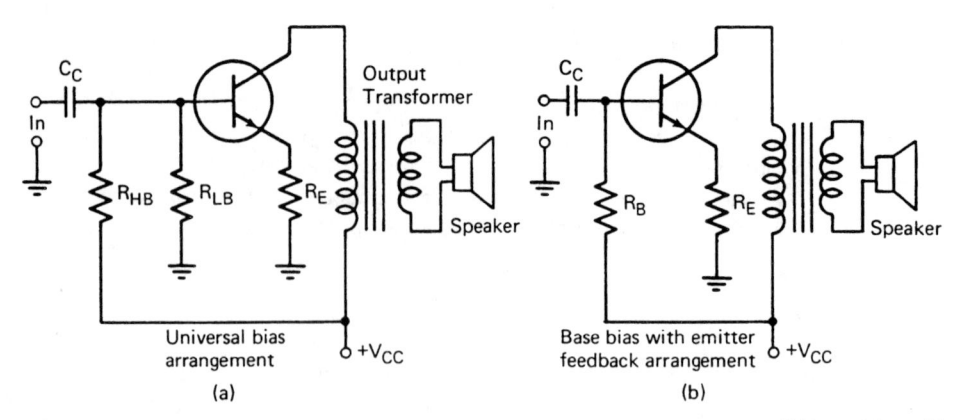

FIGURE 5-12 Single-ended power amplifiers: (a) universal-bias arrangement; (b) base bias with emitter feedback arrangement.

supply voltage. One might at first think that the high voltage at the collector would cause a much larger collector current to flow, but this does not occur because the collector current is essentially independent of the collector voltage. The collector current is determined by the biasing resistors, not by the collector load resistance or the collector voltage.

5-6 VACUUM-TUBE PUSH-PULL AUDIO AMPLIFIERS

Push-pull (P-P) *amplifiers* involve two tubes in the final power amplifier instead of one as in a *single-ended* (SE) *amplifier*. One tube amplifies the positive portion of the input signal while the other amplifies the negative portion. P-P amplifiers have many advantages over SE stages and are used extensively in small high-quality receivers and almost exclusively in high-power stereo receivers.

The input signal to a P-P amplifier must be divided into a positive and a negative phase so that when the input to one tube is positive-going, the input to the other is negative-going. The two phases must be of equal amplitude. The general process of obtaining the inverted phase is called *phase inversion*. In the following, phase inversion by use of an interstage transformer, a phase inverter, and a phase splitter is described.

Phase Inversion by Interstage Transformer

Figure 5-13 is a schematic diagram of a simple P-P amplifier that uses interstage transformer T_I to obtain phase inversion. An SE voltage amplifier $V1$ drives the primary of T_I, inducing the signal voltage into the T_I secondary. The isolation properties of the transformer prevent the plate voltage of $V1$ from passing to the T_I secondary, a function performed by the coupling capacitor in RC coupling.

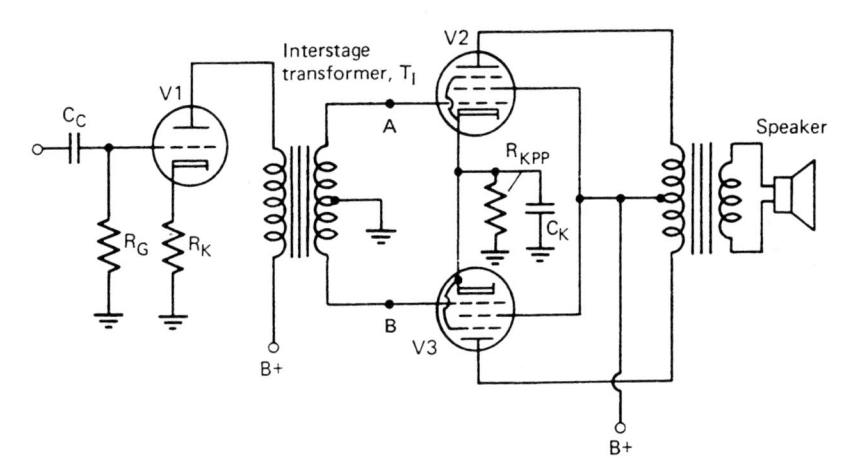

FIGURE 5-13 Push-pull power amplifier that utilizes an interstage transformer as a phase-inversion device.

Note the grounded center tap of the T_I secondary, which causes the secondary voltage to be symmetrically disposed with respect to ground. When point A is positive, point B is equally negative, and vice versa. Thus, the signal voltages at A and B are of equal amplitude and are 180° out of phase, as shown in Figure 5-14(a). This is the requirement of P-P amplifier tubes, $V2$ and $V3$.

Directing our attention to the output transformer, we note that the B+ voltage is applied to the center tap of the primary, with the B+ voltage being applied equally to the plates of $V2$ and $V3$ through the top and bottom halves of the primary. With no signal applied, equal (and small) plate currents flow from the plate of each tube through the transformer primary as shown in Figure 5-14(b). The currents flow in opposite directions so that in the transformer core, the magnetic field of one cancels the magnetic field of the other. Thus, during quiescent conditions, no net magnetic flux is present in the core.

When an input signal causes point A to go positive, the plate current of $V2$ increases while the plate current of $V3$ decreases, perhaps to zero. Then the current in the top half of the primary is much greater than the current in the bottom [Figure 5-14(c)], and a large magnetic flux is produced in the transformer core. When the input signal reverses polarity so that $V3$ conducts instead of $V2$, the current flow in the primary is through the bottom half [Figure 5-14(d)], and a strong magnetic flux is once again produced—but in the opposite direction. Thus, the magnetic flux alternates according to the input signal, and the changing flux induces the output signal in the secondary.

Cathode bias is provided by cathode resistor R_{KPP}, which may or may not be bypassed with a cathode bypass capacitor. The capacitor is often omitted so that the cathode resistor produces a small amount of degeneration, which reduces distortion at the expense of gain.

Sec. 5-6   Vacuum-Tube Push-Pull Audio Amplifiers

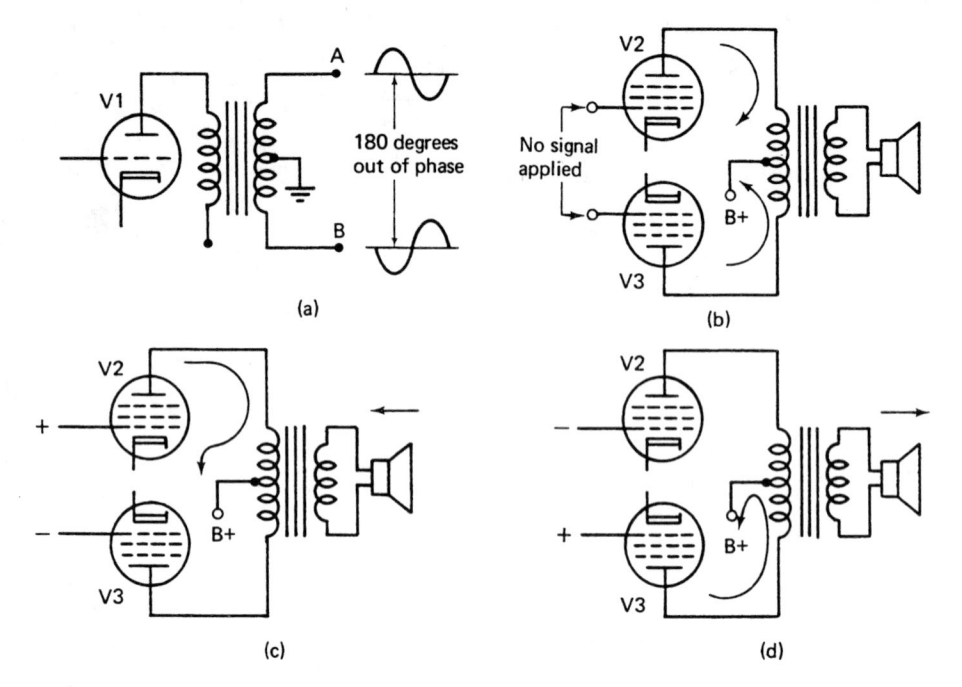

FIGURE 5-14 (a) Signals at points *A* and *B* are 180° out of phase. (b) Under no-signal conditions, the currents through the transformer primary are equal and oppositely directed. (c) When a positive voltage is applied to *V2*, the current increases as shown. (d) Current flow when the polarity of the applied signal reverses.

Output Transformer Core Saturation

In SE stages, a considerable plate current flows even under quiescent (no-signal) conditions. The current in the OT primary is always in the same direction, and therefore the magnetic flux in the core is always in the same direction. This tends to permanently magnetize the transformer core, reducing its effectiveness. In P-P amplifiers, however, very small currents flow under quiescent conditions, and, further, the small, residual currents flow in opposite directions. Consequently, residual core magnetism is not a problem, and smaller cores can be used in P-P amplifiers in comparison with SE stages of the same power rating. This is an advantage of P-P amplifiers.

Greater Efficiency of P-P Amplifiers

During quiescent conditions, appreciable plate current (about one-half the maximum value) flows in SE stages, whereas in P-P amplifiers, only small *idling currents* flow until an input signal is applied. Hence, SE stages dissipate considerable power even when at rest. The result of this, and other factors, is that the maximum theoretical efficiency of P-P stages is greater than that of SE stages. For SE stages, it is about 50% as compared to about 78% for P-P stages. Actual efficiencies obtained are, of course, less than the maximum theoretical efficiency, but the advantage of P-P over SE amplifiers is obvious.

Hum Rejection

The symmetrical and balanced nature of a P-P amplifier causes it to reject, to some extent, any extraneous signal that is common to both tubes. An immediate example is that of hum introduced by the ripple voltage of the power supply. A momentary increase of the B+ voltage (as during a ripple peak) produces a plate current increase in both tubes of approximately equal magnitudes. Since these currents flow through the output transformer primary in opposite directions, they cancel so that no hum voltage is induced in the secondary. This does not mean that no hum at all will be heard, but the P-P circuit is less sensitive to ripple-induced hum than is the SE circuit. Any hum voltage that enters the amplifier at the input will be amplified as any other signal appearing at the input.

5-7 HARMONIC DISTORTION

A pure sine wave applied to the input of an amplifier should appear as a pure sine wave also at the output. In practice, however, the signal will be somewhat distorted; the shape of the waveform will be changed by the amplifier so that it is no longer a pure sine wave. This is an undesirable feature of an amplifier.

An imperfect sine wave is the sum of voltage components of many frequencies rather than being of a single frequency. An example will make this clear. If a pure sine wave whose fundamental frequency is 1 kHz is applied to the input of an imperfect amplifier, the output voltage will consist of a large 1-kHz component plus additional components of smaller amplitude at multiples of the fundamental frequency. The output will contain voltage components not only at 1 kHz, but also at 2 kHz, 3 kHz, 4 kHz, and so on. These higher frequencies are *harmonics* of the fundamental frequency. The harmonic of order N has a frequency N times the fundamental. In the present example the third harmonic has a frequency of 3 kHz.

The amplitude of the harmonics decreases rapidly as the order (N) of the harmonics increases. The exact relationship between the amplitude of a given harmonic and the amplitude of the fundamental depends upon the type and severity of the distortion in a particular amplifier. High levels of distortion cause larger amplitudes of higher harmonics. Generally, the amplitudes decrease monotonically as the order of the harmonic increases, but this is not always the case, as we shall presently see.

The presence of the second harmonic at the output of an amplifier produces a specific type of deformity of the pure sine wave input. This is called *second harmonic distortion*. The greater the amplitude of the second harmonic, the greater is the deformity of the sine wave. Likewise, the third harmonic produces a type deformity known as *third harmonic distortion*, and this is true for all the harmonics, of which there may be a large number.

The effect of *all* the harmonics is to produce a voltage that represents the imperfection in the amplifier output. The amplitude of this voltage can be related to the amplitude of the desired fundamental frequency to obtain the percentage of total harmonic distortion, THD. This is a commonly quoted specification for commercial amplifiers. To the layman, an amplifier with a THD of 10% will not sound too bad, but to a trained ear, a THD of 1% will evoke tears. Amplifiers are available that have THD of well under 0.1%. Incidentally, when observed on a scope, a sine wave having as much as 10% THD will appear almost perfect to any but the most expert observer.

Distortion Characteristics of P-P Amplifiers

Since the plate current of the two push-pull tubes flow through the output transformer primary in opposite directions, and since the tubes are driven 180% out of phase, any even harmonic (second, fourth, sixth, etc.) distortion introduced by the P-P stage is canceled in the output transformer. The P-P amplifier output is essentially free of even-harmonic distortion. These same circuit features, however, cause the odd-harmonic distortion components to be reinforced, and they appear in the output at twice the normal amplitude. Since the amplitude of the harmonics decreases rapidly with increasing order of the harmonic, elimination of the second harmonic represents a significant improvement in performance. The thrid harmonic is troublesome, however, and in practice, places an upper limit on the dynamic range and power-output capability of a given amplifier.

5-8 SOLID-STATE TRANSFORMER-COUPLED PUSH-PULL AMPLIFIER

Basic Circuit

We introduce solid-state push-pull (P-P) amplifiers by describing the circuit of Figure 5-15. Transistor $Q1$ is a power transistor operated in the universal-bias arrangement. Interstage transformer $T1$ serves as the collector load for $Q1$ and couples the signal to P-P transistors $Q2$ and $Q3$. The secondary of $T1$ is center-tapped, and the center tap is grounded. The signal voltages appearing at the bases of $Q2$ and $Q3$ are, therefore, equal in amplitude and out of phase, as required.

When point A at the secondary of $T1$ is driven positive, point B is driven equally negative. Transistor $Q2$ is driven into conduction while transistor $Q3$ remains cut off. As $Q2$ conducts, a current flows in the top half of $T2$ as indicated by the solid arrows in the figure. When the polarity of the input signal reverses so that point B is positive, $Q3$ will be driven into conduction while $Q2$ is cut off. Current then flows in the primary of $T2$ as indicated by the dotted arrows in the figure. The changing currents in the primary of $T2$ induce a signal voltage in the secondary

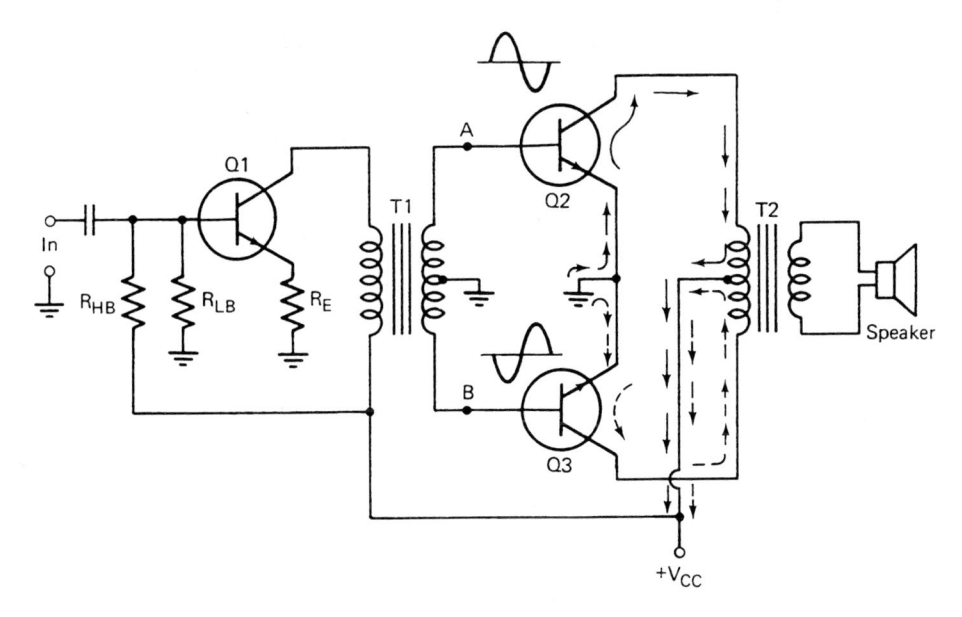

FIGURE 5-15 Transformer-coupled push-pull amplifier. This is not a practical circuit because no provision is made for reducing crossover distortion.

which is applied to the speaker. The operation of this circuit is quite similar to the vacuum-tube circuit of Sec. 5-6.

As the output transistors are driven into conduction, considerable base current flows, and therefore considerable power is dissipated in the base circuit. Transistor $Q1$ must supply this power and is the *driver* for the P-P output stage.

Crossover Distortion

Note that no bias voltage is provided for the P-P transistors. In the absence of an input signal, no collector current flows through $Q2$ or $Q3$ since neither base is forward-biased. The input signal must drive the transistors into conduction, and only the transistor receiving the positive portion of the input signal will be driven into conduction at any time. This would be acceptable, but a transistor must have a certain minimum voltage applied to the base before appreciable current begins to flow. When the input signal is instantaneously near the 0-V portion of the AC cycle, neither transistor conducts appreciably, and little signal voltage appears at the speaker. This causes *crossover distortion* because conduction ceases in one transistor before it begins in the other. There is a small interval during each "crossover" in which neither transistor is conducting. The resulting distortion is objectionable and is more pronounced at low signal levels. For this reason the amplifier in the figure is not suitable for practical application. The effect of crossover distortion is illustrated in Figure 5-16(a). The input signal is assumed to be a triangle wave.

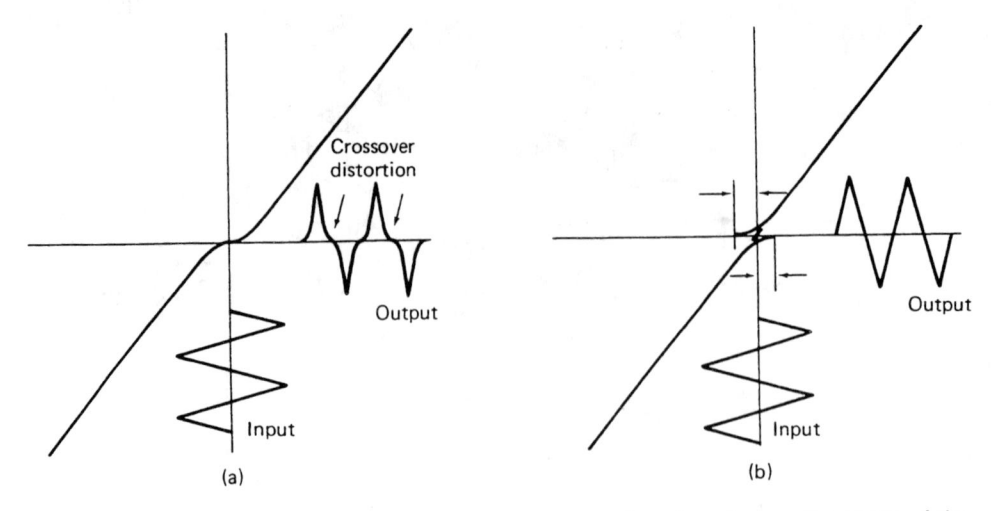

FIGURE 5-16 Effects of crossover distortion. (a) Crossover distortion changes the shape of the input triangle waveform. (b) The push-pull transistors are biased just to the point of conduction, so that the transfer curves are shifted as shown. Crossover distortion is effectively eliminated.

Crossover distortion can be reduced to negligible proportions by providing a small bias voltage to both transistors so that, in the absence of an input signal, both transistors are conducting a small amount. Then, when an input signal is applied, one transistor increases conduction smoothly while the other cuts off completely. Figure 5-16(b) shows the reduction of crossover distortion achieved by biasing the transistors slightly into the conduction region.

Practical Circuit

A widely used circuit is shown in Figure 5-17. The driver stage is identical to that of the preceding circuit, but a biasing network consisting of resistors $R4, R5, R6$, and $R7$ has been added to the P-P stage. Bias voltage for both transistors is obtained from the same network, being applied to the center tap of the $T1$ secondary, from whence it flows equally to each P-P transistor.

Resistors $R4$ and the parallel combination of $R5$ and $R6$ form a voltage divider which provides the bias voltage applied to the tap of the $T1$ secondary. Resistor $R7$ is an emitter resistor that serves both transistors, and we recognize the universal-bias arrangement. The high-base resistor is $R4$, while the low-base resistor is the parallel combination of $R5$ and $R6$. The collector load is provided by the primary of transformer $T2$.

The biasing voltage applied to the P-P transistors is critical. A slightly low voltage will cause crossover distortion. A slightly high voltage will drive the transistors far into the conduction region, causing excessive power dissipation in the

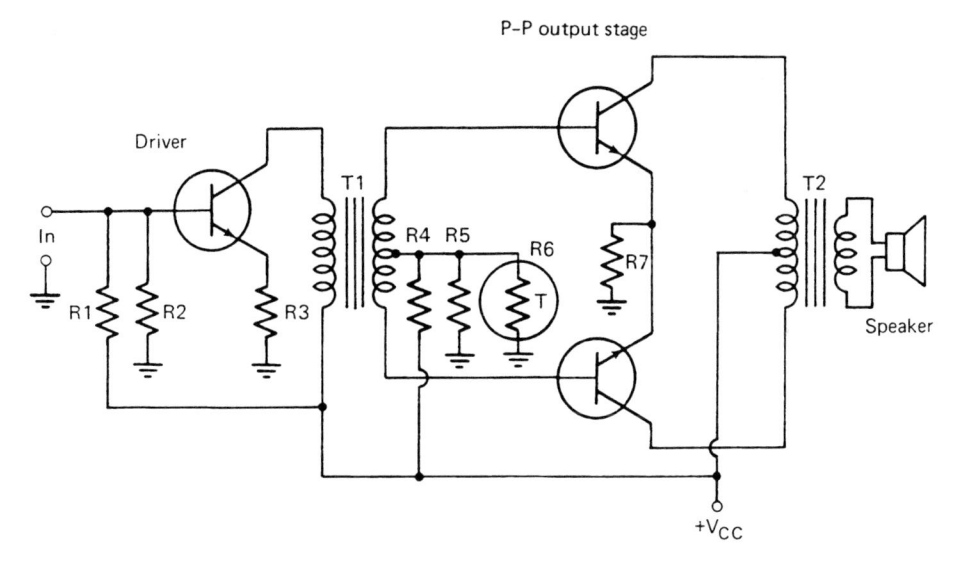

FIGURE 5-17 Practical transformer coupled push-pull amplifier with temperature-compensating resistor (thermistor) R6.

transistors. The parallel resistors R5 and R6 stabilize the bias voltage, as described in the following.

Thermistor

Resistor R6 is a *thermistor*; note the symbol. The resistance of the device depends upon the temperature, and the resistance drops as the temperature rises. The device used in this circuit has a resistance of 220 Ω at room temperature. In practice, the thermistor is located physically close to, or in contact with, the P-P transistors so that as they warm up, so does the thermistor.

The collector current in the P-P transistors tends to increase as they warm up, but as the transistors warm up, the resistance of thermistor R6 decreases. This reduces the bias voltage applied to the transistors and tends to oppose the increase in collector current.

The thermistor is placed in parallel with R5 for the following reason. Thermistors are characterized by a specified rate of change of resistance with temperature. The particular "resistance change rate" of a given thermistor may not be "exactly right" for the P-P transistors being used. By placing the thermistor in parallel with ordinary resistor R5, the rate at which the resistance of the parallel combination changes with temperature can be controlled by properly choosing R5. As R5 becomes larger, in comparison with the thermistor resistance, the change rate of the parallel equivalent resistance becomes greater. The converse is true as R5 becomes smaller. When R5 and the thermistor are equal in ohmic value, the change rate of the parallel equivalent is only 25% of the change rate of the thermistor alone.

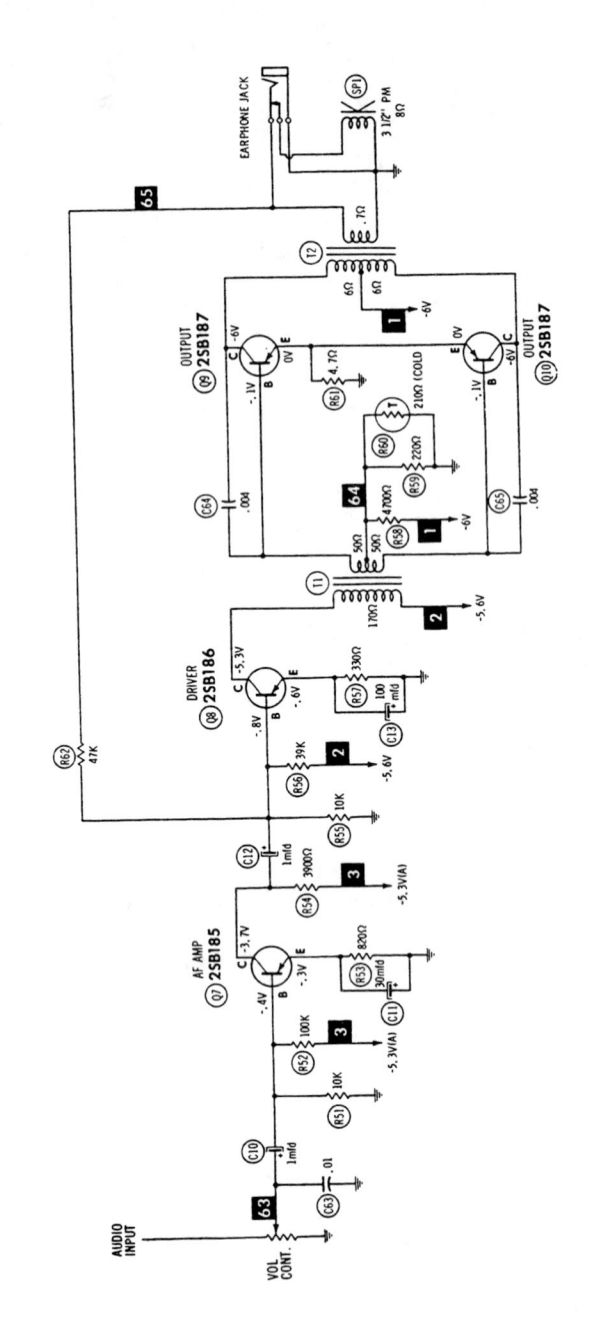

FIGURE 5-18 Complete audio section of a commercial receiver (a Photofact schematic, courtesy of Howard W. Sams & Co., Inc.)

Complete Audio Section

The complete audio section of a small, portable AM/FM receiver is shown in Figure 5-18. From the volume control, the audio signal passes through coupling capacitor $C10$ to the base of the first audio amplifier $Q7$, which is a standard universal-bias stage. Capacitor $C12$ couples the signal from the collector of $Q7$ to the base of driver transistor $Q8$. The biasing arrangement of $Q8$ is a universal bias in which the primary of transformer $T1$ forms the collector load. Resistors $R56$ and $R55$ are the high-base and low-base resistors, respectively. The function of $Q8$ is to provide sufficient power to drive the push-pull output stage consisting of $Q9$ and $Q10$.

The output stage utilizes a circuit identical to the one described above, but two features have been added. Capacitors $C64$ and $C65$ connect between the collector and the base of the output transistors to provide local negative feedback to reduce the response of the amplifier to high frequencies. Note that $R62$ connects from the high side of the speaker winding of $T2$ back to the base of the driver transistor. This connection forms a feedback loop through which a portion of the signal voltage developed across the speaker voice coil is fed back to the input of the driver stage. More is said about this connection in a later section of this section.

5-9 OUTPUT TRANSFORMER-LESS (OTL) AUDIO-OUTPUT STAGE

Because of the low impedance levels at which transistors operate, as compared with vacuum tubes, it is possible to eliminate the output transformer and drive the speaker directly without an intervening impedance matching device. Since no output transformer is used, such circuits are frequently called *OTL output stages*, and we adopt this terminology. Figure 5-19 shows a simple OTL output stage typical of those frequently encountered in small AM/FM receivers.

The driver transistor operates into the primary of interstage transformer $T1$, which features dual secondary windings. The windings are connected so that the signals delivered to the bases of $Q1$ and $Q2$ are out of phase, as required for push-pull operation. The transistors conduct alternately, as in previous push-pull amplifiers, and when $Q1$ is conducting, the output (point C) goes more negative because of the flow of electrons from the collector of $Q1$ to the emitter. When $Q2$ conducts, the output becomes less negative because electrons flow from the output (point C) through $Q2$ to ground. The signal voltage developed at point C is coupled through the large electrolytic capacitor to the speaker.

Resistors $R1$ through $R4$ form the biasing network for the two output transistors. Note that $R1$ and $R2$ are connected in series across $Q1$, while $R3$ and $R4$ are similarly connected across $Q2$. A small fraction of the voltage across $Q1$ appears at point A which is the output of the voltage divider formed by $R1$ and $R2$. The voltage at point A is delivered to the base of $Q1$ by secondary $S1$, and this voltage brings $Q1$ just into the conducting region. A similar situation exists for $Q2$, for which $R3$ and $R4$ serve as the voltage divider to deliver a small voltage to point B

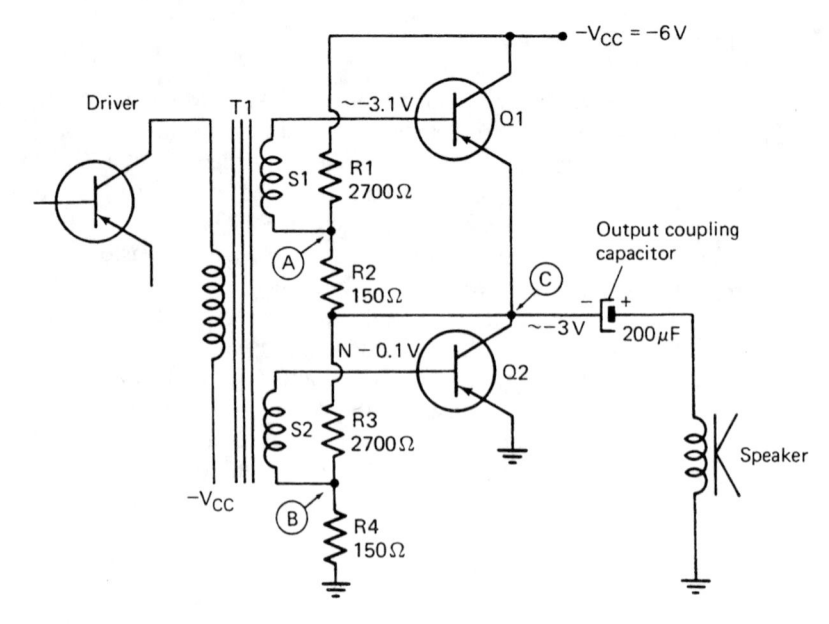

FIGURE 5-19 Output transformer-less (OTL) audio-output stage that uses a dual secondary interstage transformer.

and to the base of $Q2$. Thus, both transistors are biased just to the point of appreciable conduction in order to reduce crossover distortion.

Note that $Q1$ and $Q2$ are connected in series between ground and the power supply. Hence, the voltage at point C is very nearly one-half the power supply voltage. The output coupling capacitor blocks the flow of direct current from point C to ground through the speaker. When such an amplifier is first turned on, the output capacitor must charge up to the average DC voltage level. The charging current flows through the speaker, producing a characteristic "plop."

Temperature compensation is provided in many commercial circuits by connecting thermistors in parallel with $R2$ and $R4$, and by locating the thermistors physically close to the output transistors. An increase in temperature then causes the bias voltage applied to the transistors to be reduced.

5-10 COMPLEMENTARY-SYMMETRY AMPLIFIER

We now consider a push-pull power amplifier in which an *NPN* and a *PNP* transistor form a push-pull amplifier that does not require two signals 180° out of phase. This is possible because of the complementary nature of *NPN* and *PNP* transistors. We introduce the complementary-symmetry circuit by describing the circuit of Figure 5-20.

First, the characteristics of the transistors must be matched fairly closely. The particular type *NPN* transistor is said to be *complementary* to the *PNP* type, and the matched pair constitute *complementary transistors*.

Balanced power supplies are used for both the input and the amplifier circuits.

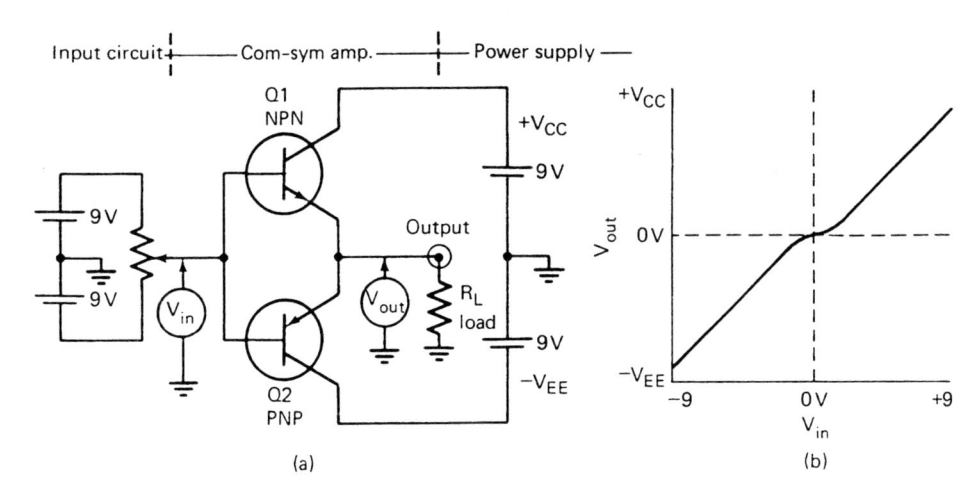

FIGURE 5-20 (a) Circuit for demonstrating the complementary-symmetry circuit operation. (b) Plot of the output voltage as a function of input voltage showing the effects of crossover distortion.

Therefore, since the transistors are in series across the power supply, the voltage at the output will be zero provided that the potentiometer at the input is at midrange. This represents the neutral point of this combination, and no current flows through R_L. Let us now see what happens as we vary the voltage applied to the bases of the transistors.

Part(b) of the figure is a graph of the output voltage V_{out} as a function of the input voltage V_{in}. As V_{in} is made more positive by moving the wiper arm upward, the NPN transistor begins to conduct and V_{out} increases. Electrons flow from ground through R_L to the emitter of $Q1$ and on to the positive terminal of the power supply, $+V_{CC}$. The output voltage V_{out} will follow very closely the voltage applied to the input V_{in}, the difference being equal to the base-emitter voltage of the NPN transistor.

As the input voltage V_{in} is made negative by moving the wiper arm of the potentiometer below the midpoint, conduction through the NPN transistors ceases as the PNP transistor begins to conduct. The voltage at the output becomes negative also, and current flow through the load resistor is reversed. Electrons now flow from the negative terminal of the power supply, V_{EE}, through PNP transistor $Q2$ and then through R_L to ground.

We observe a pronounced nonlinearity of the curve near the neutral point. This arises from the "turn-on" characteristics of the transistors in that a sizable departure of the input voltage from the neutral point must occur before conduction is substantially increased in either of the transistors. This produces crossover distortion. Fortunately, this effect can be avoided.

Practical Circuit

Figure 5-21 shows the configuration of a practical complementary-symmetry amplifier that has provisions for reducing the crossover distortion. Also, the amplifier utilizes an unbalanced power supply. Transistor $Q1$ is the driver for the comple-

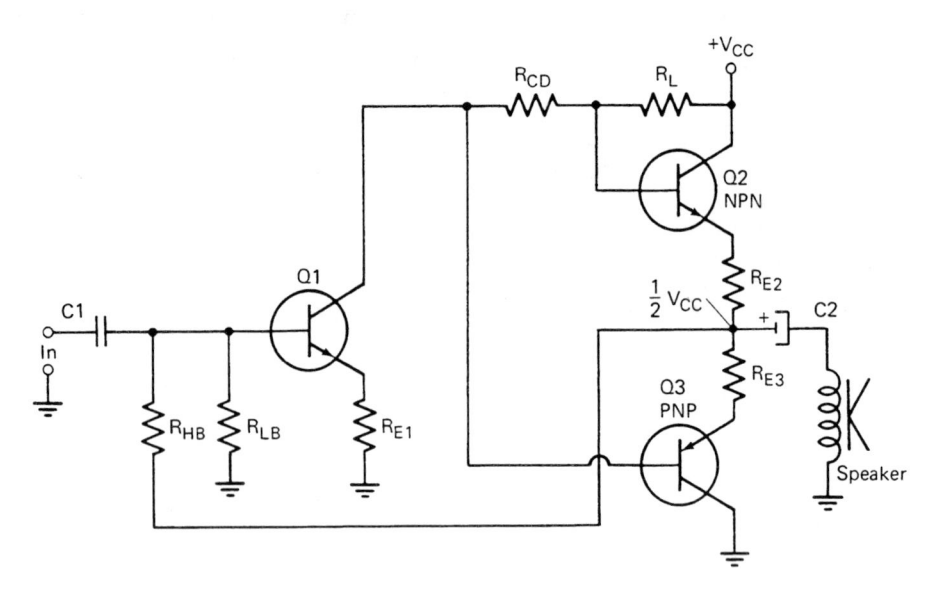

FIGURE 5-21 Practical complementary-symmetry amplifier configuration.

mentary transistors and utilizes a universal-bias arrangement. Resistor R_L is the collector load resistor for $Q1$, but note that R_{CD} is in series with R_L and is connected between the bases of the complementary transistors. The $Q1$ collector current flowing through R_{CD} develops a small voltage across R_{CD} which causes the bases of the transistors to be at slightly different potentials. The collector current of $Q1$ is such that the voltage at the midpoint of R_{CD} (inaccessible though it is) is almost exactly equal to $\frac{1}{2}V_{CC}$. Therefore, the voltage at the base of $Q2$ is slightly greater than the neutral voltage, while the voltage at the base of $Q3$ is slightly below the neutral voltage. These small voltage offsets bias the transistors $Q2$ and $Q3$ slightly into the conduction region so that the transistors are already "turned on" even under quiescent conditions. This removes the nonlinearity in the input versus output voltage curve and thereby reduces the crossover distortion.

The Q-point collector current and voltage of $Q1$ are extremely critical. Therefore, a feedback loop is arranged so that the collector current of $Q1$ is held at the proper value. This is accomplished by connecting the high-base resistor $Q2$, R_{HB}, to the output point of the complementary transistors. Any departure of the voltage at the output from the neutral point ($\frac{1}{2}V_{CC}$) will result in a correction voltage being applied to the base of $Q1$. If the output voltage begins to rise, for whatever reason, the collector current of $Q1$ increases, which, in turn, lowers the $Q1$ collector voltage. This reduces conduction in the *NPN* transistor in an effort to restore the output voltage to the neutral value.

Resistors R_{E2} and R_{E3} in the emitter leads of the output transistors are typically less than 5 Ω and provide a stabilizing influence against minor variations in the characteristics of the transistors. Capacitor $C2$ is an electrolytic capacitor, typically hundreds of microfarads, that couples the AC signal from the output of the amplifier to the speaker. The output impedance of the amplifier is very low, which makes capacitive coupling to the speaker possible, but it is not uncommon to find speakers used in such circuits of impedance values much greater than the common 3.2, 4, 8, or 16 Ω. Typical values might be as high as 60 Ω.

While the circuit described here is found in many receivers, the same basic circuit may assume many variations of such complexity that it is almost unrecognizable. Feedback loops, both positive and negative, both AC and DC, may prevail in abundance. Also, resistor R_{CD}, the crossover resistor, is frequently replaced by a diode, or a diode and a small resistor in series, to form a crossover network.

5-11 QUASI-COMPLEMENTARY AMPLIFIER

A *quasi-complementary* amplifier utilizes a complementary-symmetry stage to provide out-of-phase signals at the proper power level for driving push-pull power output transistors of the same type. A typical amplifier of this type is shown in Figure 5-22. Such amplifiers generally have a greater power output capability than the complementary-symmetry amplifier used alone.

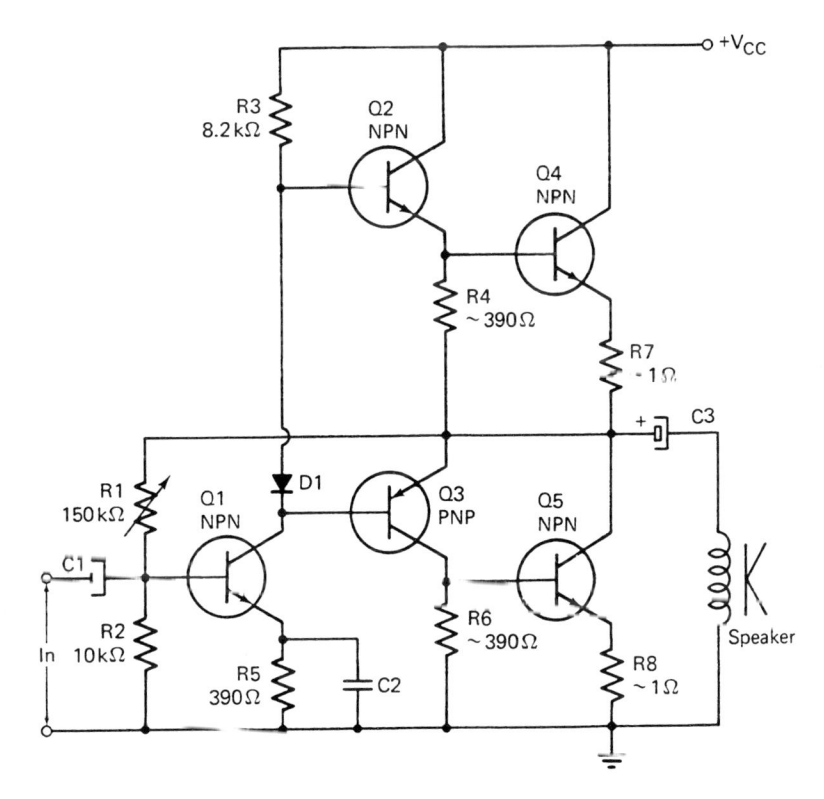

FIGURE 5-22 Quasi-complementary amplifier. The final output transistors are both of the same type (*NPN*) but are driven by a complementary pair, Q2 and Q3.

Transistor $Q1$ is the driver for the complementary pair, $Q2$ and $Q3$. The bias arrangement is universal, and negative feedback is provided to the base of $Q1$ via $R1$ from the output. The emitter resistor $R5$ is bypassed by $C2$. The $Q1$ collector load is $R3$, and diode $D1$ is the crossover diode, used to reduce crossover distortion in the complementary-symmetry pair.

The output signals developed across $R4$ and $R6$ are out of phase, since $Q2$ and $Q3$ conduct alternately in push-pull fashion. These out-of-phase signals are applied to the output transistors $Q4$ and $Q5$, whose operation is straightforward. Let us now follow the circuit action at an instant when the input signal is going positive.

When a positive-going signal is applied to the input, conduction through $Q1$ increases, causing the $Q1$ collector voltage to decrease. This less-positive voltage is applied to the bases of $Q2$ and $Q3$, and causes $Q3$ to increase conduction while $Q2$ tends to turn off. The result is that the voltage at the base of $Q5$ rises due to increased current flow through $R6$. At the same time, conduction through $R4$ decreases, reducing the voltage applied to the base of $Q4$. The effect is to turn $Q5$ on while turning $Q4$ off. This causes the voltage at the output to decrease, and this decrease is applied to the speaker by way of coupling capacitor $C3$.

5-12 COMMERCIAL AMPLIFIERS

We now describe three audio amplifier sections taken from AM or AM/FM portable receivers. The first amplifier is shown in Figure 5-23.

At first glance we note that the $Q6$-$Q7$ preamplifier is a Darlington pair. The output of the Darlington pair is capacitively coupled to driver transistor $Q8$, which powers the primary of interstage transformer $T2$. Dual secondaries drive the output transistors $Q9$ and $Q10$, which are connected in an OTL configuration. A negative

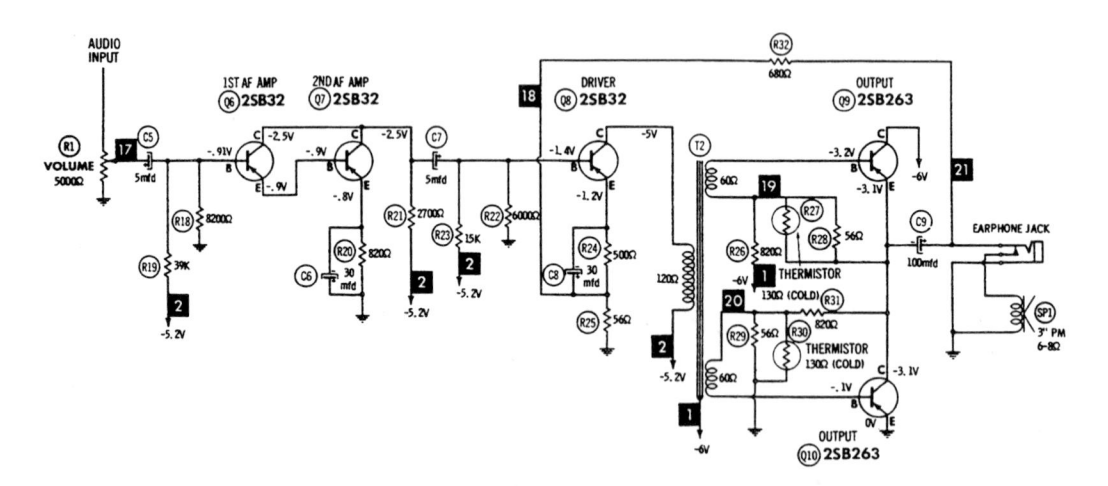

FIGURE 5-23 Complete audio section of a commercial receiver. Note the Darlington pair input stage (a Photofact schematic, courtesy of Howard W. Sams & Co., Inc.)

feedback loop connects from the output back to the emitter circuit of the driver stage.

The Darlington pair, $Q6$ and $Q7$, utilizes *PNP* germanium transistors, and is universal-biased. $R19$ and $R18$ are the high-base and low-base resistors, respectively, and the collector load is $R21$. Emitter resistor $R20$ is bypassed by capacitor $C6$. Thus, the circuit is straightforward.

Note the base-emitter voltages of the transistors in the Darlington pair, 0.01 and 0.1 V. $Q7$ conducts much more than $Q6$. This is typical of Darlington pairs. The first transistor conducts very little and consequently has a lower V_{BE} than the second transistor.

Driver transistor $Q8$ is universal-biased and features a split emitter resistance. The greater portion of the emitter resistance is bypassed, but $R25$ serves as a swamping resistor to mask variations in the AC resistance of the base-emitter junction. Also, $R25$ provides a place to connect the feedback loop involving $R32$. If $R25$ were either absent or bypassed, the emitter of $Q8$ would be at AC ground, and no feedback connection could be made to the emitter lead. The same type of transistor is used for the driver as for the Darlington pair, but note that the base-emitter voltage for the driver transistor is much greater than for either transistor of the pair. This is as expected because the driver must supply power to drive the output stage.

The output stage is straightforward. The six biasing resistors (including two thermistors) form voltage dividers across the transistors. The circuit is symmetrical in regard to the biasing network. Observe that $R26$ and the collector of $Q9$ connect to the same point and therefore are connected to each other. $R26$ and $R31$ play identical roles as high-base resistors, and the parallel network of a thermistor and ordinary resistor [($R27,R28$), ($R29,R30$)] plays the role of low-base resistor. The thermistors provide thermal stability as described earlier.

Capacitor $C9$ couples the audio signal to the speaker. An earphone jack is connected so that the speaker is disconnected when the earphone plug is inserted. Note that $R32$ is connected from the high side of the speaker to the emitter circuit of driver $Q8$. Whether the resulting feedback loop is a positive or a negative feedback loop cannot be determined directly from the schematic because the polarities of the $T2$ transformer windings are not indicated. If the transformer connections are such that the feedback is negative, the gain of the amplifier would be stabilized and the distortion characteristics of the amplifier would be improved as a result of the negative feedback. On the other hand, if the feedback is positive, the function of the positive feedback can only be to reduce the effects of negative feedback produced by other portions of the circuit. A positive feedback loop is described in the following discussion of the third commercial amplifier.

Second Commercial Amplifier

Figure 5-24 is an audio section that utilizes a direct-coupled first audio amplifier and driver. The output stage is an OTL power amplifier quite similar to those already described. The interesting features of this amplifier are the direct-coupled stages and the feedback loops incorporated.

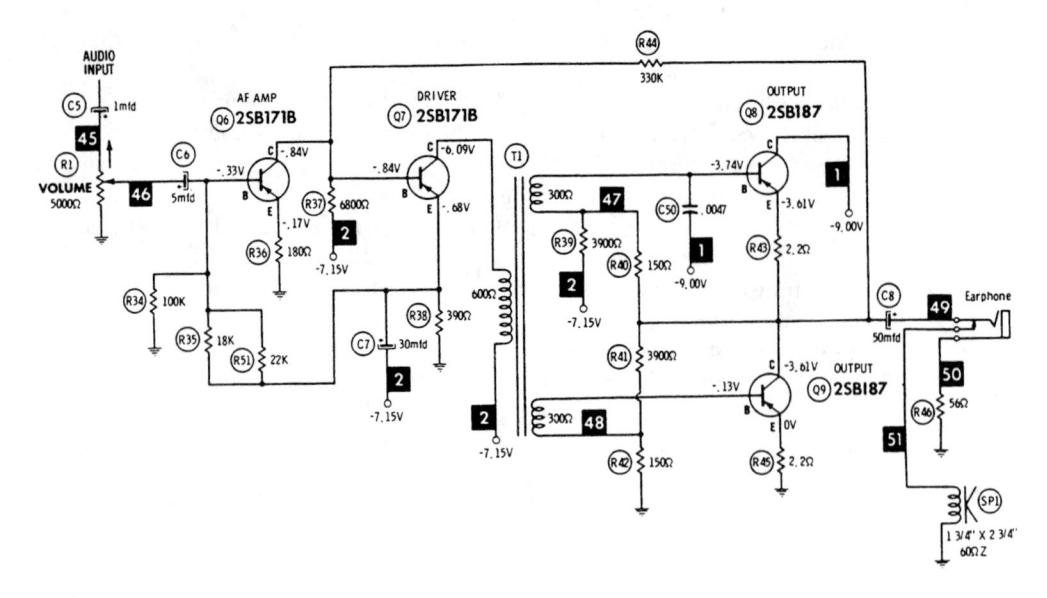

FIGURE 5-24 Complete audio section of a commercial receiver. A direct-coupled amplifier serves as the driver for the push-pull output stage (a Photofact schematic, courtesy of Howard W. Sams & Co., Inc.)

The biasing arrangement for $Q3$ is a modified universal circuit. The driving voltage for the base is obtained from a source other than the power supply, namely, from the emitter of $Q7$. The high-base resistor consists of the parallel combination of $R35$ and $R51$, and the low-base resistor is $R34$. Connecting the high-base resistance to the emitter of $Q7$ forms a DC feedback loop that tends to stabilize the DC parameters of the direct-coupled stages.

The $Q6$ collector voltage is applied to the base of $Q7$. The Q-point collector current of $Q7$ is determined by the base voltage and the emitter resistance $R38$. The primary of transformer $T1$ forms the collector load for $Q7$. The 330-kΩ resistor $R44$ forms a feedback loop from the output of the amplifier back to the base of $Q7$.

To see the stabilizing property of the feedback loop of the direct-coupled amplifier, assume that the current through $Q7$ begins to increase. The voltage at the emitter of $Q7$ will become more negative as a result of the increased current flow through $R38$. This increases the forward bias at the base of $Q6$, and the resulting increased conduction lowers the collector voltage of $Q6$ and the base voltage of $Q7$. The decrease in voltage at the base of $Q7$ tends to oppose the original increase of the $Q7$ collector current.

Capacitor $C7$ bypasses the emitter resistor $R38$ and prevents AC feedback from the $Q7$ emitter to the base of $Q6$. The capacitor holds the emitter of $Q7$ at signal ground. It does not matter that the capacitor connects to the power supply point, because the power supply point itself is a good signal ground.

Third Commercial Amplifier

The last circuit of this chapter, shown in Figure 5-25, has several interesting features and is typical of amplifiers frequently encountered. We note that all transistors except $Q8$ are *PNP* and that a positive power supply is used in conjunction with a shifted ground point.

The first AF amplifier, $Q4$, is a universal-biased emitter follower. Consequently, no load resistor appears in the collector circuit, and the output signal of the stage is taken from the emitter. $R16$ is the high-base resistor and $R15$ is the low-base resistor of the universal arrangement. In light of the emitter-follower configuration, the $Q4$ stage does not contribute any voltage gain.

The second AF amplifier, $Q5$, is biased by a universal arrangement that utilizes collector feedback produced by connecting 39-kΩ resistor $R19$ to the $Q5$ collector rather than to ground. $R18$ and $R19$ are the base-biasing resistors, and $R20$ and $R21$ are the emitter and collector resistors, respectively.

Transistors $Q6$, $Q7$, and $Q8$ form a direct-coupled, complementary-symmetry OTL output stage. $Q6$ is biased by a modified universal arrangement. The $Q6$ emitter resistor has been omitted, and high-base resistor $R23$ connects to the complementary-symmetry midpoint rather than to the power supply point (ground in this case). The collector load resistance for $Q6$ consists of the series combination of $R24$, $R26$, and the speaker impedance, with $R26$ forming the major portion of the load resistance. $R24$ reduces crossover distortion in the complementary-symmetry pair by introducing a voltage difference between the base of $Q7$ and the base of $Q8$ so that the transistors are slightly forward-biased.

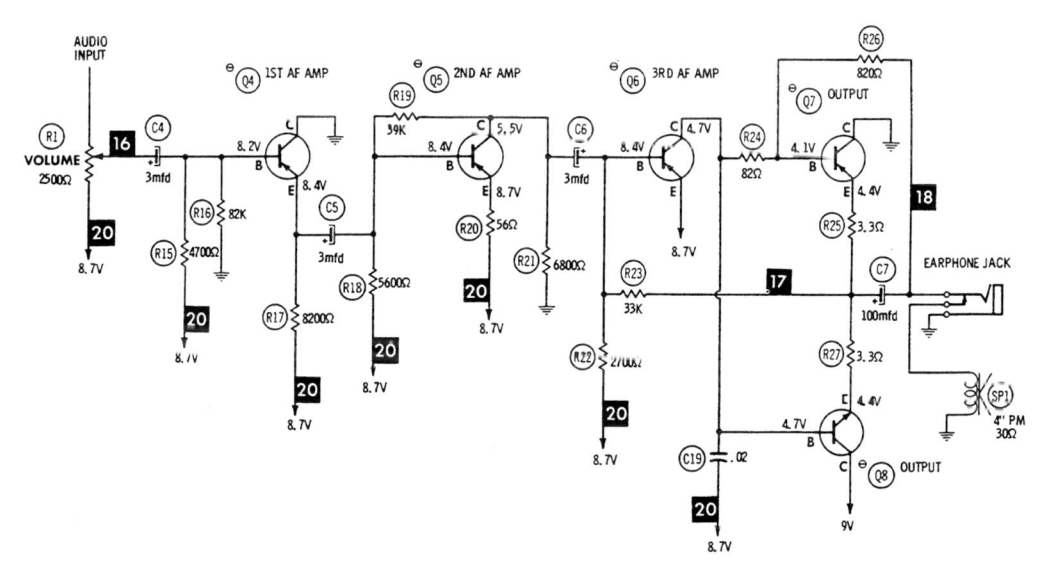

FIGURE 5-25 Complete audio section of a commercial receiver. An emitter follower serves as the first AF amplifier in this circuit (a Photofact schematic, courtesy of Howard W. Sams & Co., Inc.)

Sec. 5-12   Commercial Amplifiers   153

Note that the $Q6$ collector current flows through the speaker, or through the earphone when it is connected. The DC component of the collector current is of little significance in regard to its flowing through the speaker, but the fact that $R26$ connects to the high side of the speaker rather than directly to ground produces a positive feedback loop from the output of the amplifier back to the input of the complementary-symmetry pair. The reason for this is made clear in the following.

Observe, again, the connection of the high-base resistor $R23$ of $Q6$ to the midpoint of the complementary-symmetry pair at CTP 17. This connection gives the thermal stability required by the direct-coupled amplifier, but the connection also contributes considerable negative feedback to the signal voltage and reduces the gain of the amplifier. The undesired reduction in gain can be compensated by incorporating a positive feedback loop that tends to cancel the effects of the negative feedback loop represented by $R23$. Thus, the positive feedback loop formed by connecting $R26$ to the high side of the speaker is to counteract a portion of the negative feedback produced by the $R23$ connection. Considering the amplifier as a whole, the net feedback is negative as indeed it must be to avoid an undesirable oscillation which would result if the net feedback were positive.

Finally, we note capacitor $C19$ connected from the base of $Q8$ to ground. This is a tone-compensating capacitor that tends to remove a portion of the high frequencies from the audio in order to produce a more pleasing sound from the speaker. The fact that $C19$ connects to the positive power supply is of no consequence, since the power supply point is a signal ground. Connection of $C19$ to the supply voltage was probably done as a matter of convenience in positioning the components on the printed circuit board.

SUMMARY

1. The basic biasing arrangements of tubes, BJTs, or FETs may be cascaded to produce amplifiers to two or more stages. The coupling may be direct, capacitive (RC), or transformer coupling.

2. The total gain of a series of cascaded stages is the product of the gains of the individual stages.

3. In direct-coupled stages, any undesired deviation of the DC voltages of stage 1 will be amplified by stage 2. Thus, the DC stability of the amplifier must be assured. Negative feedback loops are frequently employed for this purpose.

4. In the common-emitter configuration of a BJT, the input signal is applied between the base and the emitter while the output signal appears between the collector and the emitter.

5. In the common-base configuration, the base is held at signal ground. The input signal is applied between the emitter and the base while the output signal appears between the collector and the base (ground).

6. In the common-collector configuration, the collector is held at signal ground. The input signal is applied to the base, and the output signal appears on the emitter. This configuration is called an emitter follower and produces, at most, a gain of 1.

7. A cathode follower is a vacuum-tube amplifier stage operated in the common-plate configuration. The input signal is applied to the grid and the output signal appears on the cathode. The stage produces, at most, a gain of 1.

8. FETs may be operated in common-gate and common-drain configurations in a manner similar to vacuum tubes and BJTS.

9. The common-base configuration of a BJT requires a biasing arrangement that includes an emitter resistor, and the emitter resistor may not be bypassed. Common-base amplifier stages have high gains (typical) and low input impedances.

10. The common-collector (emitter follower) configuration requires the presence of an emitter resistor while the collector resistor is omitted. Hence, collector feedback arrangements are not applicable to the common-collector configuration.

11. A single-ended amplifier is the opposite of a push-pull amplifier. It therefore requires only one tube or transistor. Single-ended power amplifiers typically use an output transformer as an impedance matching device between the plate or collector and the speaker to match the high impedance of the device output to the low impedance of the speaker.

12. The impedance reflected from the secondary to the primary side of a transformer equals the secondary impedance multiplied by the square of the turns ratio of the transformer.

13. A single-ended power amplifier may be obtained by replacing the plate or collector load resistor with an output transformer and speaker combination.

14. Push-pull amplifiers involve two power amplifier tubes or transistors in the output stage. These devices conduct alternately, one on the positive and the other on the negative portions of the input waveform. Push-pull amplifiers offer many advantages over single-ended amplifiers.

15. Vacuum-tube push-pull amplifiers and solid-state push-pull amplifiers that are not complementary-symmetry require two signals 180° out of phase, which may be obtained from a center-tapped transformer.

16. Push-pull amplifiers are widely used in solid-state devices. Crossover distortion must be reduced by placing a small amount of forward bias on the output transistors. Insufficient forward bias does not eliminate the distortion while excessive forward bias may cause overheating of the transistors due to excessive collector current.

17. Collector current tends to increase as a transistor heats up in operation. Therefore, temperature-compensating thermistors are frequently included to adjust the forward bias applied to the base of a transistor as the transistor warms up in operation.

18. Harmonic distortion results when an amplifier generates signal voltages at frequencies that are multiples of the fundamental frequency. Each component of the harmonic distortion (second, third, etc.) contributes a specific deformity to the signal: second harmonic distortion, third harmonic distortion, and so forth. The combined effect of all the components is the total harmonic distortion.

19. Push-pull amplifers tend to cancel even-order-harmonic distortion (second, fourth, etc.) but accentuate the odd-harmonic components.

20. The low operating impedance of transistors makes possible the design of output transformer-less (OTL) output stages that do not have an output transformer. A capacitor typically couples the output signal to the speaker. Speakers may be encountered whose impedances are as high as 40 Ω or so.

21. A complementary-symmetry amplifier utilizes a *NPN* and a *PNP* transistor to achieve push-pull operation without the requirement of a phase inverter. This is possible because of the complementary properties of *NPN* and *PNP* transistors.

22. A quasi-complementary amplifier utilizes a complementary-symmetry stage to obtain a phase inversion, and the complementary signal voltages are then applied to output transistors of the same type connected in push-pull.

QUESTIONS AND PROBLEMS

5-1. When two stages are cascaded, how is the gain of the first affected by the presence of the second?

5-2. Stage 1 of a cascaded amplifier has a gain of 8 when operated without stage 2 being connected. When stage 2 is connected, the gain of stage 1 drops from 8 to 5 due to the loading effect of stage 2. If the gain of stage 2 is 6, what is the overall gain of the cascaded amplifier?

5-3. Amplifiers that are direct coupled are capable of amplifying signals whose frequency is 0 (DC). What problem does this feature introduce into the design of such an amplifier?

5-4. A direct-coupled amplifier that uses only *NPN* transistors is subject to stacking of the collector voltages. How can this be avoided?

5-5. Explain the difference between a complementary amplifier and a complementary-symmetry amplifier.

5-6. Each transistor of a Darlington pair has a beta of 100. What is the equivalent beta of the Darlington pair?

5-7. Can all BJT biasing arrangements be used for common-base amplifiers? What is required of an arrangement for it to be acceptable for a common-base amplifier?

5-8. Must the base voltage of a common-base amplifier be 0 V DC?

5-9. Can all BJT biasing arrangements be used for emitter followers? What is required of an arrangement for it to be acceptable for an emitter follower?

5-10. (a) What is the approximate gain of an emitter follower?

(b) Is the signal inverted as it passes from input to output?

5-11. What changes are necessary to convert the basic vacuum-tube and BJT biasing arrangements to single-ended power amplifiers?

5-12. What type of tube is most often used as a power-output tube in an audio amplifier?

5-13. A small radio receiver uses a speaker whose impedance is 3.2 Ω. The output transformer has a turns ratio fo 22:1. What impedance is reflected into the plate circuit of the audio-output tube?

5-14. Suppose that the screen voltage is lost in a beam power pentode output stage. What will be the effect upon the plate current? What will be the effect upon the cathode voltage? How would the control grid voltage be affected?

5-15. As the term is applied to push-pull amplifiers, what is "phase inversion?"

5-16. Describe two methods used to achieve phase inversion in vacuum-tube circuits. Are these same methods applicable to solid-state circuits?

5-17. What causes the core of an output transformer to become magnetically saturated in a single-ended amplifer? How is this avoided in push-pull circuits?

5-18. The simplest push-pull amplifier using BJTs suffers from a particular type of distortion that is especially significant at low signal levels. What is the name of this distortion, and what can be done to minimize it?

5-19. What may occur if the transistors of a push-pull amplifier stage are biased too far into the conduction region in an effort to reduce crossover distortion?

5-20. What is a thermistor? What function may thermistors serve in a push-pull output stage?

5-21. What property of transistors makes it possible to eliminate the output transformer in certain power output stages?

5-22. What causes the characteristic "plop" when an OTL output stage is first turned on?

5-23. Why does a complementary-symmetry amplifier not require a phase inverter?

5-24. How does a quasi-complementary amplifier differ from a complementary-symmetry amplifier?

5-25. How does a crossover diode such as $D1$ in Figure 5-22 reduce crossover distortion?

5-26. Draw a voltage plot for the second AF amplifier $Q5$ in Figure 5-25.

6
RADIO-FREQUENCY AMPLIFIERS

Radio-frequency amplifiers form an important part of any receiver: in particular, the RF sections determine the receiver sensitivity and selectivity. In this chapter we describe both tube and solid-state RF amplifiers that are commonly encountered in both AM and FM receivers.

RF amplifiers almost always involve tuned circuits. Therefore, we present the important aspects of resonant circuits in the first section. Also, many RF amplifiers are designed so that the gain produced may be electronically controlled. This feature is utilized in the automatic volume control (AVC) circuitry of a receiver, known also as the automatic gain control (AGC).

In designing an amplifier that is to operate at high frequencies, many effects must be taken into account that may be ignored at audio frequencies. The interelectrode capacitance between plate and grid or collector and base, for example, gives rise to feedback from output to input that can cause the amplifier to oscillate. Noise (static) production in RF amplifiers is an extremely important consideration.

This chapter moves us a giant step closer to the understanding of a complete radio receiver. We have already studied power supplies and audio amplifiers, and after RF amplifiers are understood, there is not much left. With this bit of encouragement, let us continue.

6-1 INTRODUCTION

Strictly speaking, an RF amplifier is any amplifier that amplifies signals of higher frequencey than audio frequencies, but when applied to a superheterodyne receiver, the term *RF amplifier* refers to the amplifier (when present) that precedes the *frequency changer*. The frequency to which the antenna frequency is changed is the *intermediate frequency* (IF), and the amplifier that operates at this frequency is the *IF amplifier*. Although the frequency changer provides considerable amplification, it is not called an RF amplifier because of its more important function of changing the antenna frequency to the intermediate frequency. Both RF and IF amplifiers are described in this chapter. Frequency changers are described in Chapter 7. The relationship of these stages is shown in the block diagram of Figure 6-1.

The gain of the RF sections must be automatically controlled to produce a nearly constant signal level at the demodulator under wide signal-level variations

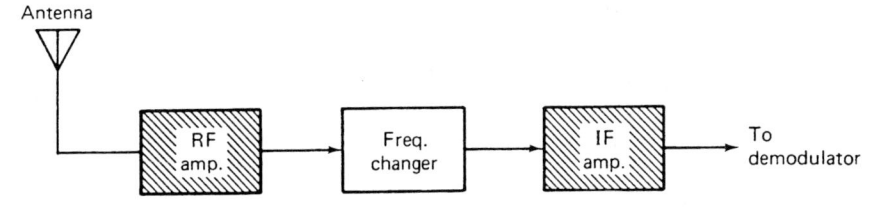

FIGURE 6-1 Partial block diagram of a superheterodyne receiver.

at the antenna. The *automatic-volume-control* (AVC) *circuits* perform this function. When the input signal is strong, the gain of the RF stages is reduced, and when the input signal is weak, the gain of the RF stages is increased. This is accomplished automatically and is not related to the manual volume control of the audio section. The gain-controlling circuitry is sometimes called *automatic-gain-control* (AGC) *circuitry*.

The RF sections determine the sensitivity and selectivity of a receiver. *Sensitivity* refers to the ability to produce an acceptable sound from weak signals. The *selectivity* is related to the tuning function and refers to the ability of a receiver to select a desired station from others of nearly the same frequency.

Tuning circuits, or resonant circuits, are extremely important to the operation of the RF and IF amplifiers of a receiver. In light of this, we begin this chapter with a brief review of the properties of tuning circuits.

6-2 RESONANT CIRCUITS

A *resonant circuit* consists of an inductor in series or in parallel with a capacitor to form either a *series* or a *parallel* resonant circuit, depending upon the connection. The resonant frequency depends upon the size of the capacitor and inductor, and may be calculated from the formula given in Figure 6-2.

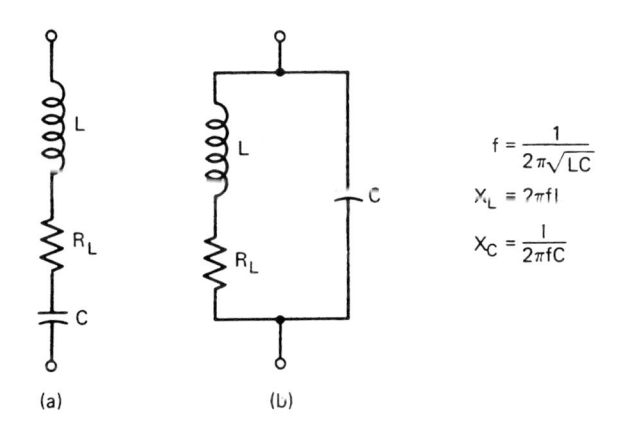

$$f = \frac{1}{2\pi\sqrt{LC}}$$

$$X_L = 2\pi f L$$

$$X_C = \frac{1}{2\pi f C}$$

FIGURE 6-2 (a) Series and (b) parallel resonant circuits. R_L is the resistance of the inductor winding.

At resonance, the impedance of a series resonant circuit drops to a minimum while the impedance of a parallel circuit reaches a maximum. Maximum current flows through a series circuit at resonance while maximum voltage is developed across a parallel circuit at the resonant frequency. Since series circuits are used less frequently, we direct the following discussion to parallel resonant circuits.

In addition to its inductance, a practical inductor also exhibits a resistance due to the resistance of the wire used to wind the inductor. The resistance is not necessarily small, and it is important in determining the sharpness of the resonance of the circuit of which the inductor is a part. At high frequencies current tends to flow only on the surface of the conductor (skin effect), causing the effective resistance to be many times the low-frequency value. In Figure 6-2 a resistor R_L is shown in series with the inductor L to present its effective resistance.

Frequency Response

Figure 6-3 illustrates a typical frequency-response curve of a resonant circuit. Maximum response occurs at the resonant frequency F_R, but a significant response occurs also for adjacent frequencies. Of interest is the *bandwidth* (BW), defined in the figure. The bandwidth is controlled by the Q, or *quality factor*, of the circuit according to the formula given in the figure.

In practical circuits, the Q of the inductor determines the Q of the circuit. The Q of an inductor is given by X_L/R_L the ratio of the inductive reactance to the effective resistance of the winding. Several factors tend to limit the Q to values from about 10 to 250. The skin effect, distributed capacitance from turn to turn of the inductor, and losses in the coil form and associated materials lower the Q.

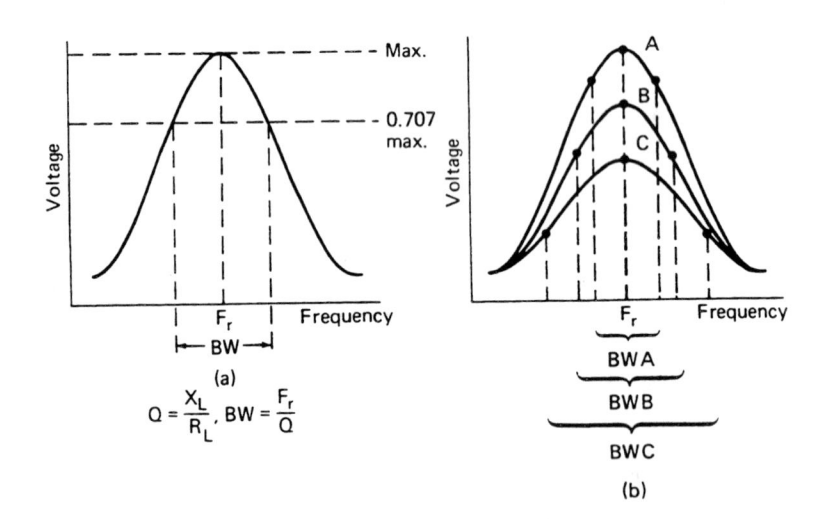

FIGURE 6-3 (a) Typical frequency-response curve; the definition of bandwidth is illustrated. (b) As the Q is increased, the peak becomes sharper and the bandwidth BW becomes smaller.

If a physical resistance is connected into the circuit, the Q will be lowered, and this provides the designer a method of reducing the Q when this is desired. The effect upon the response curve of various Q's is shown in Figure 6-3(b). As the Q is increased, the peak becomes higher while the bandwidth becomes smaller.

Parallel resonant circuits frequently appear in the output circuit of RF amplifiers, and in such cases the resonant impedance determines (in part) the gain of the amplifier. Figure 6-4 illustrates the variation of impedance as a function of frequency. Note that the resistance of the inductor. R_L, is taken into account and that it affects the maximum impedance. As R_L increases, the maximum impedance decreases. The maximum impedance also depends upon the L/C ratio. Circuits with large values of L and small values of C have higher resonant impedances. This ratio cannot be made arbitrarily large, however, because of distributed capacitance effects of practical inductors. Note further that the resonant impedance equals the reactance of the inductor or capacitor at the resonant frequency multiplied by the Q of the circuit.

Shunt Resistance

A parallel resonant circuit is frequently shunted by a resistor R_s, as shown in Figure 6-5, to lower the Q of the circuit to a new value we denote as Q_p. A formula for calculating Q_p is given in the figure. Recall that lowering the Q increases the bandwidth.

Observe in Figure 6-5 that the maximum impedance of the circuit is the parallel equivalent of the shunting resistor R_s and $Q^2 R_L$. The term $Q^2 R_L$ is the result of transforming the resistor in series with the inductor to an equivalent resistance in parallel with the inductor. The maximum impedance can never be greater than the value of the shunting resistor R_s.

We conclude this brief description of parallel resonant circuits with an example given as Figure 6-6. The references at the end of this chapter give more rigorous

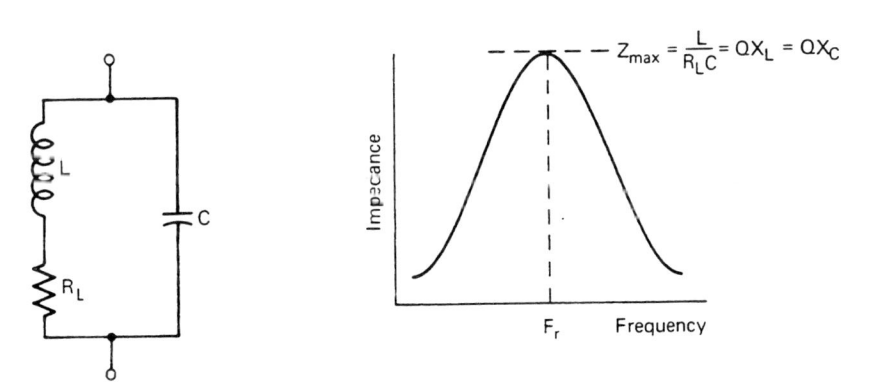

FIGURE 6-4 Variation of impedance with frequency for a parallel resonant circuit. The curve is identical to the previous frequency-response curve. Note the maximum value attained by the impedance.

Sec. 6-2   *Resonant Circuits*

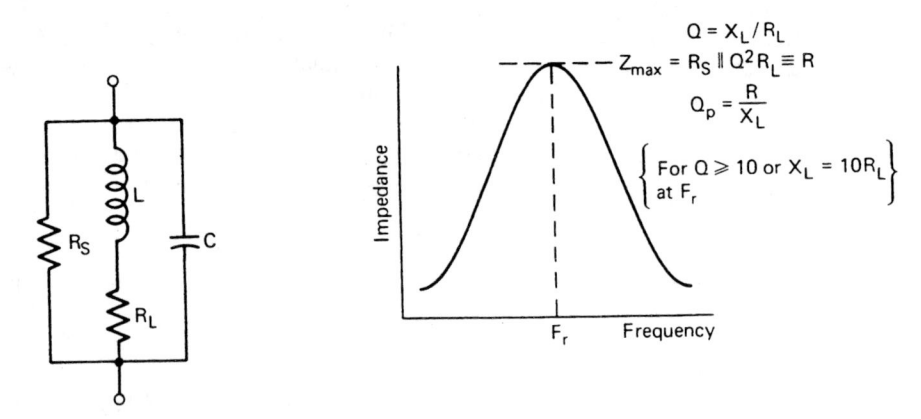

FIGURE 6-5 Resistor R_s tends to lower the Q of the resonant circuit.

and detailed descriptions. Incidentally, a parallel resonant circuit is often called a *tank circuit* because of the shape of rather large inductors in older receivers. We shall use this terminology upon occasion.

Tuned Transformers

Figure 6-7(a) illustrates a circuit in which two parallel resonant circuits tuned to the same frequency are magnetically coupled to form a tuned transformer. The frequency response depends upon the Q of the individual tuned circuits and very

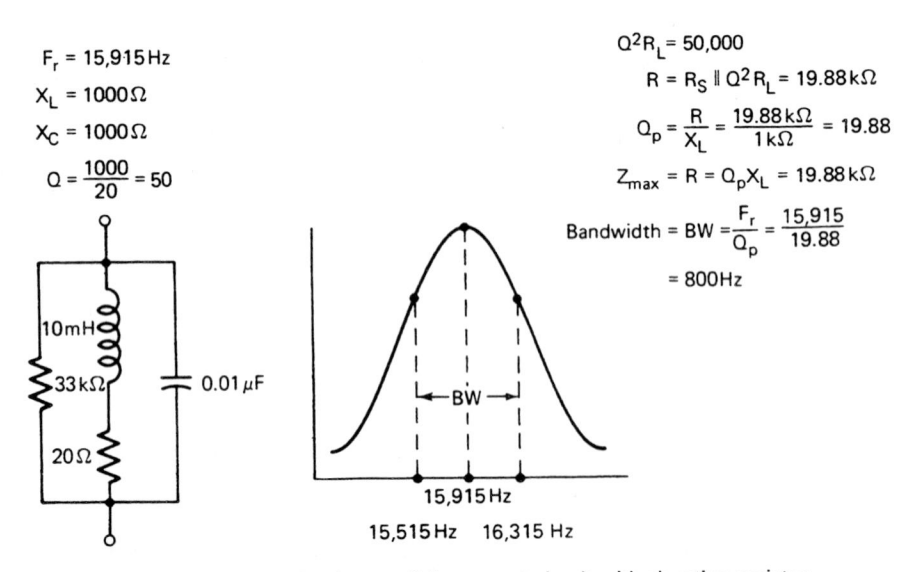

$F_r = 15{,}915\,\text{Hz}$
$X_L = 1000\,\Omega$
$X_C = 1000\,\Omega$
$Q = \dfrac{1000}{20} = 50$

$Q^2 R_L = 50{,}000$
$R = R_S \parallel Q^2 R_L = 19.88\,\text{k}\Omega$
$Q_p = \dfrac{R}{X_L} = \dfrac{19.88\,\text{k}\Omega}{1\,\text{k}\Omega} = 19.88$
$Z_{max} = R = Q_p X_L = 19.88\,\text{k}\Omega$
$\text{Bandwidth} = BW = \dfrac{F_r}{Q_p} = \dfrac{15{,}915}{19.88}$
$= 800\,\text{Hz}$

FIGURE 6-6 Analysis of a parallel resonant circuit with shunting resistor.

Radio-Frequency Amplifiers   Chap. 6

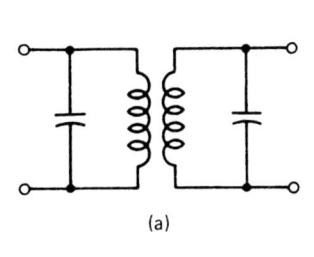

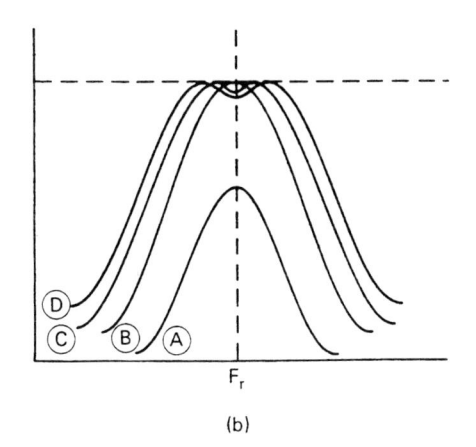

FIGURE 6-7 (a) Tuned transformer results from the magnetic coupling of two resonant circuits. (b) The frequency response depends upon the degree of coupling between the circuits. As the coupling is increased, the response broadens.

strongly upon the degree of coupling between the inductors. When the inductors are well separated (loose coupling), a frequency response is obtained as shown by curve A in Figure 6-7(b). As the inductors are moved closer together (to achieve tighter coupling), the peak of the frequency-response curve reaches a maximum at the point of "critical coupling," illustrated by curve B. A further increase in coupling causes a broadening of the response with no increase in the level of the peak. Finally, with degrees of coupling well beyond critical coupling, a *double-humped* response curve (curve D) develops that is characteristic of *overcoupled* tuned circuits.

Tuned transformers are used extensively in receivers and play important roles in determining the receiver selectivity. By controlling circuit Q's and degrees of coupling between tuned circuits, the bandwidth of the amplifiers can be controlled.

6-3 TRIODES AS RF AMPLIFIERS

Figure 6-8 shows a triode RF amplifier. Parallel resonant (tank) circuits in the grid and plate circuit tune the amplifier, and cathode biasing provides the proper bias voltage. The circuit looks good on paper, but, in practice, the interelectrode capacitances C_{gp} and C_{gk} cause the amplifier to be unstable. Feedback from the plate to the grid via C_{gp} can occur in the proper phase to cause the amplifier to oscillate and act as a signal generator rather than as an amplifier. The effect becomes worse at high frequencies, where the reactance of C_{gp} becomes smaller.

Fortunately, there are several ways to reduce the tendency of the amplifier to oscillate. If a second signal out of phase with the signal fed back through C_{gp} is fed back from the plate circuit to the grid, the two signals will cancel and the

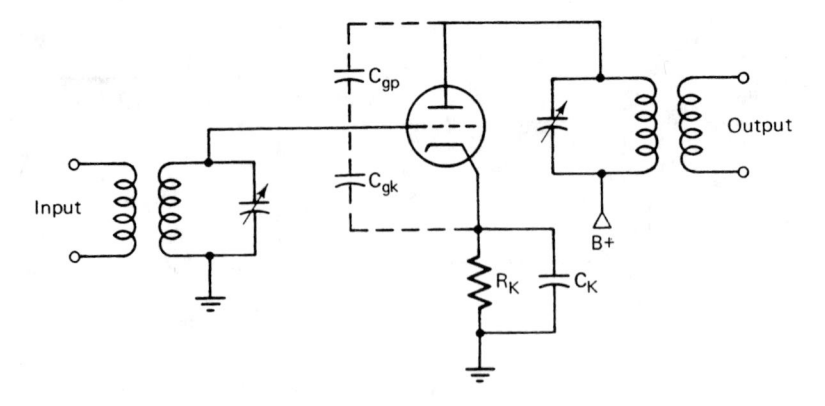

FIGURE 6-8 Simple triode RF amplifier showing the interelectrode capacitances that cause problems at high frequencies.

adverse effects of feedback through C_{gp} will be avoided. This process is called *neutralization* and is illustrated in the circuit of Figure 6-9.

The B+ voltage is applied to a tap on the plate inductor so that signal ground appears at the tap rather than at either end of the inductor. Thus, the signal at point B is out of phase with the signal at point A. The undesired feedback through C_{gp} is in phase with point A, and therefore the signal fed back to the grid via neutralizing capacitor C_N from point B is out of phase with the signal from point A. The two signals cancel at the grid, to produce no feedback at all. The result is a *neutralized* amplifier that will operate at high frequencies without going into oscillation.

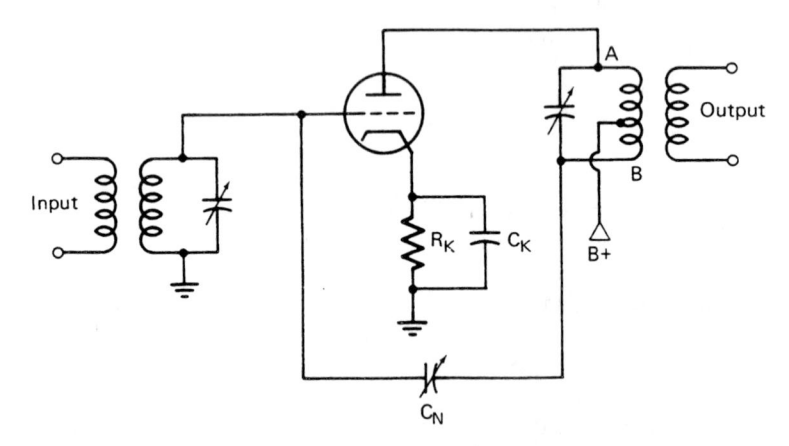

FIGURE 6-9 Neutralized triode amplifier. An out-of-phase signal is fed back from the plate circuit via C_n to neutralize or cancel the signal fed back to the grid from the plate via C_{gp} (not shown in the figure).

Miller Effect

The capacitance C_{gp} of a triode is usually on the order of a few picofarads. In an amplifier circuit, however, the effective capacitance is much larger than this because C_{gp} is amplified by the action of the tube. This is the *Miller effect*, and it may be explained as follows.

When a voltage is applied to the grid, C_{gp} must charge as the grid voltage is applied. As the grid voltage rises due to the charging of C_{gp}, the amplifying action of the tube causes the plate voltage to go down. Thus, C_{gp} charges to a much greater voltage than that applied to the grid because the plate end of C_{gp} does not remain at a fixed voltage. The result is that C_{gp} appears to be greater than it actually is by a factor approximately equal to gain of the tube.

Grounded Grid RF Amplifier

A triode operated as a grounded grid amplifier does not require neutralization, because the grid itself acts as a shield between the plate and cathode. A typical circuit is shown in Figure 6-10. The input signal causes the voltage at the cathode to vary, and since the grid voltage is held constant, a voltage variation occurs between the grid and the cathode, which controls the current flowing through the tube.

Disadvantages of the grounded grid amplifier are low input impedance and a somewhat lower gain than is achieved with neutralized, common-cathode, triode circuits. The low input impedance makes it impractical to incorporate variable

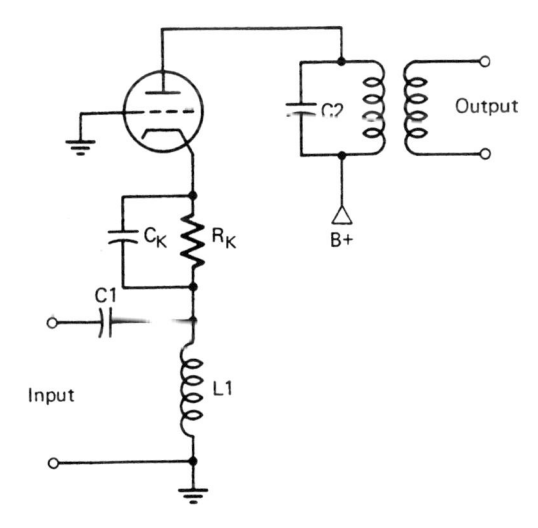

FIGURE 6-10 Grounded grid RF amplifier. The grid itself acts as a shield between the plate (the output) and the cathode (the input), so neutralization is not required.

tuning circuits into the input circuit. Grounded grid circuits are frequently encountered where the signal source is a low impedance source, such as a transmission line from an external antenna.

6-4 TRANSISTOR (BJT) RF AMPLIFIERS

Many considerations applicable to vacuum-tube RF amplifiers also apply to solid-state RF amplifiers. Resonant circuits determine the bandwidth and gain of the amplifiers, and interelectrode capacitances create stability problems. BJT RF amplifiers use methods of neutralization quite similar to the vacuum triode. In the equivalent circuit of a BJT, however, the collector-base capacitance is shunted by a resistance. This sometimes requires slightly more elaborate neutralization circuits than are required by vacuum triodes.

The same biasing arrangements described earlier are used for BJT RF amplifier stages. Additional bias considerations for RF amplifiers include selecting an operating point that results in the least noise production, and providing for operating point stability. A shift in the operating point of a BJT can result in a shift of impedance levels and changes of junction capacitance that result in detuning and loss of sensitivity.

Cutoff Frequencies

Not every type of transistor will operate at high frequencies because, at high frequencies, the alpha and the beta decrease to values well below their low-frequency value. The *gain-bandwidth product* f_t of a transistor is the frequency at which the beta is reduced to unity. Transistors normally are operated well below f_t. Audio transistors, for example, typically exhibit f_t's of 1 MHz.

The *alpha cutoff frequency* f_α is the frequency at which the alpha decreases to 0.707 of its low-frequency value; similarly, the *beta cutoff frequency* f_β is the frequency at which the beta drops to 0.707 of its low-frequency value. A relationship exists between f_β and f_t, namely, $f_\beta = f_t/\beta$, where β is the low-frequency value. The alpha cutoff frequency is typically about 1.2 times f_t, or in terms of f_α, $f_t = f_\alpha/1.2$.

For example, a transistor whose f_t is 100 MHz and whose low-frequency β is 80 has a beta cutoff f_β of 100/80 MHz = 1.25 MHz. At the same time, the alpha cutoff f_α would be around 120 MHz.

Typical Circuit

A basic common-emitter RF amplifier is shown in Figure 6-11. Tuned circuits are used both at the input and output in conjunction with transformer coupling between stages. The transistor is packaged in a metal can which is grounded by a fourth lead of the device. This is denoted by a ground symbol connected to the transistor symbol. The grounded metal can provides shielding against pickup of stray RF signals.

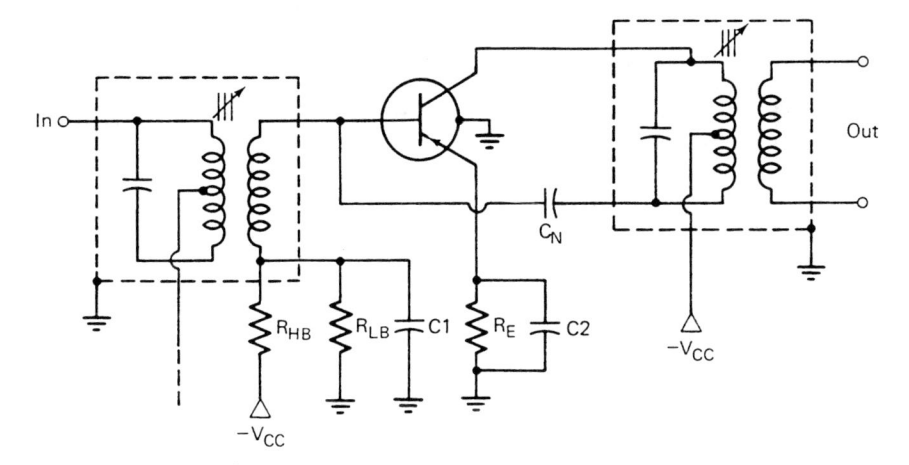

FIGURE 6-11 Typical common-emitter RF amplifier used as an IF amplifier in FM receivers.

The stage is universal biased, and the biasing resistors are labeled. The junction of R_{HB} and R_{LB} is held at AC ground by capacitor C1, and note that the DC bias voltage is applied to the base through the secondary of the coupling transformer. Emitter resistor R_E is bypassed by C2 to prevent signal degeneration.

A tapped inductor in the collector circuit provides the signal of proper phase for neutralization. Capacitor C_N is the neutralizing capacitor which feeds the neutralizing signal back to the base.

The transformer that couples the signal to the base of the transistor is not a double-tuned transformer. The low impedance of the base circuit makes it impractical to tune the secondary. Observe the dashed lines around the tuned circuits that denote metal shields to prevent coupling of the output signal back to the input. This is extremely important since both the input and output circuits are tuned to the same frequency.

Common-Base RF Amplifier

Figure 6-12 shows a common-base amplifier used as an RF amplifier in an FM receiver. The biasing is universal, and a parallel resonant circuit consisting of L1, C3, and C4 forms the collector load. The antenna signal is coupled through transformer T1 and coupling capacitor C1 to the emitter of the transistor. Capacitor C2 holds the base at signal ground.

At first, the capacitors C3 and C4 do not appear to be in parallel with L1, since the low end of these capacitors connects to ground. But the low end of L1 connects to the power supply point, which is connected to ground through a large capacitor. For high-frequency signals, the power supply appears as "signal ground," and therefore both L1 and the combination C3-C4, connect to the same signal ground, forming a parallel resonant circuit.

Capacitor C3 is a variable capacitor operator by the tuning control of the

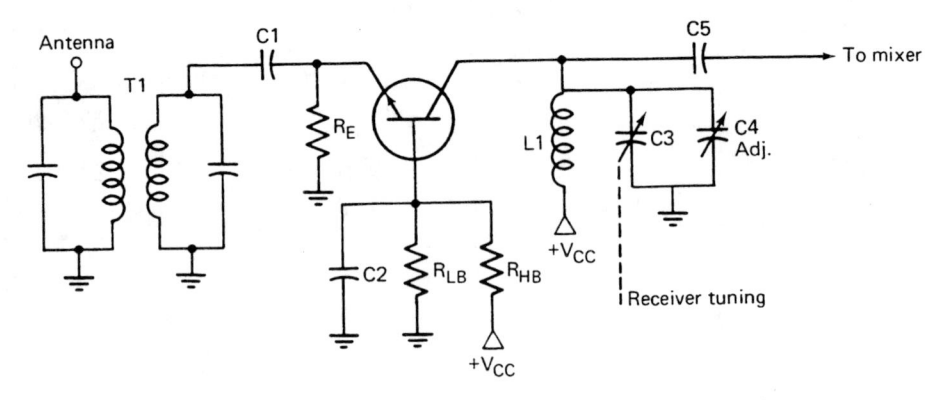

FIGURE 6-12 Common-base RF amplifier used in an FM receiver.

receiver, as indicated by the dashed line. Capacitor C4 is a small trimmer, constructed as part of C3, that is adjusted as part of the alignment procedure for the receiver. (Alignment procedures are discussed in Chapter 12.)

Another common-base amplifier is shown in Figure 6-13. A variation of a universal-bias arrangement is used that provides for DC collector feedback and AVC action. The emitter resistor is R_E, bypassed by C1. Note that the emitter current flows through the secondary of the input transformer; in the previous circuit, it did not.

Resistor R_C in the collector circuit produces a small DC voltage drop that tends to lower the collector voltage as collector current increases. The high end of R_C is held at signal ground by C3 so that only DC voltages are affected by R_C. High-base resistor R_{HB} connects to the high end of R_C to provide circuit stability as with earlier circuits involving collector feedback.

AVC action is accomplished by returning the low-base resistor R_{LB} to the AVC line. An increase in signal level causes a more negative voltage to be applied to

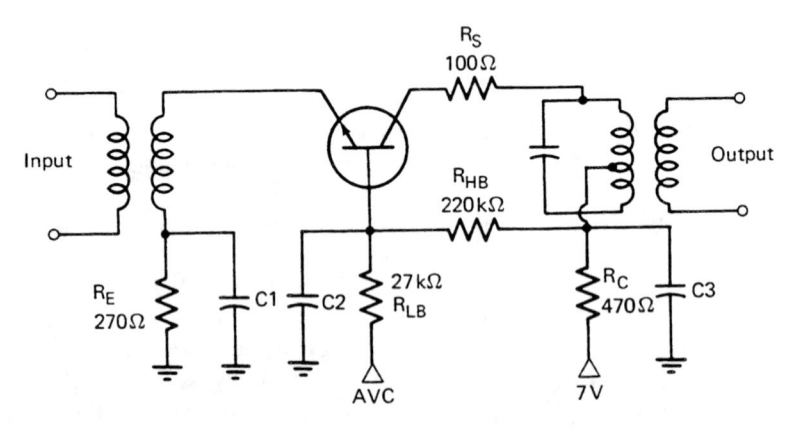

FIGURE 6-13 Common-base RF amplifier that provides for AVC action. The biasing arrangement is that of a universal bias with collector feedback.

168 Radio-Frequency Amplifiers Chap. 6

R_{LB}, which shifts the operating point of the transistor to a point of lower gain. Capacitor C2 holds the base at signal ground. An unusual feature of this circuit is the 100-Ω resistor R_S at the collector of the transistor. Under conditions of extremely large input signals (perhaps from noise pulses), it is possible for the collector voltage to drop very nearly to the voltage of the base, momentarily causing the collector-base junction capacitance to increase. An undesirable effect of this is to cause the circuit to be unstable; it tends to break into oscillation. The difficulty is avoided by including R_S, typically a few hundred ohms. We call this resistor the *collector stabilizing resistor*.

6-5 PENTODE RF AMPLIFIERS

The Vacuum Tetrode

Earlier we saw that vacuum triode RF amplifiers tend to oscillate at high frequencies because of the interelectrode capacitance between the plate and the grid. This troublesome capacitance can be reduced by a factor of 100 or more by installing a second grid in the tube between the signal grid and plate. The *screen grid*, as the second grid is called, is held at AC ground and serves as an electrostatic shield between the signal grid and the plate. The resulting tube is called a *tetrode* because it has four active elements: the cathode, signal grid, screen grid, and plate.

A high positive DC voltage must be maintained on the screen so that electrons will be drawn from the cathode toward the plate. (Recall that the screen shields the region of the cathode from the influence of the plate.) Typically, the screen voltage is equal to or slightly higher than the plate voltage, which may be as high as several hundred volts. Under these conditions, electrons from the cathode arrive at the plate with very high velocities. The impact of a high-speed electron with the metal of the plate is likely to eject one or more other electrons from the plate, a phenomenon called *secondary emission*.

Once ejected from the plate, the secondary electrons may be attracted to the screen grid because of its high positive voltage. This may cause excessive screen current, but more important, the screen current is robbed from the plate. Under certain signal conditions, the plate current may actually become smaller when, according to the voltage on the signal grid, it is supposed to become larger. This is a significant problem with tetrodes, but the problem is not so severe that tetrodes are not useful in many applications.

The Vacuum Pentode

The problem of secondary emission can be overcome by introducing yet another grid into the tube, this grid being the *supressor grid* that is located between the screen grid and the plate. It is maintained at near cathode potential, which is quite negative in comparison to the high positive voltage maintained on the plate and screen grid. The suppressor grid repels the slow-moving secondary electrons back

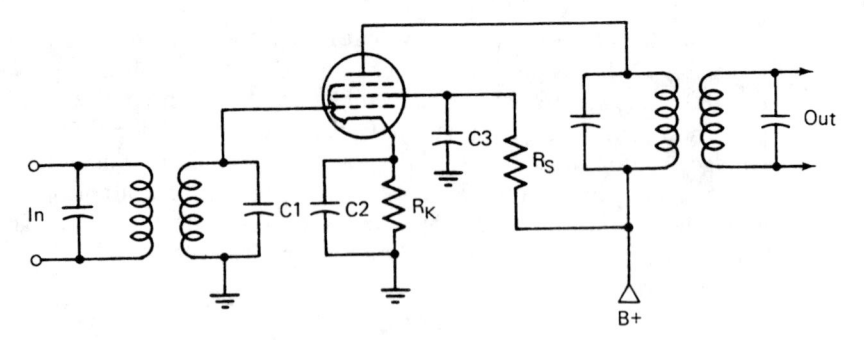

FIGURE 6-14 Pentode RF amplifier frequently used as an IF amplifier in tube-type superheterodyne receivers.

to the plate, whereas in the tetrode, they could easily be attracted to the screen. Whereas the capacitance C_{gp} of a triode is a few picofarads, typical values for pentodes are a few hundredths of a picofarad. Neutralization of pentodes is seldom required. The schematic symbol of a *pentode*, as these tubes are called, is shown in the circuit of Figure 6-14.

A pentode circuit frequently used as the IF amplifier in superheterodyne receivers is shown in Figure 6-14. Double-tuned transformers appear in both the input and output circuits to provide for a sharp tuning characteristic that sets the selectivity of the receiver. The tube is self-biased by cathode resistor R_k, which is bypassed by capacitor C_k. Screen resistor R_s and screen bypass capacitor C_s form a decoupling network that prevents the screen voltage from varying at signal frequencies.

In amplifiers that incorporate tuned circuits in both the input and the output, extreme care must be taken to avoid the slightest coupling between the output and the input. The tuned circuits are very sensitive to stray pickup, and the smallest amount of feedback will cause the amplifier to go into oscillation. As evidence of this, one may observe the shielding and careful layout of components in the IF amplifier sections of receivers.

6-6 GAIN OF RF AMPLIFIERS

A detailed discussion of the gains obtained from various RF amplifiers is beyond the scope of this text, but the following gives at least a hint of the principles involved. Recall that the transconductance g_m of an active device (tube, BJT, or FET) refers to the effectiveness of the input element (grid, base, or gate) in controlling the current flow through the device. If the input voltage is e_g, the output current variation is given by $g_m e_g$. The output current must flow through the load impedance Z_L, which causes the output voltage to vary. The variation in voltage (the output signal) is the product of the signal current $g_m e_g$ and the load impedance Z_L: $g_m e_g Z_L$. The ratio of this voltage to the input voltage e_g is the voltage gain of the amplifier,

$g_m Z_L$. Thus, the gain is dependent upon the transconductance g_m and the load impedance Z_L.

Other factors enter into the calculation of the total gain of a stage. The coupling circuits may contribute voltage gain by transformer action, and these effects must be taken into account. For our purposes, however, it is sufficient to consider that the gain is directly proportional to the transconductance and to the plate load impedance.

We saw earlier that the impedance of a parallel resonant circuit varies with frequency. Since the gain is proportional to the plate load impedance, the gain of a stage whose plate load is a paralled resonant circuit varies with frequency in the same manner as the impedance of the resonant circuit. Hence, the Q of the resonant circuit affects the gain and bandwidth of the amplifier.

The transconductance of most tubes, BJTs, and FETs used as RF amplifiers varies with the operating point (Q-point) of the device. Thus, by changing the bias voltage, the gain of the amplifier can be controlled. This idea is utilized in automatic-volume-control (AVC) circuits that are described in the next section.

6-7 AUTOMATIC VOLUME CONTROL (AVC)

In the preceding section we saw that the gain of an RF amplifier is closely related to the transconductance, which varies as the DC operating point of the active device is shifted. In tube circuits, the AVC voltage source supplies the DC grid-bias voltage in the controlled amplifiers. An increase in signal level at the demodulator causes the AVC voltage to become more negative, reducing the gain. Conversely, a decrease in signal level causes the AVC voltage to become less negative and the gain increases accordingly.

RF amplifier signal levels in receivers are small in comparison with the total dynamic range of the tube. Thus, only a small portion of the dynamic transfer curve is used at any time. The transconductance g_m is proportional to the slope of the curve, and portions of the curve that have different slopes produce different gains. From this we conclude that the transfer curve must possess considerable curvature in order for the tube to function as a controlled amplifier. This is illustrated in Figure 6-15(a).

The spacing of the grid wires affects the curvature of the transfer curve. Uniform spacing, in which the grid wire is a spiral of uniform pitch, produces a transfer curve of minimum curvature. Variable spacing produces greater curvature such as that illustrated in Figure 6-15(a). Grid wires of uniform and of variable spacing are illustrated in Figure 6-15(b).

A typical AVC circuit for an AM receiver is shown in Figure 6-16. The DC voltage component developed at the high side of the volume control is proportional to the amplitude of the incoming RF signal and serves as the source of AVC voltage. Variations of this voltage at the audio frequency are eliminated, or averaged, by the AVC filter, composed or R and C. No audio or RF signal voltages must appear on the AVC line. It is important to realize that the AVC voltage is developed by

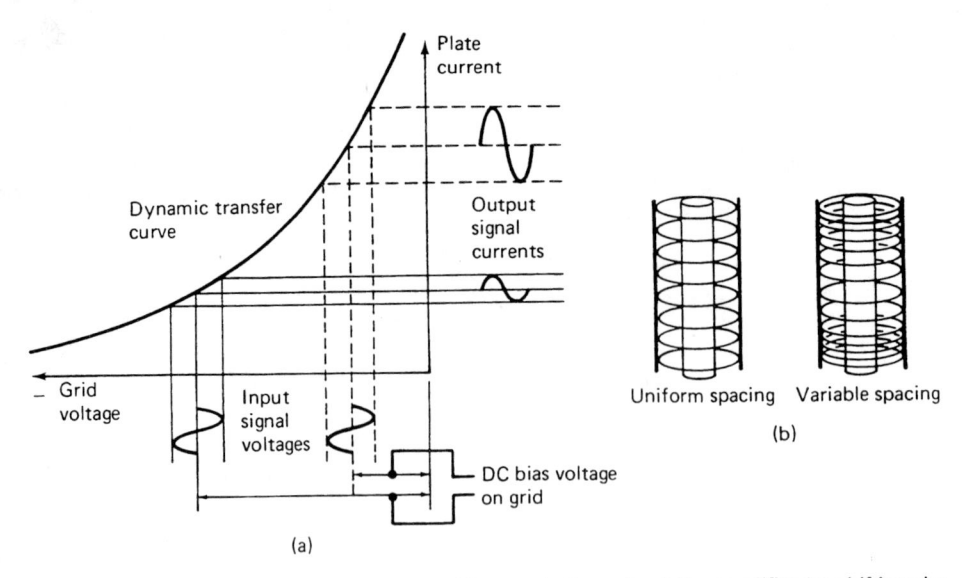

FIGURE 6-15 (a) DC bias voltage on the grid controls the gain of the amplifier by shifting the operating point on the dynamic transfer curve. (b) Uniform and variable spacing of the grid wires.

the average value of the RF signal rather than by the audio. Considerable gain-reducing AVC voltage may be developed even when no audio information (music or voice) is present.

The vacuum-tube RF amplifier circuits described earlier in this chapter may be converted to controlled-gain amplifiers by changing the grid biasing from cathode bias to AVC bias in which the grid voltage is obtained from the AVC line. This is accomplished by omitting the cathode resistor and returning the grids to the AVC line.

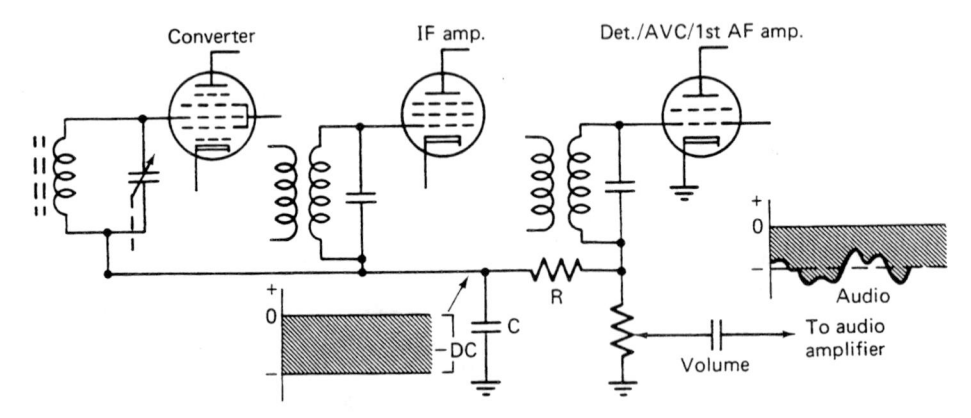

FIGURE 6-16 Typical AVC circuit for an AM receiver. Resistor R and capacitor C constitute the AVC filter which removes the audio variations from the AVC voltage.

Bipolar Transistor AVC

The transconductance of BJTs varies with the forward bias applied to the base-emitter junction. Most transistors exhibit a decrease in transconductance as the forward bias is *reduced*, but certain transistors of special design exhibit a decrease in transconductance as the forward bias is *increased*. Consequently, two modes of AVC operation are available. Systems that increase the forward bias to reduce the gain utilize *forward AVC* while systems that decrease the forward bias to reduce the gain utilize *reverse AVC*. Reverse AVC systems are more widely used. To clarify, the base voltage of an *NPN* transistor using reverse AVC must be made less positive in order to reduce the gain.

A comparison of forward and reverse AVC reveals the following. Forward AVC provides improved cross-modulation characteristics and greater signal-handling capability, since the transistor is operated well above the knee of the base-emitter conduction characteristic. On the other hand, forward AVC produces greater changes in the input and output impedances of the transistor, and also produces drastic changes in the transistor junction capacitances. These changes alter the characteristics of the associated tuned circuit, and the undesirable result is that the selectivity may change with the level of the input signal.

Reverse AVC is simpler to use because the junction capacitances are not appreciably changed by variations of the signal level. Further, the impedance levels are much higher, and they tend to increase with increasing signal level. Consequently, the tuned circuits are not as heavily loaded, and the bandwidth tends to be much more stable since the change in loading of the tuned circuits is minimal.

The AVC circuit of BJT amplifiers must be capable of supplying the base current for the controlled transistors. Since the AVC voltage is frequently derived from a point in or near the demodulator that is sensitive to loading, an AVC amplifier is frequently used to supply the current for the AVC line. In such cases the demodulator only has to supply the base current for the AVC amplifier.

6-8 JFET RF AMPLIFIERS

FETs exhibit many electrical characteristics that are quite similar to vacuum tubes. The input impedance equals or exceeds values obtained with vacuum tubes; drain current versus voltage characteristics are similar to the plate characteristics of a pentode; and the interelectrode capacitances are of the same order of magnitude as those of a triode. FETs generate only a fraction of the noise of triodes and are used as RF amplifiers in high-sensitivity receivers. The same methods of neutralization are applicable to FETs as are used for triodes, and a nonlinear transfer characteristic enables reverse AVC to be easily achieved. FETs may be connected in either the common source, common-gate, or common-drain configuration. These are analogous to the common emitter, the common base, and the emitter follower configurations, respectively, of a BJT.

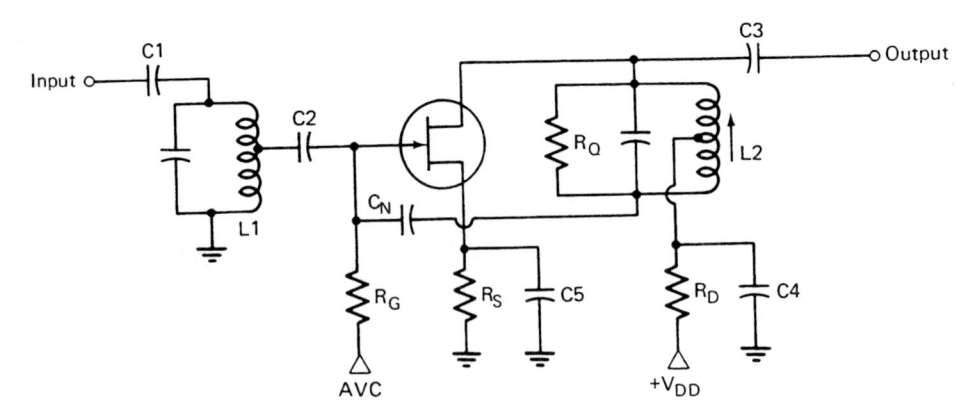

FIGURE 6-17 Typical neutralized JFET RF amplifier.

JFET RF Amplifier

A typical configuration of a JFET RF amplifier is shown in Figure 6-17. The biasing arrangement is a self-bias circuit modified so that the low end of gate resistor R_G connects to the AVC line rather than to ground. The source resistor is R_S, bypassed by C5. Resistor R_D and capacitor C4 form a filter network that causes high-frequency signals to flow to ground through C4 rather than to pass through R_D to the power supply.

The signal couples in through C1 to the $L1$ resonant circuit, from which it passes through C2 to the gate of the FET. C2 connects to a tap on $L1$ to obtain an impedance match to the gate circuit. A parallel resonant circuit forms the drain load, and C3 couples the output signal to the next stage. Resistor R_Q is in parallel with the resonant circuit to provide for a greater bandwidth by lowering the Q of the resonant circuit.

The drain supply voltage connects to a tap on $L2$ to produce an out-of-phase (with the drain) signal at the lower end of $L2$. This is the source of the neutralizing signal that is fed back to the gate via C_N.

6-9 MOSFET RF AMPLIFIERS

Stability against oscillation must be provided in MOSFET RF amplifiers because of the interelectrode caapcitance between the drain and the gate. Stability may be provided by neutralization or by slight mismatching of impedances at the input and output. Capacitance bridge neutralization is the method most commonly used.

The configuration of a typical MOSFET RF amplifier is shown in Figure 6-18. The MOSFET is self-biased by R_S and R_G, and source resistor R_S is bypassed by source bypass capacitor C_S. A parallel resonant circuit consisting of C1 and $L1$ constitutes the drain load, and the output signal is coupled to the next stage via C3.

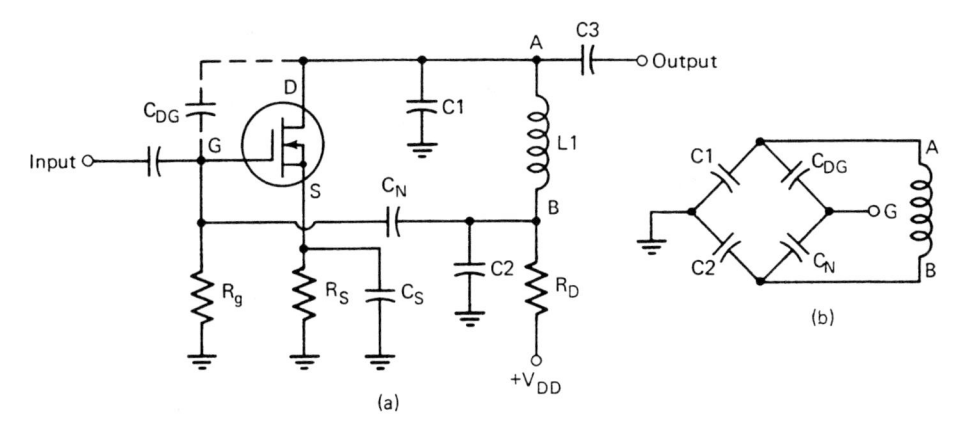

FIGURE 6-18 MOSFET RF amplifier that uses bridge neutralization.

Another method of neutralization, called bridge neutralization, is used in the circuit of Figure 6-18. Recall that when a bridge circuit is balanced (Wheatstone bridge for resistances, for example), no voltage is developed between the output points of the bridge. By making the interelectrode capacitance C_{dg} a part of a capacitance bridge, its effect can be eliminated. The bridge output terminals are connected to the source and gate so that none of the amplifier output signal appears between these terminals. Hence, the undesired coupling from output to input via C_{dg} is minimized, and the tendency of the stage to oscillate is reduced.

Capacitors $C1$, $C2$, and C_N, and interelectrode capacitance C_{DG} between the drain and the gate, form the bridge as illustrated in the figure. Resistor R_D isolates the low end of $L1$ from the signal ground of the power supply. The signal in the drain circuit is developed across $L1$, and therefore $L1$ acts as the signal source for the bridge. When the capacitors are chosen so that the bridge is balanced, no signal transfer occurs from the drain to the gate.

6-10 DUAL-GATE MOSFET AMPLIFIERS

Dual-gate MOSFETs are similar in operation to single-gate devices. The second gate, $G2$, provides additional control of drain current and is typically utilized to achieve AVC action. Two signals may be combined in a dual-gate device by applying one signal to each gate (described in Chapter 2). The second gate, when held at signal ground, tends to reduce feedback from the drain to $G1$, the signal input, so that neutralization is seldom required. Biasing voltages must be provided for each gate.

A typical circuit is shown in Figure 6-19. The input signal is provided by the antenna to the $L1$ paralled resonant circuit. Taps on $L1$ provide the proper impedance match from the antenna lead in to the signal input gate $G1$ of the MOSFET. $G1$ is biased by the voltage divider, consisting of R_H and R_L.

Gate $G2$ is used to achieve AVC action and biased by the AVC line. Resistor

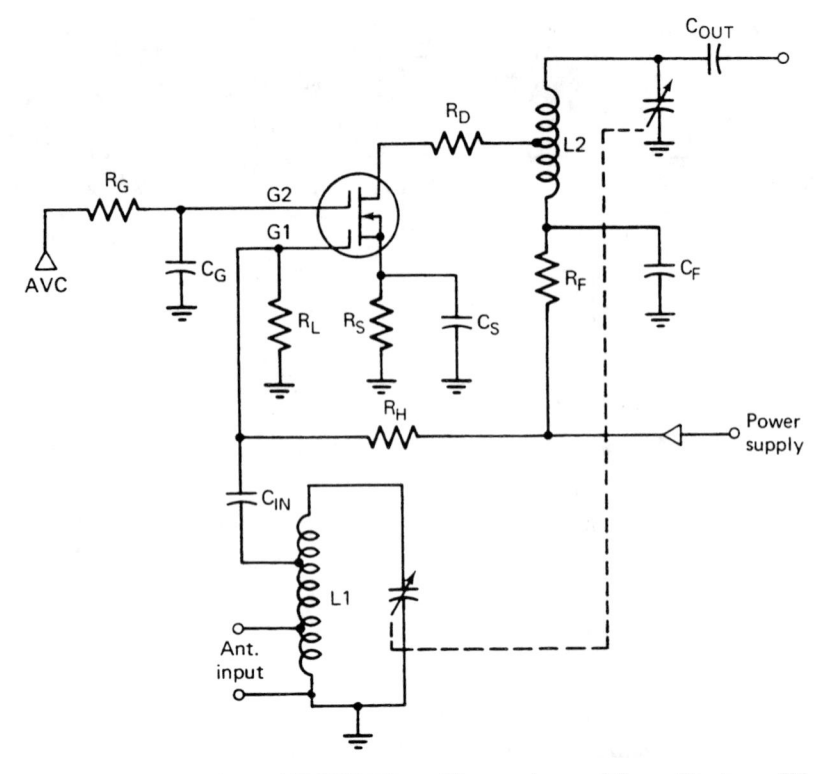

FIGURE 6-19 Dual-gate MOSFET RF amplifier used as an RF amplifier in an FM receiver.

R_G and capacitor C_G act as a filter to prevent any transfer of signals either to or from the AVC line. Since $G2$ is close to the drain of the MOSFET, some coupling will occur between the drain and $G2$.

The drain load is formed by the parallel resonant circuit ($L2$), where a tap on $L2$ provides impedance matching. R_F and C_F constitute a filter to keep the RF signals out of the power supply. Resistor R_D at the drain stabilizes the circuit against undesired oscillation. The output is coupled to the next stage by C_{out}.

AVC for MOSFET Amplifiers

The gain of a MOSFET RF amplifier depends upon the transconductance of the device, which varies with the operating point of the device and with the source-to-drain voltage. Reverse AVC involves shifting the gate bias voltage toward the negative to reduce the transconductance. Forward AVC involves an increase of the drain current which causes the drain voltage to decrease due to the additional voltage drop across a series resistor inserted into the drain supply. In the forward AVC mode, the drain current effectively controls the drain voltage, and the drain voltage alters the transconductance.

MOSFETS versus JFETs and Bipolar Transistors

The insulated gate construction of MOSFETs gives rise to an extremely small leakage current that is independent of polarity of the bias voltage. Since no current loading occurs with positive excursions of the gate voltage, MOSFETs have a much greater dynamic range than either JFETs or bipolar devices. Other advantages of MOSFETs are better cross-modulation characteristics, a low feedback capacitance that often makes neutralization unnecessary, and a typical transconductance from 1.5 to 2 times that of a JFET.

A big disadvantage of MOSFET devices is their susceptibility to damage by static electricity when out of circuit. This makes their handling more critical. While bipolar and JFET devices withstand momentary discharges without permanent damage, once the insulating material of a MOSFET is punctured, the device is permanently destroyed; no healing of the wound occurs.

6-11 NOISE

An important consideration is the amount of noise an RF amplifier contributes to the signal as it is being amplified. An amplifier with a large gain that introduces large amounts of noise is of little use. Loud pops, hums, or squeals are not classified as noise as the term is used here. The smooth, hissing static received between stations on an FM receiver is typical of the noise of which we are speaking.

The waveform of the noise "signal" is completely random, with all frequencies equally represented in *white noise*. White noise is very unpleasant to hear. A close approximation to white noise can be heard by turning the tone controls on an FM receiver to full treble with the receiver tuned between stations. *Pink noise*, on the other hand, has the high frequencies attenuated, and is quite soothing. Approximate examples of pink noise are the sound of raindrops on a tin roof during a heavy rain, and the sound of the surf at the seashore. It is an interesting exercise to try and duplicate the sound of the surf by manipulating the tone and volume controls of an FM receiver tuned off station:

Thermal Noise

A resistor acts as a noise generator because the electrons are in constant motion due to thermal agitation (heat). At any instant more electrons may be near one end of the resistor than the other, and that end of the resistor appears negative. The distribution of electrons constantly changes, and a noise voltage is developed across the resistor. A large resistance generates more noise than a small one, and a hot resistor is noisier than a cool resistor. Thermal noise is frequently called *Johnson noise*, after J. B. Johnson, who, in 1928, developed the basic theory of thermal noise.

Shot Noise

Even though the average cathode-plate current in a tube is constant, the electrons are not emitted from the cathode at a perfectly uniform rate. A minute variation of plate current exists due to the random nature of electron emission from the cathode. Variations of the plate current produce a noise voltage across the load impedance. A similar effect occurs in bipolar-junction and field-effect transistors, where variations of the collector current or the drain current produce a noise voltage across the load impedance. Noise produced by this mechanism is called *shot noise*. Shot noise increases with the level of direct current present in the device under consideration.

Partition Noise

Tube types, nsuch as tetrodes and pentodes, that provide more than one path for the cathode current to leave the tube are subject to *partition noise*. In pentodes, partition noise results from the randomly fluctuating division of current between the screen grid and the plate. A bipolar transistor is also subject to partition noise because of the division of emitter current between the collector and the base. On the other hand, triodes are not subject to partition noise, since all electrons that enter the tube must leave by way of the plate. Partition noise causes pentodes to be about six times noisier than triodes.

Flicker Noise

A type of noise predominantly associated with transistors occurs in the low audio frequencies and is called *flicker noise* or $1/f$ *noise*. The cause of this noise is not clearly understood. The noise level varies inversely with frequency; more noise is produced at lower frequencies. Generally, flicker noise becomes negligible at a frequency as high as 500 to 1000 Hz.

Semiconductor Resistance Noise

Because the semiconductor materials comprising the emitter, base, and collector of a transistor possess some amount of resistance, the materials produce a thermal noise that is characteristic of resistors. The resistance of the base is most significant in this regard. Also, random combination of thermally excited electrons and holes produces another noise component in semiconductor materials.

Noise Spectrum of a Transistor

A generalized plot of transistor noise production as a function of frequency is shown in Figure 6-20. At low frequencies, flicker noise causes an increase in the noise level, but since flicker noise is inversely proportional to frequency, the noise

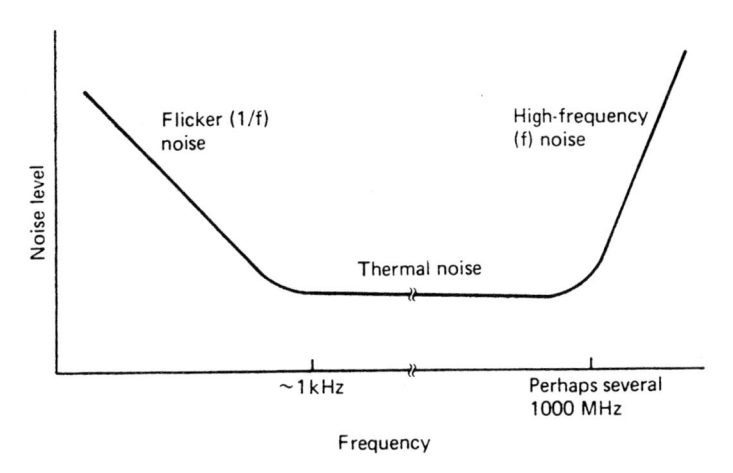

FIGURE 6-20 Generalized plot of noise level of a transistor as a function of frequency.

level drops to the "thermal noise level" that is characteristic of the midrange of a transistor's useful frequency spectrum. At high frequencies, however, the noise level once again increases, and the noise level increases as the freqeuncy increases. Consequently, the high-frequency noise is sometimes called *f noise*, since it increases with frequency.

Noise Dependence upon Bandwidth

The noise that an amplifier produces also depends upon the bandwidth of the amplifier. An amplifier designed to amplify only a narrow band of frequencies will produce less noise than a wide-bandwidth amplifier, all other things being equal.

Signal/Noise Ratio

A useful parameter for describing the relative strengths of the signal voltage and the noise voltage at a point in a system is the signal/noise ratio, S/N. In brief, the signal/noise ratio is the ratio of the signal power to the noise power at the same point:

$$S/N = \frac{\text{signal power}}{\text{noise power}}$$

If the signal voltage and the noise voltage are developed across the same resistance (or impedance), as is almost always the case, it is a simple matter to show that the signal/noise ratio is the square of the ratio of the signal voltage to the noise voltage:

$$S/N = \left(\frac{\text{signal voltage}}{\text{noise voltage}}\right)^2$$

Obviously, it is desirable that the S/N ratio should be as large as possible.

Noise Factor

Because every amplifier of a receiver contributes additional noise, it is desirable to define a parameter that gives the effect of the amplifier upon the signal/noise ratio, and it is evident that with each additional stage of amplification, the S/N ratio becomes smaller. The noise factor is defined as

$$\text{NF} = \frac{\text{input } S/N \text{ ratio}}{\text{output } S/N \text{ ratio}} = \frac{S_i/N_i}{S_o/N_o}$$

For an ideal amplifier that contributes no noise of its own, the noise factor would be 1. For practical amplifiers, the noise factor is greater than 1, which implies that the output S/N ratio is smaller than the input S/N ratio, representing a degradation of signal quality. At this point it might appear that the stages of amplification should be kept to a minimum, but this is not the case, as is explained in the following section.

The noise factor is frequently expressed in decibels (dB), and as such it is called the *noise figure*. Therefore, the noise figure F is given by

$$F = 10 \log \text{NF}$$

where the logarithm is to the base 10. For an ideal amplifier whose noise factor is 1.0, the noise factor F is 0 dB. With careful design, noise figures for RF amplifiers on the order of 2 dB can be obtained.

Noise Factor of Cascaded Stages

An important consideration is the total or equivalent noise factor resulting from connecting two or more stages in cascade. In other words, how does the equivalent noise factor depend upon the noise factor of the individual stages?

Analysis reveals the following relationship for the equivalent noise factor NF_e in terms of the noise factors of the individual stages, NF_1 and NF_2:

$$\text{NF}_e = \text{NF}_1 + \frac{\text{NF}_2 - 1}{G_1}$$

Note that the power gain G_1 of the first stage appears in this equation, and it serves to reduce the effect of the noise figure of the second stage. Typically, the equivalent noise factor for the cascaded stages is only slightly larger than the noise factor of the first stage alone. This implies that the noise factor of the first stage should be as low as possible. The noise factor of the second stage is less critical provided that the power gain G_1 of the first stage is reasonably large.

When three stages are cascaded, the resulting equivalent noise factor is

$$NF_e = NF_1 + \frac{NF_2 - 1}{G_1} + \frac{NF_3 - 1}{(G_1)(G_2)}$$

Again, note that if the power gain of the first two stages is large, the contribution to the equvialent noise factor by the third stage is negligible.

Effects of Noise

The effect of internally generated noise is to limit the maximum obtainable sensitivity of a receiver. The noise tends to cover up the weaker signals. Consequently, sensitive receivers must be carefully designed to reduce the internally generated noise or to minimize the effects of the noise. Not all noise is internally generated,

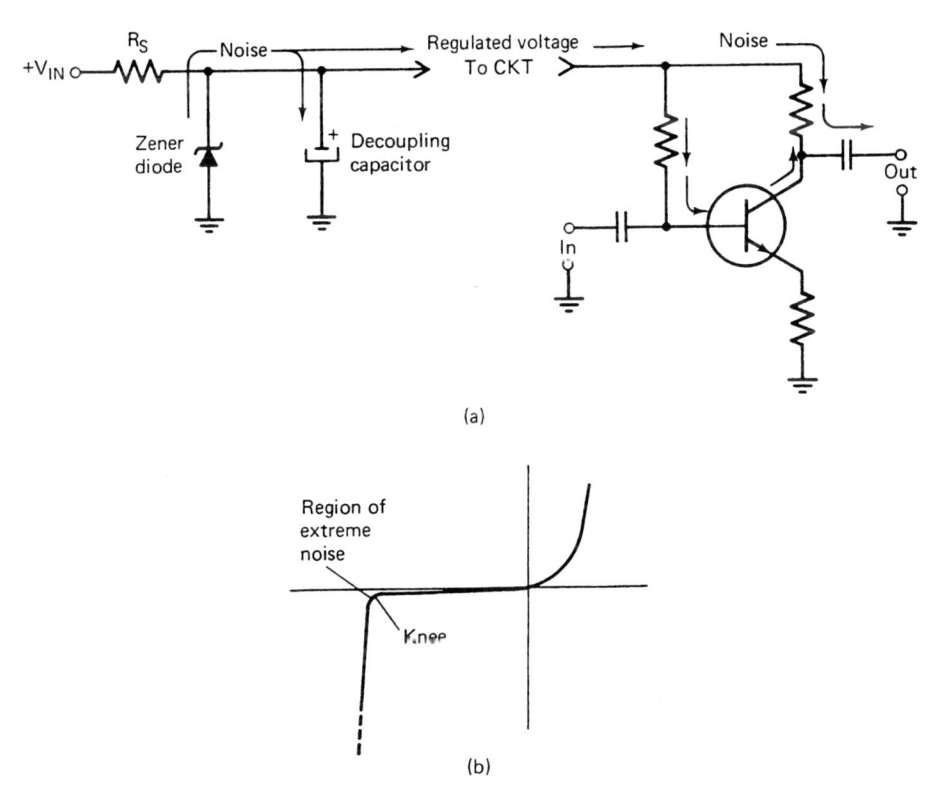

FIGURE 6-21 (a) Noise produced by a zener diode in a typical regulator circuit can cause noise to be injected into the signal path. (b) Zener diodes produce maximum noise in the vicinity of the knee of the reverse breakdown characteristic.

Sec. 6-11   Noise

however. Much of the audible noise at the speaker comes in on the antenna, being produced by atmospheric disturbances (thunderstorms), electrical machinery, and cosmic sources outside the solar system. Obviously, little can be done at the receiver to minimize the noise introduced by the antenna, because the receiver cannot distinguish between the noise and the desired signal.

The noisiest stage in most receivers is the mixer of frequency converter. The effects of mixer noise can be minimized, however, by including a low-noise RF amplifier stage ahead of the mixer. If the power gain of the RF amplifier is fairly large, the noise contribution by the mixer can be greatly reduced, in accordance with the formulas given above for the equivalent noise factor of cascaded stages. This is a major benefit of the RF amplifier—to minimize the effects of noise generated by the mixer. Likewise, since the IF amplifiers are preceded by the RF amplifier and mixer, the noise introduced by the IF stages is minimal in most receivers.

Zener Regulator Noise

As the last topic of this section we mention a particular source of noise that sometimes (as a defect) leads to extremely large noise levels in a receiver. The source is the common zener-regulator circuit shown in Figure 6-21. A zener diode is capable of generating extremely large amounts of noise that can be introduced into the signal path as indicated. The decoupling capacitor reduces the noise level by providing a low-impedance path to ground, but the zener can sometimes become such a prolific generator of noise that the capacitor fails to shunt it all to ground.

Noise generation by a zener diode is most pronounced around the knee of the reverse breakdown region. The noise diminishes on either side of the knee. A

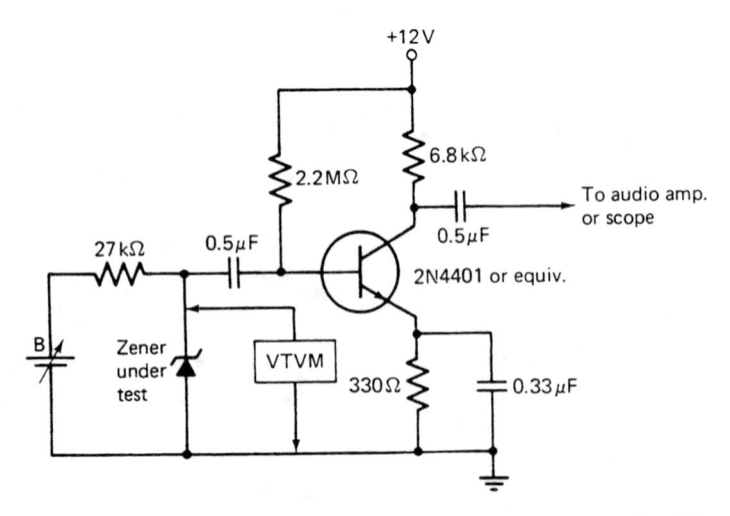

FIGURE 6-22 Demonstration circuit for investigating zener noise. The transistor stage serves as a preamplifier for driving an audio amplifier. Component values are not critical.

demonstration circuit is shown in Figure 6-22 that provides audible evidence of zener noise. Power supply B reverse-biases the zener, and the VTVM provides an indication of when the knee region is reached. The BBEF amplifier stage serves as a preamp, and the output may be applied to an audio amplifier or scope as desired. We find that considerable variation exists in the noise-generation properties of various zener diodes, and the character of the noise will change somewhat as the operating point of the zener is shifted about in the region of the knee.

SUMMARY

1. The gain and bandwidth of almost all RF amplifiers found in receivers are determined to a large degree by the properties of resonant circuits associated with the amplifiers.

2. The impedance of a parallel resonant circuit reaches a maximum at the resonant frequency. High-Q resonant circuits reach higher impedances than circuits having a lower Q.

3. The Q of a resonant circuit depends, in part, upon the amount of resistance associated with the circuit, whether the resistance is a discrete resistor or merely the resistance of the windings of the inductor.

4. At a given frequency, a large Q gives a small bandwidth, and vice versa. For a given Q, a larger bandwidth will be obtained at a higher operating frequency.

5. A resistor may be placed in parallel with an LC resonant circuit to lower the Q of the circuit and increase its bandwidth.

6. The interelectrode capacitance of a tube or transistor can cause an RF amplifier to oscillate at high operating frequencies if precautions are not taken to prevent the oscillation. Oscillation can be prevented by neutralizing or by operating the active device as a grounded grid or common-base amplifier.

7. The screen grid of a pentode effectively shields the control grid from the plate so that the tendency to oscillate at high frequencies is greatly reduced.

8. The gain of an RF amplifier is determined in part by the transconductance of the active device. The gain of the amplifier may be controlled by varying the operating point of the active device so that different values of transconductance are obtained.

9. The automatic-volume-control circuitry of a receiver serves to hold a constant audio level at the speaker against variations of the signal strength at the antenna. The automatic volume control is not related in any way to the manual control in the audio section.

10. AVC action in BJT circuits may be achieved either by forward or reverse AVC action. In reverse AVC, the most common, the forward bias applied to the base is reduced to produce a decrease in transconductance. In forward AVC, the forward bias applied to the base is increased, which, for the special transistors used, produces a decrease in transconductance.

11. Ultimately, the sensitivity of a receiver is determined or is limited by the amount of noise generated by the RF amplifiers of the receiver. The amplifier closest to the antenna is the most critical.

12. All frequencies are represented equally in white noise, but the high frequencies are attenuated in pink noise.

13. In cascaded amplifiers, the equivalent noise figure for the combination is determined primarily by the noise figure of the first stage.

14. JFETs and MOSFETs are subject to the same difficulties stemming from interelectrode capacitances as are vacuum-tube triodes and BJTs. Hence, neutralization is commonly incorporated in RF amplifiers utilizing these devices.

15. The second gate-channel configuration of a dual-gate MOSFET may be used in a gain-controlling function. In such circuits the second gate-channel tends to isolate the first gate from the drain so that feedback from the drain to the gate is minimized. Therefore, neutralization is frequently not required for RF amplifiers using dual-gate MOSFETs.

16. The Miller effect is the apparent amplification of the interelectrode capacitance associated with the active device of an inverting amplifier. The effect makes active devices more prone to oscillate at high frequencies.

QUESTIONS AND PROBLEMS

6-1. What is the difference between selectivity and sensitivity?

6-2. Describe what happens to the frequency-response curve of a double-tuned transformer as the coupling between the circuits is increased.

6-3. What difficulty results from using a triode as an RF amplifier at high frequencies? Suggest two methods to resolve the difficulty.

6-4. Describe the Miller effect.

6-5. Why does a grounded grid amplifier not require neutralization?

6-6. Why is neutralization of pentode RF amplifiers seldom, if ever, required?

6-7. Name at least two factors that affect the gain of a tuned RF amplifier that uses a pentode whose plate load consists of a parallel resonant circuit.

6-8. What is the difference between pink noise and white noise?

6-9. Give three mechanisms whereby electrical noise is generated.

6-10. Why does a triode not produce partition noise?

6-11. What is the effect of the noise generated by an RF amplifier upon the sensitivity of a receiver of which the amplifier is a part?

6-12. The gain-bandwidth product of a certain transistor is given as 1 MHz. Is it likely that this transistor would function satisfactorily as an IF amplifier at a frequency of 455 kHz? Why?

6-13. Define the alpha cutoff and the beta cutoff frequency of a transistor.

6-14. Why does a common-base RF amplifier seldom require neutralization?

6-15. What is bridge neutralization?

6-16. Why is it desirable that a receiver have an automatic-volume-control function?

6-17. Describe the AVC circuit of a small AM tube-type receiver. Is the AVC control voltage determined by the amplitude of the audio signal? Why or why not?

6-18. Does the AVC control voltage change appreciably during the transmission of a moment of silent prayer?

6-19. What is the difference between forward and reverse AVC in bipolar transistor circuits?

6-20. A *PNP* amplifier has reverse AVC. As the average level of the input signal increases, does the base become more or less positive relative to the emitter?

6-21. For what purpose is the second gate of a dual-gate MOSFET usually used in MOSFET RF amplifiers for receivers?

6-22. Calculate the resonant frequency for each of the following combinations of inductance and capacitance:

	C	L
(a)	0.01 μF	1 mH
(b)	0.001 μF	100 μH
(c)	100 pF	10 μH
(d)	1.6 pF	1.6 μH

6-23. For Prob. 6-22, calculate the inductive and capacitive reactance at the resonant frequency.

6-24. The resonant frequency of a resonant circuit is 910 kHz. Calculate the bandwidth of the circuit for the following values of Q:

(a) 20

(b) 50

(c) 100

6-25. An inductor whose reactance is 1000 Ω at the resonant frequency has an effective winding resistance of 22 Ω. What is the Q of the inductor?

6-26. If the inductor of Prob. 6-25 is resonated at 455 kHz, what will be the bandwidth of the resonant circuit?

6-27. Suppose that the inductor in Figure 6-6 is increased to 20 mH while keeping all other component values (including the resistance of the inductor) the same. Perform an analysis of the new circuit and compare the resonant frequency, circuit Q, and bandwidth of the two circuits.

6-28. Suppose that the 33-kΩ shunting resistor is removed from the circuit of Figure 6-6. What will be the effect upon the Q, resonant frequency, and bandwidth?

6-29. A circuit tuned to 455 kHz has a bandwidth of 10 kHz. What is the Q of the circuit?

6-30. If the Q of a circuit remains the same, what happens to the bandwidth as the frequency is increased? As the frequency is decreased?

6-31. If the bandwidth of a tuned circuit is 10 kHz at 550 kHz, what will the bandwidth be if the frequency is changed to 1600 kHz, if the Q remains the same?

6-32. What Q would be required for a circuit operating at 100 MHz to have a bandwidth of 200 kHz?

6-33. Why is it important that the RF amplifier of a receiver have a low noise factor and a large power gain?

6-34. Explain why the noise received at the antenna and the internally generated noise limits the ultimate sensitivity of a receiver.

6-35. Under what circumstances might additional RF gain not increase the sensitivity of a receiver?

6-36. Assume that an amplifier has a noise factor of 4. If the S/N ratio of the input signal is 8, what will be the S/N ratio of the output signal?

6-37. For an amplifier, if the input S/N ratio is 10, and if the output S/N is 6:

(a) What is the noise factor?

(b) What is the noise figure, in dB?

6-38. Suppose that the S/N ratio of a signal at the antenna terminals of a receiver is 10. If it then passes through a mixer whose noise factor is 5, what will be the S/N ratio at the output of the mixer?

6-39. If, in Prob. 6-38, the mixer is preceded by an RF amplifier whose power gain is 12 and whose noise factor is 2:

(a) What will be the noise factor of the combined RF amplifier and mixer?

(b) What will be the S/N ratio at the output of the mixer?

7
FREQUENCY CHANGERS

In this chapter we describe the heart of a superheterodyne receiver—the frequency changer. Frequently, the frequency changer is called a *mixer* or *converter*, but we reserve these words for particular types of circuits, as we shall see.

The big principle in this chapter is *heterodyning*, the process of combining two RF signals to obtain a new signal of a different frequency. This principle is widely used in all areas of electronic communications.

A brief description of basic oscillator circuits is given—an oscillator being a device that generates an RF signal. An oscillator is an essential part of every frequency changer.

You may be consoled by the fact that this is the last chapter on preliminaries before we tie everything together in the following chapter to come up with an AM superheterodyne receiver.

7-1 INTRODUCTION

The fundamental idea of a superheterodyne receiver is that the incoming station frequency is changed to a lower frequency for amplification prior to demodulation. The lower frequency is the intermediate frequency, and the frequency of each station, whether it is at the bottom or the top of the broadcast band, is converted to the same intermediate frequency. The IF frequency, as it is called, is standardized at 455 kHz for AM and 10.7 MHz for FM receivers.

The frequency changer may be the first stage of a receiver, and in such cases the antenna may be an integral part of the stage. Better AM receivers and almost all FM receivers incorporate one stage of RF amplification that precedes the frequency changer. A portion of the block diagram of a superheterodyne receiver, whether AM or FM, is shown in Figure 7-1.

7-2 PRINCIPLE OF FREQUENCY CHANGERS—HETERODYNING

When two signals of different frequencies are applied to a device having a nonlinear transfer characteristic, four frequencies appear at the output. These consist of the two original frequencies plus the sum and the difference of the original frequencies.

Antenna → RF amplifier → Frequency changer → IF amplifier

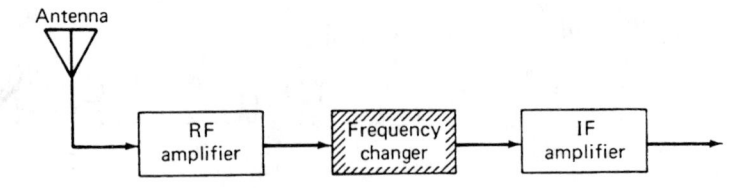

FIGURE 7-1 Portion of the block diagram of a superheterodyne receiver.

The nonlinear device is called a *mixer*, and a diode, tube, or transistor may be used as a mixer if operated in the nonlinear portion of the characteristic curves. This process, called *heterodyning*, is illustrated in Figure 7-2.

In a receiver the signal from the antenna forms one input to the mixer. The second signal is provided by an oscillator, located in the receiver, whose function is solely to provide the mixer with a signal of the proper amplitude and frequency to achieve the desired change in frequency of the antenna signal. This oscillator is called the *local oscillator*. The four frequencies that appear at the output of the mixer are (1) the antenna frequency, (2) the oscillator frequency, (3) the sum of the antenna and oscillator frequencies, and (4) the difference of the antenna and oscillator frequencies. We shall see that the difference frequency is the useful frequency at the output of the mixer.

A tuning circuit at the output of the mixer, which is also the input to the IF amplifier, selects the desired difference frequency and rejects the others. In AM receivers the oscillator frequency is arranged to always be higher than the desired station frequency by a fixed amount, typically 455 kHz. To maintain this difference in the antenna and the oscillator frequency, the oscillator and the antenna are tuned together by a ganged tuning capacitor operated by the receiver tuning control. As the antenna frequency is changed, the oscillator frequency is changed an equal amount. Therefore, the difference frequency at the mixer output will always be the same. Hence, a tuning circuit tuned to select this frequency will send this frequency and none of the others to the IF amplifier. In this manner the antenna frequency is changed to the IF frequency.

The perceptive reader may ask about the modulation (music or voice) carried by the antenna signal—a good question, since changing the frequency would be of no avail if the modulation were lost in the process. Fortunately for all super-

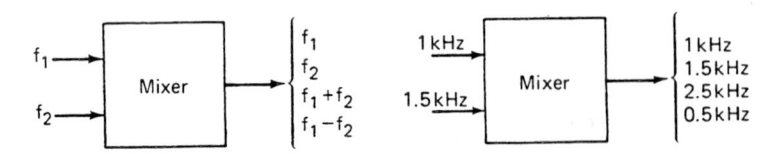

f_1 → Mixer ← f_2 → { f_1, f_2, f_1+f_2, f_1-f_2 } 1 kHz, 1.5 kHz → Mixer → { 1 kHz, 1.5 kHz, 2.5 kHz, 0.5 kHz }

FIGURE 7-2 When two different frequencies are applied to a mixer, four frequencies appear at the output. This process is called heterodyning.

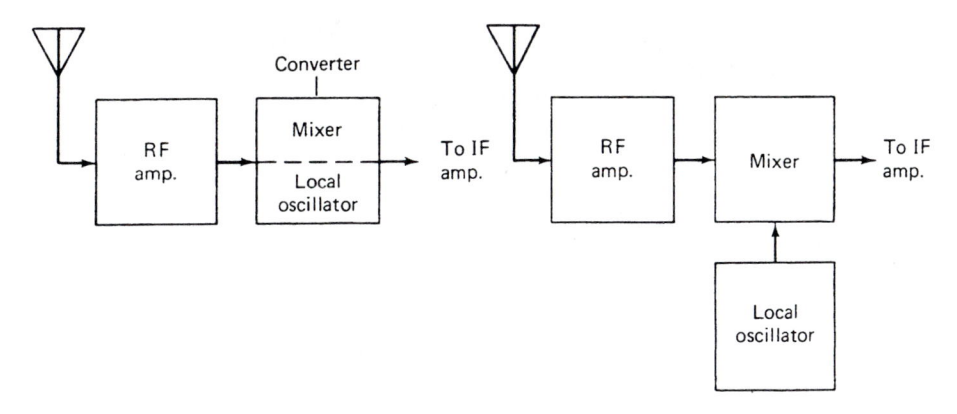

FIGURE 7-3 Combined mixer and oscillator stage is called a converter.

heterodyne receivers, the modulation is transferred to the IF frequency, whether the modulation is AM or FM. This is not a trivial question, but further discussion would take us into theoretical considerations that are best left to the engineers. We sum it up by saying, simply, that the modulation of the antenna signal is transferred to the IF signal.

Regarding terminology, the *mixer* is the element or stage in a receiver where the two signals come together in a nonlinear device to form the sum and difference frequencies. Many receivers combine the function of the mixer and local oscillator into one stage in which it is hard to separate the mixer from the oscillator. The two functions are performed simultaneously by one tube or transistor. Frequency changers of this type are called *converters* and are quite common. Receivers that do not use converters employ a completely separate oscillator stage that delivers a signal to a mixer for combining with the antenna signal. These two schemes are illustrated in Figure 7-3.

7-3 BASIC OSCILLATOR CIRCUITS

Resonant Circuits as Signal Sources

If an initial "charge" of energy is given to a parallel resonant circuit, the circuit naturally tends to oscillate, with the energy being passed back and forth between the inductor and capacitor in cyclic or periodic fashion. This continues until all the energy has been expended due to losses in the circuit. During the period of cyclic energy exchange, an alternating voltage at the resonant frequency appears across the resonant circuit. The amplitude of this signal depends upon the energy contained in the resonant circuit; the more energy, the greater the amplitude of the signal will be. Since the energy level gradually decreases due to the loss of energy from the circuit, the amplitude of the signal also gradually decreases. The result is a *damped sine wave*, illustrated in Figure 7-4.

If energy is fed to the resonant circuit at the same rate that it is lost, the energy level will remain constant, and so will the voltage developed across the circuit. In such a case the voltage appears across the resonant circuit as if the resonant circuit were actively generating the voltage. It is convenient in the analysis of oscillator circuits to regard a parallel resonant circuit as a generator of the resonant frequency signal. This is valid as long as we remember that energy must be continually supplied to the resonant circuit to sustain the oscillation.

Simple Oscillator

The Armstrong oscillator of Figure 7-5 illustrates several features of oscillators found in radio receivers. $L1$ and $C1$ form a resonant circuit that determines the frequency of oscillation. $C2$ couples the signal from the resonant circuit to the grid of the tube, and R_g is the grid resistor, which is large enough to provide grid leak biasing for the tube. Note that the plate current flows through $L2$, which is coupled to $L1$. $L2$ is the "tickler coil" that couples a portion of the output signal back to the resonant circuit to sustain oscillation. $C3$ provides a path for the RF signals to flow to ground from the power supply point.

When B+ power is first applied, a surge of electrons flows to the plate, producing a significant plate current that flows through $L2$. The current creates a magnetic field around $L2$ that couples into $L1$, producing a voltage across $L1$ that is coupled to the grid by $C2$. The voltage is of such phase that the plate current is increased, and the plate current soon reaches a maximum value because of this "positive feedback" from $L2$ to $L1$.

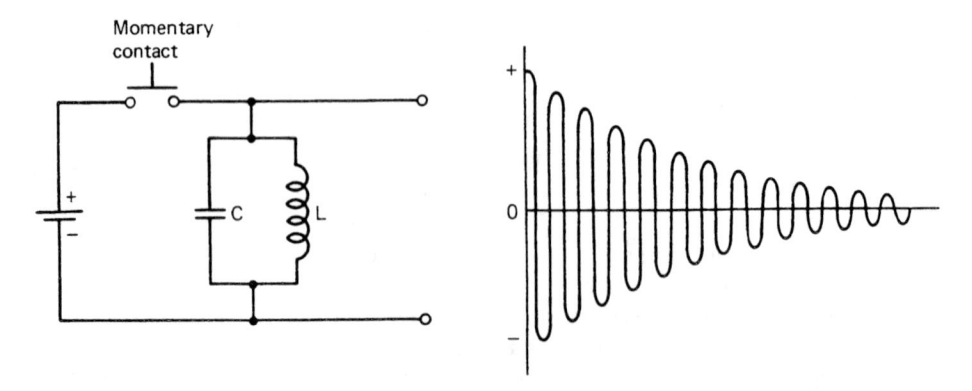

FIGURE 7-4 Damped sine wave is produced across the resonant circuit after a momentary closure of the switch.

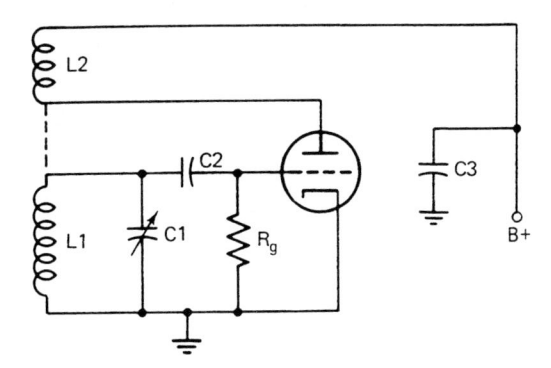

FIGURE 7-5 Armstrong oscillator. Positive feedback occurs from $L2$ to $L1$.

When the plate current is at a maximum, the magnetic field that couples into $L1$ from $L2$ is momentarily stationary, and the stationary magnetic field does not induce a voltage in $L1$. Therefore, the voltage across $L1$ begins to drop. The voltage coupled to the grid now tends to cut the plate current off, and as plate current decreases, so does the magnetic field around $L2$ and $L1$. The voltage induced in $L1$ by $L2$ now drives the resonant circuit voltage to the opposite polarity. This process continues, with $L2$ forcing $L1$ in the direction that it naturally wants to go, sustaining the oscillation.

We now point out three features of this oscillator. First, the resonant circuit determines the frequency of oscillation, and the frequency may be varied by changing the value of either $L1$ or $C1$. Second, the tube controls the application of energy to the resonant circuit. Finally, $L2$ provides positive feedback to sustain the oscillation in the resonant circuit.

Series-Fed Hartley Oscillator

Figure 7-6 illustrates an oscillator circuit that is frequently found in various forms in radio receivers. The tapped inductor $L1$ characterizes the circuit as a Hartley oscillator, and the fact that a portion of $L1$ is in series with the plate circuit makes it a *series-fed* Hartley oscillator.

Positive feedback from the plate circuit (the output) is achieved by the autotransformer action of $L1$. Plate current (actually cathode current) flowing through the bottom portion of $L1$ sets up a magnetic field that induces a large voltage in the top portion of $L1$ that is of the proper phase to produce positive feedback. The bottom portion of $L1$ may be considered as the primary of an autotransformer, while the entire $L1$ winding acts as the secondary. Components $C2$, $C3$, and R_g serve the same purpose as in the Armstrong oscillator described above.

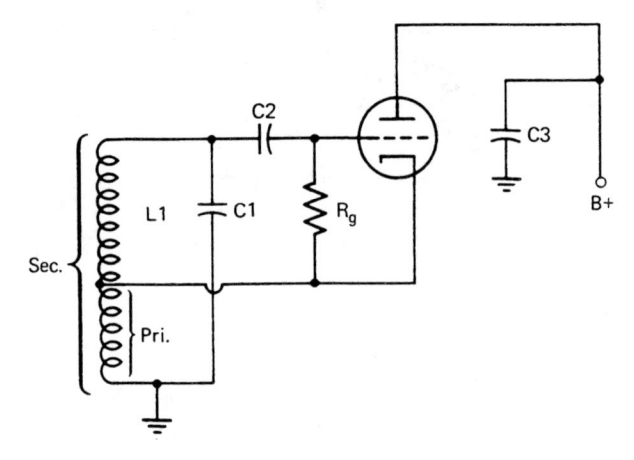

FIGURE 7-6 Series-fed Hartley oscillator.

Transistor Version. A transistor version of a series-fed Hartley oscillator is shown in Figure 7-7. A universal bias arrangement is used that consists of R_{HB}, R_{LB}, and R_E. Capacitor C3 bypasses the emitter resistor, while C4 holds the power supply point at AC ground.

This circuit differs from the vacuum-tube counterpart described earlier, in that the feedback circuit occurs in the collector lead rather than in the emitter lead. Recall that in the previous circuit the primary of the autotransformer was in the cathode lead. The effect is the same, however.

Frequency stability of the oscillator is improved by connecting the collector to a tap on L1 rather than at the end. Hence, the primary of the autotransformer is

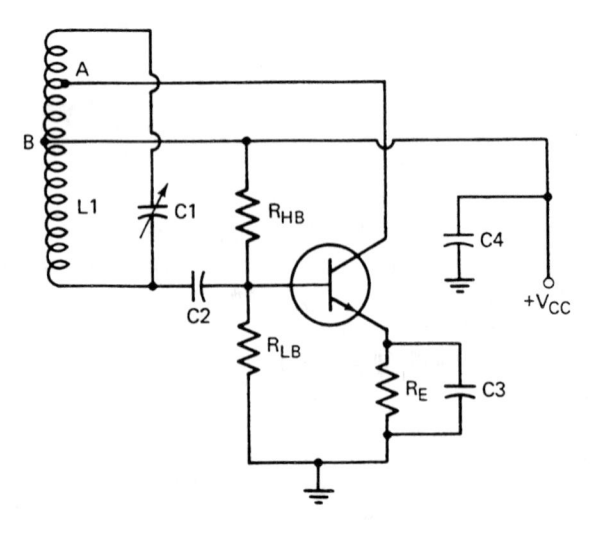

FIGURE 7-7 Transistor version of a series-fed Hartley oscillator.

Frequency Changers

only that portion of $L1$ between point A and point B. The secondary is the entire coil, as before. Note that point B is at signal ground, and observe that the collector connection is made to the high side of point B. This must be the case in order for positive feedback to occur.

Shunt-Fed Colpitts Oscillator

The tapped coil of the Hartley oscillator is undesirable at high frequencies, where the entire coil may consist of only a few turns. Location of the tap in such cases can be critical. A circuit that avoids the use of a tapped coil is the Colpitts oscillator shown in Figure 7-8. A Colpitts oscillator is characterized by a split capacitance that forms a part of the resonant circuit.

Capacitors $C1$ and $C2$, together with $L1$, form the resonant circuit. Since the capacitors are in series, the series equivalent is the capacitance that determines the resonant frequency of the circuit. Therefore, $C1$ and $C2$ have to be larger than the single capacitor used to tune a Hartley circuit.

The split capacitance provides for both ends of the resonant circuit to be ungrounded. The point between the capacitors connects to ground so that opposite ends of the resonant circuit are out of phase. Positive feedback to sustain oscillations is produced as follows.

When the base is driven more positive by the action of the resonant circuit, a larger collector current flows and causes the collector voltage to drop to a less-positive value. This negative-going surge at the collector is coupled through $C4$ back to the bottom end of the resonant circuit, and this reinforces the negative

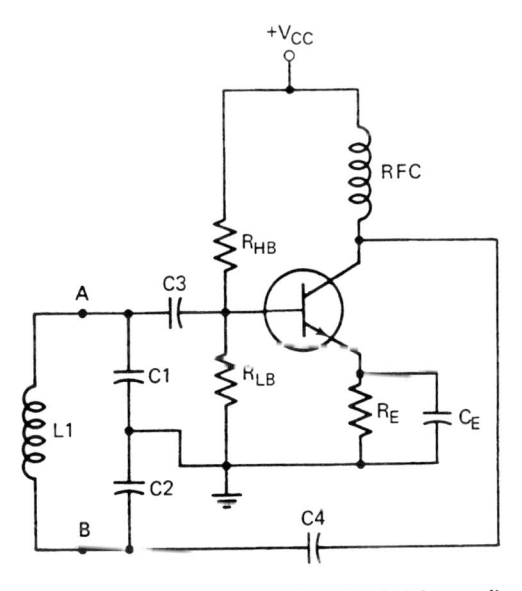

FIGURE 7-8 Transistor version of a Colpitts oscillator.

Sec. 7-3 Basic Oscillator Circuits 193

voltage that is the mirror image of the positive voltage at point A. Reinforcement of the negative voltage at B serves the same function as reinforcing the positive voltage at A. Hence, positive feedback is achieved.

7-4 SEPARATE MIXER–OSCILLATOR FREQUENCY CHANGER

The circuit of Figure 7-9 is the front end of an AM receiver that utilizes a separate oscillator and mixer as a frequency changer. Incidentally, the expression "front end" refers to the circuitry that lies on the antenna side of the input to the IF amplifier. Hence, the RF amplifier (if present), the mixer, and the oscillator (or converter) constitute the front end.

The oscillator, $Q2$, is a series-fed Hartley that utilizes a universal biasing arrangement. Inductor $L2$ and capacitors $C2$ and $C2T$ form the resonant circuit for the oscillator. $C2$ is part of a dual gang capacitor that consists of $C1$, $C1T$, $C2$, and $C2T$. $C1T$ and $C2T$ are trimmer capacitors built into the $C1$-$C2$ structure to allow adjustment of the capacitance of $C1$ and $C2$.

The collector lead of $Q2$ connects through a stabilizing resistor R_C to a tap on $L2$. Connection to the tap provides improved frequency stability. Positive feedback

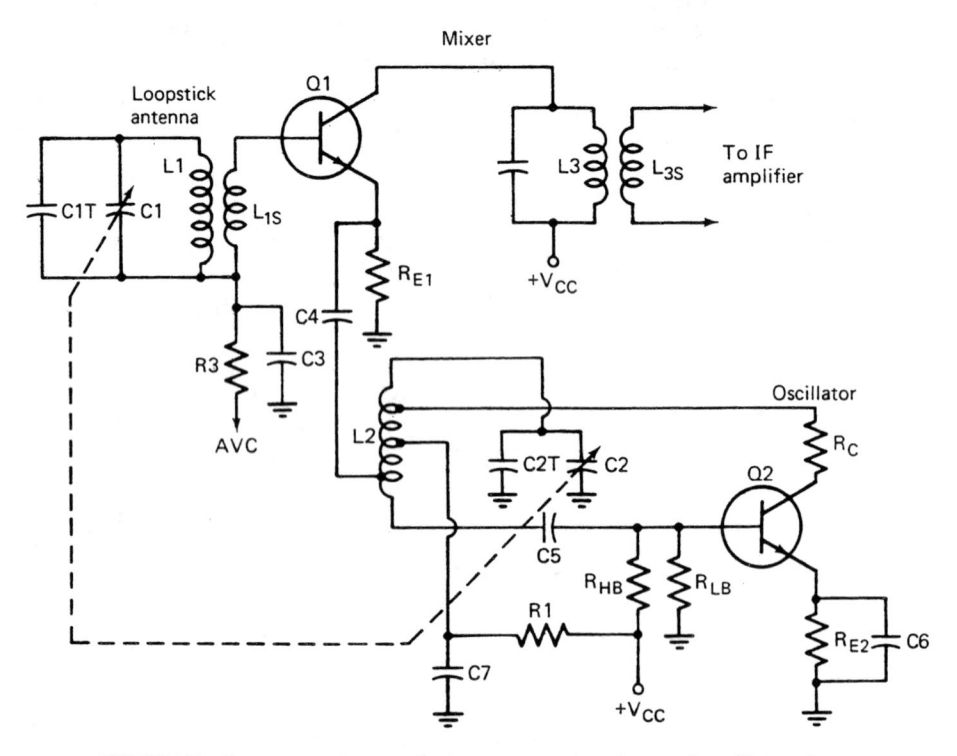

FIGURE 7-9 Frequency changer that uses separate mixer and oscillator stages.

occurs since the collector current flows through a portion of $L2$, and capacitor $C5$ couples the resonant circuit voltage to the base of oscillator transistor $Q2$. Resistor $R1$ and capacitor $C7$ form a filter for keeping the oscillator signal out of the power supply. The output of the oscillator is taken off yet another tap on $L2$ and is applied through capacitor $C4$ to the emitter of mixer transistor $Q1$.

A loopstick antenna, designated as $L1$ and $L1S$, serves as the tuning inductor for the antenna tuning circuit, $C1-L1$. Loopstick antennas are used extensively in AM receivers, and consist of a coil of wire wound on a ferrite core. The coil $L1S$ serves as a secondary of a step-down transformer, consisting of $L1$ and $L1S$, which matches the impedance of the resonant circuit to the fairly low input impedance of the transistor. Note that the antenna signal is applied to the base of the mixer transistor.

Mixer transistor $Q1$ receives the oscillator signal at the emitter and the antenna signal at the base. The base-emitter junction of the transistor serves as the nonlinear element required to accomplish mixing action. Four frequencies appear at the collector of $Q1$: the antenna and oscillator frequencies, and the sum and difference of these frequencies. The $L3$ resonant circuit in the collector lead is tuned to the difference frequency, usually 455 kHz for AM receivers, and therefore only the difference frequency is passed on to the IF amplifier.

As the receiver is tuned, $C1$ and $C2$ vary together, since they are connected to the same control shaft. When the antenna frequency changes, so does the oscillator frequency. The oscillator frequency is set a fixed amount (455 kHz) above the antenna frequency during alignment by adjusting trimmer capacitor $C2T$. This difference is maintained as the receiver is tuned over the broadcast band.

Mixer transistor $Q1$ amplifies the antenna signal as it changes the frequency. Note that the bias voltage for the base of $Q1$ is provided through $R3$ by the AVC line, so that variations in the AVC voltage (according to the strength of the incoming signal) vary the operating point of $Q1$ to control the gain. Capacitor $C3$ holds the $L1$ side of $R3$ at signal ground.

7-5 PENTAGRID CONVERTERS

Vacuum-tube AM receivers that use separate tubes (or separate sections of the same tube) for the oscillator and mixer stages are seldom encountered. A special tube has been developed, the *pentagrid converter*, that combines both oscillator and mixer functions in a single stage. The structure of this tube is shown in Figure 7-10; as the name suggests, it has five grids.

Grid $G1$ is a *signal grid*, or *control grid*, that functions as the control grid of the oscillator section of the tube. It is called the *oscillator grid*. $G2$ is a special structure, a combination of beam-forming plates and a grid that serves as the plate for an effective triode formed by the cathode, $G1$ and $G2$.

In operation, many of the electrons that are attracted toward the oscillator plate $G2$ pass on through $G2$ to the vicinity of $G3$, the RF signal grid. As with any control grid, $G3$ is biased negative relative to the cathode, and the electrons traveling toward $G3$ are repelled somewhat by the negative voltage. This results in the

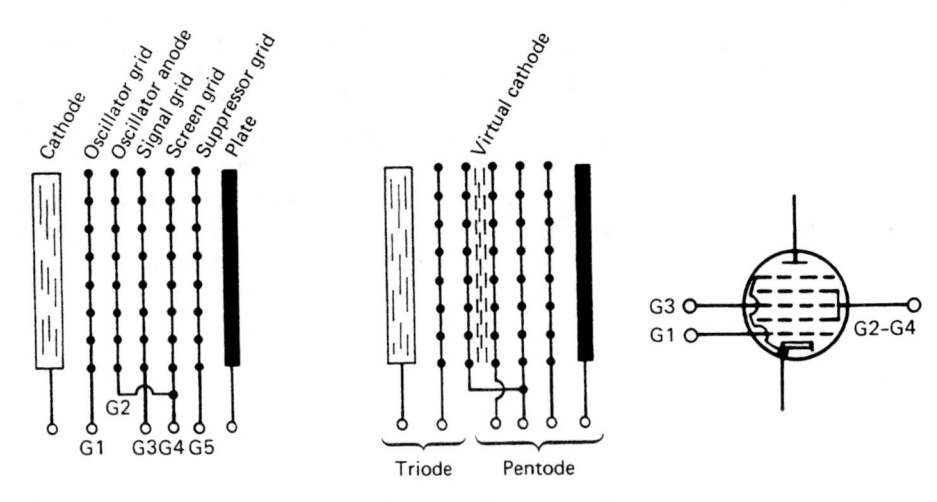

FIGURE 7-10 Structure of a pentagrid converter.

accumulation of electrons in the space between $G2$ and $G3$, and a space charge is formed that resembles, to some extent, the space charge that surrounds the cathode. The space charge between $G2$ and $G3$ is said to form a *virtual cathode*, and this virtual cathode is the source of electrons for the RF signal portion of the tube, which consists of $G3$, $G4$, $G5$, and the plate.

The signal voltage applied to $G3$ controls the flow of electrons from the virtual cathode through $G4$ and $G5$ to the plate. $G4$ acts as a screen grid and $G5$ serves as a suppressor grid. Observe that the tube resembles a triode and pentode in series.

The total current flow from the cathode to the plate is controlled by both the oscillator grid and the RF signal grid. The electron density in the virtual cathode in front of $G3$ is controlled by the oscillator grid, and the plate current depends upon both the signal grid voltage and upon the electron density in the virtual cathode. Hence, the tube effectively combines the signals appearing at the two grids.

As indicated by the tube symbol, grids $G2$ and $G4$ are connected together inside the tube. This combination of electrodes is usually held at a high DC voltage but is maintained at signal ground. Note that the suppressor grid is connected internally to the cathode.

We now examine an actual circuit that utilizes a pentagrid converter as a frequency changer in an AM receiver.

Practical Circuit

A pentagrid converter circuit that is widely used is shown in Figure 7-11. Considering that the circuit is a combination oscillator, mixer, and amplifier, it is remarkably simple. Let us first examine the oscillator section.

Oscillator tuning is provided by the parallel resonant circuit consisting of $L2$,

C2, and *C2T*. Considering the resonant circuit as a signal source, capacitor C_c couples the signal to the oscillator grid. Resistor R_G is the oscillator grid resistor and is almost always either 33 kΩ (as shown) or 22 kΩ. Positive feedback is accomplished by magnetic coupling between L_c and *L2*. The oscillator component of the cathode (and plate) current couples energy from the cathode circuit back into the oscillator resonant circuit to sustain the oscillation.

The desired station signal is selected by antenna tuning circuit *L1*, *C1*, and *C1T*. *L1* is a loopstick antenna that differs from the loopstick used for solid-state receivers in that it does not have a step-down winding. No such winding is necessary in light of the large input impedance of the tube. With the two signals applied to the respective grids, the tube functions as a combination mixer and amplifier, and the difference frequency (among others) appears at the plate of the tube. Tuned transformer *T1* then selects the difference frequency and sends it on to the IF amplifier for further amplification.

Let us now consider the biasing arrangements for the two signal grids of the tube. The oscillator grid uses grid leak biasing, and it is noteworthy that such a low ohmic value is used for the grid resistor. Typical resistances for amplifiers utilizing grid leak biasing is several megohms, but a small resistance is adequate here in light of the large amplitude of the oscillator signal. Typical voltages found at the oscillator grid are from -5 V to -10 V. This voltage will be present only if the oscillator is oscillating.

Bias voltage for the RF signal grid is obtained from the AVC line. Capacitor *C3* is the AVC filter capacitor that holds the AVC line at signal ground.

During alignment of the receiver, trimmer capacitor *C1T* is adjusted so that the antenna circuit tunes to the frequency indicated on the tuning dial. Trimmer

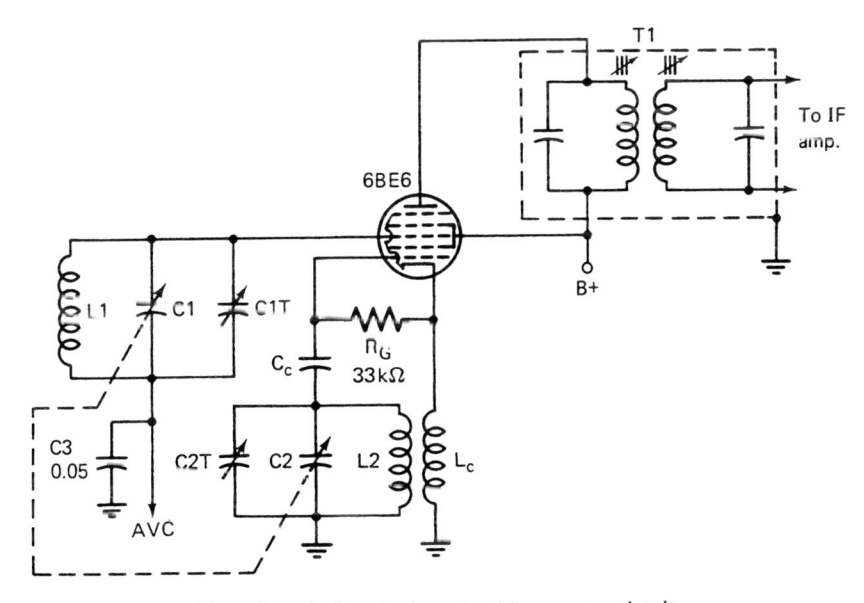

FIGURE 7-11 Practical pentagrid converter circuit.

$C2T$ is used to adjust the frequency of the oscillator so that it is exactly 455 kHz above the antenna frequency.

7-6 BIPOLAR TRANSISTOR CONVERTERS

We have seen that a pentagrid converter with its two sections combines the oscillator and mixer functions into a stage involving one tube. One version of a transistor circuit that achieves the same thing is shown in Figure 7-12. The feedback scheme resembles that of the Armstrong oscillator of a previous section.

The oscillator tank ($L2$, $C2$, $C2T$) is caused to oscillate by magnetically coupling energy from the collector circuit to the tank by way of $L2p$. The oscillator output is applied through C_c to the base circuit of $Q1$, and the oscillator voltage at the base causes collector current variations to occur at the oscillator frequency. $L2p$ functions as a tickler coil, and since the oscillator tank is tuned to the oscillator frequency, it readily responds to the oscillator component of the collector current. Hence, a strong signal at the oscillator frequency is injected into the base circuit of $Q1$.

The loopstick antenna ($L1$, $L1s$) serves as the tuning inductor for the antenna circuit. $L1s$ is a step-down winding on the loopstick that matches the tuning-circuit impedance to that of the transistor. Note that $C1$ and $C2$ are ganged capacitors, as indicated by the dashed lines.

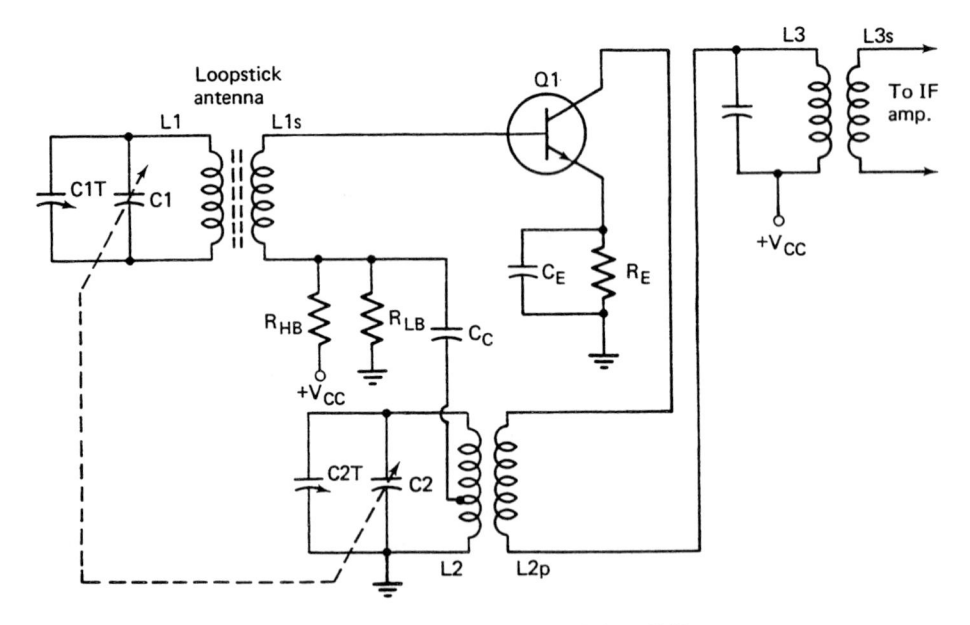

FIGURE 7-12 Converter circuit for a BJT.

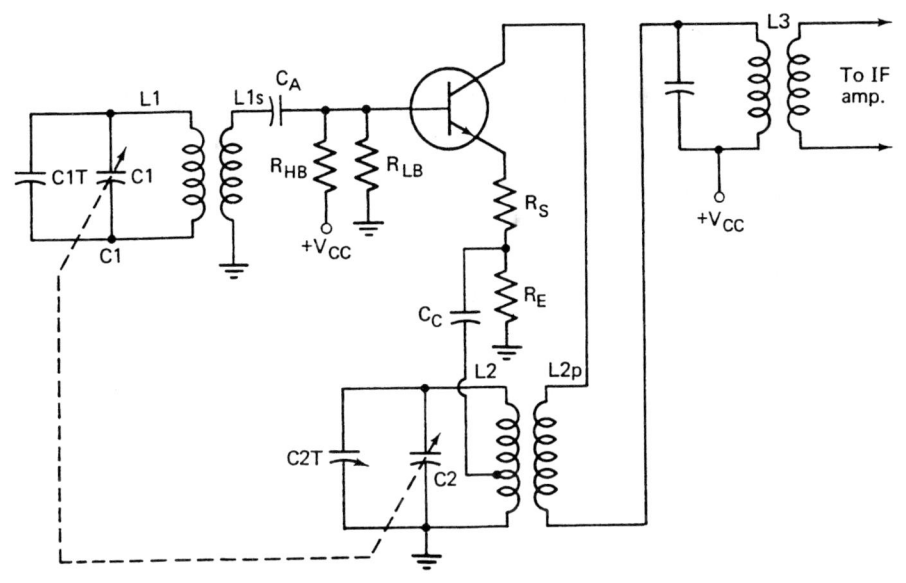

FIGURE 7-13 Variation of a transistorized converter.

The collector circuit consists of oscillator tickler $L2p$ and the $L3$ tank, connected in series. $L2p$, which is coupled to the oscillator tank, exhibits a large reactance only to the oscillator components of the collector current. Other frequency components pass through quite easily. The $L3$ circuit tunes to the 455-kHz IF frequency and passes all other frequencies. Thus, each of the two tuned circuits responds only to the frequency to which it is tuned, and each one has little effect upon the other. Therefore, the oscillation is sustained in the oscillator tank while the converted antenna frequency is passed on to the 455-kHz IF amplifier via $L3$.

A variation of the circuit described above is shown in Figure 7-13. The oscillator signal is applied to the emitter, and a different circuit is used to couple the antenna signal to the base of the transistor. The collector circuit is identical to that of the previous circuit.

Since the bottom end of $L1s$ is grounded, capacitor C_A is included to prevent the DC voltage at the base from shorting to ground. In the emitter circuit, R_s is a swamping resistor that masks variations in the AC input resistance of the emitter junction (see Sec. 4-11). No bypass capacitors are allowed, since the oscillator signal is injected at the emitter.

7-7 FREQUENCY-CHANGER CONSIDERATIONS

Now that we have seen a variety of AM frequency changers, let us consider some subtle points that will give a better appreciation of the circuits.

Oscillator Drift

When a receiver is first turned on and tuned to a station, it is highly desirable that the receiver stay tuned to the station as it gradually warms up to operating temperatures over a period of 30 minutes to an hour. It is annoying to have to retune the receiver every few minutes until it fully warms up.

The frequency stability of the local oscillator determines the stability of the receiver tuning more than any other factor. Station frequencies on AM are held to within 20 Hz of the FCC-specified frequency, and therefore the incoming signal may be considered constant in frequency. The oscillator frequency tends to vary considerably, especially during warmup, unless the circuit design includes provisions that minimize oscillator drift. Since the IF frequency is the difference between the oscillator and antenna frequencies, a change in oscillator frequency produces a corresponding change in the IF frequency. This is illustrated in the following.

If an AM station at 910 kHz is to be received, the oscillator must run at 1365 kHz to produce an IF of 455 kHz. As the receiver warms up, the oscillator tends to drift to a lower frequency. If the oscillator decreases 5 kHz, from 1365 to 1360 kHz (a change of 0.37%), the IF frequency will change from 455 to 450 kHz. Since the IF amplifier is tuned rather sharply to 455 kHz, a change in the IF frequency of 5 kHz can cause loss of volume and fidelity. A drift of 5 kHz is a very large drift, used here for purposes of illustration. In practice, the changes in oscillator frequency may amount to less than 0.1%, but the effects are still significant and must be taken into account.

Oscillator drift results from a combination of several factors. Inductances and capacitances increase in value due to thermal expansion as the operating temperature rises. The interelectrode capacitances of a tube increase as the tube warms up, and the junction capacitances of semiconductors are sensitive to temperature changes. Also, the oscillator frequency may be affected by power supply voltage variations.

Small changes in capacitance become more significant at higher frequencies. Consequently, oscillator stability is a major concern for FM receiver oscillators that typically operate from 98.8 to 118.6 MHz. An oscillator operating at 100 MHz may drift as much as 80 kHz during warmup unless care is taken to reduce this drift through careful design. A carefully designed circuit may drift 30 kHz. It is desirable, however, that the oscillator drift for FM be held to 10 kHz or less. This is generally accomplished by incorporating special temperature-compensating capacitors that decrease in value as the temperature rises. The decrease in capacitance of the compensating capacitor offsets the increase in capacitance of other components.

In addition to using temperature-compensating capacitors to improve oscillator stability, other practices include careful selection of the tube or transistor for the oscillator, shielding of the oscillator components, regulation of the oscillator power supply voltage, and locating the operating point of semiconductors so as to minimize temperature sensitivities. Another completely different approach that may be used in conjunction with the previous practices consists of using a special circuit for the specific purpose of automatically controlling the oscillator frequency. Automatic frequency control (AFC) circuits are widely used in FM receivers but are not used in AM. AFC circuits are described in Sec. 7-9.

Oscillator Pulling

If an external signal at a nearby frequency is coupled into an oscillator circuit, the frequency of the oscillator is pulled toward the external signal. The closer the two frequencies, the greater is the pulling effect. The effect depends also upon the strength of the interfering signal and upon the degree of coupling of the signal to the oscillator.

Coupling occurs in a pentagrid converter due to the interelectrode capacitance between the oscillator and the antenna grid, even though the tube is designed to minimize such coupling. The effect of oscillator pulling is to lower the IF frequency, since the difference frequency goes down as the oscillator frequency approaches the antenna frequency. The effect becomes greater at high frequencies, where the percentage difference between the oscillator and antenna frequencies becomes less for a given oscillator-antenna frequency separation. Further, at high frequencies, the reactances of interelectrode capacitances become less, which results in increased coupling.

Oscillator pulling may be minimized by providing fairly large frequency separations between the oscillator and antenna frequencies (i.e., by using a high IF frequency). This is one reason the standard IF frequency for FM is as high as 10.7 MHz. Also, oscillator pulling can be reduced by reducing coupling between the antenna and oscillator through careful circuit design and layout of components.

7-8 FREQUENCY CHANGERS FOR FM

Frequency changers used in FM receivers are more critical in design and operation than AM frequency changers because the FM broadcast band is situated at frequencies about 100 times higher than AM. Circuits that perform well at AM frequencies may not function satisfactorily at the higher frequencies. An example of such a circuit is the pentagrid converter. At high frequencies, pentagrid converters suffer a loss of conversion gain and also become susceptible to oscillator pulling. Both effects result from undesired coupling of the antenna and oscillator signals within the tube. Tubes have been designed that minimize this coupling, but even so, pentagrid converters are seldom found as frequency changers in FM receivers.

The mechanical aspects of an FM converter are quite different from lower-frequency AM circuits. Tuning capacitors are much smaller, inductors may consist of only a few turns of wire, and shielding is much more in evidence. It is not uncommon for the frequency changer section of an FM receiver to be completely enclosed in its own metal shielding compartment.

Interelectrode capacitances are more significant at high frequencies than at lower frequencies. In oscillator circuits, discrete capacitors may be omitted at points where the interelectrode capacitance of the active device provides the necessary capacitance. Feedback to sustain oscillation is frequently via an interelectrode capacitance. In such cases the feedback path will not be apparent on the schematic, since interelectrode capacitances are not indicated. The result is that common circuits, such as a Colpitts oscillator, take on different and sometimes unusual appearances.

Since FM frequency changers almost always incorporate an automatic frequency control (AFC) into the oscillator circuit, we go on to describe AFC circuitry before presenting an FM frequency changer. An FM frequency changer with AFC is described in Sec. 7-10.

7-9 AUTOMATIC FREQUENCY CONTROL (AFC)

The AFC circuit of an FM receiver controls the oscillator frequency and thereby provides stability of the receiver tuning. The effect of AFC action is twofold. First, the problem of oscillator drift is greatly reduced, since any change in oscillator frequency is opposed by the AFC circuit. Second, receiver tuning is made less critical. If the operator tunes the receiver to the vicinity of the desired station, the AFC circuitry tends to "lock in" on the station frequency.

AFC circuit operation may be understood by referring to the block diagram of Figure 7-14. The FM demodulator develops a voltage that is proportional to the "deviation" of the IF frequency from the design center value, typically 10.7 MHz. This DC control voltage is applied to a controlled reactance in the oscillator circuit and adjusts the oscillator frequency to bring the IF frequency closer to the center value.

Details of how the control voltage is developed by the FM demodulator are given in Chap. 13. At this point we address ourselves to the controlled reactance

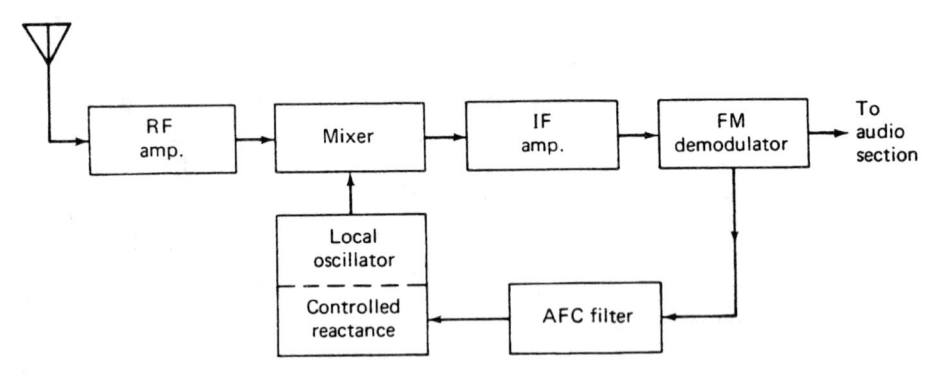

FIGURE 7-14 Partial block diagram of an FM superheterodyne receiver, showing the AFC feedback loop.

part of the oscillator circuit to see how the DC control voltage is able to tune the oscillator.

Reactance Tube

A circuit element exhibits a reactance if, when an AC voltage is applied to it, the resulting current is shifted in phase from the applied voltage. If the current leads the voltage, the reactance is capacitive, and if the current lags the voltage, the reactance is inductive. The magnitude of the reactance is determined by the magnitude of the current resulting from a given applied voltage. By controlling the current, the reactance, and hence the size of the equivalent inductor or capacitor, can be controlled or varied.

A tube may be used as a variable reactance by inserting a resistance-capacitance phase-shift network between the plate and the grid. The phase-shift network and the action of the grid produces a phase shift between the AC voltage applied to the plate and the resulting AC plate current. Thus, the tube appears as a reactance, and such a tube is called a *reactance tube*.

A DC voltage applied to the grid of a reactance tube controls the magnitude of the current resulting from the applied AC plate voltage. Therefore, the DC grid voltage can vary the size of the capacitor or inductor represented by the reactance tube. If a reactance tube is placed in parallel with the resonant circuit of an oscillator, the variable reactance exhibited by the tube will cause the oscillator frequency to vary. In short, the DC voltage applied to the reactance tube grid controls the frequency of the oscillator.

Figure 7-15 illustrates a reactance tube circuit that acts as a capacitor whose

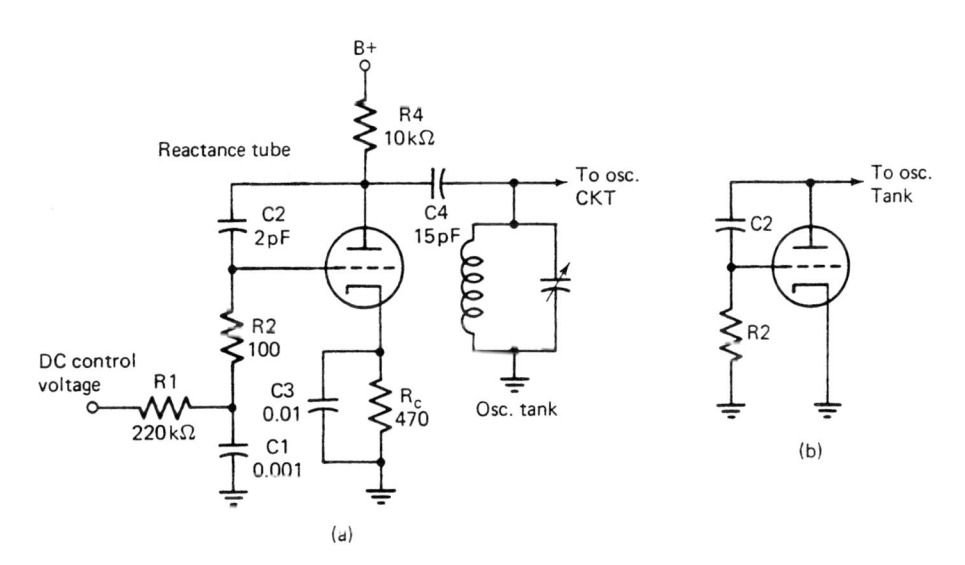

FIGURE 7-15 (a) Reactance tube circuit in which the reactance tube acts as a variable capacitor. (b) Phase-shift network between the plate and grid.

value is determined by the DC control voltage applied to $R1$. The tube is connected in parallel with the oscillator tank by $C4$ and is, therefore, in parallel with the tuning capacitor of the tank circuit. The total tuning capacitance consists of the sum of the tank and reactance tube capacitance.

Cathode biasing is used in conjunction with the input control voltage to provide grid bias for the tube. Observe that the B+ voltage is applied to the plate through the 10-kΩ resistor $R4$, which isolates the plate from the AC signal ground of the power supply. The cathode is held at signal ground by $C3$, and $C1$ places the bottom end of $R2$ at signal ground. The large value of resistor $R1$ effectively isolates the control voltage source from the grid circuit of the reactance tube. A Q-point is established by the cathode resistor and the control voltage applied to $R1$. We now consider the AC signal aspects of the tube; in particular, we are interested in what causes the tube to act as a capacitance.

The high-frequency voltage of the oscillator tank is applied to the plate of the tube through $C4$. A portion of this voltage is coupled to the grid through $C2$, but a phase shift occurs from the plate to the grid due to the series combination of $C2$ and $R2$. The reactance of $C2$ is large in comparison with $R2$, and the combination appears almost completely capacitive. Therefore, the current through the $C2$-$R2$ combination leads the voltage at the plate by 90°. The voltage at the grid is developed by the flow of current through resistor $R2$, and is, therefore, in phase with the current through $R2$. Consequently, the grid voltage leads the plate voltage by almost 90°.

The plate current is in phase with the grid voltage; as the grid voltage goes more positive, the plate current increases. Hence, the plate current leads the plate voltage because the grid voltage leads the plate voltage. Thus, the plate current leads the plate voltage by 90°. This is the characteristic of a capacitor.

Let us now see how the capacitance is varied by controlling the DC voltage on the grid. First, the AC plate current that flows as a result of the AC grid voltage depends upon the transconductance g_m of the tube. Further, g_m varies with the operating point of the tube, increasing as the DC grid voltage moves in the positive direction. When the oscillator voltage applied to the plate causes large resultant AC plate currents, the effective capacitance of the tube appears large, and vice versa. Large currents stem from large values of g_m, and vice versa. Therefore, the capacitance seems to vary with g_m, and since g_m is controlled by the average DC voltage at the grid, control of the grid voltage also controls the capacitance exhibited by the reactance tube. The capacitance increases as the DC grid voltage goes more positive.

Either triodes or pentodes may be used as reactance tubes at FM frequencies. When triodes are used, the interelectrode capacitance C_{gp} is in parallel with $C2$ (Figure 7-15), and frequently, $C2$ is omitted so that the entire capacitance is represented by C_{gp}.

Solid-state counterparts of reactance tube circuits are feasible but are seldom encountered because a simpler means is available for obtaining a voltage-dependent capacitance in the realm of solid state. The capacitance of a reverse-biased *PN* junction depends upon the applied voltage, and *PN* junction diodes have been developed specifically for this purpose. This is discussed in Sec. 7-11.

7-10 FM FREQUENCY CHANGER WITH AFC

The oscillator section of an FM frequency changer that uses a reactance tube to control the oscillator frequency is shown in Figure 7-16. The output of the oscillator couples through $C18$ to the grid circuit of the mixer tube.

The oscillator tube is $V2$, and the oscillator is a modified Colpitts that utilizes a grounded plate configuration; $C11$ holds the plate at AC signal ground. The $L5$ tank circuit tunes the oscillator, and the tank circuit voltage is coupled to the grid of the tube through $C10$. Resistor $R7$ provide grid leak biasing for the oscillator.

The parallel combination of RFC 2 and $C14$ at first appears to be a resonant circuit, but the resonant frequency is much lower than the frequency of operation of the oscillator. Therefore, the combination appears as a capacitance, since the operating frequency is much higher than the resonant frequency.

Treating the cathode circuit as a capacitor, the cathode is seen to connect to the midpoint of two series capacitances connected from grid to ground. The interelectrode capacitance C_{gp} serves as one capacitor. This split capacitance, with the cathode connected at the midpoint, is a characteristic of a Colpitts oscillator.

The reactance tube circuit is similar to the circuit described in the previous section. Isolation of the plate from the power supply is provided by RFC 1, and

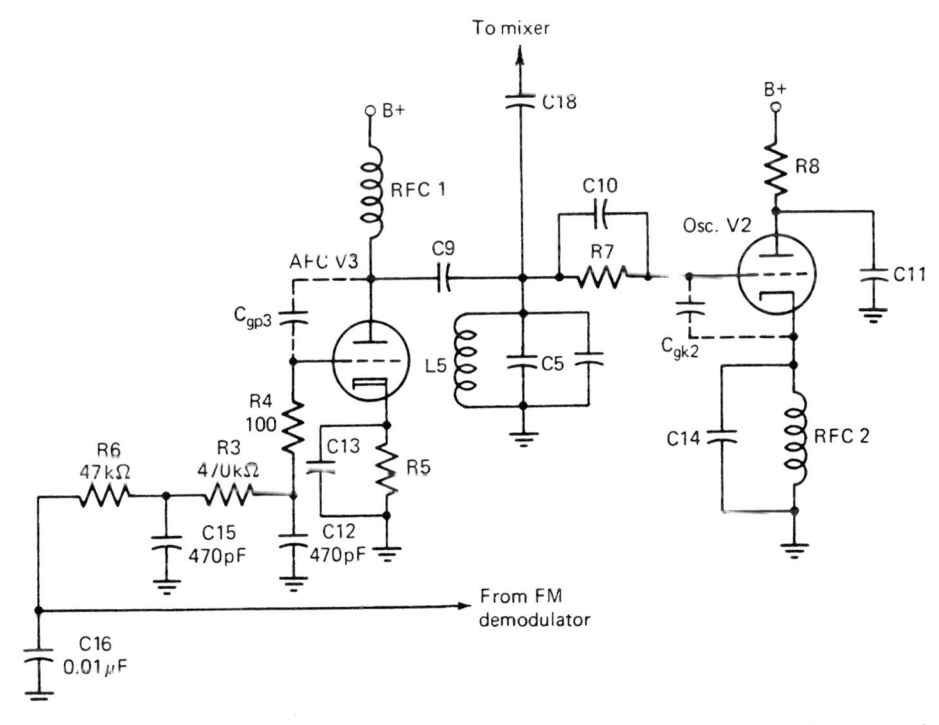

FIGURE 7-16 The oscillator section of a frequency changer that uses a reactance tube to control the frequency of the local oscillator.

the interelectrode capacitance C_{gp} plays the role of the phase-shifting capacitor that works in conjunction with R4. Capacitor C12 places the low end of R4 at signal ground. The resistor-capacitor network connecting to R3 forms a low-pass filter that removes any audio signal from the control voltage coming from the demodulator. The operation of the circuit as a variable capacitor is identical to that of the circuit described earlier. Capacitor C9 couples the plate to the tank of the oscillator.

A noteworthy aspect of this circuit is that the interelectrode capacitances of V2 and V3 are used to advantage in the circuit. These capacitors, shown as C_{gp3} and C_{gk2} in the figure, are not shown on the schematic that is a part of the service data for the receiver.

7-11 VOLTAGE-VARIABLE CAPACITORS (VVC)

Recall that a reverse-biased PN junction involves a depletion layer between conductive N and P regions of the semiconductor. The depletion layer appears as insulation between the two conductive regions, and the combination forms a capacitor whose capacitance depends upon the thickness of the depletion layer. As the thickness increases, the capacitance goes down.

The thickness of the depletion layer may be controlled by varying the reverse bias applied to the junction. As reverse bias is increased, the thickness of the layer increases and the capacitance goes down. As the reverse bias is decreased, the capacitance increases as the depletion layer becomes progressively thinner until the region of forward conduction is approached. The junction will not function as a capacitor in the forward bias region because of the large current conduction that occurs. The capacitance may be varied over a range of about 10 to 1 by controlling the magnitude of the reverse bias. Specially fabricated diodes intended for use as VVCs (sometimes called *varicaps*, *varactors*, etc.) are available over a wide range of capacitances in the picafarad (pF) range.

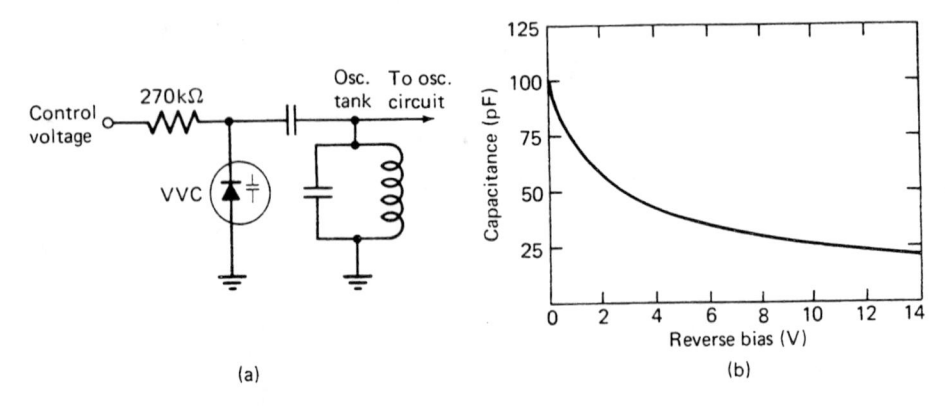

FIGURE 7-17 (a) Typical circuit for a voltage variable capacitor that is used to alter the frequency of an oscillator. (b) Typical variation of capacitance, with reverse bias voltage.

The primary use of VVCs are as voltage-variable tuning elements in such applications as controlled reactances in AFC circuits in both tube and transistor receivers. As such, VVCs perform the same function as the more cumbersome reactance tube circuit. VVCs are widely used.

A practical consideration in the use of VVCs is to prevent interaction between the DC source that supplies the control voltage and the RF circuit of which the VVC is a part. This isolation may be achieved by the use of inductors or large-value resistors. A typical circuit is shown in Figure 7-17 along with the graph of capacitance versus bias voltage for a typical VVC.

SUMMARY

1. A mixer must have a nonlinear transfer characteristic. When two frequencies are applied to the input of a mixer, the two original frequencies and their sum and difference appear at the output. This is called heterodyning.

2. The fundamental principle of a superheterodyne receiver is that the incoming station frequency is changed to a lower frequency for amplification prior to demodulation.

3. One input to the mixer of a superheterodyne receiver is provided by a local oscillator that is a part of the receiver.

4. Many receivers use frequency changers called converters that incorporate the mixer and local oscillator into one stage.

5. In oscillator circuits, the parallel resonant circuit may be considered to be a generator of a signal of the resonant frequency. Energy must be continually supplied to the resonant circuit, however, or the output will be a damped sine wave.

6. An oscillator circuit must provide positive feedback from its output to the resonant circuit. The Armstrong oscillator uses a tickler coil to magnetically couple energy into the resonant circuit. Hartley oscillators utilize tapped inductors and may be either shunt-fed or series-fed. The Colpitts oscillator utilizes a split capacitor arrangement rather than a tapped inductor.

7. The most commonly encountered transistorized converter uses an oscillator that resembles an Armstrong oscillator.

8. Almost all modern AM receivers use a loopstick antenna that serves also as the inductor for the antenna tuning circuit.

9. The local oscillator is arranged to tune at a frequency higher than the antenna frequency by an amount that remains consistent as the receiver is tuned across the band. Consequently, the difference frequency output of the mixer always occurs at the same frequency. The input to the IF amplifier is tuned to the difference frequency.

10. The most common frequency changer found in tube-type AM receivers utilizes a pentagrid converter. This tube has five grids, and, in a sense, functions as a series combination of a triode and a pentode. The triode section serves as the active device for the oscillator.

11. Oscillator drift can cause the tuned frequency of a receiver to drift considerably as the receiver warms up after turn-on. Consequently, oscillators for receivers must be care-

fully designed to minimize oscillator drift. Special temperature-compensating capacitors may be incorporated into the circuit to counteract capacitance increases in other capacitances of the circuit.

12. An uncompensated oscillator will always decrease in frequency as it warms up.

13. Oscillator pulling occurs when an external signal is coupled to the oscillator circuit. The oscillator tends to change frequency in an attempt to lock in or synchronize with the external signal. In converters, the antenna signal represents the external signal, and the oscillator is pulled toward the antenna frequency. This tends to lower the IF frequency.

14. Frequency changes for FM are more critical in design and operation than AM frequency changers.

15. FM receivers typically employ automatic-frequency-control circuits to control the frequency of the local oscillator. The AFC voltage is developed at the FM detector and is fed back to the local oscillator through an AFC filter.

16. Either a reactance tube or a voltage-variable capacitor may be used as the variable reactance of an automatic-frequency-control circuit. Voltage-variable capacitors (varicaps, varactors, etc.) have made reactance tubes obsolete, and voltage-variable capacitors are to be found in many tube-type receivers.

QUESTIONS AND PROBLEMS

7-1. What is the difference between a mixer and a converter?

7-2. What does the term "heterodyne" mean?

7-3. What is the difference between the transfer characteristic of an amplifier and a mixer?

7-4. A 3-kHz signal is mixed with a 4-kHz signal. What frequencies appear at the output of the mixer?

7-5. Is it possible for two RF signals to mix and produce a frequency in the audio-frequency range?

7-6. What function is performed by the local oscillator in the frequency changer of a radio receiver?

7-7. Explain why a parallel resonant circuit may be considered to be a signal generator when it is used in an oscillator. Under what circumstances is a damped sine wave produced?

7-8. How does the feedback that is essential to the operation of an oscillator differ from that frequently used to advantage in an amplifier?

7-9. How does an Armstrong oscillator differ from a Hartley oscillator? In what respects are they similar?

7-10. What is a distinctive feature of a Colpitts oscillator circuit?

7-11. In the frequency changer circuit of Figure 7-9, what is the purpose of capacitors $C1T$ and $C2T$?

7-12. Describe the biasing arrangements used for the oscillator and mixer transistors of the circuit of Figure 7-9.

7-13. Describe the structure of a pentagrid converter tube. Where does the virtual cathode occur? What causes the virtual cathode to form?

7-14. What signal is usually applied to the grid closest to the cathode in a pentagrid converter?

7-15. What function is served by grid $G2$ in a pentagrid converter?

7-16. What voltage is to be expected at the oscillator grid of a pentagrid converter when the oscillator section is functioning normally?

7-17. Explain how the BJT oscillator circuit of Figure 7-12 is similar to an Armstrong oscillator.

7-18. Oscillator injection is to the base circuit of the converter of Figure 7-12. To what point is the oscillator signal applied in the circuit of Figure 7-13?

7-19. Describe the effect of oscillator drift upon the tuned frequency of a receiver.

7-20. If the local oscillator of a receiver drifts to a lower frequency, does the IF frequency change to a higher or lower frequency?

7-21. Why must temperature-compensating capacitors used to correct for oscillator drift have negative temperature coefficients? That is, why must the capacitance of the units decrease with increasing temperature?

7-22. What is oscillator pulling and what may be done to minimize it?

7-23. Why does changing the grid bias voltage of a reactance tube change its effective capacitance or inductance?

7-24. The grid voltage of a certain reactance tube leads the plate voltage by 90°.

(a) What is the relationship of the plate current to the plate voltage?

(b) Does the tube appear to be a capacitance or an inductance?

7-25. Why are solid-state counterparts of reactance tubes not commonly encountered?

7-26. What is the principle of operation of a voltage-variable capacitor?

7-27. Why must a varactor diode never become forward-biased if it is to function as a voltage-variable capacitor?

7-28. Does the capacitance of a VVC increase or decrease as the reverse-bias voltage is increased?

7-29. The collector-base junction of a bipolar transistor in common amplifier circuits is reverse-biased, and a depletion layer forms between the base and the collector.

(a) Does this arrangement constitute a capacitor?

(b) What might cause the capacitance to vary?

(c) What might be the effect of this capacitance upon a parallel resonant circuit in the collector circuit?

Questions and Problems

8
PRINCIPLES OF AMPLITUDE MODULATION

Without radio waves, radio and television as we know it would not exist; at best, our electronic communications system would have to be linked by wires (or optical fibers, perhaps). This would be somewhat cumbersome for mobile communications of all sorts (for air traffic control, to say the least). In short, the physical phenomenon of electromagnetic radiation is now more or less taken for granted. Less than a century ago it was a wonder, but so was the airplane.

In the first part of this chapter we very briefly describe the transmission and reception of radio waves, and some aspects of radio waves. We give the principles involved in using radio waves to transmit information via amplitude modulation.

The concept of amplitude modulation is presented in conjunction with two simple transmission systems. AM sidebands and power relationships in an AM signal are described. The block diagram of a tuned-radio-frequency (TRF) receiver is given, and a detailed block diagram of a superheterodyne receiver is included, along with a discussion of several aspects of superheterodyne receivers.

8-1 ASPECTS OF RADIO WAVES

It is well known that radio waves are produced at the antennas of radio transmitters and that these waves carry the intelligence from the transmitting antenna to the receiving antenna. In this section we briefly describe several aspects of radio wave generation, propagation, and reception.

Let us consider the simple transmitter shown in Figure 8-1, which consists of an oscillator, an RF amplifier, and a transmission line that feeds the center of a *dipole* antenna. Electron currents are caused to flow up and down the dipole by the alternating voltage applied at the center of the dipole. Obviously, the antenna circuit, consisting of the dipole, the transmission line, and the coil to which the transmission line is attached, does not form a complete electrical circuit in the conventional sense. The result is that electrical charges develop at the ends of the dipole; the end to which electrons flow becomes negative while the other end becomes positive. The direction of electron flow alternates sinusoidally at the oscillator frequency, and therefore the polarity of the charges developed on the ends of the antenna alternates accordingly.

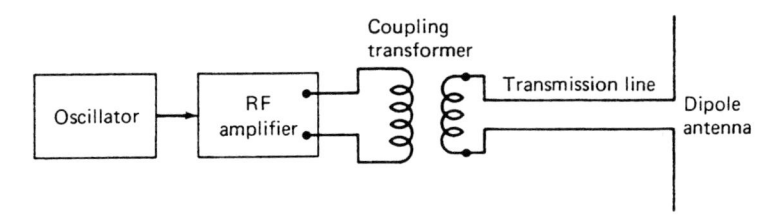

FIGURE 8-1 Simple transmitter.

The separation of charge between the top and bottom of the antenna produces an electric field in the space surrounding the antenna. Further, a magnetic field is produced by the electron currents flowing up and down the antenna. Thus, both electric and magnetic fields are produced around the antenna, and both exhibit sinusoidal variation at the oscillator frequency.

Since the oscillator drives the antenna at RF frequencies, the rapid variations cause the electric and magnetic fields to get "snapped loose" from the antenna. When this happens, the fields begin an outward journey from the antenna. The result is an *electromagnetic* (EM) *wave*, so called because the wave involves both an electric and a magnetic field. The wave moves away from the antenna at the speed of light.

For all EM waves, the directions of the electric field, the magnetic field, and the wave velocity are mutually perpendicular. The electric and magnetic fields are situated at right angles to each other in a plane perpendicular to the line of travel of the wave. Hence, EM waves are *transverse waves* that may be further described by defining the *polarization* of the wave. The polarization of an EM wave refers to the orientation of the electric field, and is typically described as being either *vertical* or *horizontal*. These types of polarization are cases of linear polarization in which the orientation of the fields remain fixed as the wave propagates through space. A more complicated type of polarization is either *circular* or *elliptical* polarization, in which the direction of the electric field constantly changes in corkscrew fashion along the length of the wave as it extends through space. Waves of vertical, horizontal, and circular polarization are illustrated in Figure 8-2.

In most cases, but not all, the orientation of the antenna determines the polarization of the waves. Vertical antennas produce vertically polarized waves, while horizontal antennas produce horizontally polarized waves. Several phenomena tend to alter the polarization as the waves propagate through the atmosphere and over the surface of the earth. Consequently, a vertically polarized wave may develop a horizontal component, a horizontally polarized wave may develop a vertical component, and a linearly polarized wave may develop components of circular polarization.

It is more than coincidental that waves emanating from a transmitting antenna travel at the speed of light. In fact, light waves are different manifestations of the same physical phenomenon; only the frequency is different. The frequency of visible light is on the order of 100,000 times as high as the highest useful radio waves (whether used for radio, TV, radar, or whatever).

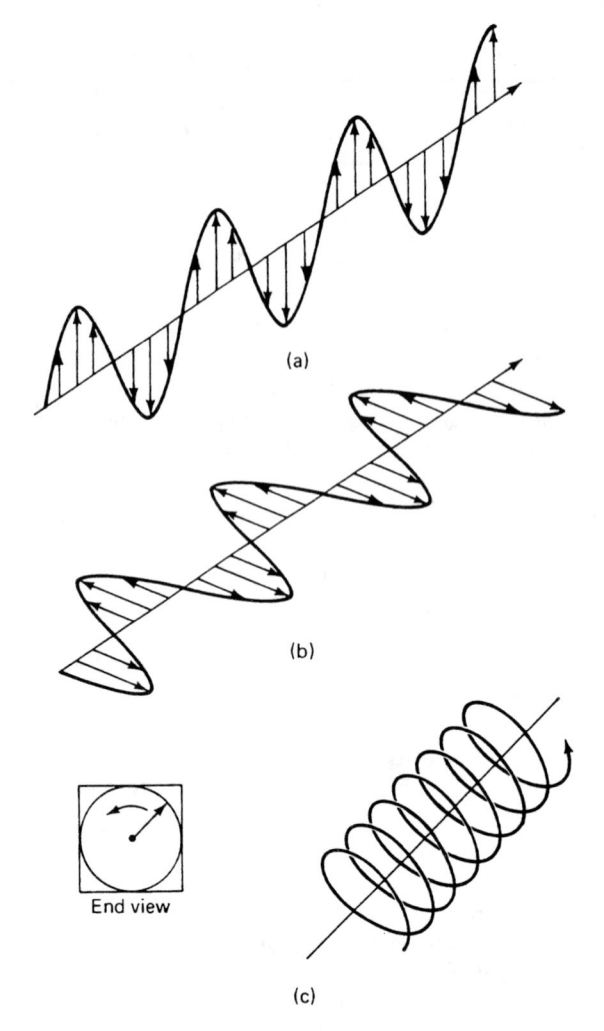

FIGURE 8-2 (a) Vertically polarized wave. (b) Horizontally polarized wave. (c) Circularly polarized wave.

Frequency and Wavelength

One complete wave is given off per cycle of the alternating current in the antenna. Since the frequency of the antenna current is determined by the oscillator, the oscillator frequency determines the number of complete waves that are produced per second by the antenna.

Because EM waves of all frequencies travel through free space at the same velocity, a given frequency gives rise to waves whose wavelength corresponds to that particular frequency. The wavelength may be defined as the distance between points of the same phase of two successive waves. Equivalently, it is the distance that a wave will travel in a time interval equal to the period of the antenna frequency. (The period is the reciprocal of the frequency.)

A simple relationship exists among the frequency, wavelength, and wave velocity:

$$f\lambda = c \quad \text{or} \quad f = \frac{c}{\lambda} \quad \text{or} \quad \lambda = \frac{c}{f}$$

where c is the velocity of light, f the frequency, and λ the wavelength. An accurate value of c is 2.9979×10^8 m/s, which we may round off to 3×10^8, or 300 million m/s. If the frequency is expressed in MHz and the wavelength in meters, the formulas above may be written

$$f_{\text{MHz}} = \frac{300}{\lambda_{\text{meters}}} \quad \text{or} \quad \lambda_{\text{meters}} = \frac{300}{f_{\text{MHz}}}$$

As examples, the wavelength of the lowest frequency (540 kHz) in the AM broadcast band is 300/0.540 = 555 m, and, for the highest frequency in the FM band (108 MHz) the wavelength is about 2.78 m.

It is sometimes of interest to calculate the distance a wave will travel in a certain length of time. This is done simply by multiplying the velocity c by the time interval in question. The reader can verify that an EM wave travels 300 m in 1 μs, or 1 mi in about 5.37 μs.

As the frequency of radio waves becomes higher, the waves act more like light waves. Waves of AM broadcast frequencies bear little resemblance to light waves, since the frequency is so very low in comparison. At FM frequencies, the waves tend more nearly to travel in straight lines, giving rise to "line-of-sight" transmission. At radar and microwave frequencies, the waves may be formed into beams much like a beam of light from a spotlight.

Reception of Radio Waves

We stated earlier that EM waves get snapped off from the transmitting antenna and travel at the speed of light to the receiving antenna. We now consider what happens at the receiving antenna when a wave passes by.

Recall that the wave involves an electric field. Further, when a charge is located in an electric field, the charge will experience a force that tends to make it move. When the electric field of the EM wave crosses the receiving antenna, it exerts a force on the free electrons within the antenna, producing electron currents. The direction of the electric field alternates due to the passing of the wave, and therefore the electron currents in the antenna alternate also, and at the frequency of the wave. Hence, electron currents in the receiving antenna alternate at the same frequency as the currents in the transmitting antenna that produced the wave.

At the transmitter, strong antenna currents produce strong transmitted waves. When a strong wave crosses the receiving antenna, strong electron currents are produced. Conversely, weak transmitting currents produce weak waves which set up weak electron currents in the receiving antenna. Thus, the currents produced in the receiving antenna represent an accurate image of the currents flowing in the transmitting antenna. Obviously, the receiving antenna current is much weaker than the transmitting current, but in other respects, the two are identical. It is for this reason that intelligence can be transmitted by radio waves.

Field Strength

The strength or intensity of a radio wave is given in terms of the strength of the electric field in units of volts per meter, or in more practical terms, in microvolts per meter. The meaning of this may be understood if we imagine a large, parallel-plate capacitor whose plates are separated a distance of 1 m. If a potential difference of 1 V is established between the two plates, the electric field strength between the plates will be 1 V/m.

If the field strength in a region is known, the signal voltage produced in a simple straight-wire receiving antenna can be computed by multiplying the field strength (in volts/meters) by the length of the antenna (in meters). This is generally of little value to a technician, however, because the important thing is the signal level made available to the input terminals of the receiver. This, of course, depends upon the type of antenna used as well as upon the field strength in the region.

Ionospheric Effects

Solar energy produces layers of ionized (electrically charged) air that surround the earth at various altitudes ranging from about 40 to 250 miles. These layers constitute the ionosphere and are distinguished broadly as being the D layer, E layer, and F layer, with sublayers distinguished within at least the E and F layers. The intensity and altitude of the layers vary with the time of day, the season of the year, and also with solar activity (sunspots, solar flares, etc.). A characteristic of the ionosphere is its variability.

The importance of the ionosphere to radio lies in the fact that the charged layers tend to reflect (refract is technically more correct) radio waves back to earth that otherwise would have been lost in space. The result is radio transmissions over distances much greater than that possible with the ground waves that remain close to the surface of the earth. This phenomenon, called *skip*, is illustrated in Figure 8-3.

The ability of the ionosphere to return waves to earth diminishes at frequencies above 30 MHz. Hence, FM stations (100 MHz) are not as susceptible to ionospheric effects. Consequently, FM stations are limited in range because of the absence of skip at FM broadcast frequencies. On the other hand, AM stations are very susceptible to ionospheric effects and commonly achieve transmissions ranging into the thousands of miles. Generally speaking, skip effects are more pronounced at night, and many AM stations that share a frequency with a station located in another geographical area are required to go off the air at sundown.

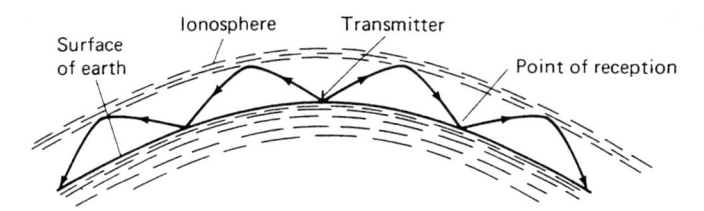

FIGURE 8-3 Ionosphere reflects radio waves back to the surface of the earth that would otherwise escape into space and be lost. This phenomenon is commonly called skip.

Atmospheric Effects—Lightning

The earliest radio transmitters utilized an electrical discharge, a spark, to excite electrical oscillations in spark-gap transmitters. In light of this, it is not surprising to find that lightning generates large amounts of RF energy that is distributed across almost the entire radio-frequency spectrum. More energy is concentrated at lower frequencies, however, and since long-distance transmission by skip is more prevalent below 30 MHz, the effects of lightning in producing radio noise is greater at low frequencies.

Antennas for AM Receivers

Early radio receivers required long wire antennas erected outside as high as practically possible. As the industry matured and portable receivers became eminent, advances were made both in antenna design and receiver sensitivity until, finally, the antenna became an inconspicuous component built into the cabinet of the receiver. Modern AM receivers almost universally employ ferrite rod antennas, which also function as the inductor for antenna tuning. Somewhat older receivers employ a loop antenna that typically is attached to the back of the cabinet. Antennas of these two types are shown in Figure 8-4.

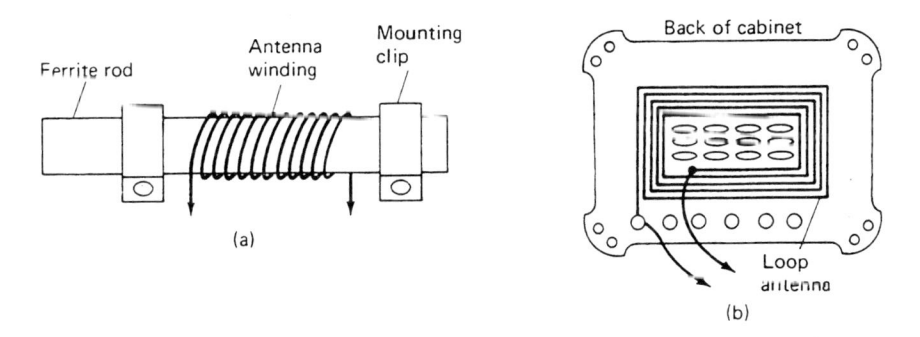

FIGURE 8-4 (a) Ferrite rod or "loopstick" antenna for AM. (b) Older type of loop antenna attached to the back of the receiver cabinet.

8-2 SIMPLE TRANSMISSION SYSTEM

Let us now consider a simple system that illustrates many concepts essential to an AM system. The system, shown in Figure 8-5, has many shortcomings, but we shall see that it is capable of transmitting information, in principle, if not in practice.

The transmitter consists of a shunt-fed Hartley oscillator whose tuning circuit is magnetically coupled to a long wire antenna. A variable voltage source provides B+ to the oscillator, and a large knob for adjusting the B+ voltage is in evidence. Note the calibration marks above the knob.

The receiver consists of a long wire antenna magnetically coupled to a parallel resonant (tank) circuit. The tank circuit is tuned to the same frequency as the transmitter oscillator. A diode, capacitor, and resistor complete the receiver except for an indicator device. We propose that a voltmeter should be used for this purpose. Let us now see how this system works.

The tapped inductor of the Hartley oscillator is familiar, and the plate capacitor feeds energy from the plate back to the grid tank to sustain oscillations. The RF choke at the plate offers an impedance which causes the plate signal to flow through the capacitor back to the grid circuit. The voltage induced in the antenna coil produces the current in the antenna, which produces the electromagnetic waves.

An important feature of this transmitter is that the strength of the oscillation may be controlled by the B+ voltage applied to the plate circuit. This feature is not unique to this circuit, however, but is common to most oscillator circuits. As the strength of the oscillation increases, the antenna current also increases, as does the strength of the EM waves radiated from the antenna. Conversely, as the B+

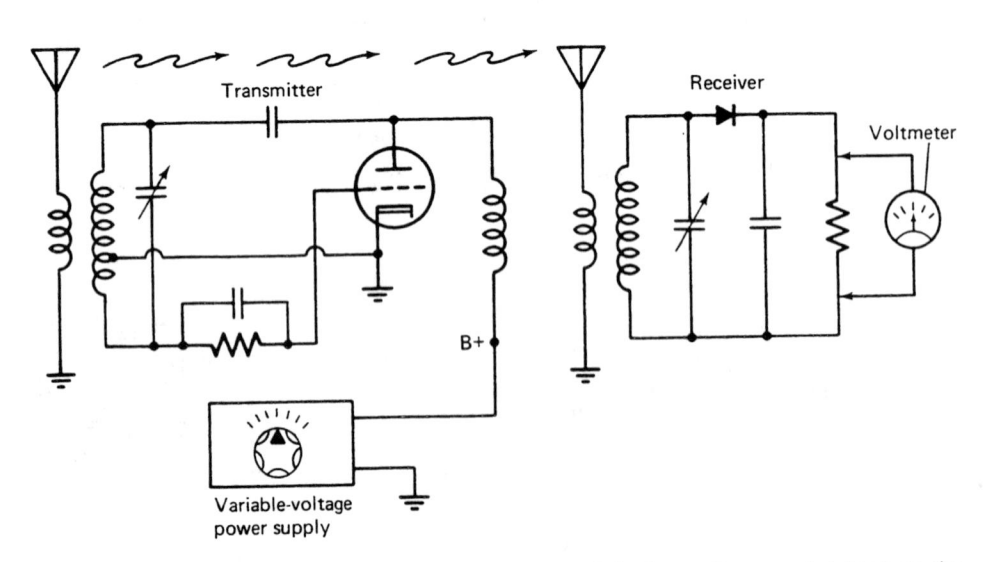

FIGURE 8-5 Simple transmission system. The setting of the B+ voltage control knob on the transmitter power supply controls the position of the needle on the voltmeter at the receiver output.

voltage is reduced, the radiated waves become weaker. Hence, the big knob on the power supply controls the strength of the waves radiated from the antenna.

At the receiver, current flows in the receiving antenna as a result of the passing waves from the transmitter. The current in the antenna coil couples energy into the tank circuit, and the antenna and tank circuit together act as a source of AC voltage. The AC voltage is applied to the diode, which acts as a rectifier and converts the AC to pulsating DC. The capacitor smooths the diode output, and the resistor completes the circuit. The voltmeter measures the DC voltage developed across the resistor.

Here is an important point. The strength of the waves passing the receiving antenna determines the strength of the AC voltage applied to the diode. Strong waves produce large AC voltages, and conversely. Further, the rectified DC voltage measured by the voltmeter depends upon the strength of the AC voltage applied to the diode. In short, the DC voltage indicated by the voltmeter is dependent upon the strength of the waves passing the receiving antenna.

Looking now at the entire system, it is evident that the position of the voltage control knob on the transmitter power supply determines the position of the voltmeter needle at the receiver. When the setting of the knob is changed, the voltmeter indication changes accordingly. This system involves the principle of amplitude modulation in a rudimentary form. The knob controls the strength or amplitude of the waves sent from the transmitter, and the amplitude of the waves passing the receiving antenna determines the indication of the voltmeter.

As a potential application of this system, suppose it is desired to transmit the temperature at a remote weather station in the vicinity of the North Pole to a data collection center located a bit farther south. The B+ knob at the transmitter could be calibrated with a temperature scale, and a corresponding scale could be devised for the voltmeter at the receiver. The operator at the weather station would read a thermometer and set the power supply knob accordingly. The voltmeter at the receiver would then indicate the temperature. Thus, the temperature information could be transmitted from the weather station to the data collection center.

We note in this system that information is transferred from the transmitter to the receiver via the RF signal that flows continually from the transmitter to the receiver as EM waves. This signal is called the *carrier*, because it "carries" the information. The amplitude of the carrier is changed or *modulated* to convey the information. We say that the carrier is *amplitude-modulated*.

8-3 VOICE TRANSMISSION SYSTEM

A transmission system capable of transmitting voice or music has several things in common with the simple system previously described, and we realize immediately that a microphone at the transmitter and a speaker or earphone at the receiver are involved. At the transmitter, the sound waves from the announcer's voice cause the microphone element to vibrate in the pattern of the sound waves, converting the pattern to an electrical signal of the same pattern. If we could see the microphone element function in extremely slow motion, we would see its position change as it vibrates back and forth in a complicated fashion under the influence of the incident sound wave.

A voice transmission system must provide for the speaker cone at the receiver to duplicate the motion of the microphone element at the transmitter. If this is arranged, the sound produced by the speaker will be a reproduction of the sound at the microphone. In essence, the position of the microphone element must control the position of the speaker cone. In this sense, a voice system is similar to the temperature transmitter of the preceding section, in which the position of the voltage knob controlled the position of the voltmeter needle. In both systems the underlying principle is that the position of one thing at the transmitter controls the position of another thing at the receiver.

If the previous system is to be upgraded to voice transmission capability, the B+ voltage applied to the oscillator must be made to vary in accordance with the electrical signal from the microphone. When this is done, the amplitude of the transmitted waves will vary in a similar manner, and the transmitted wave will be amplitude-modulated. When the B+ is high, the waves will have a large amplitude. When the B+ is low, the waves will have a correspondingly smaller amplitude. An amplitude-modulated (AM) wave is shown in Figure 8-6.

A simple system for transmitting an AM wave is shown in Figure 8-7. The output of the microphone is applied to an audio amplifier that serves as a preamplifier for a push-pull output stage. Note, however, that the B+ for the oscillator flows through the secondary of the push-pull output transformer. The voltage developed across the secondary of the transformer is either added to or subtracted from the B+ voltage of the power supply. In this manner the voltage applied to the oscillator is caused to vary in the pattern of the microphone output, and the amplitude of the transmitted waves varies in the same manner.

The receiver may be upgraded to voice capability by simply replacing the load resistor and voltmeter with an earphone as shown in Figure 8-8. The receiver will then reproduce the voice in the following manner. The waves incident upon the antenna are amplitude-modulated; they vary in strength according to the audio

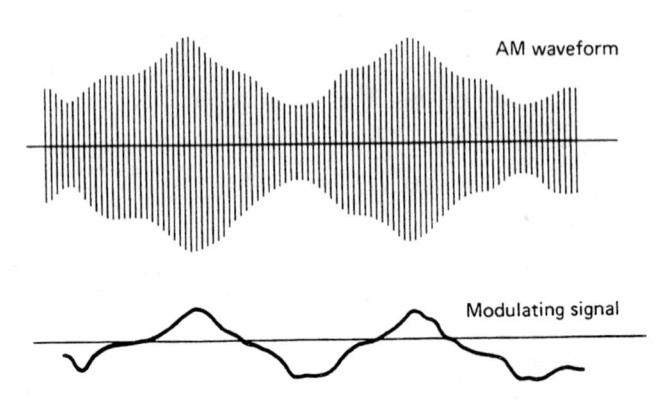

FIGURE 8-6 AM waveform. The amplitude of the signal varies with the strength of the modulating signal.

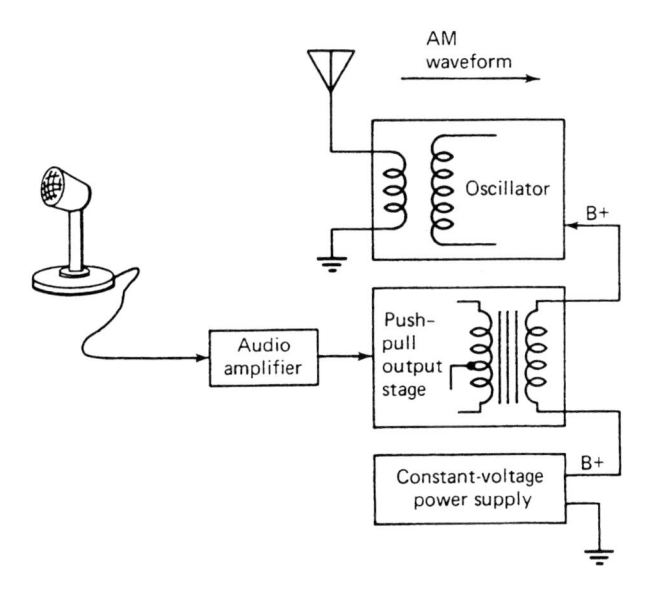

FIGURE 8-7 Simple system for transmitting an AM voice wave.

signal being transmitted. The variation in amplitude causes the voltage developed by the tank circuit and applied to the diode to vary. Since the AC input to the diode varies in amplitude, the DC output of the diode and associated filter capacitor varies accordingly. The earphone receives the output voltage of the diode and converts the electrical signal to sound.

If the diode were omitted from our simple receiver, no sound would be produced in the earphone. The RF voltage from the tuning circuit is too high in frequency for the earphone to reproduce, and even if the earphone were able to convert the RF signal to sound, it would be much too high in frequency for the ear to detect. Thus, the diode, called the *detector* or *demodulator*, is essential. After all, it is not

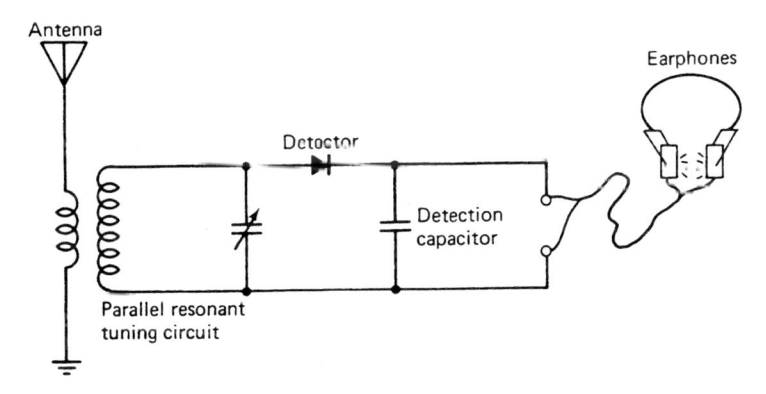

FIGURE 8-8 AM waveform modulated by voice or music may be received with this receiver.

Sec. 8-3   Voice Transmission System   219

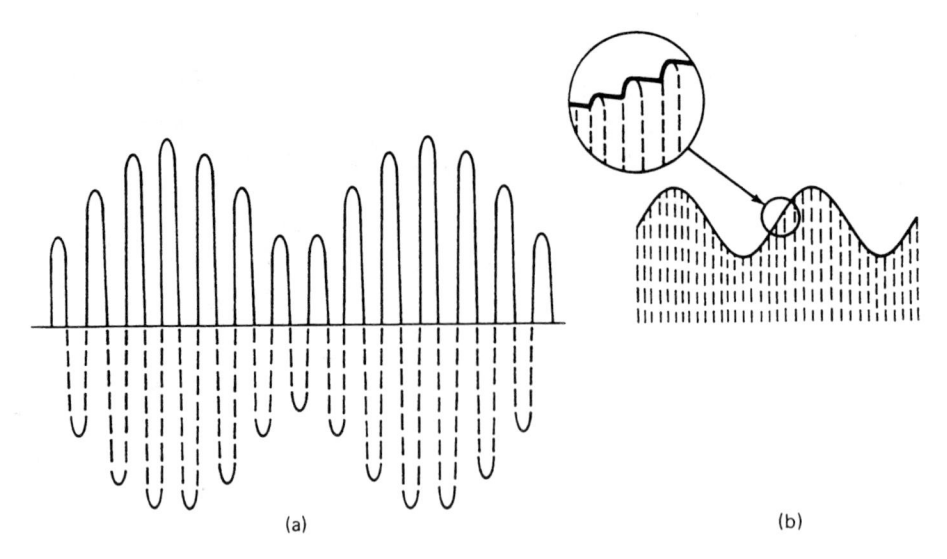

FIGURE 8-9 Details of the detection process for AM. (a) The detector diode removes the bottom (or top) portion of the input RF waveform in the fashion of a half-wave rectifier. (b) The peaks of the half-wave pulses represent the audio information. The inset shows the action of the detection capacitor in filling the gaps between the RF pulses.

the RF signal but rather the *variations* in the amplitude of the RF signal that represent the audio *intelligence* in which we are interested.

This emphasizes the concept of an RF carrier. The RF signal "carries" the audio signal via the amplitude variations. The audio signal, as carried by the RF signal, is frequently called the *modulation*.

The operation of the diode and capacitor is quite similar to the operation of a half-wave rectifier in a power supply. The capacitor acts as a filter to smooth the pulses from the diode, and in this application it is called the *detection capacitor*. The process of detection is illustrated in Figure 8-9.

8-4 PRACTICAL AM TRANSMITTER

The transmitter of the preceding section makes a good demonstration but lacks the refinement required of a useful system. We now describe the block diagram of a practical AM transmitter, shown in Figure 8-10. Only a block diagram is presented, since it is not our objective to treat transmitters in detail.

A crystal-controlled oscillator generates the carrier frequency, which is then amplified by several stages before being applied to the modulator and final amplifier combination. A crystal-controlled oscillator is required, since commercial AM transmitters must hold the frequency of the carrier to within ±20 Hz of the frequency specified by the FCC. The amplifier stages following the oscillator isolate the oscillator from the variable effects of the modulator stage, amplify the carrier to the level required by the modulator, and attenuate harmonic frequency signals

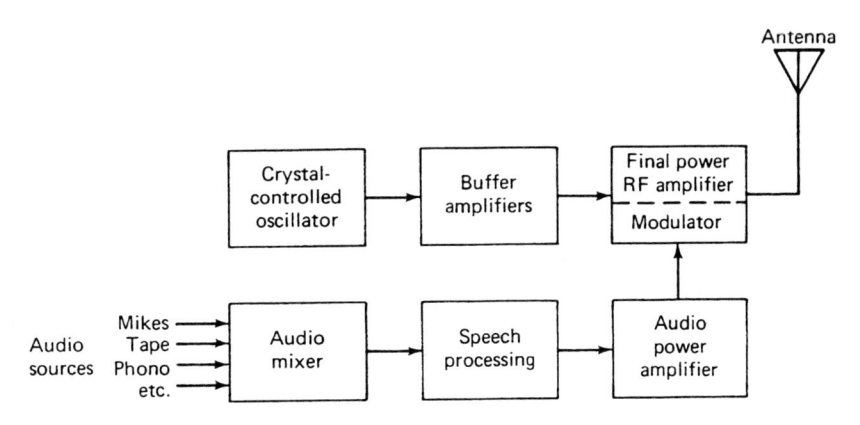

FIGURE 8-10 Block diagram of a practical AM radio transmitter.

from the oscillator. The carrier is unmodulated until it reaches the final amplifier, where it is combined with the signal from the audio section.

A typical transmitting facility has a variety of audio sources which feed into a mixer that selects the various sources and sets the volume level of each. The mixer output is applied to a "speech-processing" amplifier, which sets the frequency response and controls the dynamic range of the audio. A power amplifier then provides the audio power required by the modultor to modulate the RF carrier. Incidentally, an audio mixer is not be confused with the mixer of a frequency changer; the functions of the two are entirely different.

Characteristics of an AM Signal

At the transmitter, when the voltage produced across the secondary of the audio output transformer is of such polarity as to add to the voltage from the power supply, the voltage at the plate of the oscillator is a maximum and the radiated waves are strongest. This produces a modultion *peak* of the modulated waveform, as shown in Figure 8-11(a). On the other hand, when the secondary voltage opposes

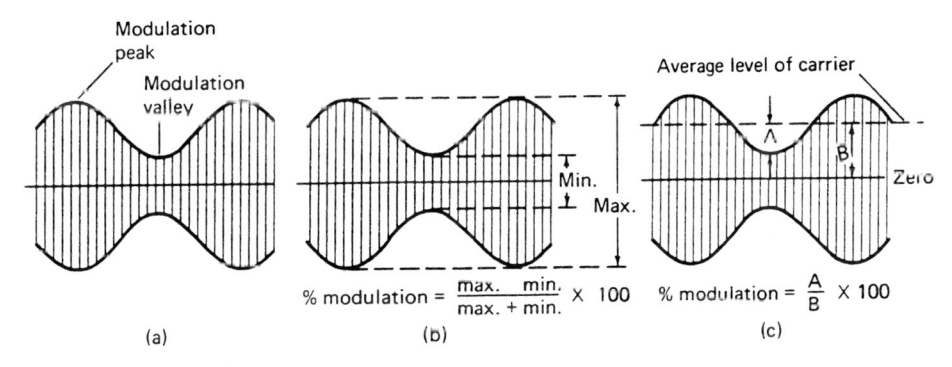

$$\% \text{ modulation} = \frac{\text{max.} - \text{min.}}{\text{max.} + \text{min.}} \times 100$$

$$\% \text{ modulation} = \frac{A}{B} \times 100$$

FIGURE 8-11 (a) Modulation peaks and valleys. (b) Definition of the percentage modulation. (c) Method of estimating the percentage modulation.

the power supply voltage, the oscillator plate voltage is a minimum, and a modulation *valley* is produced. Of course, when no sound waves strike the microphone, no voltage is developed across the secondary, and the plate voltage is constant. During this period the radiated waves are of constant amplitude.

The extent to which an RF carrier is amplitude-modulated is given by the *percentage modulation* defined by the formula and quantities defined in Figure 8-11(b). A method of estimating percentages of modulation is illustrated in Figure 8-11(c). Note that the average level of the carrier and also the zero voltage level of the carrier must first be estimated. Then the fraction A/B, expressed as a percentage, gives the percentage modulation. Various degrees of modulation are shown in Figure 8-12.

AM Sidebands

The process of amplitude modulation is similar to the process of heterodyning in the sense that both processes involve the generation of new frequencies when two original frequencies are combined. If a carrier of frequency f_c is amplitude-modulated by an audio tone of frequency f_a, the output of the modulator contains signal components of frequencies f_c, f_c+f_a, and f_c-f_a. The original carrier frequency is accompanied by a sidetone on each side of it whose frequency differs from the carrier frequency by an amount equal to the modulating frequency, f_a. This is illustrated in Figure 8-13.

The amplitudes of the three components are not the same. The sidetone amplitudes vary with the percentage modulation. When the carrier is fully modulated (100%) by the audio tone, the voltage amplitude of each sidetone reaches a maximum of one-half the carrier amplitude. When the percentage modulation is less,

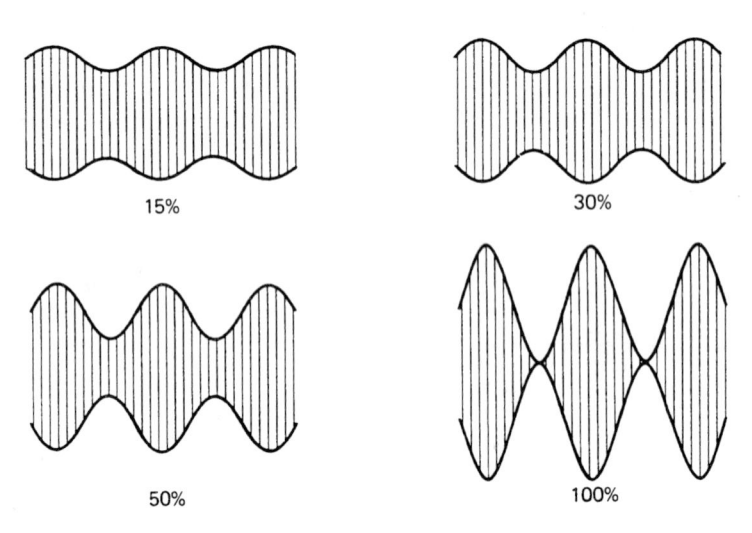

FIGURE 8-12 Waveforms illustrating various percentages of amplitude modulation.

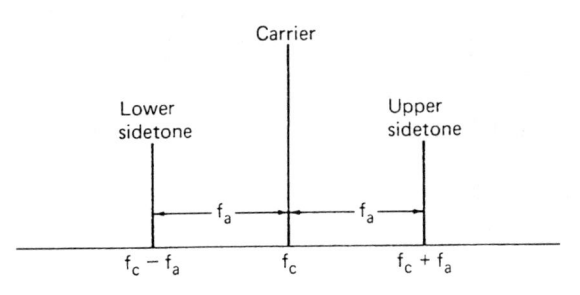

FIGURE 8-13 When a carrier is amplitude-modulated by a single audio tone, sidetones appear above and below the carrier frequency.

so is the amplitude of the sidetones. The amplitudes of the upper and lower sidetones are equal to each other.

If more than one audio tone is present in the modulating signal, there will be a corresponding number of sidetones on each side of the carrier, as illustrated in Figure 8-14. The relative amplitude of the various sidetones will be the same as the relative amplitude of the tones in the modulating signal.

A voice wave or the waveform from a musical instrument is actually the sum of signal components of many frequencies whose amplitudes depend upon the character of the sound. Equivalently, an audio waveform can be "resolved" into a band of tones, very closely spaced in frequency. When such a waveform is used to modulate a carrier, there appears in the modulator output a band of frequencies, called a *sideband*, on each side of the carrier. The high-frequency sideband is called the *upper sideband*, while the other is the *lower sideband*. The upper sideband is the mirror reflection of the lower sideband, and vice versa.

We now compare the amplitudes of the carrier and sidetones as the modulation percentage is varied from zero to 100%. At zero percent modulation, no sidetones are present, and the carrier stands alone. As the percentage modulation increases,

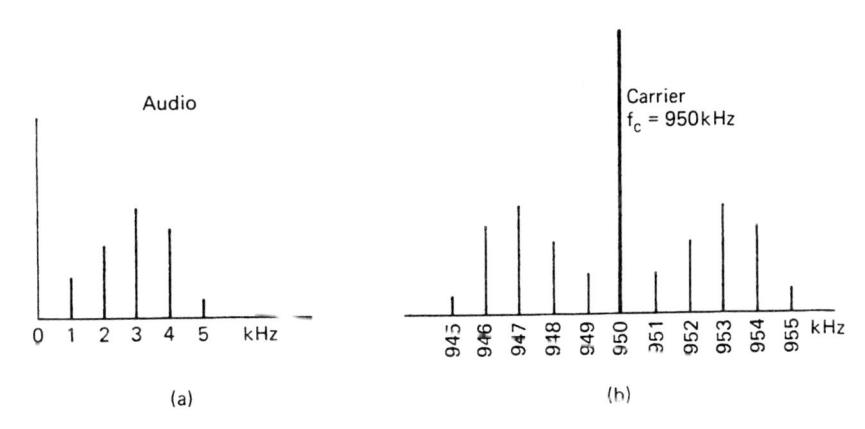

FIGURE 8-14 Each modulating tone produces a pair of sidetones in the amplitude-modulated waveform.

the sidetone amplitudes increase in direct proportion to the percentage modulation, but the amplitude of the carrier does not change. At 100% modulation, the amplitude of each sidetone is a maximum equal to one-half the carrier amplitude, but, still, the carrier amplitude remains exactly the same as if it were not being modulated at all.

At this point, recall that the RF signal voltage at the modulator output (an AM waveform) is the sum of three components: the upper sideband, lower sideband, and carrier. We have now established that the carrier component of the modulator output does not change in amplitude, and therefore any change in the amplitude of the modulator output must result from the action of the sidetones. The AM waveform at the modulator output results from alternate additive and subtractive interference of the sidetones with the carrier.

Since the sidetone frequencies are different, the sidetone components cyclically become in and out of phase with each other; sometimes they add together, other times they cancel each other completely. The frequency with which the phase change occurs is *twice* the frequency of the modulating tone. For each audio cycle, the sidetones go in and out of phase two times.

Now consider the carrier along with the sum of the two sidetone components. At those times when the sidetones are in phase, and when the sidetone sum is also in phase with the carrier, the modulator output will be a maximum. As the sidetones go out of phase, the sidetone sum approaches zero, and the modulator output falls to the level of the carrier alone. As the sidetones again come in phase, the sidetone sum is out of phase with the carrier and tends to counteract a portion of the carrier signal. Hence, at this time, the modulator output is a minimum. Continuing, the sidetones once again go out of phase, and the modulator output becomes equal to the carrier. At the last part of this cycle, the sidetones come in phase with each other and also with the carrier, producing a maximum modulator output once again. The process then repeats for the next audio cycle. This is illustrated in Figure 8-15.

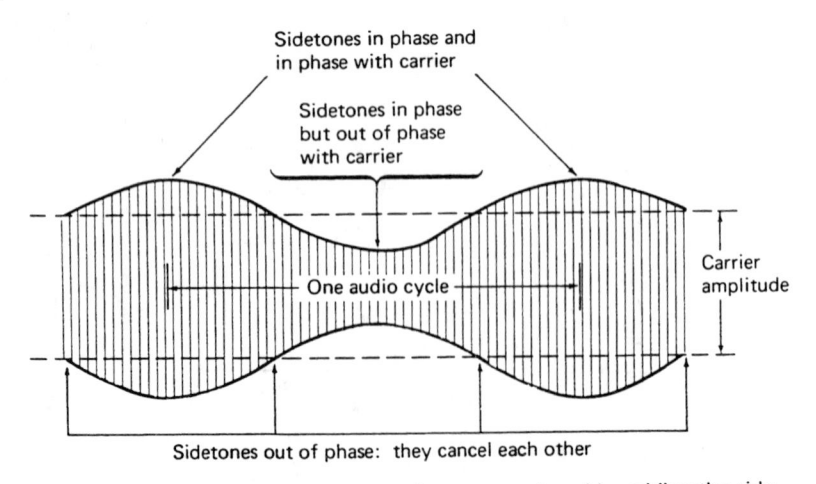

FIGURE 8-15 Modulation peaks and valleys are produced by adding the sidetone components to the constant carrier component.

Principles of Amplitude Modulation — Chap. 8

AM Power Distribution

We now consider the power contained in each sideband and in the carrier when the carrier is 100% modulated. In this case, the sideband power will be a maximum.

The power contained in a signal component is proportional to the square of the voltage amplitude of the component. At 100% modulation, the voltage amplitude of each sideband is one-half the amplitude of the carrier. Thus, the power contained in each sideband is one-fourth that of the carrier. Since there are two sidebands, the total sideband power is one-half that of the carrier, and is one-third the total transmitted power.

Suppose an AM transmitter puts out 100 W average power when unmodulated. As modulation is increased to 100%, the average power output increases to 150 W. Each sideband contains 25 W while the carrier power remains fixed at 100 W.

8-5 TUNED RADIO-FREQUENCY (TRF) RECEIVERS

Before the invention of the superheterodyne receiver, the typical commercial receiver was a TRF receiver that incorporated two or three tunable RF amplifiers in cascade prior to the detector stage, as shown in Figure 8-16. All the RF amplifiers had to be tuned simultaneously to the antenna frequency and had to be capable of operating over the entire broadcast band. Single-tuned as opposed to the more selective double-tuned transformers were used to couple the RF stages because of the practical impossibility of including a sufficient number of tuning gangs on the tuning capacitor. The selectivity varied over the band as the resonant frequency of the tuned circuits changed while the Q remained essentially constant.

Since all the RF amplifiers operated at the same frequency, extreme care had to be taken to prevent feedback between the RF amplifiers that would cause the receiver to go into oscillation. All in all, the receivers were rather unwieldly, and the advances of the superheterodyne circuit were welcome.

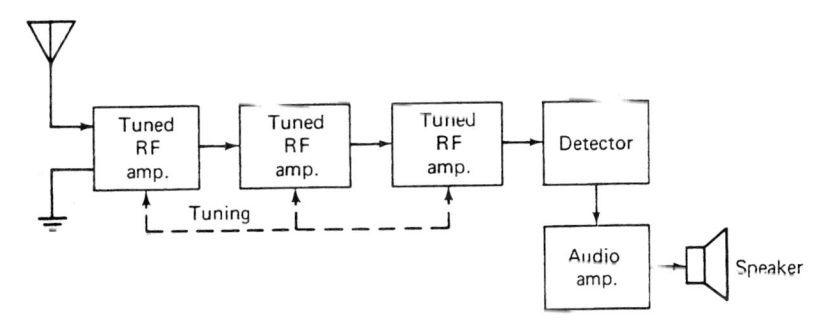

FIGURE 8-16 Block diagram of a tuned radio-frequency (TRF) receiver. (The power supply is omitted.)

8-6 SUPERHETERODYNE RECEIVERS

A block diagram of a superheterodyne receiver is shown in Figure 8-17, and the waveforms appearing at various points are indicated. Even though the circuitry of each block has been described separately in previous chapters, we shall now give a summary of each functional block. We then discuss some of the advantages, disadvantages, and other aspects of superheterodyne receivers.

In passing, we note that the superheterodyne principle, that of heterodyning to achieve frequency conversion (or frequency translation), is one of the most used concepts in communications electronics. This principle is to electronic communications as wheels are to transportation.

The antenna, which may be a loopstick enclosed within the receiver cabinet or an external long-wire antenna, is exposed to a myriad of electromagnetic waves of all frequencies. In most cases, however, the antenna or its associated circuitry is very broadly tuned so that the initial selection of the desired frequency is performed at the antenna. The signals picked up by the antenna, which may be as small as a few tens of microvolts, are delivered to the input of the RF amplifier.

The RF amplifier performs two functions: it provides the initial amplification to the signal from the antenna, and it provides the first tuning that is sharp enough to be significant in the rejection of unwanted frequencies. Even so, several station frequencies may appear at the output of the RF amplifier.

The amplification that the RF amplifier provides is important because of the noise characteristics of the mixer. A mixer tends to be noisy in operation, but the effect of the noise is much less if the amplitude of the antenna signal has been

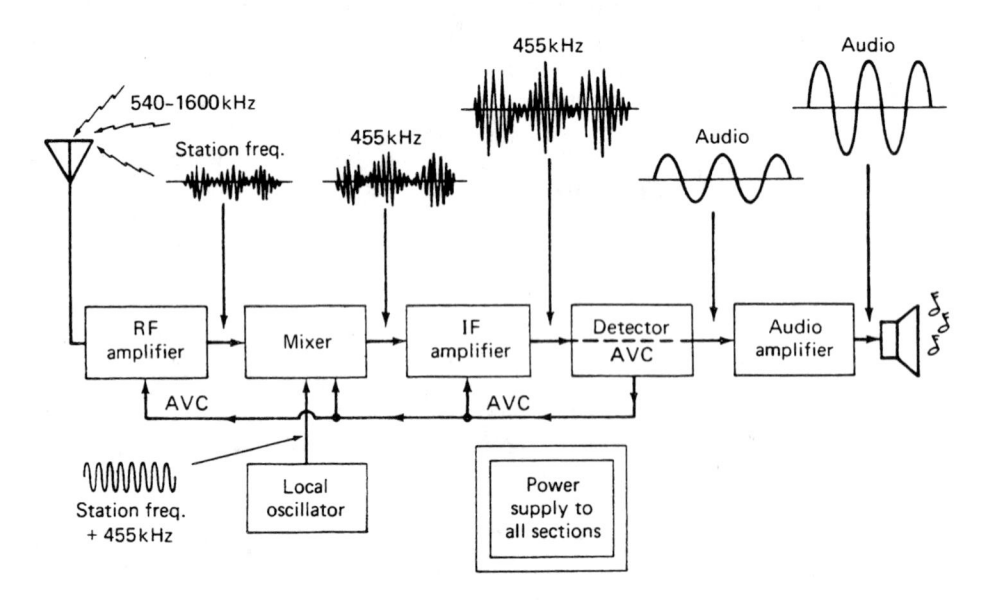

FIGURE 8-17 Block diagram of a superheterodyne AM receiver.

increased by an RF amplifier. Further, the tuning of the RF amplifier decreases the bandwidth of the signals applied to the mixer, and this reduces the total noise introduced. The RF amplifier may be omitted in lower-quality receivers.

The mixer and local oscillator work together to form the frequency changer portion of the receiver. In this stage, the antenna frequency is converted to the intermediate frequency. The local oscillator produces a pure sine wave whose frequency is 455 kHz higher than the frequency to which the antenna is tuned. This signal is sent to the mixer, where it heterodynes, or mixes, with the signal from the RF amplifier. At the output of the mixer there will appear signals of four different frequencies: (1) the frequency of the signal from the RF amplifier; (2) the frequency of the signal from the local oscillator; (3) the sum of the frequencies of the RF amplifier and local oscillator; and (4) the difference between the RF amplifier signal and the local oscillator signal. Of these four frequencies, it is the difference frequency that is of interest.

Because the local oscillator is tuned to a frequency that is higher than the RF amplifier by an amount equal to the intermediate (IF) frequency, the difference between the two is the same as the intermediate frequency. The sharply tuned circuits of the input to the IF amplifier reject the other frequencies. The amplitude of the signal applied to the input of the IF amplifier may be on the order of 20 mV.

Whereas the RF amplifier is important in determining the noise level and the sensitivity (ability to pick up weak stations) of a receiver, it is the IF amplifier that is primarily responsible for the selectivity (ability to tune sharply) of the receiver. Because the IF amplifier must operate only at one frequency (the 455-kHz intermediate frequency that is standard for AM receivers), the resonant circuits can be designed to have a very high Q, and the amplifier as a whole can have a gain of 100 or more. High-Q resonant circuits tune sharply so that unwanted frequencies of adjacent stations are rejected.

The output of the IF amplifier is applied to the demodulator (detector) that recovers the audio signal from the amplitude-modulated carrier. We have already seen that the detector is essentially a half-wave rectifier that removes either the positive or negative portion of the carrier waveform. The detection capacitor then smooths the 455-kHz pulses to form the audio.

Another function performed by the detector stage is that of providing the AVC (automatic volume control) voltage that automatically varies the gain of the preceding stages to compensate for variations in signal strength that occur at the antenna. To function properly, the detector requires an input signal from the IF amplifier whose amplitude is on the order of 1 V. If a very strong signal appears at the antenna, say 10,000 μV, then a gain of only 100 is required between antenna and detector in order for the signal to be properly detected. But if the signal is very weak, say 5 μV, a gain of 200,000 is required. The AVC circuitry (sometimes called AGC for automatic gain control) varies the gain of the RF amplifier, mixer, and IF amplifier so that essentially the same signal level is applied to the detector even when the signal level at the antenna varies over a range of 1000 to 1 or more.

The audio amplifier builds up the power level of the signal from the detector in order to drive the speaker. It also includes a manual volume control (which has nothing to do with the AVC), and it may provide a tone control so that the tone can be adjusted to accommodate individual preferences.

Considerable variation is possible in regard to the block diagram of a superheterodyne receiver. The RF amplifier may be omitted with some sacrifice in quality, as we stated earlier. The local oscillator and mixer functions may be combined in a single stage called a converter, with little if any effect upon quality. Perhaps the widest variations occur in the implementation of the AVC function. In the simplest receivers, only one stage of the IF amplifier may have controlled gain, but the better receivers will have the gain of the mixer and RF amplifier made variable also.

Finally, in light of IC technology and the popularity of FM, comparatively few AM-only receivers are built; almost all are AM-FM receivers. In these receivers, the AM functions are almost always implemented within the same chips that perform the FM functions, and almost as many designs exist as there are designers. In any event, the same basic block diagram of the superheterodyne receiver will prevail.

Local Oscillator Radiation

We described in Chapter 7 the effects of oscillator drift upon tuning stability, but there is yet another problem to be dealt with in conjunction with the local oscillator. The oscillator itself can act as a transmitter if any portion of the oscillator signal gains access to an antenna. Therefore, coupling of the oscillator to the antenna by way of discrete capacitors, interelectrode capacitance, or stray capacitance must be minimized. The FCC places restrictions on the amount of radiation that can occur from the oscillator and still be acceptable. An RF amplifier preceding the mixer reduces the radition, since the RF amplifier lengthens the path between the local oscillator and the receiving antenna. Careful circuit design and careful shielding are required to bring the radiation limits within the FCC specifications.

Antenna-Local Oscillator Tracking

The superheterodyne circuit is not without its disadvantages. One problem the designer must solve is that of tracking of the antenna and local oscillator frequencies. The two are said to track if the difference frequency remains constant as the receiver is tuned across the band. The oscillator and antenna tuning capacitors are ganged to a common shaft and tuned by a single control, but this does not assure proper tracking because the tuning ranges of the oscillator and antenna are different, and the tuning ratios are different.

Tracking may be achieved to a satisfactory degree by careful selection of L/C ratio of the antenna and oscillator tuning circuits and by using varible tuning capacitors whose plates are given a special shape so that the capacitance does not vary linearly with rotation of the shaft.

Images and Spurious Responses

A superhet is sensitive to antenna signals that differ from the local oscillator frequency by an amount equal to the IF frequency. For each setting of the local oscillator, there are two such frequencies. One is desired, while the other represents a potential source of interference. The interfering signal is called an *image* and is separated from the desired frequency by an amount equal to *twice* the IF frequency.

As an example, suppose that a receiver is tuned to a station at 600 kHz, which is near the bottom of the AM broadcast band. The local oscillator will then run at 1055 kHz, which is 455 kHz above 600 kHz. Now suppose that a strong nearby station transmits at a carrier frequency of 1510 kHz, which is near the top of the band. If the 1510-kHz signal can find its way to the mixer, it will heterodyne with the 1055-kHz signal of the local oscillator to produce a difference frequency of 455 kHz. This is the operating frequency of the IF amplifier, and the undesired 1510-kHz station will compete with or interfere with the desired station at 600 kHz.

The antenna tuning circuits favor the desired frequency since the circuits are tuned to that frequency. Conversely, the antenna circuits tend to reject the image frequencies. The image is located sufficiently far away from the desired frequency that significant interference occurs only when a very strong source is present at the image frequency. Nevertheless, image frequency interference must be considered in designing a new receiver and in selecting the IF frequency that the receiver is to use. Figure 8-18 illustrates the placement of the image frequency in relation to the oscillator and desired frequencies.

Another effect that gives rise to spurious responses stems from the fact that the signal from the local oscillator is not a perfect sine wave. The imperfections (distortion) appear as small signal voltages at frequencies that are integral multiples (harmonics) of the fundamental frequency of the local oscillator. If, for example, an oscillator running at 1055 kHz produces a distorted waveform, its output will contain signals whose frequencies are 2110 kHz, 3165 kHz, 4220 kHz, and so on. Each of these components can heterodyne with undesired station frequencies to produce 455-kHz difference frequencies that will be accepted by the IF amplifier to produce a *spurious response*. Consider the 2110-kHz signal, the second harmonic of the oscillator signal. It can heterodyne with stations at 1655 kHz or at 2565 kHz to produce a 455-kHz difference. Another example is given in Figure 8-19.

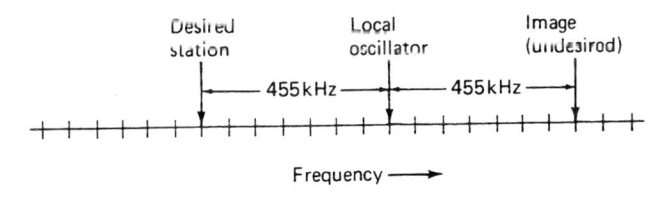

FIGURE 8-18 Superheterodyne receiver exhibits a sensitivity to stations located at the image frequency. The image frequency is related to the desired station frequency as shown.

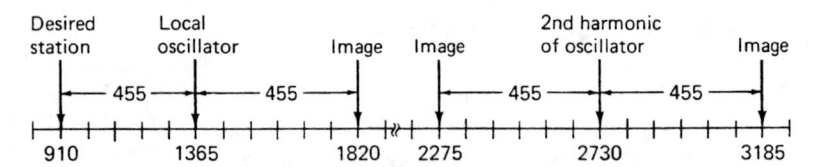

```
Desired      Local                                    2nd harmonic
station     oscillator      Image    Image           of oscillator            Image
  |←─ 455 ──→|←── 455 ──→|              |←── 455 ──→|←── 455 ──→|
├─┼─┼─┼─┼─┼─┼─┼─┼─┼─┼─┼─╫─┼─┼─┼─┼─┼─┼─┼─┼─┼─┼─┼─┼─┼─┼─┤
910           1365         1820    2275             2730           3185
```

FIGURE 8-19 Harmonics of the oscillator signal produce other undesired image frequencies. A similar situation exists for the third and higher harmonics. The image frequencies produce spurious responses in the receiver.

To summarize, antenna signals that differ from the local oscillator frequency, or any of its harmonics, by the IF frequency are subject to being admitted to the IF amplifier. The tuning circuits of the RF amplifier and of the input to the mixer attenuate the spurious responses because they are located far from the resonant frequency of the antenna circuit.

IF Amplifier

The selectivity of a superhet receiver is provided, for the most part, by the IF amplifier. Recall that the bandwidth (BW) of a tuned circuit is calculated by dividing the resonant frequency by the effective Q of the circuit. Therefore, the BW increases in proportion to the tuned frequency if the Q remains the same. Since the IF amplifier operates at a much lower frequency than the receiver front end, resonant circuits of the same Q produce a much narrower bandwidth in the IF amplifier. Further, since all antenna frequencies are converted to the same IF frequency, the bandwidth of the IF amplifier does not change as the receiver is tuned across the band. Therefore, the selectivity of the receiver is the same on the high and low ends of the band.

The IF amplifier operates at only one frequency, in contrast to the RF amplifier preceding the mixer stage, which must operate over the entire band. Single-frequency amplifiers can be designed to give much greater gains than amplifiers that must be tunable over the band.

Although considerable care is required in the design of an IF amplifier to avoid undesired feedback to preceding stages, the problems with feedback and oscillation are not as severe as with TRF receivers, because the IF and antenna frequencies are different. If a portion of the IF signal gets back to the antenna input, for example, the undesired signal will be attenuated by the antenna tuning circuits, since the frequency is different. In TRF receivers, the feedback frequency is the same as that of the antenna input.

Intermediate-Frequency Selection

Students often ask why a certain frequency is chosen for an IF frequency, so we present the following considerations. If the IF frequency is low, a smaller bandwidth results from tuned circuits of a given Q. Thus, greater selectivity is obtained by using a low IF frequency. But if the IF frequency becomes too low, the bandwidth may become too narrow. Also, the lower the IF frequency, the closer the antenna and oscillator frequency must be, and the effect of oscillator pulling and oscillator drift is greater.

Image rejection is greatest if high IF frequencies are used, because the image frequencies are located farther from the tuning frequency of the antenna. But as the IF frequency is increased, the oscillator and antenna frequencies become more widely separated, and tracking becomes a problem as the tuning ratios of the oscillator and antenna become increasingly different.

Practically speaking, selection of an IF frequency is a compromise. For AM receivers operating in the standard broadcast band, 455 kHz has been almost universally adopted. This frequency is also utilized extensively as an IF frequency in the receiver section of citizen's band transceivers.

SUMMARY

1. Current flow in an antenna produces a magnetic field in the region around the antenna. Positive and negative charge separations produce an electric field around the antenna. An electromagnetic EM wave is produced when these fields are "snapped loose" from the antenna. EM waves are of the same nature as light waves and travel at the same velocity as light waves.

2. A simple relationship exists among frequency, wavelength, and velocity of EM waves: $f\lambda = c$.

3. An EM wave travels at a velocity of about 300 million (3×10^8) m/s. An EM wave will travel 300 m in 1 μs or 1 mi in about 5.37 μs.

4. The signal level made available to the input terminals of a receiver depends upon the field strength in the region of the antenna and upon the type of antenna used.

5. Layers of ionized (electrically charged) air in the ionosphere reflect radio waves back to earth and provide for long-distance reception via "skip." Skip phenomena are more pronounced at frequencies below 30 MHz.

6. A voice or music signal may be transmitted by causing the amplitude of the carrier to vary in accordance with the amplitude of the modulating signal. Such a transmission system is an amplitude-modulated (AM) system.

7. The demodulator (or detector) of an AM receiver resembles a half-wave rectifier in a power supply.

8. When a carrier is amplitude-modulated, sidebands are produced on both sides of the carrier frequency. The upper and lower sidebands are mirror reflections of each other. Sideband amplitudes vary with the percentage modulation and never exceed one-half the carrier amplitude. The power contained in each sideband is a maximum of one-fourth that contained in the carrier.

9. The amplitude of the carrier component of an AM waveform remains constant. The peaks and valleys of the AM waveform are produced by the interference of the sidebands with the carrier component.

10. Early receivers were tuned radio-frequency (TRF) receivers. Superheterodyne receivers offered many advantages over TRF receivers.

11. Superheterodyne receivers are subject to image-frequency interference and to spurious responses caused by harmonics of the oscillator signal.

12. The selectivity of a superheterodyne receiver is determined primarily by the IF amplifier.

13. For a superheterodyne receiver to function properly, the antenna frequency and local oscillator frequency must "track."

14. Local oscillator radiation must be kept within FCC specifications.

15. An AM transmitter, when 100% modulated, will put out an average power of 1.5 times the average power output when unmodulated.

16. At 100% modulation, an AM transmitter will concentrate only one-third of the total output power into the sidebands.

QUESTIONS AND PROBLEMS

8-1. Why are radio waves often called electromagnetic waves?

8-2. What is the polarization of a radio wave?

8-3. Why are certain AM stations required to go off the air at sundown while FM stations may continue broadcasting?

8-4. What controls the strength of the oscillations in the tank circuit of the temperature transmitter described in this chapter?

8-5. What determines the strength of the oscillations in the resonant circuit of the simple receiver used in conjunction with the temperature transmitter?

8-6. Explain, by drawing appropriate waveforms, the process of AM detection.

8-7. Compare an AM detector to a half-wave power supply. How are they different? In what respects are they similar?

8-8. In what respects are the voice transmitter and the temperature transmitter similar?

8-9. Signal strengths produced by a receiving antenna vary with atmospheric and climatic conditions. What effect might this have on the simple temperature transmission system?

8-10. An AM radio station transmitting at a frequency of 980 kHz is located 25 mi from the receiver. Calculate:

(a) The wavelength of the wave, in meters and in feet.

(b) The period of the carrier frequency, in microseconds.

(c) The distance a transmitted wave will travel during one period of the carrier frequency.

(d) The time it takes for a wave to travel from the transmitter to the receiver.

(e) The number of complete waves between the transmitter and receiver at any time.

8-11. The average amplitude of an unmodulated carrier is 10 units. When modulated by a certain tone, the modulation peaks have an amplitude of 13 units while the valleys have an amplitude of 7 units. What is the percent modulation?

8-12. An AM transmitter (a CB transceiver) has an unmodulated power output of 4 W. What will the average power output be when the transmitter is 100% modulated?

8-13. A 100% modulated transmitter transmits 30 W of power.

(a) What power is contained in the upper sideband?

(b) What power is contained in the lower sideband?

(c) What is the power contained in carrier at 100% modulation?

(d) What is the power contained in the carrier when the transmitter is unmodulated?

8-14. At 100% modulation, each sideband of an AM transmitter contains 20 W of power. What is the total average power transmitted?

8-15. Describe the manner in which the sideband amplitudes vary with the percentage modulation as the percentage modulation is increased from zero to 100% if the transmitter is modulated by a single audio tone.

8-16. A 138-kHz carrier is modulated by a 2-kHz audio tone. What frequencies appear at the modulator output?

8-17. If a 400-Hz audio tone modulates a 690-kHz carrier, how many carrier cycles are involved in "carrying" one cycle of the audio?

8-18. An AM superheterodyne receiver utilizes an IF frequency of 455 kHz. The desired station is at 1240 kHz. Calculate:

(a) The frequency of the local oscillator.

(b) The frequency of the second and third harmonic of the oscillator frequency.

(c) The frequency of the spurious responses associated with the second and third harmonics.

(d) The image frequency associated with the local oscillator fundamental frequency.

8-19. An automobile receiver tuned to 1180 kHz has the local oscillator running at 1442.5 kHz. What is the IF frequency used by this receiver?

8-20. How long does it take a light wave to travel the 93 million mi (average) from the sun to the earth? How long does it take a radio wave to travel the same distance?

Questions and Problems 233

9
AM RECEIVER ANALYSIS

Previous chapters have dealt with power supplies, audio amplifiers, RF amplifiers, frequency changers, and the principles of AM and of superheterodyne receivers. We are now ready to investigate the circuitry of complete receivers, to see how the various stages tie together to make a functional unit.

Several complete schematics are fully described in this chapter. Both solid-state and vacuum-tube receivers are included, but the emphasis, obviously, is on solid state.

9-1 PRELIMINARY CONSIDERATIONS

Schematic Diagrams

At first, the schematic diagram of a complete receiver may appear hopelessly complex, an extensive jumble of components that bear little resemblance to the separate circuits already studied. Fortunately, the initial intimidation quickly passes as we learn to "divide and conquer," to break the schematic down into sections that correspond to the block diagram. The first objective in studying a schematic is to identify the signal path that begins with the antenna, passes through the various stages, and finally ends at the speaker.

Almost always, the signal flow on schematic diagrams is from left to right, beginning at the upper left and then extending downward to additional "lines" as necessary. Hence, the antenna circuit is typically found at the upper left corner of the circuitry, and the speaker appears near the lower right. The power supply circuitry usually appears near the bottom.

Within stages, the signal usually flows from the base to the collector, but this is not the case for the less frequently encountered emitter-follower and common-base circuits. The output is from the emitter of an emitter follower. For a common-base circuit, the input is to the emitter while the output appears at the collector. Good schematics have the circuits arranged so that the signal path is easy to follow.

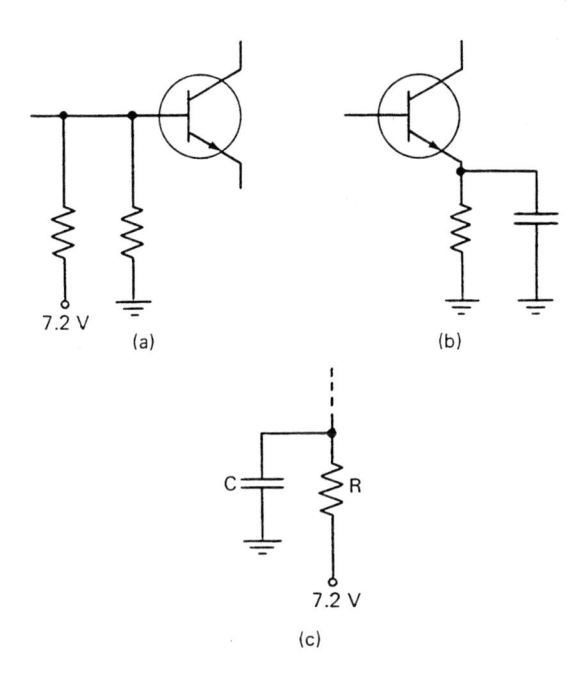

FIGURE 9-1 (a) Voltage divider of a universal bias arrangement. (b) Emitter resistor and emitter bypass capacitor. (c) Decoupling network.

Three Familiar Networks

The networks shown in Figure 9-1 are commonly encountered in receivers, so that it is more convenient to refer to each combination rather than to the individual components. Part (a) is the voltage-divider portion of a universal bias arrangement as it usually appears on schematics. Part (b) is an emitter resistor and its bypass capacitor. Part (c) is a decoupling network, described below. Quick recognition of these and other networks will help unclutter a schematic.

Decoupling

When several stages are brought together and powered by a common power supply, it is important that the signal of one stage not be *coupled* to the other stages by way of the power supply. In other words, the stages must be *decoupled*. This is especially important if the stages are cascaded as they are in radio receivers. If cascaded stages are not decoupled, the output signal can be fed back to an input stage where it is again amplified; the result is an oscillation. The coupled stages can act as an oscillator, and the receiver will emit various screams, howls, or whistles along with the desired audio signal, or it may just scream and howl without producing any useful signal at all.

Sec. 9-1   Preliminary Considerations

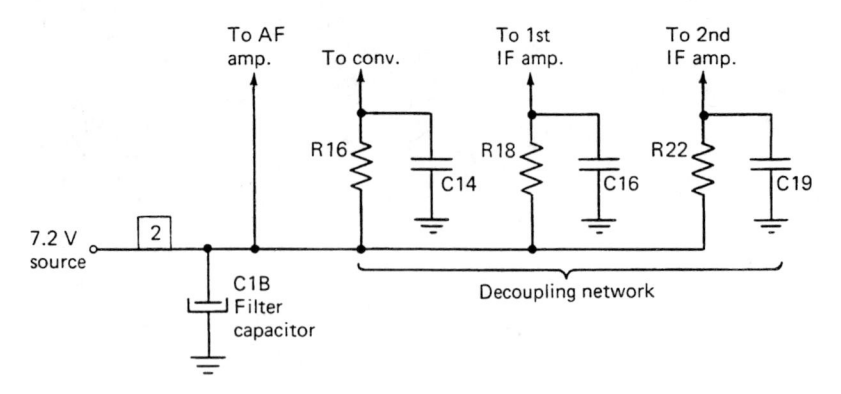

FIGURE 9-2 Decoupling networks in the receiver of Figure 9-3.

The decoupling network of Figure 9-1(c) is inserted between the power supply and the stage being decoupled. The resistor presents an impedance to AC signals coming from the stage, and the capacitor shunts the signals to ground. Put differently, the capacitor forces the stage side of the resistor to be an AC ground. The distribution of the 7.2-V source of the receiver of Figure 9-3 is shown in Figure 9-2. Three decoupling networks are in evidence.

9-2 AM RECEIVER; POWER SUPPLY

The complete schematic of a solid-state receiver is shown in Figure 9-3. The power supply is a transformerless supply. An optional clock assembly connects to the AC line, and a tap on the clock motor winding provides power for lamp $M1$, which illuminates the clock. A luminescent panel, located behind the tuning dial, illuminates the tuning indicator and the scale. Resistor $R39$ limits the current flow through the panel, which glows when the receiver is switched on.

A line filter capacitor $C26$ is connected from the AC line to ground to remove high-frequency noise that may be present on the AC line. This capacitor also absorbs sharp voltage transients that frequently appear on the AC line and prevents these transients from reaching the rectifier diode $X1$.

Power Supply Circuitry

The power supply is a half-wave system with two B+ sources. The 88-V source powers the audio output stage, and the 7.2-V source provides power for all other stages.

Diode $X1$ is connected directly to the AC line and conducts on the positive half-cycles of the AC line voltage. Fusible resistor $R40$ is in series with $X1$ and performs three different functions. It produces a voltage drop, which lowers the B+ to 88 V at CTP 1. It limits the surge current through diode $X1$ when the

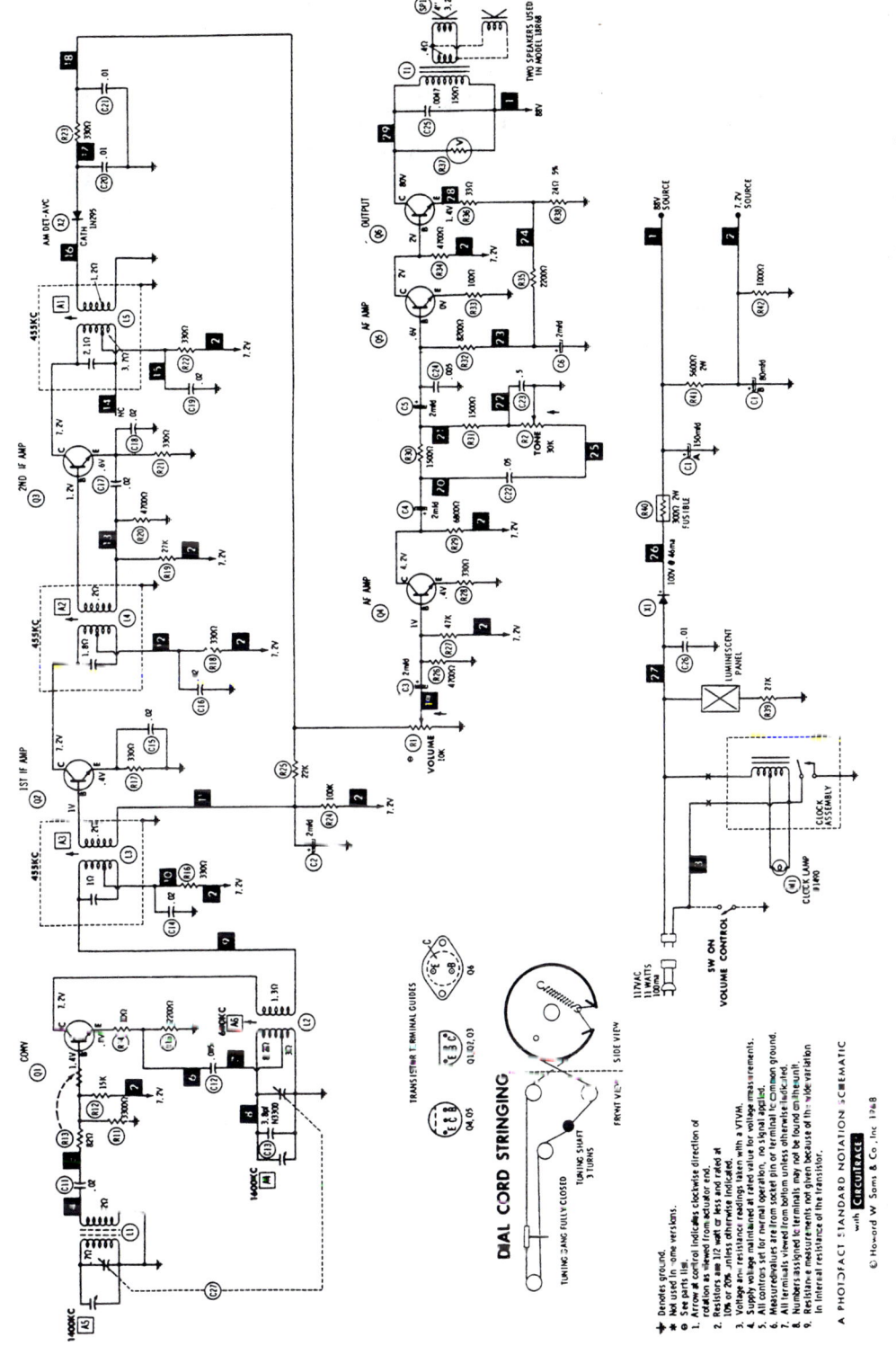

FIGURE 9-3 Typical AM receiver with line-operated power supply and direct-coupled, single-ended audio amplifier (a Photofact schematic, courtesy of Howard W. Sams & Co., Inc.)

receiver is first turned on and capacitor C1A charges. Finally, it acts as a fuse. If excessive current is demanded from the power supply because of a faulty component, R40 will heat up, open, and break the circuit.

Electrolytic capacitor C1A is the input filter capacitor and removes most of the ripple from the pulsating DC voltage coming from the half-wave rectifier. The voltage that appears on C1A forms the 88-V source designated CircuiTrace point 1 (CTP 1), which is the power supply for the audio-output stage.

Power for the other stages is obtained from the 88-V source through resistor R41, which serves as a voltage-dropping resistor to reduce the voltage at CTP 2 to 7.2 V. The electrolytic capacitor C1B provides a further reduction of 60-Hz ripple, and also provides for decoupling. Resistor R42 is a bleeder resistor that helps to stabilize the 7.2-V source.

9-3 DIRECT-COUPLED, SINGLE-ENDED AUDIO AMPLIFIER

The audio amplifier section of this receiver, shown in Figure 9-4, consists of three stages. The first audio amplifier, Q4, is an *RC*-coupled voltage amplifier, while the second and third stages, involving Q5 and Q6, form a direct-coupled power amplifier which drives the speaker. A tone control network is present in the coupling circuit between the first and second stages. Incidentally, we shall refer to variable (by user) tone control components as being tone *controls* as opposed to fixed tone-*compensation* components that are not variable and have no front panel control.

First AF Amplifier

The audio signal first appears at the detector diode X2, where the audio is produced by demodulating the 455-kHz IF signal. The audio signal enters the audio section at volume control R1 and is then coupled through C3 to the base of Q4.

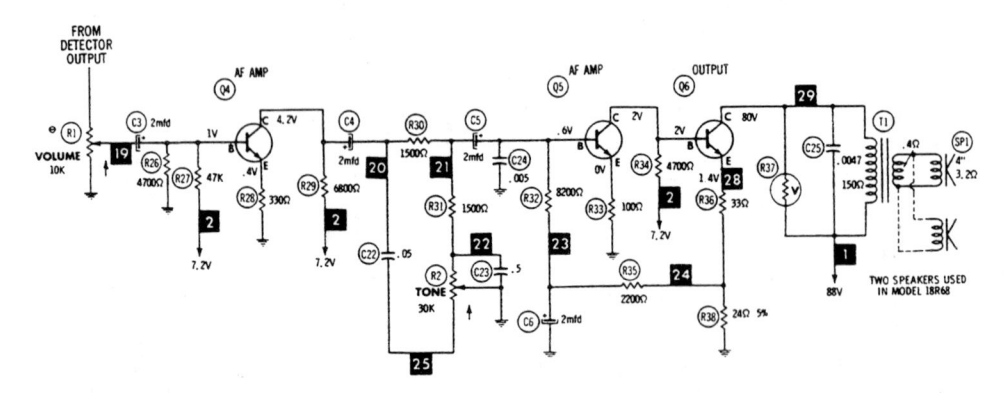

FIGURE 9-4 Three-stage audio section with tone control. The driver is direct-coupled to the single-ended output stage.

238   AM Receiver Analysis   Chap. 9

The $Q4$ stage is a universal-biased voltage amplifier described in Sec. 4-8. The high-base resistor is $R27$ and the low-base resistor is $R26$. The 7.2-V source provides both biasing and collector voltages for this stage. Emitter resistor $R28$ provides thermal stability and stabilizes the gain of the stage against variations in the transistor parameters. Resistor $R29$ is the collector load resistor across which the output voltage of the stage is developed as the collector current varies in accordance with the input signal.

Tone Compensation

The speaker in this receiver is relatively small and reproduces the high-frequency components of the audio signal (the "highs") with greater efficiency than it reproduces the "lows." To avoid an unnatural, "tinny" sound, it is necessary to attenuate the highs somewhat to obtain a proper and pleasing balance between the lows and the highs.

Fixed tone compensation is provided by capacitors $C24$ and $C25$. Capacitor $C24$ shunts a portion of the highs from the base of $Q5$ to ground. Capacitor $C25$ is connected across the audio output transformer $T1$ and allows high-frequency components to bypass the output transformer.

Tone Control

Figure 9-5 shows the tone control circuitry included between the collector of $Q4$ and the base of $Q5$. Capacitors $C4$ and $C5$ isolate the tone control circuitry from the DC voltages present at $Q4$ and $Q5$, respectively. Resistor $R30$ effectively divides the tone control circuit into two parts. The first part consists of capacitor $C22$ and resistance $R2A$. The second part is composed of $R31$, $C23$, and $R2B$.

All audio signals appearing at point A are attenuated somewhat by the shunting effect of $C22$ and $R2A$. Since $C22$ exhibits a smaller reactance at higher frequencies,

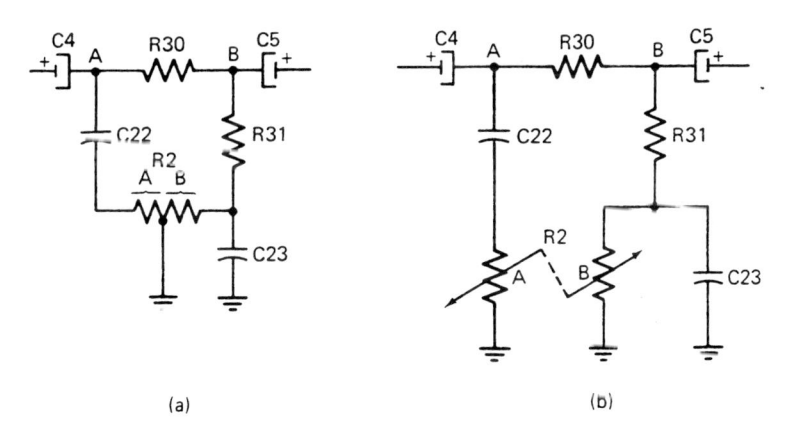

FIGURE 9-5 Tone control circuit drawn to illustrate operation.

more highs than lows are shunted to ground, and therefore the highs are attenuated more than the lows. The impedance seen by the highs looking from point A to ground will be primarily determined by R2A since the reactance of C22 becomes small in comparison at high frequencies. Therefore, R24 controls high-frequency attenuation and thereby functions as a tone control.

The second part of the tone control circuit provides a path from point B to ground through resistor R31 in series with the parallel combination of C23 and R2B. Capacitor C23 passes highs to ground more easily than lows, and the net effect of R21 and C23 is to attenuate the highs.

Now consider that R2B provides a shunt for C23. As R2B is reduced in value, C23 is shunted, and lows as well as highs pass to ground more easily. The net effect is twofold. As R2B decreases, frequency discrimination is diminished and lows and highs are attenuated an equal amount. Second, as R2B decreases, an appreciable attenuation of all audio frequencies occurs, because of the net reduction of the impedance between point B and ground. This is desirable for the following reason.

The volume should not change as the tone control is varied. Only the tone should change. As high-frequency attenuation is reduced by increasing resistance R2A, the overall volume level would increase, except as R2A increases, R2B decreases, and a reduction of R2B serves to reduce the overall volume level as described above. Taken together, the two parts of the tone control augment each other to form an effective tone control that does not influence the volume.

Direct-Coupled Power Amplifier

Transistors Q5 and Q6 form a direct-coupled amplifier that incorporates a feedback loop to provide DC stability. The Q5 stage is a base bias with emitter feedback arrangement in which the positive voltage for the Q5 base is obtained from the Q6 emitter circuit via base resistors R32 and R35. DC feedback is obtained by connecting the base resistor of Q5 to the emitter of Q6, as we shall see.

Direct-coupled amplifiers are especially sensitive to thermally induced "Q-point stray," since any undesirable shift in the operating point of the first transistor will be amplified by the second and subsequent transistors. Unless measures are taken to provide the circuit with a certain "natural stability," amplified deviations may result in distortion, loss of volume, or even component failure.

The Q-point power dissipation of Q6 in Figure 9-4 is almost 2 W. The transistor heats up during operation, and a rise in temperature almost always produces a corresponding increase in the DC beta of the transistor. This causes an increase in collector current, which, in turn, increases the power dissipation and makes the problem even worse. Fortunately, it is neither difficult nor expensive to prevent this *thermal runaway*.

As described in Sec. 4-4, any resistance in the emitter lead tends to stabilize the collector current. But a far greater stability is achieved by connecting R32 and R35, the base resistors of Q5, to the voltage divider in the emitter lead of Q6 rather than to the positive supply. If the Q6 collector current increases, the emitter voltage will rise and increase the positive voltage appearing at CTP 24. This increase will couple through $R32$ and $R35$ to the base of $Q5$ and increase the collector current of $Q5$.

Going further, the larger collector current of $Q5$ will cause the voltage at the collector of $Q5$ to decrease due to the IR drop across the collector load resistor. The collector of $Q5$ is directly coupled to the base of $Q6$, and as the collector voltage of $Q5$ decreases, the forward bias at the base of $Q6$ decreases. This reduces the collector current in $Q6$, counteracting the initial increase.

AF Signal Feedback

Since the emitter resistances of $Q5$ and $Q6$ are unbypassed, they produce negative signal feedback as well as DC feedback. This reduces the gain of the amplifier but provides stability of the gain against variations of the transistor parameters.

The negative feedback path of $R32$ and $R35$ does not affect the gain of the amplifier because of capacitor $C6$. This capacitor shunts to ground any audio signal that might reach CTP 23. By breaking the $Q5$ base resistance into two parts ($R32$ and $R35$) and inserting $C6$, AF signal isolation is achieved while still providing DC feedback for stabilizing the DC voltage levels.

Output Circuit

Only capacitor $C25$ and resistor $R37$ deserve comment in the collector circuit of $Q6$. Capacitor $C25$ has been mentioned as a tone-compensation capacitor, but it serves another purpose along with *voltage-dependent resistor $R37$*.

To understand the purpose of these components, consider what happens to the collector voltage of $Q6$ as the collector current is cut off very sharply due to a very strong and rapidly changing input signal or a noise "spike" that reaches the base of $Q6$. As the current flowing in the primary of $T1$ decreases very rapidly, the collapsing magnetic field surrounding $T1$ induces a voltage in the primary that attempts to maintain the current at its previous level. This is the property of an inductor, and the self-induced voltage across the primary of $T1$ is of such polarity that it adds to the 88-V source voltage at CTP 1. The result is that a very high voltage can be generated at the collector of $Q6$ that might exceed the breakdown voltage of the transistor, resulting in its destruction.

The operating characteristic of a voltage-dependent resistor ($R37$) is that its resistance decreases sharply when the voltage across it reaches a certain level. In the presence of the voltage spikes across the transformer primary, $R37$ conducts heavily and attenuates the voltage surge.

The voltage peaks described above are very short in duration. Since short pulses act much like high frequencies, capacitor $C25$ also absorbs or significantly attenuates the pulse. Thus, the $Q6$ collector voltage is prevented from reaching levels that might destroy the transistor.

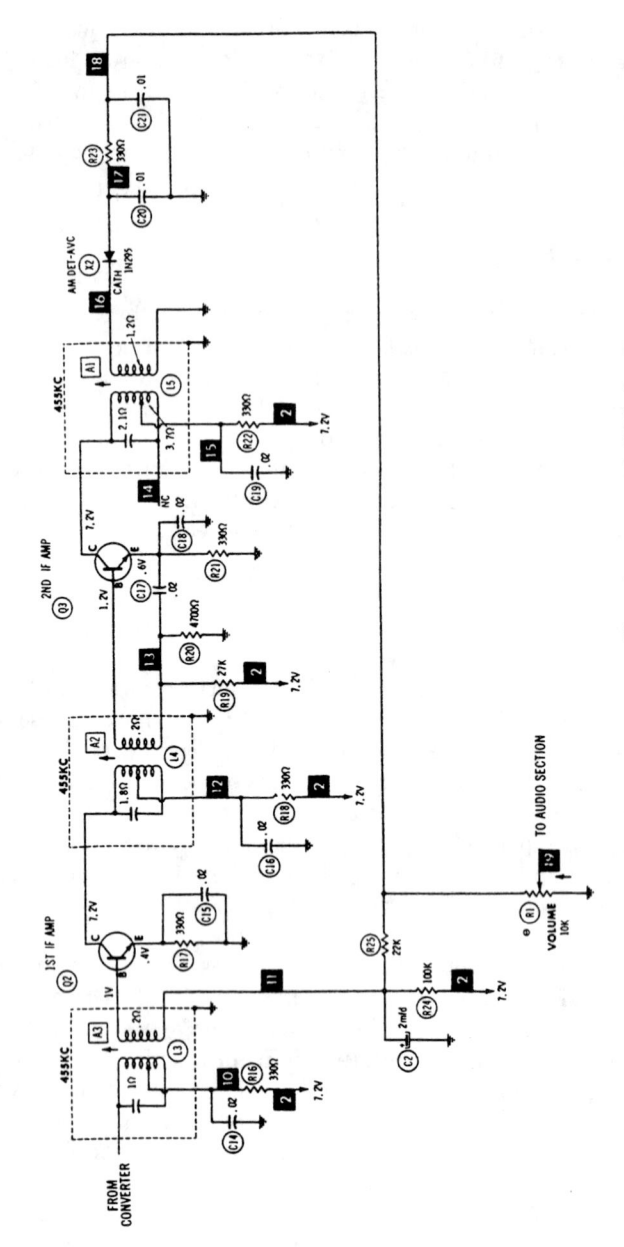

FIGURE 9-6 Two-stage IF amplifier, detector, and AGC circuitry.

9-4 IF AMPLIFIER, DETECTOR, AVC

The IF amplifier, shown in Figure 9-6, is a two-stage amplifier, typical of IF amplifiers found in solid-state receivers. The first stage is gain-controlled by the AVC circuit. The output of the IF amplifier is applied to the detector diode $X2$, which recovers the audio signal and produces the AVC voltage. Since the bias for the first IF stage is derived from the AVC line, we now describe the detector and see how the AVC voltage is developed.

Detector and AVC

The detector and AVC circuit is redrawn in Figure 9-7 to show the similarity to a half-wave rectifier. The filter is obvious ($C20$, $R23$, $C21$), and the volume control resistor forms the detector load. $R25$ samples the voltage at the volume control and applies it to AVC filter capacitor $C2$, which averages the audio voltage to a slowly varying voltage proportional to the amplitude of the incoming IF signal. Pull-up resistor $R24$, connected to the 7.2-V source, pulls the AVC line to a low positive potential on the order of 1 V. Thus, the AVC voltage, being slightly positive, can be used as the bias voltage for first IF amplifier, $Q2$.

The detector output is negative-going. When a strong IF signal is present, the input to $R25$ goes more negative, reducing the positive voltage of the AVC line. This, in turn, reduces the bias at $Q2$ and thereby reduces the gain of the first IF amplifier.

First IF Amplifier

The first IF amplifier (Figure 9-6) is a typical transformer-coupled tuned amplifier. The biasing arrangement is similar to a universal bias. The high-base resistor is $R24$ and the low base resistor is $R25$. The connection of $R25$ to the AVC line rather

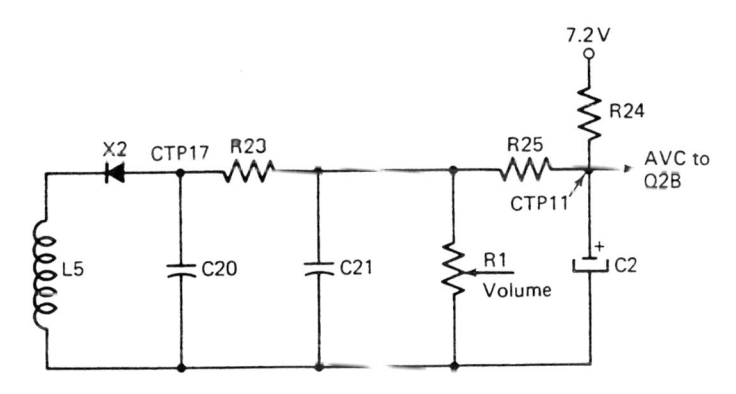

FIGURE 9-7 Detector and AVC circuit drawn to show similarity to half-wave power supply.

than to ground represents the departure from a pure universal-bias arrangement. Emitter resistor $R17$ is bypassed by $C15$, and the collector load is formed by the tuned primary of $L4$.

The first IF transformer, $L3$, couples the signal from the converter to the base of $Q2$. The tuned circuit selects the 455-kHz difference frequency at the converter output. Note that only the primary of $L3$ is tuned. Tuning of the secondary is impractical because the low impedance of the base circuit "wrecks the Q" of the tuned circuit.

The output voltage is developed across the tuned primary of $L4$ and is magnetically coupled to the $L4$ secondary. The B+ supply voltage appearing at CTP 12 is applied to a tap on the $L4$ primary winding to match the load impedance to the collector of $Q2$. Resistor $R18$ and capacitor $C16$ provide decoupling to prevent the IF signal from reaching the 7.2-V source.

Second IF Amplifier

The second IF amplifier (Figure 9-6) is typical of a transformer-coupled tuned amplifier that has no AVC provisions. The biasing arrangement is pure universal, with the emitter resistance bypassed. Capacitor $C17$ provides the return path to ground for IF signal voltages in the base-emitter circuit of $Q3$. Recall that the $Q3$ emitter is held at signal ground by $C18$.

The collector circuit is similar to the circuit of the first IF amplifier. The B+ voltage is applied to a tap on the $L5$ primary, and a decoupling network ($C19$, $R22$) keeps the IF signal out of the power supply.

In passing, note that the three IF transformers are shielded with the shields grounded. They are tuned by a ferrite slug, and even though the circuit within each is the same, they are not identical, as indicated by the DC coil resistances on the schematic.

Converter Stage

The converter of this receiver, shown in Figure 9-8, is almost identical to a converter described in Sec. 7-5. A universal arrangement biases $Q1$, and the output circuit features two windings in series, tuned to different frequencies. The converter stage mixes the antenna signal, applied to the base, and the oscillator signal, injected at the emitter. Tuning of the antenna and oscillator is accomplished with a dual-gang tuning capacitor that includes trimmer capacitors $A4$ and $A5$ in parallel with the main tuning gangs.

The oscillator tank features a temperature-compensation capacitor $C13$ rated 3.8 pF and N3300. This capacitance decreases (as denoted by the N) 3300 parts per million for each degree Celsius rise in temperature. A 1°C change in temperature reduces the capacitances by 0.013 pF. The decreasing capacitance offsets increasing capacitances in other parts of the circuit.

A "step-down" loopstick antenna applies the antenna signal via $C11$ and $R13$ to the base of $Q1$. The function of $R13$ is hard to discern from the schematic alone,

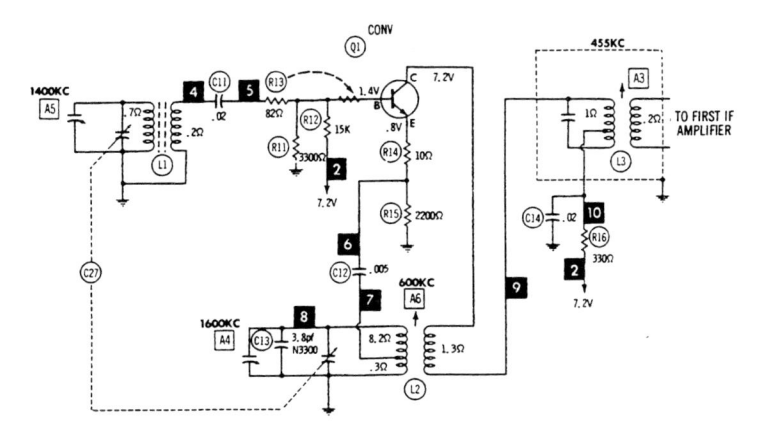

FIGURE 9-8 Typical converter.

but it is probably included to reduce any tendency to oscillate at the antenna frequency.

9-5 VARIATION: BRIDGE RECTIFIER; PUSH-PULL AUDIO AMPLIFIER

The schematic of the next receiver to be considered is shown in Figure 9-9. The converter, IF amplifier, detector, and AVC stages are almost identical to those of the previous receiver. Only the power supply and the audio section are appreciably different.

An examination of the base-emitter voltages of the transistors reveals that all are approximately 0.3 or less. This indicates that the transistors are germanium, and the symbols and negative collectors indicate the transistors to be *PNP*.

Power Supply

The AC line of this receiver connects only to the primary of the power transformer *T*1, to the on-off switch, and to the clock when the clock is included. The AC line does not connect to the chassis, and hence the chassis is not "hot." Power transformer *T*1 is a step-down transformer that steps the line voltage down to 11.2 V for application to the bridge rectifier.

The bridge rectifier consists of four diodes connected in the bridge configuration and encapsulated into a single unit, *X*1. Only four leads or terminals are in evidence, two of which are for the AC input while the other two are the positive and negative output. The positive terminal of the bridge is grounded, and the negative terminal constitutes the output. The power supply output is negative so as to be compatible with the *PNP* transistors.

A three-section filter provides three different source voltages, designated as CTP 1, CTP 2, and CTP 3. Electrolytic capacitors *C*1, *C*2, and *C*3 provide filtering

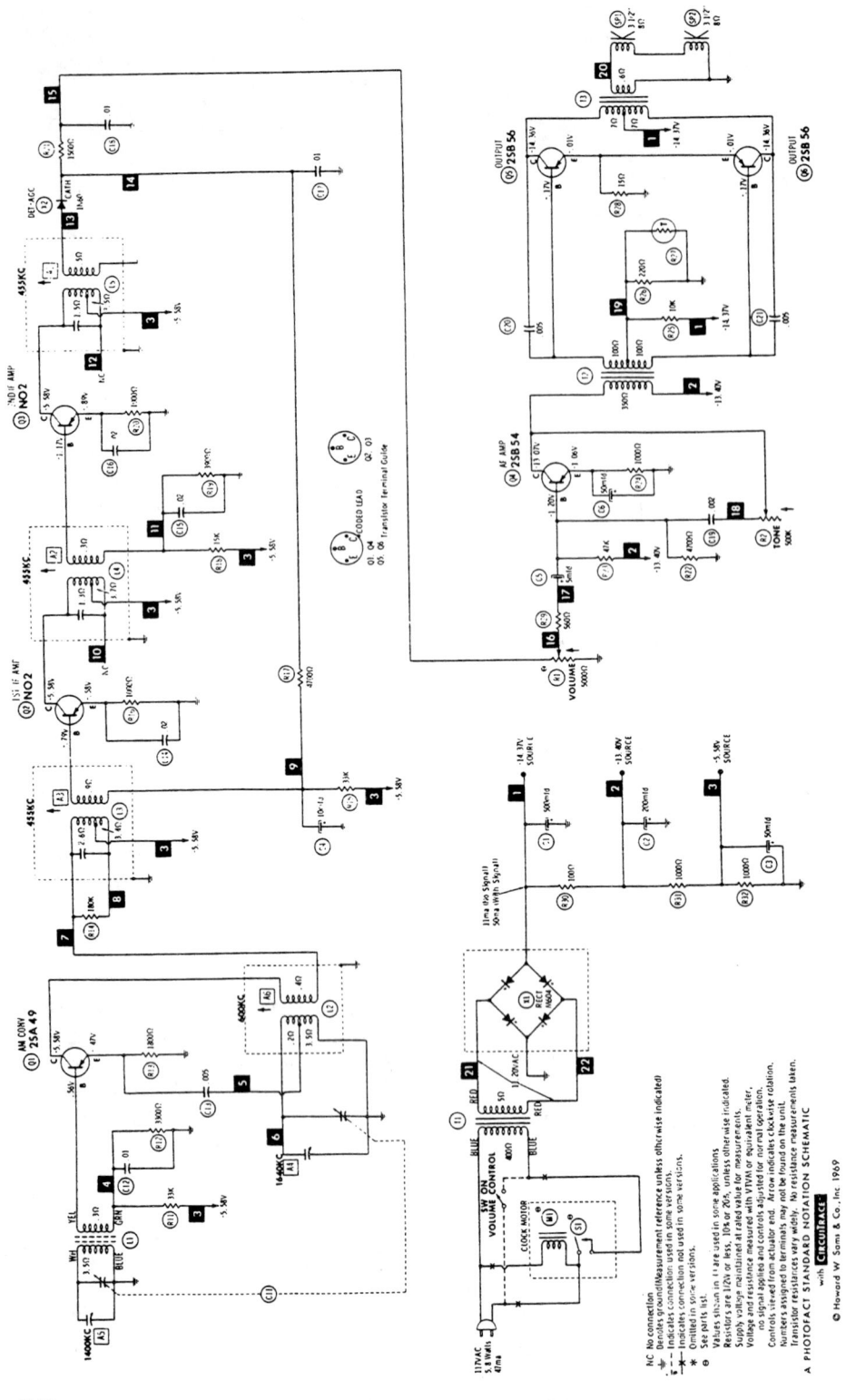

FIGURE 9-9 Typical AM receiver with bridge power supply and push-pull output stage (a Photofact schematic, courtesy of Howard W. Sams & Co., Inc.).

and decoupling. Observe the polarity of the filter capacitors and note that the positive plates of the capacitors connect to ground. Resistors $R30$ and $R31$ are filter resistors that separate filter capacitors $C1$, and $C2$, and $C3$. $R32$ is a bleeder resistor that stabilizes the voltages appearing at CTP 2 and CTP 3.

Audio Section

Transistor $Q4$ is the first audio amplifier and driver for the audio-output stage. With the exception of the tone network, $R2$ and $C19$, it is a standard universal-biased amplifier with transformer coupled output. $R23$ is the high-base resistor; the low-base resistor is $R22$. Emitter resistor $R24$ is bypassed by $C6$ to prevent degeneration of the audio signal. Capacitor $C5$ prevents the bias voltage at the base of $Q4$ from leaking to ground through the volume control resistor $R1$.

Tone Control

The tone control ($R2$ and $C19$) functions by controlling the amount of high-frequency feedback from the collector to the base of $Q4$. The signals appearing at the collector and the base are 180° out of phase, and a signal fed back from the collector to the base will produce negative feedback and a reduction in gain. More highs are fed back than lows, because of the reactance of $C19$, and this results in an enhancement of the low frequencies. Tone control resistor $R2$ controls the amount of high-frequency feedback and, therefore, controls the high-frequency gain of $Q4$.

Audio Output Stage

The audio-output stage is a standard transformer-coupled push-pull power amplifier, as described in Sec. 5-8. Thermal stability is achieved by including a thermistor $R27$ in parallel with $R26$, located physically close to output transistors $Q5$ and $Q6$. As $Q5$ and $Q6$ heat up during operation, $R27$ also warms up, and as the temperature of $R27$ rises, its resistance goes down. This reduces the bias voltage applied to the output transistors and prevents the rise in collector current that normally occurs as the output transistors heat up. Capacitors $C20$ and $C21$ are tone-compensation capacitors that provide negative feedback to the bases of $Q5$ and $Q6$, thereby reducing the high-frequency gain of the amplifier.

9-6 VARIATION: TRANSFORMER POWER SUPPLY; OTL PUSH-PULL AUDIO AMPLIFIER

The receiver shown in Figure 9-10 uses a transformer input to a half-wave power supply and an audio output stage that is "output transformer-less" (OTL). An interesting feature of the power supply is that the winding for the clock motor is

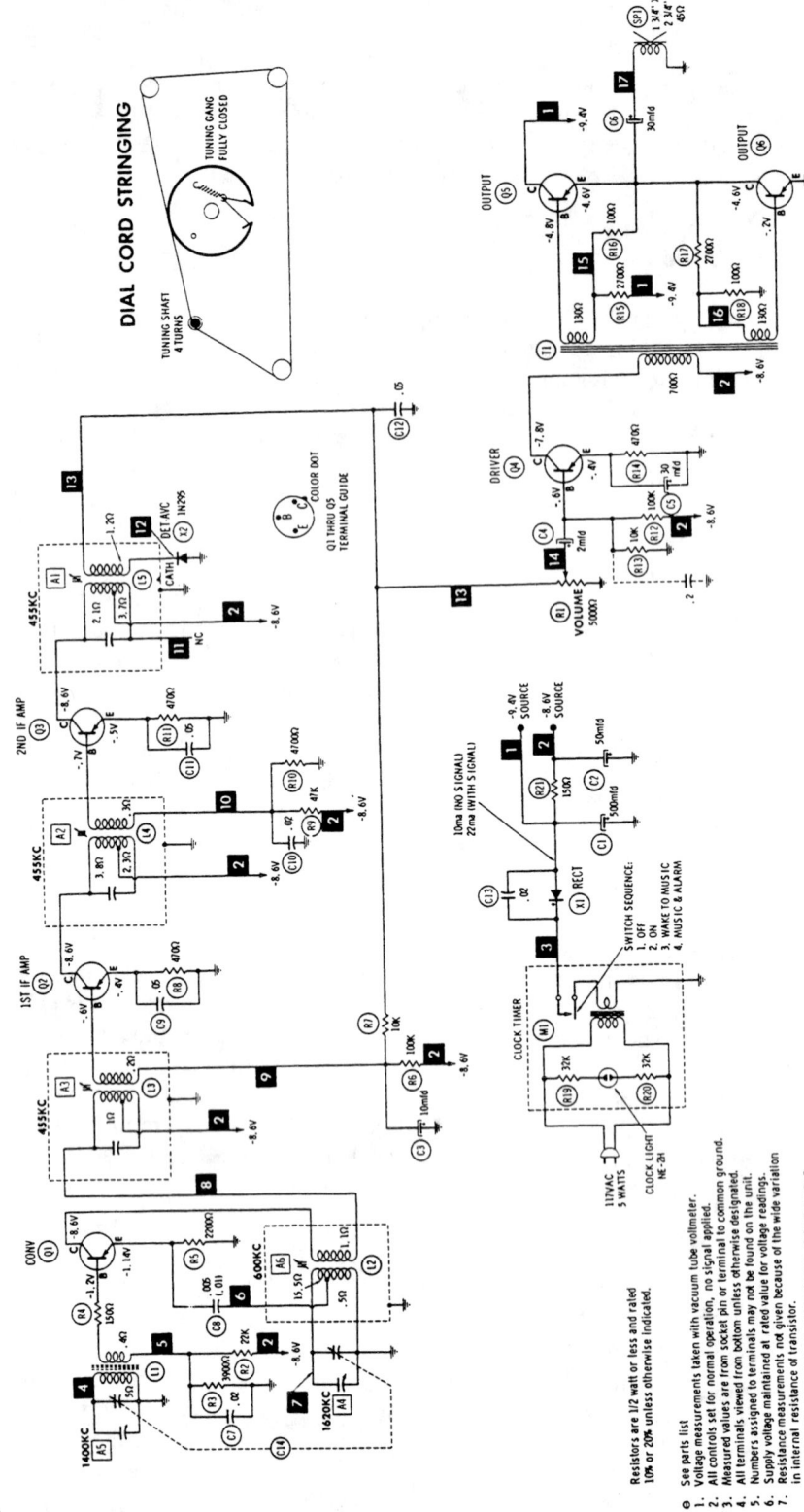

FIGURE 9-10 AM receiver with half-wave power supply and OTL audio output stage (a Photofact schematic, courtesy of Howard W. Sams & Co., Inc.).

used as the primary of a transformer. An extra winding on the magnetic core of the clock serves as a secondary and provides low-voltage AC for the receiver. The audio output transformer is replaced by a capacitor which couples the audio output to the speaker.

This receiver employs circuit configurations for the converter, IF amplifier, AGC, and driver stages discussed previously, and the reader should be at ease with these circuits. In this receiver the detector is relocated slightly, but the operation is the same as for the previous circuits.

Power Supply

The AC input circuit contains a neon lamp, an input transformer that is part of the clock, and the receiver on-off switch connected in series with the high side of the clock-transformer secondary. The neon lamp NE-2H in series with two current-limiting resistors $R19$ and $R20$ is placed across the AC line to provide illumination for the dial of the clock. The AC line is connected across the clock motor winding so that the clock runs continuously.

The on-off switch completes the circuit between the low-voltage secondary and rectifier diode $X1$. When the switch is open, the extra winding has no effect on the operation of the clock. When closed, the switch completes the circuit to $X1$, and the receiver comes on.

The B− circuitry (the power supply output is negative) is composed of rectifier diode $X1$ with transient protection capacitor $C13$, input filter capacitor $C1$, filter resistor $R21$, and filtering-decoupling capacitor $C2$. Diode $X1$ is connected so that it conducts on negative portions of the AC input signal, producing a negative source voltage. Capacitor $C13$ absorbs sharp voltage spikes that sometimes appear on the AC line that otherwise might destroy the rectifier diode. The filter network is standard.

Detector

The detector functions in the same manner as before, but the diode has been moved to the low side of the last IF transformer ($L5$) secondary, as shown redrawn in Figure 9-11. The diode conducts when the voltage across the secondary of $L5$ is such that CTP 12 is negative. Electron flow in the detector circuit is from CTP 12 through $X2$ to ground.

9-7 VARIATION: A COMPLICATED POWER SUPPLY

The next receiver to be examined is shown in Figure 9-12. With the exception of the power supply, the circuitry is quite similar to that of receivers described earlier. Note, however, that the biasing arrangement used for the converter, IF amplifiers, and first AF amplifier is the emitter bias with two supplies.

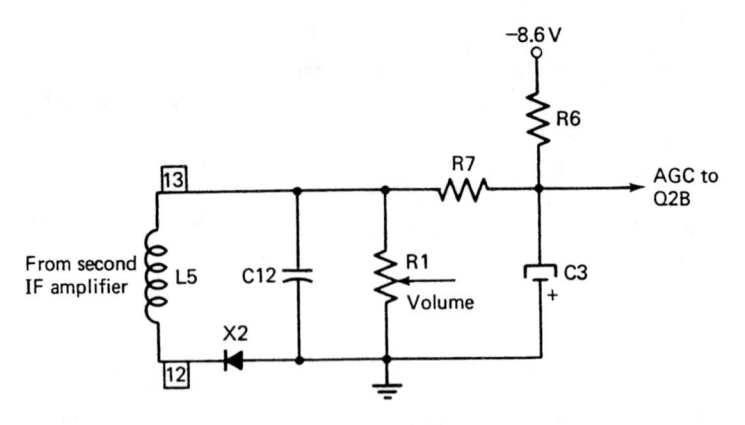

FIGURE 9-11 Detector and AGC circuit redrawn to show similarity to half-wave power supply.

To simplify the power supply, it is redrawn in Figure 9-13 with only the essential components shown. The dial lamp, clock assembly, and all filter capacitors have been omitted for the sake of clarity. The secondary side has been drawn upside down from the way it appears on the schematic in order to have the circuit appear more nearly standard. Our objective is to identify the paths of electron flow.

Rectifier $X1$ conducts only when CTP 18 is positive relative to CTP 1. The electrons flow from the CTP 1 end of the secondary through $R18$, $R19$, $R17$, $R16$, and $X1$ to CTP 18 as indicated in the figure. The lines to the right of the source points indicate electron flow through the various stages of the receiver circuitry.

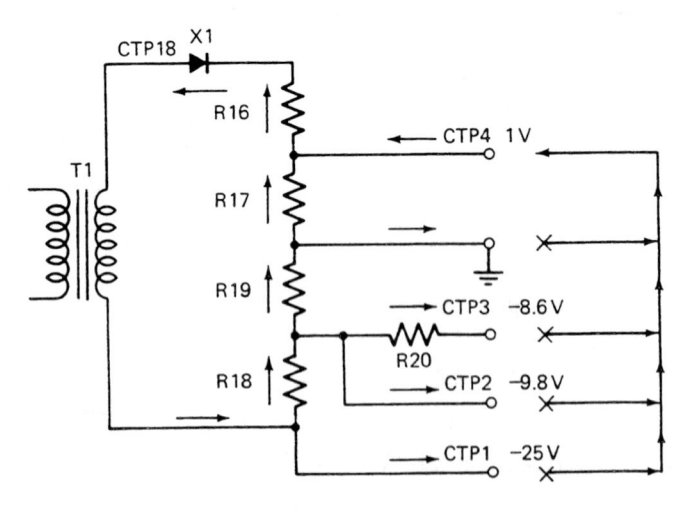

FIGURE 9-12 Power supply redrawn to show paths of electron flow. The filtering-decoupling capacitors and dial lamp have been omitted. Arrows indicate the direction of electron flow. The lines to the right of the source points indicate electron flow through the various stages of the receiver.

AM Receiver Analysis   Chap. 9

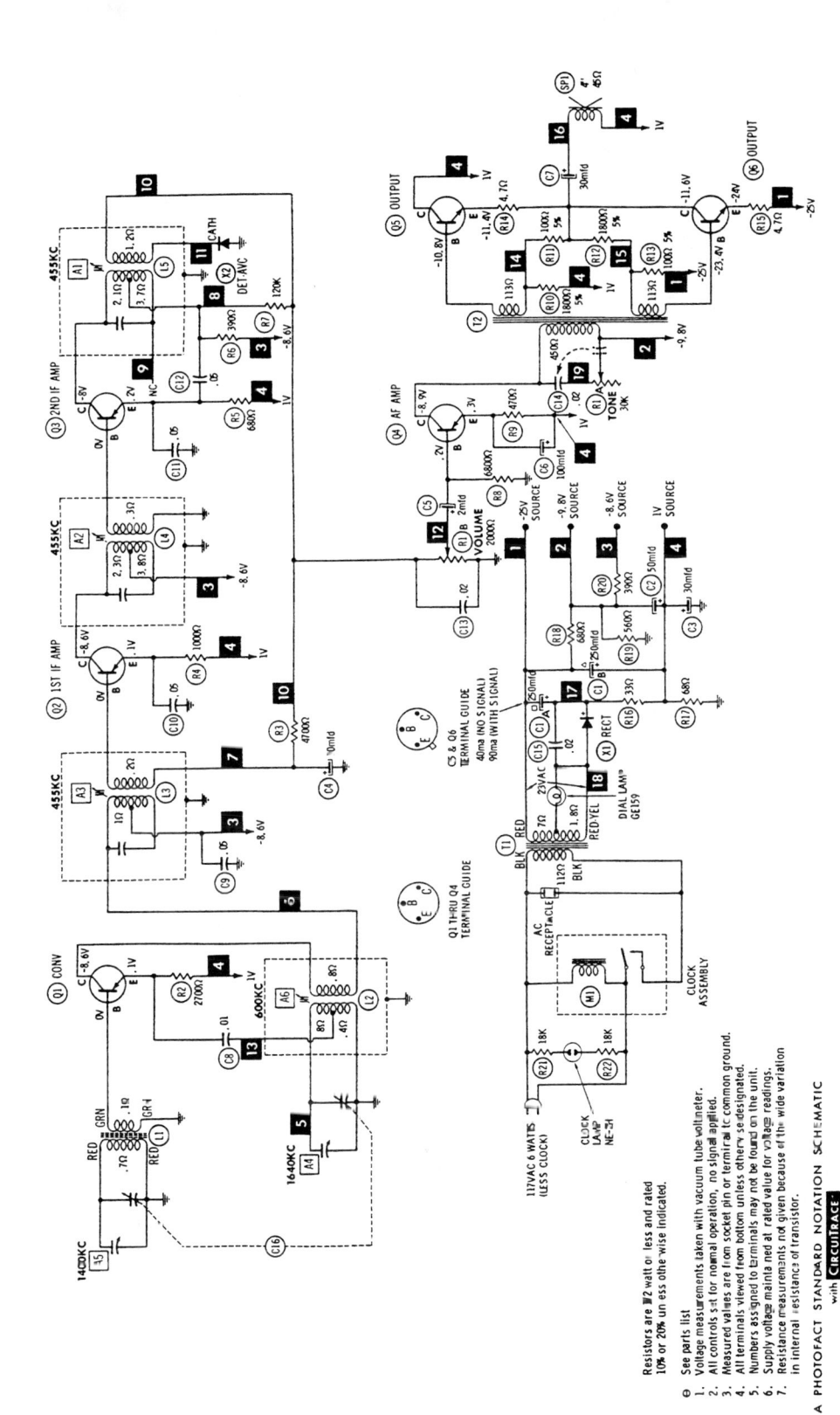

FIGURE 9-13 AM receiver with an unusual power supply (a Photofact schematic, courtesy of Howard W. Sams & Co., Inc.).

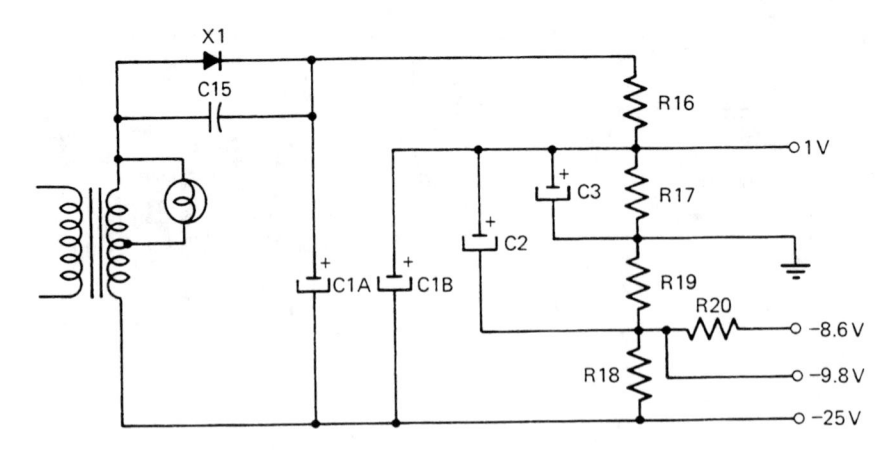

FIGURE 9-14 Power supply redrawn with filtering-decoupling capacitors and dial lamp included. Note that all decoupling capacitors return to the +1-V source.

The circuit is now recognizable as a half-wave rectifier circuit with a voltage divider attached to the output. Note that a central point of the divider is selected to be ground, and this causes points on opposite sides of the ground point to have voltages of opposite sign. Any point on the voltage divider could be ground, but by selecting this particular point, the 1-V positive source is obtained along with the negative sources.

When the filtering and decoupling capacitors are added, along with the dial lamp and transient suppression capacitor $C15$, the circuit will appear as in Figure 9-14. Capacitors $C1A$ and $C1B$ are the filter capacitors for smoothing the 60-Hz ripple, whereas $C2$ and $C3$ are mainly for decoupling. Note that $C2$ and $C3$ do not connect to ground, and note that the −8.6-V source has no decoupling capacitor shown on the power supply schematic.

The 1-V source is connected to the various circuits of the receiver where a ground is typically encountered, and in fact, the 1-V source acts as a "signal ground." Decoupling capacitors $C1B$, $C2$, and $C3$ are all returned to the 1-V source. The −8.6-V source is used only for the converter and IF amplifiers, and the decoupling filter for this source is $C9$, located on the primary side of $L3$ in the converter output circuit.

9-8 VARIATION: QUASI-COMPLEMENTARY AF AMPLIFIER WITH DARLINGTON PAIRS

The receiver of Figure 9-15 has two noteworthy stages, the power supply and the audio amplifier. The power supply is very simple. The entire audio amplifier is direct-coupled and is a bit complex.

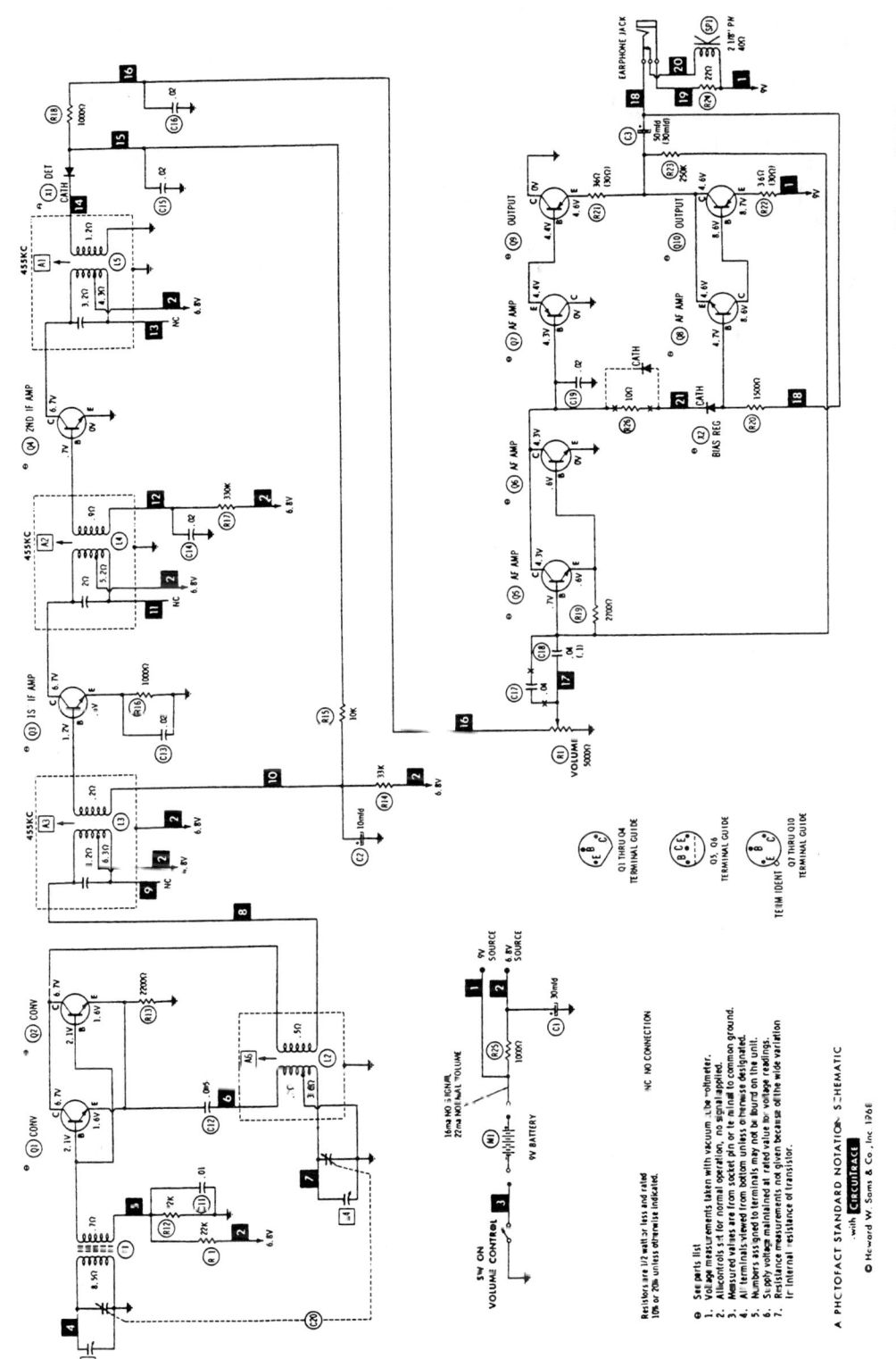

FIGURE 9-15 AM receiver with a battery power supply and a direct-coupled audio amplifier (a Photofact schematic, courtesy of Howard W. Sams & Co., Inc.).

253

Power Supply

The power supply consists of only four components. A 9-V "transistor battery" supplies the power when the switch is closed. The 9-V source CTP 1 is taken directly from the positive terminal of the battery. Resistor $R25$ drops the voltage from 9 V to 6.8 V, the source designated as CTP 2.

The circuit provides an example in which decoupling capacitors are required even when there is no requirement for filtering out ripple. The battery produces perfect DC, and ripple is not a problem. Resistor $R25$ and capacitor $C1$ form a decoupling network that serves to decouple the sources CTP 1 and CTP 2.

Audio Amplifier

The audio section consists of a Darlington first audio amplifier which is direct coupled to a "quasi-complementary" audio output stage (Figure 9-15). The output stage is OTL and operates into a 40-Ω speaker. Thermal stability is provided by a DC feedback loop from the midpoint of the series connected outputs to the base of input transistor $Q5$. A second feedback loop from the amplifier output to the power amplifier input provides a small amount of positive feedback. This prevents excessive negative AC feedback from seriously diminishing the amplifier gain.

A Darlington pair is composed of two direct-coupled transistors connected so that the combination acts much as a single transistor of very high beta, as described in Sec. 5-2. The high beta gives high input impedances and large power gains.

Darlington pair $Q5$-$Q6$ is the first audio amplifier and driver for the audio power amplifier. The biasing scheme for the $Q5$-$Q6$ combination is that of simple base bias, where the base resistor $R23$ connects to the output midpoint. This connection provides both AC and DC negative feedback from the amplifier output to the input. Resistor $R19$ reduces the current flow through silicon transistor $Q5$ to a very low level, as indicated by the small base-to-emitter voltage. The effect of this is to maintain the current flow through $Q6$ at a moderate level also.

Figure 9-16 shows a simplification of the collector circuit of $Q5$-$Q6$. The collector load resistor is $R20$, and CTP 18 is the positive supply voltage. The collector of $Q5$-$Q6$ is direct-coupled to the bases of $Q7$ and $Q8$, and the $Q5$-$Q6$ collector voltage constitutes the bias voltage for $Q7$ and $Q8$.

Diode $X2$ and resistor $R26$ (or a second diode in some versions) produce a small voltage difference between the bases of $Q7$ and $Q8$. This provides a small forward bias to these transistors, which brings them just into the conduction region. This reduces crossover distortion, and $X2$ and $R26$ form the crossover network.

The positive voltage for the collector circuit of $Q5$ and $Q6$ is obtained from CTP 18, and this point derives its positive voltage from the battery by way of the speaker winding. Note that the speaker connects to the positive terminal of the battery. The voltage at CTP 18 is a combination of the battery voltage and the output signal voltage developed across the speaker. Therefore, the amplifier output couples through $R20$ and the crossover network to the bases of $Q7$ and $Q8$. This constitutes a feedback loop for AC signals, and the feedback is positive. The purpose of this loop is described below.

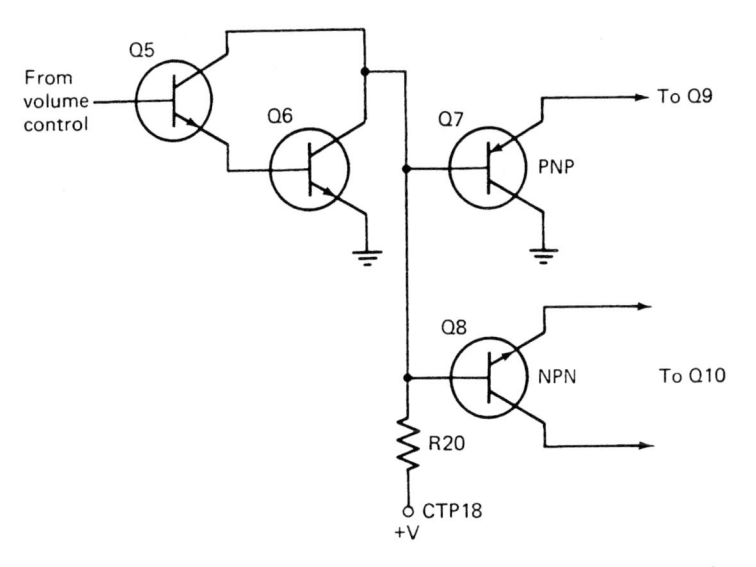

FIGURE 9-16 Simplified collector circuit of Q5-Q6. The crossover network has been omitted. Transistors Q7 and Q8 form a complementary pair.

PNP transistors Q7 and Q9 are connected in the standard Darlington configuration as demonstrated in Figure 9-17. A positive-going input signal tends to turn the pair off. No phase inversion occurs from the base of Q7 to the emitter of Q9. These two transistors form the top half of the series-connected output stage.

The bottom half is composed of *NPN* transistor Q8 and *PNP* output transistor Q10. This is a Darlington pair which uses a different configuration because of the use of an *NPN-PNP* pair, but the configuration has the same beta-multiplying properties as the pair Q7-Q9. Since the input transistor Q8 is *NPN*, the pair acts as an *NPN* transistor in that a positive-going input signal causes the pair to conduct.

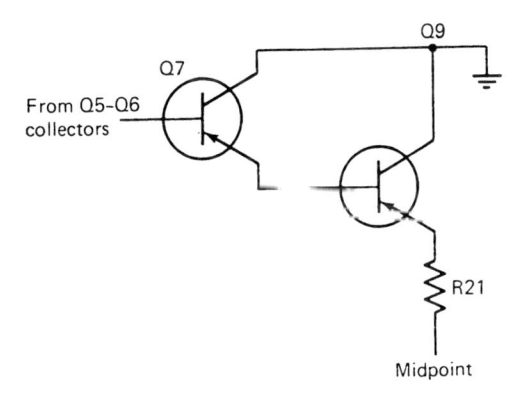

FIGURE 9-17 Darlington pair Q7-Q9 redrawn for comparison with the Darlington configuration as usually shown.

Sec. 9-8   *Variation: Quasi-Complementary AF Amplifier with Darlington Pairs*   **255**

To see the action of the output stage, assume that the *Q5-Q6* collector voltage is simultaneously positive-going, as indicated in Figure 9-18. The signal applied to the bases of *Q7* and *Q8* will also be positive-going. Since the bottom half of the series outputs has an *NPN* input transistor, the bottom pair will conduct more heavily, causing the midpoint voltage to go more positive. When the polarity of the applied signal reverses and becomes negative-going, the top half will conduct more heavily, and the mid-point voltage will fall toward zero. This rise and fall of the midpoint voltage constitutes the output of the amplifier.

An automatic phase inversion is accomplished in this circuit by transistors *Q7* and *Q8*, which control the output transistors. One is *NPN*, the other is *PNP*, and the circuit action is similar to the input of a complementary-symmetry amplifier. Transistors *Q7* and *Q8* form a complementary pair, but the output transistors are both *PNP*. Hence, this is a *quasi-complementary* amplifier.

Capacitor *C3* couples the output of the amplifier to the speaker as in previously described OTL output stages, but in this circuit the other terminal of the speaker goes to the 9-V supply rather than to ground. This is a bit disturbing until it is noted that only the battery separates CTP 1 from ground. The internal resistance of the battery is small compared to the impedance of the speaker, and therefore

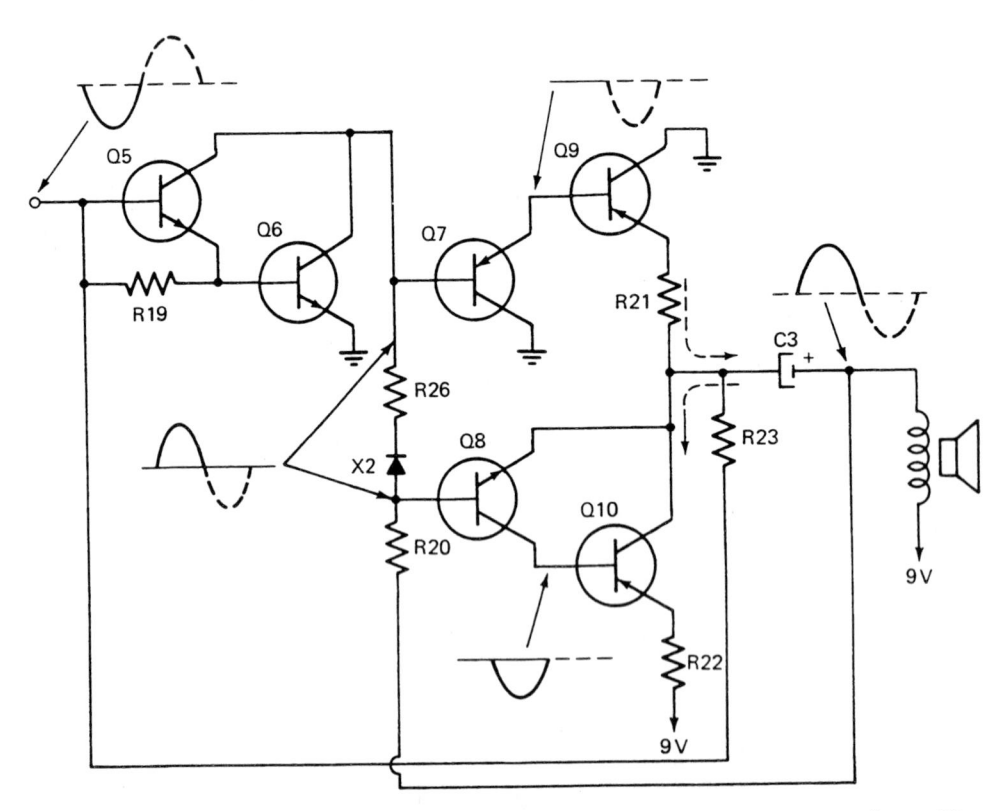

FIGURE 9-18 Signal-phase relationships and current flow for quasi-complimentary audio amplifier.

256   AM Receiver Analysis   Chap. 9

CTP 1 acts as a ground for AC signals even though the DC voltage at that point is 9 V.

This receiver has provisions for an earphone, as indicated by the earphone jack in series with the speaker. When the earphone plug is inserted into the jack, the speaker is disconnected and $R24$ is placed in series with the earphone. This reduces the output volume to a comfortable listening level.

DC Feedback Loop

For purposes of stabilizing the DC levels of the amplifier, the base bias resistor of $Q5$ is connected to the midpoint of the output transistors rather than to the positive supply. This connection provides negative DC feedback to the base of $Q5$, and any improper deviation of the output midpoint voltage will be corrected as a consequence of this feedback loop. This connection also provides a considerable negative feedback for AC signals as well as DC, and this effect reduces the gain of the amplifier. The AC feedback is partially compensated by positive feedback via CTP 18.

To understand the operation of the DC feedback loop, assume that the midpoint voltage deviates in the positive direction. This will couple through $R23$ to produce an increase in the forward bias of $Q5$, resulting in a decrease in voltage at the $Q5$-$Q6$ collector which will couple to the inputs of the complementary $Q7$ and $Q8$. A decrease in voltage at the $Q7$-$Q8$ inputs causes the midpoint voltages to go down, counteracting the initial rise. The feedback voltage opposes the original deviation and stabilizes the midpoint voltage.

AC Feedback Loop

The second feedback loop in this amplifier is an AC or signal feedback loop. A portion of the output signal developed across the speaker is fed back through $R20$ to the $Q7$ and $Q8$ inputs. These inputs are in phase with the output, and thus the feedback is positive.

This might arouse some concern because of the association of positive feedback with oscillators. While this particular loop provides a small amount of positive feedback, it serves only to cancel a part of the negative feedback present in the amplifier as a whole. The net feedback must be negative or the amplifier would indeed oscillate.

Negative feedback is produced by the loop of $R23$ to the base of $Q5$, and by resistors $R21$ and $R22$ in the emitter leads of the output transistors. The negative feedback is desirable from the point of view of DC stability, but if excessive negative feedback is present, the gain of the amplifier can be seriously reduced.

Only AC or signal feedback occurs from the speaker through $R20$ even though a DC voltage is present on the loop. Capacitor $C3$ isolated CTP 18 from the midpoint of the outputs, so that only the AC voltage appearing across the speaker constitutes the feedback signal. The fact that it rides on the 9-V DC voltage is of no consequence.

Converter, IF, Detector, and AGC

The schematic of these stages is shown in Figure 9-15. With only two exceptions, the circuits are almost identical to those found in the receivers already described. The first exception is that no emitter resistance is used for the second IF amplifier, $Q4$. The biasing arrangement is that of a simple base bias where $R17$ is the $Q4$ base resistor.

The second exception is the use of two parallel transistors, $Q1$ and $Q2$, for the converter. This is unusual, and, frankly we can only speculate as to why this is done. Connecting transistors in parallel lowers the input impedance but effectively doubles the collector current. Since the converter output is by way of magnetic coupling in $L2$ and $L3$, a larger current and variation thereof would enhance the magnetic coupling, since magnetic field strengths are related to the magnitudes of the currents producing the magnetic fields.

Even though the manufacturer's reasoning for including two parallel transistors where only one is commonly found is unknown, the operation of the circuit is the same as if only one transistor were used. Likewise, servicing procedures are the same.

9-9 A VACUUM-TUBE RECEIVER

Power Supply

The power supply of the receiver of Figure 9-19 is quite similar to the power supply described in Sec. 3-9. This receiver, however, has a clock and a timed appliance outlet present in some versions, and the location of the off-on switch depends upon whether the particular version includes the clock. When the clock is present, the switch is included as part of the clock mechanism. Otherwise, the switch is mounted on the volume control. In either case, the switch interrupts the side of the AC line that connects to the receiver chasis.

As is typical of these receivers, the tube filaments are connected in a "series filament string," which is connected directly across the AC line. The order in which the tubes are connected in the series string is important because the AC filament voltage tends to introduce hum into the signal path due to the capacitive coupling between the filaments and the cathodes. The detector/first AF tube is most sensitive to hum and is connected on the end of the filament string where the AC voltage is lowest. The converter tube is the next-most sensitive and appears next in the string, followed by the IF amplifier and the audio output tube. The rectifier is least affected and, consequently, is connected into the filament string where the AC voltage is highest.

Filter capacitor $C1A$ provides initial smoothing of the positive pulses appearing at the rectifier cathode. The partially filtered voltage is then applied to the tap on the primary of the audio output transformer $T1$. The 300-Ω section of the transformer serves its normal function, but the 12-Ω "hum-bucking winding" reduces the level of hum appearing in the speaker as a result of the poorly filtered voltage

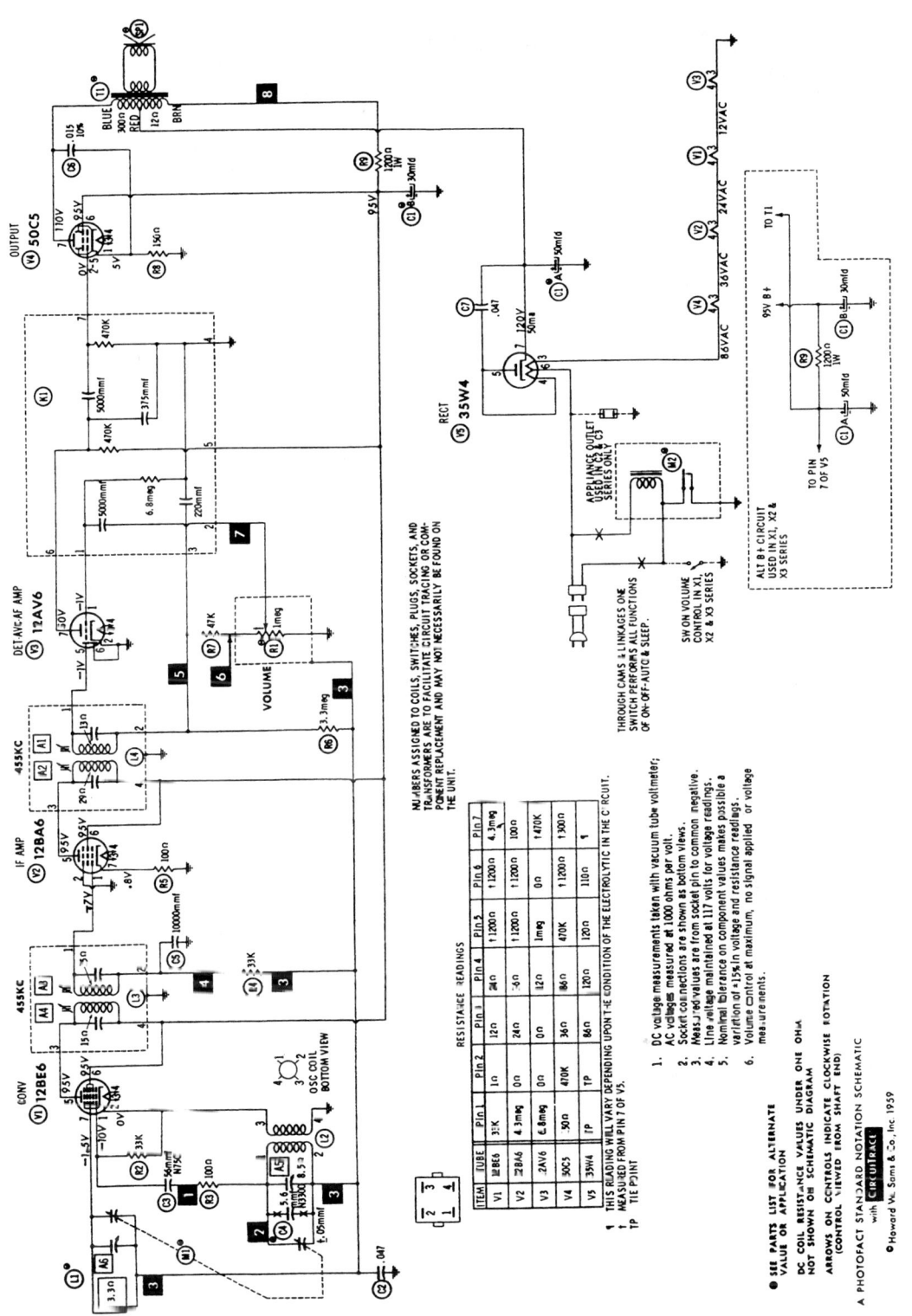

FIGURE 9-19 Typical five-tube superheterodyne receiver (a Photofact schematic, courtesy of Howard W. Sams & Co., Inc.)

at the rectifier cathode. The hum winding is actually in series with filter resistor R19, which leads to the 95-V source point at filter capacitor C1B.

Current flow in the transformer primary is in opposing directions. The audio output plate current flows through the 300-Ω top section, but the plate and screen current from the rest of the receiver flows through the 12-Ω bottom section. The direction of these currents is such that the magnetic fields tend to cancel.

Current flow through the tubes increases with increasing supply voltage during a ripple peak. This variation of plate current at the ripple frequency tends to produce 60-Hz hum in the speaker. When the hum-bucking winding is present, the increase in audio output plate current flowing through the top section of the transformer primary is partially offset by increased current flow from the rest of the receiver that flows through the bottom section of the primary. The result is reduction in ripple-induced hum.

Capacitor $C7$, connected between the plate and cathode of the rectifier, prevents the high-voltage transients that sometimes appear on the AC line from being applied directly across the rectifier tube. These very brief pulses, which may arise from nearby electric motors and other industrial machinery, may have amplitudes of several hundred volts. Because the pulses are of short duration, the filter capacitor absorbs them with little, if any, adverse effect.

Converter, IF Amplifier, and Detector/AVC

The converter and IF amplifier of this receiver are almost identical to the pentagrid converter described in Sec. 7-5 and the IF amplifier described in Sec. 6-5. Therefore, for the sake of brevity, we proceed directly to the detector stage.

The detector functions exactly as a half-wave rectifier, but its resemblance to a half-wave rectifier is obscured by the way it is drawn on most schematics and by combining the detector and first AF amplifier functions in a single tube. The detector circuit is redrawn in Figure 9-20.

The 12AV6 includes two diode plates in addition to the triode section. The same cathode serves both the triode and the diodes. Typically, one of the plates is used as the plate of the half-wave rectifier that constitutes the detector, while the other is rendered electrically inert by connecting it to the cathode. Since the cathode is connected to ground, no interaction between the detector and AF amplifier occurs as a result of using a common cathode for the two sections.

Recall that a 455-kHz amplitude-modulated waveform is developed across the secondary of $L4$ and is applied to the detector diode. On the positive half of the input waveform, the detector diode conducts and produces electron flow, as indicated by the arrow in the figure. No conduction occurs when the diode is driven negative by the negative portion of the input waveform. The input signal is rectified. Resistor $R7$ and volume control resistor $R1$ constitute the detector load, and the output voltage of the detector is developed across these resistors. The amplitude modulation of the signal applied to the detector diode causes the electron current to vary accordingly, producing the audio signal.

The 220-pF detection capacitor connected across the $R1$-$R7$ combination serves the same function as the filter capacitor in a power supply. It smooths the pulses

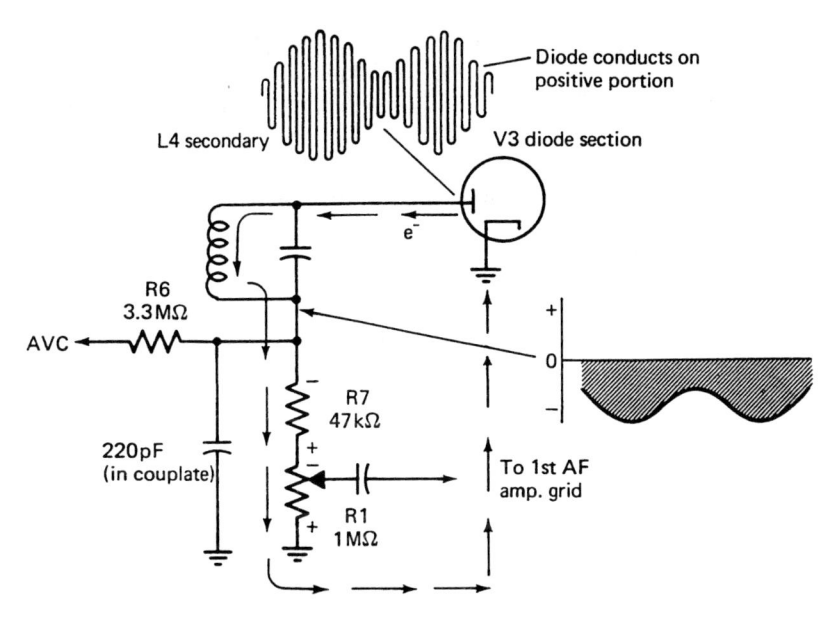

FIGURE 9-20 Detector circuit of the receiver of Figure 9-19. The arrows indicate the path of electron flow.

of the detector output to form the audio signal. This capacitor is physically located in the couplate, and at first glance it may not be apparent that it is the filter for the detector circuit. The couplate is a small encapsulated printed circuit that includes the components shown inside the dashed-line box between V3 and V4 on the schematic.

Another way to view the function of the capacitor is to consider that it shunts to ground the 455-kHz voltage component that appears at the top of R7. This clarifies at least one function of R7—to provide a resistance that reduces the 455-kHz signal voltage reaching the volume control. The 455-kHz component passes to ground through the detection capacitor rather than through the 47-kΩ resistor. The result is that very little of the 455-kHz component is passed to the audio amplifier.

AVC Source

The detector circuit is arranged so that the output is negative-going; as the input signal to the detector increases in amplitude, the detector output goes more negative. This is not significant insofar as the audio signal is concerned, but it is important to the AVC circuit.

Recall that the transconductance of the converter and IF amplifier decreases as the grid bias becomes more negative, resulting in lower gains. A large signal at the detector coming from a strong station causes the detector output to go more negative. A portion of this voltage is taken from the detector via R6, the AVC

Sec. 9-9   A Vacuum-Tube Receiver   261

filter resistor, and is applied to the AVC filter capacitor $C2$. The $R6$-$C2$ time constant is long relative to audio frequencies, so that all audio signals are removed from the AVC line. Hence, a voltage appears on the AVC line that is proportional to the average strength of the IF signal applied to the detector, and the AVC voltage goes more negative as the signal level increases. This voltage is applied to the converter and IF amplifier tubes to provide the automatic-gain-control function.

Audio Section

The first audio amplifier is shown, redrawn, in Figure 9-21. It is simple, but the fact that every resistor and capacitor is included in the couplate gives the circuit a less familiar appearance both on the schematic and in the physical circuit. Two things are worthy of note. First, grid leak bias is used since it is desirable to ground the cathode to avoid interaction between the detector and amplifier functions of the cathode. The 6.8-MΩ resistor R_{G1} is typical of the large resistances of this application. Second, any IF signal that appears on the plate of the first AF is shunted to ground through the 375-pF capacitor C_F. Also, this capacitor tends to reduce the high-frequency (treble) content of the audio signal, since a portion of the "highs" are shunted to ground.

The audio output stage is a typical single-ended amplifier using the 50C5 beam power pentode. Note the symbols used inside the tube to denote the beam-forming plates, which are internally connected to the cathode.

Capacitor $C6$ is the tone compensation capacitor, which reduces the treble content of the audio to obtain a more pleasing sound from the small speaker. Output transformer $T1$ matches the relatively high impedance of the tube to the lower impedance of the speaker. The primary of $T1$ appears a bit unusual, because of the hum-bucking winding described earlier, but its operation is straightforward.

9-10 VT RECEIVER WITH RF AMPLIFIER AND PP AUDIO AMPLIFIER

In this section we describe a receiver that utilizes an RF amplifier ahead of the converter to provide better sensitivity and noise characteristics. In the audio section a push-pull audio output stage is used to obtain improved performance. The schematic is shown in Figure 9-22.

The power supply is a half-wave supply that utilizes a selenium rectifier $M1$ as the rectifying element. Selenium rectifiers are now obsolete, having been replaced by smaller, more reliable silicon diodes, but they are frequently encountered in older receivers. Resistor $R9$ is a fusible resistor that acts both as a surge-limiting resistor and as a fuse. If the rectifier $M1$ were to short circuit, raw AC would be applied to capacitor $C1A$, and this would soon cause it to short, also. This would produce a short circuit directly across the AC line. In such a case the fusible resistor heats up and opens the circuit.

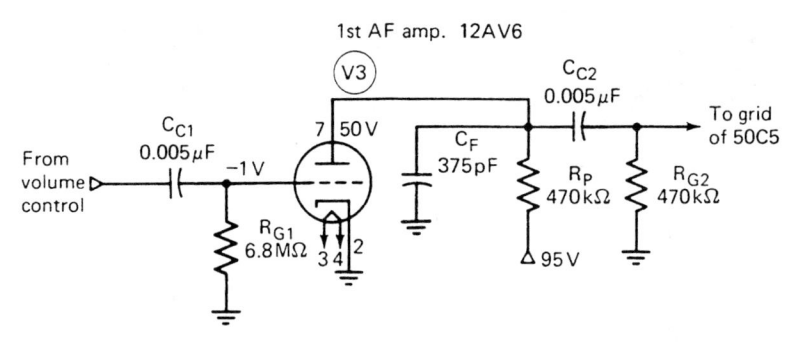

FIGURE 9-21 First audio amplifier of the receiver of Figure 9-19.

RF Amplifier, Converter, and IF Amplifier

The input signal to the RF amplifier is provided by the antenna $L1$, which is tuned by one section of the three-gang tuning capacitor $M2$. The RF amplifier tube, a pentode, is biased by AVC voltage applied to the signal grid, and, additionally, by cathode resistor $R3$. The gain of the RF amplifier is varied by the AVC circuit, as is the gain of the converter and IF amplifier.

The only unusual feature of the converter is the capacitive winding associated with the oscillator coil. What at first appears to be (and is) an open winding is actually a few turns of wire that act as a capacitor; it couples the oscillator signal to the oscillator grid of the pentagrid converter.

The IF amplifier stage is standard, but an IF filter is built into the second IF transformer can. This filter serves the same purpose as the detection capacitor but offers improved performance.

Detector—AVC

The operation of this circuit is quite similar to the detector of the previous receiver even though it looks more complicated. Electron flow is from the detector diode (pin 5 of $V4$) down through the $L5$ secondary, through the 50-kΩ IF filter resistor, then through $R7$ to ground. The audio signal is developed at the top of $R7$ and is coupled to the volume control circuit by $C8$.

The AVC voltage is also derived from the top of $R7$ at CTP 5. Resistor $R5$ and capacitor $C3$ form the AVC filter so that the voltage at CTP 2 does not vary at audio frequencies.

Tone Compensation at the Volume Control

Tone compensation at the volume control is desirable for two reasons. One is that the human ear, at low volume levels, responds less to low audio frequencies (lows) than to higher frequencies (highs). Second, the small speakers of table-model

Sec. 9-10 VT Receiver with RF Amplifier and PP Audio Amplifier 263

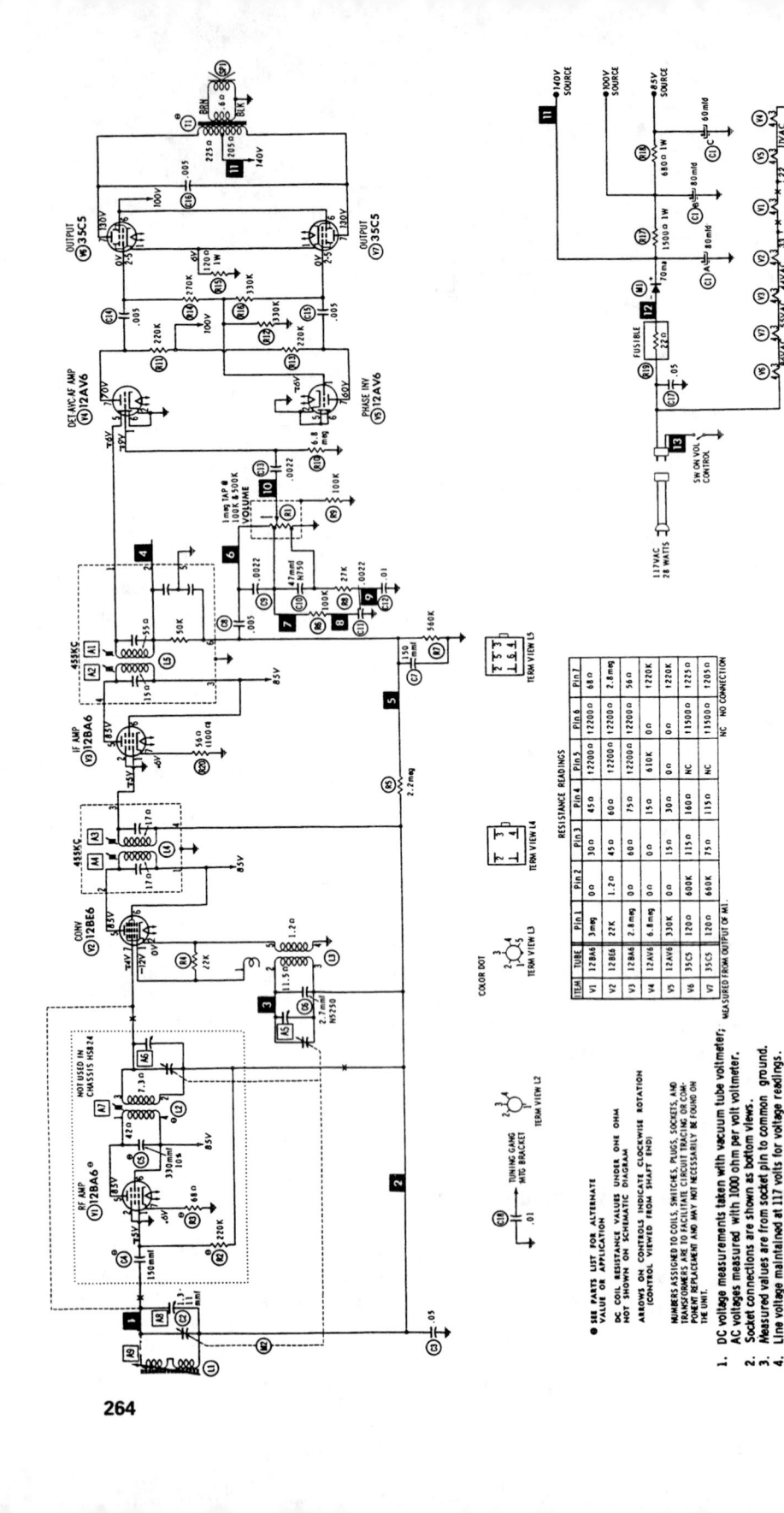

FIGURE 9-22 Receiver that utilizes an RF amplifier and push-pull audio amplifier (a Photofact schematic, courtesy of Howard W. Sams & Co., Inc.)

receivers tend to reproduce the lows less efficiently than the highs, rendering the sound with a more treble or "tinny" quality. Therefore, without tone compensation, as the volume is reduced the low-frequency bass notes tend to drop out first, causing the treble quality of the sound to increase. If a balanced tone is to be maintained at all volume levels, a means of reducing the highs must be provided at low volume levels. A reduction of the highs results in the same effect as an increase in the lows.

A capacitor exhibits less reactance to high frequencies because capacitive reactance is inversely proportional to frequency. Therefore, a capacitor connected from an audio line to ground will have a greater shunting effect on high frequencies than on low frequencies. At high frequencies, the reactance becomes very low so that most high-frequency components are shunted to ground.

If a resistance is placed in series with the capacitor, the shunting impedance of the resistor-capacitor combination will never be less than the ohmic value of the resistor. Hence, at low frequencies, the impedance will be high, because of the capacitive reactance, but as the frequency increases, the capacitive rectance decreases and the impedance decreases. The impedance of the combination, however, will never drop below the value of the resistor.

In Figure 9-23, two *R-C* combinations are tapped onto the volume control element, one at *B* and one at *D*. When the wiper arm is below a tap, a portion of the highs is lost via the *R-C* combination connected to the tap. If the wiper arm is below two taps, an even greater portion of the highs is lost. This provides the equivalent of a bass boost at low volume levels.

Incidentally, referring to the schematic of Figure 9-22, we note a 100-kΩ resistor connected from the volume control housing to ground. This is to provide a degree of shielding of the volume control element against stray signal pickup without having the volume control become "hot," as it would be if it were connected directly to

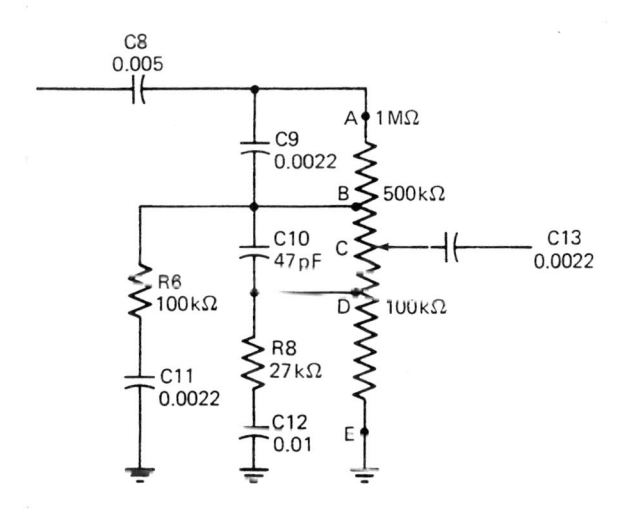

FIGURE 9-23 Tone-compensation circuit of the receiver of Figure 9-22.

Sec. 9-10 VT Receiver with RF Amplifier and PP Audio Amplifier 265 chassis ground. This is for reasons of safety in the event that the volume control knob is lost and the control shaft is exposed.

Audio Section

The audio output stage is a push-pull power amplifier that utilizes two beam power pentodes. However, no interstage transformer is used. Instead, an additional stage called a *phase inverter* (V5) is inserted into the signal path to V7, and this causes the signal at V7 to be inverted relative to the signal at V6, as is required for push-pull operation.

From the volume control, the audio signal passes through C13 to the grid of the first AF amplifier, V4. From the plate of V4 the signal passes through C14 to the grid of V6 and also to a voltage divider consisting of R14 and R12, as is shown more clearly in Figure 9-24. Only a fraction of the voltage at point A appears at point B, so that only a much smaller signal is applied to the grid of phase inverter V5. But V5 amplifies the signal applied to its grid, and the net effect is that the signal at the plate of V5 is restored to the amplitude of the signal at point A. However, the signal is inverted relative to the signal at point A, as required.

To ensure that V5 does not contribute any net gain, negative feedback is used around V5 to stabilize the stage at unity gain. This is described in more detail in the following.

As stated above, the voltage at point B is applied to the grid of V5, but point B is driven both by point A (through R14) and by point C (through R16). Further, the voltages at points A and C are out of phase. Hence, when point A drives B in the positive direction, the signal from point C tries to drive point B in the negative

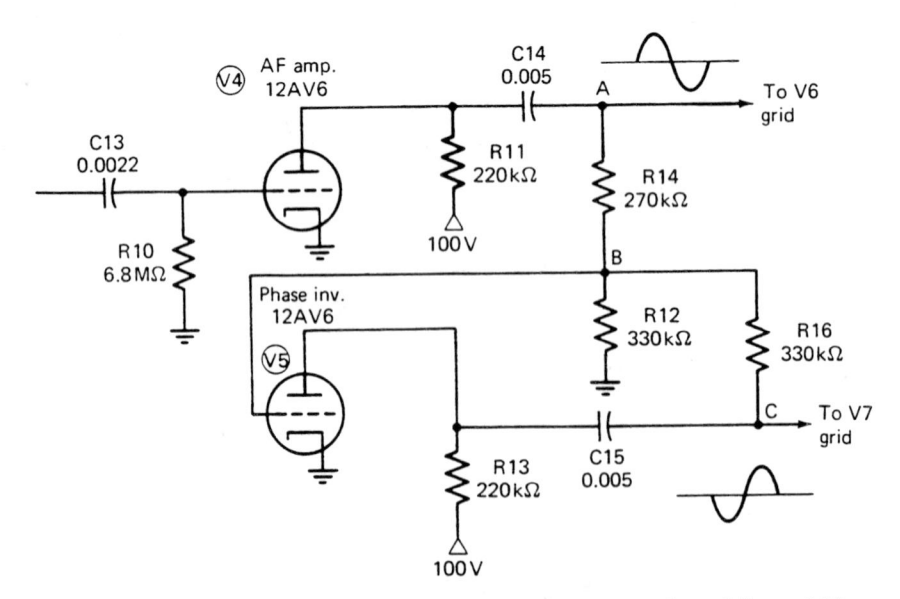

FIGURE 9-24 First AF amplifier and phase inverter of the receiver of Figure 9-22.

direction. The result is that point C cancels most of the signal arriving at B from point A. In other words, $R16$ introduces negative feedback at point B.

Comparing $R14$ and $R16$, we note that $R14$ is smaller (270 kΩ versus 330 kΩ). Therefore, point A will have more effect on point B than point C has on B. Roughly 0.55 of the voltage at A is applied to B while about half (0.5) of the voltage at C is applied to point B. Recalling that the signals at points A and C are to be equal (though out of phase), we see that only a small fraction of the voltage at point A remains uncanceled at B. A more exact analysis than is implied here reveals that only about 7% of the voltage at A appears at B. Consequently, when $V5$ contributes a gain of about 14.5, the voltage at point C, the grid at $V7$, will equal that at point A, the grid of $V6$. This is the desired result.

Normal tolerances in resistors and tubes introduce an uncertainty in the actual gain a given amplifier stage will produce. In the phase-inverter application, gain variations in the phase inverter stage result in an undesirable signal unbalance between the grids of the P-P tubes. The negative feedback scheme tends to stabilize the output of the phase inverter against variations in the gain of the stage. If the phase-inverter output becomes too large, the amount of signal feedback is also increased, and this reduces the input to the phase inverter so that the output is brought back in line. In the present circuit, for example, a 25% increase in gain of the phase inverter stage produces only a 4% increase in the voltage applied to the grid of the P-P tube. Thus, the negative feedback gives rise to greatly increased stability against signal unbalance between the P-P tubes.

Output Stage

The P-P output stage, beyond the phase inverter, is typical. The cathodes are tied together, and bias voltage is provided by $R15$. Screen voltage is taken from the 100-V source of the power supply so that the screen voltage is less than the plate voltage. Capacitor $C16$ is a tone-compensation capacitor that gives a deeper quality to the sound from the speaker.

SUMMARY

1. Three networks appear frequently in radio receivers and should be recognized immediately. They are (1) the voltage divider of a universal biased stage, (2) an emitter resistor and its bypass capacitor, and (3) a decoupling resistor and the associated capacitor to ground.

2. A decoupling network is used to prevent the signals of one stage from being passed to another by way of the power supply connections.

3. In solid-state receivers that are not battery operated, a step-down transformer is typically used to provide low-voltage AC for application to either a half-wave or a full-wave rectifier. Bridge rectifiers are widely used.

4. Power supplies for the older five-tube AM receivers are almost always half-wave supplies that are transformerless. The highest B+ voltage is on the order of 120 V. Tube filaments in a series string are connected directly to the AC line, and almost always, one side of the AC line will connect through the on-off switch directly to the chassis. Therefore, the chassis is "hot," and care is due in handling the set.

5. Power supplies in solid-state receivers may be rather elaborate in order to provide biasing voltages for the various stages of the receiver.

6. The line filter capacitor prevents undesired RF signals from entering the receiver via the AC line.

7. The antenna for an AM receiver is typically a ferrite rod (loopstick) antenna. Older receivers may have a loop antenna attached to the back of the cabinet. In both cases, the winding of the antenna forms the inductance for the antenna tuning circuit.

8. A solid-state converter typically utilizes one transistor which serves as both mixer and local oscillator. The collector circuit incorporates two tuned transformers in series, one to sustain oscillations in the oscillator resonant circuit, and the other to couple the output of the converter to the IF amplifier.

9. The converter of tube-type AM receivers is almost always a pentagrid converter.

10. An RF amplifier may precede the converter in more elaborate receivers.

11. The IF section of a discrete-transistor receiver (as opposed to one that uses ICs) almost always uses two stages of amplification, with the first stage having its gain controlled by the AVC network. Tuned transformers provide coupling between stages. Input to the IF section is from the converter, and its output is applied to the detector (demodulator). The standard IF frequency for AM is 455 kHz.

12. The sharpness of the tuning circuits in the IF amplifier enhances the selectivity of a receiver.

13. Tube-type receivers typically have only one stage of IF amplification.

14. The detector circuit of an AM receiver resembles a half-wave rectifier. In addition to its function as a demodulator of the AM waveform, the detector also produces a slowly varying DC voltage for the AVC.

15. The AVC control voltage depends upon the average amplitude of the IF signal rather than upon the level of the audio signal at the output of the detector.

16. The stages in the audio section of a solid-state receiver may be *RC*-coupled, transformer-coupled, or direct-coupled. Push-pull output stages are more commonly encountered than single-ended output stages.

17. The output stage may use a transformer as an impedance matching device between the output transistors and the speaker, or an OTL (output-transformer-less) circuit may be used.

18. Audio sections usually have either two or three stages of amplification. The first stage is a voltage amplifier, while the second stage is the driver for the output stage. If only two stages are used, the first audio amplifier also serves as the driver.

19. Phase inversion for push-pull amplifiers may be achieved via an extra stage, a phase splitter, a transformer, or a pair of complementary transistors.

20. All or any portion of the audio section may be direct-coupled. When direct coupling is used, negative feedback is used to stabilize the DC voltage levels.

21. The audio section of tube-type AM receivers typically involves one stage of voltage amplification ahead of the audio output stage. Grid leak (or contact) bias is used for the first AF amplifier. Cathode biasing is used for the output stage, which almost invariably utilizes beam power pentodes.

22. Small AM receivers often utilize small speakers. Tone compensation is required to avoid excessive high-frequency response, which would give a "tinny" quality to the sound.

23. As the volume is reduced, the low-frequency components tend to drop out first, causing the tone to shift toward the treble. This may be avoided by tone-compensation capacitors connected to taps on the resistance element of the volume control.

QUESTIONS AND PROBLEMS

Refer to Figure 9-3 for Probs. 9-1 to 9-7.

9-1. What is the purpose of $R40$ in the power supply?

9-2. What type of transistor is $Q4$?

(a) *PNP* or *NPN*?

(b) Silicon or germanium?

9-3. What is the purpose of $R19$ and $R20$ in the second IF amplifier?

9-4. What is the purpose of $C19$ and $R22$ in the second IF amplifier?

9-5. What function is performed by $C26$ in the power supply?

9-6. What is the purpose of $C6$ in audio amplifier?

9-7. What type of device is $R37$, which is connected across the output transformer primary? What is its function?

Refer to Figure 9-9 for Probs. 9-8 to 9-12.

9-8. What type of device is $R27$ in the audio output stage? What is its function?

9-9. In the power supply, what does the dotted line around $X1$ indicate?

9-10. Why is the positive terminal of the bridge rectifier connected to ground?

9-11. What is the purpose of $C19$ that connects to the base of $Q4$, the first audio amplifier?

9-12. What is the purpose of $C6$ connected in parallel with $R24$ in the first audio amplifier?

Refer to Figure 9-10 for Probs. 9-13 and 9-14.

9-13. What type of transformer arrangement is used to obtain the low-voltage AC for the low-voltage power supply?

9-14. Referring to transformer $T1$ in the audio section:

  (a) What is the reason for having two secondary coils instead of one secondary coil that is center-tapped?

  (b) Assuming that the two secondary coils are identical, what determines that the signals applied to the bases of the output transistors will be 180° out of phase?

Refer to Figure 9-12 for Probs. 9-15 through 9-17.

9.15. Referring to the power supply used in this receiver, is it half-wave or full-wave?

9-16. The power supply uses a single-diode $X1$ as a rectifier, yet both a positive and a negative voltage are developed for use by the stages of the receiver. Explain how this is accomplished.

9-17. What is the purpose of $C9$, located near the converter stage?

Refer to Figure 9-15 for Probs. 9-18 and 9-19.

9-18. What is the purpose of $R26$ and diode $X2$ in the collector circuit of $Q5$ and $Q6$?

9-19. What is the name applied to the circuit configuration of $Q5$ and $Q6$?

9-20. What is the purpose of $C12$ in the base circuitry of the converter of Figure 9-9?

9-21. The AVC circuitry used in Figures 9-3 and 9-9 is similar, yet not identical. What is the difference?

9-22. If the signal level increases in the IF amplifier of Figure 9-9, does the AVC voltage developed at CTP 9 go more positive, or does it go more negative? Why?

9-23. What is the purpose of $R2$ and $R3$ in the converter of Figure 9-10?

9-24. What is the purpose of $C10$ in the first IF amplifier of Figure 9-12?

9-25. If a strong but unmodulated carrier is present at the detector input of Figure 9-12, what will be the polarity of the voltage developed at the high side of the volume control, CTP 10?

9-26. What is the purpose of $C2$ near the first IF amplifier in Figure 9-15?

9-27. If the IF signal applied to the detector in Figure 9-15 increases, does the AGC voltage at CTP 10 go more negative or more positive?

9-28. What is the purpose of $C15$ and $C16$ near the detector in Figure 9-15?

9-29. What is the approximate collector current of $Q4$, the first audio amplifier of Figure 9-3?

9-30. What current flows through $R41$ of the power supply in Figure 9-7?

9-31. Compare the antenna input circuitry to the converter of Figure 9-7 and 9-3. The low side of the loopstick secondary is grounded in Figure 9-3 but is connected to a resistor network in Figure 9-7. How can the secondary be grounded in one circuit and not grounded in the other?

9-32. From a consideration of the base voltage of $Q8$ and the DC voltage present at CTP 18 in Figure 9-15, estimate the current flowing in the collector circuit of the $Q5$-$Q6$ Darlington pair.

9-33. Identify the biasing arrangement of the second IF amplifier in Figure 9-15.

9-34. Why can the audio output stage of a receiver tolerate a power supply voltage with more ripple than can be tolerated by the other stages?

9-35. What method is used to bias the oscillator grid in the pentagrid converters in the vacuum-tube receivers described in this chapter? What method is used to bias the antenna signal grid?

9-36. Describe an easy way to determine if the local oscillator of a pentagrid converter is working.

9-37. When a receiver is not tuned to a station, will a 455-kHz IF signal be present at the input to the IF amplifier?

9-38. What is the purpose of the 375-pF capacitor in the couplate of Figure 9-19?

9-39. What values are used for AVC filter capacitors in Figure 9-19 and 9-22? What values are used for AVC filter resistors?

9-40. If the voltage at the cathode of a cathode-biased amplifier is known, how can the plate current be determined?

9-41. What are the two functions of a fusible resistor?

9-42. What is the function of the resistor-capacitor network attached to the taps on the volume control of Figure 9-22?

9-43. Describe the manner in which negative feedback is used to ensure that signals of equal amplitude are applied to the two push-pull output tubes in the circuit of Figure 9-22.

10
INTRODUCTION TO TROUBLESHOOTING

This chapter marks the beginning of our treatment of electronic troubleshooting. In the previous chapters we have investigated the circuitry of AM receivers, and, now, with a good understanding of complete AM receivers, we can address ourselves to the servicing and repair of these receivers. The theory and description of FM and FM stereo receivers are presented in subsequent chapters.

This chapter is devoted to discussions of a general nature pertaining to such items as tools, equipment, replacement parts, and failure mechanisms of tubes and transistors. A procedure is given for using an ohmmeter to determine the type (NPN or PNP) and terminal connections (emitter, base, and collector) of a BJT. The basic troubleshooting procedure, including the techniques of signal injection and signal tracing, is described in Chap. 11.

10-1 TOOLS FOR ELECTRONIC SERVICING

Efficient servicing requires many tools for cabinet disassembly, component replacement, and general use. Figure 10-1 is a list of recommended tools for servicing radio receivers. Most servicing jobs can be completed with only a few of the listed items, but many of the tools help the technician to save time (and money) and to do the job better. An ample and organized supply of tools connotes professionalism and commands respect.

Service Data

Good service data are essential for efficient servicing. Circuits vary widely. A schematic of each receiver encountered should be available and should include voltage measurements, component replacement data, layout sheets, and alignment instructions. Layout sheets are photographs or drawings that show the location of each component, test point, and alignment adjustment.

Most manufacturers publish service data for their receivers. If a technician plans to limit his servicing activities to receivers of two or three manufacturers, the service data of the manufacturer are probably the most economical to acquire. If the intent, however, is to service any set that comes along, service information much wider

1. Screwdriver assortment (flat blade and Phillips)	20. Contact cleaner
2. Socket wrenches	21. Freeze spray
3. Nut drivers	22. Cement for plastics
4. Small open-end wrenches	23. Epoxy cement
5. Allen wrenches	24. Alcohol
	25. Lubricating oils assortment
6. Regular and miniature needle-nose pliers	26. Hacksaw
	27. Pocket knife
7. Regular and miniature diagonal cutters	28. Wire strippers
	29. Drill
8. Miniature end cutters	30. Small bench vice
9. Wire pliers (large)	31. Alignment tools
10. Soldering gun, 100/140 W	32. Electrical tape
11. Soldering iron, 25 W	33. Clip leads
12. Soldering aids	34. Cheater cord
13. 60/40 solder for electronics	
14. Desoldering device	35. Penlight
15. Heat sinks	36. Flashlight
	37. Magnifying lense
16. Steel-wire brush	
17. Toothbrush	38. Small plastic containers
18. Small paintbrush	39. Miscellaneous hardware
19. Cleaning cloth	40. Sandpaper, steel wool, files

FIGURE 10-1 List of recommended tools for servicing radio receivers.

in scope is needed. The standard source of service information is the Photofacts published by Howard W. Sams & Co., Inc., of Indianapolis, Indiana. The Photofact folders contain schematics, layout sheets, alignment instructions, parts lists and substitutes, and, when appropriate, instructions for disassembly. They are obtainable directly from the publisher or from almost any electronic component supply house.

Receiver Disassembly

Screws encountered in receivers may be hex, slot, or Phillips head of almost any size, and they may not be readily accessible. Thus, it is desirable to have an assortment of screwdrivers and nutdrivers available. In some cases a socket wrench or an open-end wrench is required, because the location of a screw or bolt behind other components may prevent access to the screw with ordinary screwdrivers. Screws located at the bottom of a deep hole can be very difficult to replace if screw-holding drivers are not available. Alternatively, a bit of putty (or chewing gum) on the tip of a screwdriver will hold a screw to enable you to get it started. When screws of different sizes are involved, a colored mark from a felttip pen on the screw and hole from which it came helps to get the item back in the proper place.

Items removed should be meticulously placed in a container that is always kept with the receiver. This is especially important if the receiver must be "put back" while a part is ordered. This simple practice can save untold heartache, because nothing is as frustrating and time-consuming as to have repaired a receiver only to find that when the cabinet is about to be reassembled, the screws are hopelessly lost.

If the line cord is connected to the receiver through an interlock that is part of the back panel, a "cheater cord" is required to apply power to the chassis. Jumper wires may be required for connecting speakers, switches, antennas, and so forth, that may be left in the cabinet during the servicing procedure.

Sometimes the knobs on the front panel controls can be very difficult to remove. If a screwdriver is used to pry the knob loose, the cabinet may be damaged or, even worse, the knob may be broken. An old rag or a handkerchief slipped under and wrapped around the knob to apply pressure uniformly minimizes the chances of damage to the knob or the cabinet.

Removing and Replacing Components

Several types of cutters and pliers are necessary for removing and installing parts. Components in tube-type equipment are generally large enough to be worked with regular-size tools. A pair of needle-nose pliers is needed for manipulating lead wires, and a pair of diagonal cutters is required for cutting lead wires in tight places close to soldered terminals. A regular-size pair of diagonal cutters is also useful for breaking up a tube socket located on a printed circuit (PC) board to facilitate removal.

Solid-state sets typically utilize smaller components, and special, miniature cutters and pliers are required. The smaller components must be handled carefully. Full-size cutters and pliers often lead to damaged components, and furthermore the components are frequently so closely spaced that only miniature tools are able to reach the part. Crowded PC boards sometimes necessitate the use of a special type of cutter, called an *end cutter*. The cutting portion is located on the tip of the jaw, and it can be used to cut component leads while holding the handles of the cutter nearly perpendicular to the board.

Most semiconductor devices (diodes, transistors, integrated circuits) are soldered directly onto the circuit board, and it is necessary to prevent excessive heating of the device during the soldering operations required to remove or replace suspected devices. Whenever possible, a heat sink should be attached between the solder joint and the body of the device to absorb heat that otherwise would be transferred through the leads to the semiconductor materials. Commercial heat sinks may be used, or a pair of needle-nose pliers may be used to grip the lead being soldered. When the device fits close to a PC board, it may be impractical to attach a heat sink; in such cases, the soldering operation must be done quickly. A brief cooling-off period should be allowed between successive connections to the device.

Soldering and Desoldering Devices

Several types of soldering devices are required to handle the wide range of component sizes and connections found in various receivers. Tube-type receivers (whether hand-wired or PC) generally require a soldering gun with multiple heat settings.

A 100/140-W gun will cover most jobs. The 100-W setting is sufficient for removing most components, whether mounted on PC board or terminal strips. Ground connections to the chassis and shields require more heat, because of the additional metal that must be heated; the 140-W setting of the gun is usually appropriate here.

A soldering gun is awkward and tends to get too hot for the small components of solid-state receivers. A 25-W pencil iron is more appropriate. Excessive heat applied to a foil on a PC board can cause the foil to come loose from the board, a highly undesirable situation.

PC boards pose yet another problem. Resistors and capacitors can be removed one lead at a time. But with transformers, IF transformers, or integrated circuits, three or more leads have to be heated and removed simultaneously. All the while, consideration must be given to the amount of heat transferred to the device being removed. In such cases a desoldering device is required to remove the solder from each lead.

Several types of desoldering devices are available. Copper braid can be used to absorb the solder from each lead connection. A soldering gun or iron is used to melt the solder, and then the braid is placed in contact with the solder to remove it from the joint. Suction devices are available which suck the molten solder from the joint. The suction devices appear as hand-operated squeeze bulbs or as spring-loaded devices. Desoldering irons are available that contain the source of heat in conjunction with a hollow tip fitted with a squeeze bulb. The bulb is squeezed to remove the air, the joint is heated until the solder is melted, and the bulb is then released to "sip the solder" from the joint.

Light Sources

The workbench should be well illuminated by ambient light, and a localized light source is desirable for concentrating light on the receiver. An incandescent or fluorescent lamp mounted on a swivel base is ideal for this purpose. A magnifying glass is sometimes helpful when working on solid-state receivers.

It is sometimes difficult to find the same point on both the foil side and the component side of a PC board. For example, a certain solder joint might be known to be the base connection of a certain transistor. But when the technician looks at the component side of the board, several transistors may be in the area, and it may not be immediately obvious which transistor corresponds to the joint in question. A light source placed behind the foil side of the board causes the PC conductors and solder joints to appear as shadows. Most PC boards are translucent to some extent, but the foils and solder joints are completely opaque. This simple trick makes circuit tracing much easier.

A small penlight is also available in locating components. By placing the penlight behind the foil side, only a very small section of the board is illuminated. The component in question can then be readily identified. Going in the other direction, if the penlight is held above a component, the shadow of the component will be visible from the foil side of the board.

Even with good light, the circuitry may still be hard to see because receivers tend to collect dust and dirt on the PC board. A small paintbrush and a toothbrush are ideal for cleaning the PC board before beginning work.

Alignment Tools

A receiver is aligned by adjusting the tuned circuits of the receiver to the proper frequency. The tuned circuits have either variable capacitors or variable inductors. Variable capacitors usually have a slotted screw that varies the distance between the plates of the capacitor. Metal screwdrivers should not be used to make these adjustments because the stray capacitance of the blade affects the resonant frequency of the tuned circuit. Plastic or fiberglass screwdrivers, or a screwdriver with a plastic shaft and a small metal tip, are required.

Variable inductors have movable ferrite cores that screw in or out to change the inductance. These cores may be slotted for (nonmetal) screwdriver adjustment, or they may be hollow for a hex-head "diddle stick." Alignment tools come in many shapes and sizes, and a technician should have a good assortment available. Use of the wrong tool or careless use of the right tool can lead to "busted cores" of inductors, and the attendant frustration and loss of time need no elaboration.

Chemicals

Several chemicals are necessary for repairing and cleaning various parts of a receiver. Contact cleaner in a spray can is used to clean switch contacts and volume/tone control potentiometers. Freeze spray, a refrigerant packaged in a pressurized container, is useful for locating temperature-sensitive intermittent components. It is also useful for cooling solid-state components prior to soldering operations in order to minimize heating of the device. A good epoxy glue and a glue for repairing plastic parts are useful for mending broken knobs, cracks in plastic cabinets, torn speaker cones, broken PC boards, and so on. A lightweight oil is sometimes needed to oil moving parts.

Miscellaneous Items

A few miscellaneous items are needed for various tasks. A drill and bits may be needed for drilling new mounting holes for replacement transformers, speakers, and so on. A heat source (e.g., a hair dryer) is useful for finding intermittent components that break down due to heat. A box of assorted screws, nuts, bolts, and so forth, can save many headaches when a small item wanders away.

Isolation Transformer

In many receivers, the chassis or ground foil is connected directly to the AC line. This produces a particular hazard to the technician and to the equipment because some items of test equipment have ground leads that connect to the ground of the AC line. If a grounded ground lead is connected to a chassis that is "hot," a short circuit of the AC line will result, and a heavy arc will occur until a fuse blows or a circuitbreaker opens. Or, if the technician touches the ground lead and the chassis at the same time, a serious electrical shock will result.

Two things should be done to avoid this situation. First, establish a common ground for all pieces of test equipment that must plug into the AC line. Second, use an isolation transformer (see Sec. 3-1) with each and every receiver that is to be serviced. The importance of this cannot be overemphasized.

Safety Glasses

Safety glasses or goggles should be worn when servicing any type of equipment. Technicians frequently press a soldering gun against a contact to get a good thermal contact to the joint being unsoldered. When the solder melts, the gun may slip off the contact, letting it spring back to its original position, slinging molten solder into the eyes of the unsuspecting technician. Another sad fact of life is that capacitors sometimes explode with impressive violence. Power supply filter capacitors and line filter capacitors are the most likely candidates, and they may go suddenly, without warning. Considering that the technican frequently works with his or her face very close to the set, eye protection should be a mandatory requirement of any service bench.

10-2 EQUIPMENT CONSIDERATIONS

Many receivers with simple or obvious problems may be serviced with only simple hand tools, but more sets with more difficult problems can be handled in less time if the service bench is well equipped. In fact, many problems will be unsolvable if proper equipment is not available. Economic considerations dictate what equipment is justifiable. A person setting up a service business to be operated full-time can justify a larger investment in equipment than a person who engages in servicing only as a sideline.

Many types of test equipment are available for troubleshooting a defective receiver, and these types may be classified as basic measuring instruments (volt meters, ammeters, ohmmeters), signal generators, signal tracers (signal tracers as such or oscilloscopes), component checkers, and parts-substitution devices. Also, suitable power supplies must be available for providing power to auto radios or other battery operated equipment when the batteries are not present. A list of recommended equipment for a well-equipped service bench is shown in Figure 10-2. In the following sections we will describe a few aspects of commonly used test equipment.

Multimeters: EVM vs. VOM

A multimeter is a multirange device for measuring voltage, current, and resistance. The desired mode (volts, amps, or ohms) is selected by the user. The least sophisticated multimeter contains no internal amplifiers; power from the circuit under

> 1. High-input-impedance voltmeter (VTVM or TVM)
> 2. Volt-ohm-milliammeter (VOM)
> 3. Signal generators (AM, FM, audio)
> 4. Oscilloscope
> 5. Signal tracer
> 6. Sweep/marker generator
> 7. Stereo generator (for FM multiplex)
> 8. Component checkers
> 9. Frequency counter
> 10. Power supply for battery substitute

FIGURE 10-2 Recommended equipment for servicing radio receivers.

test drives the meter movement. This type of multimeter is usually called a VOM, for volt-ohm-milliammeter.

Electronic multimeters (EVMs) incorporate amplifier circuits between the test prods and display device so that the amplifier (as opposed to the circuit under test) drives the display device. Earlier EVMs consisted of the vacuum-tube voltmeters (VTVMs), which were followed by solid-state versions of the same device. In these instruments, the display device was a swinging-needle meter movement. The advantage of the meters with amplifiers is that of high input impedance.

Modern EVMs may have either digital or swinging-needle displays plus other features such as automatic range selection and overload protection. Although digital displays are great, they are not without at least one disadvantage. When a voltage is measured that is not steady, the display often flickers incoherently, whereas a needle would drift smoothly with the voltage. For some applications, the needles are more informative! The choice of display, however, is a minor point. What is important is the high input impedance of the EVM as compared to the lower impedance of the VOM. This is illustrated in the following.

A typical EVM has an input impedance of 11 MΩ or higher, and the input impedance remains the same on all ranges. The input impedance of a VOM may be determined from the "ohms per volt" sensitivity of the meter given by the manufacturer. A typical sensitivity is 20,000 Ω/V. The input impedance is different for different ranges, and is given by the ohms/volt sensitivity multiplied by the maximum voltage of a particular range. If a 20-kΩ/V meter is switched to a 5-V range, the input impedance will be 5 $\times$ 20 kΩ = 100 kΩ. If the meter is switched to a 100-V range, the input impedance will be 100 $\times$ 20 kΩ = 2 MΩ. Note that the input impedance becomes higher as higher ranges are selected.

Figure 10-3 illustrates the effect of meter impedance upon a voltage measurement in a simple circuit. Two 500-kΩ resistors are connected in series, with 10 V applied to the combination. A voltage of 5 V is developed across each resistor, but significantly different voltage may be measured if the wrong type of meter is used.

An EVM of input impedance 11 MΩ and a VOM of input impedance 200 kΩ are used to measure the voltage developed across one of the resistors. When the meter is connected, the resistance of the meter is placed in parallel with the resistance across which the voltage is being measured. This causes the voltage distribution in the circuit to change.

Figure 10-3(b) shows the equivalent circuit with the EVM connected. The large impedance of the meter produces only a small effect; the measurement is in error only about 2%. In Figure 10-3(c), however, the 200-kΩ resistance of the VOM changes the voltage distribution drastically; the reading is in error by 56%! Such a discrepancy in a practical circuit might cause a technician to form incorrect conclusions.

The effect is less pronounced in circuits involving lower impedances, as dem-

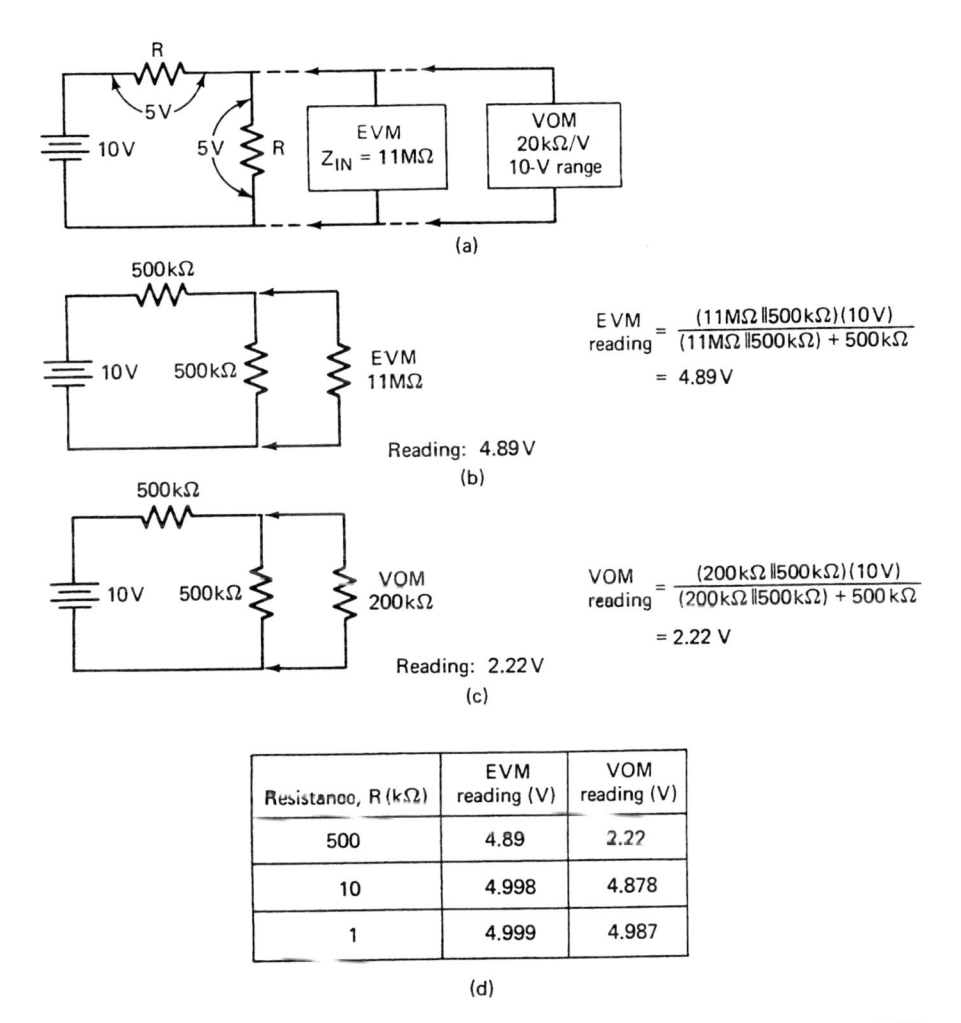

$$\text{EVM reading} = \frac{(11M\Omega \| 500k\Omega)(10V)}{(11M\Omega \| 500k\Omega) + 500k\Omega} = 4.89 \text{ V}$$

$$\text{VOM reading} = \frac{(200k\Omega \| 500k\Omega)(10V)}{(200k\Omega \| 500k\Omega) + 500k\Omega} = 2.22 \text{ V}$$

Resistance, R (kΩ)	EVM reading (V)	VOM reading (V)
500	4.89	2.22
10	4.998	4.878
1	4.999	4.987

(d)

FIGURE 10-3 A voltmeter connected to a circuit changes the voltage in the circuit. A VOM affects the circuit more than an EVM.

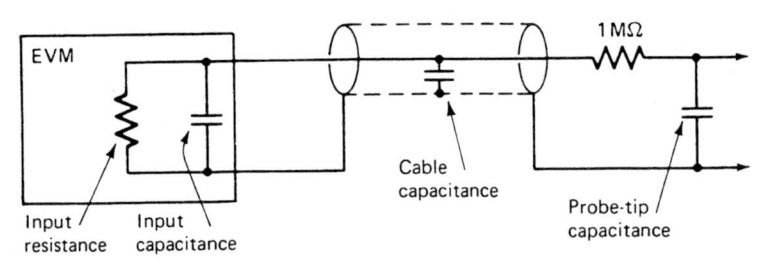

FIGURE 10-4 Simplified equivalent circuit of an EVM, showing the various capacitances involved.

onstrated in Figure 10-3(d), in which the resistors have been lowered to 10 and 1 kΩ respectively. When the resistors are 1 kΩ, no significant difference exists in the readings.

Another important consideration is the shunt capacitance of the meter. A simplified equivalent circuit of an EVM is shown in Figure 10-4, in which a capacitance appears in parallel with the input resistance. This capacitance results from distributed capacitance between the lead wires and associated input circuitry of the meter. In high-frequency circuits, the shunt capacitance can shift the resonant frequency of tuned circuits, or, if a meter is used that has a very large shunt capacitance (such as a VOM), the circuit may become completely inoperative.

For example, the oscillator in a tube-type receiver may be checked for oscillation by measuring the DC voltage present at the grid of the oscillator tube (frequently part of a pentagrid converter). The presence of a negative voltage indicates that the oscillator is working because the voltage results from grid leak biasing and will only be present when a fairly large signal drives the oscillator grid. The fairly low shunt capacitance of an EVM will alter the frequency slightly, but the oscillator will continue to operate. On the other hand, if a simple VOM is connected, the shunt capacitance and circuit loading will probably stop the oscillator altogether, and no voltage will be measured.

A typical EVM has a large series resistor located in the probe near the tip to minimize the effects of distributed capacitance associated with the lead wires going to the meter. A VOM has no such resistor. In short, the effect of connecting a VOM to a high-frequency circuit is so drastic that it should never be done. An EVM is far superior.

Resistance measurements may be made with either an EVM or VOM. As with any ohmmeter used to check semiconductor devices, the polarity of the test leads should be determined, and the test voltage of the ohmmeter should be checked to see that it is not large enough to destroy transistors and diodes.

A typical EVM has no provision for current measurement, whereas a VOM usually will measure DC currents up to a fraction of an amp. This is one advantage of a VOM over an EVM. Another advantage is that a VOM is isolated from ground so that it can be used to measure voltages across resistors when neither end of the resistor connects to ground. An example would be a plate load resistor. A VOM may be connected directly across the resistor so that the voltage across the resistor

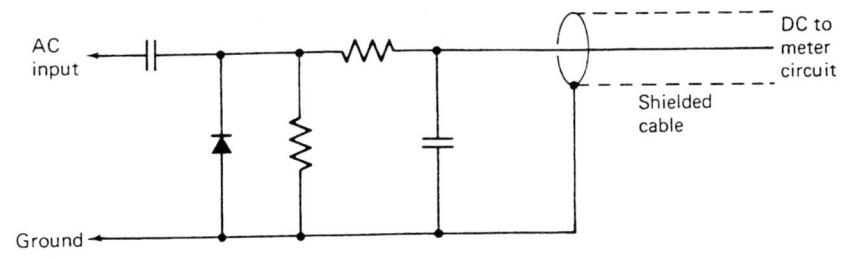

FIGURE 10-5 Circuit configuration of a typical RF probe for an EVM. Resistance and capacitance values must be chosen to achieve compatibility with a particular instrument.

is read directly. With an EVM, the voltage must be determined at both ends of the resistor and then subtracted to get the voltage across the resistor.

RF Probe for EVM

Most EVMs require a special RF probe for measuring high-frequency voltages. The probe is actually a detector that converts the RF signal into a proportional DC voltage that is sent to and measured by the meter. The schematic diagram of an RF probe is shown in Figure 10-5. Resistance and capacitance values must be selected for a particular EVM so as to achieve calibration. Typical RF probes allow voltage measurements at frequencies up to about 100 MHz. Beyond that, special instruments are required that are not commonly found around typical service shops.

Signal Generator

A signal generator is used to apply an appropriate signal to a stage to determine if the stage is performing properly. (This is covered in detail in Sec. 11-2.) Three different types of signals at various frequencies and amplitudes are necessary in order to cover the requirements of AM and FM receivers: audio, AM, and FM. The audio signal can be any frequency from about 50 to 10,000 Hz, but a frequency in the range 400 to 1000 Hz is preferred. An AM signal should be available at frequencies of 262.5 kHz, 455 kHz, and 535 to 1605 kHz. An FM signal should be available at 10.7 MHz and over the range from 88 to 108 MHz. In most cases, an AM generator can be used in place of the FM generator by altering procedures somewhat. This is discussed in Chapter 15.

The various signals may come from separate generators or from a combination generator. The output signal should be variable in amplitude, and the percentage modulation of AM and FM generators should be variable.

In addition to troubleshooting, AM and FM generators are used for alignment of receivers, as described in Chapters 12 and 15. Consequently, the frequency calibration should be fairly accurate. An AM generator can be used for FM alignment with some sacrifice in precision. A sweep-marker generator should be available for FM alignment (Chapter 18). For FM stereo, a special FM stereo generator should be used.

Component Checker

Special equipment for testing components makes work easier on occasions, but these instruments are not essential, since other methods can be used to check most components. Vacuum-tube and transistor checkers are quite common, and equipment is available to check capacitors and inductors. Resistors may be tested with the ohmmeter function of an EVM or VOM. In any case, a suspected component may be checked indirectly by substitution of a component known to be good. However, this involves some risk for the good component if another circuit defect caused the failure of the original component.

Use of EVM as a Component Checker

An EVM (ohmmeter function) can be used to check resistors, coils, transformers, and capacitors. Resistors should show the proper resistance, and the tolerance of the resistor should be taken into account. For example, if a given resistor has a tolerance of 10%, the resistor is assumed good if the measured resistance lies in the range 90 to 110% of the marked value. A 220-Ω 10% tolerance resistor might measure anywhere from 198 to 242 Ω and be good. In practice, resistors are frequently found to have changed value while in the circuit. This is more likely to occur if the resistor runs hot (or warm) or if it is subjected to large voltages.

Inductors may be checked for continuity with an ohmmeter. The DC resistance of coils is frequently given in the service data, and any significant departure from the correct value indicates a defect. The resistance values are usually rather low, on the order of a few ohms.

The primary and secondary windings of a transformer may be checked in the same manner as a coil. Iron core transformers and inductors can be checked for leakage to the core by an ohmmeter set on a high range. Also, leakage from the primary to the secondary can be checked. Either defect may cause problems in a given circuit.

Capacitors can be checked for leakage with an ohmmeter, but the ohmmeter must charge the capacitor to the supply voltage of the ohmmeter before the final resistance reading can be made. When first connected the needle will deflect upscale toward 0 Ω. It will then gradually settle back to a larger resistance value as the capacitor charges. The final resistance will depend upon the type of capacitor being tested and, to some extent, upon the capacitance.

Small disc ceramic capacitors with fairly low working voltages should show a resistance of several hundred megohms or more, while disc capacitors with high working voltages (400 V or greater) should indicate almost an infinite resistance. Large electrolytic filter capacitors are usually assumed to be good if the resistance is more than at least 50 kΩ and preferably 100 kΩ.

The kick of the needle when the ohmmeter is connected is smaller for smaller capacitances. With a typical ohmmeter set on the $R \times 1\text{-M}\Omega$ range, the kick is just noticeable with a 100-pF capacitor. A 0.0047-μF capacitor kicks the needle to about 40 MΩ before it proceeds to (almost) infinity. A 0.1-μF capacitor causes the needle to rise to about 5 MΩ before it quickly settles to about 1000 MΩ. With practice, a technician can get a good idea of the capacitance of an unknown capacitor by observing the kick of the needle when the ohmmeter is first connected. The capacitor must be discharged each time the test is made or erroneous conclusions will be drawn.

Sometimes a capacitor will check *good* (show a large resistance) at low voltages but will break down and leak when higher voltages are applied. Figure 10-6 illustrates a method of using an EVM (preferred) or a VOM and a high-voltage power supply to determine the resistance of a capacitor at high voltages. The power supply established a voltage V_s across the series combination of the meter and the capacitor. The voltage indicated on the meter, V_m, is related to the resistance of the capacitor R_c by the formula given in the figure. Large values of R_c give small meter readings, and vice versa.

With the EVM on a high-voltage scale, connection is made to the high voltage. As the capacitor charges, the meter will show a deflection that gradually subsides to a small value (near zero) as the capacitor becomes fully charged. The meter is then switched to a low-voltage scale to measure the final voltage shown on the meter. Any voltage indication indicates leakage in the capacitor. Care should be taken to avoid application of voltages higher than the working voltage of the capacitor.

In an actual servicing situation, the B+ voltage of a receiver can be used as the power supply for high-voltage testing of capacitors. Also, it frequently happens that the suspected capacitor connects in the circuit to a high-voltage source, and in such cases, it is only necessary to disconnect the other end of the capacitor and attach the voltmeter.

The resistance of the voltmeter must be known in order to use the formula of Figure 10-6 for calculating the resistance of a leaky capacitor. For VOMs this is obtained by multiplying the ohms/volt sensitivity of the meter by the maximum reading of the range used. For EVMs it is typically 1 MΩ in all ranges.

Capacitor leakage resistance $= R_c = \left(\dfrac{V_s}{V_m} - 1\right) R_m$

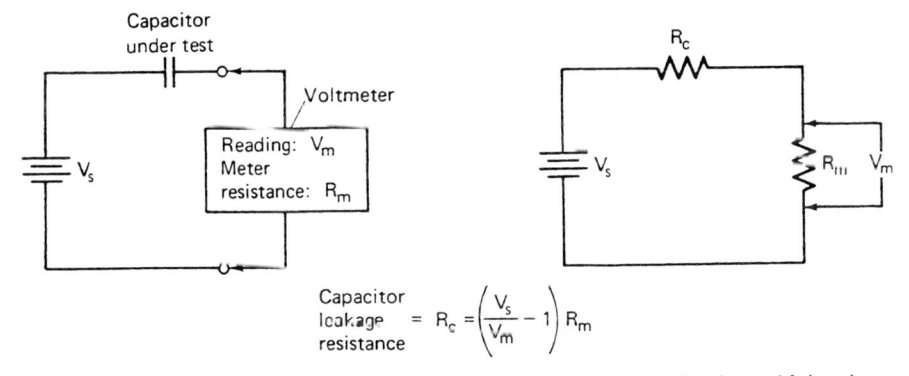

FIGURE 10-6 Leakage resistance of a capacitor can be determined using a high-voltage power supply and a voltmeter. Note that the internal resistance of the meter must be known.

Oscilloscope

The sophistication and quality of a particular "scope" determine how useful it will be in analyzing the various waveforms of a receiver. Desirable features include a calibrated vertical section, a triggered sweep horizontal section, a bandwidth from DC to 10 or 15 MHz, and provisions for a low-capacitance probe for use in high-frequency circuits. Scopes are available that are suitable for servicing in a price range that is affordable by most service organizations. Service scopes are less sophisticated than more expensive laboratory scopes found in educational institutions and industrial enterprises.

In addition to displaying the waveform of a signal, a scope with a calibrated vertical section may be used as a voltmeter for DC and high-frequency signals. If the vertical section is uncalibrated, only approximate determinations of signal amplitude can be made.

Another important consideration is the frequency response of the scope, typically being from DC up to a frequency generally determined by the expense of the scope. Least expensive scopes may have a bandwidth of 1 or 2 MHz, with successive steps being to 5 MHz, to 10 MHz, and to 15 MHz and higher for progressively more expensive instruments.

In practice, a scope will respond to frequencies higher than the upper-bandwidth limit, but the calibration becomes invalid due to the loss of sensitivity. Nevertheless, the scope is useful at these higher frequencies. For example, a 10-MHz scope will respond to the 10.7-MHz FM IF carrier, and to some degree, so will a 5-MHz scope. The presence of the 10.7-MHz signal could be confirmed with a 5-MHz scope by observing an illuminated band across the CRT screen even though the scope is unable to display a stable, discernible sine wave.

Most scopes of recent manufacture have *triggered sweep* horizontal sections, as opposed to the older *recurrent sweep* scopes. In recurrent sweep scopes, the "dot" executes the trace-retrace cycle at a frequency determined by the sweep frequency controls together with the synchronizing circuits that adjust the frequency slightly to produce a stable display. In triggered sweep scopes, the dot moves at a controlled *speed* during the trace, with the number of traces per second determined by the *sweep rate* and by the number of trigger pulses that arrive per second. Triggered sweep scopes can be used to determine the frequency of a signal fairly accurately, and in general, they are easier to use than the recurrent sweep types.

When the probe of a scope is connected to a circuit, the circuit is disturbed by the presence of the probe in much the same way that a circuit is affected by connection of an EVM. Of particular importance at high frequencies is the capacitance introduced by the lead wires between the scope and the probe. This capacitance is large enough to completely alter the operation of a tuned circuit to which the probe is connected.

Special, low-capacitance probes must be used at high frequencies so that the circuit under test is disturbed a minimal amount. A large-value resistance is connected in series with the input to isolate the probe tip from the capacitance of the leads and from the input capacitance of the scope.

The large series resistance causes an attenuation of the input signal, usually by a factor of 10. Hence, low-capacitance probes are 10X probes, and an input voltage to a 10X probe must be 10 times the input to a 1X probe to produce similar displays. At the same time, the 10X probe can be applied to voltages 10 times greater than to a 1X probe without fear of damage to the input circuitry of the scope.

Signal Tracer

A signal tracer is essentially a high-gain audio amplifier that is usually equipped with an RF demodulator probe. The output is an audible signal from the speaker of the signal tracer. Many signal tracers include a *magic eye* to provide a visual indication of signal amplitude, and in such cases the speaker may be switched off if desired.

A signal tracer is used much like a scope, with obvious shortcomings in comparison to a scope. Only rough approximations of signal amplitude can be made, and distortion content of the signal is hard to discern. Also, the signal tracer responds only to audio information, but the RF demodulator (AM) probe permits tracing of an IF signal provided that the signal is amplitude-modulated. The probe recovers the audio component, which is then heard in the speaker.

The same considerations must be given to the input capacitance of a signal tracer as for a scope or voltmeter. Connecting the probe to a tuned circuit changes the resonant frequency of the circuit, perhaps enough to render the circuit inoperative.

A signal tracer will not respond to FM signals, even when the RF probe is used because the probe responds only to AM. Therefore, a signal tracer is useless in troubleshooting the IF section of an FM receiver.

By and large, a signal tracer is most useful in troubleshooting audio circuits, and, perhaps, IF amplifiers in AM receivers. A signal tracer is of little if any value in the front end of an AM receiver or in the radio-frequency sections of an FM receiver. In almost all cases, the same objectives can be accomplished with signal-injection techniques. The one exception, perhaps, is the case in which excessive noise is being generated somewhere in a receiver. A signal tracer could possibly isolate the source of noise; in this case, signal injection would be of no value.

10-3 FAILURE MECHANISMS IN TUBES

Tubes fail to function properly for many reasons, the most obvious fault being that of a burned-out or broken heater. Heater continuity can be checked with an ohmmeter, or the proper voltage can be applied directly to the heater pins on the tube, in which case the heater should light.

After a tube has been in service for an extended period (the length of which depends upon the application), the ability of the cathode to emit electrons decreases to the point where the tube no longer functions properly. In short, not enough electrons are given off by the cathode. The *cathode emission* can be temporarily increased by supplying the heater with a higher-than-normal voltage, but this is not an economical solution in radio receivers. The practical solution is to replace the tube.

A tube sometimes becomes *gassy* as a result of the presence of excessive air molecules in the vacuum region. This condition is indicated by a blue glow inside the tube during operation, but the presence of a blue glow does not necessarily mean that the tube is gassy. Tubes operated either at high voltages or at high frequencies will sometimes emit a blue glow from the vicinity of the plate or screen grid when the tube is perfectly good. Substitution and observation of a tube known to be good usually indicates whether the suspected tube is really defective.

Occasionally, the elements within a tube will short, with shorts being most likely to occur between the grid and the cathode or between the heater and the cathode. The short may be present only at operating temperatures, and such high-temperature shorts may not be detected by a tube checker. Again, substitution of a tube known to be good is the most practical servicing procedure.

Microphonics

The elements within a tube sometimes become susceptible to mechanical vibrations, and it is possible for the mechanical vibrations to be transferred to the electrical signals flowing through the tube. This can give rise to a wide variety of squeals, howls, and various other noises. A tube with this problem is said to be *microphonic*.

A tube can be checked for microphonics by tapping it lightly with a screwdriver to determine if it is particularly sensitive to vibration. Tubes in older receivers with tube sockets that are worn and dirty normally produce a noise when jarred or moved around, but a microphonic tube will be extremely sensitive. We once observed a tube so sensitive to vibration that the ticking of a wristwatch placed in contact with the tube was reproduced with good volume in the speaker. Microphonic tubes may occur in any section of a receiver; they are not limited to the audio section.

Tube Sockets

It is fitting at this point to mention the tube socket as a potential cause of difficulty with a tube. The pins of the tube must make good electrical contact with the connectors of the tube socket. In older receivers the socket connectors may have become dirty, corroded, or bent so that good contact is not achieved between the pins of the tube and the connectors of the socket. A problem with a tube socket may be overlooked unless the technician makes it a point to remember the importance of the socket. If inspection reveals an unusual amount of dust, deformed contacts, or a general dirty appearance, the socket should be considered suspect. Another symptom of a dirty socket is the production of loud, banging noises in the speaker when the tube is wiggled in the socket. The tube should seat distinctly as the socket connectors grip the pins.

10-4 FAILURE MECHANISMS IN TRANSISTORS

Transistors sometimes fail in operation since there are several things that can happen to a transistor, resulting in complete failure of the device or in degraded characteristics. Transistors are subject to more "anomalies" than vacuum tubes, and it is not uncommon for a transistor to check *good* by ohmmeter and by transistor checker and still fail to function in the circuit.

Transistor junctions fail by becoming open, shorted, or leaky, as revealed by an ohmmeter or transistor checker. An open junction will read infinite resistance with both polarities of the ohmmeter. A shorted junction will read a resistance much lower than normal for a good junction, perhaps near zero on the ohmmeter scale. A leaky junction will read less than infinity in the reverse direction. We note, however, that germanium transistors tend to leak much more than silicon devices, and a large, high-power germanium transistor can exhibit an impressive amount of leakage and still be good. On the other hand, any silicon transistor that exhibits a measurable leakage must be considered suspect. We now describe, in brief, a few of the mechanisms that cause transistors to fail.

The external surface of the semiconductor is rather susceptible to contamination by moisture and other contaminants. A contaminated surface may give rise to leakage currents among the emitter, base, and collector regions. These surface conduction paths are essentially in parallel with the junctions. Surface effects are minimized in *passivated* devices, in which the external surface of the semiconductor crystal is coated with a thin layer of silicon dioxide (glass) or other material that protects the junctions from contamination.

Contaminants within the semiconductor material, under the influence of temperature and applied voltage, may give rise to a gradually deteriorating junction that eventually becomes leaky or shorts out completely. A nonuniform thickness of the base region, resulting from irregular impurity diffusion during manufacture, may cause high current densities in the thin areas that produce localized "hot spots" in the base region. The high temperatures of the hot spots may produce further diffusion of the impurities, which ultimately shorts the emitter to the collector.

Transistors frequently fail because the contacts to the semiconductor material become open or intermittent. The contacts may be sensitive to vibration, temperature, mechanical stress, or applied voltage.

Transistors sometimes become noisy in operation, and such a transistor in a receiver may generate so much static that the desired signal is almost obscured. All transistors contribute a certain amount of noise to the signal, but excessive noise is generally a symptom of a defective junction, a contaminated device, or a bad internal contact. The noise generation is frequently rather sensitive to temperature changes, with the noise decreasing as the device is cooled. This provides a possible means of locating a noisy transistor in a receiver.

Heat may be regarded as "Public Enemy No. 1" for semiconductor devices. Since the diffusion processes used to fabricate the transistor are controlled by temperature, it is easy to visualize that high temperatures due to excessive current flow or improper heat sinking would allow these diffusion processes to continue. In short, high temperatures limit the useful life of any solid-state device.

10-5 OHMMETER IDENTIFICATION AND TESTING OF BJTs

An ohmmeter may be used to determine whether a BJT is *NPN* or *PNP* and which lead is the base, the emitter, and the collector. No prior knowledge of the transistor is required except that it is a bipolar junction type. But first, two precautionary items should be mentioned.

Some ohmmeters, in the low-resistance ranges, can deliver enough current to damage the transistor under test. Therefore, only high-resistance ranges ($R \times 1$ kΩ, for example) should be used. Second, the polarity of the voltage delivered to the ohmmeter test prods must be known. On simple VOM (volt-ohm-milliampere) meters, it is almost never the lead marked (+) or colored red. The polarity can be determined with a DC voltmeter by noting the direction of voltmeter needle deflection when the voltmeter is connected to the ohmmeter.

It is a simple matter to check a diode (a *PN* junction) with an ohmmeter. The resistance should read almost infinite in one direction and much lower in the other. The actual resistance indicated in the forward direction depends upon the particular ohmmeter being used. A typical forward resistance might be 15,000 Ω. The absolute value is not important, but the resistance of a *PN* junction known to be good should be measured and the resistance memorized for that particular meter and for the particular scale used. Almost all *PN* junctions will measure this same resistance if the junction is good. We shall refer to this resistance as being the *ohmmeter junction resistance*, or OJR. Silicon and germanium diodes may give different OJRs on any given ohmmeter, and when this is the case, the difference should be noted.

Procedure

To identify the type (*NPN* or *PNP*) of a BJT and the lead connections, proceed as follows. Mentally label the leads of the transistor 1, 2, and 3 for purposes of reference. Attach the positive ohmmeter prod to lead 1. Connect the other prod to lead 2, and note the reading. It should be either infinity or OJR. Then disconnect the prod from lead 2 and connect it to lead 3. Does the ohmmeter indicate the same resistance as it did for lead 2? If it does, lead 1 might be the base, but at this point we cannot be sure. If the readings were not the same, we can be sure that lead 1 is not the base.

If the resistance of lead 2 and 3 were both infinity, reverse the prods and repeat the test. The negative prod now connects to lead 1. If the resistance at leads 2 and 3 are now the same, lead 1 is the base. If they are different, lead 1 is not the base. We now assume, for our purpose, that lead 1 was not the base.

Connect the positive ohmmeter prod to lead 2 and check the resistance at leads 1 and 3. If the resistance is OJR at both leads, lead 2 is the base. If the resistances are different, lead 2 is not the base. If the resistance is infinity at both, we are not sure. Reverse the ohmmeter prods and repeat the test. The same reading (OJR) at both leads 1 and 3 confirms that lead 2 is the base. If the resistances are different, lead 2 is not the base.

At this point we will either have found the base or we will have eliminated two or three possibilities. If we have not found it, it must be the remaining lead 3. Thus, we have now found the base, one way or the other.

Once the base has been located, we can determine if the transistor is *PNP* or *NPN*. Note the polarity of the prod, which when connected to the base yields OJR when the other prod is connected to either of the other two leads. If the polarity of the base prod is positive, the transistor is *NPN*. If it is negative, it is *PNP*.

Collector Identification

We must now identify which of the remaining leads is the emitter and which is the collector. If the transistor is *NPN*, connect the positive prod to the lead you suspect to be the collector, and connect the negative lead to the emitter. (Reverse this for *PNP*.) The resistance should read infinity for a good transistor. We now connect a high resistance (1-MΩ) between the collector and the base and note the resulting ohmmeter reading. If the resistance goes rather sharply toward OJR, our guess about the collector and the emitter is probably correct. If it is sluggish so that the indicated resistance is not too far from infinity, the collector and emitter are probably reversed.

Trick

Here is a helpful trick. It is inconvenient to have to hunt for a 1-MΩ resistor to connect between the base and the collector, so short a moistened finger between the base and the collector. Practice with a known transistor will reveal the proper amount of moisture and pressure needed to establish contact between the leads of the transistor. With the collector and emitter properly identified, the ohmmeter will quite promptly go to OJR. With the elements reversed, a large resistance will be indicated, perhaps 10 times the OJR.

If in following the procedure described above the ohmmeter ever goes to zero resistance, the junction between the two leads is shorted and is therefore defective. If a junction reads infinity in both directions, the junction is open. Whenever a resistance reading between infinity and OJR is obtained, first ascertain that no fingers are touching the ohmmeter leads, and then suspect the junction to be leaky.

A word of caution: a transistor that checks good according to the procedure above might still be bad. The procedure is a DC check that determines if the junctions are intact. A transistor might have degraded junctions that could check good using the ohmmeter but might not function in an actual circuit. But if a transistor checks out as bad in the test above, it is definitely bad.

10-6 CHOOSING REPLACEMENTS FOR DEFECTIVE PARTS

After the defective part has been located, several things must be taken into account when selecting the replacement. It should be made of the same type of material and should have the same value, voltage rating, power rating, tolerance, and temperature coefficient as the original. When the replacement matches the original in all aspects, the replacement is an *exact replacement*. An exact replacement is not always required, and, for that matter, may not always be readily available. Some variation from the original part is permissible, and this is the topic of the following sections.

Resistors

Four factors are significant when considering a replacement for a defective resistor: the type of material of which the resistor is made, tolerance, ohmic value, and power rating.

Common types of resistors are, in the order most frequently encountered, carbon composition, wire wound, and metal film. Carbon and wire-wound resistors are most common in receivers, while metal film resistors are used primarily in precision applications such as measuring instruments or low-noise circuitry. Carbon composition resistors generate more noise and are larger in size for the same power rating than wire-wound resistors, but they exhibit minimal inductance and are inexpensive. In low-noise circuits such as audio preamps and RF amplifiers, carbon composition resistors should not be used to replace either wire-wound or metal film resistors.

Wire-wound resistors exhibit considerable inductance in addition to the resistance and should not be used to replace carbon composition resistors in high-frequency circuits such as IF amplifiers, oscillators, mixers, and RF amplifiers. In most circuits, however, the different types of resistors can be interchanged without difficulty, although the physical size may not permit replacing a wire-wound resistor with a carbon composition of equal power rating. Circuits that are not sensitive to the type of resistor used are power supply circuits, high-level audio circuits, and high-frequency circuits where the high-frequency signals are not applied to the resistor.

Resistor tolerances are typically 5, 10, or 20%, with 10% being the most common in receivers. A replacement resistor should have a tolerance equal to or better (smaller) than the original. A 5% resistor will replace a 10% resistor, and a 10% resistor will replace a 20% resistor. A 20% resistor should not be used as a replacement for an original with a lower tolerance unless an ohmmeter is used to verify that the resistance of the 20% resistor is actually within the limits of the lower tolerance.

Most receivers use resistors whose ohmic values are standard in the industry. The replacement resistor should be of the same ohmic value, although it is possible to arrange series and/or parallel combinations of resistors to obtain a correct value when the needed value is not readily available. For example, two 100-Ω resistors in parallel produce an equivalent resistance of 50 Ω, and two 3.3-kΩ resistors in series will replace a 6.8-kΩ resistor. Combinations of resistors should be checked with an ohmmeter to be sure they are within tolerance for the particular application.

The power-handling capability of the replacement must be equal to or better than the original. A larger rating can be used to replace a smaller rating, but never the other way around. A 1-W resistor may replace a 1-W or a ½-W resistor, but it is not acceptable as a replacement for a 2-W resistor. In resistor combinations, if equal power dissipations occur in all resistors, the total power-handling capability is equal to the sum of the individual capabilities if adequate ventilation is provided. For example, two 100-Ω 2-W resistors may be connected in series to produce a 200-Ω 4-W resistor, or the two resistors may be connected in parallel to produce a 50-Ω 4-W resistor.

Control Potentiometers

A variable resistor (*potentiometer* or "pot") should be replaced with a part having the same resistance value and *taper* as the original. By taper is meant the rate of change of resistance with degree of rotation of the control shaft, and a linear taper produces the same change in resistance per degree of rotation over the entire range of the resistance element. An audio taper is not linear; it produces a greater change in resistance per degree of rotation near the clockwise limit or high-volume end of the range. Such a taper is desirable for audio volume controls because the human ear requires greater changes at high volume levels to produce the same perceived changes in volume.

Resistance elements made of a carbon composite material are most common in receivers, and replacement components should also be of this type. A wire-wound resistor should be not be substituted for a composition type because the variation of resistance is not smooth in wire-wound resistors. Small, discrete jumps in resistance occur as the wiper moves from turn to turn of the resistance element. On the other hand, a composite can be substituted for a wire-wound provided that the power dissipation capability of the two are equal. A composite type is generally larger than a wire-wound type of the same ohmic value and power rating.

Typical volume control pots for vacuum-tube receivers have a resistance of 1 MΩ, while volume controls in solid-state sets are typically about 10 kΩ. The difference results from the fact that solid-state sets operate at much lower impedance levels than tube sets.

Of major concern is whether the replacement is physically compatible with the application. Space for the control may be limited; a shaft of certain length and diameter must accommodate the knob; the control frequently operates in conjunction with a switch; and a given control may be intended either for front mounting or for mounting on a PC board. In many cases an exact replacement must be obtained from the receiver manufacturer, but it is often possible to adapt a general replacement potentiometer to the particular application.

Capacitors

Five things must be considered when selecting a replacement capacitor. These are: the capacitance, the dielectric, the tolerance, the voltage rating (working voltage), and the temperature coefficient.

A replacement capacitor should generally have the same type of dielectric as the original, although this is not a strict requirement. Critical applications for capacitors are those involving high frequencies and resonant circuits, and the replacement should duplicate the original as closely as possible. Less critical applications include bypass capacitors, coupling capacitors in audio circuits, and various filter capacitors.

The same considerations apply to the tolerance of capacitors as apply to resistors. The one difference that is perhaps worthy of mention is that electrolytic capacitors are sometimes given an unsymmetrical tolerance such as -10%, $+50\%$. This usually is of little concern to the technician, however, because of the noncritical applications to which electrolytics are applied.

The voltage rating of the replacement should be equal to or greater than the original. A capacitor with a large working voltage may be substituted for a capacitor with a lower working voltage, but usually the higher the working voltage, the larger is the size of the capacitor for the same capacitance. High-voltage capacitors have greater plate separation (thicker dielectrics), which requires that plates with larger surface areas be used to produce the desired capacitance. Physical-size considerations may limit the working voltage of the replacement to that of the original.

Temperature Coefficient

Three temperature-coefficient designations are given to disc ceramic capacitors: N, P, and NPO. NPO capacitors have *neither positive* nor *negative* temperature coefficients; the coefficient is zero. P-type capacitors have a *positive* coefficient, and N-type have a *negative* coefficient. The complete coefficient is given by the letter followed by a number, such as N750 or P150. The number gives the parts per million change of capacitance per degree Celsius change in temperature. In selecting a replacement part, a capacitor with exactly the same temperature coefficient as the original should be selected.

Transformers

In addition to obvious concerns such as physical size and method of mounting, a replacement transformer must have windings (and appropriate taps), turns ratios, current capabilities, and total power rating so that it can perform the function of the original. Extra, unused windings on a replacement may be left open if care is taken to insulate the connections to the windings. The ends of unused windings should *not* be shorted together.

General replacement transformers are available to the service industry, and the manufacturers of such transformers publish cross-reference and replacement guides that enable the service technician to determine the replacement for a given part number in the receiver. Great care must be taken to ensure compatibility of the replacement, or extensive and perhaps catastrophic damage may be done to the receiver in the event that the windings are improperly connected or develop the wrong voltage.

Solid-State Devices

Manufacturers of replacement semiconductors publish extensive cross-reference and replacement guides that provide replacement information on almost any transistor that may be encountered. In the event no replacement can be found from the replacement lines of semiconductors, the exact replacement can be ordered from the manufacturer of the receiver; this is seldom required, however.

The better-known manufacturers of replacement semiconductors include General Electric (GE), International Rectifier (IR), Motorola (HEP), Sylvania (ECG), and RCA (SK). The initials given in parentheses by each manufacturer refer to the first part of the type designations of the replacement devices. For example, typical RCA devices are SK3018 and SK3049; a Sylvania device, for purposes of illustration, is the ECG123.

10-7 TIPS ON REPLACING DEFECTIVE PARTS

Sometimes the installation of a replacement part is not as simple as it may appear. Modifications may have to be made to the replacement part or to the way it is mounted in order to get it situated in a satisfactory manner.

Volume Controls

Replacement volume or tone controls will probably have to be cut and shaped to fit. First cut the shaft of the replacement to the correct length by measuring and marking, as shown in Figure 10-7. During the cutting operation, be sure that the hacksaw does not tend to rotate the shaft in relation to the body of the control, because damage may result if the shaft is forced to rotate beyond the stops built into the device. Clamp the vise to the shaft, *not* to the body of the control.

After cutting the shaft to length, mark the length and depth of the index for the knob as shown in the figure. Then make the cut for the index using either a

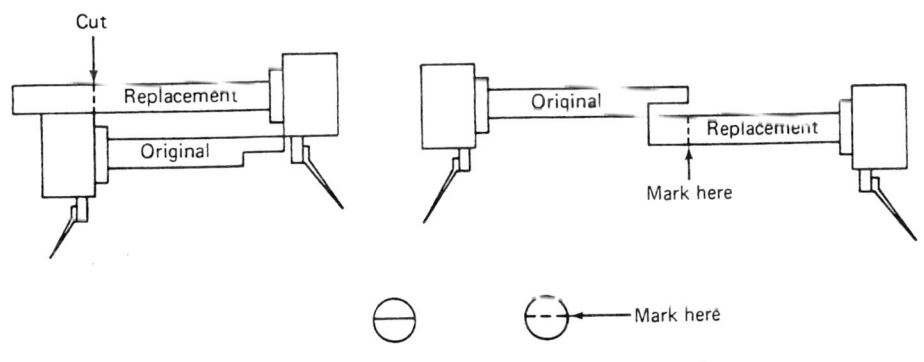

FIGURE 10-7 Sizing replacement volume control.

hacksaw or a file. Try the knob to ascertain the proper fit before installing the control.

Top-Side Exchange of Components

Occasionally, it is rather difficult to get to the back side of the PC board to unsolder a component. In such a case the replacement may be exchanged from the top. Cut the leads of the defective component as close to the component as possible, and straighten the portion of the leads left on the PC board. Form small loops at the ends of the leads of the replacement component, and then slip these loops over the original leads that were left on the PC board by the defective component. Do not use excessive heat in the soldering operation, since the solder holding the leads of the original component might melt and let go. Transistors can be replaced in the same manner, but this technique, illustrated in Figure 10-8, should be used only if a great deal of work is involved in getting to the other side of the board.

Heat Sinks

When replacing power transistors, due consideration must be given to the heat-sinking properties of the mounting arrangement. Duplicate the original assembly as closely as possible, paying particular attention to any insulators that may be involved. If the original used an insulator, be sure to use an insulator with the replacement. If the original did not use an insulator, do not use an insulator with the replacement even though an insulator might be included in the package that contained the replacement.

Silicone compounds have been developed that provide good thermal conductivity while exhibiting good electrical insulation properties. This silicone grease is applied to mating surfaces to improve the transfer of heat from the body of a power transistor to the chassis. In many applications it is essential that the silicone grease be used. If omitted, the transistor will run much hotter, resulting in a much shorter useful life for the device.

Another factor that affects the heat transfer between a transistor and a chassis is the pressure with which the transistor and chassis are forced together by the mounting screws. The mounting screws or bolts should be tightened firmly, but not so tightly that the transistor or mounting hardware is damaged. Common sense must be used in this regard.

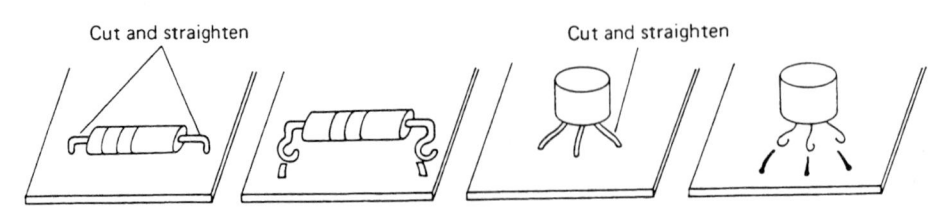

FIGURE 10-8 Installation of a replacement component from the component side of a PC board.

Bridge with insulated wire Bridge with solder

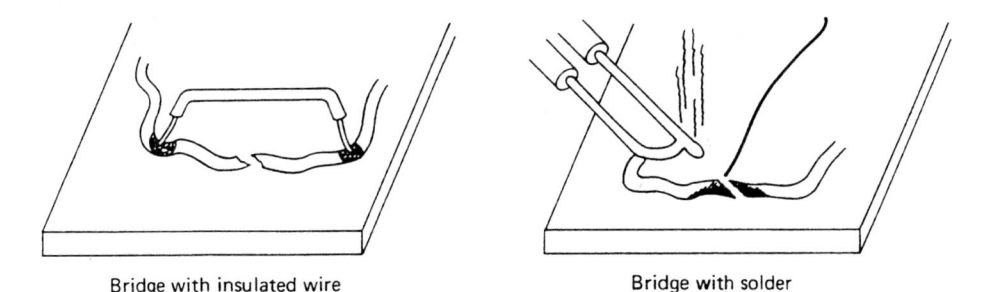

FIGURE 10-9 Two methods for repairing a broken foil on a printed circuit board.

Forming Leads

Plastic tab transistors frequently must have the leads formed during the process of installation. Do not bend the leads without supporting the leads between the point of the bend and the body of the transistor. It is an easy matter to break one of the leads internally.

The suggested replacement for a defective transistor may not be physically identical to the original, even though the two are electrically identical. Frequently, the basing diagram of the replacement differs from the original, requiring that the leads of the replacement be crossed in order to be compatible with the circuit board. Insulation should be used on one lead, and care should be exercised in bending the leads.

Broken PC Boards

Sometimes portable receivers get dropped, and this frequently results in a broken PC board. The printed circuit foils can be repaired either by soldering at the break or by using short pieces of hookup wire to bridge the break as in Figure 10-9. Frequently, a protective coating must be scraped from the foils before soldering can be done successfully. If the break in the board is severe, use a good grade of epoxy glue to give the board physical strength.

SUMMARY

1. Electronics servicing is expedited by a good assortment of hand tools, good service data, and a well-lighted service bench.

2. A desoldering device may be required to remove a component from a PC board if the component has several terminals that cannot be removed one at a time.

3. A source of light placed behind a PC board will allow PC conductors to be discerned as silhouettes.

4. It is important to use the proper alignment tools for making alignment adjustments.

5. An assortment of chemicals should be available: contact cleaner, freeze spray, alcohol, epoxy glue, plastic glue, lubricating oils, and so forth.
6. Every service bench should have an isolation transformer.
7. An EVM is perhaps the most important and most used item on a service bench. With an EVM, checks can be made of resistors, capacitors, inductors, transformers, tube filaments, and solid-state devices.
8. The high impedance of an EVM is necessary to minimize disturbance of the circuit being tested. Lower-impedance VOMs are restricted in application.
9. Whenever a scope, EVM, or signal tracer is connected to a high-frequency circuit, the capacitance of the test probe and associated circuitry disturbs, to some extent, the circuit under test. Unless special low-capacitance probes are used, the circuit may be rendered completely inoperative when the probe is connected.
10. A voltmeter and a DC power supply can be used to determine the leakage resistance of a capacitor at high voltages.
11. One of the most useful test instruments is the oscilloscope. The usefulness is generally determined by the quality and sophistication of the instrument.
12. A service bench should have an audio generator and an AM-modulated RF generator.
13. A signal tracer is essentially a high-gain audio amplifier, usually equipped with an RF demodulator (AM) probe.
14. Most common tube failures are due to broken heaters, loss of cathode emission, shorted elements, gas within the envelope, and microphonics.
15. The tube socket must not be overlooked as a possible cause of a defect.
16. Common transistor failures are due to open leads to the semiconductor materials, shorted junctions, or leakage from the collector to the base or from the collector to the emitter.
17. The junctions of a transistor may check *good* even though the high-frequency properties of the transistor may be seriously degraded. In such case, a DC check of the device may show it to be good, although it fails to perform in-circuit.
18. Germanium transistors typically show much more leakage than silicon devices. Any silicon device that shows appreciable leakage should be considered suspect.
19. Heat is "Public Enemy No. 1" for semiconductors.
20. An ohmmeter can be used to determine whether a transistor is *NPN* or *PNP*.
21. An ohmmeter can be used to identify the base, emitter, and collector leads of a transistor.
22. In choosing a replacement resistor, consideration must be given to type, tolerance, ohmic value, and power rating.
23. Potentiometers should be replaced by units having the same taper as the original.
24. In choosing a replacement capacitor, consideration must be given to the capacitance, tolerance, type of dielectric, voltage rating, and temperature coefficient.

25. Manufacturers of replacement semiconductors publish extensive cross-reference guides to assist in the selection of a replacement component.

26. Minor modification of replacement components is sometimes required to achieve physical compatibility with the particular application.

QUESTIONS AND PROBLEMS

10-1. Explain how a flashlight can be used to facilitate circuit tracing by rendering PC conductors "visible" from the component side of the board.

10-2. Give a procedure for removing a knob that is stuck on a control shaft.

10-3. Why is it important to have the proper alignment tools? What is wrong with using metal screwdrivers for making alignment adjustment?

10-4. What is an isolation transformer and for what is it used?

10-5. Describe the importance of eye protection for a technician. How can one suddenly come to have an eyeful of hot solder?

10-6. For what purpose is an assortment of colored felt tip pens useful when a receiver is being disassembled?

10-7. What is the typical input impedance of an EVM? How does this compare with the input impedance of a 20-kΩ/V VOM set on a 5-V range?

10-8. Explain why an EVM and a VOM might give different readings in a DC circuit containing large resistances.

10-9. What might be the effect of connecting a VOM to the oscillator grid of a pentagrid converter?

10-10. Explain what properties of resistors, capacitors, inductors, and transformers can be checked with an EVM. What properties of these components cannot be checked with an EVM?

10-11. Describe in detail how the leakage resistance of a capacitor can be measured with a voltmeter.

10-12. What are the important aspects of an oscilloscope that should be considered when the purchase of a scope is contemplated?

10-13. How does a 10X probe differ from a 1X probe?

10-14. In what section of a receiver is a signal tracer most useful? If a scope is available, is a signal tracer really needed?

10-15. Suppose that voltage measurements made from the component side of a PC board indicate that a certain transistor is defective, and that the foil side of the board can be accessed only with great difficulty. Describe a procedure for installing the replacement transistor.

10-16. Describe two techniques for repairing a hairline crack in a PC board.

10-17. Describe a procedure for removing a multiterminal component if all PC terminal connections must be removed simultaneously.

10-18. What is a microphonic tube?

10-19. How can a gassy tube be recognized? Is this a conclusive test?

10-20. Comment upon the pros and cons of investing in a stock of a wide variety of tubes rather than investing the same money in a tube checker.

10-21. What parameters of a transistor cannot be checked with an EVM?

10-22. The test lead marked (+) and colored red on most VOMs is actually the negative terminal in the ohmmeter mode:

    (a) How could this be ascertained for a given VOM?

    (b) What might be the result of a transistor identification test if the ohmmeter polarities are unknowingly reversed?

10-23. Experience has shown that transistors that check *good* on a transistor checker sometimes fail to function in circuit. Should this fact be considered when the purchase of a transistor checker is contemplated? What alternative to the purchase of the transistor checker might be suggested?

10-24. What is the procedure for using freeze spray to locate a noisy transistor?

10-25. A large power transistor is to be replaced. The original is smeared with a white, sticky, greaselike compound. What is this material, and what is its purpose?

10-26. Can freeze spray be squirted on the glass envelope to determine if a tube is microphonic? Comment on the possible hazard produced by squirting the freeze spray on the hot glass envelope of the tube.

10-27. What is meant by the "taper" of a potentiometer used for a volume control?

10-28. Describe the things that must be considered when selecting a replacement

    (a) Resistor.

    (b) Capacitor.

    (c) Inductor.

    (d) Transistor.

10-29. Describe the construction of a wire-wound resistor.

10-30. Why should wire-wound resistors not be used in high-frequency circuits?

10-31. In a leakage resistance test of a capacitor, using a 100-V power supply and an 11-MΩ VTVM, the voltmeter reading was 10 V. Calculate the leakage resistance of the capacitor.

10-32. Following Prob. 10-31, suppose that the voltmeter read 20V, 30V, . . . 100V. Compute the leakage resistances and construct a graph of leakage resistance versus voltmeter reading.

10-33. To what does the bandwidth of an oscilloscope refer?

10-34. List three ways in which tubes fail.

10-35. Draw a diagram to show how a high-resistance leakage path between the collector and the base of a transistor tends to bias the transistor into conduction in a manner similar to the base-bias arrangement described in Chap. 4.

10-36. When the resistance of a *PN* junction is measured with an ohmmeter, what is the significance of a resistance of $0\ \Omega$?

11
BASIC TROUBLESHOOTING PROCEDURE

In this chapter we describe the fundamental procedures for troubleshooting electronic devices. The techniques of signal injection, signal tracing, and voltage-resistance analysis are presented within the framework of the basic troubleshooting procedure.

We emphasize troubleshooting by procedure, being methodical, proceeding according to a plan. Each test or each measurement should move the technician one step closer to the conclusion. Procedures must be flexible, however, and more than one procedure may be applicable to a given symptom. In this text we suggest what we think is the best procedure for each of several symptoms, but another qualified and experienced technician may have distinctly different suggestions as to the order in which checks are made and whether to use signal tracing or signal injection, for example.

Finally, we must comment that in practice, troubleshooting is never as easy, logical, and forthright as a textbook always makes it sound. But then again, a textbook also cannot provide the feeling of warm, enveloping joy that sets in when a stubborn, frustrating defect is finally located.

11-1 BASIC TROUBLESHOOTING PROCEDURE

The objective, when servicing a defective receiver, is to find and repair the defect in the least time consistent with high-quality workmanship and a professional approach. A systematic approach to localizing the defect is needed so that only a few components have to be checked. This involves a process of elimination. The various sections and stages are subjected to various tests to determine which are not functioning properly. These tests can be grouped into a series of logical steps called the *basic troubleshooting procedure* (BTP). The BTP is general in nature and applies to the majority of symptoms encountered. The BTP is outlined in Figure 11-1, and each step is explained in the following section.

Questioning of the Owner

The owner can be a valuable source of information if encouraged to describe the problem. Ask tactful questions to get a precise description. For example, does the receiver make a buzzing sound, a hum, a scratchy sound, a howl, a squeal, a putt-

10-34. List three ways in which tubes fail.

10-35. Draw a diagram to show how a high-resistance leakage path between the collector and the base of a transistor tends to bias the transistor into conduction in a manner similar to the base-bias arrangement described in Chap. 4.

10-36. When the resistance of a *PN* junction is measured with an ohmmeter, what is the significance of a resistance of 0 Ω?

11
BASIC TROUBLESHOOTING PROCEDURE

In this chapter we describe the fundamental procedures for troubleshooting electronic devices. The techniques of signal injection, signal tracing, and voltage-resistance analysis are presented within the framework of the basic troubleshooting procedure.

We emphasize troubleshooting by procedure, being methodical, proceeding according to a plan. Each test or each measurement should move the technician one step closer to the conclusion. Procedures must be flexible, however, and more than one procedure may be applicable to a given symptom. In this text we suggest what we think is the best procedure for each of several symptoms, but another qualified and experienced technician may have distinctly different suggestions as to the order in which checks are made and whether to use signal tracing or signal injection, for example.

Finally, we must comment that in practice, troubleshooting is never as easy, logical, and forthright as a textbook always makes it sound. But then again, a textbook also cannot provide the feeling of warm, enveloping joy that sets in when a stubborn, frustrating defect is finally located.

11-1 BASIC TROUBLESHOOTING PROCEDURE

The objective, when servicing a defective receiver, is to find and repair the defect in the least time consistent with high-quality workmanship and a professional approach. A systematic approach to localizing the defect is needed so that only a few components have to be checked. This involves a process of elimination. The various sections and stages are subjected to various tests to determine which are not functioning properly. These tests can be grouped into a series of logical steps called the *basic troubleshooting procedure* (BTP). The BTP is general in nature and applies to the majority of symptoms encountered. The BTP is outlined in Figure 11-1, and each step is explained in the following section.

Questioning of the Owner

The owner can be a valuable source of information if encouraged to describe the problem. Ask tactful questions to get a precise description. For example, does the receiver make a buzzing sound, a hum, a scratchy sound, a howl, a squeal, a putt-

If all these questions receive a positive response, the receiver should be completely reassembled and checked out again to ensure that the reassembly process did not introduce any new problems. This "after-assembly" check is very important, because many things can happen during final assembly. If any of the questions brings a negative response, the new problem must be corrected. When any of the first six items is answered "no," the alignment of the receiver should be checked as described in Chap. 12.

After the receiver is reassembled, it should be played for at least 30 minutes with an application of item 8 every once in a while. When satisfactory performance is established, the receiver is ready to be cleaned up for delivery. Also, the charges for the repair should be calculated at this time.

Cleanup

After the receiver is approved for delivery, the cabinet should be cleaned and polished. This creates a good impression with the customer, who will reason that if the technician does a good job on the outside, he must have done a good job on the inside. The work area should also be cleared and tools returned to their proper place. The next job will go faster if the work area is not cluttered from the previous job. A neat work area also reflects a well-organized technician who takes pride in his work. All these factors create a positive impression with the customer.

11-2 SIGNAL INJECTION

Signal injection consists of applying a signal generator to the input of a stage and determining if the proper signal appears at the output, as indicated by an *indicator*. The speaker is the indicator commonly used in a receiver, although a scope, signal tracer, or other signal-detecting device can be used. Signal-injection procedures are useful in locating the defective section or stage of a receiver. The procedure is illustrated in the following section.

Audio Section

The block diagram and signal-injection chart of Figure 11-2 illustrates the process of using signal injection to localize the problem in an audio amplifier. The test signal is applied at various points, and at each point the question of whether the test signal is heard at the speaker is asked. The conclusions drawn and subsequent action is indicated on the signal-injection chart.

If the test signal applied to the volume control (point A) is heard in the speaker with proper volume, the audio section is working, and the trouble is located in the sections that preceed the volume control. If the signal is not heard at the speaker, the defect must be in the audio section. We proceed to localize the defect by moving the test signal input closer to the speaker.

If the test signal applied to point B produces a sound in the speaker, the coupling

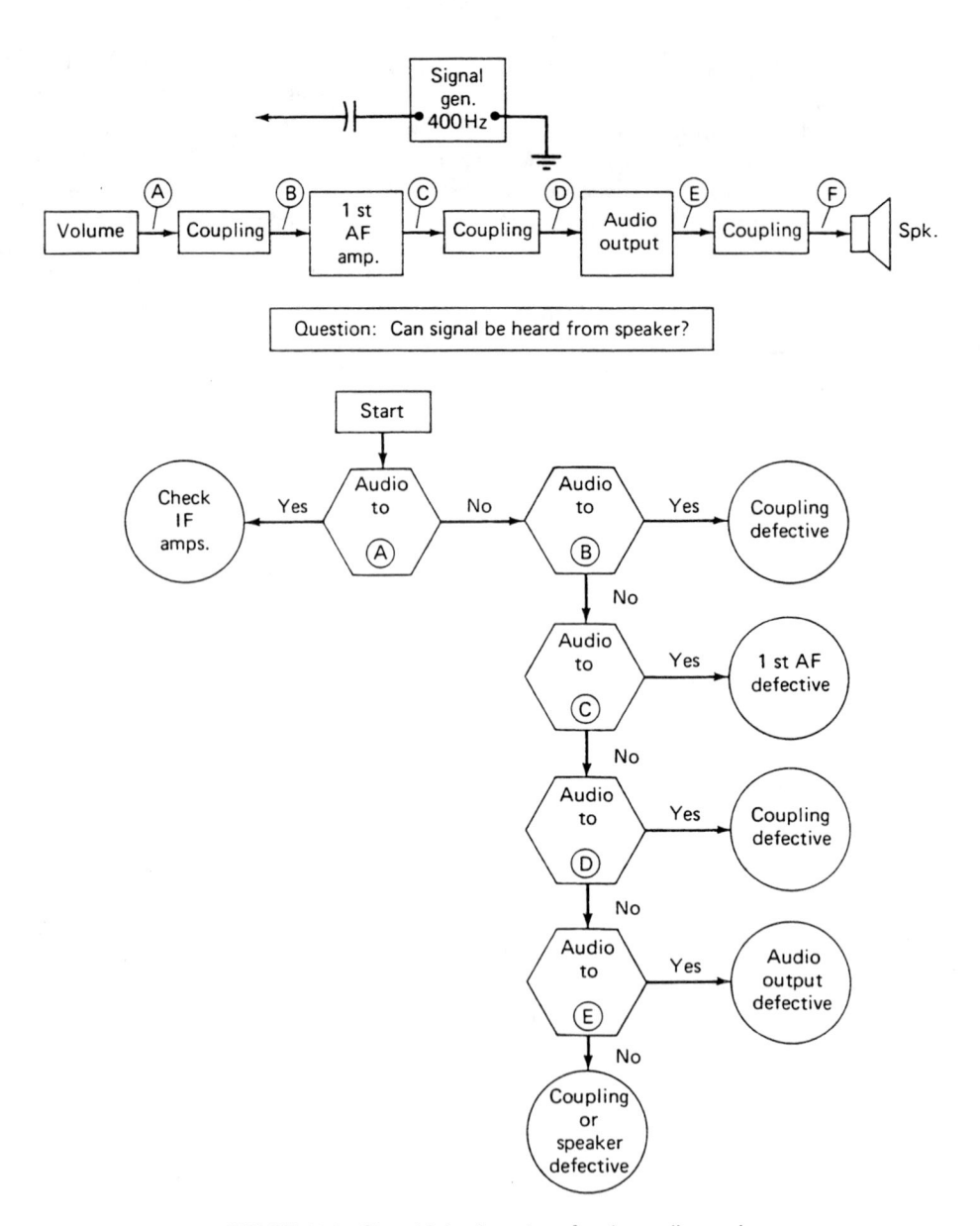

FIGURE 11-2 Signal-injection chart for the audio section.

device or network between the volume control and the input to the first audio amplifier is defective. If no sound is produced, the defect lies still closer to the speaker, and we move the test signal input to point *C*. A sound at the speaker implies that the first audio amplifier is defective, but if no sound is produced, the test signal is moved to point *D*, the input to the power audio amplifier.

This process is continued until the signal is heard at the speaker. By noting

> 1. Question the owner.
> 2. Verify the complaint.
> 3. Perform sensual examination of receiver.
> 4. Perform isolation procedures to isolate the defective section, the defective stage, and, finally, the defective component.
> 5. Replace or repair the defective component.
> 6. Aircheck for normal operation.
> 7. Clean up the bench and do paperwork.

FIGURE 11-1 Major steps in the basic troubleshooting procedure.

putt? How long has the receiver been like this? When did it first happen (during a thunderstorm? when first turned on?)? Does it do it all the time (the problem may be intermittent)? If the set is new, it may have brought a problem with it from the factory.

During the initial conference with the owner, any obvious defects such as broken knobs or a frayed line cord should be brought to the owner's attention to determine the extent to which these "optional repairs" are desired. An agreement should be reached as to the maximum amount of money the owner wishes to invest in the repair, or an estimate should be given before the repairs are rendered. Be sure the owner understands that a charge will be made for any estimate that is based upon a detailed circuit analysis even if repairs are not performed.

Verification of the Complaint

The receiver should be turned on, before it is disassembled, to verify that the receiver does indeed act as the customer said. Sometimes, in transporting the set from home, new defects occur, or the original problem cures itself. Problems that cure themselves are usually due to lose or dirty connections, and such problems tend to return. This initial "turn-on" should be done with the owner present if possible, to avoid incurring the blame for any new problems that may have developed. Other problems overlooked by the owner may be present, and these should be brought to his attention. It frequently happens that time and other involvements do not allow the initial turn-on to be done in the presence of the owner, but it should be done before disassembling the receiver so that any damage that occurs during disassembly can be readily identified. A wire pulled loose or broken during disassembly can very easily go unnoticed.

Sensual Examination

As the receiver is disassembled, inspect the circuitry very carefully. Look for broken wires or components, charred components, missing tubes, and so forth. After the set is turned on, smell for unusual odors, look for a whisp of smoke, feel suspected components for overheating or the absence of heat. (Remember, power resistors and power transistors are supposed to be warm.) Ascertain that all vacuum tubes are lit up.

Listen carefully to the symptom and try to isolate it to a particular stage or even to a particular part. Some symptoms are caused only by certain parts, and if the part can be identified quickly, much time will be saved. This process of localizing a defective part by the symptom it produces is known as *symptom diagnosis*.

Isolation Procedures

If the defect is not found during the steps describe above, isolation procedures must be employed to locate the defect. These procedures consist of symptom diagnosis, signal injection, signal tracing, voltage analysis, resistance analysis, and parts checking or parts substitution. The usefulness of symptom diagnosis, identifying a defect solely from the symptom it produces, increases with a technician's experience and is described in Chapter 12. The other procedures are described in later sections of this chapter. Systematic application of the procedures, plus a bit of logical thinking and careful analysis, will, in the majority of cases, lead to the defect.

Replacement or Repair of Defective Part

In electronic servicing, the consensus is that locating the defect is the major portion of the battle. Replacing components is usually a routine matter, although there are many things to consider. The extent to which components can be repaired depends on the type of component and the skill of the technician.

Air Check

When repairs have been completed, the receiver should be subjected to a thorough check before it is reinstalled into the cabinet. The checklist should include the following items:

1. Is the volume level normal and sufficient?
2. Is the sound clear, distinct, and free of unusual sounds?
3. Does the receiver exhibit a normal sensitivity?
4. Does the dial indicator function properly? Does the pointer cover the entire band?
5. Do the stations come in at the right place on the dial?
6. Do all controls function properly?
7. Are all indicator and dial lamps illuminated?
8. Does the receiver operate satisfactorily when jarred or shaken?

where the signal appears, the defective stage or component can be identified. It should be noted that as one proceeds toward the speaker, larger test-signal amplitudes are required to drive the speaker to usable volume. The signal generator may not provide sufficient power to drive the speaker from point E or F. Hence, it is important to know the limitations of the equipment being used.

IF Amplifier

If the signal is heard at the speaker when the test signal is applied to the volume control, the defect lies somewhere between the volume control and the antenna. The procedure now consists of applying a modulated RF signal (at the IF frequency, typically 455 kHz) to the input of the IF amplifier, which is the same point as the output of the mixer. The generator is set to produce an amplitude-modulated test signal at the IF frequency of the receiver. When the modulated signal is applied to the mixer output, the modulating tone should be heard in the speaker if the IF amplifier is working properly. If the tone is not heard, the signal is being lost somewhere between the mixer output and the volume control.

A block diagram and signal-injection chart of the IF amplifier and detector is shown in Figure 11-3. The test-signal injection point is moved from the mixer output toward the volume control until the signal is heard in the speaker. The suspected stage or component is the one immediately before the point at which the signal first appears. If the signal could be heard when the test signal was applied to the output of the mixer (point G), the *front end* (RF amplifier, mixer, oscillator) is defective.

Front End

The block diagram and injection chart of the front end is shown in Figure 11-4. The generator is used to apply a modulated RF signal of IF frequency to the input of the mixer. If no signal is heard at the speaker, the mixer is defective. If a signal is heard, we change the generator to a frequency in the AM broadcast band and tune the receiver to the same frequency. If no signal is heard, the oscillator stage is defective.

We reach this conclusion by remembering that the oscillator must produce an unmodulated signal that is heterodyned with the antenna signal to produce the frequency conversion of the antenna signal to the IF frequency. Since the mixer will amplify a signal at the IF frequency, we know it is working. But since it will not pass a signal in the broadcast band, we reason that the broadcast signal is not being converted to the IF frequency. The stage to suspect immediately is the oscillator.

By using signal-injection procedures, we can isolate the defect to a section (audio, IF, front end), to a stage (first IF amplifier, second IF amplifier, detector), and sometimes even to the defective part. If we can isolate a part, it can be checked by the appropriate means to confirm the defect, and, hopefully, the troubleshooting job will be done. If, however, the defect is only isolated to a stage, voltage and

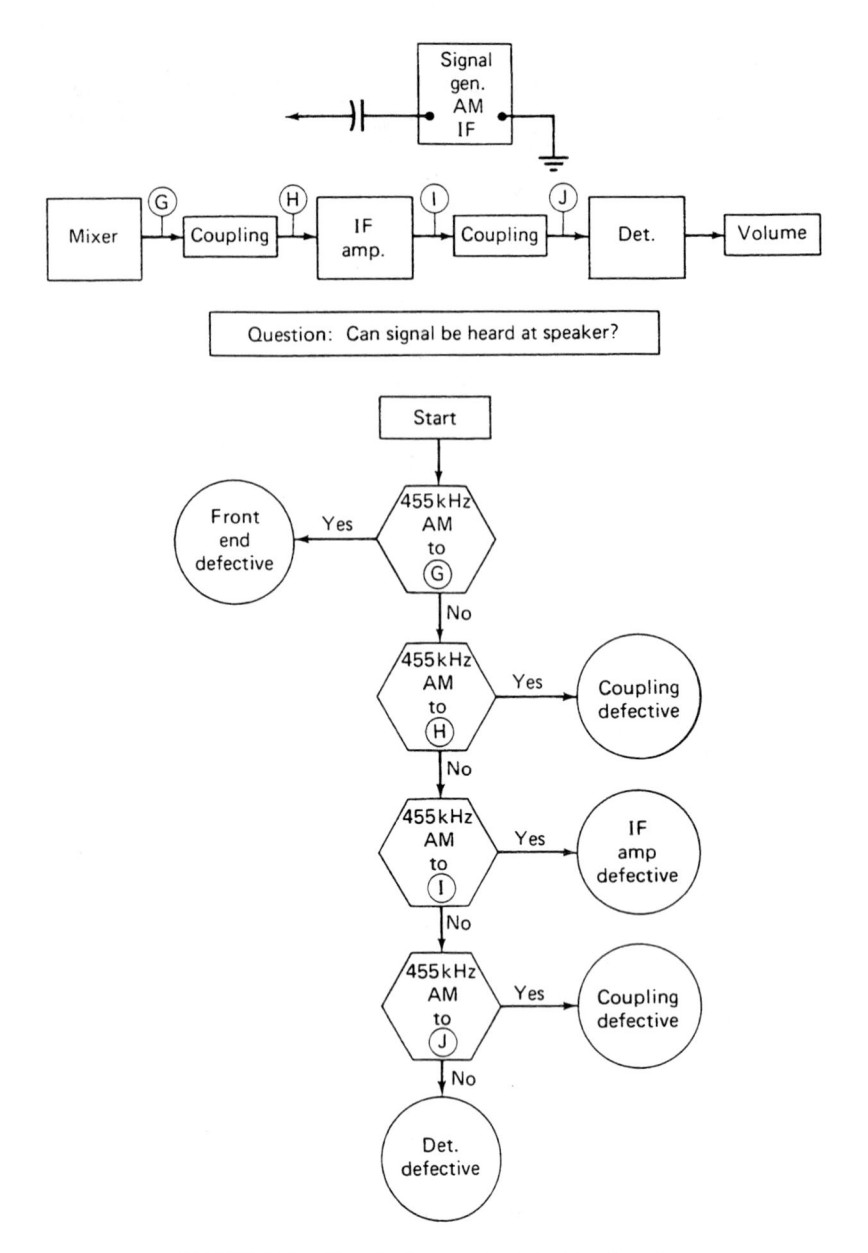

FIGURE 11-3 Signal-injection chart for the IF section.

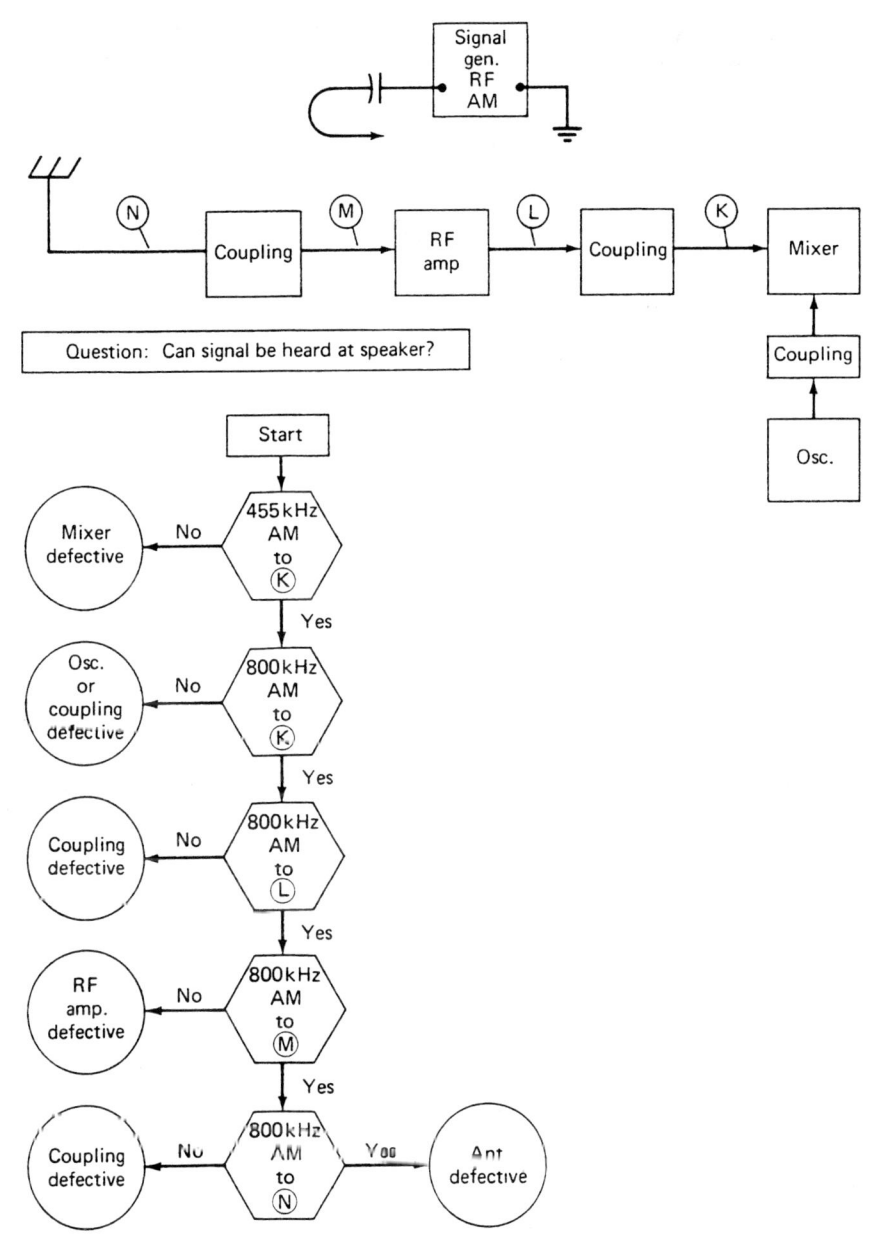

FIGURE 11-4 Signal-injection chart for the front end.

Signal Injection 307

resistance analysis must be used to isolate the defective part. Details of this procedure are given later, but now we turn our attention to signal tracing.

11-3 SIGNAL TRACING

Signal tracing consists of following the signal as it progresses through the various stages of a receiver. If the receiver is defective, the signal will be lost at the point of the defect. A signal tracer or an oscilloscope can be used in this capacity, and the indicator is the audible output from the signal tracer or the waveform displayed on the scope. A block diagram of a receiver and a signal-tracing chart are shown in Figure 11-5. The appropriate question is: Can the proper signal be seen or heard? The procedure begins at the volume control with the receiver tuned to a station and with the volume control set at maximum volume.

If the front end and IF sections are functioning properly, an audio signal should be present if the receiver is tuned to a station. If no audio signal is present, try tuning the receiver to a known station. If no signal appears, the defect is in the front end or IF amplifier. If an audio signal is present, the defect is in the audio amplifier. The defect may be further localized by moving the tracer probe to point B in Figure 11-5(a) and then to points C, D, E, and F until the signal is lost. The signal-tracing chart indicates the conclusions to be drawn.

If the audio signal is not found at the volume control, we move back to the output of the mixer to see if the signal is present at that point [point G in Figure 11-5(a)]. This is the input to the IF amplifier, and the signal that should appear is a modulated (by voice or music) RF signal of the IF frequency. If a signal tracer is being used, a detector probe must be used to demodulate the signal in order to produce an audible output. If a scope is being used, a faster sweep rate might be in order. Once again, the receiver must be tuned to a station for the IF frequency to appear.

If the signal is present at the output of the mixer, it is followed through the IF section as indicated on the block diagram and on the signal-tracing chart. The point at which the signal is lost identifies the area of the defect.

If the IF signal is not present at the mixer output, and if the receiver is tuned to a station, the defect must lie somewhere in the front end. In this area, signal-tracing procedures must be carried out with great care, and the success of this method depends largely upon the quality and limitations of available equipment. First, the signal tracer cannot be used to check the oscillator because the oscillator signal is not modulated. Therefore, the output from the detector probe of the signal tracer will be a DC voltage which will not produce an audible signal in the speaker of the signal tracer.

An oscilloscope may be used to determine if the oscillator is functioning, but care must be used in the method of coupling the scope to the oscillator. The scope probe may *load down* the oscillator until it stops oscillating, only to begin again when the probe is removed. A low-capacitance probe must be used if direct connection is to be made, but frequently it is possible to pick up the oscillator signal by merely attaching a short piece of wire to the scope probe and holding it near

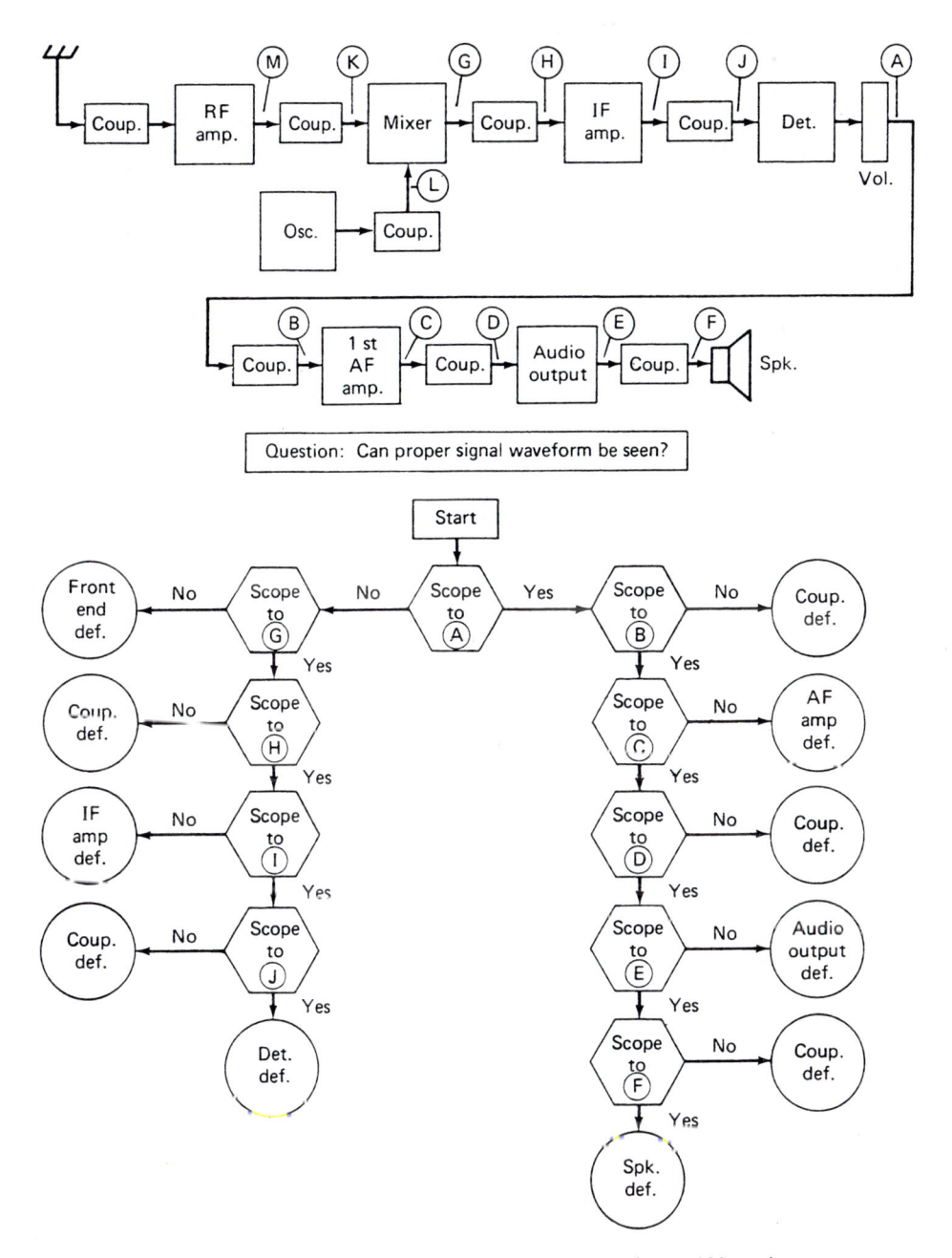

FIGURE 11-5 Signal tracing with an oscilloscope for an AM receiver.

Sec. 11-3 — Signal Tracing

the oscillator tuning coil. A sine wave should be observed if the oscillator is functioning.

At the output of the RF amplifier (point *M*), neither the scope nor the signal tracer may be useful, because of the low signal level present at this point. Unless the sensitivity of the scope or tracer is extremely good, there will be no indication of a signal even if it is present. Second, the effect of the instrument upon the tuned circuits must be considered. It is very important that the limitations of the equipment be known. If the equipment cannot provide a usable indication at the oscillator output or RF amplifier output, the problem must be isolated by other procedures, such as signal injection or voltage and resistance analyses.

11-4 VOLTAGE-RESISTANCE ANALYSIS

Voltage-resistance (V-R) analysis consists of comparing the voltage or resistance values given on the schematic with those measured in a defective receiver and interpreting any differences that exist. The voltages shown on the schematic are nominal values and normally vary somewhat as a result of component tolerances or variations in line voltage. Any significant differences found in these comparisons are analyzed to determine which part or parts are causing the problem. The suspected part is then checked to confirm the defect, or a good part is substituted to see if normal operation is restored.

Basic Circuit

This process can best be understood by applying it to a few basic circuits. We begin with the simple series resistor circuit shown in Figure 11-6. If *R3* were to open, the voltages shown in chart (a) would be found. The circuit is open, and no current

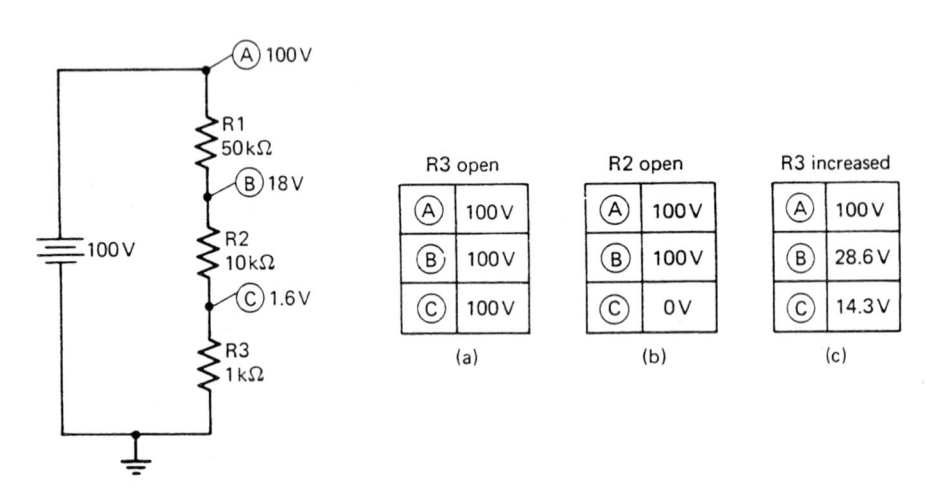

FIGURE 11-6 V-R analysis of a simple circuit.

flows. But when the EVM is connected to point B, the circuit is completed through the voltmeter. The resistance of the EVM (11 MΩ) is so much larger than $R1$, however, that almost all the supply voltage is developed across the meter. Very little voltage appears across $R1$, because of the small current that flows through the voltmeter. The same reasoning accounts for the high-voltage reading obtained when the voltmeter is connected to point C. The abnormally high voltage at B and C causes us to suspect that $R3$ is open. An ohmmeter can be used to measure $R3$ to confirm our suspicion.

Chart (b) of the figure shows the readings found when $R2$ is open. With the EVM placed at point B, the circuit is completed through the meter and a reading close to the supply voltage is obtained. At point C, no voltage is found, because the circuit is not complete even with the meter connected. This indicates a voltage drop of 100 V across $R2$, a situation that could exist only if the resistance of $R2$ were many times greater than that of $R1$ or $R3$. Also, there is no voltage drop across $R3$, which indicates that no current is flowing in the circuit. This causes suspicion to fall on $R2$, and the next step is to check $R2$ with an ohmmeter.

In many cases a resistor will not open completely but will increase in value to many times its original resistance. The effect of this is shown in chart (c), where $R3$ has increased from 1 kΩ to 10 kΩ. As $R3$ increases in value, the voltage from point B to ground also increases. An increase in either $R2$ or $R3$ could cause an increase in voltage at point B. But the voltage at point C also has increased, and this can occur only if $R3$ has increased. We reason further that if $R2$ had increased in value, the voltage at point C would have decreased. From this we conclude that $R3$ is the defective resistor, and the suspicion is confirmed by checking $R3$ with the ohmmeter.

V-R Analysis of a Simple Power Supply

Now that the basic procedure has been established, we turn our attention to the simple half-wave power supply shown in Figure 11-7. Note that the circuitry in the receiver is represented as resistances connected to source points A, B, and C. The receiver circuitry causes the series voltage-divider arrangement to be much more complicated than it appears to be at first glance.

We begin by assuming that resistor $R3$ is open. Measured voltages are given in chart (a) of the figure. Note that the voltage at point B does not rise to 145 V. Current flow through $R2_{CKT}$ maintains the voltage at B at an only slightly increased value of 115 V. The voltages at points A and B increase slightly because, with $R3$ open, less current flows through $R1$ and $R2$. Even when $R3$ is open, a complete circuit exists from point B to ground through $R2_{CKT}$.

Since 0 V is measured at point C, a voltage of 115 V is lost across $R3$. In light of the slightly increased voltage at point B, we reason that excessive current is not flowing through and therefore the only thing that could produce a large voltage across $R3$ is for $R3$ to have increased to a value much greater than $R1$, $R2$, or $R3_{CKT}$. This leads us to believe that $R3$ is open, a suspicion that is confirmed by measuring the resistance of $R3$ with an ohmmeter.

Before $R3$ is replaced, $C3$ and $R3_{CKT}$ should be checked for a decrease in

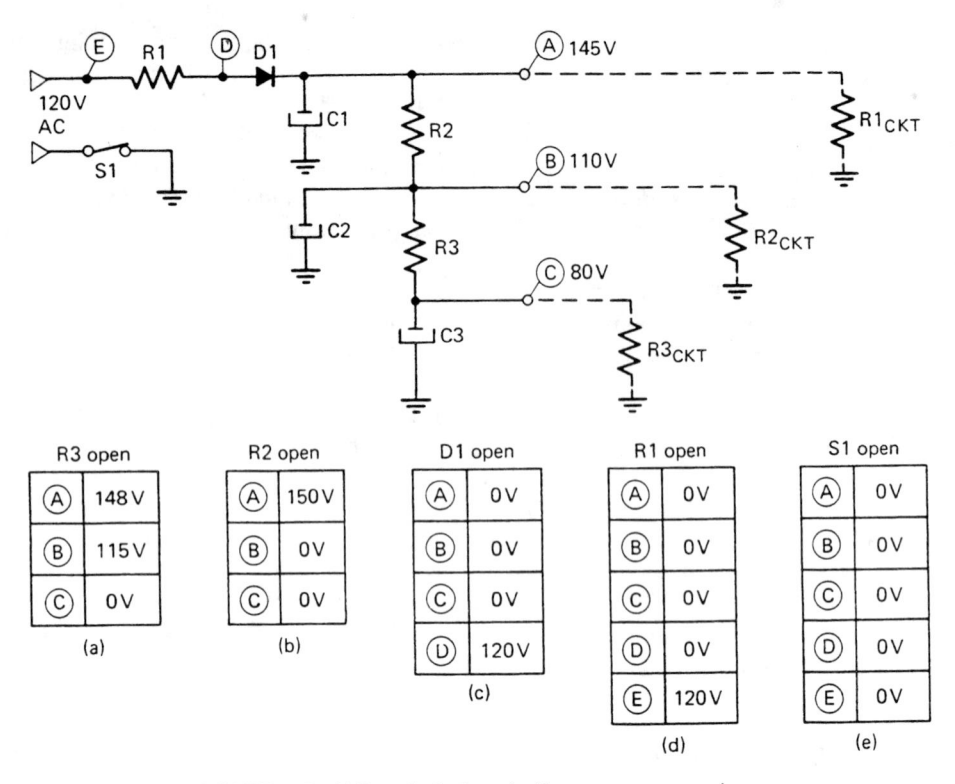

FIGURE 11-7 V-R analysis for a half-wave power supply.

resistance that could have caused excessive current through $R3$, subsequently causing it to open. This is important, especially if the voltage at C is zero. The resistance from point C to ground should measure at least 50 kΩ in a tube receiver or about 10 kΩ in a solid-state receiver. If a lower reading is obtained, $C3$ is disconnected and checked out of circuit. If defective, it is replaced; if it is good, the problem is in the circuitry represented by $R3_{CKT}$. The procedures for finding a short circuit in the receiver circuitry are given below.

Short Circuit in the Receiver Circuitry

While replacing a defective rectifier in a vacuum-tube receiver, it is a simple matter to make an ohmmeter check of the resistance from the cathode of the rectifier to ground to check for short circuits on the power supply line. The resistance reading should be at least 50 kΩ, obtained after the filter capacitors have been given time to charge due to the ohmmeter voltage. In solid-state receivers, however, ohmmeter readings are not as reliable because of the PN junctions that may or may not be forward-biased by the ohmmeter voltage. Therefore, the procedures for finding a short circuit are different for the two types of receivers.

For a tube-type receiver, with the ohmmeter connected to the rectifier cathode, systematically disconnect sections of the receiver circuitry from the B+ line until the ohmmeter returns to a high-resistance reading. The principle, here, is to divide and conquer. Possible defects are shorted electrolytic filter capacitors, a shorted tone-compensation capacitor connected from plate to cathode of the output tube, leakage from the primary winding to the core of the audio output transformer, defective tube sockets, shorted decoupling or screen bypass capacitors, and a variety of random things such as component leads touching the chassis or bits of wire or solder lodged between conductors.

For solid-state receivers, a short circuit that is severe enough to burn out the rectifier will produce an ohmmeter reading that is quite low, perhaps 100 Ω or less. In this case, the ohmmeter can be used in a manner similar to that used for tube-type receivers. The resistance from rectifier cathode to ground is monitored while various sections of the receiver are disconnected.

For short circuits that are less severe, an approach that is usually successful is to apply power to the receiver with the new rectifier installed and then make voltage measurements to locate the low-resistance path to ground. In typical receivers, the power supply is divided into two or more sections, the sections being separated by resistors. If an unusually large voltage drop occurs across a given resistor, it is likely that the short circuit lies on the low-voltage side of the resistor.

When a low voltage is discovered in a particular section, components or sections of circuitry can be disconnected until the low voltage again goes high. We should remember, however, that excessive conduction through a transistor having shorted junctions can pull the supply voltage down. In this case, voltage measurements made around the transistor should reveal the defect so that the procedure of disconnecting components can be circumvented.

In regard to disconnecting components, it is easier said than done, and at best, the procedure is time consuming. Often it is possible to open a circuit by removing one lead of a resistor, but in other cases, there may be no simple way to disconnect sections of a circuit. Of course, the foils on the circuit board may be cut, but this procedure is recommended only as a last resort. Whenever possible, the short circuit should be located by voltage measurements and careful analysis of the voltages that are obtained.

11-5 VOLTAGE-RESISTANCE ANALYSIS OF TUBE AMPLIFIERS

The type of analysis described above can also be applied to amplifier stages. A typical triode amplifier with normal voltage readings is shown in Figure 11-8. Since it is an easy matter to replace the tube of a suspected stage with a tube known to be good, the voltage readings are concerned primarily with finding defective components in the associated circuitry. We now examine the effect upon circuit voltages of several possible defects.

Chart (a) shows the voltages expected if the cathode resistor R_k opens. No current will flow through the tube, and there will be no IR drop across the plate load resistor. Consequently, the plate voltage V_p will be found to equal the power

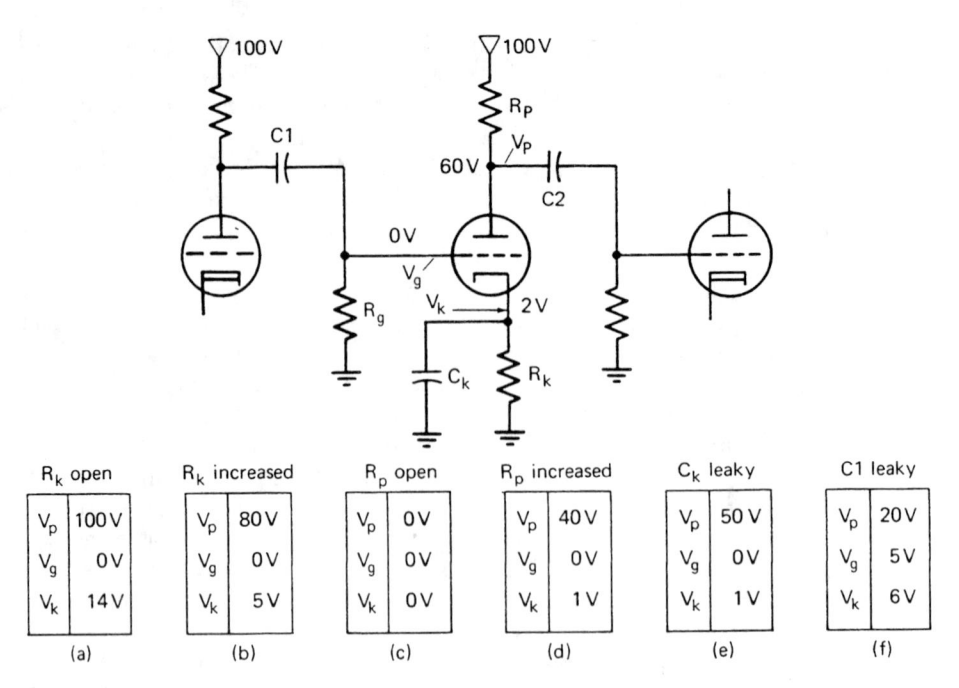

	R_k open	R_k increased	R_p open	R_p increased	C_k leaky	C1 leaky
V_p	100 V	80 V	0 V	40 V	50 V	20 V
V_g	0 V	0 V	0 V	0 V	0 V	5 V
V_k	14 V	5 V	0 V	1 V	1 V	6 V
	(a)	(b)	(c)	(d)	(e)	(f)

FIGURE 11-8 Voltage analysis for triode amplifier stage.

supply voltage. The grid voltage will not change noticeably from the normal value, but the cathode voltage will measure much higher than normal, for the following reason.

The plate draws electrons from the cathode, and the loss of electrons tends to make the cathode positive. In normal operation, electrons flow through the cathode resistor to the cathode at the same rate that the plate draws electrons from the cathode, and the cathode voltage stabilizes at a small positive voltage. But when the cathode resistor opens, the cathode becomes "starved" for electrons and rises to a positive voltage that is nearly equivalent to the power supply voltage. When the EVM is connected to the cathode, however, electrons flow to the cathode through the meter, and the meter resistance acts like an extremely large cathode resistor. The result is that the cathode voltage is pulled down from the power supply voltage to a value typically from 6 to 10 times the normal cathode voltage. The action of the tube under the influence of the large "cathode resistor" (the meter resistance) and resultant bias voltage tends also to limit the level to which the cathode voltage will rise.

A large voltage at the cathode can be caused by excessive current flow through the tube and cathode resistor. In the present case, however, the high voltage at the plate indicates that little, if any, current is flowing through the tube, since there is no *IR* drop across the plate load resistor. The grid voltage is normal and therefore cannot be responsible for having cut the tube off, and we immediately suspect that the cathode resistor is either open or has increased in value. The suspicion can be checked by removing power from the set and measuring the resistance of R_k.

If R_k is not completely open but has increased in value, the plate voltage will lie between the normal value and the power supply voltage. Also, the cathode voltage will be higher than normal but will not be as high as for the case when R_k is completely open. This is depicted in chart (b).

If the plate load resistor R_p should open, the voltages indicated in chart (c) may occur. Both the plate and cathode voltage will be zero, since no complete circuit exists between the power supply (B+) and the plate of the tube. Obviously, no current will flow through the tube, and no current will flow through the cathode resistor to produce a cathode voltage. The grid voltage will not be affected appreciably if R_p opens. Any time a very low or zero voltage is found at the plate of a tube, it is wise to check the voltage at the power supply source point that supplies the voltage to the tube in question. The defect might lie in the power supply circuitry rather than in the plate circuit.

If the plate load resistor should merely increase in value rather than become completely open, the plate voltage will assume a value lower than normal but greater than zero, as shown in chart (d). Lowering the plate voltage reduces the plate current, and, consequently, the cathode voltage. The decrease in cathode voltage provides the clue that causes suspicion to fall upon the plate load resistor. Reduced cathode voltage implies reduced current conduction through the tube. This, plus the fact that the plate voltage is low, indicates an increase in R_p, since a decrease in current would cause the late voltage to rise if R_p remains at the correct value.

If the cathode capacitor C_k develops excessive leakage, the cathode voltage will decrease because of the decreased resistance from cathode to ground [chart (e)]. The reduced cathode voltage manifests itself as a reduced bias voltage for the tube, and the current conduction through the tube will increase. The increased plate current results in a lower plate voltage, because of the increased IR drop across the plate load resistor. Here we have a low voltage both at the plate and at the cathode, and this symptom is similar to that of an increased R_p. An ohmmeter check of R_p will quickly eliminate it as a possible cause of the improper voltages. The cathode circuit then becomes the prime contender for the defect, and an ohmmeter check should reveal the defective component.

A lowered resistance between cathode and ground can be caused if R_k changes resistance to a smaller value. The symptoms would be the same as for a leaky cathode bypass capacitor, but an ohmmeter will identify the defective component.

Leakage in coupling capacitor $C1$ may result in the voltages shown in chart (f). The grid voltage V_g is drive positive since the coupling capacitor fails to completely block the high positive voltage at the plate of the preceding tube. The leakage resistance of C and the grid resistor R_g form a voltage divider that applies a fraction of the plate voltage of the preceding tube to the grid of the tube in question. The positive voltage applied to the grid causes the tube conduction to increase, resulting in a larger current flow through R_p and R_k. This causes an increase in voltage at the cathode but a decrease in voltage at the plate, as a result of the larger IR drop across R_p. The most likely cause of a positive voltage to appear on the grid is for $C1$ to be leaky.

No change in DC voltages would result in the present circuit if $C1$, $C2$, or C_k

opened, but there would be a drastic change in AC signal levels. These possibilities are discussed in Chapter 12, together with the rather weird symptoms that can result if the grid resistor R_g opens.

Pentode Amplifier

Much of the foregoing analysis applies to a pentode amplifier, which involves a screen grid and associated components. The screen resistor and the screen bypass capacitor C_s may become defective, and we shall see the effect of such a failure in the following. Figure 11-9 illustrates a typical pentode amplifier with normal voltage readings.

If the cathode resistor R_k opens, the voltages given in chart (a) may be obtained. The plate voltage will rise to that of the power supply, and the cathode voltage will rise to several times its normal value while the grid voltage remains near the normal value. The screen circuit action will be similar to that of the plate; the screen voltage will rise to the power supply voltage, since screen current will be reduced to zero and the *IR* drop occurring across R_s will disappear. Similar effects, but to a lesser degree, will occur if R_k increases in value rather than becoming fully open.

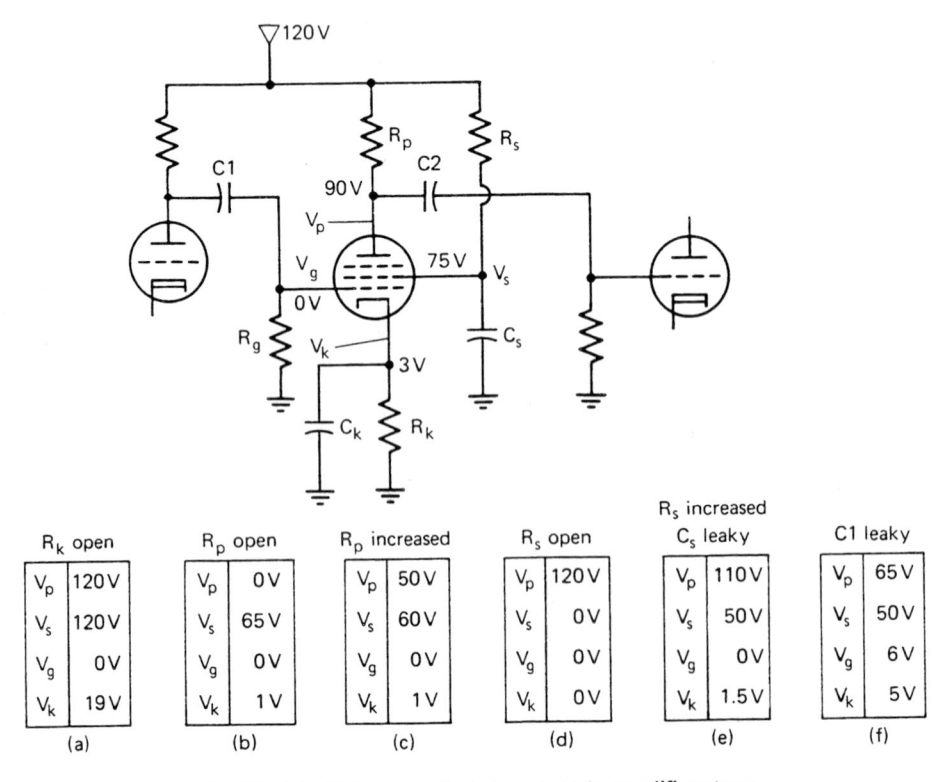

R_k open		R_p open		R_p increased		R_s open		R_s increased C_s leaky		C1 leaky	
V_p	120 V	V_p	0 V	V_p	50 V	V_p	120 V	V_p	110 V	V_p	65 V
V_s	120 V	V_s	65 V	V_s	60 V	V_s	0 V	V_s	50 V	V_s	50 V
V_g	0 V	V_g	0 V	V_g	0 V	V_g	0 V	V_g	0 V	V_g	6 V
V_k	19 V	V_k	1 V	V_k	1 V	V_k	0 V	V_k	1.5 V	V_k	5 V
(a)		(b)		(c)		(d)		(e)		(f)	

FIGURE 11-9 Voltage analysis for pentode amplifier stage.

Chart (b) applies to an open plate load resistor. The plate and grid voltages are zero as before, but the cathode voltage does not decrease to zero as it did in previous circuits. It will be lower than normal, but the screen current still flows through the resistor, and therefore a voltage is developed across it. With 0 V on the plate, the screen grid current increases because the plate no longer draws electrons away from the screen. The increased screen current depresses the screen voltage, as a result of a larger *IR* drop across the screen resistor R_s.

Chart (d) indicates possible voltage measurements with the screen resistor open. The plate voltage rises to the power supply voltage, which indicates that little, if any, plate current is flowing. The screen voltage drops to zero for obvious reasons. Recall that in a pentode the electrons are drawn from the cathode largely because of the voltage on the screen. Loss of screen voltage therefore results in a greatly reduced cathode current, which, in turn, results in a negligible *IR* drop across the cathode resistor. Consequently, the cathode voltage will measure nearly zero.

If R_s opens, the screen grid is left "floating," in that any electrons it may collect will be trapped here. This gradually builds up a large negative voltage that effectively shields the cathode from the attractive influence of the plate. Thus, an open screen grid circuit reduces plate current essentially to zero.

The effect of a leaky screen capacitor C_s depends upon the severity of the leakage. The leakage will cause increased current flow through the screen resistor, which results in lower screen voltages [chart(e)]. This, in turn, will result in decreased current flow through the tube, giving an elevated plate voltage and a depressed cathode voltage. If the leakage increases to the point where C_s is practically short-circuited, the screen voltage will be nearly zero and a very large current will flow through R_s, which may result in its overheating and eventual destruction. The current flow into the rest of the B+ power supply may be sufficient to pull the entire B+ supply voltage down to lower than normal values.

11-6 VOLTAGE-RESISTANCE ANALYSIS OF TRANSISTOR CIRCUITS

The effects of defects in transistor circuits can be analyzed in much the same way as vacuum-tube circuits, but a significant difference exists between the two insofar as mechanical aspects are concerned. Vacuum tubes are always "socketed," whereas transistors are almost always soldered to a printed circuit board. A tube known to be good can be substituted into a vacuum-tube circuit in order to eliminate the tube as a possible defect. On the other hand, it is sometimes rather difficult to remove a transistor, and it is desirable to determine the status of a transistor without removing it from the circuit. Because of this, voltage charts are included in the following sections to demonstrate the voltage changes that may occur when the transistor itself is defective. In the majority of cases, the defect will be found to be in the transistor rather than in the associated circuitry.

The charts shown represent voltage levels for a negative ground power supply, and the charts can be applied to *PNP* circuits by changing the polarities of the indicated voltages. The important thing to observe in each chart is the direction of the voltage change; the size of the change is less significant, since transistor circuits vary considerably. The same defect may give rise to different voltage readings in different circuits, but the following analysis illustrates the principles involved.

Before proceeding, let us comment briefly on the use of an ohmmeter in transistor circuits. Many resistance measurements can be made in vacuum-tube circuits without removing the tube or disconnecting the component, since the elements of a tube are isolated electrically when the tube is inoperative. Such is not the case for transistors. The ohmmeter battery voltage may be great enough to forward bias the transistor junctions, leading to great confusion in interpreting the ohmmeter readings. Junction resistances are extremely variable, and because of this, resistance measurements are less valuable in transistor than in vacuum-tube circuits.

Universal Bias with Collector Feedback

Figure 11-10 shows a universal-biased stage modified so as to provide collector feedback. Voltages given in the boxes indicate the measurements that would be obtained if a positive ground were used. This is equivalent to connecting the common lead of the voltmeter to the positive power supply point and then measuring other voltages relative to that point.

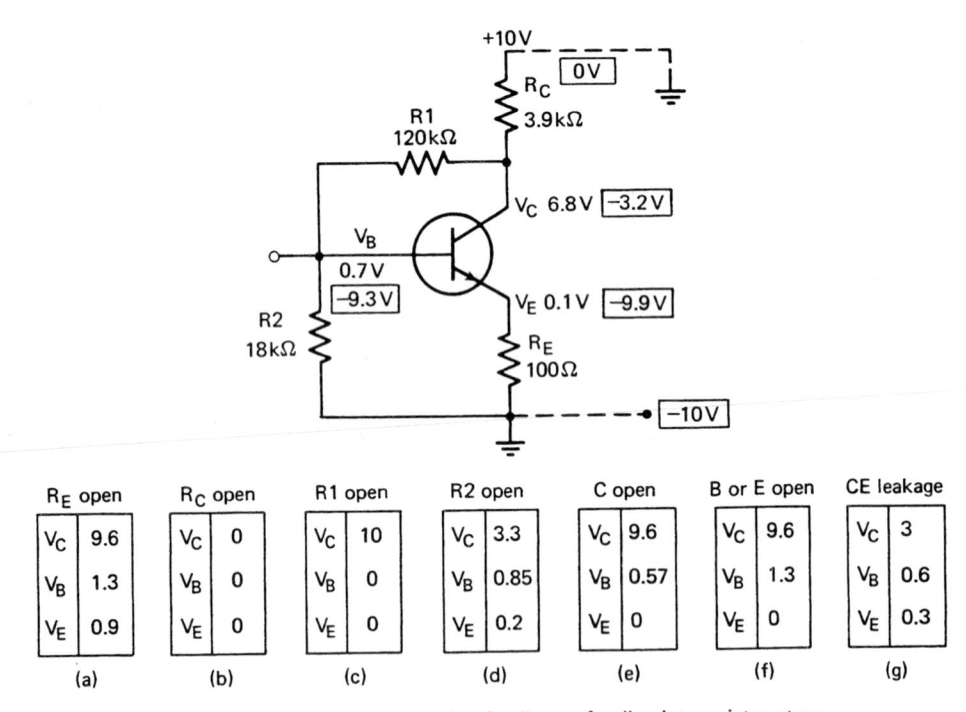

FIGURE 11-10 Voltage analysis of collector feedback transistor stage.

	R_E open	R_C open	R_1 open	R_2 open	C open	B or E open	CE leakage
V_C	9.6	0	10	3.3	9.6	9.6	3
V_B	1.3	0	0	0.85	0.57	1.3	0.6
V_E	0.9	0	0	0.2	0	0	0.3
	(a)	(b)	(c)	(d)	(e)	(f)	(g)

318 Basic Troubleshooting Procedure Chap. 11

If R_E opens, the collector current will drop to zero and the collector voltage will rise almost to the power supply voltage. Current flow through R_C will not be zero, however, because resistors $R1$ and $R2$ still form a complete circuit to ground. A small voltage drop occurs across R_C, resulting in the collector voltage being slightly lower than the power supply voltage. With R_E open, the voltages that appear at the base and emitter of the transistor are determined by the voltage distribution across the series combination of R_C, $R1$ and $R2$, since the transistor becomes completely inactive. These voltages are given in chart (a) of the figure.

If collector load resistor R_C opens, the power supply voltage will be removed from the entire circuit. Consequently, the voltage at all three terminals of the transistors will be zero, as indicated in chart (b).

Chart (c) shows what happens if $R1$ opens. The forward bias will be removed from the base and the collector current will go to zero. The collector voltage will rise to the power supply voltage while the emitter and base voltages drop to zero.

If $R2$ opens, the base voltage will rise somewhat, since the voltage divider formed by $R1$ and $R2$ becomes open at ground. The forward bias at the base is increased, which results in an increased collector current. The collector voltage decreases under the influence of the increased current, as shown in chart (d). In such a case the transistor is frequently driven into saturation, as evidenced by the very low collector voltage.

The effects of defects in the transistor itself are shown in charts (e) through (g). If the collector-base junction opens, the base-emitter junction will remain forward-biased and the base current will increase. At the same time, the emitter current and emitter voltage will decrease, since the base current is small in comparison with the normal collector current. The effect of the increased base current is to cause a larger voltage drop across $R1$, making the base voltage decrease slightly. With almost 0.6 V between the base and emitter, the transistor should be conducting, but the high collector voltage indicates that it is not. A check of the transistor is in order.

If either the base or emitter leads open inside the transistor, no collector current will flow, and the collector voltage will increase. The base voltage may increase slightly (or a lot, depending upon the circuit) and the emitter voltage will go to zero. In chart (f) we note that 1.3 V appear between the base and the emitter, yet the transistor is not conducting, since the collector voltage is at the supply point. We immediately suspect the transistor.

Leakage from the collector to the emitter is a common defect in transistors and causes much larger-than-normal collector currents for a given base-emitter voltage. The emitter voltage may rise as a result of the larger than-normal conduction, so that the B-E voltage is much smaller than normal. At the same time, the collector voltage will be pulled down by the increased collector current. In chart (g), the transistor is near saturation (as indicated by the low collector voltage) even though the B-E voltage is only about 0.3 V. A silicon transistor with only 0.3 V between the base and the emitter should be almost completely cut off. We therefore suspect the transistor to be "leaky," and should confirm this suspicion by checking the device.

+10 V

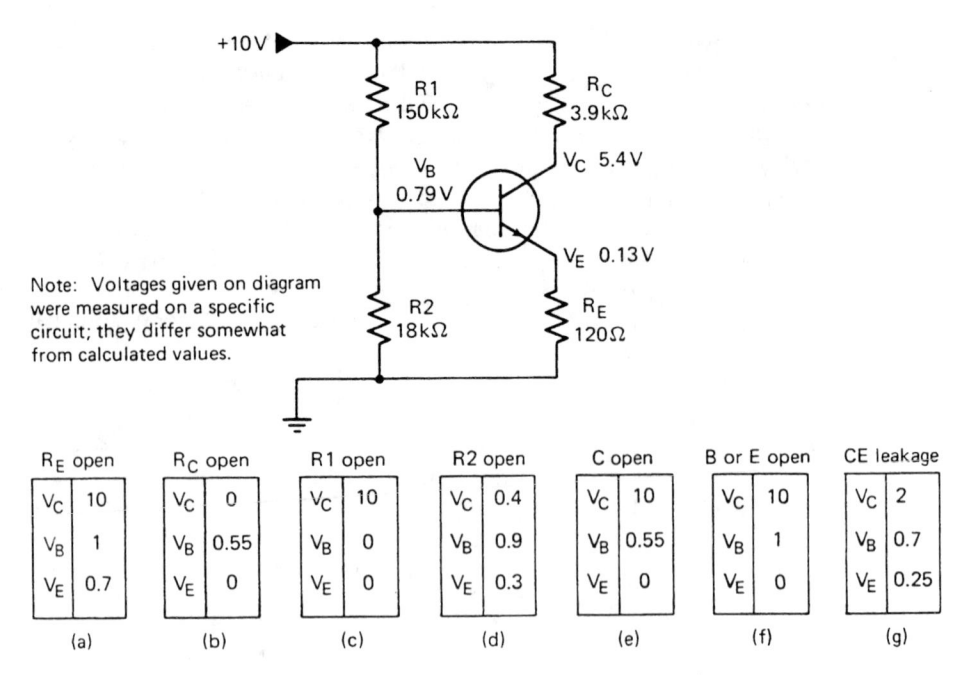

R1 150 kΩ
Rc 3.9 kΩ
VB 0.79 V
Vc 5.4 V
VE 0.13 V
R2 18 kΩ
RE 120 Ω

Note: Voltages given on diagram were measured on a specific circuit; they differ somewhat from calculated values.

RE open	Rc open	R1 open	R2 open	C open	B or E open	CE leakage
Vc 10	Vc 0	Vc 10	Vc 0.4	Vc 10	Vc 10	Vc 2
VB 1	VB 0.55	VB 0	VB 0.9	VB 0.55	VB 1	VB 0.7
VE 0.7	VE 0	VE 0	VE 0.3	VE 0	VE 0	VE 0.25
(a)	(b)	(c)	(d)	(e)	(f)	(g)

FIGURE 11-11 Voltage analysis of universal-bias transistor stage.

Universal Bias

Figure 11-11 illustrates a universal-biased stage with normal voltages given along with voltage charts for common defects. Let us change our approach slightly from the preceding. Rather than take a given defect and see what voltages would result, let us assume that we are measuring voltages and forming conclusions or suspicions as to what the defect might be.

Suppose, first, that we measure the collector voltage, finding it to equal the power supply voltage. This tells us that the transistor is not conducting. To determine whether a bias voltage is present at the base, we measure the base voltage and find that the voltage is 0.55 V, slightly lower than normal. Moving on to the emitter, we find it at 0 V (very nearly). What defect could cause these voltages?

Experience tells us to suspect the transistor first, and then other components. The base-emitter voltage (0.55) is large enough so that a silicon transistor should exhibit appreciable conduction, but we have none in the present circuit. The fact that the base voltage is nearly normal indicates that the base resistors are probably good, but we note that the base voltage is slightly low, which could be caused by increased base current. What defect can cause an increase in base current at the same time the collector current is zero? This points to an open in the collector circuit, and since the power supply voltage appears at the collector, the collector resistor R_C must be conductive. The only other possibility is that the collector circuit inside the transistor is open. Subsequent testing of the transistor should confirm our suspicion.

Proceeding to another defect, suppose we find the collector voltage to be 10 V, the base to be 1.1 V, and the emitter to be 0 V. What does this mean? Very quickly we note the large voltage between the base and the emitter, 1.1 V. A junction must be conducting very heavily to develop such a large voltage, but the collector voltage tells us that the transistor is not conducting at all. We immediately conclude that the B-E junction of the transistor is open. We note, in passing, that the base emitter voltage is one of the best indicators of a transistor's well-being that is available to a technician.

As a final exercise for this circuit, suppose that the collector voltage is found to be rather low, say 0.4 V. At the same time, the base voltage measures 0.9 V and the emitter measures about 0.28 V. What can cause the low collector voltage? The collector voltage will be lowered by either excessive collector current or an increase in value of R_C. Since the emitter voltage also is increased, we suspect an increased collector current. Observe that the base voltage is rather high, but note that the emitter voltage also has risen, so that the base-emitter voltage remains at a reasonable value (0.62 V). We assume from this that the B-E junction of the transistor is good. We discount the possibility of C-E leakage, since the B-E voltage and increased conduction seem compatible. Something is apparently driving the base voltage upward. An increased $R2$ or a decreased $R1$ would do this, but a careful examination of the PC board reveals a questionable solder connection where $R2$ connects to ground. Resoldering the connection restores the circuit to normal operation.

In retrospect, the faulty connection at $R2$ opened the lower part of the voltage divider ($R1$, $R2$), causing the base voltage to rise. This increased the collector current, decreasing the collector voltage while increasing the voltage at the emitter. Such defects in solder connections are fairly common.

Base Bias with Emitter Feedback

The third and final circuit is a base bias with emitter feedback stage shown in Figure 11-12 with normal voltages. The circuit differs from the two previously described, since the base is not tied to ground through a resistor. We therefore briefly account for the voltages developed when various defects occur. The voltages are given in charts accompanying the circuit.

First, if R_E opens, no part of the circuit will be grounded, and all voltages in the circuit will rise to the power supply level as shown in chart (a). If the collector resistor R_C opens, the circuit will be reduced to a series combination of $R1$, the B-E junction, and R_E. Since $R1$ is rather large compared to the other elements in the series, most of the power supply voltage will be dropped across the resistor. Only a small current will flow, and the base-emitter voltage of the transistor will be correspondingly small. The voltage at the emitter will be negligible, since the IR voltage developed across the emitter resistor is very small. The collector will assume a voltage nearly equal to that of the base, since it is left "floating" when R_C opens. These voltages are shown in chart (b).

Chart (c) shows the effect of an open base resistor, $R1$. Forward bias is lost at the B-E junction, and conduction through the transistor ceases. Therefore, the

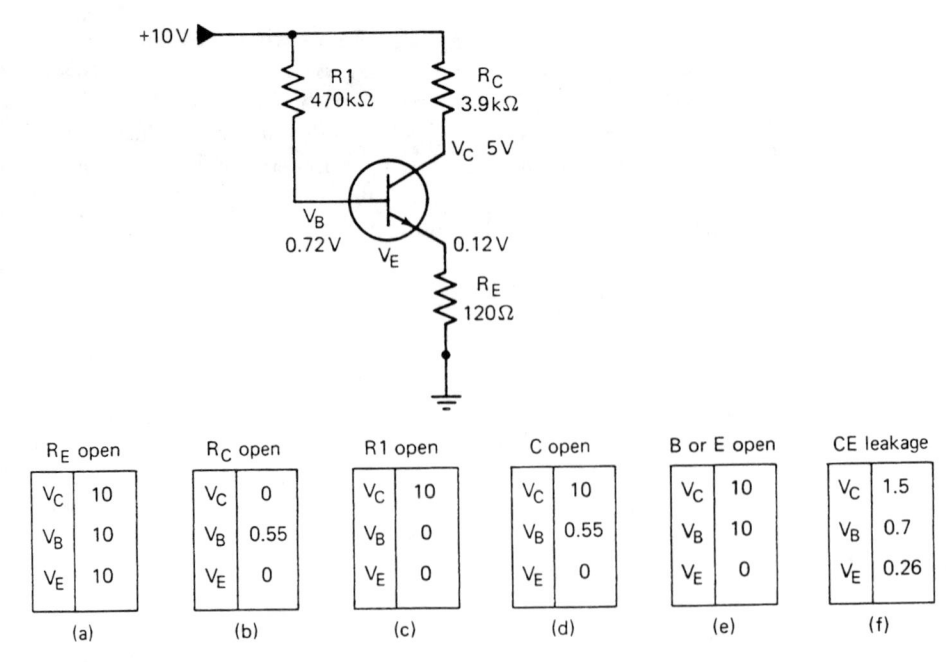

FIGURE 11-12 Voltage analysis of base bias with emitter feedback transistor stage.

	R_E open		R_C open		R1 open		C open		B or E open		CE leakage
V_C	10	V_C	0	V_C	10	V_C	10	V_C	10	V_C	1.5
V_B	10	V_B	0.55	V_B	0	V_B	0.55	V_B	10	V_B	0.7
V_E	10	V_E	0	V_E	0	V_E	0	V_E	0	V_E	0.26
(a)		(b)		(c)		(d)		(e)		(f)	

collector voltage rises to the power supply voltage while the base and emitter voltages go to zero.

Charts (d) and (e) illustrate the effect of open leads inside the transistor. We leave this part for the reader to analyze. The effect of leakage in the transistor, shown in chart (f), is to produce a collector current that is much larger than normal for the given B-E voltage. The result is shown in the chart; the collector voltage is low, while the base and emitter voltages are high.

11-7 ANALYSIS OF AN IC STAGE

We may assume that an IC stage is defective when a good signal is present at the input and a faulty signal appears at the output. The faulty signal may be distorted, or it may have the wrong amplitude, such as no amplitude at all.

Our first inclination is to assume that the IC is bad, and in many cases this assumption will be correct. However, it often happens that some other defect in the circuit causes the IC not to operate properly even though the IC is good. Our objective is to make as many measurements as possible to confirm that the IC is definitely bad before we go to the trouble of removing it from the circuit.

Service data will provide DC voltages that should appear at each pin of the IC, as in Figure 11-13. It is a simple matter to measure these voltages and make comparisons. Less obvious, however, is the conclusion to be drawn when a voltage is found in error. The defect may be inside the IC, or, just as likely, it may be in circuitry external to the IC.

First, ascertain that power supply voltage is being applied to the IC. Locate the power supply pin and measure its voltage (15 V on pin 9 in Figure 11-13). It is best to make the measurements from the top side of the IC if possible, touching the voltmeter prod to the pin itself rather than simply to the solder connection. Sometimes a poor connection or open will exist between the pin and the solder pad, and it is not unusual to find the pin tucked under the IC package rather than penetrating into the solder pad. Tucked-under pins are very difficult to see unless you specifically look for them.

When the power supply connection is verified, measure the voltage on the ground pin of the IC. It should be zero, obviously, but if it is not, a bad connection exists somewhere along the ground line. In Figure 11-13, zero volts should be measured at pins 4 and 8 and on the tab.

If no defect is found with the power supply and ground, we then examine the voltage on the other pins of the IC. Our first concern is whether the voltage that is supposed to appear on a particular pin comes from inside the IC or is applied to the pin by an external circuit? Because a DC voltage cannot pass through a capacitor, any pin isolated by capacitors must be driven by circuitry inside the IC. Such is the case for pins, 1, 3, 6, 7, and 10 of Figure 11-13. Therefore, if no voltage appears at pin 3, for example, we may conclude either that pin 3 is not being driven to 7 V, or that C7 is very leaky and is shorting pin 3 to ground.

Check the voltage on the other pins of the IC before beginning any desoldering for out-of-circuit checks of the capacitors or other components. When appropriate,

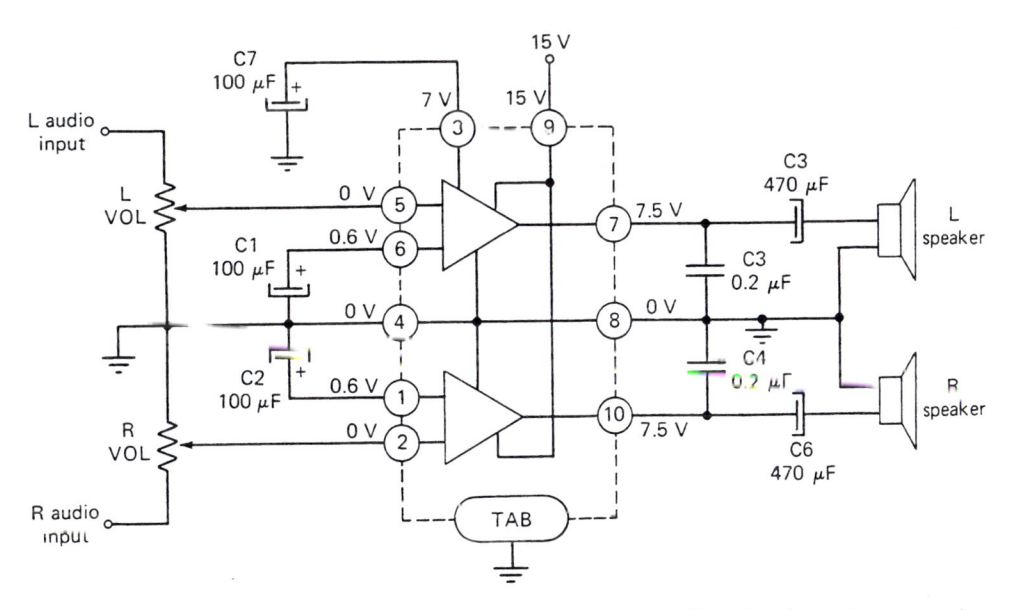

FIGURE 11-13 Power IC audio amplifier for an FM stereo receiver. Note that the voltage appearing at each pin of the IC is given.

Analysis of an IC Stage

compare voltages from one channel to the other, such as at pins 7 and 10 or 6 and 1. Capacitor C7 is evidently common to both channels. If it is defective, the effect on each channel should be the same. However, if C6 or C4 were shorted, we would expect a lower voltage at pin 10 than at pin 7. Pay particular attention to the voltages at the inputs to the IC because a wrong voltage here will give a corresponding wrong voltage at the output.

In most cases, when troubleshooting an IC stage, we are dealing with an unknown entity—the IC itself. We will seldom have a detailed description of the circuitry inside the IC. Therefore, we cannot be sure how a defect at one pin will affect the voltage at another pin. Conversely, it is not always possible to analyze a set of wrong voltages and pinpoint the defective component. Hence, when there is nothing left to do, remove the IC.

With the IC removed, it is likely that an ohmmeter or capacitor checker can be used to check for short circuits and defective components. In Figure 11-13, capacitors C1, C2, and C7 can be checked directly, while C3, C4, C5, and C6 can be checked for leakage. If no defects are found, we then suspect, with some confidence, that the IC is defective. A replacement should be installed, and, hopefully, the problem will be solved.

Sometimes, as everyone knows, installing a new IC will not solve the problem. When this occurs, even after having checked the circuitry around the IC, we start over. Recheck each voltage as before, but now the checks should include the stages on both sides of the stage we initially suspected. All available techniques should be used—signal tracing, signal injection, and so forth.

11-8 GENERATOR COUPLING TO RECEIVER TEST POINTS

A bit of caution is due when injecting a test signal at the plate of a tube or, to a lesser extent, at the collector of a transistor. It must be remembered that a high DC voltage is present at the plate of a tube, and sometimes this voltage can be several hundred volts. Indiscriminate connection of the generator lead directly to the high voltage may result in damage to the output circuitry of the generator.

Figure 11-14 shows a DC blocking capacitor connected in series with the gen-

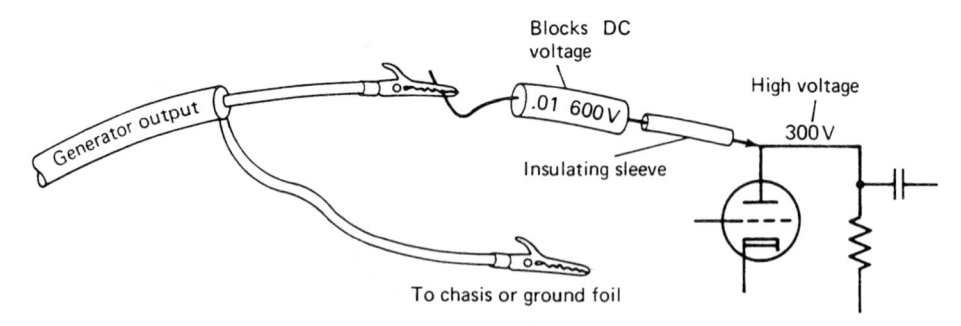

FIGURE 11-14 Capacitor in series with the generator lead blocks the DC voltage at the plate of the tube.

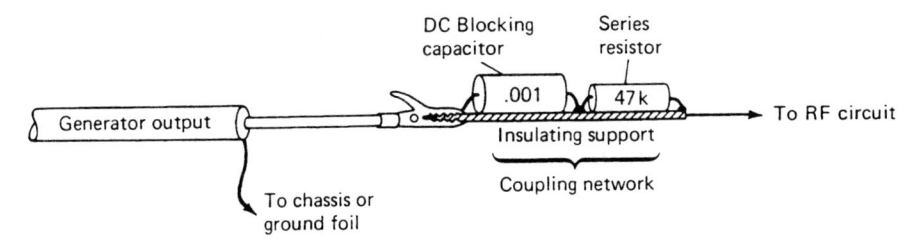

FIGURE 11-15 Coupling network to prevent the capacitance of the generator cable from detuning the RF circuits to which the generator is connected.

erator lead to block the DC voltage. For most vacuum-tube circuits, a 0.01-µF 600-V capacitor is adequate. Smaller capacitances (0.001 µF) may be used at the IF and RF frequencies in these sections of the receiver.

Generally speaking, the lower impedances of solid-state circuits require larger capacitances to avoid attenuation of the generator signal. For audio injection, electrolytic capacitors on the order of 1 µF are suitable. Remembering that electrolytic capacitors are polarized, pay attention to the polarity of the capacitor when connecting the generator lead. As in vacuum-tube circuits, smaller capacitances can be used in RF sections. Frequently, the service data will recommend that the capacitance be connected in series with the generator lead.

In some cases, when working with the RF sections of a receiver, the capacitance of the generator cable will seriously detune the circuit to which the generator is connected. The effect may be reduced by connecting a fairly large resistor in series with the generator lead, as shown in Figure 11-15. Experience will dictate the optimum resistance for a given generator, but usually a resistance of 10 to 100 kΩ will suffice, larger resistances being required for vacuum-tube circuits because of the higher impedance levels involved. A technician should prepare a set of coupling networks in advance so that they will be readily available when needed.

SUMMARY

1. The major steps in the basic troubleshooting procedures are: complaint verification, sensual examination, isolation procedures, replacement of defective parts, aircheck, cleanup, and paperwork.

2. A defective receiver should be turned on prior to disassembly (in the presence of the owner, if possible) to verify the complaint and so that any problem introduced by disassembly can be recognized.

3. As a technician gains experience, he or she will become increasingly able to identify defective components from the symptoms they produce.

4. The three general procedures for isolating a defect are (1) signal injection, (2) signal tracing, and (3) voltage-resistance analysis.

5. When a receiver has been repaired, it is important to aircheck the receiver both before and after the receiver is reassembled.

6. After a receiver is approved for return to the owner, the cabinet should be cleaned and polished.

7. Signal injection consists of applying a signal from a generator to the input of a stage and determining if the proper signal appears at the output.

8. If an audio test signal applied to the slider of the volume control is heard with normal volume and fidelity at the speaker, the audio section is functioning normally and the defect (if any) lies in the RF sections between the volume control and antenna.

9. An amplitude-modulated RF generator is required for signal injection in the RF, mixer, and IF stages of a receiver. Accurate frequency calibration is important.

10. Signal-injection procedures are preferred over signal tracing in the front end of a receiver due to the low signal levels involved.

11. An oscilloscope may be used as a signal tracer; indeed, an oscilloscope is preferred.

12. In all signal-tracing procedures, a signal must be present for the signal tracer to trace. A receiver must be tuned to a station, or a signal generator must supply the signal to the input of the stage under test.

13. Normally, signal tracing begins at the volume control of a receiver with the receiver tuned to a known station.

14. Signal-tracing procedures applied to the front end of a receiver are likely to be futile unless the signal tracer or scope has extremely good sensitivity.

15. Voltage-resistance analysis consists of comparing the voltage or resistance values given on the schematic with those measured in a defective receiver and interpreting any difference that exists.

16. V-R analysis is used as the final step in almost all service jobs to locate the faulty component after the defective stages have been isolated by signal tracing or by signal injection.

17. Resistance analysis is less valuable in solid-state receivers because of the variable nature of the measured resistance of *PN* junctions.

18. A resistance-analysis procedure may involve the systematic disconnection of several components in the event of a serious short circuit in the power supply.

19. A voltage analysis, properly interpreted, can yield much useful information as to whether a tube or transistor stage is working properly.

20. When a silicon (germanium) transistor is functioning properly as an amplifier, a voltage on the order of 0.7 (0.3) V will appear between the base and the emitter.

21. Higher-than-normal cathode or emitter voltages usually indicate excessive current conduction through the tube or transistor.

22. Higher-than-normal plate or collector voltages usually indicate a lower-than-normal current conduction through the tube or transistor.

23. In most cases the usefulness of a V-R analysis and the benefit derived from the analysis is dependent upon how well one understands the operation of the basic biasing arrangements.

24. Leakage from the collector to the emitter is a common defect in transistors and causes much larger than normal collector currents for a given base-emitter voltage.

25. A capacitor should be connected in series with the output lead of a signal generator to protect the output circuitry of the generator from the high DC voltages that may be encountered in the plate or collector circuits of a receiver.

26. Before hastily removing the IC from an IC stage that malfunctions, carefully measure the voltages around the IC to see if a defective component other than the IC itself is causing the malfunction.

27. In stereo receivers, comparisons of voltages between channels can often lead to the defective component.

28. After an IC is removed, ohmmeter checks of capacitors and other components around the IC can be made under open-circuit conditions.

QUESTIONS AND PROBLEMS

11-1. What are the seven major steps in the basic troubleshooting procedure?

11-2. What are some of the things a technician should look for when a receiver is first examined?

11-3. Should the initial turn-on of a defective receiver be done before or after the receiver is disassembled?

11-4. What is "symptom diagnosis?"

11-5. Briefly describe the three general procedures for isolating a defective component or stage in a malfunctioning receiver.

11-6. Why is it important to aircheck the receiver after it is reassembled?

11-7. Why should the charges for the repair and servicing of a receiver be figured immediately after the job is completed?

11-8. What are some of the things to be checked during the final aircheck of a receiver?

11-9. Why is it important to clean and polish the cabinet of a receiver that has been repaired? Why not let the owner do this?

11-10. Describe the procedure for using signal injection to isolate a defect in the audio section of a receiver. What type of test signal is employed?

11-11. As one moves toward the speaker during a signal-injection procedure, the amplitude of the generator signal must be increased. Why?

11-12. What type of signal generator is required for signal injection in the IF stages?

11-13. What will be heard at the speaker of a good receiver if a 455-kHz unmodulated sine wave is injected into the IF amplifier?

11-14. What precautions are in order when a signal is to be injected at the plate of a tube?

11-15. Explain how signal injection can be used to determine if the local oscillator of a receiver is functioning properly.

11-16. What advantages does an oscilloscope have over a signal tracer? If a good-quality oscilloscope is available, should one invest in a signal tracer?

11-17. What are some points to consider in deciding whether to invest in a signal tracer or to invest in an oscilloscope?

11-18. Will a 455-kHz IF signal be present in a receiver when the receiver is not tuned to a station? How does this affect signal-tracing procedures?

11-19. What type of probe is required in order to use a signal tracer in the IF section of a receiver?

11-20. Without making any direct connection to the receiver, how can a scope be used to determine if the oscillator of a receiver is operating?

11-21. Why may signal-tracing procedures fail to be productive in the front end of a receiver?

11-22. Can a signal tracer with an RF demodulator probe be used to determine if the local oscillator of a receiver is working? Why or why not?

11-23. How would the voltage at points A, B, and C be affected in Figure 11-7 if $R3$ merely increased in value rather than to become completely open?

11-24. In Figure 11-7, suppose that $R3_{CKT}$ opens so that no current flows through $R3$. What would be the voltage at point C as compared to point B?

11-25. In Figure 11-7, suppose that $C2$ becomes leaky, but not so that catastrophic failure of any components occurs. How would the voltages at points, A, B, and C be affected?

11-26. Describe the procedure for finding a short circuit in the power supply wiring of a receiver.

11-27. A rectifier diode in a receiver is found to be open. What checks of the circuitry should be made before a new rectifier is installed?

11-28. In Figure 11-8, suppose that $C1$ is suspected to be leaky when the high voltage is applied. How could this suspicion be confirmed?

11-29. An increase in voltage at the cathode of a tube can be caused by at least three different malfunctions. What are they?

11-30. Suggest checks to confirm each malfunction of Prob. 11-29.

11-31. Give three malfunctions that can cause the plate voltage to be too low.

11-32. Suggest checks to confirm each malfunction of Prob. 11-31.

11-33. What malfunction can cause the plate voltage to be too high?

11-34. What malfunction can cause the control grid voltage to be too positive?

11-35. When the control grid becomes too positive, what will be the effect upon the plate and cathode voltages?

11-36. What happens to the plate current, plate voltage, and cathode voltage when the screen grid voltage drops to zero in a pentode?

11-37. Why are resistance measurements less valuable in solid-state circuitry than in vacuum-tube circuitry?

11-38. As the collector current increases abnormally in a transistor circuit, what happens to:

    (a) The collector voltage?

    (b) The emitter voltage?

11-39. Suppose that the emitter resistor increases in value. What is the effect upon the:

    (a) Collector voltage?

    (b) Collector current?

    (c) Base voltage?

11-40. What voltage measurements or other indications might lead one to believe that the collector current of a transistor is excessively large?

11-41. What are the indications exhibited by a circuit in which the collector current is too small?

11-42. What voltage measurements might indicate that a transistor is defective?

11-43. In determining the status of an in-circuit transistor, what is one of the most revealing voltage checks that can be made?

11-44. Compare the voltage-analysis procedure for tubes with that of transistors. Are they quite similar?

11-45. Briefly describe the steps that should be taken in troubleshooting an IC stage that is defective. What should be done prior to removing the IC?

11-46. Give two reasons why a check-by-substitution is not the recommended first step in troubleshooting an IC stage.

12
TROUBLESHOOTING AM RECEIVERS

This chapter presents detailed descriptions of how the basic troubleshooting procedure is applied to servicing AM receivers. The object is to go as directly as possible from the symptom to the defect. We describe specific symptoms, give the most likely cause, and indicate an isolation procedure that should identify the defect. The procedures are flexible in that more than one procedure may be applicable to a given symptom. The best procedure for a given situation depends upon the equipment that is available and upon the experience and preference of the technician.

In troubleshooting, each check made and each measurement performed should lead the technician one step closer to finding the problem. There should be method in the madness. A technician is in trouble when procedures are abandoned and the troubleshooting degenerates to random measurements, part substitutions, alignment adjustments, and a variety of other hit-or-miss efforts that border on the frantic. So make every effort to be systematic and methodical at all stages of the troubleshooting program.

12-1 INTRODUCTION

The overall procedures for servicing an AM receiver are basically the same whether the receiver is a solid-state or a tube-type receiver. The techniques of signal injection and signal tracing are essentially the same, and many voltage analysis procedures are similar. However, when we get down to details, several differences exist in the methods used for the two types of receivers.

Tubes have filaments; transistors and ICs do not. Therefore, the first objective in servicing a tube set is to get the filaments lighted. Two procedures for troubleshooting a series filament string are given in a following section.

Tubes are always installed in sockets, whereas in radio receivers, transistors and ICs seldom are. It is a simple matter to substitute a tube known to be good for one that is suspected to be bad. In solid-state receivers, many tests are made in order to determine the condition of the transistors or ICs without having to remove them from the circuit. In-circuit transistor checkers are available, but their performance in many circuits is questionable.

An ohmmeter is quite useful in troubleshooting tube-type receivers because the elements within the tube are isolated from each other. Resistance data are typically included in the service information for tube-type receivers. For solid-state devices, however, the small voltage supplied by the ohmmeter can bias the junctions into conduction. This confuses the readings obtained because the reading will vary depending upon the particular type of ohmmeter being used. For this reason resistance data are seldom included in the service information for solid-state receivers.

The ohmmeter section of the newer digital multimeters often has a special low-voltage ohmmeter function in which the voltage appearing at the ohmmeter test prods is not large enough to bias a junction into conduction. To some degree, at least, this avoids the problem of conducting junctions.

In many cases, transistors that check good out of the circuit with an ohmmeter or transistor checker will not function properly in the circuit. All ohmmeters and many transistor checkers only test the DC parameters of a transistor, and a transistor may have severely degraded AC parameters even though the DC parameters check good. After other possible causes of a problem are eliminated, the transistor involved should be replaced.

In any electronic servicing endeavor, care should be taken not to accidentally short circuit the terminals of any device with voltmeter test prods, oscilloscope probes, or other instrument. This is especially true for solid-state circuits, where a momentary short can cause much damage. In this regard, tube-type circuits are somewhat more forgiving. Voltmeter test prods should be sharpened to a point to avoid having them slip from a terminal.

12-2 SYMPTOM-FUNCTION DIAGNOSIS

As technicians gain experience, they find that most symptoms are related to specific stages or components in a receiver. A list of the most commonly encountered symptoms and the most probable cause of each is given in Figure 12-1 for tube-type receivers and in Figure 12-2 for solid-state receivers. For most symptoms, there are only a few possible causes. The ability to observe a symptom and then recognize the receiver function that is not being performed satisfactorily is called *symptom-function diagnosis*. This ability to identify defective components by merely listening to a defective receiver is acquired through experience and a thorough understanding of receiver operation. The economic benefits of this ability need little elaboration.

A technician should maximize the benefits of his experience, first, by learning as much electronic theory as possible, and then by making notes at the service bench for future review and reference. When a receiver having unusual symptoms is repaired, try to understand why the defect produced the observed symptoms. Try to learn something beneficial from each receiver serviced.

12-3 FILAMENTS DO NOT LIGHT

With unlighted filaments, none of the tubes will operate, and no sound will come from the speaker. Since the filament string and switch form a series circuit across the AC line, it is a simple matter to use a voltmeter to measure AC voltages along

Symptom	Probable cause
1. Dead receiver; filaments do not light.	Power supply: AC input, filament string.
2. Filaments light; no sound from speaker.	B+ portion of power supply; audio output stage.
3. Low 60-Hz hum; no static with volume control fully clockwise.	AF amplifier; detector; IF amplifier; B+ portion of power supply.
4. Low 60-Hz hum; lots of static with volume at maximum.	RF amplifier; converter; IF amplifier.
5. Loud 60-Hz hum with volume control at minimum setting.	Filter capacitor; heater-cathode leakage in the audio section; leakage on the B+ line.
6. Distorted audio.	AF amplifier; audio output tube; B+ portion of power supply.
7. Loss of sensitivity.	RF amplifier; converter; IF amplifier: low B+ voltage; misalignment.
8. Insufficient volume, sensitivity normal.	Audio section; low B+ voltage.
9. Oscillations.	Decoupling capacitors; open ground connections; RF amplifier; converter; IF amplifier; misalignment.
10. Noisy reception.	Dirty tube sockets; breaks in PC board; leaky IF cans.
11. Intermittent operation.	Depends upon symptom when malfunction occurs. Trouble is often due to breaks in PC board.

FIGURE 12-1 List of commonly encountered symptoms and the probable cause of each for tube-type receives.

the string to locate the problem. We look for an open circuit in the switch, a tube filament, or the circuit wiring.

On the schematic of Figure 12-3, for example, assume that the $V1$ filament is open. The chart shows the voltages measured and indicates a voltage drop of 117 V across $V1$. In general, a voltage drop equal to the line voltage will be found across the open component. Be on the lookout for dirty tube sockets, bent pins on tubes, deformed socket connectors, or tubes located in the wrong sockets.

An alternative procedure is to use an ohmmeter to look for the open circuit. This is illustrated in Figure 12-4.

Heater-cathode shorts along the filament string may cause some unusual symptoms. Certain filaments may be lighted while others are out—even in a series filament string! In Figure 12-3, if $V1$ develops a heater-cathode short, $V5$, $V4$, and $V2$ will light up much brighter than normal, while $V1$ and $V3$ will be out.

If $V4$ develops such a short, the filament in $V5$ will probably be burned out as the result of excessive current flow. If, after determining that $V5$ has an open filament, a replacement is installed, it will probably light up with excessive brilliance and go out again. Such actions should cause the technician to suspect a short circuit in $V4$ even though the filament of $V4$ may be good.

Symptom	Probable cause
1. Dead receiver.	Speaker circuit; power supply; audio-output stage; earphone jack.
2. Turn-on plop; very little static.	AF amplifiers; IF amplifiers; detector.
3. Turn-on plop; lots of static.	Front end; IF amplifiers.
4. 60-Hz hum with volume at minimum setting.	Power supply filter capacitors; audio output stage.
5. Distorted audio.	Audio section; power supply.
6. Loss of sensitivity.	Front end; IF amplifiers; power supply.
7. Insufficient volume; sensitivity normal.	Audio section; power supply.
8. Oscillations.	Decoupling or neutralizing capacitors; front end; IF amplifier; alignment.
9. Noisy reception.	Defective transistor; poor circuit connections on printed circuit board.
10. Intermittent.	Depends upon symptom when malfunction occurs; usually problem is a transistor or poor printed circuit connection.

FIGURE 12-2 List of common symptoms and probable cause for a solid-state receiver.

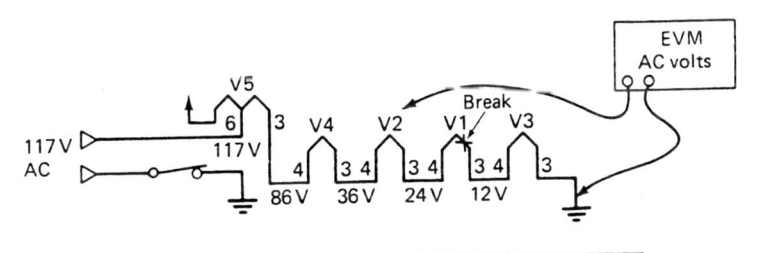

Tube pin	Measured (V)	Normal (V)
V5, P6	117	117
P3	117	86
V4, P4	117	86
P3	117	36
V2, P4	117	36
P3	117	24
V1, P4	117	24
P3	0	12
V3, P4	0	12
P3	0	0

FIGURE 12-3 AC voltage analysis of a series filament string.

Sec. 12-3 *Filaments do not Light* **333**

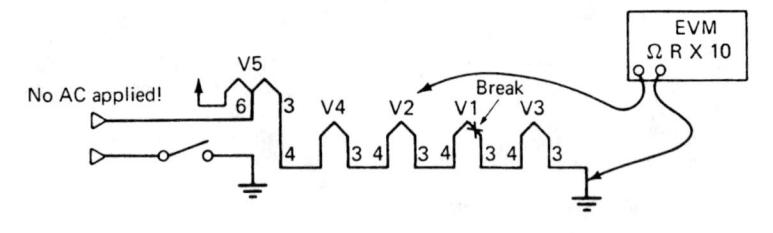

Tube pin	Measured (Ω)	Normal (Ω)
V5, P6	∞	110
P3	∞	86
V4, P4	∞	86
P3	∞	36
V2, P4	∞	36
P3	∞	24
V1, P4	∞	24
P3	12	12
V3, P4	12	12
P3	0	0

FIGURE 12-4 Resistance analysis of a series filament string.

12-4 DEAD RECEIVER

A receiver is *dead* if no sound of any kind comes from the speaker. We immediately suspect the power supply, the audio output stage, or the speaker itself.

Check the speaker by placing an ohmmeter across the speaker terminals. A small pop or click should be heard in the speaker as the ohmmeter is connected, and the resistance indicated should be on the order of the DC resistance of the speaker, typically 8 Ω or less.

If no click or pop is heard when the ohmmeter is connected, connect a test speaker in parallel with the suspected speaker and check for normal operation. If proper operation is restored, the speaker is defective.

Almost all solid-state receivers emit a characteristic *plop* when first turned on, the plop being caused by the initial flow of current through the output transistor. If the plop is not heard, and if no hum, hiss, or static is audible, a defect in the power supply or audio output stage is indicated. We check the power supply first, as illustrated by the following example.

EXAMPLE 12-1: In this example we assume that the receiver of Figure 12-5 is dead. After the back is removed, a substitute speaker is connected across the speaker terminals, but no sound is heard when the set is turned on. It is then turned off, and an ohmmeter check of the audio output transformer secondary shows 0.7 Ω. From this it is concluded that the speaker circuit is normal. The chassis is pulled.

A check of the B+ voltages shows 105 V at CTP 1 (90 V) and 0 V (9.6 V) at CTP 2. The absence of voltage at CTP 2 causes $R26$ to be suspected to be open. An ohmmeter check of $R26$ in-circuit shows 500 kΩ, which confirms our suspicion. To check for a possible short circuit, the resistance from CTP 2 to ground is measured and found to be 680 Ω, the

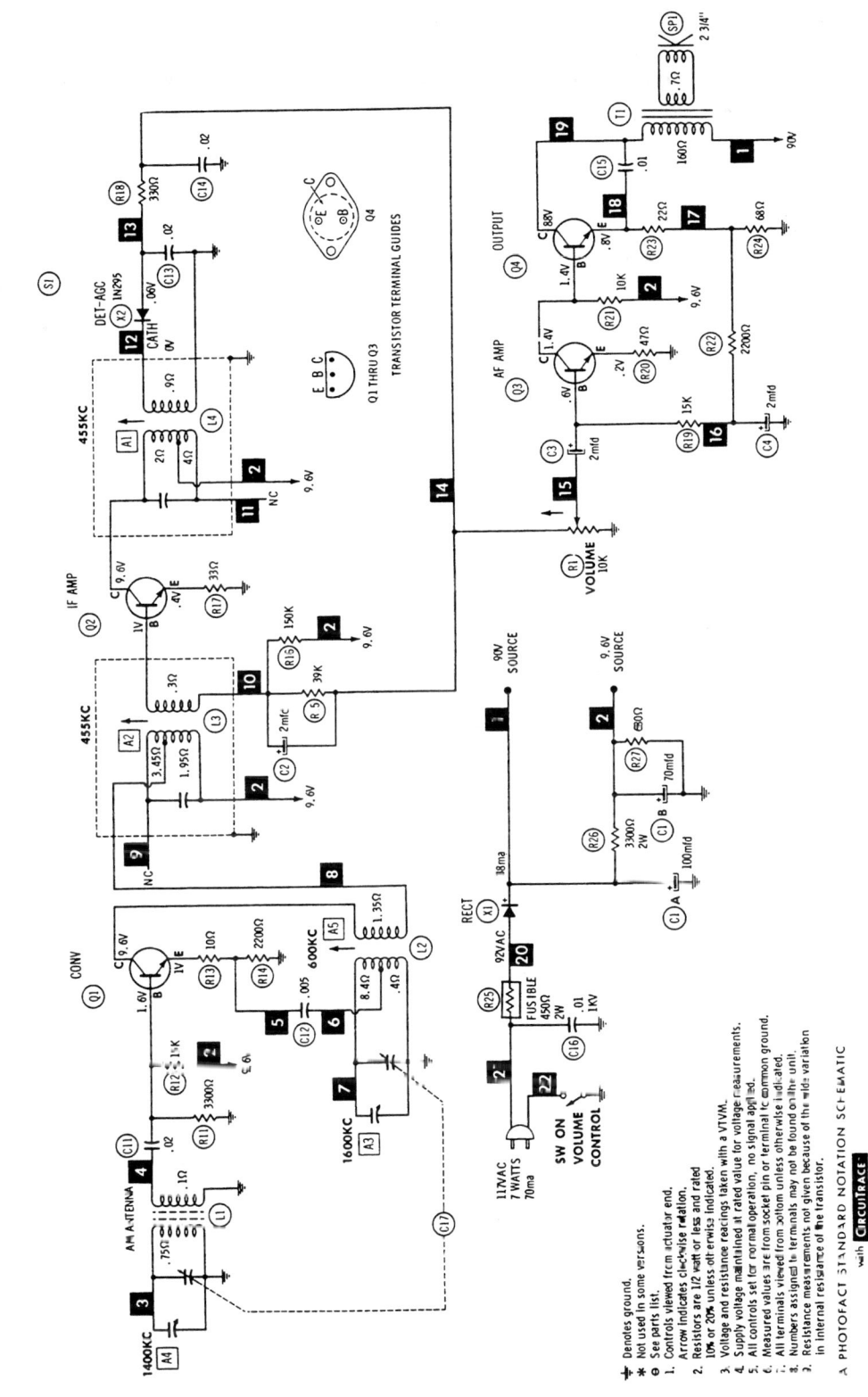

FIGURE 12-5 Receiver having a standard converter, a single IF stage, and a direct-coupled, single-ended audio-output stage (a Photofact schematic, courtesy of Howard W. Sams & Co., Inc.)

335

resistance of R27. This is normal. After checking R26 out-of-circuit to confirm the defect, a replacement is installed and normal operation is restored.

Power Supply OK

When the power supply voltages are found to be normal but no sound comes from the speaker, we turn our attention to the audio section, as illustrated in the following example.

EXAMPLE 12-2: In this example we assume that the B+ voltages of Example 12-1 are found to be slightly higher than normal: 94 V (90 V) at CTP 1 and 11 V (9.6 V) at CTP 2. The audio output stage is checked, and since direct coupling is used, the driver is checked also. The voltages are found to be:

Q3	Measured	Normal	Q4	Measured	Normal
C	11	1.4	C	94	88
B	0	0.6	B	11	1.4
E	0	0.8	E	0	0.8

From the excessive forward bias on the B-E junction of Q4 it is evident that either the base or the emitter must be open. Q4 is checked in-circuit with an ohmmeter, and unusual resistances are obtained. With both polarities of the ohmmeter, a resistance of about 10 kΩ is obtained from the base to the emitter, and a slightly higher resistance between the base and the collector. The transistor is removed and checked out-of-circuit, and the base is found to be open. An infinite resistance is indicated from the base to each of the other terminals, with both ohmmeter polarities.

Since Q3 is direct-coupled to Q4, it is checked in-circuit also, and the base appears to be open. It is removed from the circuit and rechecked to confirm the defect. The resistors around the stages are quickly checked and are found to be normal. Normal operation of the receiver is restored when transistors Q3 and Q4 are replaced with suggested replacements given in the service data.

In retrospect, if Q4 had been replaced without checking Q3, the new Q4 would have been quickly destroyed by excessive current resulting from excessive voltage at the Q4 base. It is very important to check all transistors of a defective direct-coupled amplifier when one stage is found to be bad. It is likely that, in the example above, the initial failure of Q3 caused the subsequent failure of Q4.

Alternate Procedure for Tube-Type Receivers

After the speaker is tested with an ohmmeter and found to be good, it is a simple matter to substitute the audio output tube before going on to check the power supply. This can be done without having to remove the chassis from the cabinet. If the receiver still does not operate, remove the rectifier tube and measure the resistance from the cathode terminal (pin 7 of a 35W4) to ground to check for shorts on the B+ line. In most cases the cathode terminal can be contacted from the top of the tube socket by carefully inserting a voltmeter probe or small wire into the socket hole for the cathode pin of the tube. If a satisfactory resistance is obtained, substitute a known-to-be-good rectifier tube and check again for normal operation. If normal operation is not restored, or if a short was found on the B+ line, the chassis must be pulled for further testing.

Another Dead Receiver with an Audio Problem

The following example shows that the failure of a single component in the audio section of a solid-state receiver can produce complete silence in the speaker even though the power supply and output transistors are good.

EXAMPLE 12-3: This example refers to the receiver of Figure 12-6, and we assume that the receiver is dead. The speaker and B+ circuits check normal, so the voltages around the audio output transistors, $Q5$ and $Q6$, are checked. They are found to be:

Q5	Measured	Normal	Q6	Measured	Normal
C	−10	−10	C	−10	−10
B	−0.5	−0.28	B	−0.5	−0.28
E	−0.45	−0.12	E	−0.45	−0.12

The base voltages are higher than normal, which indicates either collector–base leakage, an absence of base current, or a fault in the bias network, $R24$, $R25$, and $R26$. The emitter voltage is higher than normal, and this indicates either an increase in emitter current or an increase of resistor $R27$. Since the forward bias on the base–emitter junction is only 0.05 V, and since no "turn-on plop" was heard at the speaker, the possibility of excessive conduction is ruled out. We now suspect that emitter resistor $R27$ is open. It is checked in-circuit, and the measured resistance is several thousand ohms. It is removed and rechecked to confirm the defect. A replacement is installed, and normal operation of the receiver is restored.

Leakage from the collector to the base in the output transistors could have caused $R27$ to overheat and increase in value. The voltages around $Q5$ and $Q6$ are checked after replacing $R27$. The voltages are found to be normal, and we assume that the transistors are good.

In this example the bias network was not a prime suspect, since the absence of the turn-on plop indicated a lack of conduction. Further, if the bias network had been at fault, a larger base–emitter voltage would have been found.

12-5 TURN-ON PLOP IS PRESENT; LITTLE STATIC

The turn-on plop in solid-state receivers or a low-level hum in the speaker of tube-type receivers indicates that the speaker circuit, audio output stage, and power supply are probably functioning normally. The absence of static indicates a probable defect in the AF amplifiers, detector, or IF amplifier. However, a reduced power supply voltage could cause improper operation of the AF or IF stages, reducing the level of static that is present. Therefore, we check the power supply voltages first, and then use signal injection or signal tracing to isolate the defective stage.

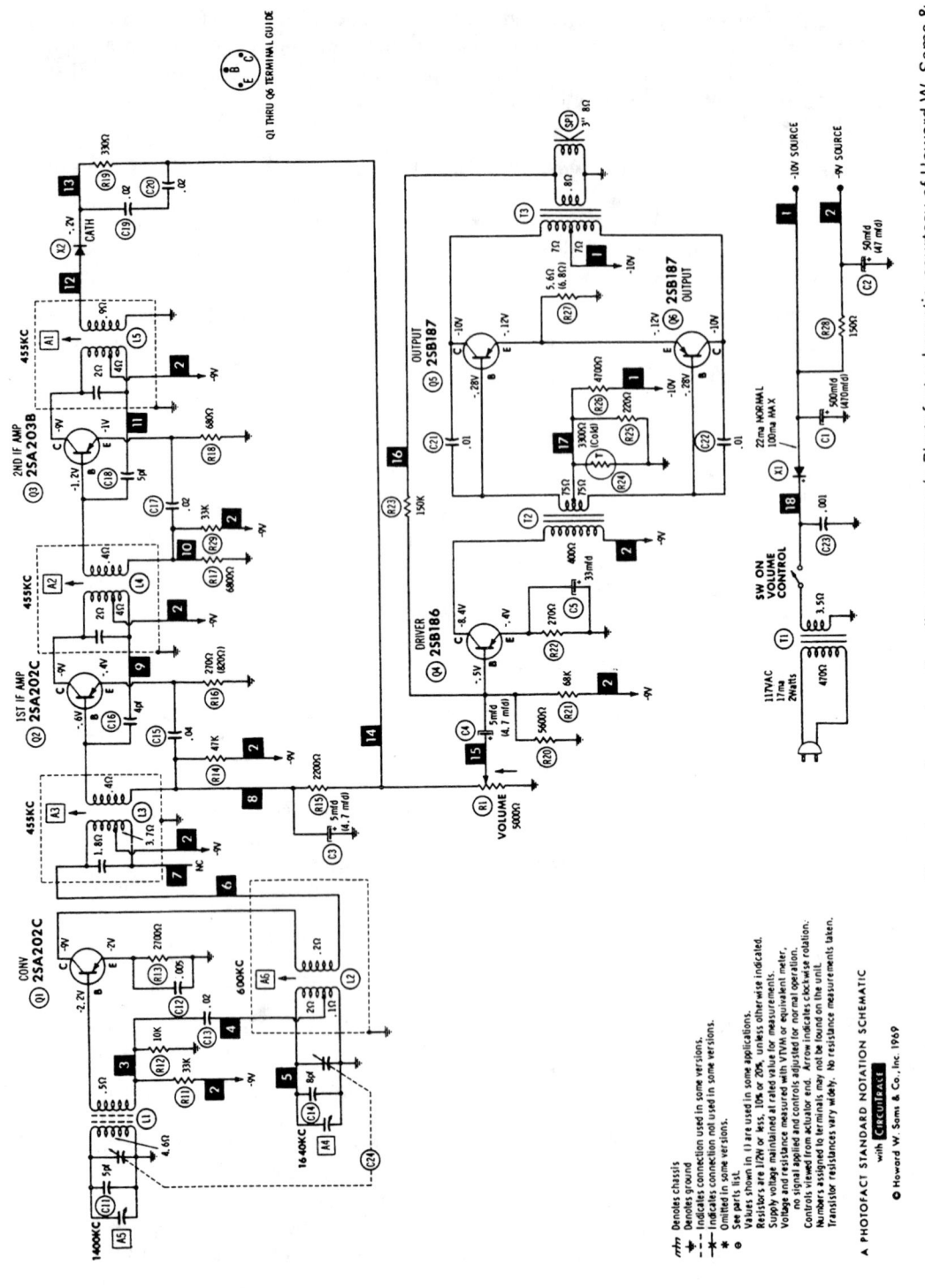

FIGURE 12-6 Receiver having a two-stage IF section and a push-pull audio-output stage (a Photofact schematic, courtesy of Howard W. Sams & Co., Inc.)

EXAMPLE 12-4: Suppose that the receiver of Figure 12-6 exhibits this symptom. The B+ voltages are checked and found to be normal. With the volume control fully clockwise, a small metal screwdriver is used to tap the wiper arm of the volume control. Nothing is heard at the speaker—an abnormal response. Since the output transistors are probably good, the voltages around driver transistor Q4 are checked instead of continuing with the signal-injection process. The voltages are found to be:

Q4	Measured	Normal
C	0	−8.4
B	−0.2	−0.5
E	−0.09	−0.4

The absence of collector voltage causes suspicion to fall upon the *T*2 primary as being open. The voltages are measured on both sides of the *T*2 primary, and −9 V is found on the B+ side while 0 V is found on the collector side. The resistance across the winding is measured in-circuit and is found to be very large. The transformer is removed from the circuit, and the winding is found to be open. Resistances in parallel with the winding produced the "less-than-infinity" indication of the ohmmeter when the winding was checked in-circuit.

The insulation is carefully removed from around the primary terminals of the transformer, but no break of the conductor can be found. We therefore conclude that a repair is impossible, and a new transformer must be ordered. The old transformer is placed in the container that holds the screws removed from the cabinet, and the disassembled receiver is carefully placed on a shelf to await the new part. The container holding the screws and transformer is taped to the cabinet so that it will not become separated from the receiver.

Alternate Procedure for Tube-Type Receivers

Before removing the chassis, replace or check the first AF/detector, IF amplifier, and rectifier tubes. If normal operation is not restored, remove the chassis from the cabinet and measure the power supply voltages. If the power supply voltages are normal, inject a signal at the high side of the volume control (with the control set for maximum volume) to check the audio section. This can be done with an audio signal generator, or an alternative is to tap the volume control with the tip of a conductive screwdriver held so that it contacts at least a finger of the technician. The human body is a good source of 60-Hz voltage, since the body picks up the energy radiated from the AC line.

If no sound comes from the speaker when the signal is injected at the volume control, the audio section is defective, and the signal injection process is continued toward the speaker in order to isolate the defect.

12-6 TURN-ON PLOP WITH LOTS OF STATIC

Static is electronic noise that is greatly amplified and reproduced in the speaker. When a defective receiver emits static from the speaker, the implication is that the power supply, audio section, and at least part of the IF section are functioning normally. Hence, the problem is most likely in the front-end section, in the RF amplifier or converter. However, reduced power supply voltage could cause the local oscillator to be so far off frequency that no stations can be received, and in some cases a defective IF amplifier by itself will produce these same symptoms.

The following procedure is appropriate. While removing the chassis from the cabinet, carefully inspect the antenna circuitry, looking for breaks in the typically small wires associated with the loopstick antenna. Measure the power supply voltages to be sure they are normal. Confirm that the audio section is functioning by injecting a signal at the volume control. If no defect is found thus far, inject a 455-kHz AM signal into the IF amplifier at the output of the converter, with care being taken to keep the amplitude of the injected signal very low. If the modulation tone is heard at the speaker, the IF amplifier is functioning normally, and we then suspect the front end.

A 455-kHz signal applied to the input of the converter will be heard in the speaker because the IF amplifier will respond to the 455-kHz signal at the converter output. This will occur even if the oscillator is not working. However, if an 800-kHz AM signal is applied to the converter, a frequency change must occur in order for the IF amplifier to accept the signal. This requires that the oscillator be functioning and that the receiver be tuned to 800 kHz so that the oscillator will be running at the proper frequency to convert the 800-kHz test signal. This provides a convenient method of testing the converter stage, as illustrated in the following example.

EXAMPLE 12-5: The receiver of Figure 12-7 is assumed to emit static from the speaker but not receive stations. No obvious defects are found with the antenna circuit, so the power supply voltages are measured and are found to be normal. A screwdriver applied to the volume control produces a low 60-Hz hum in the speaker. We prepare for a signal-injection procedure.

An AM signal generator is used to inject a 455-kHz AM signal through a 0.01-μF capacitor to the collector of $Q1$, and a tone is heard from the speaker. The generator is moved to the base of $Q1$, and the signal is still heard. The generator frequency is changed to 800 kHz, the receiver is tuned back and forth around 800 kHz, but nothing is forthcoming from the speaker. We now suspect that the oscillator is not working.

The DC voltages around oscillator transistor $Q2$ are measured and found to be:

$Q2$	Measured	Normal
C	−2.75	−4
B	−1.3	−1.2
E	−2.5	−1.1

The large emitter voltage indicates that the transistor is conducting heavily, but, on the other hand, the base–emitter junction of the transistor is reverse-biased. The transistor appears to have a collector–emitter short. An ohmmeter placed across the transistor reads almost 0 Ω, so the transistor is removed and checked out-of-circuit to confirm the defect. A new transistor is installed and the receiver then operates normally. The sensitivity appears normal and the stations are received at the proper place on the dial, so an alignment is deemed unnecessary. If the sensitivity is questionable, or if the stations come in at the wrong place, a complete alignment is required.

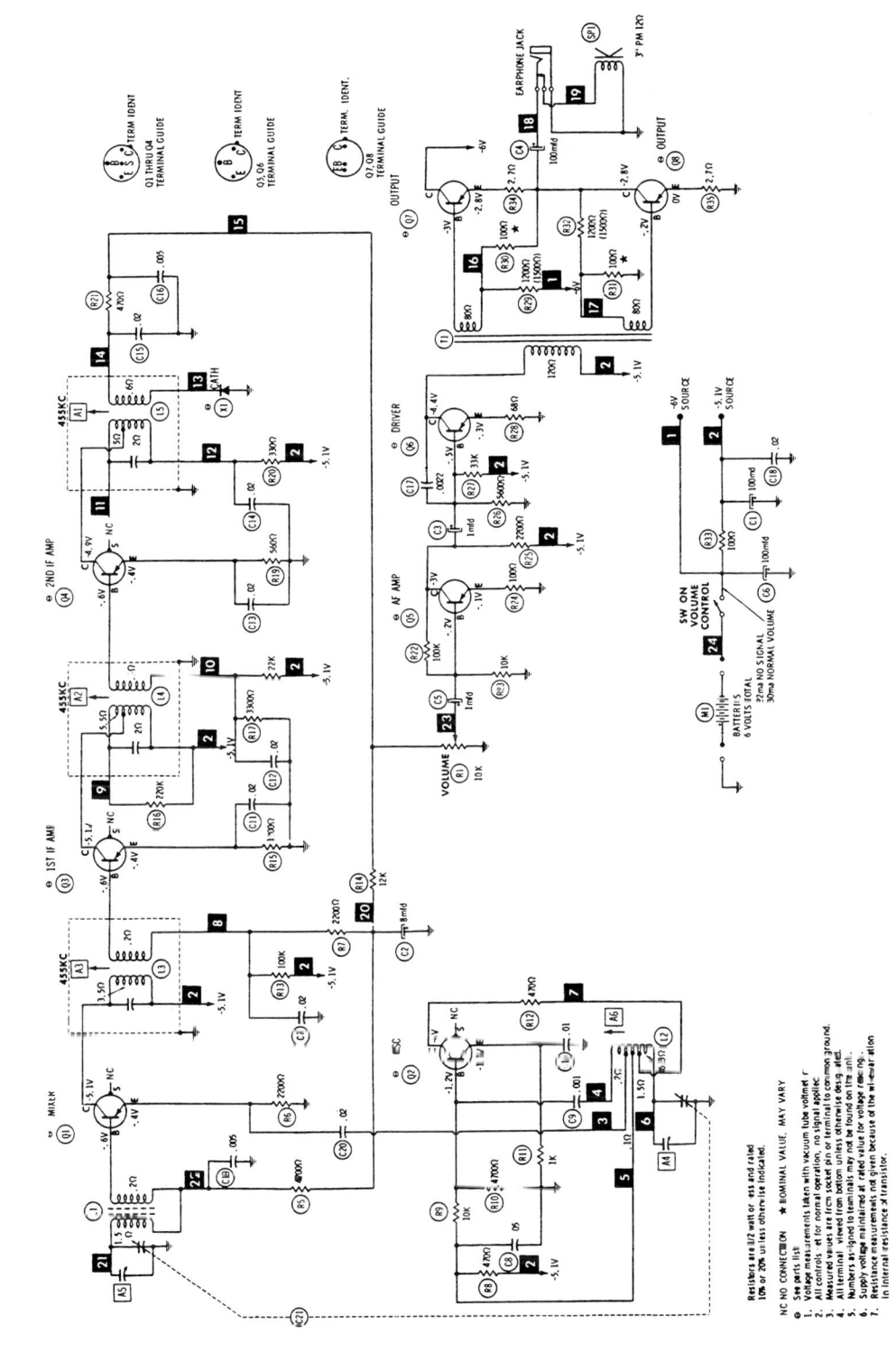

FIGURE 12-7 Receiver having a separate stage for the local oscillator.

EXAMPLE 12-6: Refer to the receiver of Figure 12-6, which we now assume emits considerable static but refuses to pick up a station. No obvious defects are discovered around the antenna circuit, and the B+ voltages check *normal*. A screwdriver on the volume control produces a normal response. An AM signal generator is used to inject a 455-kHz signal through a 0.01-µF capacitor to the collector of converter transistor $Q1$. A tone is heard, and the generator is then moved to the base of the converter. The tone is again audible. The generator frequency is changed to 800 kHz, and the receiver is tuned back and forth through 800 kHz. Nothing is heard from the speaker. Obviously, the oscillator is not working, so we proceed to check the DC voltages around the converter. All voltages are found to be normal.

At this point we reason that the defect is of such a nature that the DC voltages are not affected. An ohmmeter is used to check the CTP 5 side of the oscillator coil, and the resistance is normal. The printed circuit board is carefully inspected for cracks and cold solder joints, but nothing is found. Capacitors $C13$ and $C12$ are bridged with new units, but still there is no change.

We are now running out of things to check, so we remove the converter transistor and check it with both an ohmmeter and a transistor checker. It checks *good*. Realizing that the ohmmeter and the particular transistor tester we are using makes only DC checks of a transistor, we find the replacement for the transistor and install it in the circuit "just in case." Normal operation is obtained when the receiver is turned on.

Something had happened to the converter transistor to degrade its characteristics enough to keep the oscillator from oscillating. This is fairly common; a transistor will check out normally on an ohmmeter or transistor checker and yet will not operate in the circuit. This is especially true for oscillator circuits. When all other possibilities are eliminated, replace a suspected transistor even if it checks *good*.

Considerations for Tube-Type Receivers

Obviously, because tubes are easily replaced, the IF amplifier, converter, and RF amplifier tubes should be substituted before removing the receiver from the chassis. This should be standard practice for tube-type receivers.

Whether or not the oscillator is oscillating can be determined by measuring the voltage at the oscillator grid of the pentagrid converter. If the oscillator is running, a negative voltage of several volts will be present at the oscillator grid. If this voltage is not present, the oscillator is probably defective. An alternate procedure is to hold a scope probe near the oscillator coil or converter tube. The scope should pick up the oscillator signal, even with no direct connection made to the circuit.

Because tube-type receivers are likely to be rather old, we should ensure that dust and grime have not accumulated between the plates of the tuning capacitors and produced a short circuit. Also, corrosion of the windings in the various coils can lead to shorted turns and open windings, either of which will affect the operation of the receiver.

12-7 LOUD RIPPLE HUM WITH VOLUME CONTROL COUNTERCLOCKWISE

A loud 60-Hz hum from the speaker that does not vary appreciably with the setting of the volume control indicates excessive ripple on the power supply line. This ripple may arise either from open or leaky filter capacitors or from leakage somewhere on the power supply line that causes excessive current to flow in the power supply. As the current increases, the ripple voltage becomes greater.

A test capacitor is useful in troubleshooting power supply filters and consists of an electrolytic capacitor of about 1000 μF rated at 25 or 50 V. (For tube sets, the capacitor should be about 40 μF and rated at 450 V.) The leads should be insulated with spaghetti so that the technician can handle the capacitor safely in live circuits. Proper polarity must be observed, and the polarity should be clearly marked to avoid confusion.

With the chassis pulled and with power applied, bridge the test capacitor across the input filter capacitor of the power supply. No solder connection is necessary; the leads of the test capacitor can be hand-held to make the proper contact. Be careful to observe the proper polarity. If the hum disappears when the test capacitor is connected, the input capacitor is either open or has decreased in value and must be replaced. If the hum is still present, remove power from the receiver, disconnect the input filter capacitor, and substitute the test capacitor. If the hum disappears, the input filter is leaky and must be replaced. If no change in hum occurs, the trouble is elsewhere. Reconnect the input filter, and then perform the same checks for the other filter capacitors.

Early in the checkout procedure, one should feel the audio output transistors to determine if they are excessively warm. The transistors should be warm, but if they are uncomfortable to the touch, they are probably leaky. In such a case the transistors are removed and checked. All resistors around the output stage are checked in-circuit (when possible) to determine that none has changed value. All direct-coupled transistors are checked in circuit and are removed when necessary for further checking.

If the output transistors and the output transformer appear to be at normal operating temperatures, the filter capacitors in the power supply are bridged with test capacitors. If the hum remains, the B+ line is checked for leakage with an ohmmeter. Normal resistance from the B+ line to ground should be about 10 kΩ, but in some sets, resistances as low as 500 Ω may be normal due to the presence of bleeder resistors. Check the schematic carefully to determine the normal resistance that should be found. Components are then disconnected one at a time until the resistance reading increases. This is done in the same manner as is done with the vacuum-tube sets.

EXAMPLE 12-7: The receiver of Figure 12-6 exhibits a loud 60-Hz hum even when the volume control is at minimum. Output transistor $Q6$ is much hotter than $Q5$ and emitter resistor $R27$ appears to be discolored. An ohmmeter is used to measure the collector–emitter (C-E) resistance of $Q5$ and $Q6$. A reading of 50 Ω is obtained for $Q6$, and 64 Ω for $Q5$. $Q6$ is removed and still has 50 Ω C-E resistance when checked out-of-circuit. $Q5$ is also removed, and exhibits a high C-E resistance that is typical of germanium power transistors on the ohmmeter we are using. When $Q5$ was checked while it was in the circuit, the low resistance of $Q6$ and the output transformer primary (14 Ω) were in parallel with $Q5$, giving the 64-Ω reading. Emitter resistor $Q27$ is checked and found to be 8 Ω (5.6), and is replaced along with $Q6$. Normal operation is restored.

On inexpensive and low power receivers, it is not necessary to replace push-pull output transistors with matched pairs. On more powerful stereo receivers, when one output transistor of a push-pull amplifier must be replaced, replace both output transistors with a matched pair of replacement semiconductors. This ensures the cleanest possible sound with minimum crossover distortion. In a stereo receiver, when only one channel has a defective output stage which is replaced with a matched pair, it is not necessary to change the transistors in the good channel.

12-8 DISTORTED AUDIO

In most cases, distorted audio results from a defect in the audio amplifier section as opposed to a defect in the IF amplifier or front end. The distortion may be caused by a defective power audio IC, transistor, or tube, or it may be caused by a defect in the biasing circuitry of discrete transistor or tube amplifiers. Distortion will result when one transistor of a push-pull pair becomes defective while the other continues to operate. Another possibility, of course, is that the speaker itself may be defective, having a warped cone that causes the voice coil to rub against the magnet core. Substitution of a good speaker is the best test of a speaker suspected to be defective.

Signal injection or signal tracing may be used to isolate the defective stage, beginning at the output and working back toward the volume control. V-R analysis is then used to locate the defective component. When direct-coupled stages are involved, a voltage analysis may prove futile, and an out-of-circuit check of each transistor may be necessary.

EXAMPLE 12-8: The receiver of Figure 12-7 is used to illustrate the procedure. The speaker is substituted with a test speaker, and the distortion remains. A scope is used to check the waveform at the bases of $Q7$ and $Q8$, after tuning the receiver to a local station. A normal audio waveform is found at both bases, but at the output point, CTP 18, the signal is severely clipped. The voltages around $Q7$ and $Q8$ are checked and found to be:

Q7	Measured	Normal	Q8	Measured	Normal
C	−6	−6	C	−3.3	−2.8
B	−4	−3	B	−0.6	−0.2
E	−3.3	−2.8	E	0	0

Since the transistors are germanium, the base–emitter voltage on $Q8$ is much too high and should cause $Q8$ to saturate. However, the collector voltage on $Q8$ is higher than normal, which indicates a decrease in conduction. This would result if the base or emitter of $Q8$ were open. An ohmmeter check of $Q8$ reveals a very high resistance between the base and emitter in both directions. $Q8$ is removed, and an out-of-circuit check confirms that the emitter is open. Replacement of $Q8$ with a new transistor restores normal operation.

EXAMPLE 12-9: Refer now to the receiver of Figure 12-8, which we assume to have distorted audio. The speaker is substituted with no improvement. A DC voltage analysis reveals that the voltages are far from normal throughout the entire direct-coupled amplifier. Signal tracing or signal injection would be useless, since a DC problem has thrown the bias voltages of all transistors way off. At the same time, the feedback loop from output to input confuses the issue so that no conclusions can be drawn as to where the trouble might be.

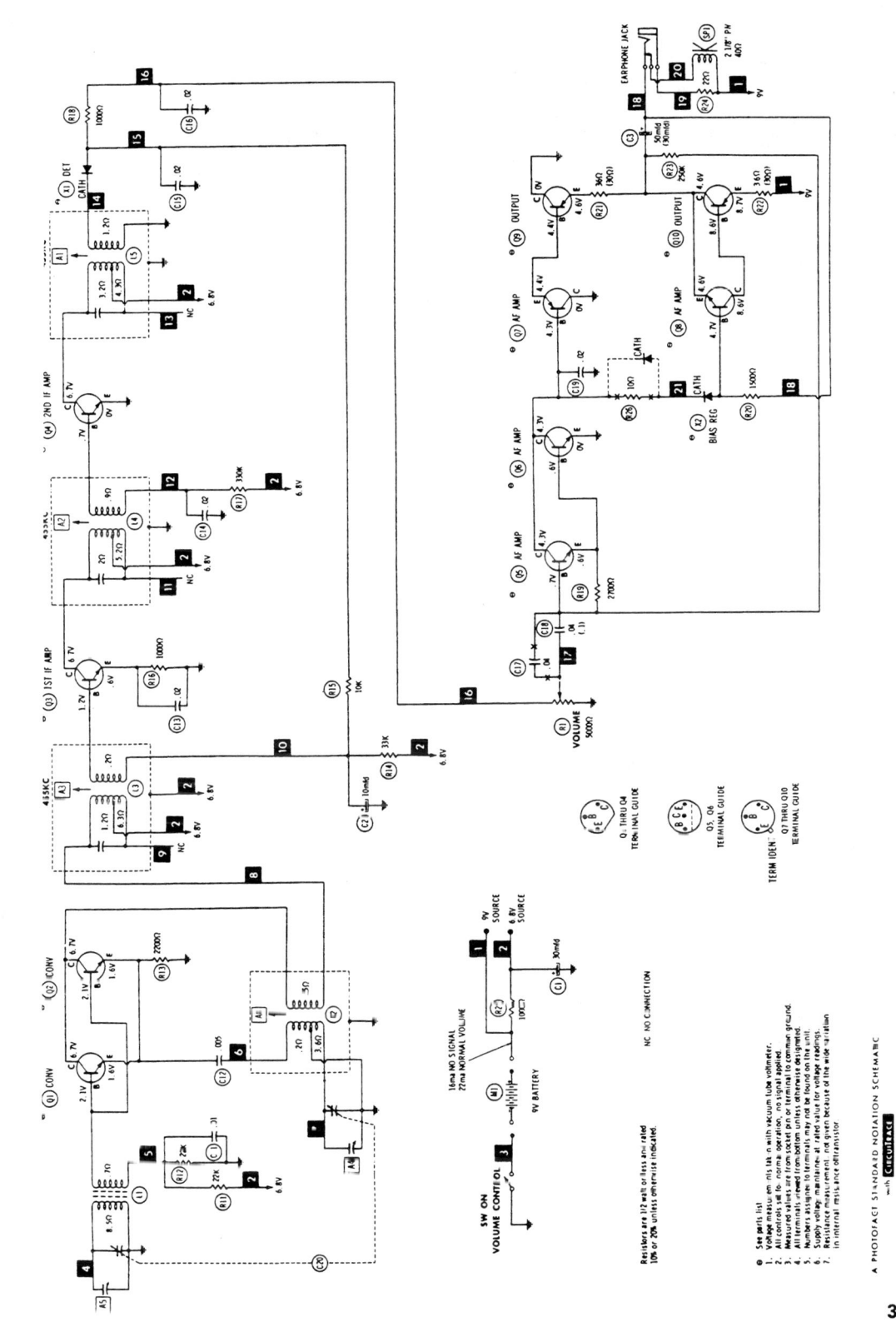

FIGURE 12-8 Receiver having a direct-coupled, OTL output stage. The other stages are typical.

345

Either of two approaches can be used at this point. Each transistor can be checked in-circuit and removed if necessary to confirm the check, or a variable voltage power supply can be used to clamp one of the voltages in the amplifier to the proper voltage. For purposes of illustration, we choose the latter procedure.

Examining the circuit, we note that the $Q5$-$Q6$ collector connection divides the amplifier in half, separating the preamp and output stages. It also appears that no damage will be done if we use an external power supply to pull the voltage at that point from its low value (about 1.5 V) up to the normal value of 4.3 V. We set the external supply to 4.3 V, connect the negative lead of the supply to ground, turn on the receiver, and then connect the positive lead of the external supply to the collector of $Q6$. We note that the external supply voltage drops slightly, so we adjust it back to 4.3. Quickly, now, we measure the voltages around $Q7$, $Q8$, $Q9$, and $Q10$, and find them to be close to normal. We remove the external supply.

We now conclude that the problem is in the Darlington pair preamplifier. $Q5$ and $Q6$ are checked in-circuit, and $Q6$ shows a low resistance from its collector to emitter. To be sure, both $Q5$ and $Q6$ are removed from out-of-circuit testing, and only $Q6$ is found to be defective, the defect being excessive collector–emitter leakage.

Since clamping the $Q6$ collector voltage produced the proper voltages around the other transistors of the output stage, it is not necessary to test these transistors. We replace $Q5$, install a new $Q6$, and find that normal operation is restored. This makes us very happy, but we allow the receiver to play for about 1 hour to be sure that everything holds together.

Comment

Substituting an external supply in a direct-coupled amplifier circuit can save valuable time when done properly. However, careful analysis of the circuit is required to determine a point that will provide a valuable division of stages and also cause no harm to befall the transistors in the circuit. If the technician is unsure of where to apply the external supply voltage, or if he is unsure as to whether damage may result, the external supply should not be used. Instead, all transistors should be checked to determine which (if any) are defective, and then the rest of the circuit should be checked component by component in the hope that something turns up.

This provides us an opportunity to speak of the advantages of an in-depth understanding of circuit operation. If the circuit is understood, the effects of connecting the external supply can be ascertained; otherwise, it is impossible. Direct-coupled amplifiers are notorious for being difficult to service because everything is coupled together. Sometimes, the only way to make any progress at all is to use an external supply to force a voltage to its proper value and then see what happens elsewhere.

12-9 LOSS OF SENSITIVITY

A receiver that displays a loss of sensitivity will function normally on strong, local stations but will be very weak or inoperative on weak, distant stations that are normally received. A loss of gain in the front end or IF amplifier is usually responsible, but a reduced power supply voltage can produce the same effect. Loss of gain can be caused by weak tubes, defective transistors, or ICs that do not function normally, misalignment, a bias change, or degeneration produced by an open bypass capacitor in the emitter circuit of transistors or in the cathode circuit of tubes. If the same power supply line is used for the audio section, and if the audio is normal, it is unlikely that the problem is caused by a reduced power supply voltage.

The troubleshooting procedure begins by removing the back of the cabinet or case and inspecting the antenna circuitry for an open circuit or a cracked core in a loopstick antenna. For tube-type receivers, the IF, converter, and RF amplifier tubes should be either substituted or tested. If nothing is found, remove the chassis and measure the power supply voltages to be sure that the loss of sensitivity is not caused by a low voltage. Continuing, bridge a test capacitor across decoupling capacitors and bypass capacitors to ensure that an open capacitor is not causing degeneration by introducing undesired negative feedback.

If the defect remains, a signal generator is used to inject signals at various points in the receiver to determine the relative gain of each stage. The exact procedure to be followed at this point depends upon the type or sophistication of the available generator and the technician's familiarity with the generator.

Generally speaking, life is made easier if the generator has a calibrated RF output, but a calibrated generator is not essential provided that the technician is familiar with the typical control settings necessary to produce a given speaker output when the signal is injected at various points in the receiver. This information is gathered by injecting signals into properly operating receivers of various types—tube, solid state, with and without RF amplifiers, etc.—and recording the results for future reference. Then, in a troubleshooting situation, if the generator output has to be increased significantly to obtain the standard output, a loss of gain is indicated for the stage involved. Figure 12-9 shows the settings for a hypothetical generator.

Standard Output

Since the human ear is not a sensitive indicator of the speaker output—and we admit that this depends upon the ear involved—an AC voltmeter should be connected across the speaker terminals to serve as an indicator. If we take 50 mW to be our standard output, which produces a "good" volume level in typical, small speakers, we calculate the rms AC voltage at the speaker terminals to be 0.63 V for an 8-Ω speaker. For any other speaker impedance, the voltage is calculated by taking the square root of (0.05 times the impedance). For a 4-Ω speaker, the voltage is 0.45 V. At a given service bench, any "standard" output may be adopted as long as it is used consistently.

Modulation

If the generator has a modulation control, it is essential that the same percentage of modulation be used consistently, since this directly affects the speaker output. Many servicing procedures recommend 30% modulation; we do not recommend percentages greater than about 50%.

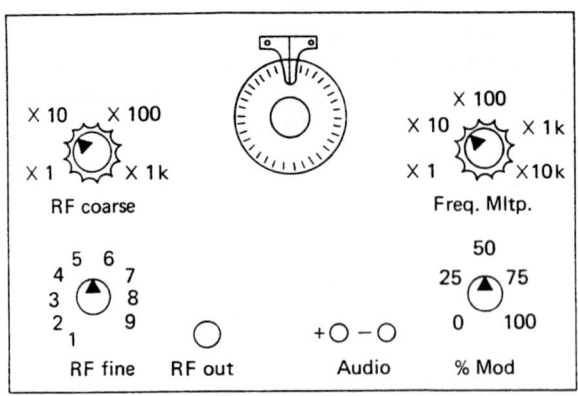

Standard output, 50 mW; receiver volume control, max.

Tube receiver: no RF amp.

Signal (kHz)	Injection point	Coarse	Fine
455	Grid IF amp.	X100	4
455	Grid conv.	X10	8
800	Grid conv.	X10	2
800	Loop	X10	5

Tube receiver: with RF amp.

Signal (kHz)	Injection point	Coarse	Fine
455	Grid IF amp.	X100	5
455	Grid conv.	X10	8
800	Grid conv.	X10	4
800	Grid RF amp.	X1	8
800	Loop	X10	1

Tube receiver: no RF amp.

Signal (kHz)	Injection point	Coarse	Fine
455	Grid IF amp.	X100	4
455	Grid conv.	X10	8
800	Grid conv.	X10	2
800	Loop	X10	5

Tube receiver: with RF amp.

Signal (kHz)	Injection point	Coarse	Fine
455	Grid IF amp.	X100	5
455	Grid conv.	X10	8
800	Grid conv.	X10	4
800	Grid RF amp.	X1	8
800	Loop	X10	1

FIGURE 12-9 Settings for a hypothetical generator to produce a standard output for solid-state and vacuum-tube receivers.

Procedure

The general procedure to be followed when the average generator settings are known is as follows. With an AC voltmeter connected across the speaker terminals, the generator is coupled through a 0.1-μF capacitor to the input of the IF amplifier, with the generator set to produce a 455-kHz AM signal. The volume control is set for maximum, and the generator output is adjusted so that the voltmeter indicates the standard output.

If the generator setting is higher than normal, a problem exists in the IF or detector stages. The generator is then moved to the input of the mixer so that all the IF transformers can be checked for misalignment. If adjustment of the tuned circuits results in a significant increase in speaker output, the problem was most likely due to misalignment, and a complete alignment procedure is in order. If the adjustments produce little change in the speaker output, the IF and detector stages are given a V-R analysis to determine the problem.

If the IF amplifier checks normal in the procedure above, the generator is moved to (or kept at) the input to the mixer (still at 455 kHz) and is adjusted for standard output at the speaker. If the generator setting exceeds the normal, the IF transformer that represents the mixer load is checked for proper adjustment, just to be sure. If the adjustment does not restore proper operation, the mixer is subjected to a V-R analysis to determine the defect.

If a standard output is achieved with normal generator settings when a 455-kHz signal is applied, the generator frequency is then changed to 1000 kHz and the receiver is tuned to the same frequency. This will test the frequency-changing ability of the mixer. If the generator setting is greater than normal, the front end is aligned according to the alignment instructions, and the sensitivity is rechecked. If greater-than-normal settings are still required, the oscillator becomes suspect, and a V-R analysis is in order.

If the mixer seems to operate normally, the defect must then be in the RF amplifier or antenna circuit. The generator signal is applied to the input to the RF amplifier, being careful to tune the receiver to the frequency of the generator, and the process is repeated. After the defect has been corrected (assuming success), the receiver is given a complete alignment.

Using an Unknown Generator

If the generator settings for standard output are unknown, a slightly different procedure is used. Beginning with the last IF amplifier stage, apply a small 455 kHz signal to the output of the stage (to plate or collector) and note the response in the speaker. Then move the generator to the input to the same stage and note the response, which should be louder if the stage has gain. If a substantial increase in signal output is not obtained, the stage is considered suspect.

This procedure is continued all the way back to the input of the mixer. However, in going from the output of the mixer to the input, a large increase in signal level will not be observed even if the circuit is operating normally because the tuned circuits of the mixer are tuned to a frequency significantly different from that of the generator (455 kHz). But if there is no change in signal level or if it decreases, the mixer is then considered suspect.

In this procedure it is important that the output of the generator be kept as small as possible, for two reasons. The first is that the AVC circuitry of the receiver will attempt to hold the signal level constant for large signals, thereby masking the gain of the stage being investigated. The second is that the ear is more sensitive to amplitude variations of small signals than it is to amplitude variations of very large signals.

As always, when investigating the IF amplifier, mixer, and RF amplifier, the alignment of the tuned circuits should be kept in mind. An IF transformer that is just slightly out-of-tune can have a very large effect upon the performance of the receiver.

Examples

The following examples illustrate the procedure to be used when the generator settings for a standard output are known. We assume that the generator of Figure 12-9 is used. For each example we assume that the receiver will receive only strong, local stations, and that the antenna has been inspected and found good. For tube-type receivers, we assume that the tubes have been substituted or checked. Further, we assume that the power supply voltages are normal and that bridging any bypass capacitors had no effect. In preparation for the signal-injection process, a voltmeter set to measure AC volts is connected across the speaker.

Our first three examples pertain to a tube-type receiver, shown in Figure 12-10. This receiver is described in Sec. 9-9.

EXAMPLE 12-10: Continuing the procedure described in part above, the generator is connected through a 0.1-μF capacitor to the IF amplifier grid (pin 1, $V2$) and adjusted for 455-kHz AM, 50% modulation. With the volume control set for maximum control, the generator output is adjusted to give 0.45 V rms across the speaker. The resulting generator setting is X100-10 (X100-4 is normal, from Figure 12-11), and this indicates a lack of sensitivity. The following IF transformer adjustments ($A1$ and $A2$) are checked and found to be tuned for a maximum response (peaked) at the IF frequency, which is normal. The generator is disconnected, and the voltages around the IF amplifier ($V2$) are measured and found to be:

12BA6	Pin	Measured	Normal
Plate	5	95	95
Screen	6	95	95
Grid	1	−0.4	−0.7
Cathode	7	3	0.8

These readings cause us to suspect that $R5$ has increased in value, giving rise to the high-cathode voltage. It is checked with an ohmmeter and found to be 500 Ω. We now see that the lack of sensitivity was caused by $V2$ being biased almost to the cutoff point where the transconductance of the tube is very low.

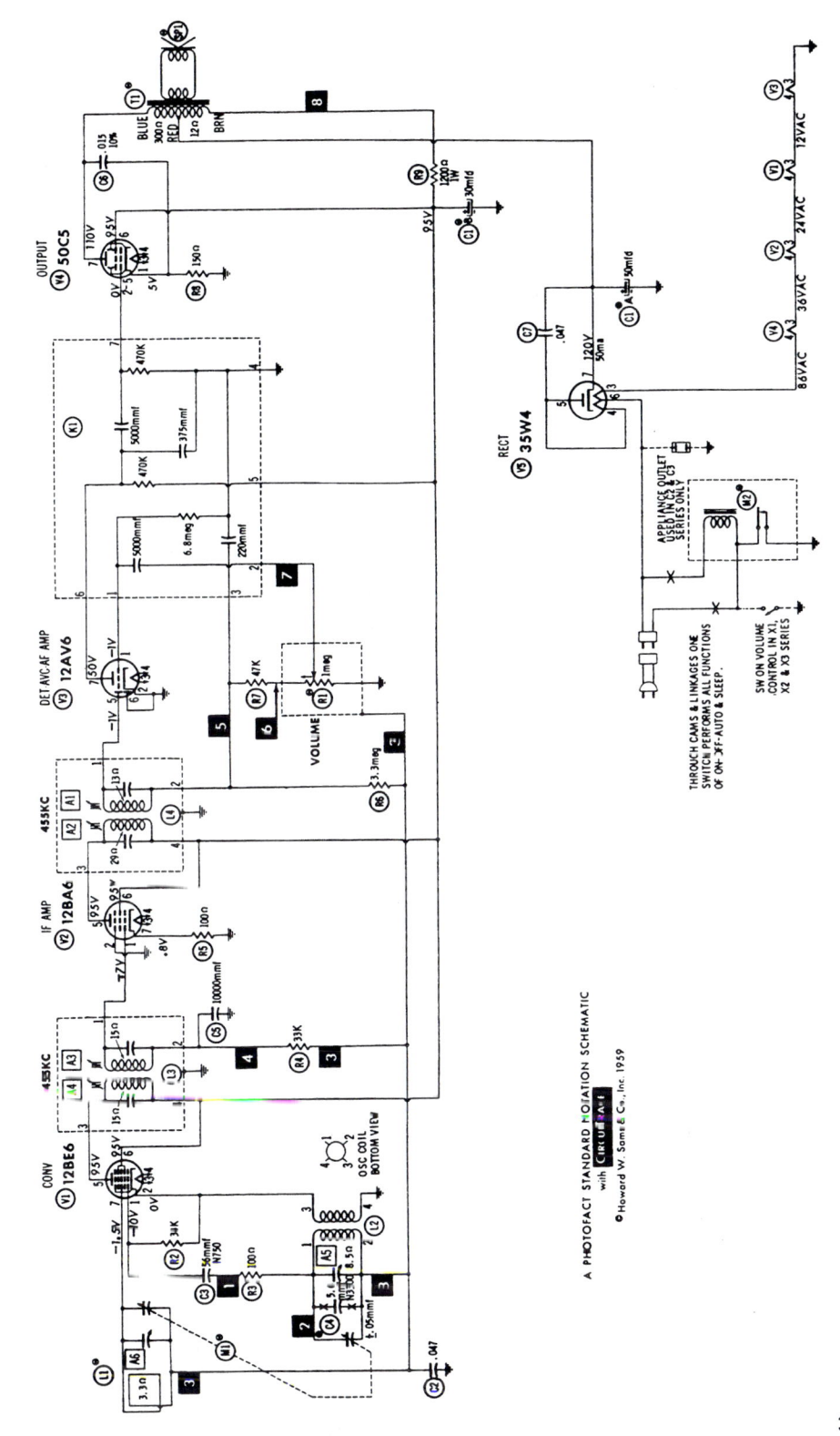

FIGURE 12-10 Typical five-tube AM receiver (a Photofact schematic, courtesy of Howard W. Sams & Co., Inc.)

EXAMPLE 12-11: With the same receiver (Figure 12-10), let us assume that the generator settings at the IF amplifier grid were X100-4, which is normal. The generator is then moved to the antenna grid of the converter (pin 7, $V1$) and readjusted for the standard output. The generator setting obtained is X100-5 [X10-8], and this indicates that there is a problem in the mixer stage. We check the adjustments ($A3$ and $A4$) of the first IF transformer, and to our great delight, a significant increase in signal output is obtained. The generator is again adjusted for standard output, and the setting now is X10-7, which indicates normal operation.

In this case the lack of sensitivity was caused by out-of-tune resonant circuits in the first IF transformer. This is not unusual, since the tuned frequency will drift as the receiver ages, is exposed to vibration due to its own speaker, or is subjected to heat, cold, and humidity. After the problem is solved, the entire receiver should be given the benefit of a careful alignment.

EXAMPLE 12-12: Continuing with the same receiver, let us suppose that a different result was obtained when the generator was connected to the antenna grid of the converter. Assume that the generator setting to produce standard output was normal (X10-8). In such a case the generator frequency is then changed to 1000 kHz and the receiver is tuned to the same frequency. The setting for standard output is found to be X10-10 (X10-2), which indicates that a problem exists in the antenna or oscillator circuit. An injection loop (see Sec. 12-14) is formed and the front end is aligned according to the instructions given in the service data. Adjusting the antenna trimmer $A5$ produces a significant increase in signal output, restoring the normal sensitivity of the receiver. Once again, a detuned resonant circuit was the culprit.

EXAMPLE 12-13: We assume that the solid-state receiver of Figure 12-7 exhibits a loss of sensitivity. Further, we assume that a generator is to be used whose settings for standard output are known and are given in Figure 12-9. The B+ voltages are found to be normal.

With a 455-kHz AM signal injected at the base of the second IF amplifier $Q4$, the receiver produces standard output when the generator setting is X1K-5 (X100-9), which is too high. Emitter bypass capacitor $C13$ is bridged with a test unit from the $Q4$ emitter to ground, and the generator setting has to be reduced to X100-8, which is normal, $C13$ is replaced, but normal operation is not restored. The generator setting once again is X1K-5 for standard output.

Something funny is going on, so we carefully inspect the PC board around $C13$, and a hairline crack is found on the ground side of $C13$. Solder-bridging the crack restores normal operation, and the original $C13$ is reinstalled.

Sensitivity problems can be very frustrating if the generator settings for standard output are not known. Therefore, we emphasize the importance of taking the time to obtain the average settings using receivers of various types that are in good working order.

EXAMPLE 12-14: In this example we assume that we must use a generator for which the typical settings are unknown. We refer again to the receiver of Figure 12-7, for which the B+ voltages are found to be normal.

An AM generator at 455 kHz is connected to the collector of the second IF amplifier $Q4$. The generator output is adjusted so that a barely audible output is obtained at the speaker, with the volume control set at maximum volume. The generator is then moved to the base of the transistor, and the output from the speaker increases substantially, indicating that the stage has considerable gain.

The generator output is reduced until the speaker output is barely audible, and the adjustment $A1$ of IF transformer $L2$ is found to be peaked. The generator is then moved to the collector of the first IF amplifier $Q3$, and the tone drops out at the speaker. The generator output is increased slightly, and the tone reappears. This is normal since the IF transformer between the collector of $Q3$ and the base of $Q4$ is a step-down transformer.

The generator is then applied to the $Q3$ base and a noticeable increase in volume is observed. The generator output is again reduced, and adjustment $A2$ is found to be slightly off the peak. The generator is moved to the collector of mixer $Q1$, and the generator output has to be increased because of the step-down action of the IF transformer. Moving the generator to the base of the mixer causes loss of the signal at the speaker because the base circuit of the mixer is tuned to broadcast frequencies, not 455 kHz, the frequency of the generator. Hence, the generator output must be increased to bring the signal back in.

Adjustment $A3$ is found to have little effect upon the signal amplitude, and this causes suspicion to fall upon either the transformer or $Q1$. The DC voltages around $Q1$ are found to be:

$Q1$	Measured	Normal
C	−5	−5.1
B	−0.7	−0.6
E	−0.6	−0.4

The base-emitter bias voltage on $Q1$ is reduced to 0.1 V instead of the normal 0.2 V. This would reduce the gain, and could be responsible for the loss of sensitivity. An ohmmeter is used to measure the $L3$ primary resistance, which is 3.4 Ω. A visual inspection of the solder connections turns up nothing that looks suspicious, so $Q1$ is removed and is found to have excessive collector–emitter leakage. This accounts for the reduced gain, and also accounts for the broad tuning response of $A3$ in that defective transistor $Q1$ loaded the tuned circuit, destroying the Q. $Q1$ is replaced with a new unit, $A3$ is peaked, and normal operation is restored.

We note, in passing, that the oscillator function in this receiver is performed by another transistor, $Q2$. When a defect occurs in a converter transistor, which performs both oscillator and mixer functions, the oscillator is apt to stop oscillating. In such a case the lack of oscillator voltage is likely to lead to the defect.

12-10 INSUFFICIENT VOLUME, SENSITIVITY NORMAL

A receiver that picks up all available stations but is lacking in volume at the speaker is described as having normal sensitivity with insufficient volume. The audio amplifiers are usually responsible for this symptom, although reduced B+ voltages can produce it also.

The preferred procedure is that of signal tracing. A scope is better suited to this task than a signal tracer, but either should work. There should be no loss of signal across coupling capacitors, but a drop in signal level normally occurs across interstage and audio-output transformers. These transformers are step-down transformers used for impedance matching purposes. A loss of 30 to 50% of the signal voltage is normal. We note here that solid-state OTL output stages contribute little if any voltage gain, since they are power amplifiers concerned primarily with providing a large current capability.

EXAMPLE 12-15: Referring to the receiver of Figure 12-5, the B+ voltages check normal. When a station is tuned in, an audio signal is found at the high side of the volume control, and the amplitude of the signal is about 0.8 V p-p. The scope probe is moved to the wiper arm of the pot, and the signal level is significantly reduced. An ohmmeter check of the volume control pot is made, being careful to place the negative prod of the ohmmeter at the high side of the control. This reverse-biases detector diode $X2$ and IF amplifier $Q2$ so that better isolation of the volume control is obtained. The control appears to be normal.

DC voltages measured around the audio transistors $Q3$ and $Q4$ are normal so that the transistors are not prime suspects. No bypass capacitors are present that could open and cause degeneration, but we note that $C4$ is included to prevent AC feedback from the emitter of the output transistor to the base of the input transistor. If $C4$ opened, degeneration and a reduction of the voltage at the wiper arm of the pot would occur.

Since no signal voltage is supposed to appear at CTP 16, we apply the scope probe to that point to see if $C4$ is doing its job. As we suspected, a signal voltage is present, and we now know what is going on. An AC signal voltage is being fed back from $Q4$ via $R22$ and $R19$ that is canceling a portion of the signal applied to the base of $Q3$. The negative feedback signal also couples through $C3$ to the wiper of the volume control, killing the signal there. We confirm this suspicion by bridging $C4$ with a good unit, and normal operation is restored. $C4$ is replaced, and normal response at the speaker is obtained.

EXAMPLE 12-16: This example illustrates how the presence of a signal voltage, where no signal voltage is supposed to be, can lead us directly to the faulty component. In the receiver of Figure 12-6, which we are troubleshooting for a lack of volume at the speaker, a signal voltage is found at the emitter of the first audio amplifier $Q4$. But the emitter resistor is bypassed by $C5$, and there should be no signal voltage present at the emitter. Therefore, $C5$ must be open or a fault in the PC wiring has removed $C5$ from the circuit. Bridging the capacitor restores normal operation, so we conclude that a loss of capacitance has occurred, and replace the faulty $C5$. Normal operation is restored.

12-11 RECEIVER OSCILLATIONS: MOTORBOATING, SQUEALS, HOWLS

A receiver that generates unusual sounds such as squeals and howls or a low-frequency putt-putt (motorboating) is oscillating. The oscillation may be either at RF or audio frequencies. A defect in the receiver develops a positive feedback loop that couples that output of a stage back to the input of a previous stage. The coupling may be through the power supply or the AVC line, or coupling may occur by radiated signals from one stage to another.

There are two basic types of receivers oscillations: motorboating and squealing. Motorboating is usually produced by the audio section as a result of an open power supply filter capacitor or an open decoupling capacitor. Squeals, howls, and whistles are usually produced in the IF amplifier or front end due to open decoupling capacitors, open neutralizing capacitors, bad ground connections, or misalignment. A bad ground connection to a transformer shield can cause undesired signal radiation and oscillation.

Procedure

After the chassis is pulled, bridge the decoupling capacitors in the power supply and AVC circuit with test capacitors. If the oscillation persists, determine the effect of volume control variations on the oscillation. If it has no effect, the oscillation is coming from the audio section. If the speaker response varies with volume setting, the IF amplifier or front end is responsible for the oscillation.

All ground connections in the suspected section should be inspected carefully and resoldered. In the IF and front end, pay particular attention to the grounds of tuned circuit shields.

With an oscillating front end or IF amplifier, bridge any neutralizing capacitors that may be present, and check their circuit connections very carefully. If the oscillation remains, we must determine exactly which stages are oscillating so that they can be checked out more carefully.

If a scope is available, start at the last IF can and move back, stage by stage, until the oscillation signal is no longer present. This locates the input stage to the oscillating section. If a scope is not available, a 0.1-μF capacitor can be used to kill the oscillation by connecting it from various signal points to ground. It can therefore be used to isolate the oscillating stages. Once again, start at the last IF can and proceed back toward the antenna. The stage on the speaker side of the point where the capacitor does not kill the oscillation is the input stage to the oscillating section.

Once the oscillating stages are isolated, the tuned circuits at the input and output are adjusted to see if the oscillation will stop. If so, a complete alignment is then performed. If the oscillation still persists, the circuit connections are checked very carefully for cold solder joints or other abnormality.

EXAMPLE 12-17: To illustrate, we assume that a high-pitched whistle comes from the speaker of the receiver of Figure 12-6. The power supply filter capacitors, $C1$ and $C2$, and the AVC filter capacitor $C3$ are bridged with no results. The ground connections for the IF transformer shields are inspected and resoldered. Several other ground connections are resoldered "just in case." Neutralizing capacitors $C16$ and $C18$ are bridged, but the problem remains.

The high-pitched nature of the whistle leads us to believe the oscillation is at RF frequencies. Audio oscillations tend to be of a rougher "texture" than what is heard. To confirm this, we connect a 0.1-μF capacitor momentarily from the base of the second IF amplifier $Q3$ to ground. The sound disappears. Hence, $Q3$ is involved in the oscillation, but more than one stage may be participating. The capacitor is moved to the base of the first IF amplifier $Q2$, and the sounds are still present. This focuses our attention upon $Q3$.

The IF transformer tuning adjustments $A1$ and $A2$ are turned one-half turn in each direction with no effect. The circuitry around $Q3$ is carefully inspected with a magnifying glass, but nothing is found. Finally, we remove $Q3$ and replace it with a new transistor, and the receiver then operates normally.

We can only speculate as to why some transistors function normally for a period of time and then choose to oscillate. Something inside the transistor changes with age; perhaps the junction capacitance increases, or leakage develops. The point is that something indeed goes wrong, and the transistor must be replaced. Incidentally, the transistor would check *good* by ohmmeter and by transistor checker, so beware.

12-12 NOISY RECEPTION

Noisy reception is characterized by abnormal pops, clicks, and crackles coming from the speaker along with the station information. Frequent causes of this type of problem are noisy tubes or transistors, leaky transistors, leaky IF transformers, noisy resistors or capacitors, and intermittent PC board connections.

The volume control can be used to determine whether the problem is in the audio section. If the noise goes away when the volume is turned down, the source of the noise is on the antenna side of the volume control. If the noise is indeed coming from the audio section, the audio tubes of a tube-type set should be substituted, or freeze spray should be used to cool the audio transistors in a solid-state set. The noise generation of a transistor is greatly reduced when it is cooled. Diodes, especially zener diodes, are also capable of producing large quantities of noise.

If the noise disappears when the volume control is turned fully counterclockwise, the IF or front-end sections are responsible for the difficulty. Tubes should be substituted or transistors sprayed with freeze spray, as appropriate. If the problem is not located, we turn to the IF transformers.

In tube sets, which operate at higher voltages than solid-state sets, an IF transformer may have leakage from the capacitor on the primary to the capacitor on the secondary. The insulation between the capacitors breaks down and allows the B+ voltage on the primary to be applied sporadically to the secondary. This can be quickly checked with a voltmeter. Any positive voltage on the secondary side confirms leakage, and the transformer must be replaced. Leakage problems are rare in the IF transformers of solid-state sets, but the possibility should not be overlooked.

When the noise is not due to a tube or IF can, a scope or signal tracer can be used to determine where the noise originates. If the problem is known to be in the audio section, the scope is used to check the noise level at the various signal points to determine the source. Remember, a small amount of noise is normal. Look for spikes in the normal noise.

Many receivers exhibit noisy reception only when handled or adjusted. This is usually due to dirty tube sockets, dirty control potentiometers, or poor printed circuit connections. If a scraping, scratching noise is heard when the volume or tone control is adjusted the adjusted, the control should be cleaned. This is done by spraying contact cleaner into the control while rotating the shaft back and forth several times.

A similar procedure is followed to clean a noisy tuning gang that produces a scratching sound in the speaker as the receiver is tuned. Carefully spray contact cleaner between the plates and rotate the shaft. If the noise persists, a thin knife or razor blade is very carefully used to separate the plates.

Dirty tube sockets are cleaned by removing the tube and spraying contact cleaner into the pin connectors and reinserting the tube. If the noise persists, it is necessary to clean the tube pins with a knife or fine sandpaper, and, perhaps, to bend the socket connectors closer together. A pointed object inserted between the connector and socket base can be used to bend the contacts so that they grip the tube pins more tightly.

Poor printed circuit connections may be located by using an insulated tool to apply pressure to various points to determine the most sensitive area of the board. A magnifying glass and a good light source may be used to inspect the board for hairline cracks or cold solder joints.

12-13 INTERMITTENT PROBLEMS

A receiver that malfunctions repeatedly after a period of normal operation, or that malfunctions for short or long periods at random times, is said to be intermittent. The difficulties in servicing such problems are legendary.

Intermittent defects may be classified as being either thermal or mechanical. Mechanical defects include dirty tube sockets, bad solder connections, and breaks in printed circuit (PC) wiring. Thermal defects include components that break down, change value, become leaky, and so on, with a rise in temperature.

The symptoms that occur can be used to determine the most likely cause of the problems, as in the preceding sections. The tubes of suspected stages should be replaced, and the tube sockets should be cleaned. The conditions that cause the problem to develop can also provide a clue to the defect. Problems that occur when the receiver is moved or jarred are usually of mechanical origin, while problems that occur only after the receiver has operated for a while are usually thermal problems. The isolation procedure used to locate the defect will vary depending upon the nature of the problem.

Mechanical Defects

If a receiver malfunctions irregularly, sometimes when first turned on, we suspect a mechanical defect and may proceed as follows. First, wiggle each tube gently in the socket and note the effect. Remove one or two tubes and examine the tube sockets. They should be clean and should not have bent connectors. Replace the tubes of the stages that the symptom causes you to suspect. Tap each tube lightly with the handle of a screwdriver, and tap various areas of the chassis to see if mechanical vibration has any effect.

Wiggle the front panel controls, paying attention to the on-off switch if the receiver goes completely dead. If nothing develops from these preliminary checks, remove the receiver from the cabinet and systematically examine every inch of the chassis and printed circuit board under a good light and magnifying glass. Look for anything unusual—burned or charred components, bad solder connections, short circuits, cracks in the PC board, solder bridges between conductors, strands of fine wire, and so on. Pay attention to ground connections made to the chassis. Solder connections to the chassis are frequently "cold" because of the large amount of heat required to heat the chassis metal, and screwed-down connectors frequently develop poor or intermittent contact.

Note the effect of pressing against different areas of the PC board. Sometimes an intermittent can be made to come and go by slightly deforming the board. In such a case find the area of the board that requires the least pressure to affect the defect. The defect will be close by.

When a bad solder connection is suspected in an area of the board but the exact connection cannot be located, the shotgun approach of reheating all connections in the area may be justified.

Thermal Problems

When the problem occurs only after the receiver has been in operation for a while, the defect is likely to be temperature-sensitive. Spray cans of pressurized coolant, which we call *freeze spray*, can sometimes be used to cool components one at a time until normal operation is restored. Freeze spray should never be sprayed directly on tubes, and its effectiveness on large electrolytic capacitors is limited. When cooling a component restores normal operation, reheat the component with a (sparingly used) soldering iron to see if the problem recurs. If the problem can be made to come and go by heating and cooling a single component, the component should be replaced.

If the problem does not recur when the suspected component is heated, heat a nearby component. It is difficult to control the coolant as it is sprayed into the receiver, and in most cases several components will be cooled at the same time.

In the event that a receiver acts up only after operating for several hours, a hair dryer or similar device may be used to heat the receiver in an attempt to bring on the problem more quickly. Since ventilation is improved when the receiver is out of the cabinet, some defects may never occur until the receiver is reinstalled. In such cases the receiver should be covered appropriately to reduce heat loss.

Special Problems

In the event that the defect remains momentarily but occurs quite frequently, test equipment may be connected to the receiver and left with it until the problem develops. A power supply problem may be confirmed by connecting a voltmeter and then noting whether the power supply voltage changes during the malfunction. Signal-injection and signal-tracing procedures can also be implemented in this fashion, but this procedure is time-consuming and may tie up equipment for long periods.

In some cases, when the malfunction occurs, simply touching a test lead to a point may restore normal operation. Try connecting the test instrument before applying power to the receiver, or turn it off and on to see what happens.

A weakness of solid-state devices is the point of attachment of lead wires to the semiconductor material. These contacts can become completely open or merely intermittent. Intermittent contacts may be sensitive to vibration, temperature changes, voltage, or voltage transients. The following paragraphs describe infrequent but not unusual experiences.

It once became apparent that a pass transistor of a voltage-regulator circuit had an open between the base and emitter. The base–emitter voltage was 11 V in the forward direction, so the conclusion was fairly obvious. The transistor was removed and checked out-of-circuit, and was found to be good. Suspecting that an error had been made in measuring the base–emitter voltage, the transistor was reinstalled and the voltage remeasured. Once again, 11 V appeared between the base and emitter. The transistor was promptly replaced, and normal operation of the regulator was restored. As a curiosity, the transistor was again checked out-of-circuit—good—and was then thrown out the back door. The device was malfunctioning when a high voltage (12 V) was applied, but would check out normally with the low voltage of the ohmmeter.

Another receiver would play 5 minutes and then be off for 10, play 5 more minutes, and go off again, and so on. Freeze spray applied to the first AF amplifier would cause the receiver to come on, and a soldering iron lightly applied to heat the transistor would cause the problem to return. The receiver would operate until the transistor reached a certain temperature at which the transistor opened up. It then would cool down and start to function again. The temperature-sensitive transistor was replaced.

A technician, faced with a stubborn intermittent in a TV set, was delighted to find that he had just acquired the power to make the problem go away by waving his hands over the set. One difficulty, however, was that the problem returned when he removed his hands. The fault turned out to be a glass-encapsulated diode that had become sensitive to light. It would work in the dark, but not in the light. The technician's hand was throwing a shadow on the diode, causing it to function normally.

A receiver once came in that, at turn-on, would sometimes come on normally and sometimes would only exhibit a hum with a little static. Later, it was discovered that the receiver could be caused to come on by touching, ever so lightly, any portion of the signal path up to the second IF amplifier with a small metal screwdriver. When a signal generator was connected to the RF and IF sections in an attempt to locate the problem by signal injection, the receiver would begin to function normally. A voltmeter prod would also turn the thing on. It was discovered, however, that touching anything from the collector of the second IF amplifier on to the speaker had no effect on the problem. The second IF amplifier was replaced, and normal operation was restored. The faulty transistor had an intermittent that was sensitive to voltage transients.

12-14 AM RECEIVER ALIGNMENT

The purpose of alignment is to adjust the tuned circuits of a receiver to the proper frequencies. Simply stated, alignment is accomplished by using a signal generator to inject the correct frequencies (and waveform) into a receiver, and adjusting the tuned circuits to obtain the proper response on an indicator. For AM alignment, the indicator is usually a voltmeter connected either across the speaker or to the AVC line. The adjustments are made to obtain a maximum output.

The alignment procedure is divided into two parts: (1) the IF amplifier, and (2) the front end. The IF alignment is performed by injecting an AM signal at the IF frequency (usually 455 kHz) at the mixer input and adjusting the IF tuned circuits for maximum response as shown in block diagram form in Figure 12-11. It is extremely important that the generator output be kept as low as possible because, if a large signal is injected, the action of the AVC circuit will mask the effect of "peaking" the IF cans, and the alignment will be poor at best. Use only the output necessary to provide a useful indication—the smaller the better.

After the IF alignment is completed, the front end is aligned by using the generator and an injection loop to radiate a signal to the receiving antenna. An injection loop consists of 8 or 9 turns of wire formed into a loop about 6 in. in diameter as shown in Figure 12-12. The object of the front-end alignment is to adjust the antenna and oscillator circuits so that they differ by an amount equal to the IF frequency. Also, the circuits are adjusted for proper tracking (i.e., so that station frequencies are received at the proper point on the dial).

Front-end alignment begins by setting both the generator and the receiver to 1600 kHz. The oscillator adjustment is then adjusted to obtain maximum output, which will occur when the oscillator frequency is greater than the generator frequency an amount equal to the IF frequency. The generator and receiver are then set at 1400 kHz (according to the dial on the receiver), and the antenna adjustment is made to achieve maximum response. Next, the generator and receiver are set to 600 kHz, and the antenna adjustment is adjusted for maximum response while rocking the tuning control back and forth to maintain the receiver tuning at the generator frequency. These steps—at 1600, 1400 and 600 kHz—are repeated several times to achieve the best possible result.

The purpose of the procedure of the preceding paragraph is to adjust the antenna and oscillator tuning capacitances so that the best tracking is achieved. A receiver

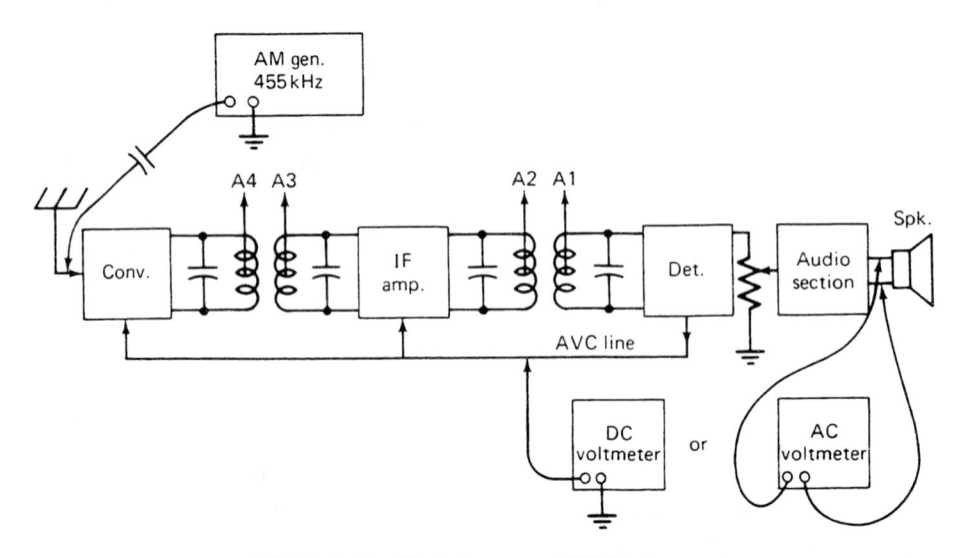

FIGURE 12-11 Block diagram of AM IF alignment.

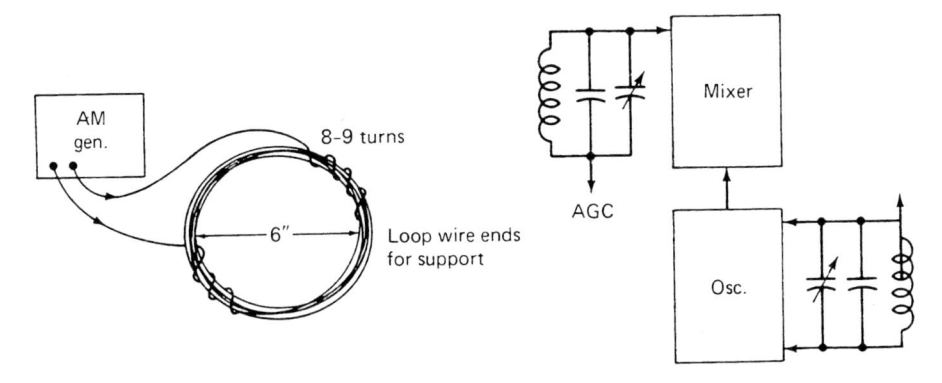

FIGURE 12-12 Injection loop for front-end alignment.

tracks when the oscillator and antenna frequencies change at the same rate as the receiver is tuned across the band. When a receiver is tracking properly, the stations will come in at the proper place on the dial, over all the band. When stations come in at the proper place on one end of the band but not at the other, the receiver is not tracking properly, and further adjustment is necessary.

Generator Accuracy

An important consideration is the accuracy of the frequency coming from the generator. If the generator is off-frequency, the receiver may not perform very well after the alignment process. Fortunately, it is a simple matter to check the accuracy of the generator frequency.

Tune a working receiver to the known frequency of a station in the area. With an injection loop, feed in an unmodulated signal from the generator and vary the generator frequency to obtain a *beat note* in the receiver. The generator is adjusted for *zero beat*, and at that point the generator frequency is the same as the station frequency. If the generator setting is higher than the frequency of the station frequency, the generator output is too low by an amount equal to the difference in the station frequency and generator dial indication at zero beat.

Alignment Instructions

Whenever a receiver is to be aligned, the alignment instructions for that particular receiver should be followed to the letter. The instructions are contained in service data provided by the manufacturer, and alignment instructions are included in the service data published by Howard W. Sams & Co., Inc. These instructions state what frequency to use, where it should be injected, where the indicator should be connected, and what response to adjust for. A typical set of alignment instructions from Sams service data is shown in Figure 12-13 for the receiver of Figure 12-14.

ALIGNMENT INSTRUCTIONS

Volume control should be at maximum position. Output of signal generator should be no higher than necessary to obtain an output reading. A6 and A7 are not used in Chassis HS-624.

	SIGNAL GENERATOR COUPLING	SIGNAL GENERATOR FREQUENCY	RADIO DIAL SETTING	OUTPUT METER	ADJUST	REMARKS
1.	High side thru, .1 µf to pin 7 (grid) of Converter, Low side to common.	455 kHz 400 Hz Mod.	Tuning gang fully open	Across voice coil	A1, A2, A3, A4	Adjust for maximum output.
2.	High side thru, .1 µf to pin 7 (grid) of Converter, Low side to common.	1620 kHz	Tuning gang fully open	Across voice coil	A5	Adjust for maximum output.
3.	High side thru, .1 µf to pin 1 (grid) of RF Amp. Low side to common.	1400 kHz	1400 kHz Signal	Across voice coil	A6	Adjust for maximum output.
4.	High side thru, .1 µf to pin 1 (grid) of RF Amp. Low side to common.	600 kHz	600 kHz Signal	Across voice coil	A7	Adjust for maximum output. Repeat steps 2, 3 and 4.
5.	Fashion loop of several turns of wire and radiate signal into loop of receiver. Adjust for maximum output.	1400 kHz	1400 kHz Signal	Across voice coil	A8	Adjust for maximum output with chassis in cabinet.
6.	Fashion loop of several turns of wire and radiate signal into loop of receiver. Adjust for maximum output.	600 kHz	600 kHz Signal	Across voice coil	A9	Adjust for maximum output with chassis in cabinet.

FIGURE 12-13 Alignment instructions for the receiver of Fig. 12-14.

Typical Procedure

Step 1 of Figure 12-13 gives the IF alignment procedure. An AM signal at 455 kHz is injected through a 0.1-µF capacitor to the mixer grid. An AC voltmeter is connected across the speaker terminals to serve as an indicator. The tuning gang is fully open (tuned to the top of the band) to minimize interference from the

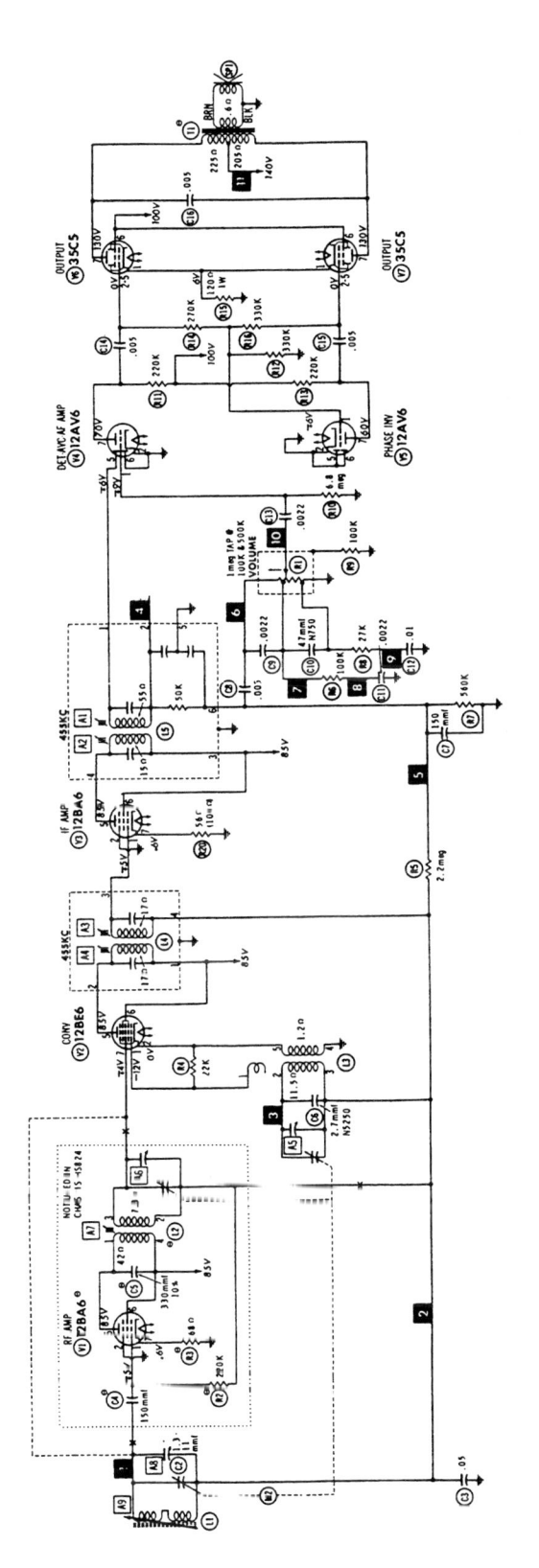

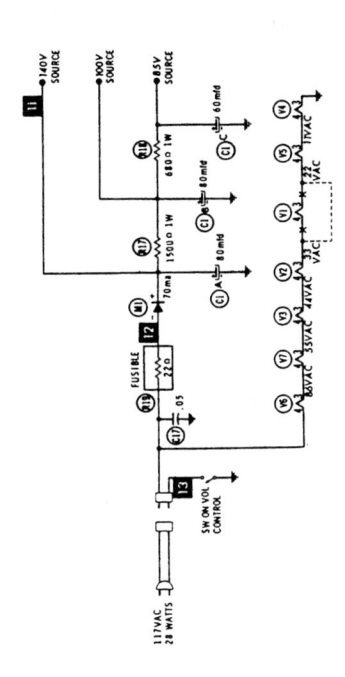

FIGURE 12-14 AM receiver with RF amplifier and push-pull audio-output stage (a Photofact schematic, courtesy of Howard W. Sams & Co., Inc.)

oscillator. The generator output is adjusted to give the minimum usable reading on the lowest range (usually about 0.2 V rms on the 1.5-V scale).

With the "setup" as described, the IF tuned circuits are adjusted by tuning $A1$, $A2$, $A3$, and $A4$ to produce maximum deflection on the meter. If one or more of the adjustments cause the output to increase drastically, the generator output should be reduced to a low level to avoid AVC effects. The adjustments should be made several times to ensure optimum performance.

Step 2 of Figure 12-13 is to adjust the oscillator frequency so that proper tracking is achieved. With the tuning gang still fully open, the generator frequency is set to 1620 kHz. Oscillator trimmer $A5$ is then adjusted until the tone is heard at maximum volume from the speaker.

Steps 3 and 4 are to adjust the RF amplifier plate load to the proper frequency. The generator is connected through a 0.1-μF capacitor to the grid of the RF amplifier, and the generator and receiver are set to 1400 kHz. Trimmer capacitor $A6$ is then adjusted to maximum output. The generator and receiver are then set to 600 kHz, and inductance $A7$ is adjusted for maximum output. Steps 2, 3, and 4 are then repeated several times to achieve equal output levels at 600 and 1400 kHz, and to have the receiver dial indicate as closely as possible to the correct frequency.

Steps 4 and 5 provide for the antenna circuit to be adjusted to the correct frequency in the same manner as the plate load of the RF amplifier. Note that an injection loop is used for injecting the signal into the receiver.

Precautions

Sometimes a tuning slug or core of an inductor is encountered that will not turn with moderate pressure. Never try to force the slug, or the rather fragile ferrite material may crack. Put a small amount of penetrating oil into the opening of the inductor and allow several minutes for it to penetrate. Then try to turn the slug again. If it fails to turn, the tuned part will have to be replaced.

An adjustment that reaches the limit of the adjustment before maximum output is obtained may have a cracked slug. Carefully remove the slug and inspect it thoroughly. A cracked slug must be replaced before a proper alignment can be achieved. Never use worn alignment tools, because they can cause the slugs to crack.

SUMMARY

1. In servicing a receiver that emits no sound whatever, the first priority is to obtain some sound—static, a hum, or otherwise—from the speaker. The problem could be in the power supply, audio section, or in the speaker itself.
2. Almost all solid-state receivers emit a "plop" when first turned on due to the initial flow of current through the output transistors.

3. When the turn-on plop is present, the power supply audio output stage, and speaker circuit are probably functioning normally. If static is present in the speaker, the problem is probably in the receiver front end. If no static is heard, the problem may be in the IF amplifier, detector, or AF amplifiers.

4. A low 60-Hz hum heard at the speaker indicates that the B++ power supply and audio output stage are probably good. The defect may lie in the first AF stage, in the low-voltage section of the power supply, or in the detector or IF amplifier.

5. A loud 60-Hz hum that does not vary with the volume control may be caused by an open filter capacitor in the power supply or by leakage on the B+ line that places a heavy current load on the power supply.

6. If an audio signal injected at the volume control is reproduced at the speaker, it can be reasonably assumed that the power supply and the audio section are functioning normally.

7. The base-emitter voltage V_{BE} of a properly operating transistor will be on the order of 0.7 V for silicon (0.3 V for germanium) transistors. This provides a quick means of performing an initial check of a transistor without having to remove the transistor from the circuit.

8. In making voltage measurements in solid-state receivers, care must be exercised not to let the voltmeter probe slip and cause an accidental short circuit.

9. The DC parameters of a transistor may check good, even though the high-frequency properties are seriously degraded.

10. Leakage from collector to base is a fairly common defect in transistors and tends to produce a larger-than-normal collector current.

11. Meaningful resistance measurements can sometimes be made between the terminals of a transistor in-circuit, but careful consideration must be given to circuit resistances in parallel with the measured resistance.

12. When a defective semiconductor is found and replaced in a direct-coupled amplifier, it is essential that all other semiconductors in the direct-coupled section be checked prior to applying power to the circuit.

13. The low impedances of transistor circuits makes the volume control less sensitive to 60-Hz hum introduced by touching the wiper with a finger. An audio generator may be required for signal injection in order to get a reliable test.

14. An external power supply may be used to clamp a point of a direct-coupled amplifier to a certain voltage to isolate the defect. Extreme care must be exercised when using this procedure.

15. Static originates in the RF sections of a receiver, primarily being introduced at the antenna. The presence or absence of static at the speaker can provide clues as to whether the RF sections are operating.

16. A loss of sensitivity may result from a defect in the RF amplifier, converter, or IF amplifier. Also, a loss of sensitivity may occur if the B+ power supply voltage is too low or if the receiver is misaligned.

17. IF transformer windings sometimes become open as a result of corrosion and deteri-

oration of the windings. Leakage paths which may also develop from primary to secondary, give rise to noisy operation.

18. Generally, when a problem is found to exist in the IF/RF sections of a receiver, a signal-injection procedure is the most convenient way to further isolate the problem.

19. Whenever a tuning adjustment is encountered that fails to tune properly, the circuit containing the adjustment is not likely to be functioning normally. Therefore, a "failure to tune" can sometimes lead one to the defect.

20. A technician should be alert to the fact that owners sometimes "tighten the screws" in the IF transformers. This is certain to misalign the IF section and may result in internal damage to the transformers.

21. Insufficient volume with normal sensitivity is usually caused by a fault in the audio section or by the portion of the power supply that drives the audio section.

22. A signal voltage present at a point that normally is at signal ground indicates an open bypass capacitor or an open decoupling capacitor.

23. Receiver oscillations can be caused by open decoupling capacitors, open ground connections, improper alignment, or open neutralizing capacitors.

24. Intermittent defects in transistors may be sensitive to mechanical vibration, heat or cold, and voltage or current effects.

25. Intermittent problems are the most difficult to troubleshoot because they often seem to go away just as the troubleshooting procedure is begun. Judicious tapping, pulling, pushing, and pressing may help to locate a mechanical intermittent, and freeze spray or a source of heat may reveal a thermal intermittent.

26. The first and foremost requirement in servicing a difficult problem in a solid-state receiver is to ascertain that all semiconductors are good and are properly installed. Untold heartache can be caused if the emitter and collector leads are reversed because the transistor will almost work when connected backward. Further, new devices may be damaged after installation.

27. Receiver alignment should be performed with only sufficient generator signal amplitude to provide a useful indication in order to avoid the complicating influence of the AVC network.

28. The IF stages are aligned first, reducing generator amplitude as tuned circuits are peaked.

29. An injection loop may be used to radiate a test signal to the antenna of the receiver.

30. The manufacturer's alignment instructions, when available, should always be followed to the letter to assure proper tracking of the oscillator and antenna tuning circuits in the face of a possible subtle peculiarity of a particular receiver.

The following items apply to vacuum-tube receivers.

31. The immediate objective, when faced with a dead receiver, is to get all filaments lighted. Either a voltage or a resistance analysis may be used to located the defect that may be in the line cord, switch, or other part of the AC input circuit.

32. Unusual symptoms associated with series filament strings may be caused by heater-

cathode shorts. These may occur only after the receiver has been operating for several minutes, or longer.

33. Distorted audio is generally caused by a defect in the audio section. Grid bias voltages should be checked carefully. A leaky coupling capacitor can cause a positive voltage to appear on the grid, resulting in distortion.

34. Receivers having a push-pull audio output stage may render distorted audio at an imperceptibly reduced volume level if one side of the push-pull stage becomes inoperative.

35. When the oscillator is running normally, a negative voltage of several volts will be present at the oscillator grid. Absence of this voltage indicates that the oscillator is not working properly.

36. A 455-kHz IF signal introduced at the antenna signal grid of a pentagrid converter will be heard at the speaker even if the oscillator is not working. A test signal in the broadcast band, however, injected at the same point, will not be heard unless (1) the receiver is tuned to the signal frequency, and (2) the oscillator is working.

QUESTIONS AND PROBLEMS

12-1. What causes the "plop" that is heard when most solid-state receivers are turned on, and what is the significance of the plop in regard to troubleshooting?

12-2. What may be the result of accidentally letting the test prod of a VTVM momentarily short the base of a transistor to the collector? What may be done to minimize this possibility?

12-3. The voltage on the base of a certain transistor is +2 V. The voltage on the emitter is 1.3 V. What conclusion can be formed about the state of the transistor?

12-4. The base-emitter voltage of a given transistor in the IF stage is found to be 0.65 V. Is the transistor necessarily "good?" Why or why not?

12-5. The base-emitter voltage of a certain transistor is found to be 4 V. Is the transistor necessarily bad? Might it be "good" at high frequencies? Why or why not?

12-6. A universal biased RC-coupled voltage amplifier is found to have a normal base voltage, a collector voltage that is too low, and an emitter voltage that is too high. What is probably wrong with the transistor?

12-7. Suppose that the collector-base junction of a transistor opens. What effect would this have on the measured

(a) Collector voltage?

(b) Emitter voltage?

(c) Base voltage?

12-8. What is the significance of static in the speaker of a solid-state receiver?

12-9. Briefly outline a troubleshooting procedure for a receiver that exhibits a turn-on plop and has considerable static in the speaker.

12-10. Suppose that the detector diode opens. What symptoms are likely to be produced?

12-11. What factors may indicate that the audio output transistors are conducting excessively?

12-12. What is a matched pair? Under what circumstances should matched pairs be used?

12-13. In troubleshooting a direct-coupled amplifier, why is it important to test all semiconductors after one has been found to be defective?

12-14. What factors or defects may cause distorted audio in a solid-state set? (Do not forget crossover distortion.)

12-15. Why are direct-coupled amplifiers more difficult to troubleshoot?

12-16. Describe the procedure for troubleshooting a receiver for loss of sensitivity.

12-17. What special technique may be used in troubleshooting a direct-coupled amplifier? What precautions are necessary?

12-18. Could a defect in the AVC network cause a loss of sensitivity? How?

12-19. Give a procedure for troubleshooting a receiver that is oscillating.

12-20. How does the speaker output during an IF-stage oscillation differ from the output caused by an audio oscillation?

12-21. What symptoms are likely to be produced by weak batteries in a portable receiver?

12-22. What are two common causes of noisy reception in solid-state sets?

12-23. How can an oscillating IF stage be confirmed?

12-24. How can a 0.1-μF capacitor be used to localize a source of noise in a receiver?

12-25. Briefly explain the procedure for using freeze spray to identify a noisy transistor. Are all transistors temperature-sensitive?

12-26. What steps may be taken in troubleshooting an intermittent solid-state receiver?

12-27. Suppose that a small transistor portable is encountered for which no service data are available. If the battery is missing and if the polarity is not marked, there may be some question as to whether the ground foil should be positive or negative. Explain how electrolytic capacitors in the receiver can reveal the power supply polarity.

12-28. Describe a test for determining if the local oscillator is working, using

  (a) A signal generator.

  (b) An oscilloscope.

12-29. Sometimes replacement audio-output transistors perform satisfactorily with the important exception that they overheat due to excessive collector current flow caused by slightly different bias voltage requirements. In such case, minor modifications of the bias network may be in order.

  (a) Discuss the advisability of modifying the bias network.

(b) What resistances should be changed in Figure 12-7 to reduce the collector current in the output transistors?

(c) Same as part(b) for Figure 12-6.

12-30. A scope indicates excessive crossover distortion in the output stage of Figure 13-5. What components should be checked, or perhaps modified, to get rid of the distortion? Describe the procedure you would use.

12-31. What is a quick check of:

(a) A coupling capacitor in the audio section of a receiver?

(b) A filter capacitor suspected to be open in a power supply?

(c) A filter capacitor suspected to be open?

12-32. Suppose that leakage develops on the B+ line until a heavy current load is placed on a power supply. What symptoms are likely to be produced at the speaker?

12-33. Briefly describe the procedures for locating the defect that causes an undesired oscillation in a receiver.

12-34. What will be the symptom produced when an IF amplifier is misaligned?

12-35. IF static is present (or absent) at the speaker when the volume control is tuned to maximum volume. What deductions can be made about the IF amplifiers?

12-36. Suppose that an open circuit occurs in the detector circuit of a receiver. Is it likely that significant static will be present at the speaker?

12-37. Which section of a receiver is aligned first? Where is the generator connected, and what frequency is selected?

12-38. Why is it necessary to keep the generator output signal as small as possible?

12-39. To align the IF amplifier, the generator may be connected to the output of the converter. In such case, however, the primary of the first IF transformer cannot be tuned to the proper frequency. Why is the tuning of the primary not possible, and where should the generator be connected?

12-40. Where are the trimmer capacitors for the antenna and oscillator tuning physically located?

12-41. What is an injection loop, and how may one be constructed?

12-42. Briefly describe the procedure for aligning the front end of a receiver.

12-43. How can the accuracy of an RF signal generator be checked?

12-44. What percentage modulation is usually recommended for receiver alignment?

12-45. Comment upon the importance of using the manufacturer's alignment instructions rather than the generalized instructions given here. Is it likely that the two will be significantly different?

12-46. What consideration must be given to the AVC circuit of a receiver during alignment?

12-47. What would be the effect upon the selectivity of a receiver if, during alignment of the IF amplifier, the signal generator frequency was slowly drifting downward from the 455-kHz IF frequency? What would be the effect upon the sensitivity?

The following apply to vacuum-tube receivers.

12-48. Describe the procedure for finding an open filament in a series filament string using:

    (a) Voltage analysis.

    (b) Resistance analysis.

12-49. What unusual symptoms may be caused by a heater-cathode short in a filament string?

12-50. What is an easy way to inject an audio test signal at the volume control? Explain how this test effectively divides the receiver in half, isolating the defect either to the RF or to the audio sections of the receiver.

12-51. Generally, a DC voltage is present on the antenna loop or loopstick of small tube-type receivers. What is the source of this voltage?

12-52. What effect do the tuning circuits associated with the mixer grid have upon a 455-kHz signal injected at that point?

13
PRINCIPLES OF FM RECEIVERS

This chapter marks the beginning of our study of frequency-modulation receivers, and in this chapter we describe the FM system, the operation of FM detectors [Foster-Seeley discriminator, radio detector, quadrature, and phase-locked-loop (PLL) detectors]. In block diagram form, we present an FM superheterodyne receiver and typical block diagrams of combination AM-FM receivers. FM-only receivers are rarely encountered, and, therefore, the two chapters following this are devoted to a detailed description of AM-FM receivers.

While many communications services use frequency-modulation systems, we shall refer only to the commercial FM broadcast band, even though much of what follows is equally applicable to FM systems in other areas.

13-1 INTRODUCTION TO FM

In an FM system the frequency of the carrier is caused to deviate from the FCC specified center frequency an amount determined by the instantaneous amplitude of the modulating signal. When the instantaneous voltage level of the modulating signal is positive, the FM modulator causes the carrier to be higher than the center frequency. Conversely, when the instantaneous voltage level of the modulating signal is negative, the carrier is caused to be lower than the center frequency. The transmitted signal waveform of an FM transmitter is constant in amplitude, but the frequency varies or deviates in accordance with the modulating signal. A comparison of AM and FM waveforms is shown in Figure 13-1(a).

The manner in which the FM carrier varies in frequency is shown in Figure 13-1(b). Starting an audio cycle at A, we note that the carrier is at the center frequency F_C. As the audio voltage rises to point B, the carrier moves to a proportionately higher frequency. A maximum is reached at point C, and then the carrier frequency decreases as the audio voltage proceeds toward the negative peak at point G. The cycle is completed via H and I.

Production of an FM Waveform

One method for generating an FM waveform is illustrated in Figure 13-2. A varactor diode, which acts as a variable capacitor, is coupled by capacitor $C2$ to the resonant circuit of an oscillator that produces the high-frequency carrier. The varactor ca-

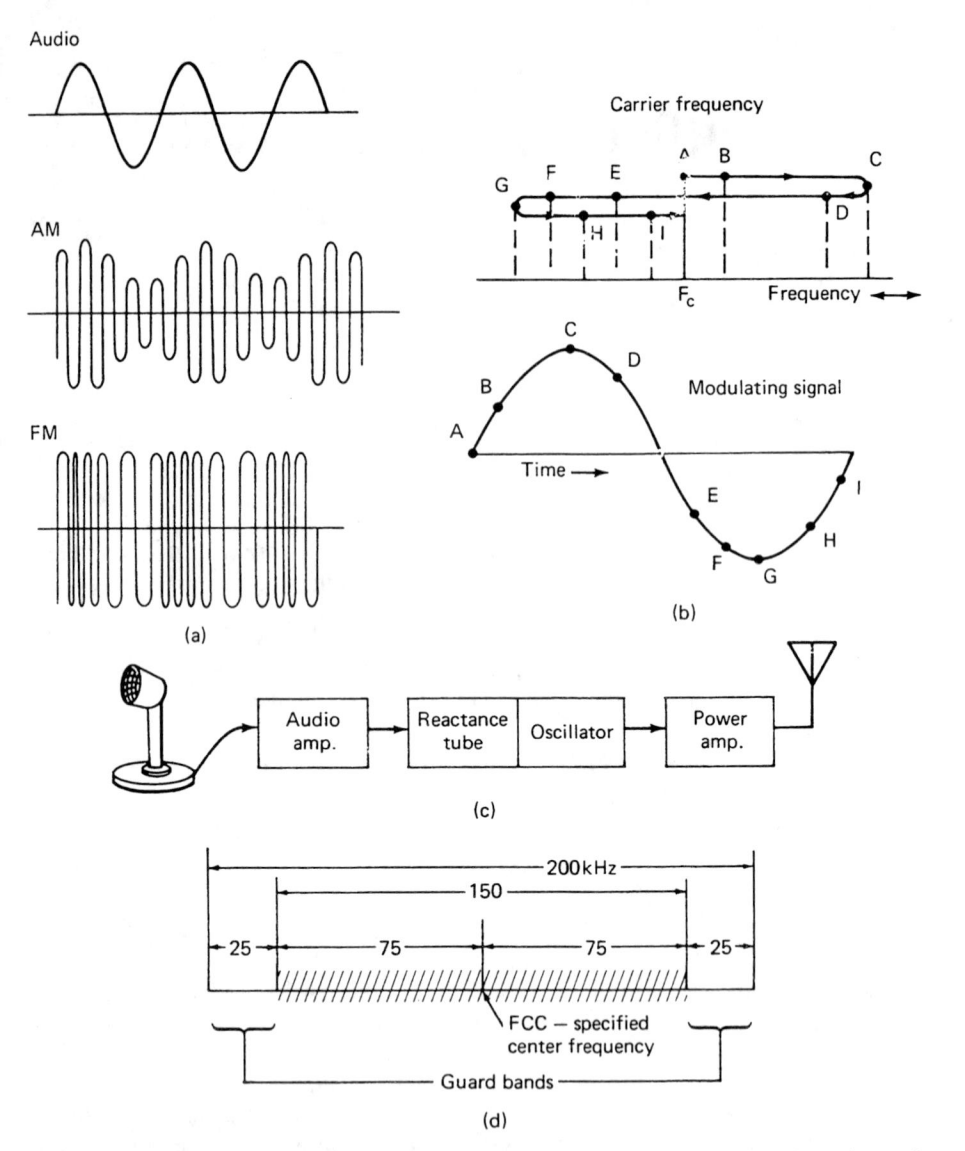

FIGURE 13-1 (a) Comparison of AM and FM waveforms. (b) Illustration showing how the carrier frequency varies according to the modulating signal. (c) Block diagram of a simple FM transmitter. (d) Frequencies relevant to a commercial broadcast channel.

pacitance C_v appears essentially in parallel with $C1$ because the reactance of $C2$ at the oscillator frequency is small enough to be ignored. Hence, the capacitance that resonates with $L1$ is the parallel equivalent of $C1$ and C_v, and a variation of either $C1$ or C_v will cause the oscillator frequency to vary.

The varactor diode is kept reverse-biased by a DC voltage applied through $R3$ by the voltage divider consisting of $R1$ and $R2$. The large ohmic value of $R3$ isolates

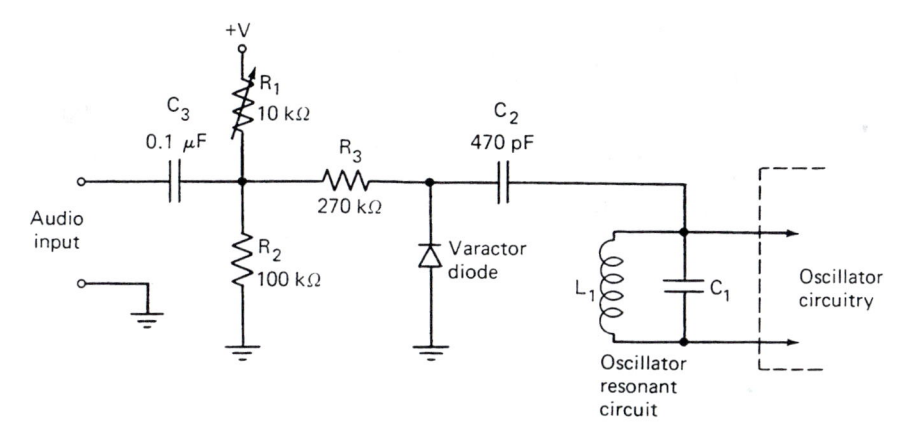

FIGURE 13-2 The audio input signal varies the capacitance of the varactor diode, which is connected in parallel with the oscillator resonant circuit. Changes in capacitance produce changes in frequency of the oscillator, producing an FM waveform.

the bias network and the audio circuitry from the varactor and oscillator resonant circuit, which operate at high frequencies. But since virtually no DC or audio-frequency currents flow through $R3$ (because the varactor is reverse-biased and the reactance of $C2$ is large at audio frequencies), the audio signal voltage lost across $R3$ is negligible.

An audio signal voltage applied to the varactor via $C3$ and $R3$ causes the varactor capacitance to vary, and this causes the oscillator frequency to vary also. When the audio signal produces an instantaneous increase in voltage across the varactor, its capacitance decreases and the oscillator frequency goes up. When the audio signal produces a decrease in voltage across the varactor, its capacitance increases and the oscillator frequency goes down. Thus, the audio signal causes the frequency of the oscillator to vary above and below its center frequency. This is FM.

The *deviation* is the instantaneous difference in frequency between the transmitted and center frequencies. A maximum allowed deviation of ±75 kHz has been established by the FCC for broadcast FM. Accordingly, a deviation of 75 kHz represents 100% modulation. This is an arbitrary choice, a compromise between high-fidelity transmission capability, on the one hand, and efficient use of radio-frequency spectrum space, on the other. Modulation percentages less than 100% are calculated by expressing the ratio of the given deviation to 75 kHz as a percentage. Hence, a deviation of 45 kHz is 60% modulation, and so forth.

The FM broadcast band ranges from 88 to 108 MHz and consists of 100 channels, each 200 kHz wide. Channel center frequencies are situated at odd tenths of a megahertz from 88.1 to 107.9 MHz. Guard bands are established for each channel as shown in Figure 13-1(d).

Noise Reduction by Amplitude Limiting

Since intelligence is transmitted via carrier-frequency variations in an FM system, the amplitude of the signal is unimportant, assuming that sufficient amplitude is available. Consequently, the FM signal can be "limited" at the receiver to remove

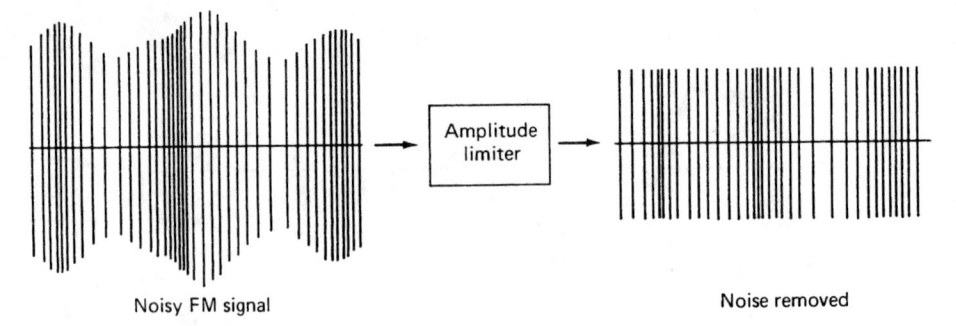

FIGURE 13-3 Noise can be removed from an FM signal by an amplitude limiter. Note that the frequency variations remain.

most noise prior to FM detection. As shown in Figure 13-3, the frequency variations remain after the amplitude limiting has occurred.

Although most noise is represented by amplitude variations that may be removed by a limiter stage, sharp noise pulses can produce a momentary shift in phase of the FM carrier. The FM detector will respond to this phase shift and produce a pop or a click at the speaker. Hence, amplitude limiting cannot give perfectly noise-free reception.

FM Propagation

The propagation characteristics of broadcast-band FM signals are significantly different from AM because FM frequencies are roughly 100 times as high as AM frequencies. This accounts for the fact that FM reception distances are only slightly greater than line of sight. Further, the high frequencies are not reflected back to earth by the ionosphere; long-distance transmission via "skip" does not occur except in rare circumstances. We emphasize that it is the much higher frequencies of the FM broadcast band that produces the different propagation characteristics. It has nothing to do with the type of modulation used.

Antennas for FM

FM broadcast waves are horizontally polarized (as are TV signals), so that horizontal, TV-type antennas are used for external FM antennas. Many wide-band TV antennas are designed to be suitable also for FM reception. The same flat, twin-lead ribbon cable is used to connect an FM antenna to the receiver as is used for television, and receivers are frequently provided with two terminals for connecting an external antenna lead-in.

Other table-model receivers may provide only a single terminal for connecting a short piece of wire to serve as an antenna. Portable receivers employ telescoping antennas that are extended to enhance FM reception. Receivers powered from the AC line frequently connect a small high-voltage capacitor to one side of the AC line to pick off the FM signals that appear on the line. Loopstick or ferrite core antennas are not used as FM antennas.

Audio Fidelity of FM Transmissions

In comparison with AM, FM reception is more nearly noise-free, owing to the amplitude limiting that removes most of the noise. Mere noise-free reception is not necessarily high fidelity, however, since fidelity is related to the audio bandwidth of the transmitted signals. For FM, audio frequencies from 50 to 15,000 Hz are transmitted. The highest audio frequency an AM station can transmit without exceeding the 10-kHz bandwidth is 5000 Hz. Consequently, FM reception is of much higher fidelity, not because of any inherent quality of FM over AM, but because of the greater bandwidth allotted to FM by the FCC.

Modulation Index

A parameter that is useful in theoretical discussions of FM is the modulation index M. By definition, $M = \Delta F/f_a$, where ΔF is the deviation produced by an audio tone whose frequency is f_a.

For example, a 2-kHz audio tone of sufficient amplitude to produce a 30-kHz deviation gives rise to a modulation index of 15:

$$M = \frac{\Delta F}{f_a} = \frac{30 \text{ kHz}}{2 \text{ kHz}} = 15$$

If the amplitude of the audio tone is increased so that a 60-kHz deviation results, then

$$M = \frac{60 \text{ kHz}}{2 \text{ kHz}} = 30$$

Note that the modulation index M depends upon the amplitude of the audio tone because it is the amplitude of the audio tone that determines the deviation ΔF. As the audio amplitude increases, so does the modulation index M (for a given f_a).

13-2 FM SIDEBANDS

The sidebands produced by frequency modulating a carrier are significantly different from the sidebands resulting from amplitude modulation. Both the amplitude and frequency of the modulating signal affects the sideband structure. In theory, an FM waveform resulting from single-tone modulation contains an infinite number of sidebands. In practice, however, the sideband components far from the center frequency are small in amplitude and may be ignored. This makes it possible to contain an FM transmission within a 200-kHz channel. Sideband components having an amplitude of 1% or more of the unmodulated carrier amplitude are called *significant sidebands*.

Sideband Frequencies

If a carrier of center frequency F_c is modulated by a single audio tone of frequency f_a, upper sidebands appear at F_c plus integral multiples of f_a, and lower sidebands appear at F_c minus integral multiples of f_a. For example, if a 100-kHz carrier is FM modulated by a 2-kHz audio tone, upper sideband components are 102, 104, 106, . . . kHz, and lower sideband components are 98, 96, 94, . . . kHz. Thus, the sidebands are symmetrically located relative to the center frequency. In this respect AM and FM sidebands are similar, but in AM there are only two sidebands. Figure 13-4 illustrates sideband frequency spacing as it depends upon the frequency of the modulating signal. Neither amplitude nor phase relationships are considered in Figure 13-4.

Sideband Amplitudes

The amplitude of each pair of FM sidebands is given by Bessel functions, a set of functions that comprise solutions to Bessel's differential equation. Rather than to embark upon a detailed analysis of this, we present two examples in Figure 13-5 that indicate the general nature of FM sidebands as they vary with modulating frequency and deviation. More detailed discussions of FM sideband amplitudes may be found in the references at the end of the chapter.

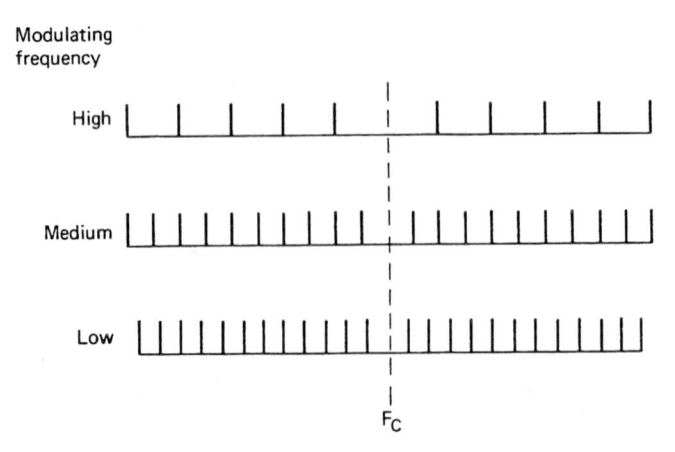

FIGURE 13-4 Sideband frequency spacing as it depends upon the modulating frequency.

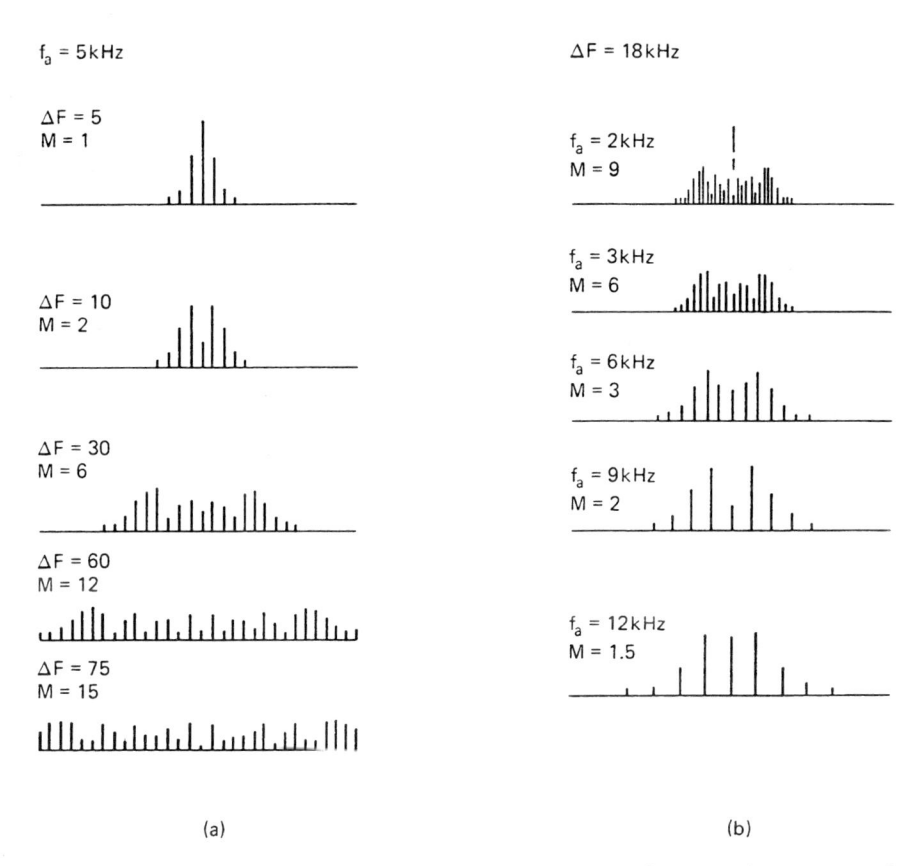

FIGURE 13-5 Typical FM sideband structures. (a) Modulating frequency is constant at 5 kHz and the amplitude is gradually increased. (b) Amplitude is held constant while the modulating frequency is increased.

In part (a) of the figure, the modulating frequency is 5 kHz for all cases. From top to bottom, the amplitude is increased so that deviations of 5, 10, 30, 60, and 75 kHz result. Note that the sideband spacing remains constant and more sidebands appear with increasing deviation.

In part (b) the amplitude is held constant while the modulating frequency is varied. Modulating frequencies are, from top to bottom, 2, 3, 6, 9, and 12 kHz. As the frequency increases, sideband spacing becomes greater while fewer sidebands, larger in amplitude, appear.

Note that the center-frequency component does not remain constant in amplitude. For certain values of M, the carrier component disappears entirely. In that case, all the transmitted power is contained in the sidebands. At all times, however, the average power output of an FM transmitter is constant and does not vary with modulation.

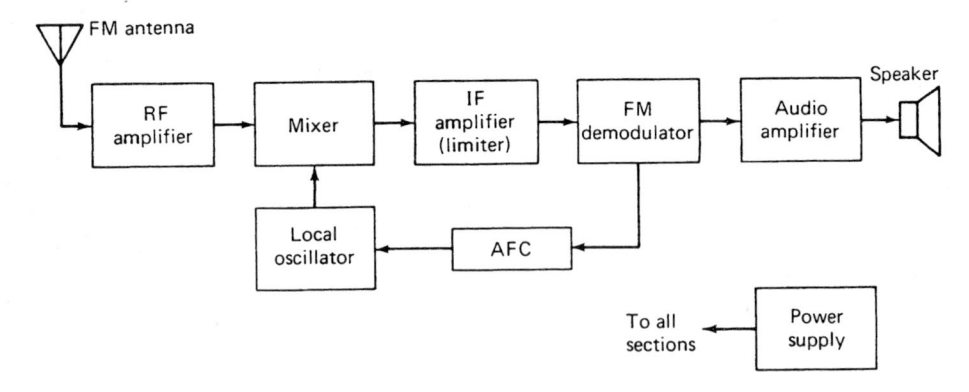

FIGURE 13-6 Block diagram of a superheterodyne receiver for FM. The IF amplifier may consist of two or more stages and typically includes provisions for amplitude limiting.

13-3 SUPERHETERODYNE RECEIVERS FOR FM

The block diagram of an FM superheterodyne receiver, shown in Figure 13-6, is essentially the same as that of an AM receiver. An RF amplifier is almost always included in FM receivers. The IF amplifiers may serve also as amplitude limiters, and the IF frequency is almost always 10.7 MHz. An FM detector converts the frequency deviations into an audio signal. The most frequently encountered FM detector is the ratio detector, but Foster-Seeley discriminators are not unusual. An automatic-frequency-control loop is almost always present that utilizes a signal developed at the FM detector to control the frequency of the local oscillator in the frequency changer. The audio amplifier stage is usually quite similar to those of AM receivers.

Combination AM-FM receivers are more desirable than FM-only receivers, and such receivers share common stages between the AM and FM sections. A block diagram of a typical receiver is shown in Figure 13-7, and this combination of stages is frequently encountered. The first and second IF amplifiers are designed to operate at both the FM IF frequency of 10.7 MHz and the AM IF frequency of 455 kHz. The audio stage serves both sections, and the power supply is switched between the FM and AM front ends.

The common IF stages employ series-connected parallel resonant circuits tuned to 10.7 MHz and 455 kHz. The two IF frequencies are widely separated so that the AM tuned circuits do not affect FM operation, and vice versa. Thus, the stages function at both frequencies and require no switches for changing from one to the other. This is described in detail in Chapter 14.

13-4 LIMITERS

The purpose of a *limiter* is to remove amplitude variations of the input signal and provide an output of constant amplitude while preserving any frequency variations that may exist in the input signal. In an FM receiver, the limiter immediately

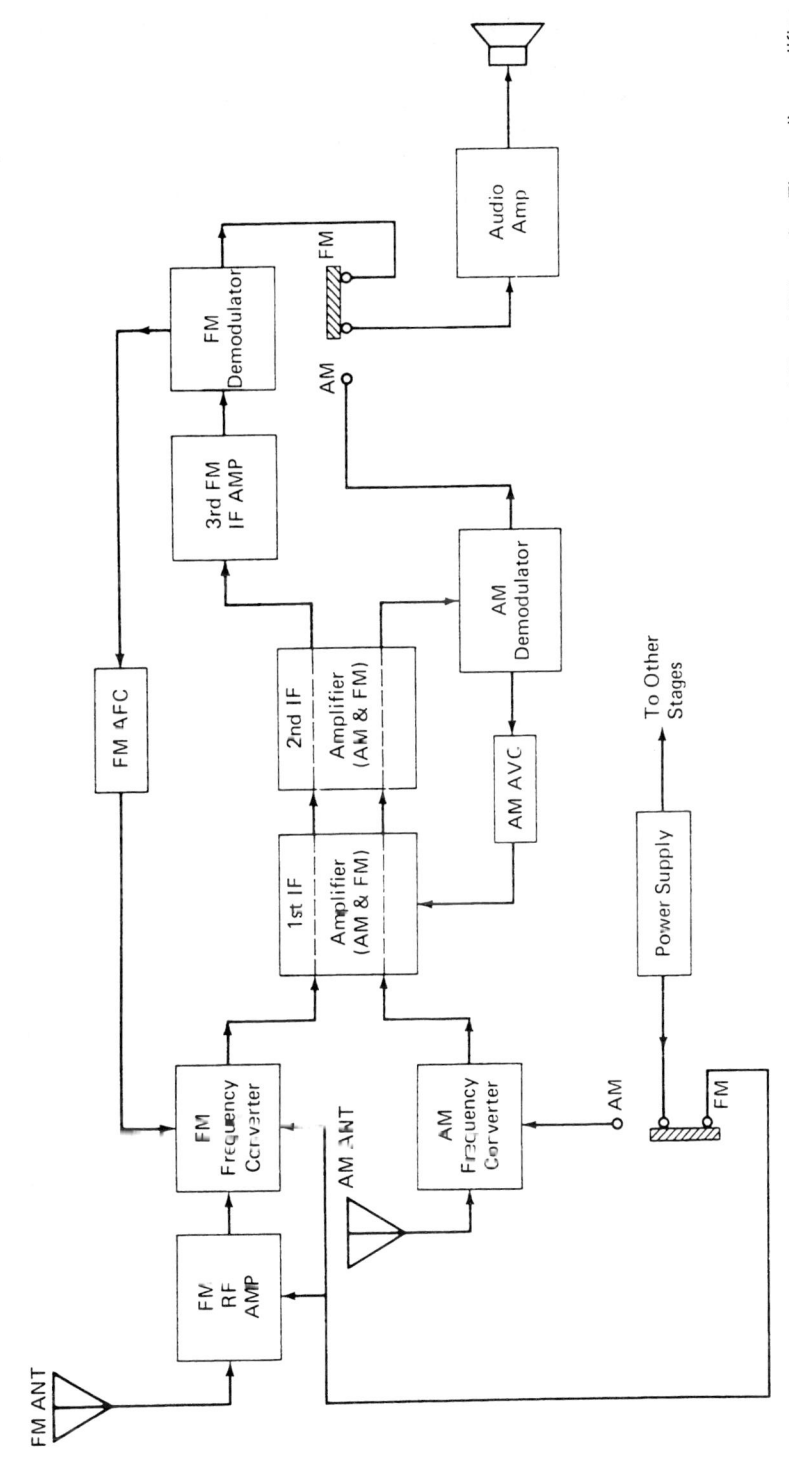

FIGURE 13-7 Block diagram of a typical AM-FM receiver. Two stages of the IF amplifier are used in both the AM and FM modes. The audio amplifier and power supply also serve in both modes.

precedes the FM detector. Since a ratio detector provides its own limiting action, an additional limiter is optional, but a limiter will always precede a Foster-Seeley discriminator. Increased use of ratio detectors causes limiters to be encountered less often now than when the predominant FM detector was a discriminator.

A signal amplitude greater than a minimum, *threshold* amplitude is required for complete limiting action to occur. If the input signal to the limiter drops below the limiting threshold, limiting action will be lost. Limiters are typically driven by a signal voltage of several volts.

Sharp-Cutoff Pentode Limiter

The plate current of a sharp-cutoff pentode such as a 6AU6 will become zero when the grid is only a few volts negative. On the other hand, if the tube is operated with rather low plate and screen gird potentials (50 V), the plate current will reach a maximum (or saturate) quite readily as the grid voltage becomes less negative. This sharp-cutoff and quick saturation characteristic can provide limiting action if the input signal is large enough to drive the tube from one extreme to the other.

A typical pentode limiter circuit is shown in Figure 13-8. A low voltage for the plate and screen is provided by the $R1$-$R2$ voltage divider. Grid leak biasing is provided by R_G and C_G. Input to the stage is from a preceding FM IF amplifier, and the plate operates directly into the primary of the discriminator transformer.

Solid-State Limiters

Limiting action in a bipolar transistor stage can be achieved by operating the stage as an overdriven amplifier. The stage may be overdriven by reasonably small signal amplitudes if the collector voltage is very low, as illustrated in Figure 13-9(a). On

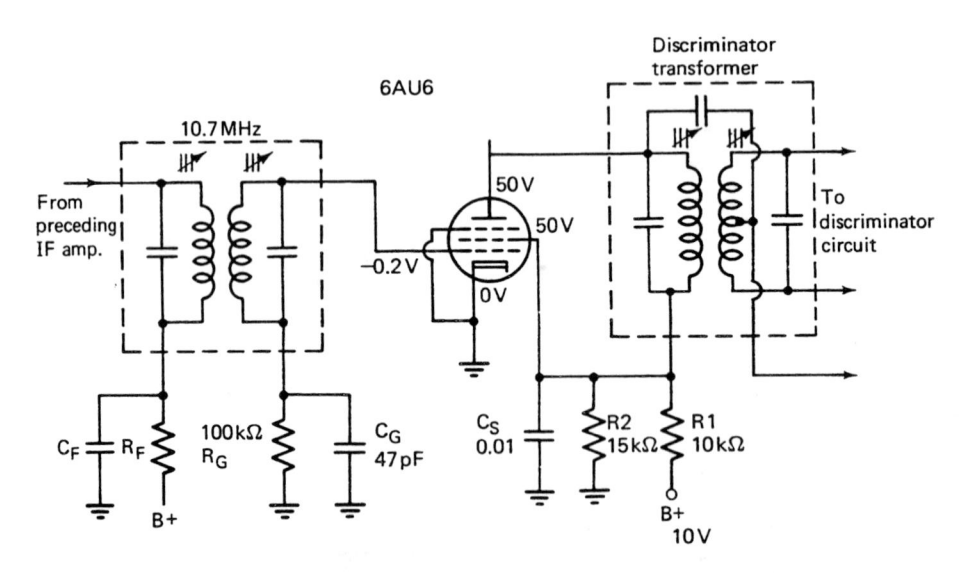

FIGURE 13-8 Typical amplitude limiter utilizing a sharp cutoff pentode.

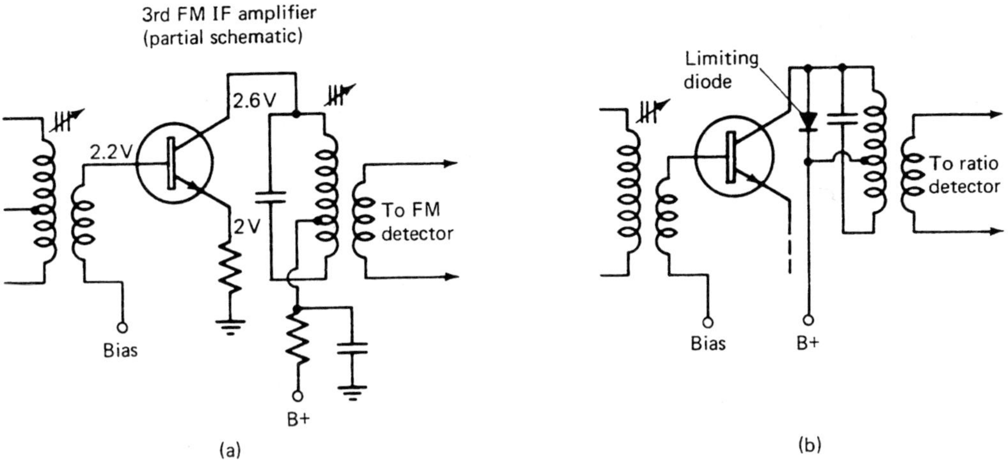

FIGURE 13-9 Two forms of solid-state limiters. (a) The low collector-base voltage suggests the limiting function of this stage as an overdriven amplifier. (b) Limiting diode across resonant circuit of an FM IF amplifier provides limiting action ahead of ratio detector.

positive input peaks, the transistor saturates, and on negative peaks the transistor cuts off completely.

Another trick of the solid-state trade is to connect a diode across a portion of a resonant circuit as shown in Figure 13-9(b). At high signal levels the diode will conduct, the junction capacitance will change, the circuit Q will be lowered, and limiting will occur to some degree. We shall see both types of these limiters in the receivers of subsequent chapters.

13-5 PHASORS—VECTOR REPRESENTATION OF SINE WAVES

An explanation of phase-shift FM detectors demands that phase relationships between two or more sine waves be illustrated. Therefore, we present this brief description of phasor representation of sine waves before proceeding to FM detectors.

A sine wave represents the sinusoidal increase and decrease in voltage or current in a circuit. If we attach an arrow to a rotator as in Figure 13-10, the up-and-down movement of the tip of the arrow is sinusoidal. We ignore the to-and-fro horizontal movement and consider only the vertical height of the tip of the arrow.

The length of the arrow determines the amplitude of the resulting sinusoidal motion, and the rate of rotation—the revolutions per second—determines the frequency. We may relate this to a voltage by assuming that the voltage becomes more positive as the tip of the arrow moves upward. Zero voltage corresponds to a horizontal arrow, and negative voltages are represented when the arrow points downward.

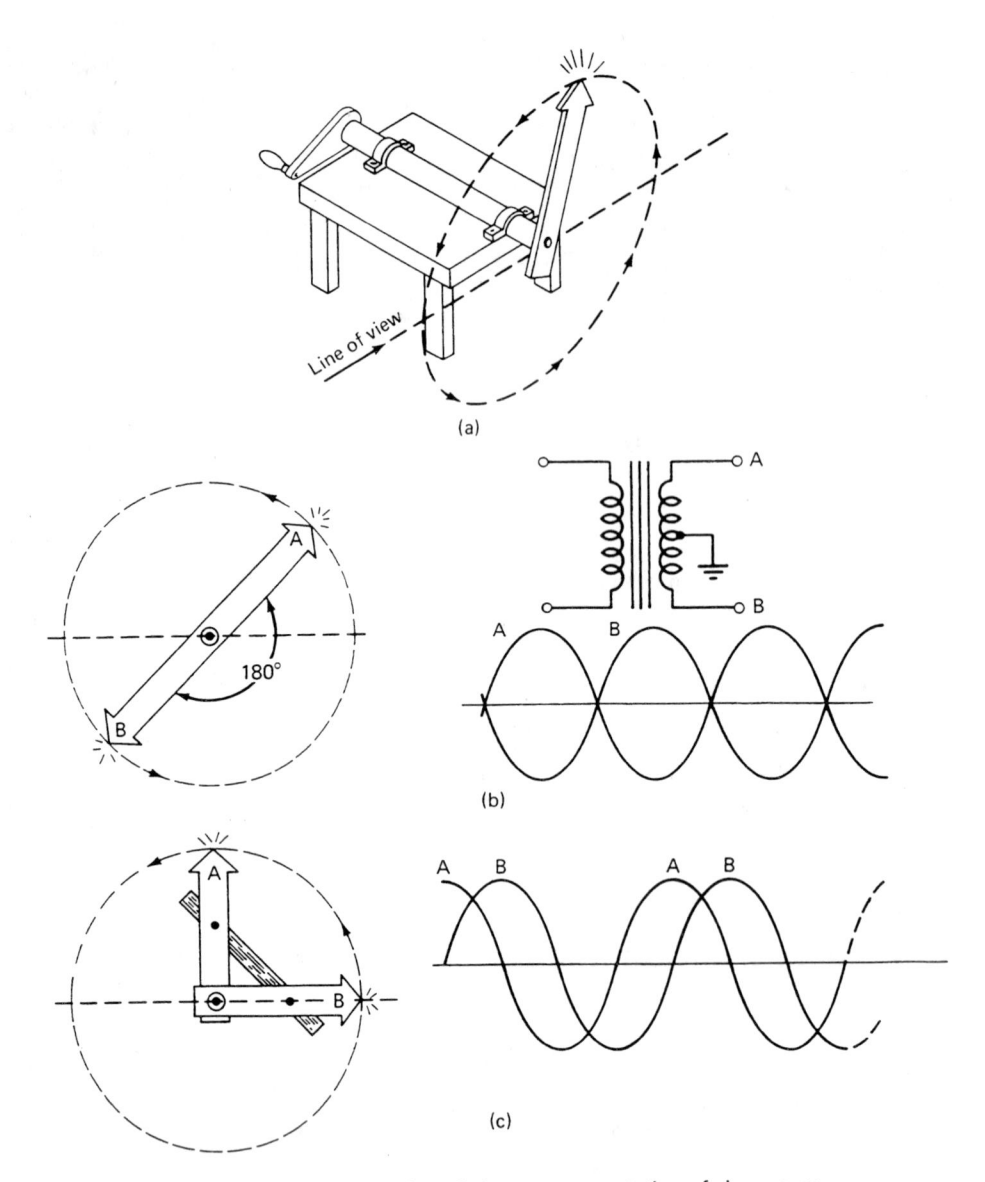

FIGURE 13-10 Illustration of phasor representation of sine waves.

Voltages 180° out of phase may be represented on the phasor machine (the arrows are called *phasors* or *vectors*) by attaching the arrows as shown in part (b) of the figure. Then, when turned, the tip of arrow *A* will reach a maximum height when *B* is at a minimum. An electrical analogy is the voltage at opposite ends of a transformer winding.

Part (c) of the figure represents sinusoidal voltages 90° out of phase, with voltage *A* leading voltage *B*. The arrows are held at 90° to each other by the brace, and

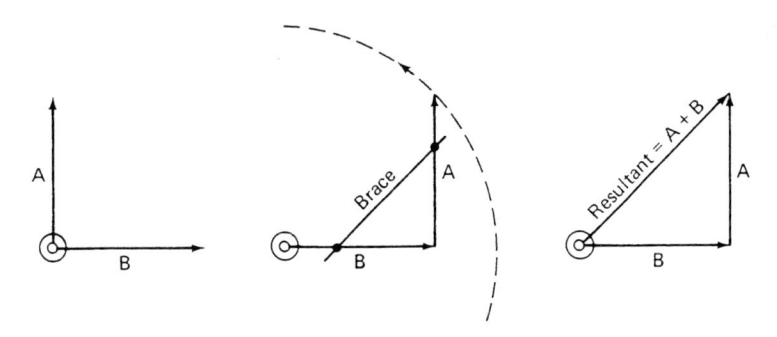

FIGURE 13-11 Addition of two sine waves 90° out of phase produces a resultant sine wave of greater amplitude than either of the originals, as illustrated here with phasors.

the entire framework rotates together in a counterclockwise direction. When A is at a maximum, B is at zero—exactly like sine waves 90° out of phase, as indicated.

Thus, we see that phasors (or vectors) can depict phase relationships between sine waves. We emphasize now that these sine waves were *not added*, since the tails of each phasor attach at the point of rotation (i.e., the axle of the phasor machine).

Phasor Addition of Sine Waves

Phasor addition of sine waves is achieved by connecting the tail of one to the tip of the other—the order is unimportant—maintaining the exact angular relationship between the two phasors. Figure 13-11 illustrates the addition of two sine waves 90° out of phase. A brace is included to hold the phasors in place relative to each other.

We see, then, that the addition of two sine waves produces another sine wave. The amplitude of the *resultant* may be greater or less than either of the originals, depending upon the phase angle between the originals. Further, the phase of the resultant is different from the originals. As illustrated, the head-to-tail combination of A and B may be replaced by a resultant C drawn from the center of rotation as shown. Figure 13-12 illustrates the addition of phasors of various amplitudes and phases. Note that the process is applicable to more than two phasors.

In this discussion we assumed all sine waves to be of the same frequency so that the arrows could be "braced" relative to each other. When the frequencies of all phasors are not the same, the phasors rotate relative to each other and the situation can easily get out of hand, requiring a phasor machine of enormous mechanical complexity.

13-6 FOSTER-SEELEY DISCRIMINATOR

A Foster-Seeley discriminator is shown in Figure 13-13. The function of this circuit is to convert changes in frequency to changes in voltages. It is therefore a frequency-to-voltage converter, and it may serve as an FM detector.

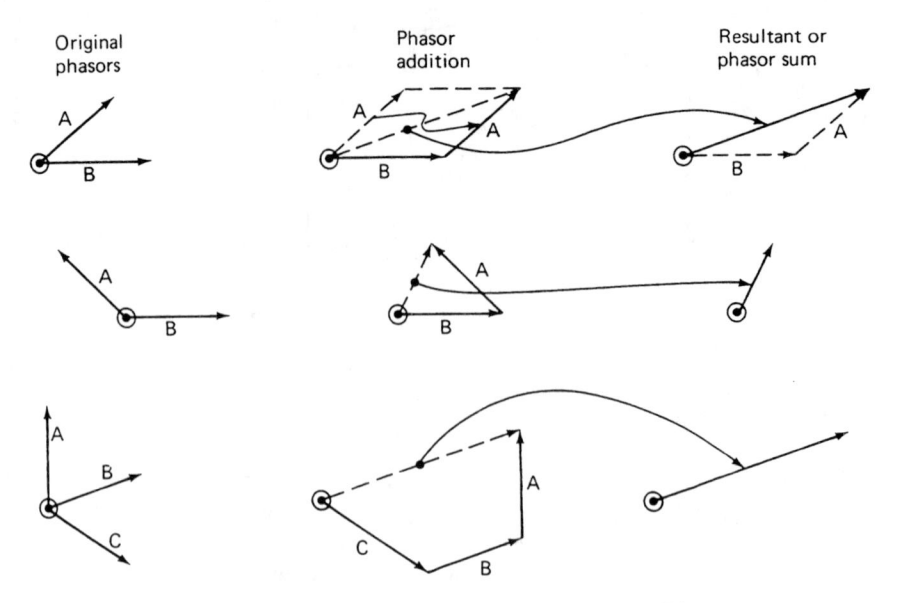

FIGURE 13-12 Three illustrations of phasor addition.

Very much in evidence is the discriminator transformer with a capacitor connecting the primary and secondary circuits. Both primary and secondary are tuned to the FM IF center frequency of 10.7 MHz. Note that discriminator diodes $D1$ and $D2$ are connected in the same direction. We now point out a property of loosely coupled tuned circuits that is essential to the operation of both the discriminator and ratio detector.

Phase Relationship of Primary-Secondary Voltages

In Figure 13-14 the two resonant circuits are loosely coupled and tuned to the same frequency f_c. The low end of the primary is at RF ground, and the center of the secondary is grounded. A driving voltage of frequency f_c is applied at point A.

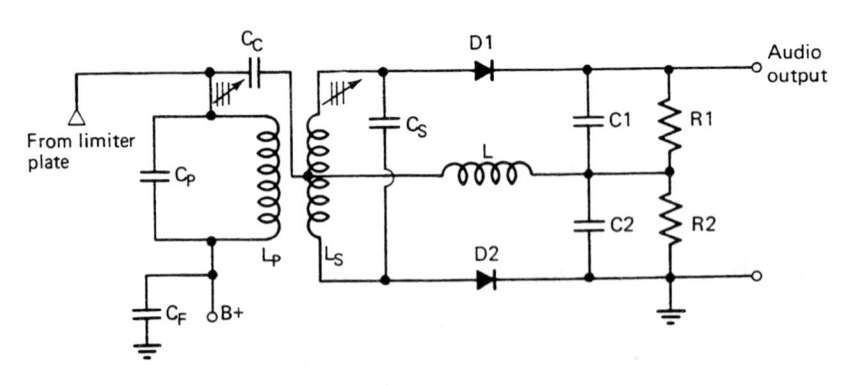

FIGURE 13-13 Foster-Seeley discriminator.

Principles of FM Receivers   Chap. 13

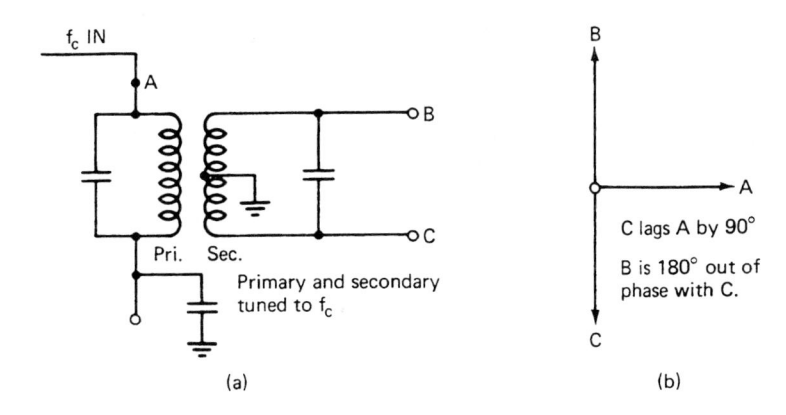

FIGURE 13-14 (a) Tuned transformer with the input signal of the same frequency as that to which the primary and secondary are tuned. The circuits are loosely coupled. (b) Phasors indicating the phase relationships of the voltages appearing at points A, B, and C.

Under these conditions the voltage induced in the secondary will be 90° out of phase with the voltage at point A.

The voltages at points B and C, on opposite ends of the secondary winding, are 180° out of phase with each other. In a phasor diagram, the phasors are drawn in opposite directions, as shown in part (b) of the figure. When the driving voltage at A is of the same frequency f_c as that to which both the primary and secondary are tuned, the voltages at B and C will differ in phase from the voltage at A by 90°, as indicated.

A mechanical analog consisting of two equivalent pendulums suspended as in Figure 13-15 demonstrates the 90° phase difference between the two coupled res-

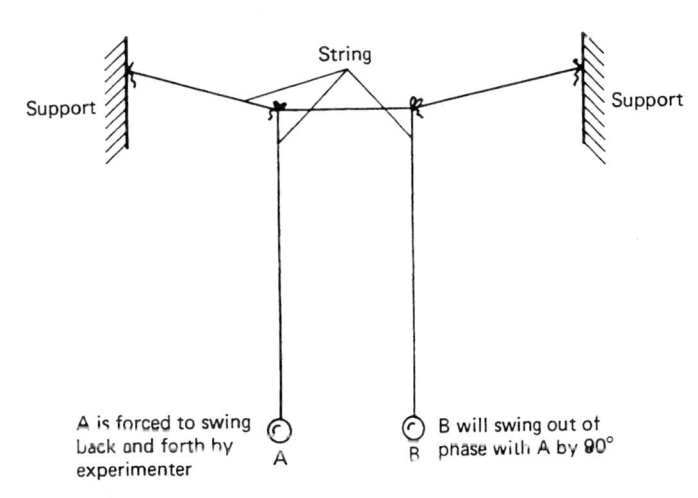

FIGURE 13-15 Two identical pendulums, loosely coupled, can be used to illustrate phase relationships in resonant circuits.

Sec. 13-6   *Foster-Seeley Discriminator*   **385**

onant circuits. Pendulum A is caused to swing steadily at its natural frequency. Pendulum B is driven by A via the common support string, and will lag A in phase by 90°. This simple demonstration illustrates a universal principle of loosely coupled resonant systems. It is amusing, and sometimes even works.

We now consider what happens when the driving frequency f_{in} is higher than f_c, the resonant frequency of the tuned circuits. First, the primary circuit will oscillate at the driven frequency since it is more-or-less forced to do so. Coupling to the secondary, however, is loose, and the reluctance of the secondary to oscillate at the higher-than-resonant frequency produces a greater than 90° phase shift from primary to secondary. The phasor diagram for this case is shown in Figure 13-16(a). The secondary circuit, not wishing to oscillate so rapidly, lags farther behind than when the driving frequency is equal to the resonant frequency. Remembering the phasor machine in reference to Figure 13-16(a). We see that points B and C will reach a maximum later than before.

When the driving frequency is lower than f_c, the secondary circuit wants to oscillate faster than the incoming signal, and the secondary voltage lags the primary by less than 90°. The secondary tends to "catch up" with the primary voltage, and so moves ahead of its natural, 90° behind, position. This gives rise to the phasor diagram of Figure 13-16(b).

In summary, the secondary voltage lags the primary voltage by 90° when f_{in} equals f_c. When f_{in} is greater than f_c, the secondary voltage lags by more than 90°. When f_{in} is less than f_c, the secondary voltage lags by less than 90°. The amount of phase shift is proportional to the difference between f_{in} and f_c.

The above can be demonstrated by slightly changing the length of pendulum A in Figure 13-15. When driving pendulum A is longer than B, the phase angle should decrease, and when A is shorter than B the phase angle should increase.

Phasor Addition of Primary and Secondary Voltages, $f_{in} = f_c$

Figure 13-17(a) shows the transformer portion of a discriminator. Capacitor C_C couples the primary voltage to the center of the secondary so that the primary voltage is applied equally to diodes $D1$ and $D2$. But we must not forget that the

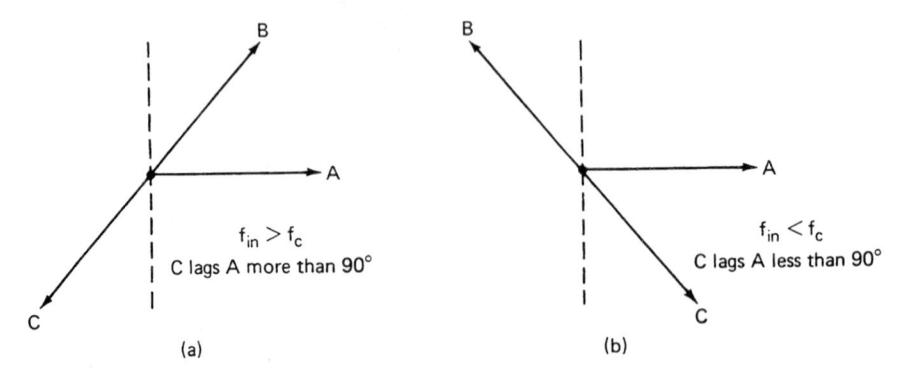

FIGURE 13-16 When the input frequency to the circuit of Fig. 13-14 varies either above or below the tuned frequency of the circuits, the primary voltage and secondary voltage become more or less out of phase, as shown.

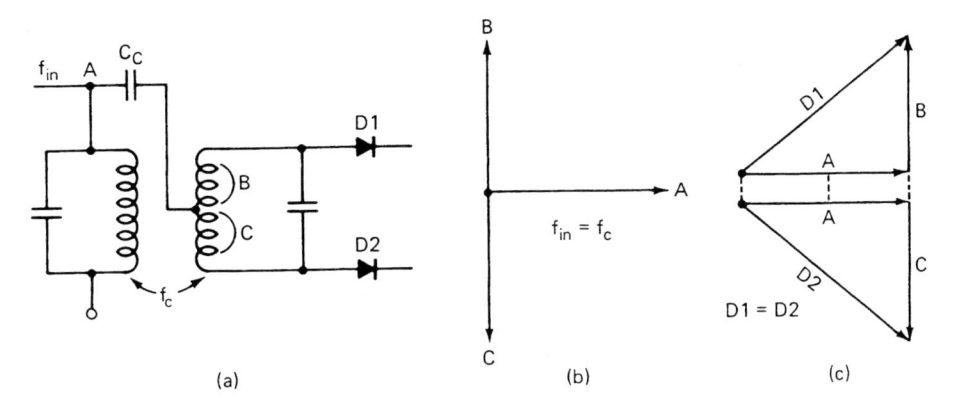

FIGURE 13-17 (a) Portion of a Foster-Seeley discriminator. When the input frequency equals the tuned frequency of the resonant circuits, the phase relationships are as shown in part (b). Equal voltages are applied to the diodes as illustrated in part (c).

transformer action provides an additional voltage component to the diodes. This is the magnetically induced component that shifts in phase as described above. Phasor voltage B is developed in the top portion of the secondary, and phasor voltage C in the bottom. Thus, the voltage appearing at each diode consists of two parts: the voltage from the primary via C_C, and the induced voltage, B or C.

The voltage applied to $D1$ is the phasor sum of A and B, shown at the top of Figure 13-17(c) as taken from the "unadded" phasor diagram of Figure 13-17(b). The voltage applied to $D2$ is the phasor sum of A and C, as shown in the lower diagram of Figure 13-17(c). Note that when f_{in} equals f_c, equal voltages are applied to the diodes.

$f_{in} > f_c$

Referring now to Figure 13-18(a) we see that B and C have shifted in phase relative to A since f_{in} is greater than f_c. When A and B are added to obtain the voltage applied to $D1$, we find the resultant voltage to be greater than before. On the other hand, when A and C are added to obtain the voltages at $D2$, the resultant is less than before. Thus, when f_{in} does not equal f_c, unequal voltages are applied to the diodes. Further, when f_{in} is greater than f_c, the greater voltage will be applied to $D1$.

$f_{in} < f_c$

Figure 13-18(b) shows the phase shift, phasor addition, and resulting diode voltages when f_{in} is less than f_c. Note that unequal voltages are applied to the diodes, with $D2$ receiving a larger voltage than $D1$.

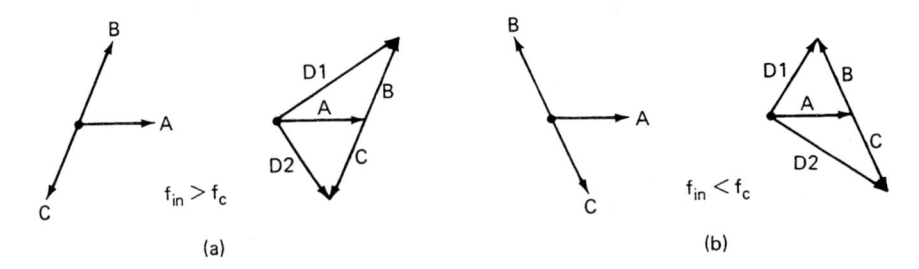

FIGURE 13-18 When the input frequency F_{in} varies relative to f_c, the phase relationships change and cause different resultant voltage amplitudes to be applied to the diodes, $D1$ and $D2$.

Summarizing, when f_{in} equals f_c, the diode voltages are equal. When f_{in} is greater than f_c, $D1$ receives a greater voltage than $D2$. When f_{in} is less than f_c, $D2$ receives a greater voltage than $D1$. This action forms the heart of the operation of a discriminator and also of a ratio detector.

Discriminator Operation

We now refer to the schematic of the discriminator secondary circuit of Figure 13-19. Diodes $D1$ and $D2$ serve essentially as half-wave rectifiers to rectify the 10.7-MHz phasor sum applied to each diode. Resistors $R1$ and $R2$ serve as load resistors for the respective diodes. Electron flow resulting from the rectification is as indicated in the figure. Capacitors $C1$ and $C2$ smooth the pulses from the diodes, and inductor L prevents the 10.7-MHz voltage at the secondary center-tap from being shunted to ground through $C2$.

Note carefully the polarity of voltages $V1$ and $V2$. The output voltage is $V1$ minus $V2$. When f_{in} equals f_c, equal voltages will be applied to the diodes and $V1$ will equal $V2$. Consequently, the output, $V1$-$V2$, will be zero (when $f_{in} = f_c$).

When the input frequency rises above f_c, the phase shift of the secondary voltage

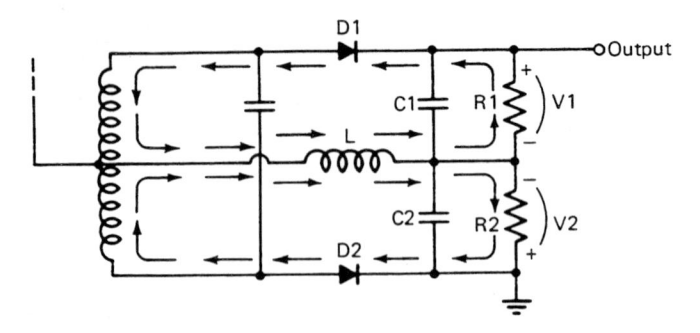

FIGURE 13-19 Secondary circuit of a Foster-Seeley discriminator showing the paths of electron flow through $D1$ and $D2$.

causes the voltage at $D1$ to increase while the voltage at $D2$ decreases. Hence, $V1$ will be greater than $V2$, and the output $V1$-$V2$ will be positive (when $f_{in} > f_c$).

When the input frequency falls below f_c, the phase shift in the opposite direction causes $V2$ to become greater than $V1$, and the output will be negative (when $f_{in} < f_c$).

If the input voltage is frequency-modulated by a sinusoidal modulating tone, the sinusoidal variation of f_{in} will produce a sinusoidal voltage at the output that is an exact replica of the original modulating tone. Thus, the discriminator is an FM detector.

Discriminator Sensitivity to AM

If the amplitude of the input signal to the discriminator increases or decreases so that the new amplitude is a number x times the original amplitude, the new values of $V1$ and $V2$ will be $xV1$ and $xV2$, and the new output voltage will be $x(V1$-$V2)$, or x times the original output. Thus, the output voltage is directly dependent upon the input amplitude. Therefore, a discriminator has no natural immunity to AM riding on the input FM signal. For this reason a limiter always precedes a discriminator in an FM receiver in order to remove the AM, which otherwise would contribute noise to the output.

An example will make this clear. Suppose that f_{in} and V_{in} are such that $V1 = 3$ and $V2 = 1$. The output is $V1 - V2$ or $3 - 1 = 2$. If V_{in} is now doubled so that x in the above is 2, the new $V1 = 6$ and the new $V2 = 2$. The new output becomes $6 - 2 = 4$, which is twice the original output.

Variation

A commonly encountered variation of the basic discriminator circuit is shown in Figure 13-20. Capacitors $C1$ and $C2$ of Figure 13-13 have been replaced by capacitor $C12$. Since no capacitor now connects to the $R1$-$R2$ junction, inductor L of Figure

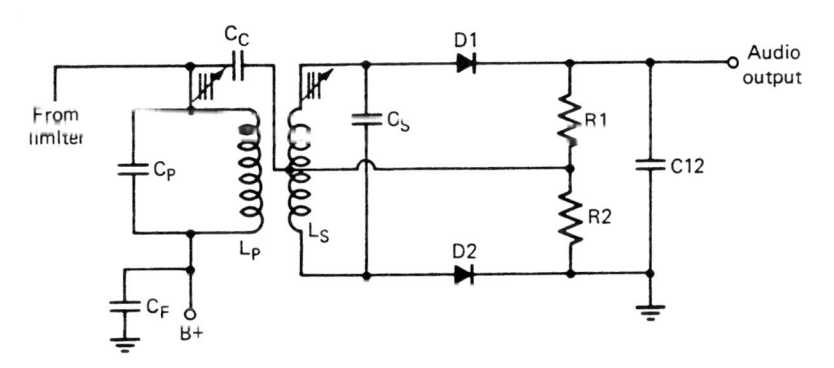

FIGURE 13-20 Frequently encountered variation of the Foster-Seeley discriminator.

13-13 may be omitted. The parallel equivalent of $R1$ and R then serves the same function. The operation of this circuit is the same as for the previous circuit.

13-7 RATIO DETECTOR

The ratio detector is by far the most popular FM detector in FM receivers, and a basic circuit is shown in Figure 13-21. Prominent features include a *tertiary winding* L_T that is part of the transformer, diodes connected in opposite polarity, and a large stabilizing capacitor C_R. We shall see that the circuit operation is similar in many respects to the discriminator. Note, however, that the output is taken from the tertiary winding.

Both the primary and secondary are tuned to the center frequency f_c, as in the discriminator, and the same shifting in phase of the secondary voltage occurs as before. The primary voltage is applied to the secondary center tap by L_T, the tertiary winding, which consists of a few turns of wire coupled tightly to the low end of L_P. Since L_T is tightly coupled to L_P, no phase shift occurs from L_P to L_T; the tertiary voltage is in phase with the primary voltage. Thus, a voltage is applied to the center of the secondary that is in phase with the primary voltage—as in the discriminator.

It is apparent, then, that voltages are applied to the diodes in the same manner as in the discriminator, and the voltages vary with f_{in} as before. Thus, this aspect of the ratio detector needs no further explanation.

Referring to the partial circuit of Figure 13-22(a) when f_{in} equals f_c, diodes $D1$ and $D2$ conduct equally. Electron flow to the left through $D1$ is equal to the flow to the right through $D2$. Consequently, no current flows in the tertiary winding, and C_A remains uncharged. Current only flows around the loop, as shown in the figure, with equal voltage drops across $R1$ and $R2$.

In part (b) of the figure, we assume that f_{in} is greater than f_c, and further, we assume that the transformer connections and winding polarities are such that when f_{in} is greater than f_c, diode $D2$ conducts more than $D1$. Note that a current (equal

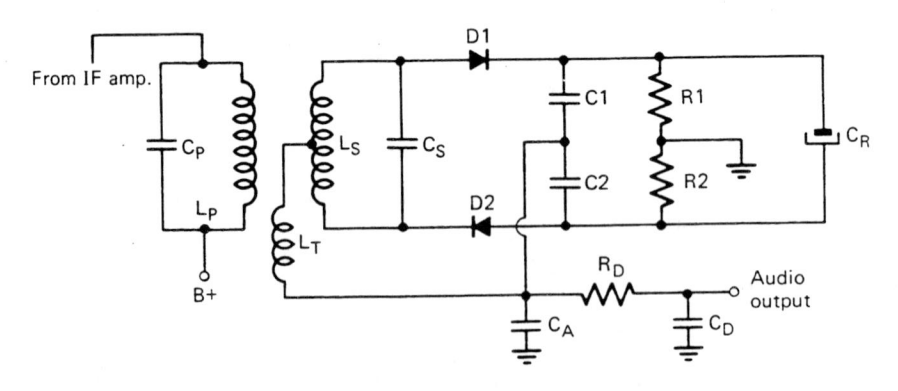

FIGURE 13-21 Balanced ratio detector.

Principles of FM Receivers   Chap. 13

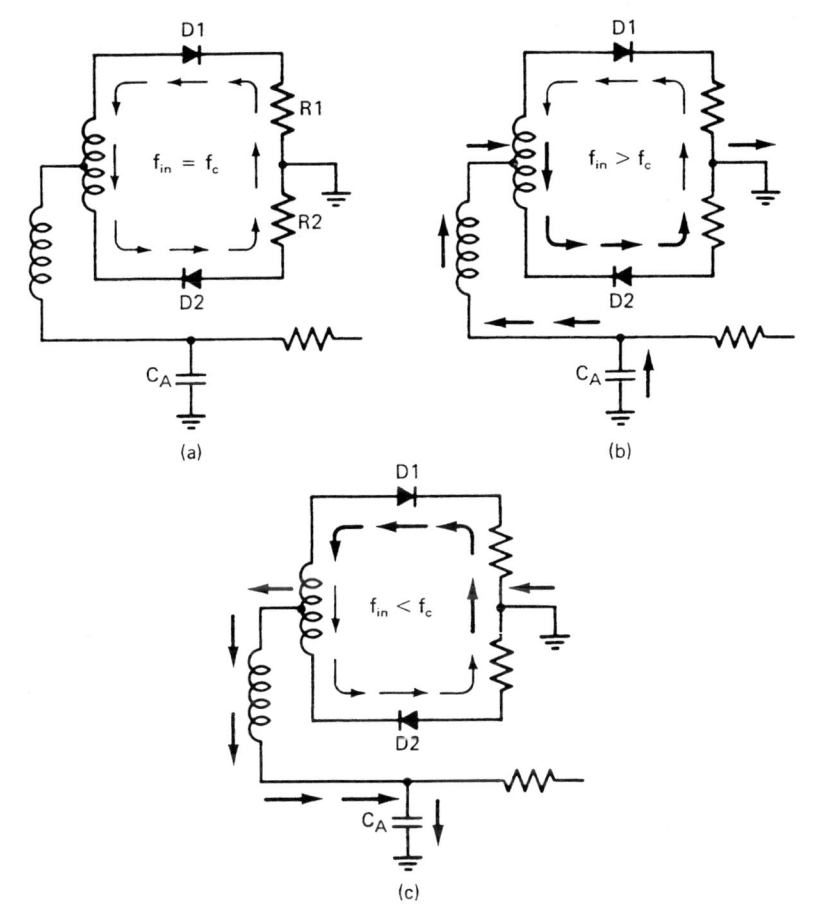

FIGURE 13-22 As the input frequency f_{in} varies above and below f_c, the different voltages applied to the diodes cause a net current to flow in the tertiary winding.

to the difference in $D1$ and $D2$ currents) flows in the tertiary winding and also in the ground connection. Capacitor C_A is charged positively, and the voltage developed on C_A represents the output of the ratio detector.

In part (c) the situation is reversed. Diode $D1$ conducts more than $D2$, and C_A becomes charged negatively as f_{in} falls below f_c. Thus, the output voltage depends upon the value of f_{in} relative to f_c.

We now consider the complete secondary circuit of the ratio detector of Figure 13-23. Capacitors $C1$ and $C2$ complete the circuit for the 10.7-MHz voltage developed across the tertiary winding so that the voltage at the low end of the tertiary winding is applied to the $C1$–$C2$ sides of diodes $D1$ and $D2$. Capacitor C_R is the stabilizing capacitor that is largely responsible for the AM-limiting properties of a ratio detector.

When an input signal is applied, diodes $D1$ and $D2$ conduct (perhaps unequally), and voltages $V1$ and $V2$ are developed across $R1$ and $R2$, respectively. The polarity

Sec. 13-7 Ratio Detector 391

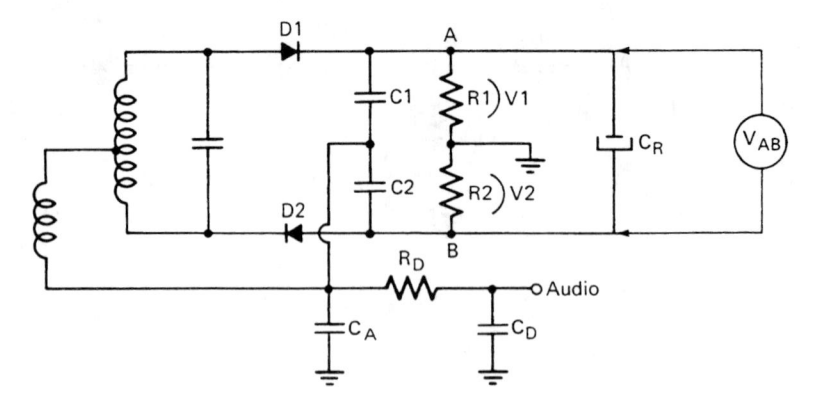

FIGURE 13-23 Complete secondary side of a ratio detector. Capacitors C_1 and C_2 provide a return path for the 10.7-MHz voltage developed across the tertiary winding and applied to the diodes. C_R is the stabilizing capacitor and C_A is the capacitor across which the audio signal is developed. R_D and C_D form a deemphasis network to reduce the ratio detector response at high frequencies.

of V1 and V2 is such that they add to give V_{AB}, the voltage between points A and B in Figure 13-23. The average amplitude of the input signal determines the value of V_{AB}. Capacitor C_R charges to voltage V_{AB}.

Since C_R is an electrolytic whose capacitance is a few microfarads, its effect is to hold V_{AB} constant. Now, V_{AB} = V1 + V2, so that when f_{in} changes to increase V1, V2 must necessarily decrease. Note that as f_{in} varies, the ratio of V1 and V2 varies accordingly, hence, the name *ratio detector*. The voltage division across R1 and R2 occurs in conjunction with the charging and discharging of C_A, developing the output signal.

Ratio-Detector Response

A typical plot of ratio-detector output as a function of frequency is shown in Figure 13-24. The curve is called an *S-curve* because of its shape. The output of a discriminator is, for all practical purposes, identical to that of a ratio detector.

The central, linear region is the active or useful portion of the curve. The output is zero when the input frequency equals the center frequency, typically 10.7 MHz. As the input frequency increases, the output becomes positive and increases with frequency to the positive peak of the S-curve. At this point the frequency is at the upper limit of the IF amplifier bandwidth, and, if the frequency is further increased, the input amplitude falls off quite sharply. This produces the downward turn of the response, forming the upper portion of the S-curve. The IF response is indicated in Figure 13-24.

A similar response is obtained when the input frequency falls below the center frequency. The output becomes increasingly negative as the frequency becomes lower. The negative peak occurs at the lower limit of the IF amplifier bandwidth.

The S-curve may be inverted by reversing the direction of the ratio-detector

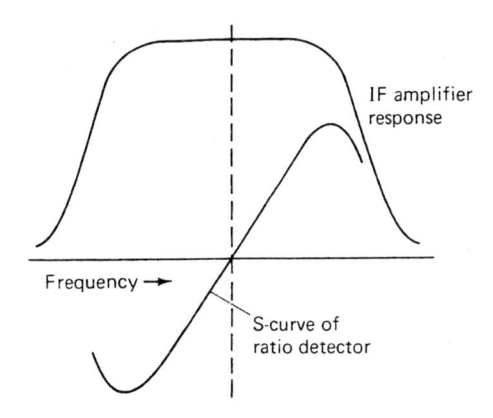

FIGURE 13-24 S-curve response of ratio detector or discriminator. The IF amplifier response is also indicated.

diodes or by reversing the polarity of a winding in the ratio-detector transformer. The S-curve polarity displayed in a given servicing situation will also depend upon the characteristics of the sweep generator employed. In a servicing situation, the polarity is unimportant. In the receiver the S-curve polarity is frequently dictated by AFC requirements. Receivers in which the ratio-detector output becomes more negative with increasing frequency are not uncommon.

Limiting Action of Ratio Detector

The voltage V_{AB}, held constant by C_R, is responsible for the limiting properties of the ratio detector. Limiting action results from the variable loading of the high-Q resonant circuits of the ratio detector transformer. Assume that an increase occurs in the amplitude of the input signal. Diode conduction increases, increasing the loading on the tuned circuits, lowering the Q. The lower Q of the primary winding reduces the gain of the previous stage. This, in turn, reduces the input voltage to the diodes, counteracting the increase in amplitude.

As C_R charges slowly to a higher voltage, diode conduction decreases, and the circuit Q's return to their normal value. The effect of the slowly varying voltage across C_R is to slow the response of the ratio detector to an increase in input amplitude to a point where the coupling capacitors attenuate the slowly varying voltage to the point of extinction.

If the input amplitude drops, diode conduction decreases, circuit loading decreases, and the circuit Q's become higher than normal. The decrease in amplitude is opposed until C_R slowly adjusts to the new amplitude. Thus, variable loading of the tuned circuits produced by the effect of C_R upon diode conduction gives the ratio detector its immunity to amplitude variations of the input voltage.

Ratio detector circuits can sometimes become "overstabilized" so that when an increase in amplitude suddenly occurs, the output of the ratio detector actually drops momentarily. This can be avoided by connecting compensation resistors as

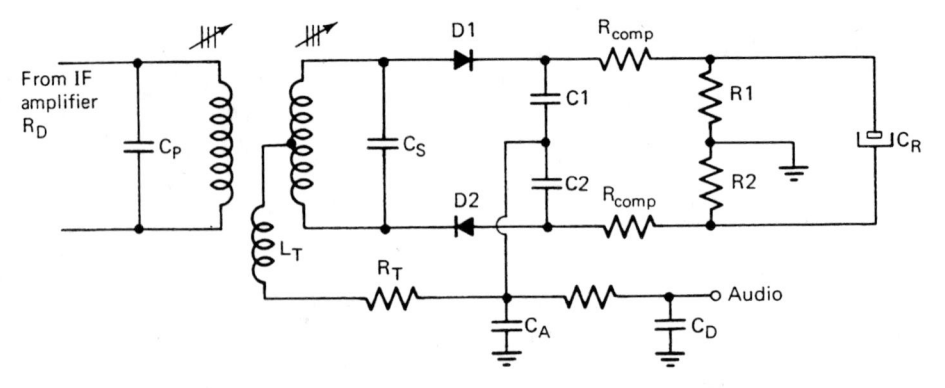

FIGURE 13-25 Ratio detector that includes compensation resistors to avoid overstabilization and a resistor in series with the tertiary winding to limit diode current at high signal levels.

shown in Figure 13-25. The compensation resistors then become a part of the diode load, and then, only the $R1$–$R2$ portion of the load is stabilized. Typical compensation resistors are on the order of 1 to 1.5 kΩ. Resistor R_T of Figure 13-25, in series with the tertiary winding, limits maximum diode current and reduces AM sensitivity at high signal levels.

Variation

An unbalanced ratio detector is shown in Figure 13-26. Capacitors $C1$ and $C2$ are eliminated since the tertiary voltage is returned to the diodes via C_A and C_R. Resistors $R1$ and $R2$ are replaced by $R12$, and no compensation resistors are included. Its operation should be obvious to the reader. Naturally, this circuit cannot match the more complicated circuit insofar as fidelity and limiting properties are concerned.

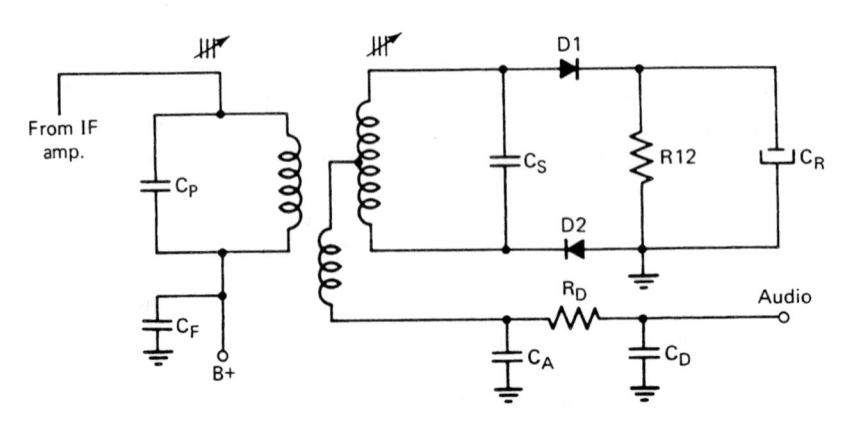

FIGURE 13-26 Unbalanced ratio detector.

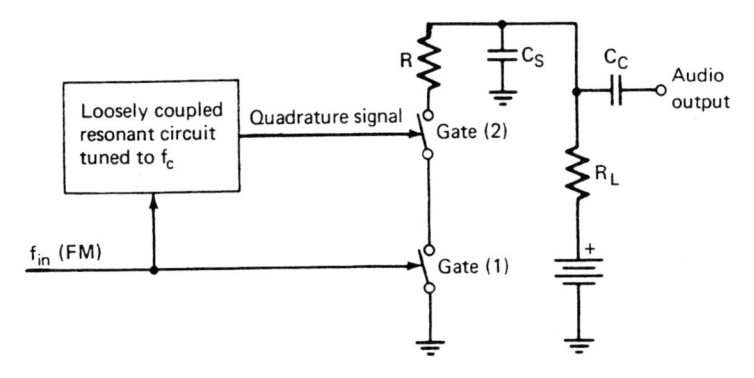

FIGURE 13-27 General scheme of a quadrature detector. The phase of the quadrature signal varies relative to the input signal as the input signal varies above and below the resonant frequency f_c. Conduction through R_L varies as the opening and closing of the gates become more or less in phase.

13-8 QUADRATURE DETECTORS

A quadrature detector is a phase-shift type of detector because its operation depends upon the shifting phase of a resonant circuit voltage relative to a loosely coupled driving voltage. In Sec. 13-6 we saw that the phase of the voltage of a resonant circuit at resonance lags the driving voltage by 90°. The resonant circuit voltage is said to be "in quadrature" with the driving voltage. Further, we saw that the phase lag varies as the driving frequency varies above and below the natural frequency of the resonant circuit.

The general scheme of a quadrature detector is shown in Figure 13-27. The driving and quadrature signals drive series-connected gates as shown. The gates (indicated as switches) are capable of being cut completely on or off by a small positive or negative voltage, respectively. Both gates must be on for conduction through R_L to occur.

When both gates are on, electrons flow from ground through R to C_s and tend to discharge C_s. When either gate is off, no conduction through R occurs, and C_s charges to a higher voltage due to electron conduction from C_s through R_L to the positive supply. The actual voltage developed on C_c depends upon the fraction of the time that both gates are simultaneously on. Capacitor C_c couples the variation in voltage across C_s to the output; this is the audio signal, and the net effect is to convert variations in frequency to variations in voltage. This is the essence of FM demodulation.

Figure 13-28 illustrates the action of the quadrature detector when the input is equal to, is less than, and is greater than the resonant circuit frequency f_c. Square waves are shown, due to the off-on controlling characteristics of the gates. When f_{in} equals f_c, the input and quadrature signals are 90° out of phase, and the resulting conduction is indicated. A change in f_{in} causes the phase of the quadrature signal

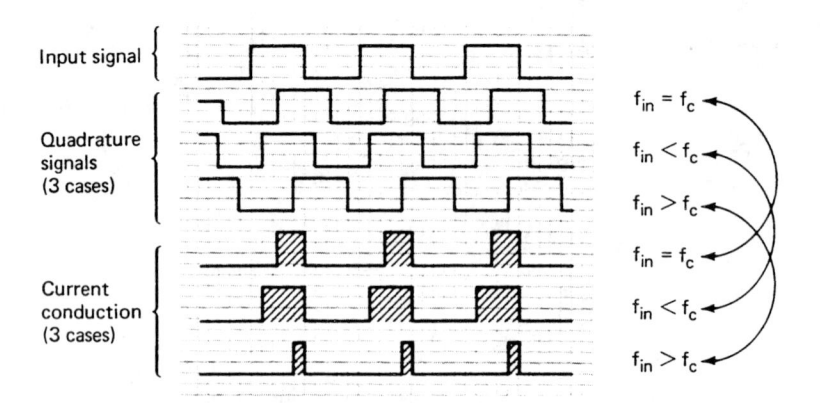

FIGURE 13-28 Quadrature signal varies in phase relative to the input signal. The three cases are illustrated, and the current flow for each case is indicated.

to shift, altering the conduction as shown. Thus, conduction occurs as a series of pulses whose width varies according to the relationship of f_{in} to f_c.

A quadrature detector commonly found in the sound (FM) section of television receivers is shown in Figure 13-29. The tube, a 6BN6, is specially constructed to achieve "gating" action of the quadrature and signal grids. The quadrature resonant circuit is caused to oscillate by weak coupling to the input signal, which occurs within the tube. Resistor R at the plate enhances this coupling. An FM signal applied to the input is demodulated, and the audio output appears atop R_L.

Solid-State Quadrature Detectors

Integrated circuit (IC) quadrature detectors are sometimes encountered in solid-state AM-FM receivers. Inner details of the IC "chip" are known only to (perhaps) the circuit designer, but we can be sure that the gating arrangement is used in

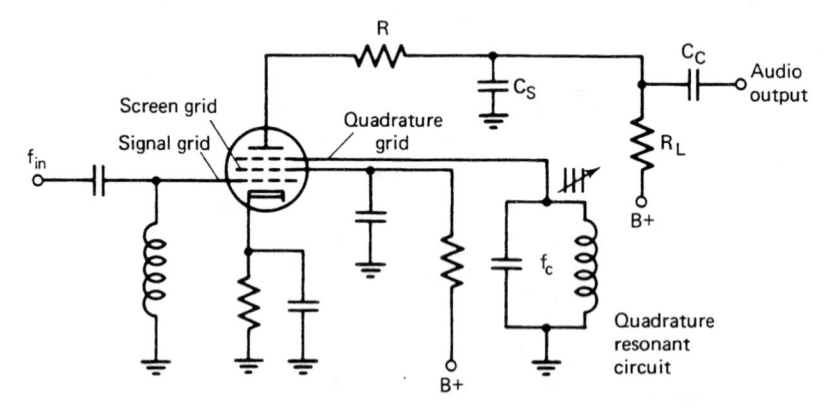

FIGURE 13-29 Quadrature grid detector commonly found in the sound (FM) section of a TV receiver.

```
                Slowly varying
                correction voltage
                       |
f_in   ┌─────────┐   ┌─────────┐   ┌──────────┐
──────▶│ Phase   │──▶│Low-pass │──▶│ Voltage  │──●──▶ f_out = f_in
       │detector │   │ filter  │   │controlled│
       └─────────┘   └─────────┘   │oscillator│
            ▲                      └──────────┘
            │                            │
            └────────────────────────────┘
```

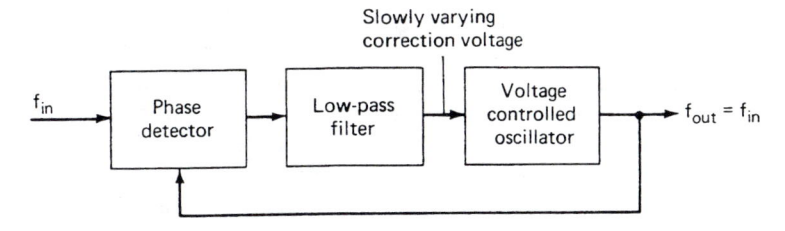

FIGURE 13-30 Basic phase-locked-loop arrangement.

some fashion. Such circuits are characterized by an FM IF input to the chip, an external (to the chip) quadrature resonant circuit, and an audio output directly from the chip. We shall see one of these in Figure 18-3.

13-9 PHASE-LOCKED-LOOP (PLL) FM DETECTOR

Even though PLL FM detectors are seldom encountered in commercial AM-FM receivers, we describe a PLL circuit in this section and show how it can be used as an FM detector. PLL circuits are frequently encountered in FM stereo decoders, and that application of PLLs is described in Chap. 17.

The block diagram of a PLL is shown in Figure 13-30. A *voltage-controlled oscillator* (VCO) is a stable oscillator whose frequency is controlled by a DC (or slowly varying) voltage applied to the control input. In the PLL circuit, a portion of the output of the VCO is fed back to a phase detector which also receives an input signal whose frequency f_{in} is of the same order of magnitude as the output frequency f_{out} of the VCO. The phase detector compares the frequency (or phase) of f_{in} and f_{out}, and generates a correction voltage whose amplitude and polarity depend upon the relationship between f_{in} and f_{out}. The correction voltage is applied via a low-pass filter to the control input of the VCO and tends to alter the VCO frequency until f_{out} is exactly equal to f_{in}. The low-pass filter assures that only slowly varying voltages are applied to the control input of the VCO.

The use of a PLL as an FM detector is shown in Figure 13-31, where the interest lies in the correction voltage rather than in the VCO output. With zero correction

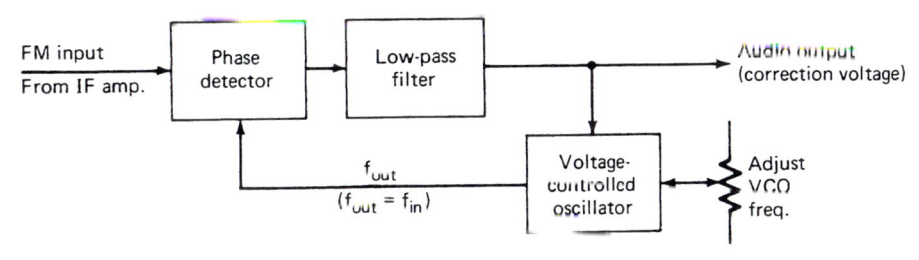

FIGURE 13-31 PLL used as an FM detector. The correction voltage to the VCO is the audio output signal.

voltage applied to the VCO, it is adjusted to run at the FM center frequency of 10.7 MHz. Thus, when f_{in} = 10.7 MHz, the phase detector gives a zero voltage output, the correction voltage is zero, and everything is stable. Suppose that f_{in} increases in frequency. The phase detector now finds f_{out} too low, and it generates a positive correction voltage in order to raise f_{out} to the same frequency as f_{in}. A positive voltage now appears at the output.

If f_{in} falls below 10.7 MHz, the frequency to which the VCO is adjusted, the phase detector will find f_{out} too high, and a negative control voltage will be generated to reduce f_{out} to the same frequency as f_{in}. It is now clear that if frequency-modulated input is applied to the phase detector, the correction voltage will represent the audio output. Thus, the circuit serves as an FM detector.

In Chap. 17 we describe the application of a PLL to FM stereo decoding, and we illustrate how the VCO frequency can be a multiple of f_{in}. This is a very useful feature of PLL circuits.

13-10 PREEMPHASIS AND DEEMPHASIS

It is found that the high-frequency portions of the audio spectrum contribute more noise to FM reception than the lower frequencies. The highs, therefore, tend to have a lower signal/noise ratio than the lows. The noise contribution of the high-frequency region can be reduced by transmitting the highs at increased relative volume levels and then reducing the level the same amount at the receiver. The

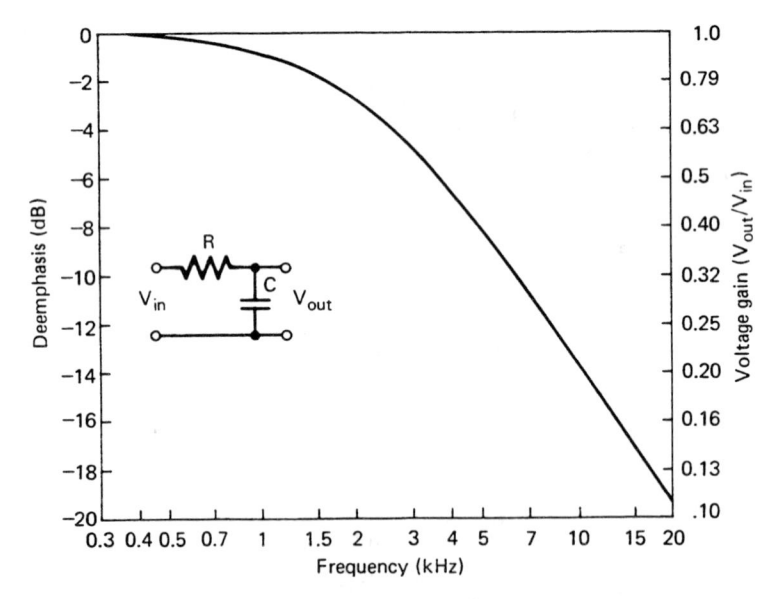

FIGURE 13-32 Deemphasis curve for a 75-μs time constant, as is standard for commercial FM broadcast receivers.

boosting of the highs at the transmitter is called *preemphasis*, and the reduction of the highs at the receiver is called *deemphasis*.

SUMMARY

1. The amplitude of the transmitted wave from an FM transmitter is always constant. The modulating signal causes the transmitted frequency to deviate above and below the center frequency. The amplitude of the modulating signal determines the extent of the deviation, while the modulating frequency determines the rate at which the deviations occur.

2. The deviation established by the FCC as corresponding to 100% modulation for commercial FM broadcasting is ±75 kHz. The FM broadcast band ranges from 88 to 108 MHz and consists of 100 channels, each 200 kHz wide.

3. Amplitude limiting in an FM system provides a means of removing much of the noise that must be tolerated in an AM system. A ratio detector has inherrent AM rejection properties, whereas a discriminator has none.

4. The high fidelity of FM transmissions is due in part to the wide bandwidth (50 Hz to 15 kHz) of the transmitted signal.

5. The modulation index M is the ratio of the deviation to the modulating frequency. Since the deviation depends upon the amplitude of the modulating signal, the modulation index depends upon both the amplitude and frequency of the modulating signal. The FM sideband structure depends upon the modulation index, which can range from zero to 1500 in commercial FM broadcasting.

6. The sideband frequency spacing is determined by the frequency of the modulating signal.

7. Superheterodyne receivers for FM almost always include an RF amplifier, and the IF stages may be designed to provide amplitude limiting. An FM detector is required, and the detector typically produces a control voltage for automatic frequency control of the local oscillator.

8. With the possible exception of the more elaborate receivers, most FM receivers are combined AM-FM receivers that may be operated in either the AM or FM mode. Such receivers typically employ two or more "common stages" that operate in both the AM and FM mode.

9. The purpose of a limiter is to remove amplitude variations from an FM signal while preserving the frequency variations. Typical limiters include a sharp cutoff pentode operated with low plate and screen voltages, a BJT operated as an overdriven amplifier, and a limiting diode connected in parallel with a resonant circuit of an IF amplifier stage.

10. Phase relationships of sine waves are easily pictured by the use of phasors. The length of a phasor represents the amplitude of the sine wave, and its angular position on a phasor diagram indicates its phase relationship to other phasors on the diagram.

11. Phasor addition of sine waves is achieved by connecting the tail of one to the tip of the other while maintaining the exact angular relationship between the two. The phasor sum (or the resultant) is then the phasor drawn from the tail of the first to the tip of the second.

12. The sum of two sine waves is another sine wave, whose amplitude and phase is generally different from either of the two originals.

13. Phase-shift FM detectors depend upon the shifting in phase of the secondary voltage of a tuned transformer as the input frequency varies above and below the resonant frequency of the tuned circuits of the transformer. The discriminator, ratio detector, and quadrature detector are phase-shift FM detectors.

14. When the input frequency to a tuned transformer (loosely coupled) is the same as the resonant frequency of the transformer, the secondary voltage will lag the primary voltage by 90°. When the input frequency rises above the resonant frequency, the phase lag becomes more than 90°, and when the input frequency falls below the resonant frequency, the phase lag becomes less than 90°.

15. The frequency deviations of the input signal to a discriminator or to a ratio detector cause unequal conduction in the diodes. The unequal conduction leads to the development of the audio signal.

16. The stabilizing capacitor of a ratio detector is primarily responsible for its immunity to AM. AM rejection is achieved by variation in the impedance reflected into the primary winding of the ratio detector transformer.

17. The output voltage of a discriminator or of a ratio detector is a function of the input frequency; the output versus frequency characteristic is the S-curve.

18. Quadrature detectors depend upon series-connected gates to produce variations in conduction as the phases of the voltages driving the gates vary with the input frequency to the detector. One gate is driven directly by the input frequency, while the driving voltage for the other gate is obtained from a resonant circuit loosely coupled to and driven by the input signal.

19. A phase-locked loop consists of a phase detector, a low-pass filter, and a voltage-controlled oscillator. The correction voltage produced by the phase detector and low-pass filter maintains the VCO frequency at the frequency of the input signal.

20. A PLL can be used as an FM detector. As such, the correction voltage applied to the VCO becomes the audio output.

21. Preemphasis and deemphasis are used in commercial FM broadcasting to improve the signal-noise ratio at high frequencies. The standard time constant is 75 µs.

QUESTIONS AND PROBLEMS

13-1. Draw waveforms to illustrate the difference between AM and FM.

13-2. Define deviation. What is the maximum deviation allowed by the FCC for broadcast FM?

13-3. Describe an FM broadcast channel. How wide are the guard bands? How many channels are in the FM band? What is the spacing between channels?

13-4. Draw diagrams to show how amplitude limiting can reduce noise due to amplitude variations in an FM waveform.

13-5. Which FM detector *must* be preceded by a limiter? For which is a limiter optional?

13-6. Suppose that the amplitude of a 1-kHz tone that is modulating an FM transmitter is such that a deviation of 30 kHz is produced.

(a) What is the percentage modulation?

(b) What is the modulation index?

(c) What is the total frequency swing of the transmitted waveform?

(d) If the amplitude of the modulating signal were doubled, what would be the new percentage modulation, modulation index, and deviation?

13-7. What accounts for the fact that the propagation characteristics of broadcast band FM signals are significantly different from the propagation characteristics of broadcast-band AM signals?

13-8. Briefly describe FM antennas.

13-9. What factors cause FM transmissions to be of higher fidelity than AM transmissions? (Consider only broadcast-band transmissions.)

13-10. Suppose that the center frequency of an FM channel is 101.5 MHz and that a modulating tone of 1 kHz is present. What would be the frequencies at which the first three upper sidebands appear? Would each sideband necessarily be present?

13-11. Suppose that the modulating frequency is gradually increased. What happens to the spacing of the sidebands?

13-12. Do amplitude variations of the modulating signal affect the sideband spacing?

13-13. In FM transmissions, does the center-frequency component of the transmitted waveform remain constant in amplitude? Compare this with an AM system.

13-14. In general, describe the change in the sideband structure of an FM signal if a modulating tone of constant frequency is gradually increased in amplitude.

13-15. In general, describe the change in the sideband structure if the amplitude of the modulating tone is held constant while the frequency is gradually increased.

13-16. What stages of a superheterodyne receiver for FM are different from a superheterodyne receiver for AM?

13-17. What is the function of an AFC circuit?

13-18. Why is an AVC (or AGC) circuit not required for an FM receiver?

13-19. What is the standard IF frequency for FM receivers? How many times greater is this than the standard IF frequency for AM?

13-20. Describe two different block diagrams of AM-FM receivers.

13-21. Why is a low plate voltage provided for a sharp-cutoff pentode operated as a limiter?

13-22. What type of biasing arrangement is typically used for pentode limiters?

13-23. How does the action of the limiter stage of an FM IF amplifier complicate the alignment procedure for the IF amplifier? Can the speaker response be used as an indication of optimum alignment?

13-24. How can a bipolar transistor IF amplifier stage be made to operate as a limiter?

13-25. Describe the operation of a limiting diode. Where is the diode usually connected?

13-26. Describe the representation of sine waves by phasors.

13-27. On a phasor diagram, how are phasors added?

13-28. Two sine waves 64° out of phase are added together. Is the result a sine wave? How does the amplitude of the resultant compare with the originals? Draw a diagram to illustrate.

13-29. Describe the phase relationship between the primary and secondary voltages of a loosely coupled tuned transformer when the input frequency is

(a) Equal to

(b) Higher than

(c) Lower than

the frequency to which the transformer is tuned.

13-30. Practice until you can draw from memory the discriminator circuits of Figures 13-13 and 13-20.

13-31. Explain in detail, by drawing phasor diagrams, how the voltages applied to the diodes of a discriminator vary with the frequency of the input signal.

13-32. Same as Prob. 13-31, but for a ratio detector.

13-33. What is the purpose of the tertiary winding of a ratio detector? What component of a discriminator does it replace? Describe the coupling of the tertiary winding to the primary of the ratio detector transformer.

13-34. Explain why the current flow through the tertiary winding equals the difference in current flow through the diodes.

13-35. When referring to a schematic diagram, how can a discriminator be quickly distinguished from a ratio detector?

13-36. What is the purpose of the stabilizing capacitor of a ratio detector?

13-37. What is the purpose of the compensating resistors of a ratio detector?

13-38. Practice until you can draw from memory the ratio detector circuit of Figure 13-25.

13-39. Briefly describe the principle of operation of a quadrature detector.

13-40. Briefly describe the principle of operation of a phase-locked-loop circuit used as an FM detector.

13-41. How does the combination of preemphasis and deemphasis reduce the noise at high audio frequencies in an FM system?

14
AM-FM RECEIVER ANALYSIS

FM receivers invariably are designed to include an AM section so that the receiver may be switched to either the AM or FM mode. Such receivers are AM-FM receivers. They range in complexity and sophistication from the small battery-operated portable to the larger units that form the heart of a home music system.

Invariably, all but the smaller AM-FM receivers will include the circuitry for receiving FM stereo, which is described in a later chapter. In this chapter we describe the circuitry of nonstereo AM-FM receivers.

14-1 TYPICAL AM-FM FRONT END

The FM front end of a typical AM-FM receiver is shown in Figure 14-1 and the AM converter of the same receiver is shown in Figure 14-2. The AM converter is standard and is exactly like the converter described in Sec. 7-6 and shown in Figure 7-13. Therefore, no additional description should be required at this point. We proceed directly to the FM section.

The FM front end consists of a common-base RF amplifier and a common-base converter. The RF amplifier is a controlled-gain amplifier whose bias is derived, in part, from the FM RF AGC circuit. The oscillator is provided with a varactor diode, which serves as the voltage-controlled capacitor for the automatic frequency control.

FM RF Amplifier

A switched 9.2-V supply (CTP 3) provides power for both the RF amplifier and converter. Supply voltage is applied to the RF amplifier through decoupling network $C17$ and $R57$. The biasing arrangement for $Q1$ is that of base bias with emitter feedback, but the base driving voltage is derived partially from the AGC circuit in order to vary the bias voltage at the base and thereby control the gain of the amplifier. Base resistor $R50$ connects to the positive supply, which is standard, but the AGC line connecting to CTP 4 is a negative source which tends to reduce the

403

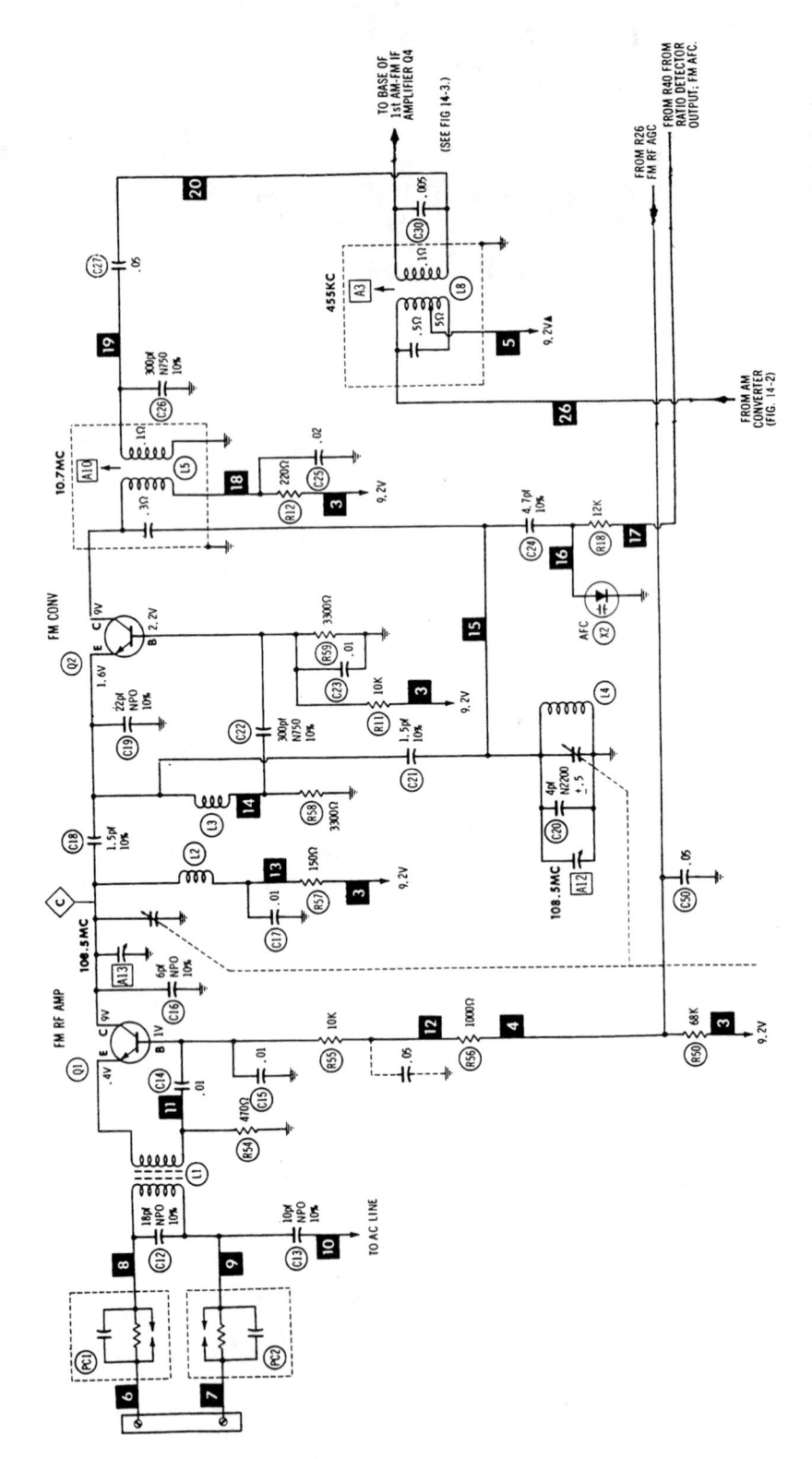

FIGURE 14-1 FM RF amplifier and FM converter section of an AM-FM receiver (a Photofact schematic, courtesy of Howard W. Sams & Co, Inc.)

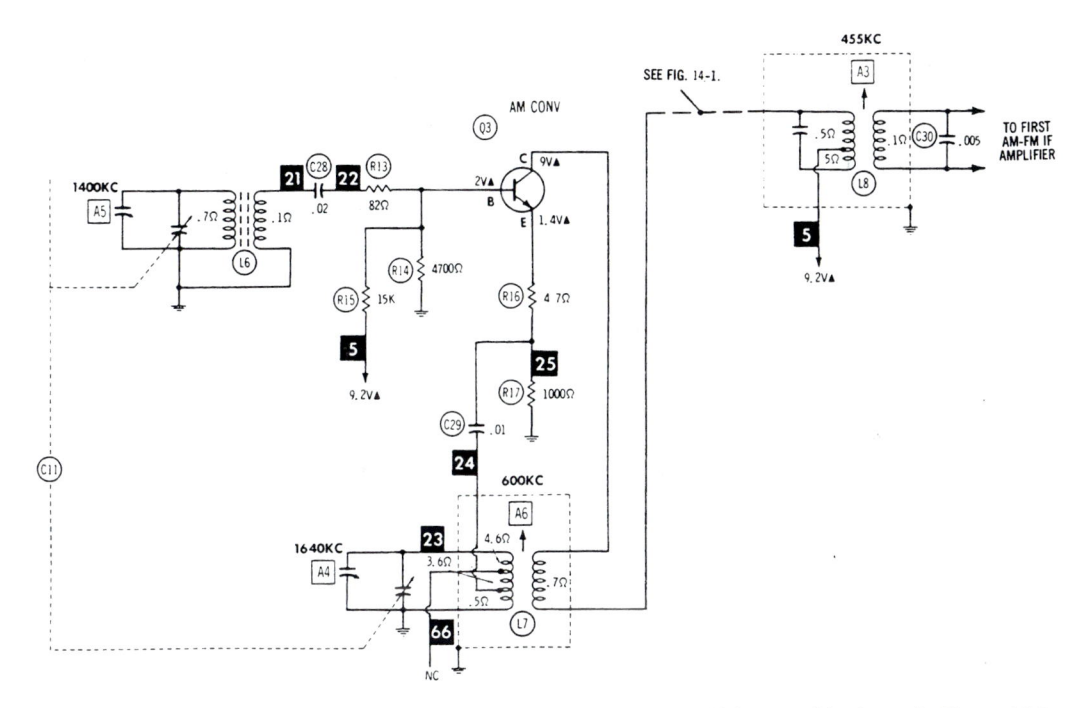

FIGURE 14-2 AM converter section of the AM-FM receiver whose FM front end is shown in Figure 14-1.

net positive voltage applied to the base through $R56$ and $R56$. $C50$ is the AGC filter capacitor. This AGC source is shown in Figure 14-3 near $Q5$ at CTP 34, and will be described in detail in Sec. 14-2. Capacitor $C15$ holds the base of the RF amplifier at signal ground, and $C14$ serves as an emitter bypass capacitor to bypass emitter resistor $R54$ in order to place the low side of the $L1$ secondary at signal ground.

External antenna connections for a 300-Ω twin-lead are provided at CTPs 6 and 7, with PC1 and PC2 providing protection against static charges induced on the antenna during electrical storms. A charged cloud passing overhead can induce large static voltages on the antenna even if no discharge of lightning occurs. It is hoped that such charges will be harmlessly dissipated in the RC network or by the spark gap in parallel. The spark gap is composed of two conductors closely spaced so that arc-over will occur before the voltage becomes great enough to puncture the capacitors. Of course, such a network could not be expected to successfully dissipate a direct hit of lightning on the antenna.

Capacitor $C12$ broadly tunes the $L1$ primary to the FM band. The low input impedance of the RF amplifier loads the tuned circuit, lowering the Q, so that the entire band is passed.

Capacitor $C13$ at CTP 10 provides an alternative FM antenna by coupling the FM input circuit to the AC line. In strong signal areas, an external antenna is probably unnecessary.

Typical AM-FM Front End

Complex Coupling

Complex coupling is used between the RF amplifier output and the input to the converter. A parallel resonant circuit is formed at the RF amplifier collector by $L2$ and $C16$, and $A13$, and one section of ganged $C11$. $C19$, $L3$, and $C22$ form a filter circuit that has a very low impedance to ground at the 10.7-MHz IF frequency, and this filter helps to prevent the converter from oscillating at the IF frequency.

FM Converter

A universal bias arrangement is utilized for the converter transistor, composed of base resistors $R11$ and $R59$ and emitter resistor $R58$. The base is held at signal ground by $C23$. Collector voltage is applied through the primary of $L5$.

The oscillator component of the collector current of $Q2$ passes through the capacitor inside the $L5$ shield and passes via CTP 15 to the oscillator tuning circuit $L4$. The output voltage of resonant circuit $L4$ is coupled through $C21$ to be injected at the emitter of $Q2$. Since no phase inversion occurs between the emitter and collector of a common-base amplifier, the oscillator signal feedback from the collector to the oscillator tuned circuit is of the proper phase to sustain the oscillation.

The AFC network consists of $C24$, *varactor* diode $X2$, and isolating resistor $R18$. The control voltage for the varactor $X2$ is supplied from the ratio detector, and it will be zero when the IF frequency is properly centered, as determined by the oscillator frequency. If the oscillator drifts to a lower frequency, the IF will also decrease as a result, and a negative voltage developed at the ratio detector will be applied to the varactor via $R18$. This increases the width of the depletion layer of the junction and reduces the junction capacitance. Since the varactor capacitance is in parallel with the oscillator tuning circuit, the reduction of the varactor capacitance will raise the oscillator frequency to its proper level.

At this point we mention that the polarity of the ratio detector output can be determined by examining the direction in which the varactor AFC diode is connected. That is, as the FM IF frequency goes up, does the ratio detector output become positive or does it go negative?

It is impossible to determine this from the schematic of the ratio detector because the polarity is controlled by the phasing of the ratio detector transformer windings. Recall, however, that an increase in reverse bias across the varactor reduces the capacitance and tends to raise the oscillator frequency. Then assume the oscillator is "too low," and note the correction voltage that must be applied to the varactor in order to increase the reverse bias. This gives the output polarity of the ratio detector because the correction voltage comes from the ratio detector.

14-2 TYPICAL AM-FM IF SECTION

The IF section of the receiver whose front end is shown in Figures 14-1 and 14-2 is shown in Figure 14-3. In the FM mode, $Q4$, $Q5$, and $Q6$ form a three-stage FM IF amplifier, but in the AM mode only $Q4$ and $Q5$ are used in a two-stage AM IF amplifier.

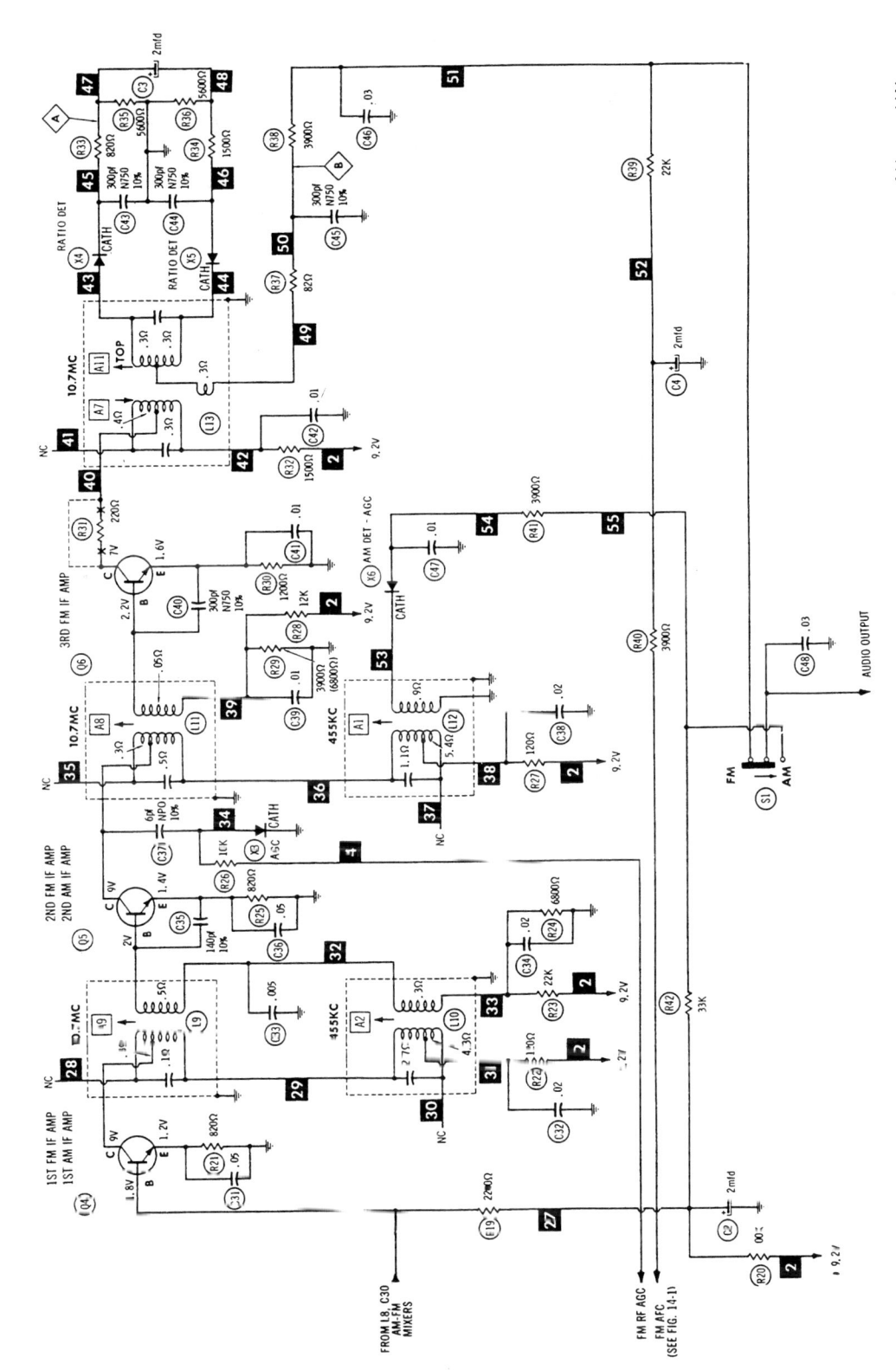

FIGURE 14-3 AM-FM IF section and AM and FM detectors of the receiver of Figures 14-1 and 14-2 (a Photofact schematic, courtesy of Howard W. Sams & Co., Inc.)

The stages $Q4$ and $Q5$ are able to function in either the AM or FM mode because of the great difference in frequency between the AM (455 kHz) and FM (10.7 MHz) intermediate frequencies. Cascaded resonant circuits in the collector circuits of $Q4$ and $Q5$ respond to one frequency but reject the other.

For example, when $Q4$ is operating in the AM mode, the AM IF signal at 455 kHz easily passes through the $L9$ tuned circuit, which is tuned to 10.7 MHz. The $L10$ circuit, however, responds to the 455-kHz signal and couples it to the base of $Q5$, as if the $L9$ secondary were not present. In the FM mode, the $L9$ circuit at the $Q4$ collector responds to the 10.7-MHz FM signal and couples it to the base of $Q5$. The presence of the $L10$ circuit has little effect upon the FM signal. The switched power supply for the AM and FM front ends prevents both signals from appearing at the same time.

An automatic gain control (AGC) voltage is developed by diode $X3$ in the collector circuit of $Q5$ for application to the FM RF amplifier. The AM detector circuit follows $Q5$, and the output of the detector is fed to the function switch. The AM AGC voltage is developed by the detector and is applied to the first AM IF amplifier.

A ratio detector follows the third FM IF amplifier, and the output of the ratio detector is used to form the AFC control voltage. The audio is applied to the function switch and passes on to the audio amplifier in the FM mode. None of the B+ sources for the IF amplifier stages are switched.

First AM-FM IF Amplifier

An unswitched 9.2-V B+ supply powers $Q4$ which serves as the first IF amplifier for both AM and FM functions. Resistor $R20$ at CTP 27 serves as the high-base resistor of a modified universal arrangement. In the AM mode, the AM AGC becomes active and modifies the voltage appearing at CTP 27, which drives the series base resistor $R19$ which limits the base current. In the FM mode, the AM AGC becomes inactive for lack of signal input to the AM detector, and the series combination of $R42$, $R41$, diode $X6$, and the secondary of $L12$ serve as the low-base resistor for the universal arrangement. Note that in this capacity, AM detector $X6$ is forward biased so that DC conduction occurs. The diode, however, plays no significant role other than to develop a "knee voltage" across it since it is forward-biased.

Signal input to $Q4$ is from the AM or FM mixer and is applied directly to the base. In the AM mode, $Q4$ develops the 455-kHz IF signal across $L10$, and it is then applied to the base of $Q5$ via $L9$. $C33$ appears in parallel with the secondary of $L10$ and appears to tune the secondary to 455 kHz.

In the FM mode, the output of $Q4$ is developed across $L9$, which couples the 10.7-MHz signal directly to the base of $Q5$. Capacitor $C33$ returns the IF signal to ground. Capacitor $C35$ appears to form a series resonant circuit with the secondary $L9$. The resonant rise in voltage across the capacitor is applied to the $Q5$ base.

Second IF Amplifier

Transistor $Q5$ serves as the second IF amplifier for both AM and FM. A standard universal bias arrangement is used with the familiar series-connected tuned circuits at the input and output of the stage. The operation of this stage should be obvious, so we proceed to describe the FM RF AGC source, consisting of $C37$, $R26$, and AGC diode $X3$.

To see how the AGC source develops a negative voltage that is proportional to the signal strength at the $Q5$ collector, consider the circuit when the $Q5$ collector goes to a positive peak of one cycle of the 10.7-MHz FM IF frequency. Electrons will be drawn off the top plate of $C37$, and at the same time, electrons will be drawn from ground through $X3$ to the bottom plate of $C37$. When the positive peak at the $Q5$ collector "relaxes," electrons flow back onto the top plate of $C37$, and electrons flow from the bottom plate toward the diode. But the electrons cannot pass through the diode to ground; they are trapped. Their only avenue of escape is through $R26$ onto the AGC line. Thus, CTP 34 appears as a negative source whose output is controlled by the strength of the IF signal at the $Q5$ collector.

The negative voltage of the AGC line is smoothed by $C50$ of Figure 14-1 and is applied to the biasing network of the FM RF amplifier. As the FM IF signal increases in strength, the negative voltage increases, and the bias applied to the RF amplifier is reduced. This, in turn, reduces the transconductance and, therefore, the gain.

AM Detector

The AM detector is straightforward. We should mention, however, that the volume control, which is not shown in the figure, serves as the major portion of the detector load. Resistor $R42$ is the AGC filter resistor and $C2$, an electrolytic, is the AGC filter capacitor.

Third FM IF Amplifier and Ratio Detector

These circuits also are "standard" and have few features worthy of comment at this point. Capacitor $C40$ appears to be series-resonant with $L11$ secondary, and we note that $R31$ is a stabilizing resistor included in the $Q6$ collector circuit of some versions of the receiver.

Inside the shield designation of the ratio detector transformer, note the arrows underneath the adjustment designations $A7$ and $A11$. One points up and the other points down. This signifies that the top slug in the transformer core adjusts the secondary while the bottom slug adjusts the primary. This tidbit is very useful during an alignment procedure.

The ratio detector is a balanced type, very similar to those described earlier. The only feature not previously encountered is the IF signal return path from the tertiary winding to the diodes. In this circuit it is through $R37$ and $C45$ to ground, and then from ground to the junction of capacitors $C43$ and $C44$.

The FM AFC voltage is taken off through $R39$ at CTP 52, and is filtered by $C4$ and $R40$. It then passes to the varactor diode at CTP 16 of Figure 14-1. Note that no capacitors appear in series with the ratio detector output. This must be the case since the AFC voltage is of necessity a DC voltage. The full schematic shows that a DC blocking capacitor follows the volume control, however.

14-3 COMMON AM CONVERTER—FM FIRST IF AMPLIFIER

The receiver of Figure 14-4 utilizes a simple switching network to convert the first FM IF amplifier to an AM converter. Also, it incorporates diode overload protection instead of FM AGC, and the base-collector junction capacitance of a transistor is used as the voltage-variable capacitor in the AFC network for the FM oscillator. At first glance the circuit appears rather complicated, having an abundance of tuned circuits, but we shall see that it is similar in many respects to the receiver of the previous discussion.

FM Front End

The biasing arrangement for the RF amplifier is that of base bias with emitter feedback. The power supply is inverted; the collector is grounded while the emitter is supplied a negative voltage.

Complex coupling is used between the RF output and the converter input, as usual. A new feature, however, is the overload diode connected across the parallel resonant circuit at the RF output. If the signal level becomes sufficiently large to bias the diode into conduction, the resulting conduction will load the resonant circuit and lower the Q. This, in turn, lowers the collector load impedance so that the gain of the stage is reduced. In this capacity, the diode $X2$ is called *RF overload diode*.

A similar diode appears across the 10.7-MHz resonant circuit at the output of the converter consisting of the $L5$ primary and $C33$. The principle of operation is the same, and $X3$ is called *IF overload diode*.

The oscillator tuned circuit is the $L4$ parallel resonant circuit. Feedback to sustain the oscillation is coupled from the collector to the circuit by $C33$. The oscillator signal is injected at the emitter of the converter by $C28$. A 10.7-MHz trap for the IF signal is formed by $L3$, $C27$, and $C29$ to avoid oscillation of the converter stage at the IF frequency.

The AFC function is provided by $Q3$, which is biased by a variation of the universal-bias-with-collector-feedback arrangement. The high-base resistance is $R14$, which connects to the collector side of collector load resistor $R15$ in order to provide collector feedback. The low base resistance is composed of $R13$, which goes to ground, and the series combination of $R12$ and $R11$, which goes to the output of the ratio detector. Note that the base is held at signal ground by $C16$ and $C31$.

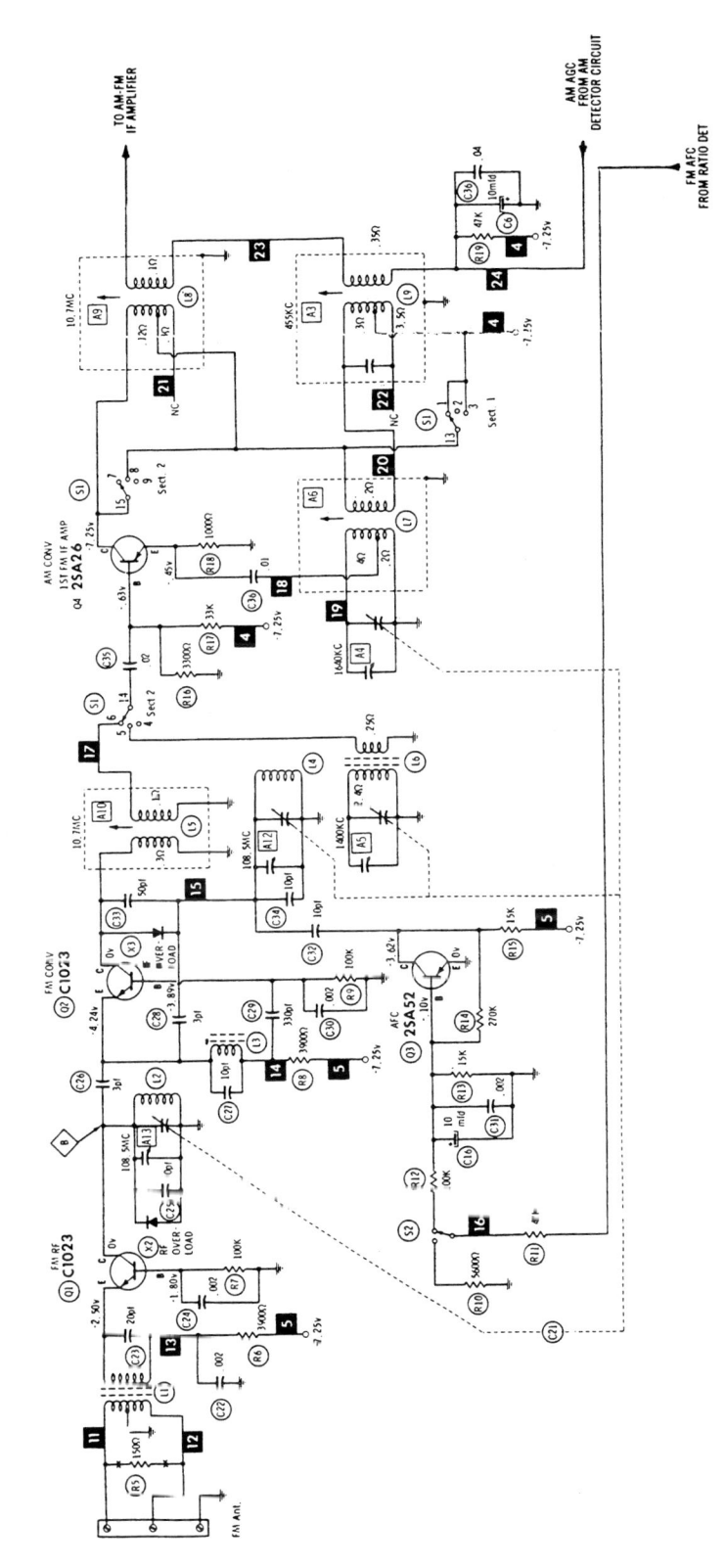

FIGURE 14-4 Front end of an AM-FM receiver in which the first FM IF amplifier serves as the AM converter (a Photofact schematic, courtesy of Howard W. Sams & Co., Inc.)

Observe, now, that the transistor $Q3$ is in parallel with the oscillator resonant circuit, coupled by $C32$. In particular, the base-collector junction capacitance of $Q3$ is in parallel with the resonant circuit. Therefore, if the junction capacitance varies, the frequency of the oscillator will be changed.

The base-collector junction capacitance varies with the voltage difference between the base and the collector, because this voltage determines the width of the depletion layer between the base and the collector region. As the collector voltage rises relative to the base, the capacitance decreases because the depletion layer becomes wider. This decrease causes the oscillator frequency to increase.

The collector voltage is caused to vary by increasing or decreasing the collector current to produce a variation of the voltage drop across the collector load resistor. Thus, large collector currents give low collector voltages and large capacitances. Small collector currents produce small capacitances, by the same reasoning. Hence, by controlling the base voltage, and, in turn, the collector current, the base-collector junction capacitance can be controlled.

Suppose, for example, that the oscillator frequency is too low. The IF will also be below the center frequency, and let us suppose the ratio detector output polarity is such that a low frequency produces a positive output. If this positive output voltage is applied to the base of $Q3$ of the present circuit, the collector current would decrease (note the *PNP* transistor). This, in turn, would decrease the capacitance and tend to bring the oscillator frequency back to its proper value.

Switch $S2$ at the base of $Q3$ is the AFC defeat switch. When disconnected from $R12$, the control voltage is removed from the base of $Q3$. AFC action is then lost. $R10$ then provides the proper load for the ratio detector when the base circuit of $Q3$ is disconnected.

The AFC defeat switch is desirable for the following reasons. First, if an attempt is made to tune a weak station that is rather closely located to a much stronger station, the AFC circuit will pull the receiver away from the weak station to the stronger station. It may be impossible to tune the weak station, or the tuning may shift from one to the other periodically. Second, the AFC action tends to cause the station "to follow" the tuning dial so that accurate tuning may not result. If the AFC is temporarily defeated, the station can be accurately tuned, followed by reactivation of the AFC function.

First FM IF Amplifier

With the function switch in the position shown (Figure 14-4), $Q4$ functions as the first FM IF amplifier. The FM IF signal is coupled from $L5$ through $C35$ to the $Q4$ base. The emitter is essentially at signal ground because the AM oscillator coil $L7$ exhibits a low impedance at 10.7 MHz. Tuned circuit $L8$ is the 10.7-MHz collector load which couples the output of $Q4$ to the next stage. The low end of $L8$ returns to a power supply connection which is a signal ground.

In the AM mode, $Q4$ receives the AM RF signal from the loopstick antenna $L6$ by way of pins 5 and 14 of $S1$. In the collector circuit, $S1$ pins 15 and 8 short out the $L8$ FM transformer. Farther down, $S1$ section 1 pins 13 and 2 remove the short from AM circuits $L7$ and $L9$ placing them in operation. This forms the standard converter circuit.

14-4 COMPLEX COUPLING IN IF SECTIONS

A complex-coupled AM-FM IF section is shown in Figure 14-5. The FM signal is developed across $L7$ by transistor $Q5$, and is coupled by $C45$ to the $L8$ primary. The secondary of $L8$ then couples the signal to $Q6$. This type of coupling offers excellent selectivity. The secondary of $L7$ provides a neutralizing signal which is coupled to the base of $Q5$ by $C43$.

A similar coupling arrangement follows second FM IF amplifier $Q6$. Also, AGC voltage for the FM RF amplifier is developed in the collector circuit of $Q6$ by AGC diodes $X6$ and $X7$, which form a voltage doubler whose output is negative going. Capacitor $C54$ couples a small portion of the IF signal to the diodes. With the exception of the voltage-doubler circuit, the operation of this circuit is identical to that of the AGC circuit in Figure 14-3.

The AM input to $Q6$ also uses the coupling circuit described above, but the AM output of $Q6$ is more conventional. Observe the overload diodes $X3$ and $X5$ in the AM coupling circuit to $Q6$.

Ceramic and Crystal Filters

The IF section of Figure 14-6 uses crystal filters instead of parallel resonant circuits to define the pass band of the IF amplifier and thereby determine the selectivity of the receiver. The ceramic filters $Y201$ and $Y202$ of the FM section have a resonant frequency of 10.7 MHz and will only pass a narrow band of frequencies around this frequency. These units provide excellent selectivity and greatly simplify the FM IF alignment procedure. The ceramic filters have no adjustment, and the resonant frequency does not change with age.

The first AM IF amplifier tuned circuit employs a 455-kHz crystal filter to couple the signal from the mixer to the first AM IF amplifier. The operation of the crystal filter is essentially the same as the ceramic filter, and it provides excellent selectivity.

Since adequate selectivity is provided by the crystal or ceramic filters, an RC-coupling network is employed between IF amplifiers $Q202$ and $Q203$. Thus, the expense of tuned transformers is avoided by substituting less expensive resistors and capacitors. RC-coupled IF amplifiers are not frequently encountered because the RC-coupling network provides no tuning whatever.

14-5 TUNING METER

Many of the more elaborate AM-FM receivers incorporate tuning meters to assist the user in obtaining precise tuning. A wide variety of tuning meter circuits are in use. One such circuit is illustrated in Figure 14-7.

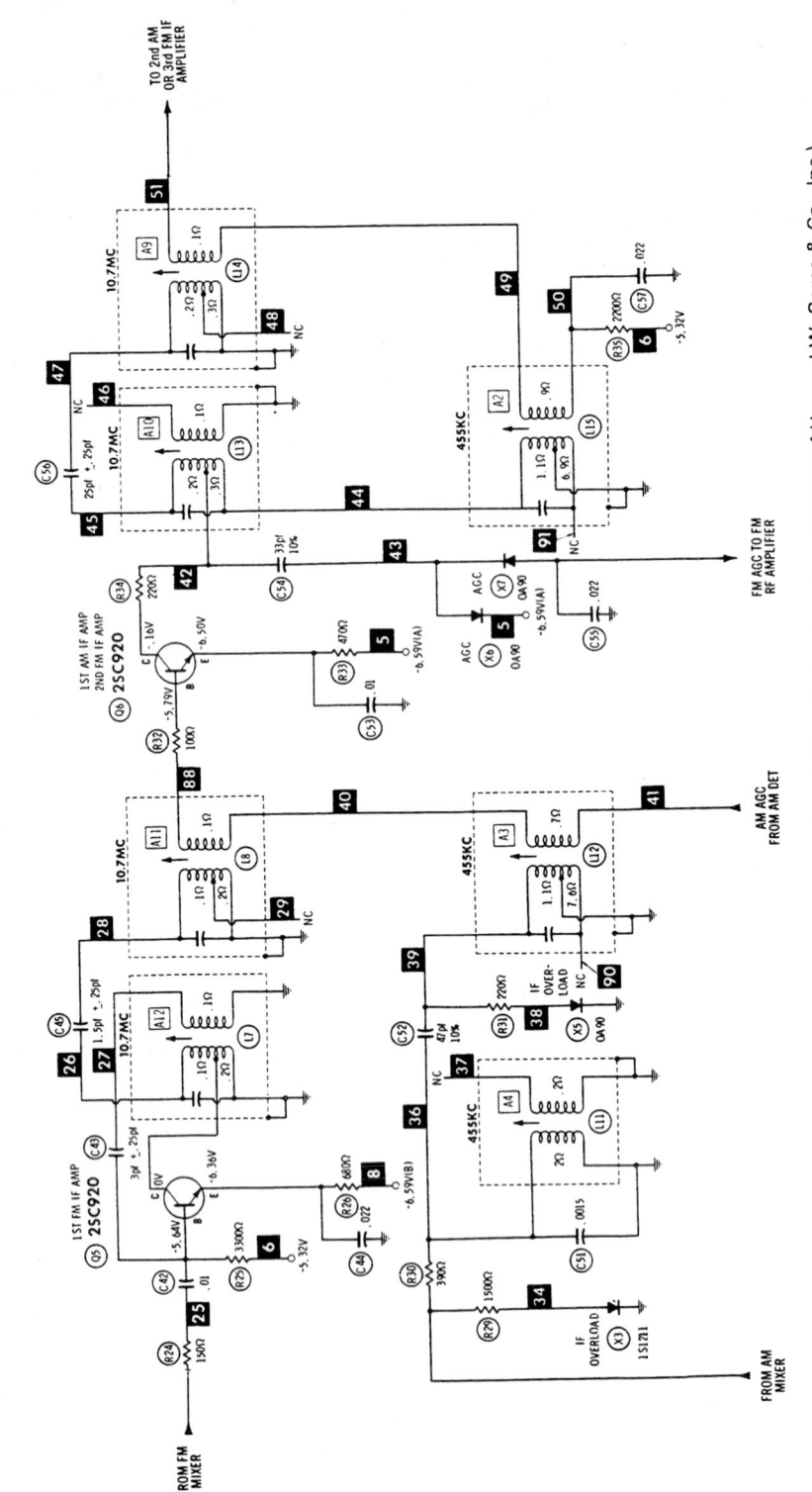

FIGURE 14-5 Complex-coupled IF section of an AM-FM receiver (a Photofact schematic, courtesy of Howard W. Sams & Co., Inc.)

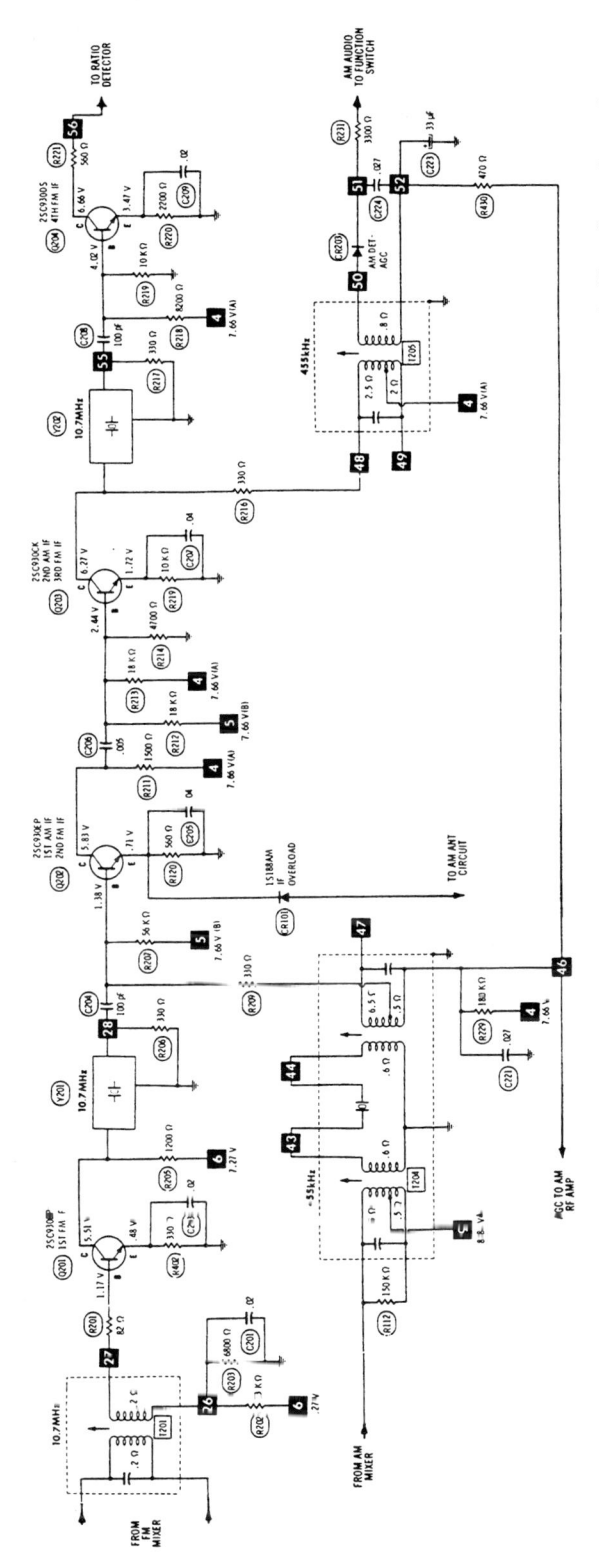

FIGURE 14-6 IF section of an AM-FM receiver in which ceramic and crystal filters define the pass band of the IF section (a Photofact schematic, courtesy of Howard W. Sams & Co., Inc.)

415

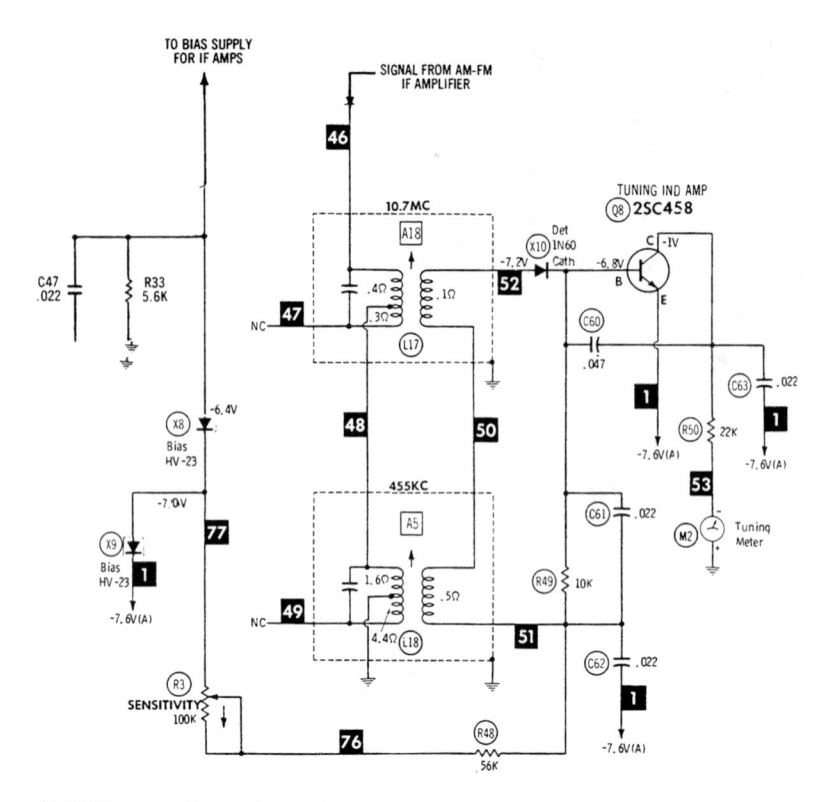

FIGURE 14-7 Circuit for driving the tuning meter of an AM-FM receiver. Many variations of such circuits exist.

The current for driving the meter is provided by the collector current of $Q8$, the tuning indicator amplifier. In brief, the IF signal for either the AM or FM is sampled and rectified in order to produce a DC voltage proportional to the strength of the IF signal. This voltage is then applied to the base of the indicator amplifier, and the base voltage controls the current flowing through the meter. The maximum current will flow through the meter when the input signal is strongest, and therefore tuning is performed to obtain maximum meter indication.

In Figure 14-7 the AM and FM signals are applied to series-connected resonant circuits. Either the AM or FM circuit will respond and apply the sample of the IF signal to the diode $X10$. Resistor $R49$ is the load for the diode, and a DC voltage will be developed across $R49$ in proportion to the strength of the input signal. Capacitor $C61$ smooths the diode output, and $C62$ removes any audio variations that may be present.

A -7.0-V bias voltage is applied to CTP 77, and sensitivity adjustment $R3$ varies the voltage of the $Q8$ base circuit so that with no input signal, the meter indicates zero.

When an input signal is applied to $X10$, it will conduct in proportion to the strength of the input signal and a voltage will be developed across $R49$ of such polarity that the end near CTP 51 becomes negative while the end near the base becomes positive. The voltage across $R49$ increases the forward bias at the base of $Q8$ and causes the collector current of $Q8$ to increase. This causes the meter to deflect upscale.

Capacitors $C60$ and $C63$ help to remove the audio variations from the tuning indication. $C60$ provides negative AC feedback from collector to base, and $C63$ shunts audio signals to ground.

14-6 TWO POWER SUPPLIES FOR AM-FM RECEIVERS

Power supplies found in AM-FM receivers vary in complexity from simple half-wave circuits to full-wave circuits with active filters and zener-regulated sources. In this section we present two power supplies that are typical of what may be encountered.

Unswitched Power Supply

A more complicated half-wave supply is shown in Figure 14-8. Note that no switching is used in changing from the AM to the FM mode. The receiver uses a common AM converter–FM IF amplifier that enables it to use an unswitched power supply.

A similar power supply was described in Sec. 9-7 in Figures 9-12 and 9-14. At that point the suggestion was made that complicated power supplies be redrawn, omitting the coupling capacitors and other nonessential components, in order to determine the direction of electron flow. The present power supply is shown redrawn in Figure 14-9 to show the paths of electron flow. The operation of such a supply is not immediately apparent.

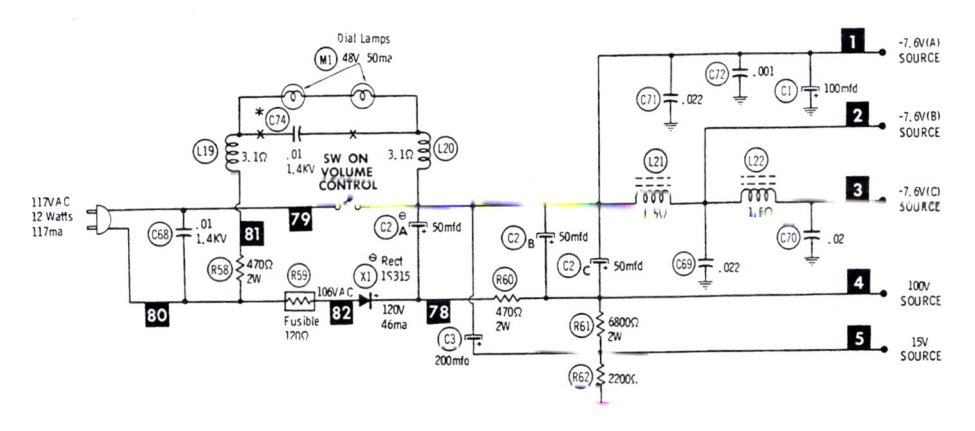

FIGURE 14-8 More complicated power supply that requires a bit of examination to understand. This supply is redrawn in Figure 14-9.

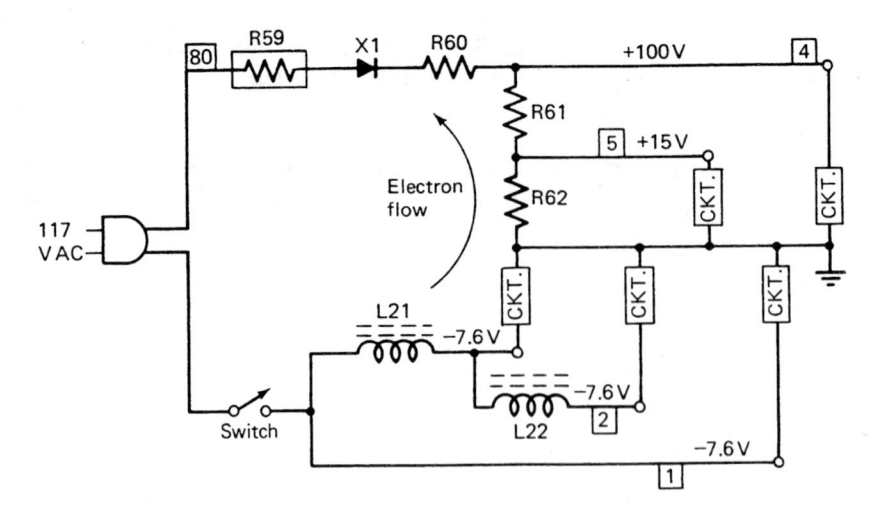

FIGURE 14-9 Essential elements of the power supply of Figure 14-8. The filter capacitors have been omitted. Note that circuit resistances constitute part of the series-connected voltage divider consisting of R60, R61, R62, and circuit resistances connected to the −7.6-V lines.

Full-Wave Supply with Center-Tapped Transformer

Figure 14-10 is a full-wave power supply with both positive and negative sources. It also includes an active filter in which a transistor is used to "amplify" the capacitance of a filter capacitor.

Two sets of rectifiers are connected to the same center-tapped secondary to provide one positive source of 14.74 V and several negative sources. The direction of the diodes determines whether the output will be positive or negative. The positive source at CTP 4 is obvious in its operation and deserves no further comment.

Rectifier diodes $X1a$ and $X1b$ provide the main negative source, which is filtered by electrolytic capacitor C2 and designated CTP 1. This source is not highly filtered.

From that point, a voltage is applied to transistor $Q14$, which functions as an active filter. The transistor is biased almost to saturation by base resistor $R72$. Note the electrolytic capacitor C1 connected to the base of the transistor. Next we discuss the principle of this type of active filter.

When a capacitor is connected to the base of a transistor, as is done in this circuit, the effective value of the capacitor connected to the base is *multiplied* by the *beta* of the transistor. That is, if the active filter were replaced by a single capacitor, the replacement capacitor would have to be beta times as large as the given capacitor in order to provide the same filtering action. It is easy to see that such an active filter can provide extremely large effective capacitances.

The variety of sources connected to the filter output is not unusual until we arrive at diode $X2$ leading to CTP 6. This diode is forward-biased and produces a controlled voltage drop of one "junction voltage" from CTP 5 to CTP 6.

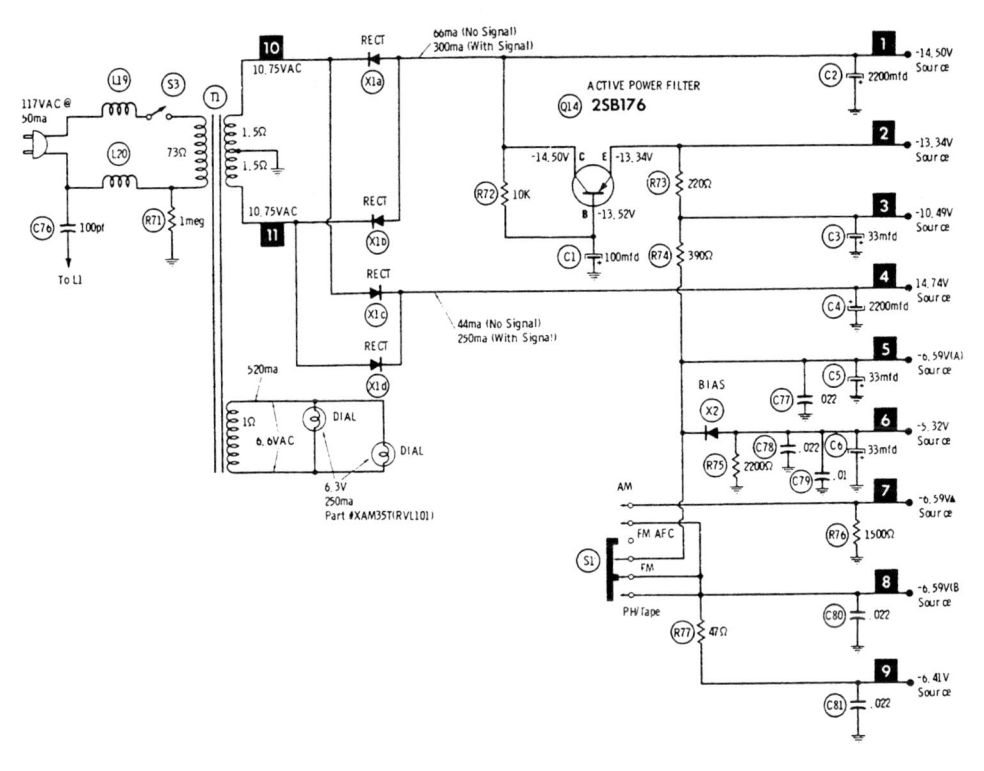

FIGURE 14-10 Power supply that utilizes an active power filter (a Photofact schematic, courtesy of Howard W. Sams & Co., Inc.)

A section of the function switch is also shown on the power supply to depict how certain sources are switched in order to select the various functions of the unit.

14-7 TUBE-TYPE FM FRONT END

In this and the following sections we describe typical circuits found in the older tube-type AM-FM receivers. You may observe that many features of tube and solid-state circuits are similar. A typical tube-type FM front end is shown in Figure 14-11. The AM, IF, and detector circuitry for the same receiver is shown in Figure 14-13. We note the following in a quick overview of the circuit.

The dashed line surrounding the circuit indicates that the circuitry is enclosed in metal shielding. The FM RF amplifier is a grounded grid triode, and a 300-Ω balanced antenna input is provided. A triode is also used for the FM converter whose output frequency is the standard 10.7 MHz. Both triode sections are enclosed within the same glass envelope.

The powdered iron sleeve on the filament line (the insulated filament supply

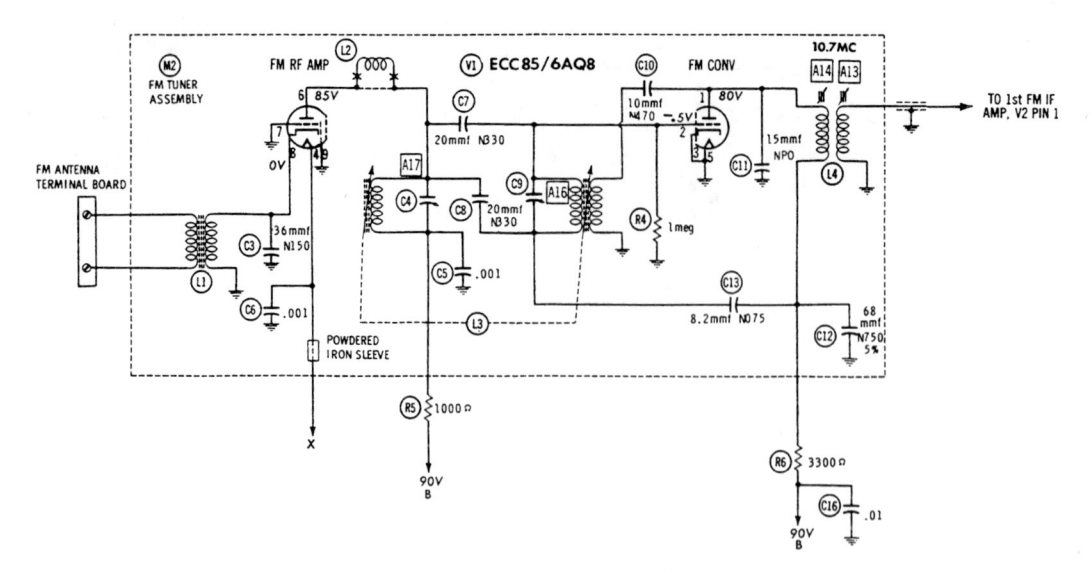

FIGURE 14-11 FM amplifier and FM converter sections of a typical tube-type AM-FM receiver (a Photofact schematic, courtesy of Howard W. Sams & Co., Inc.)

lead passes through the sleeve), acts as an inductor, and with C6 forms a filter circuit to prevent the RF or oscillator signals from being coupled to other stages by the filament line.

The same switched B+ is applied to both stages. The function switch will supply the B+ to the FM front end when the switch is in either the FM or FM stereo position.

FM RF Amplifier

B+ voltage is applied to the plate circuit of the RF amplifier through a decoupling network consisting of R5 and C5. The parallel resonant circuit A17 forms the plate load and is tuned to the antenna frequency. Variable tuning is achieved by varying the inductances of both the RF plate load and the oscillator coil. A mechanical assembly moves powered iron slugs into and out of the coils. Capacitor C4 is a trimmer capacitor that is adjusted during the alignment procedure. In some versions of the receiver, inductor L2 is included to improve the stability of the RF amplifier.

In the cathode circuit, a transformer L1 with a broadly tuned secondary couples the antenna signal to the cathode of the tube. The low input impedance of the grounded grid amplifier makes it impractical to provide variable tuning for the secondary of L1. Capacitor C3 causes the secondary to be very broadly tuned over the entire FM band from 88 to 108 MHz.

The coupling network between the RF amplifier and the converter involves a balanced bridge to avoid coupling the RF signal to the oscillator coil. The bridge is redrawn in Figure 14-12 and note that the oscillator coil appears across the output of the bridge. Thus, the RF voltage applied to the oscillator coil is minimized. Recall from Sec. 7-7 that when an external signal is coupled into an oscillator circuit, the oscillator frequency tends to "pull" toward the external frequency. Note that the grid-cathode capacitance C_{gk} of the converter forms an active part of the bridge.

FM Converter

This type of FM converter is called an *autodyne* converter or, frequently, a *self-oscillating mixer*. Feedback at the oscillator frequency occurs from the plate through $C10$ to the primary of the oscillator coil $A16$. The secondary of the oscillator coil is resonated by the combination $C7$, $C8$, and trimmer $C9$.

Resonant circuit $L4$ is tuned to the IF frequency of 10.7 MHz. Capacitors $C11$ and $C12$ resonate with the primary of $L4$. Bridge neutralization of the converter at the IF frequency is provided by $C11$, $C12$, $C13$, and C_{gp} of the tube in a manner quite similar to the bridge neutralization circuit of Sec. 6-9. Observe that this circuit involves two bridges, one at the input and one at the output, and many circuit components are active in both bridges. An in-depth understanding of such a circuit can only be obtained through a mathematical analysis of each bridge.

Resistor $R6$ and capacitor $C16$ form a decoupling network that prevents the high-frequency signals from entering the power supply and being coupled to the other stages of the receiver.

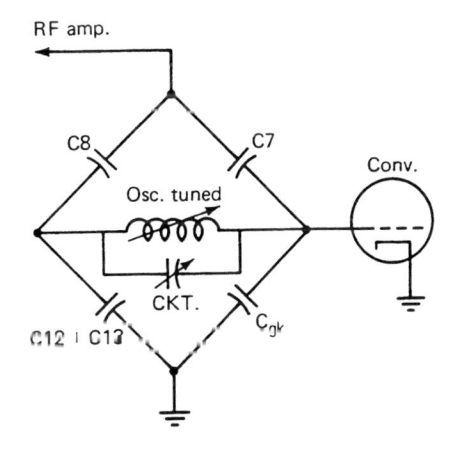

FIGURE 14-12 Bridge circuit formed by the components that couple the signal from the RF amplifier to the converter of Fig. 14-11. The bridge reduces the amount of antenna frequency voltage applied to the oscillator tuned circuit.

14-8 AM FRONT-END AND AM-FM IF SECTIONS

The AM converter and AM-FM IF sections that accompany the preceding FM front end are shown in Figure 14-13. A standard pentagrid converter serves as the AM converter. Vacuum-tube $V4$ serves as an IF amplifier for both the AM and FM modes The AM output of $V4$ is coupled to the AM detector $X1$ by $L10$. The output of the detector then goes via the function switch to the audio section.

Either the AM or FM front end is placed in operation by switching the power supply sources. At the AM converter tube, the half-circle symbol beside the 105-V source indicates that the source is switched. Voltage will be applied to the AM converter only when the function switch is in the AM position.

The output of the AM converter is applied to the grid of $V4$ via the first IF can $L8$ and inductor $A11$. The tuned circuit $L5$ of the FM IF section does not exhibit an appreciable reactance to the 455-kHz AM IF. Therefore, the AM IF signal passes through the inductor portion of the $A11$ resonant circuit with ease. $V4$ serves as the IF amplifier for the AM section.

The 455-kHz AM output signal of $V4$ passes easily through FM resonant circuit $A10$ to the second AM IF transformer $L10$. The output of $L10$ is applied to the AM detector.

The AM detector is typical of those found in AM receivers. Resistor $R22$ and the capacitors inside the shield of the second IF transformer form an IF filter, and $R23$ forms the detector load cross which the audio signal voltage is produced. Capacitor $C29$ couples the audio signal to the function switch and ultimately to the audio amplifier section.

Resistor $R10$ is the AVC (or AGC) filter resistor, and $C19$ is the AVC filter capacitor. In the AM mode, AVC voltage is applied to the control grid of $V4$. In the FM mode, the AVC line is grounded so that only the self-bias developed by $R11$ is effective at $V4$.

All in all, if one can bring himself to ignore the FM circuitry, the AM section of this receiver is rather simple. We admit, however, that the function switching arrangement lends its share of complications.

FM IF and FM Detector

The input to the first FM IF amplifier $V2$ comes from the FM front end of Figure 14-11. This stage is standard in all respects. Its output is coupled through $L5$ to the grid of $V4$, which, in the FM mode, operates at 10.7 MHz as the second FM IF amplifier.

In the FM mode the AM IF tuned circuits have little effect on the 10.7-MHz signal. The capacitors used to tune the 455-kHz resonant circuits are so large that the 10.7-MHz FM IF passes through quite easily. Consequently, we may consider the AM IF resonant circuits as "dead shorts" when considering the FM IF frequency.

With this established the operation of $V4$ in the FM mode is obvious. Its output

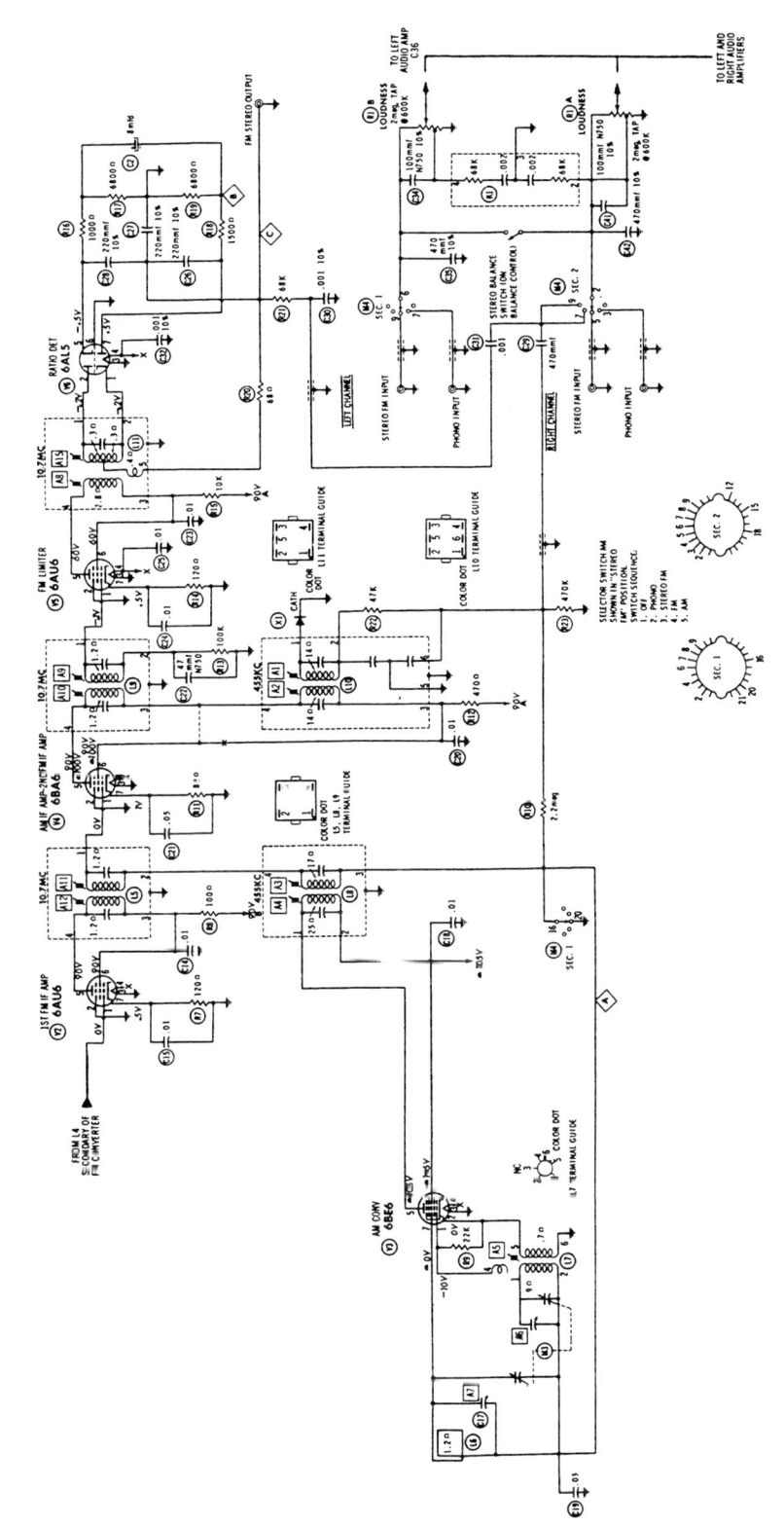

FIGURE 14-13 AM converter, AM-FM IF section, AM detector, and ratio detector of the receiver whose FM RF section is shown in Figure 14-11.

is coupled through $L9$ to limiter stage $V5$. A similar limiter circuit was described in Sec. 13-4. Grid leak biasing is provided by $R13$ and $C22$, while a component of self-bias (cathode bias) is provided by $R14$ and $C24$. The characteristically low plate and screen voltages of the limiter are produced by $R15$ in series with the B+ supply. $C23$ returns RF voltages of the screen and plate circuits to ground. The limiter output is applied to the ratio detector.

Recall that fairly large signal levels appear at the limiter and ratio detector. Because of this, and due to the capacitance between the cathodes and filaments, capacitors, $C25$ and $C32$ are connected from the filament to ground near $V5$ and $V6$. This prevents the large signals from being coupled to other tubes by the filament line.

We note in passing that the B+ voltage to the limiter is not switched off when the function switch is placed in the AM mode. Consequently, a small current flows through the tube, but this is of little consequence.

Ratio Detector

The ratio detector is almost exactly like the balanced ratio detector described in detail in Sec. 13-7. Therefore, our description at this point will be brief.

Electrolytic capacitor $C2$ is the stabilizing capacitor that gives the ratio detector its amplitude-limiting properties. Resistors $R17$ and $R19$ form the major portion of the diode load resistance, but compensation resistors $R16$ and $R18$ are also included. Recall that the compensation resistors are included to ensure that the circuit is not overstabilized by $C2$. Capacitors $C26$ and $C28$ provide a return path for the IF signal voltage applied to the diodes.

Capacitor $C27$ is usually drawn from point (C) to ground, but in any event, it is the capacitor across which the audio voltage is developed. Resistor $R20$ in series with the tertiary winding limits the maximum diode current and prevents an increase in sensitivity to AM at high signal levels. Point (C) represents the output of the ratio detector. $R21$ and $C30$ form a deemphasis network to restore the high-frequency audio components to their proper amplitudes following preemphasis at the transmitter. An output jack labeled "FM stereo output" is connected to point (C) and is intended for use with an "add-on" FM stereo decoder. FM stereo is described at length in Chapters 16 and 18.

Function Switch

The inputs to the stereo audio amplifiers are selected by the function switch $M4$. Since the AM and FM signals are applied only to the right channel section of $M4$, a shorting switch is provided on the balance control to tie the inputs of the two amplifiers together. When a stereo record is played on the changer or when an FM multiplex adapter is used, the switch is left open.

14-9 AM-FM FRONT END COMMON AM CONVERTER— FIRST FM IF AMPLIFIER

A more complicated AM-FM front end is shown in Figure 14-14 and the remainder of the IF circuitry is shown in Figure 14-15. A pentode is used for the FM RF amplifier, and three triodes constructed in the same glass envelope are used for the FM mixer, oscillator, and AFC stages. A complicated switching arrangement is used to allow *V*3 to function as the FM converter and as the first FM IF amplifier.

FM RF Amplifier

The pentode RF amplifier receives the antenna signal through an impedance matching network consisting of *C*2, *L*1, and *C*3, which matches the 300-Ω antenna impedance to the input impedance of the RF amplifier. AGC voltage from the limiter grid leak circuit is applied through *R*1 to the RF amplifier grid.

Complex coupling is used at the output of the RF amplifier to increase selectivity and to match impedance levels between the RF output and mixer input. The input to the grid of the mixer appears as a fairly low impedance to the RF signal.

FM Oscillator and AFC

The FM oscillator uses a triode connected as an Armstrong oscillator. Feedback is by transformer coupling from the plate to the oscillator tuning circuit. Switched B+ is applied through decoupling network *R*8 and *C*15. Grid leak bias is developed for the oscillator by *C*13 and *R*7.

Automatic frequency control (AFC) of the oscillator is provided by reactance tube *V*2B. The plate of the reactance tube is coupled to the tuning circuit of the oscillator through C_{gp} of the oscillator tube and through *C*13. A fixed bias is provided for the reactance tube cathode by the voltage divider formed by *R*5 and the cathode resistor. The grid-plate capacitance of the reactance tube serves as the capacitor that is usually connected between the plate and grid as part of the phase-shift network. The control voltage comes from the FM detector (a discriminator in this case) via *R*4 and *R*5. Capacitors *C*11 and *C*12 provide decoupling, and *R*3 is a part of the phase shift network. Recall that a detailed description of reactance tubes was given in Sec. 7-9.

The DC voltage output of the discriminator depends upon the IF frequency applied to the discriminator. When the IF carrier is at the center frequency of 10.7 MHz, the DC output of the discriminator is zero as shown in Figure 14-16. If the FM oscillator drifts to a higher frequency, the IF carrier frequency will also increase to a frequency greater than 10.7 MHz. From the S-curve of Figure 14-16 it is evident that this will produce a positive voltage at the discriminator output.

This positive voltage is then coupled through a filter network to the control grid of the reactance tube, resulting in an increase of the capacitance exhibited by

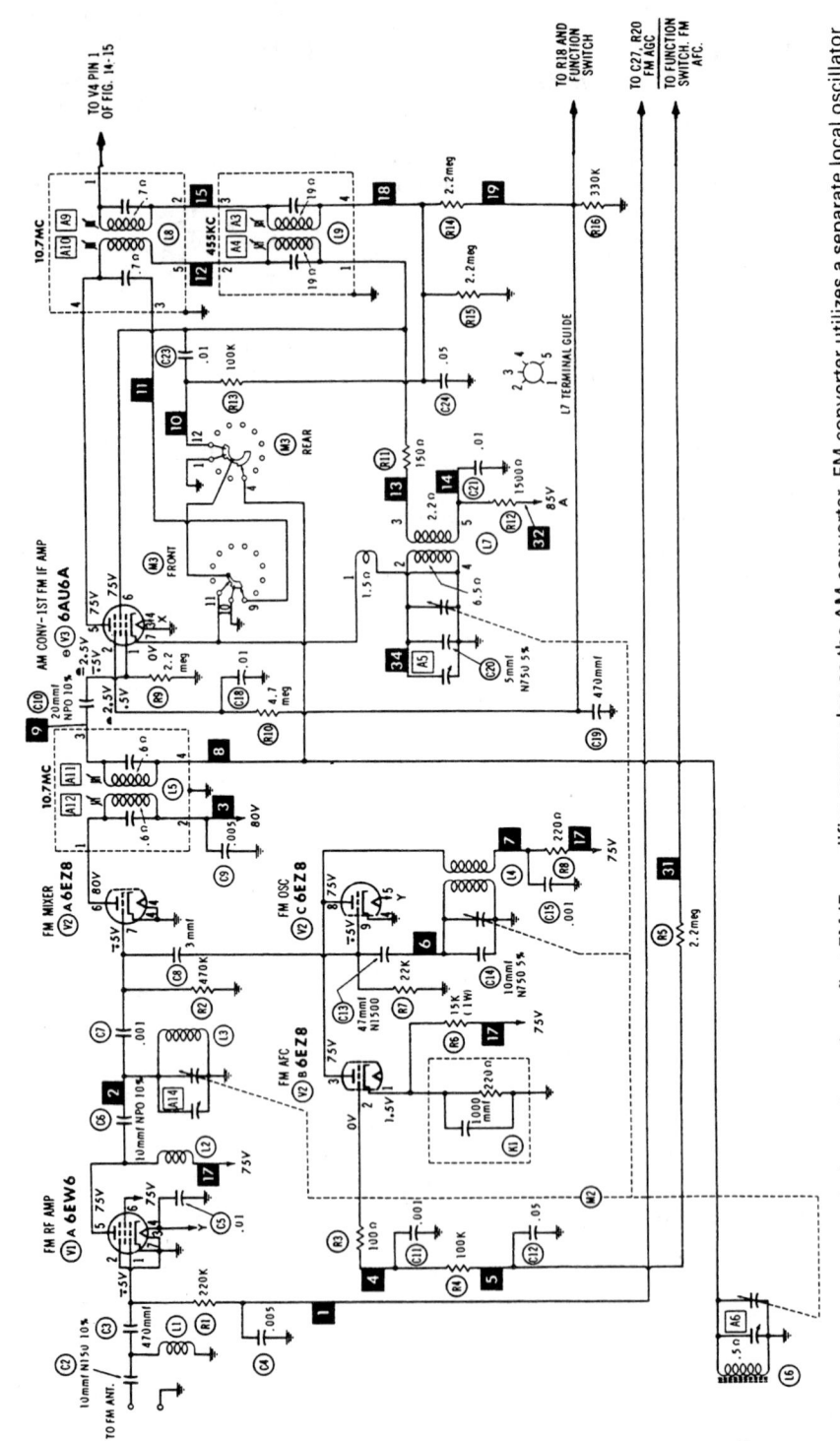

FIGURE 14-14 Front end of a receiver whose first FM IF amplifier serves also as the AM converter. FM converter utilizes a separate local oscillator stage that incorporates an automatic frequency control circuit, *V2B* (a Photofact schematic, courtesy of Howard W. Sams & Co., Inc.).

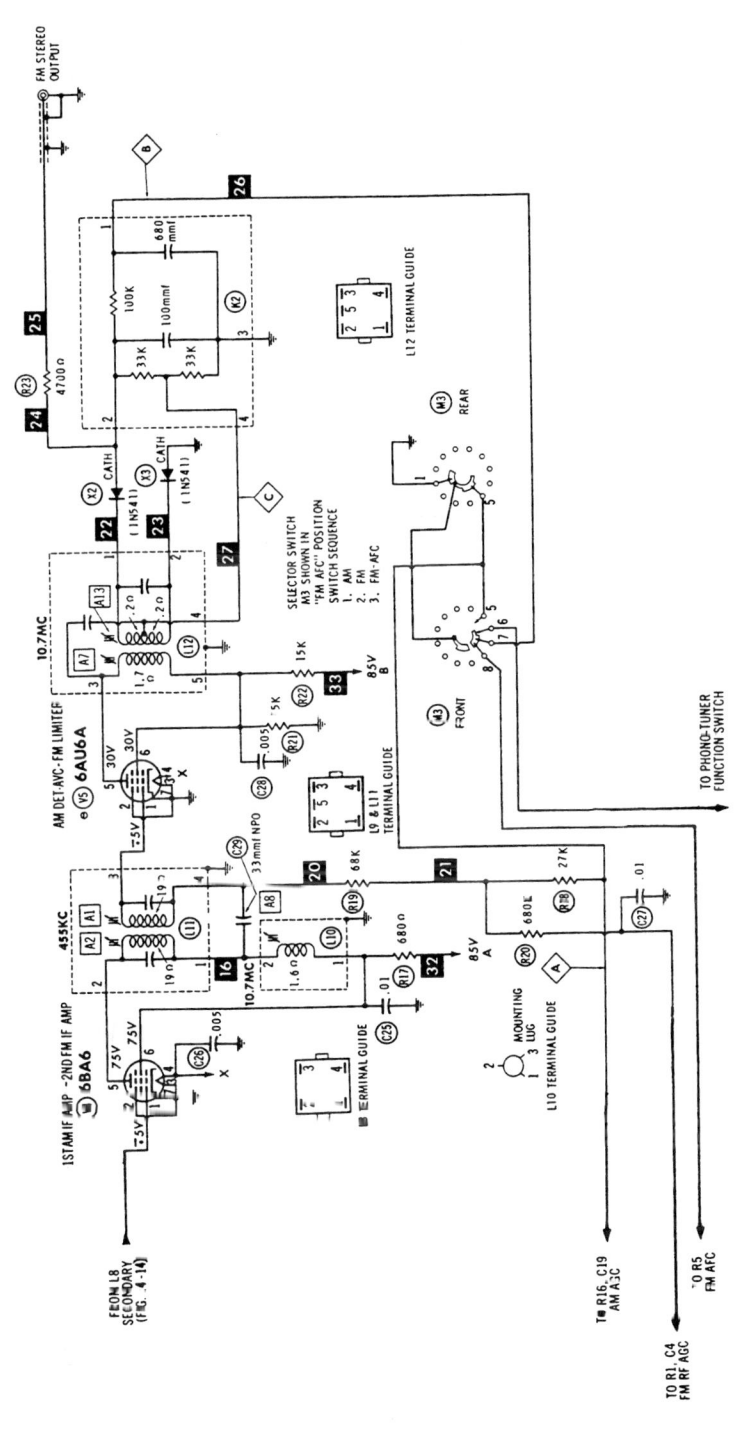

FIGURE 14-15 IF amplifier and detector circuitry of the receiver whose front end was shown in Figure 14-14. Note that the V5 control grid serves as the AM cetector diode.

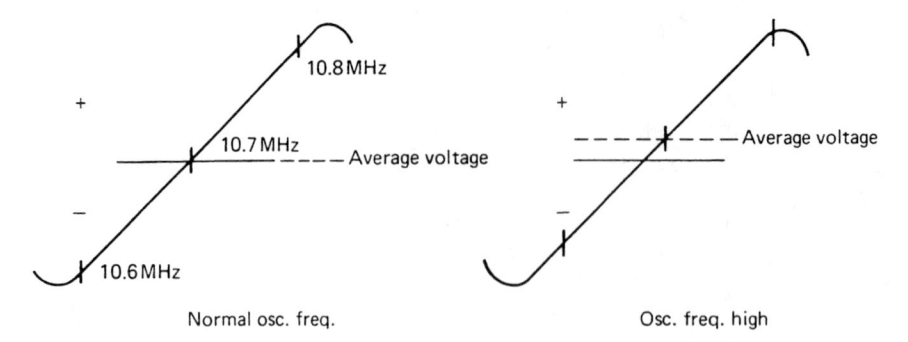

FIGURE 14-16 When the oscillator frequency becomes higher than normal, the FM detector produces a net DC voltage which may be used as a control voltage for the AFC circuit. (This figure refers to the circuit of Figure 14-15.)

the tube. This increase in capacitance, since it is coupled to the oscillator tuning circuit, causes a reduction of the oscillator frequency. This, in turn, brings the IF carrier center frequency back to its proper value of 10.7 MHz.

FM Mixer

The mixer stage is simple. The antenna signal is applied through $C7$, and the oscillator signal is injected through $C8$. The mixer output circuit is tuned to the difference frequency of 10.7 MHz, and the mixer output signal is applied to $V3$, which, in the FM mode, serves as the FM IF amplifier.

AM Converter; First FM IF Amplifier

The AM-to-FM switching arrangement that enables $V3$ to function as the first FM IF amplifier and also as the AM converter makes the circuit configuration difficult to recognize. Therefore, we have redrawn the circuit as it appears in the FM mode in Figure 14-17 and as it appears in the AM mode in Figure 14-18. It is suggested that the reader relate the redrawn circuits back to Figure 14-14, paying particular attention to the switch connections.

In the FM mode, $V3$ receives plate voltage from the unswitched 84-V/AB+ source through decoupling network $R12$ and $C21$, and through $L7$, $R11$, $L9$, and $L8$. The capacitor connected to pin 3 of $L8$ tunes the 10.7-MHz primary of $L8$, and is connected to ground in the FM mode. The cathode is also connected to ground by function switch $M3$ in order to remove the AM oscillator coil $L7$ from the FM circuit. In the grid circuit, CTP 8 is grounded to remove the AM antenna. The AM AGC function is removed by grounding the low end of $R10$ that connects to the suppressor grid of the tube. In order to place the low end of $L9$ at signal ground and to remove the AM AGC function from the FM second IF amplifier, $C23$ and $R13$ are grounded by $M3$ at CTP 10.

After the circuit is redrawn (Figure 14-17), its configuration becomes recognizable if we remember that the AM tuned circuits appear as short circuits to the FM signal. The stage is a pentode RF amplifier and should require no further elaboration.

AM Converter

In the AM mode, all the function switch ground connections are opened so that the antenna $L6$, AM oscillator coil $L7$, and the AM AGC line are placed in the circuit. The 10.7-MHz tuned circuits appear as short circuits to the AM signal, and their presence is of no consequence.

Feedback for the AM oscillator is provided by the primary of $L7$ to the oscillator tuned circuit. The oscillator signal is inductively coupled by the 1.5-Ω winding of $L7$ to the cathode of $V3$. The station signal, received by AM antenna $L6$, is coupled through $L5$ and $C10$ to the grid. In the plate circuit, $L9$ picks out the 455-kHz difference signal and couples it through the $L8$ secondary to the grid of $V4$, the AM IF amplifier (Figure 14-15). AM AGC is applied through $R10$ to the suppressor grid of $V3$, and is applied through $R14$ to the control grid of $V4$, the AM IF amplifier.

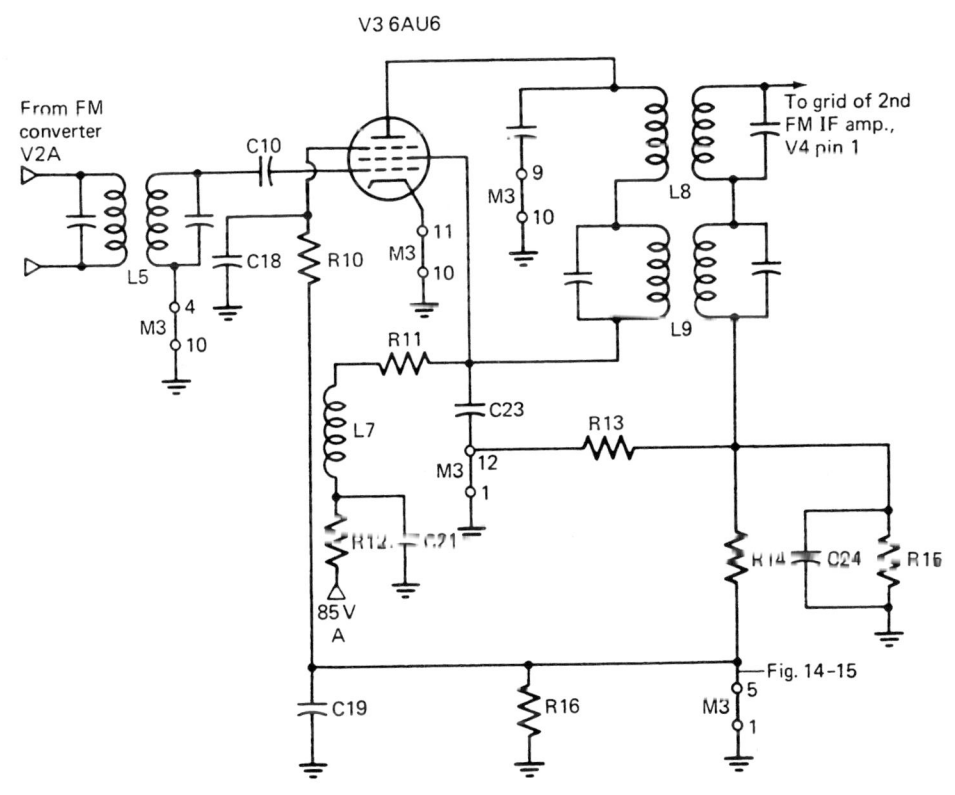

FIGURE 14-17 First FM IF amplifier redrawn from the schematic of Figure 14-14.

Sec. 14-9   *AM-FM Front End Common AM Converter—First FM IF Amplifier*

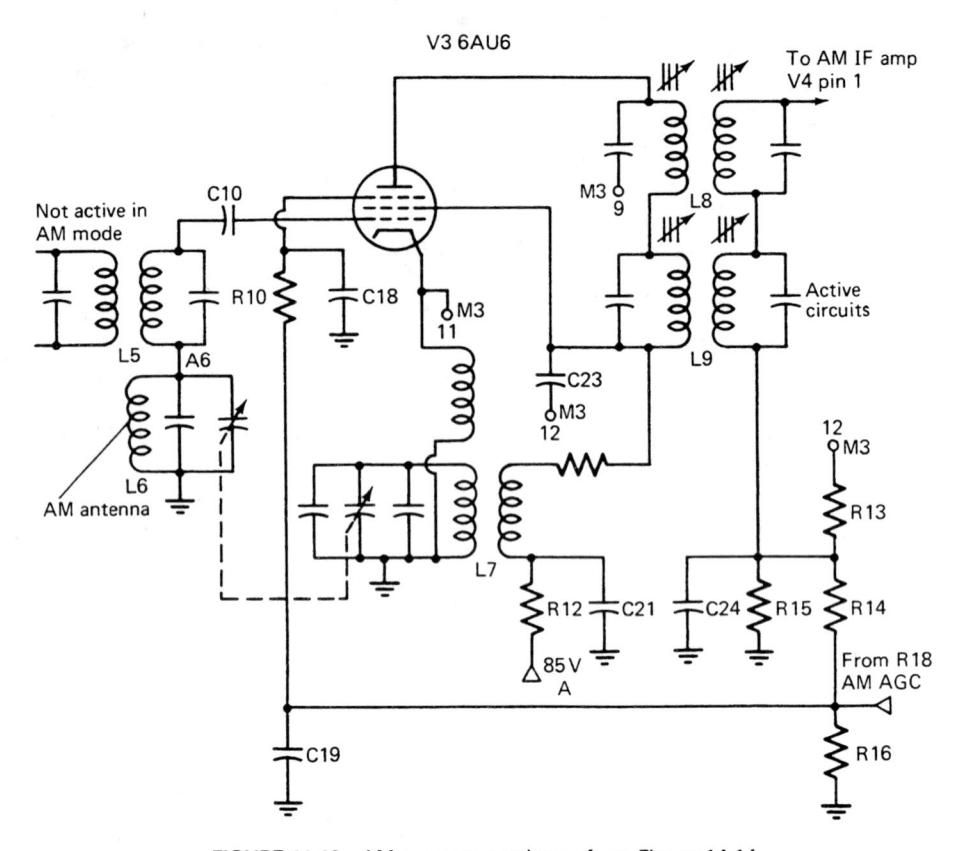

FIGURE 14-18 AM converter redrawn from Figure 14-14.

AM Detector; FM limiter; F-S Discriminator

Figure 14-15 is from the same receiver as Figure 14-14. It contains a common AM-FM IF amplifier stage, a common AM detector-FM limiter stage, and a Foster-Seeley discriminator. A portion of the function switch arrangement is shown below the discriminator. The operation of $V4$ is the same as in the previous FM IF amplifier, and it therefore needs no further discussion.

AM Detector

In the AM mode, the 455-kHz AM signal is coupled through $L11$ to the AM detector circuitry. The grid of $V5$ serves as the plate of the AM detector diode. The detector load consists of $R19$, $R18$, and $R16$ of Figure 14-14. IF filtering action is provided by $C29$ and $C19$ of Figure 14-15. The audio signal is developed at test point $<A>$ and is then passed on to the audio amplifier section via the function switch $M3$. AM AGC voltage is developed by $R14$ and $C24$ for the AM IF amplifier $V4$, and is developed by $R10$ and $C18$ for the AM converter $V3$ (Figure 14-14).

FM Limiter

In the FM mode the 10.7-MHz FM IF signal is complex-coupled from $L10$ through $C29$ to the limiter grid. Grid leak bias is provided by $C29$, $R19$, and $R18$. In the FM mode the low end of $R18$ is connected to ground. FM RF AGC is developed from the grid leak bias voltage by $R20$ and $C4$ (Figure 14-14). Reduced plate voltage is provided for the limiter by voltage divider $R22$ and $R21$, with $C28$ used for decoupling. The grid leak bias and low plate voltage providing limiting action as described in Chap. 13.

Discriminator

The limiter output is coupled to the discriminator as described in Chap. 13. The 100-K resistor and 680-pF capacitor near test point $<B>$ form the deemphasis network. The DC voltage at $<B>$ is used to control the AFC tube in the front end, and the audio component is applied to the audio amplifier through the ever-present function switch that complicates the entire arrangement. A separate output (CTP 25) is available for connecting an FM stereo adaptor for receiving FM stereo transmissions.

14-10 RATIO DETECTOR WITH TUNING INDICATOR

Figure 14-19 shows the IF amplifier and ratio detector of a receiver having a tuning indicator. The IF amplifier circuitry is straightforward, and a standard, balanced ratio detector is used as the FM demodulator. We turn our attention to the tuning indicator.

The DC voltage at pin 7 of the 6AL5 (test point $<B>$) is negative-going and is proportional to the strength of the input signal to the ratio detector. Naturally, this voltage will be greatest when the receiver is exactly on station. Therefore, this voltage is used as the control voltage for the tuning indicator, described in the following.

The tuning indication is provided by a fluorescent screen upon which impinges an electron beam, or electron ray. A *ray control electrode* deflects the electron ray to produce a shadow of varying extent on the screen. Varying the voltage of the ray control electrode alters the extent of the shadow, which serves as the indication. A filter network to remove audio-frequency variations from the control voltage is composed of $R20$, $C29$, $R21$, and $C30$. The control voltage is applied to pin 1, the control grid, of the tuning indicator.

Detailed descriptions of several types of tuning indicators are given in the references. Such indicators are seldom if ever found in solid-state receivers (meters are preferred) because of the high voltages involved. The plate (pin 9), control grid (pin 1), and cathode form a triode amplifier which increases the voltage variation applied to the ray control electrode. $R22$ is the plate load resistor of this amplifier. The grid shown connected to the cathode is actually a beam-forming

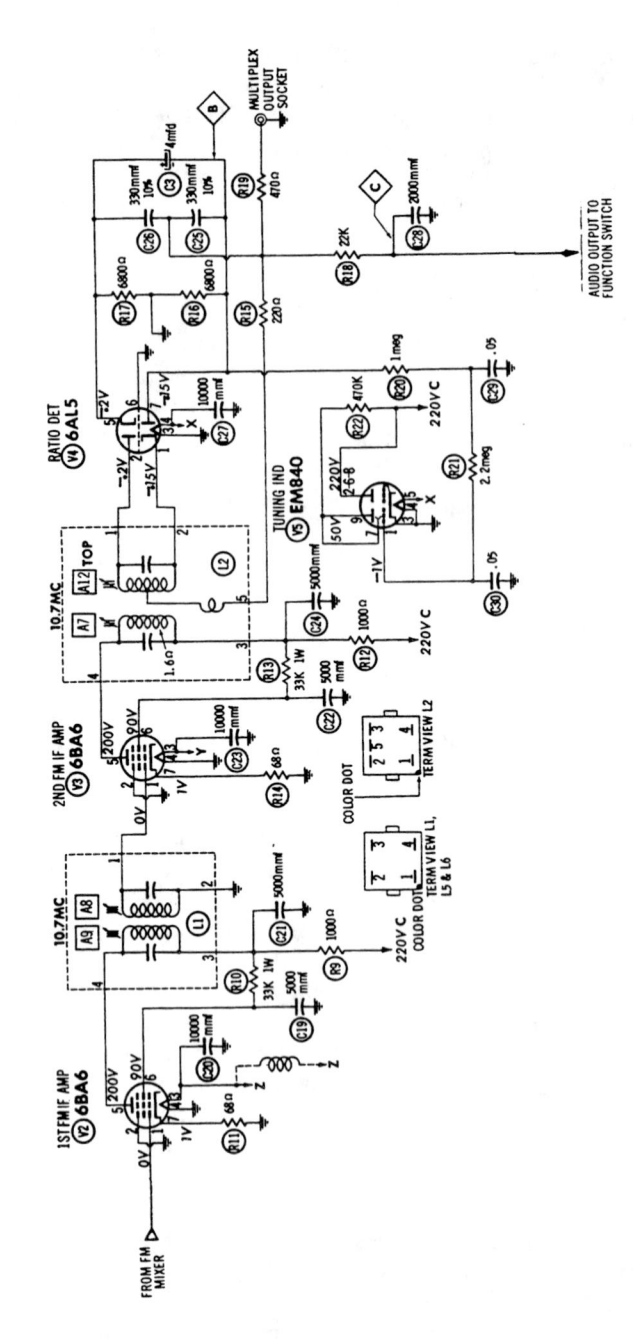

FIGURE 14-19 FM IF amplifier and FM detector of a receiver having a tuning indicator (a Photofact schematic, courtesy of Howard W. Sams & Co., Inc.).

plate that limits the extent of the electron beam. The plate connected to pins 2, 6, and 8 is the fluorescent screen.

SUMMARY

1. FM receivers almost always utilize an RF amplifier. The RF amplifier and converter sections are frequently constructed on a separate subchassis to form the FM tuner.

2. Several stages of an FM receiver may be also used in conjunction with the AM section of the receiver. Common stages may be the power supply, audio amplifier, one or more IF amplifiers, and, sometimes, an FM IF stage that serves as the AM converter.

3. The AM converter circuits typically found in AM-FM receivers are very nearly identical to those of AM only receivers. However, when the first FM IF amplifier is used as the AM converter, the switching function can make the circuit appear rather complicated.

4. Common-base RF amplifiers and FM converters are commonly encountered in the FM front end of AM-FM receivers.

5. FM receivers almost always provide terminals for connecting an external antenna.

6. Bridge circuits are sometimes employed to minimize the amount of antenna signal that is applied to the oscillator tuning circuit, which, otherwise, would cause oscillator pulling.

7. The input circuits to the FM front end may be capacitively coupled to the AC line in order to use the AC as the FM antenna.

8. The FM IF section may involve as many as three stages, IF amplifiers and limiters. The AM IF section seldom involves more than two stages and frequently uses only one stage of IF amplification.

9. In the IF sections, resonant circuits for the AM and FM modes are commonly connected in series. The tuned frequency of the circuits are sufficiently far apart (455 kHz versus 10.7 MHz) that little interaction occurs.

10. The FM section will almost always involve an AFC line running from the FM detector to a voltage-variable reactance in the FM local oscillator. The FM section may also incorporate an FM AGC function.

11. Frequently, a portion of the IF signal is taken off, prior to limiting, for rectification and use as a control voltage for the FM AGC or for a tuning indicator.

12. Overload diodes, connected in parallel with tuned circuits in the IF amplifier or FM front end, are commonly used to prevent overloading of the receiver by very strong signals.

13. Complex coupling networks are frequently encountered in AM-FM receivers.

14. Ceramic or crystal filters may be used in an IF section to define the bandwidth and the selectivity of the receiver. The use of ceramic filters normally eliminates several IF transformers, and untuned amplifiers may be used in conjunction with the filters.

15. More elaborate AM-FM receivers may incorporate a tuning meter. The meter requires additional driving circuitry, typically, a meter amplifier.

16. Older AM-FM receivers may include a stereo amplifier intended for the stereo output of a built-in record changer. Newer receivers having stereo amplifiers almost always have FM stereo reception capability. these receivers are discussed in a later chapter.

17. The FM detector of some AM-FM receivers has a "stereo multiplex" or "MPX" output for attaching a separate stereo decoder for FM stereo. This is now an outdated practice.

18. Switched power supplies are frequently encountered in AM-FM receivers.

19. The function switch that switches between the AM and FM modes lends considerable complication to the schematic of an AM-FM receiver.

20. The filament string of an FM receiver typically contains components to prevent coupling of various stages via the filament line.

21. The control grid of an FM IF amplifier tube may be used as a detector diode in the AM mode.

QUESTIONS AND PROBLEMS

14-1. In regard to the FM RF amplifier $Q1$ of Figure 14-1:

(a) Describe the bias arrangement.

(b) What is the purpose of $C14$ and $C15$?

(c) What is the purpose of $C13$ near CTP 10?

(d) What is the purpose of the spark gaps associated with PC1 and PC2?

14-2 In regard to the FM converter of Figure 14-1:

(a) What biasing arrangement is used?

(b) What is the purpose of $C23$ Of $C20$? Of $C24$?

(c) Would you expect the DC voltage at CTP 16 at $X2$ to be positive or negative relative to ground?

(d) Identify the service adjustments associated with the FM RF amplifier and converter.

14-3. In regard to the AM converter of Figure 14-2:

(a) Identify the biasing arrangement.

(b) What is the purpose of $C28$?

(c) Describe the mechanical construction of $L6$.

(d) How many service adjustments are associated with the converter, not counting $L8$?

14-4. Refer to the IF section of Figure 14-3 for the following:
  (a) Why are 455-kHz AM IF signals able to pass through the tuned primary circuit of $L9$ and $L11$?
  (b) What are the functions of the capacitors $C33$ and $C34$ in the base circuit of $Q5$?
  (c) Calculate the capacitive reactance of $C33$ at 455 kHz and at 10.7 MHz.
  (d) Is the voltage developed at CTP 34 due to $X3$ and an input signal coupled to $X3$ via $C37$ positive or negative?
  (e) What is the purpose of $R33$ and $R34$ in the ratio detector?
  (f) What is the purpose of $C43$ and $C44$ (in conjunction with $R37$ and $C45$)?

14-5. Redraw the AM converter first FM IF amplifier stage ($Q4$) in Figure 14-4 as it appears when operating in the AM mode. Do not include components that function only in the FM mode.

14-6. In regard to Figure 14-4:
  (a) Explain the operation of the AFC circuit ($Q3$).
  (b) As the $Q3$ base voltage is made more positive, does the FM oscillator frequency increase or decrease?
  (c) What is the purpose of $S2$ in the vase circuit of $Q3$?
  (d) Why is it desirable to have an AFC defeat switch?

14-7. Refer to the IF section of Figure 14-5 for the following:
  (a) What is the purpose of the secondary of $L11$?
  (b) What use is made of the $L7$ secondary?
  (c) What use is made of the $L13$ secondary?
  (d) What is the purpose of $C45$ and $C56$ (near the top of the figure)?
  (e) Explain the operation of the FM AGC voltage doubler consisting of $C54$, $X6$, $X7$, and $C55$.

14-8. Refer to Figure 14-6 for the following items:
  (a) Why is the two-stage amplifier consisting of $Q202$ and $Q203$ able to operate at either 455 kHz or 10.7 MHz?
  (b) What is the likely purpose of $R112$ near 455-kHz transformer $T204$?
  (c) Compare the number of service adjustments in Figure 14-6 to the partial IF section of Figure 14-5.

14-9. Explain the operation of the tuning meter circuit of Figure 14-7.

14-10. At first, diodes $X9$ and $X8$ (Figure 14-7) appear to form a voltage doubler, but this is not the case. What is the purpose of these diodes? Are they silicon or germanium?

14-11. Describe the operation of the active power filter of Figure 14-10.

Questions and Problems 435

14-12. Redraw the rectifier circuit of Figure 14-10 and show that diodes $X1a$, $X1b$, $X1c$, and $X1d$ are connected in a standard bridge rectifier configuration.

14-13. What is the purpose of $C76$, $L19$, and $L20$ near the line plug of Figure 14-10?

14-14. In Figure 14-11:

(a) What is the purpose of the powdered iron sleeve?

(b) Why is the antenna input circuit not tuned with a variable component?

(c) Why is $C8$ connected to the low side of resonant circuit $A16$?

(d) What is the name of the FM converter?

(e) Why is $R6$ necessary since the FM converter would probably work as well with 90 V on the plate as with 80?

14-15. In Figure 14-13:

(a) On the schematic, what designates that the power supply for the AM converter is switched?

(b) Why are the tuned secondaries of $L5$ and $L8$ connected in series?

(c) What is the purpose of the $C22$ and $R13$ combination associated with $V5$, the FM limiter?

(d) Why is the voltage at the limiter plate (pin 5 of $V5$) reduced to 60 V by $R15$?

(e) What approximate plate current flows through the limiter?

14-16. What is the purpose of the capacitors $C25$ and $C32$ connected to the filaments of $V5$ and $V6$ of Figure 14-13?

14-17. Give the function of each component of the ratio detector of Figure 14-13. Include $R21$ and $C30$.

14-18. Compare the FM RF amplifier of Figure 14-14 with that of Figure 14-11. Is either amplifier neutralized? Why or why not?

14-19. What is the purpose of $R3$ in the AFC circuit ($V2B$) of Figure 14-14?

14-20. In Figure 14-15, what type of FM detector is used?

14-21. What is the purpose of $R21$ and $R22$ in the plate circuit of $V5$ in Figure 14-15?

15
TROUBLESHOOTING AM-FM RECEIVERS

In this chapter we present troubleshooting procedures for AM-FM receivers. Our emphasis is on the FM sections since procedures given in Chapter 12 are used for AM.

Since many of the symptoms exhibited by the FM sections are caused by the same type of defect that produces a similar symptom in AM, with which you are now familiar, we encourage you to read each symptom, refer to the appropriate schematic, and then try to anticipate what procedure should be used and what are the most likely causes of the symptom. This approach will make this chapter much more interesting.

15-1 INTRODUCTION

The addition of FM stages makes a receiver appear more complex, but the same basic troubleshooting procedures developed in previous chapters are applicable. The AM, audio, and power supply sections are treated using the procedures developed in Chapter 12, and many of the same procedures are applicable to the FM sections if certain precautions are observed.

Signal tracing in the FM IF section, which typically operates at 10.7 MHz, requires the use of a scope with a low-capacitance probe. The low-capacitance probe minimizes circuit loading. A signal tracer will respond only to the AM noise content and not to the FM signal. Hence, signal tracers are of little use.

Signal injection must be done carefully to avoid loading the tuned circuits with the capacitance of the generator leads and the stray capacitance associated with the DC blocking capacitor commonly placed in series with the high side of the generator. Signals can be injected into the front-end stages by placing the generator lead close to the injection point without making physical contact with the circuit. If it is necessary to physically connect the generator lead to the circuit, a 100-kΩ resistor should be placed in series with the generator "hot" lead at the end of the lead next to the receiver. This resistance reduces the effect of generator lead capacitance upon the resonant frequency of the tuned circuits.

Signal injection procedures can often be performed successfully in an FM receiver by utilizing an AM signal generator. This results from two factors. Most AM signal generators produce a small amount of FM in the course of amplitude

437

modulation, and most FM receivers do not have perfect limiting action. An AM waveform will produce a small response at the FM detector. This combination of imperfections usually results in a usable tone being produced at the speaker.

If the available AM generator does not go as high in frequency as 88 MHz, the bottom of the FM broadcast band, it may be possible to use the second or third harmonic of the generator to inject a signal into the front end of an FM receiver. If the generator only goes to 30 MHz, for example, the third harmonic of 30 MHz is 90 MHz, and will probably be usable.

Since the AM, audio, and power supply section can be checked by procedures already described, only one comment will be made regarding these stages. The use of the same stage in an AM-FM receiver for both AM and FM functions makes it unnecessary to check the stage if the stage operates normally in either the AM or the FM mode. Hence, the AM-FM function switch can be used to eliminate certain "common stages" as being defective. For example, if the FM functions normally, we know immediately that the power supply and audio output stages are working. Figure 15-1 is a list of symptoms relating to the AM, audio, and power supply sections.

The FM and common stages may cause various symptoms as shown in Figure 15-2. Procedures for troubleshooting each of these symptoms are described in the following sections of this chapter. Examples are given to illustrate the application of the procedures, and for each example, the AM IF amplifier and the first and second FM IF amplifiers are considered to be the common stages. For receivers with different organizations, the possible causes for some symptoms will be different. We trust these differences will be obvious to the reader.

The procedures and examples given are for solid-state sets. When a tube set is encountered, the first step is to check or replace suspected tubes; then follow the

Symptom	Probable cause or defective section
1. Dead receiver.	Power supply, audio output stage, speaker.
2. FM normal; no AM.	AM converter; AM detector; power supply.
3. Distorted audio.	Audio amplifiers; power supply.
4. 60-Hz hum.	Power supply filters; leakage on B+ line.
5. FM normal, insufficient AM sensitivity.	AM converter; AM IF alignment.
6. FM normal, oscillations in AM mode.	AM converter, AM IF section misaligned.
7. Motorboating.	Open decoupling capacitor (possibly filter capacitor in power supply).
8. FM normal, AM intermittent.	Mechanical or thermal intermittent in the AM sections.

FIGURE 15-1 List of symptoms relating to the AM, audio, and power supply sections of an AM-FM receiver.

Symptom	Probable cause or defective section
1. AM normal, no FM.	FM front end; FM IF stage; FM detector; power supply for FM sections.
2. No AM or FM, pop or hum from speaker.	Common IF stages; power supply; audio amplifiers.
3. AM normal, insufficient FM sensitivity.	FM front end; FM IF section; misalignment.
4. Insufficient AM and FM sensitivity.	Common IF amplifier.
5. AM normal, oscillation on FM.	FM front end; FM IF misalignment; neutralizing capacitors.
6. AM normal, FM distorted.	Misalignment, FM detector.
7. AM normal, noisy FM.	FM front end; FM IF stage.

FIGURE 15-2 List of symptoms relating to the FM and common stages of an AM-FM receiver.

same procedure as given in the following, making the necessary adaptation to vacuum-tube circuitry.

The examples are related to a single receiver because of the size and complexity of typical schematics of AM FM receivers. The schematic has been broken up into sections and is presented in the following pages as Figures 15-3 to 15-6.

15-2 AM NORMAL; NO FM

A receiver that works normally in the AM mode but is completely inoperative on FM usually has a defect in one of the FM stages or the FM B+ power supply. The FM stages that could cause this problem are the FM RF amplifier, the FM converter, the third FM IF amplifier and the FM detector, usually a ratio detector. The FM IF amplifiers that also serve in the AM mode are assumed normal since the AM section functions normally. If lots of hiss is present at the speaker, the FM front end is the most likely cause. On the other hand, if little hiss is present, the third FM IF amplifier or the FM detector is the most likely cause.

In most receivers the physical layout of the FM front end makes it difficult to gain access to the FM RF amplifier, mixer, and oscillator. Metal shielding usually surrounds the circuits that operate at or above the FM broadcast band, but limited access may be obtained by removing parts of the shielding compartment. Great care should be taken not to move or deform the coils around the front end. The resonant frequency of the tuned circuits may be altered so that the receiver fails to work properly even after the original defect is located and repaired. Be careful!

The checkout procedure given assumes that the level of hiss from the speaker is insufficient to draw a definite conclusion about the status of the IF stages. The

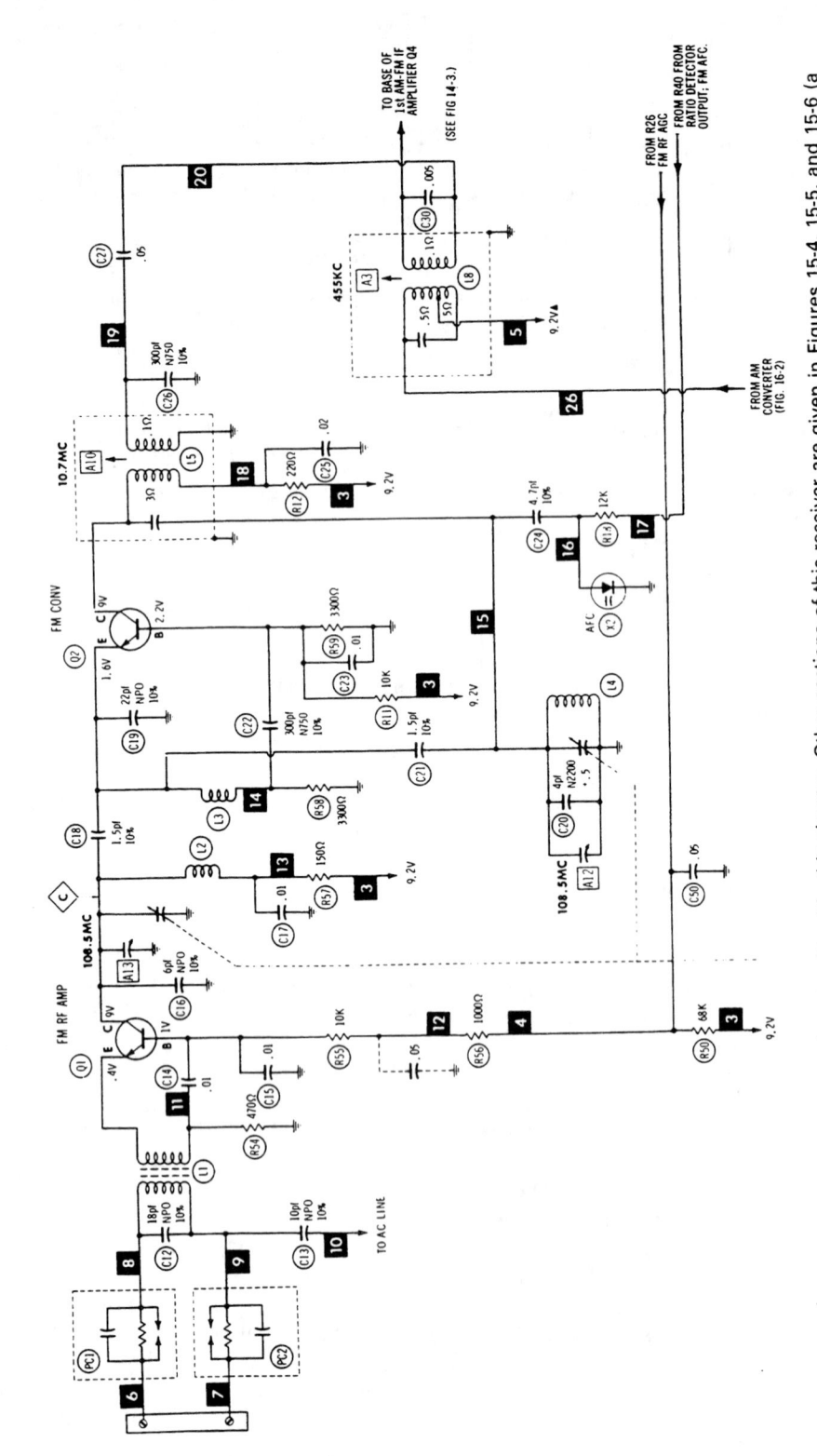

FIGURE 15-3 FM front end of the receiver described in this chapter. Other sections of this receiver are given in Figures 15-4, 15-5, and 15-6 (a Photofact schematic, courtesy of Howard W. Sams & Co., Inc.).

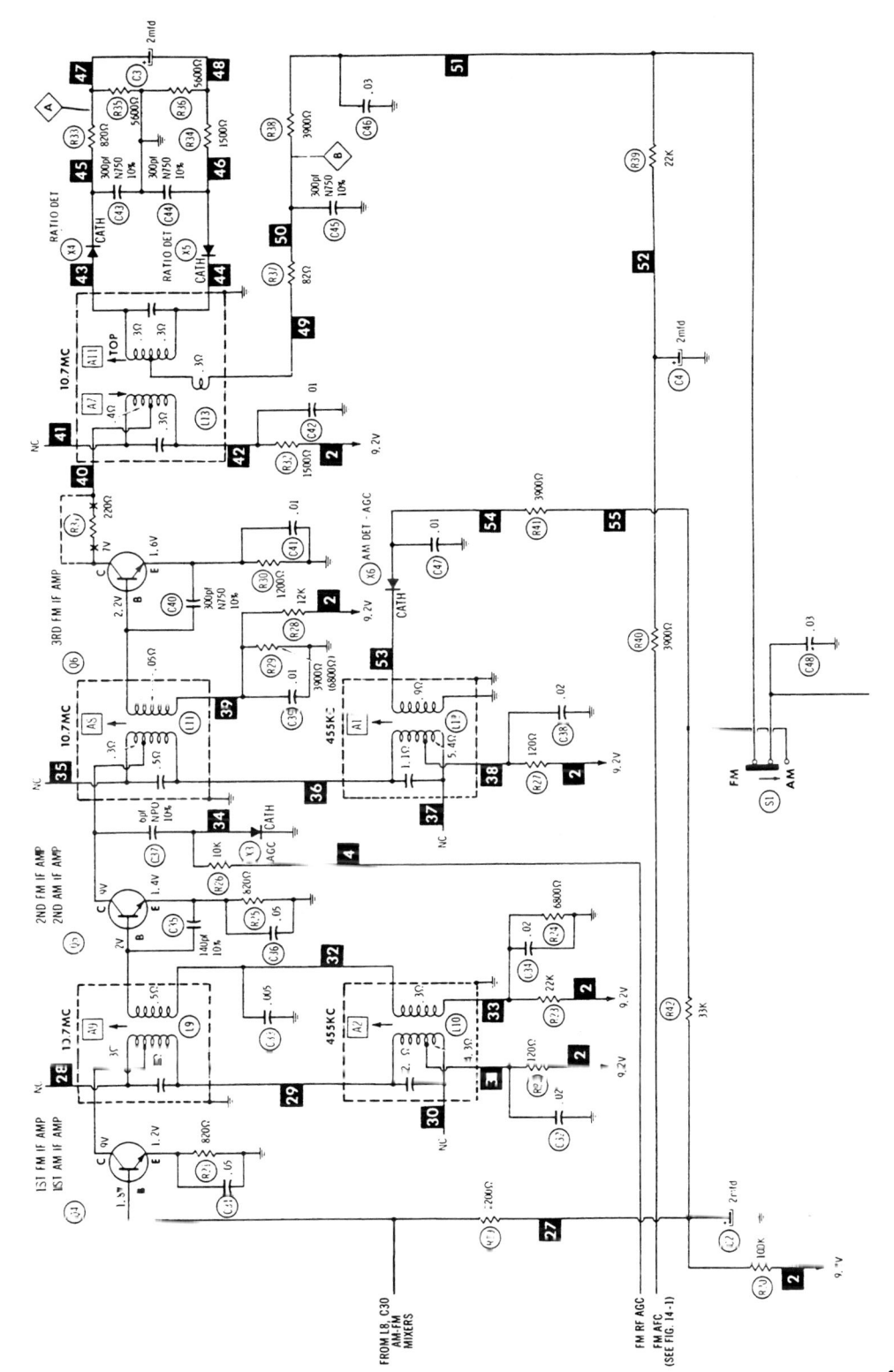

FIGURE 15-4 IF amplifier and detector sections.

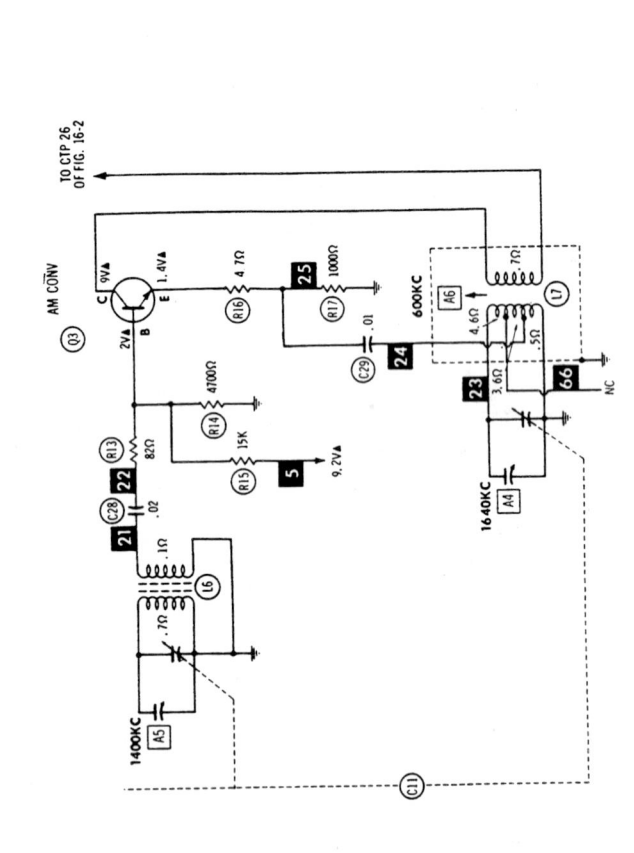

FIGURE 15-5 AM converter and notations given on the schematic.

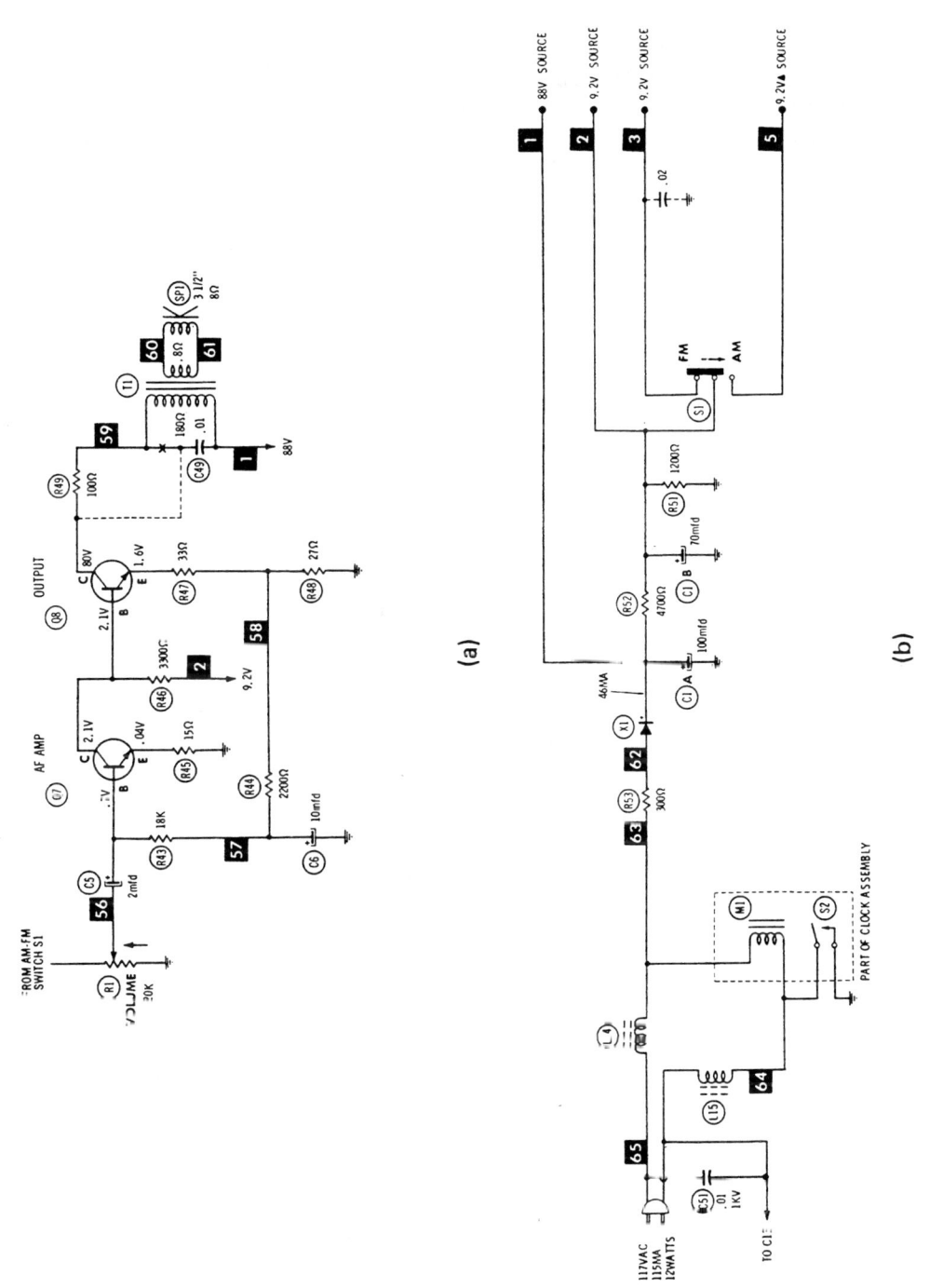

FIGURE 15-6 (a) Audio section and (b) power supply of the receiver whose other sections were shown Figures 15-3, 15-4, and 15-5.

FM B+ voltage is checked first at the FM front end. This B+ source is usually switched by the function switch, and the absence of FM B+ could be due to a defective switch. If the B+ checks normal, the third FM IF amplifier and FM detector are checked by signal injection or by signal tracing with a scope. If the third FM IF amplifier checks out normally, the front end is checked by signal injection to locate the defective stage. V-R analysis or part substitution is then used to determine the defective component.

EXAMPLE 15-1: To illustrate this procedure, we refer to the receiver of Figure 15-3, assuming that the AM functions normally, but that the FM produces only a hiss from the speaker. The chassis has been pulled, and the B+ voltage at CTP 3 (the B+ supply for the FM front end) is present when the function switch is in the FM position. A scope is used to check for a signal at test point <B> (at CTP 50 near the ratio detector of Figure 15-4), and no output is observed from the ratio detector. The scope is moved to the collector of the third FM IF amplifier Q6, and no signal is observed. The scope is moved to the base of Q6, and a large noise signal is present. Tuning the receiver to a known station reveals an FM signal at the base of Q6.

The DC voltage levels around Q6 are measured and found to be:

Q6	Measured	Normal
C	4.5	7
B	2.3	2.2
E	3.8	1.6

The increased emitter voltage, the lack of B-E forward bias, and the reduced collector voltage indicate that Q6 has excessive C-E leakage. The transistor is removed, checked, and found to be defective as suspected. Installation of a new Q6 restores normal operation of the receiver.

The FM sensitivity, selectivity, tracking, and clarity should be examined carefully whenever a component is changed in the FM IF amplifier or FM front end. If the reception is not normal in all aspects, an FM alignment is necessary. FM alignment procedures are described at the end of the chapter.

EXAMPLE 15-2: As a variation of the preceding example, we now assume that no signal is found at the base of Q6, and we continue the isolation procedure. An AM signal generator is used to inject a 10.7-MHz AM signal at CTP 19 (FM converter output, Figure 15-3), and a 10-kΩ resistor is attached to the generator output lead to minimize circuit loading. A tone is heard from the speaker which indicates that the FM IF amplifier is working. The generator is moved to test point <C> (output of RF amplifier) and, again, a tone is heard from the speaker. No gain is noticed, but the input to the converter Q2 is not tuned to 10.7 MHz, so this is normal.

With the generator still applied to test point <C>, the generator frequency and the receiver are both tuned to 90 MHz. No tone is heard at the speaker, so the receiver tuning is rocked back and forth around 90 MHz. The tone is still not heard, and since the 90-MHz signal does not pass through the converter, we conclude that the oscillator portion of the converter is not working.

The DC voltage levels around converter transistor $Q2$ are measured and found to be:

$Q2$	Measured	Normal
C	9	9
B	2.3	2.2
E	3	1.6

These voltages indicate collector–emitter leakage in the transistor. It is removed and checked to confirm the suspicion. Replacement of $Q2$ with a new unit restores normal operation. Tracking, sensitivity, selectivity, and clarity are normal, so no alignment is necessary.

15-3 NO AM OR FM; TURN-ON PLOP (OR HUM) PRESENT

The characteristic "plop" of the speaker at turn-on, or a hum from a tube-type receiver, indicates that the B+ and audio output stage are functioning normally. For both the AM and the FM to be inoperative, a common IF stage, audio amplifier stage, or a common B+ source point must be defective.

The checkout procedure for this symptom is as follows. The B+ voltage for the audio and common IF stages is checked first. If normal, signal injection of signal tracing is used to check the audio amplifier stages. Then the common IF stages are checked for a defective stage. When the defective stage is located, a V-R analysis or parts substitution is used to locate the defective component. The common IF amplifier stages can be checked either in the AM or FM position.

EXAMPLE 15-3: We now assume that the receiver of Figure 15-3 exhibits this symptom. The receiver is placed in the AM mode, and the B+ voltage at CTP 2 (power supply source point for the IF amplifier stage) is normal. An audio signal injected at the output of the AM detector (CTP 55 of Figure 15-4) by touching it with a small metal screwdriver produces a low 60-Hz hum at the speaker. This indicates that the audio amplifiers are working.

A scope is connected to the collector of the second IF amplifier $Q5$, and no signal is observed. The scope is moved to the base of $Q5$, and a noise signal is present. Tuning the receiver to a known station produces an AM signal at the base of $Q5$. The DC voltage levels are measured around $Q5$ and found to be:

$Q5$	Measured	Normal
C	0	9
B	1.7	2
E	1	1.4

In light of the lack of collector voltage, there must be an open in the collector circuit. The ohmmeter is connected across the IF transformer primary (between CTP 35 and CTP 36), and a very large resistance is indicated. The large resistance is also obtained when measured from the $Q5$ collector to CTP 36. The $L11$ primary is apparently open, so the transformer is removed from the circuit.

The shield is carefully removed from around the transformer windings, and a visual inspection reveals a broken wire on the terminal that connected to CTP 36. A careful repair is made, and an ohmmeter reveals normal resistances for the various portions of the coils. The shield is replaced, and the unit is soldered back into the circuit. Normal operation is restored.

15-4 AM NORMAL; INSUFFICIENT FM SENSITIVITY

A receiver that functions normally in the AM position but which receives only strong, local FM stations most likely has a defect in the FM IF or FM front end. In most cases, the defect will be in the FM RF amplifier.

It should also be noted that in rare cases a defect in a common AM-FM stage can produce this symptom. The common stage may operate normally at the AM IF frequency, yet fail to operate at the higher FM IF frequency. Since this possibility is unlikely, the common IF stages are not initially considered as a possible cause. However, the overall sensitivity of the FM IF amplifier is checked before attempting to check the FM front end.

The checkout procedure is basically the same as that used for insufficient AM sensitivity in Chap. 13, with one important exception. Do not "tweak" or alter the adjustments of the IF amplifier because FM alignment is more complex than simply tuning for a maximum.

After the B+ voltages are checked, signal injection is used to determine if each FM stage has adequate gain. If a generator with known output levels is used, the FM IF is checked first by injecting a signal at the first FM IF amplifier. If the generator output required to produce standard output at the speaker is higher than normal, the third FM IF amplifier is checked next. If the IF amplifier sections appear to be operating normally, we proceed to check the RF amplifier and mixer at the receiver front end.

When the generator output is unknown, we begin the signal-injection process at the collector of the third FM IF amplifier and proceed toward the antenna, checking the gain of each stage. If a stage shows little or no gain, V-R analysis is used to determine the defective component. If no noticeable defect is uncovered with the signal-injection process, an FM alignment is performed with care taken to note any adjustment that does not seem to tune or function properly. In some cases the realignment will cure the lack of sensitivity, and no circuit defects will be discovered. In other cases an adjustment that does not tune properly will lead to the defect.

EXAMPLE 15-4: We now demonstrate a typical procedure, referring to the receiver of Figure 15-3. The B+ voltages are found to be normal, and a signal generator with unknown output levels is used for the signal-injection procedures.

A 10.7-MHz, 50% modulated AM signal is injected at the collector of the third FM IF amplifier $Q6$ (Figure 15-4), and with maximum generator output, a tone is barely audible at the speaker. The generator is moved to the base of $Q6$, and a much louder tone is produced. The generator output is reduced, and the generator is moved to the base of $Q5$, the second IF amplifier. A loud tone is again emitted from the speaker. The generator is reduced by the same amount as before, and is moved to the base of the first IF amplifier, Q4, where a loud tone is again obtained. Thus, the IF stages seem to have about the same amount of gain, so the front end is suspected as the cause of the problem.

Since experience has taught us that the RF amplifier is the prime suspect, a thorough checkout of the mixer and oscillator is delayed. The generator frequency is set to 90 MHz and is applied to test point <C> (Figure 15-3), the collector of the RF amplifier. With the receiver also tuned to 90 MHz, the generator output is adjusted to produce a low-level tone at the speaker. The generator is then moved to the emitter of RF amplifier Q1, and no tone is heard at the speaker, even when the receiver is rocked about 90 MHz on the tuning dial. The generator output has to be increased greatly to produce an audible tone. This indicates that the FM RF amp is not amplifying, but is actually causing a signal loss.

The DC voltage levels around the RF amplifier are now measured and found to be:

Q1	Measured	Normal
C	9.2	9
B	2	1
E	0	0.4

Excessive forward bias of the base-emitter junction and zero voltage at the emitter implies that Q1 has an open base or emitter. Q1 is removed, and the suspicion is confirmed with an ohmmeter. Normal operation is obtained when the new unit is installed.

EXAMPLE 15-5: Suppose that a generator is available for which the output settings required to produce the standard output of a receiver are known. We assume the output settings given in Figure 15-7 are applicable to an FM receiver with an RF amplifier and three stages of IF amplification.

A voltmeter is connected across the 8-Ω speaker, and the volume control is set at maximum. The generator is connected to the secondary side of the first IF transformer (CTP 19 of Figure 15-3), and the generator output is adjusted to produce 0.9 V rms across the speaker. (Note that 0.9 V across 8 Ω gives 100 mW, the standard output for the generator settings.) The resulting generator settings are X10-7, which indicates that the FM IF stages have normal gain.

The generator is moved to test point <C> (RF amplifier collector) and is readjusted for 0.9 V at the speaker. At 90 MHz, with the receiver tuned to the signal, the generator

For 100 mW at speaker

Signal	Injection point	Coarse	Fine
10.7 MHz	Base 1st FM IF	X10	0
90 MHz	Input to mixer	X10	2
90 MHz	Input to RF amp.	X1	6
90 MHz	Ant. terminals	X1	7

FIGURE 15-7 Settings of a typical generator required to produce the standard output at the speaker of an FM receiver.

settings are X100-2, when the normal settings are X10-2. This indicates that the mixer is actually causing a loss of signal! The generator is moved to the emitter of the mixer, $Q2$, and the generator setting has to be reduced to X10-3, which is very close to normal.

Evidently, coupling capacitor $C18$ is open. Careful inspection reveals no breaks in the printed circuit connections, and no suspicious solder joints are found. $C18$ is bridged with a 3-pF test capacitor by soldering the test capacitor from $L2$ to $L3$, and the receiver exhibits near-normal sensitivity, but sensitivity is still lacking. A 1.5-pF exact replacement for $C18$ is installed, and normal operation is restored.

15-5 OSCILLATIONS ON FM; AM NORMAL

Squeals or howls in the FM mode when a station is being received indicates that one or more of the FM stages are oscillating. In FM receivers the oscillations will only be noticeable when the receiver is tuned to a station. The oscillation could be coming from the FM RF amplifier, mixer, or third FM IF stage. Possible causes of the oscillation are open decoupling capacitors, open neutralizing capacitors, open grounds, or misalignment. Since AM reception is normal, it is unlikely that a common AM-FM IF stage is responsible for the difficulty. The same procedures applicable to AM oscillations (Chapter 13) are also applicable to FM. The checkout procedure should proceed as follows.

The power supply and AGC (AVC, if you prefer) filter capacitors should be bridged first. If the oscillation persists, the circuit connections in the FM IF amplifier and front end are inspected, with suspicious connections being resoldered. Next, bridge the neutralizing capacitors. If nothing good comes from all this, a scope or shunting capacitor is used to determine which stages are oscillating, and the oscillating stage is thoroughly rechecked. The input and output tuned circuits are also adjusted as in AM receivers, but a complete realignment will be necessary if the tuned circuits are altered.

EXAMPLE 15-6: We assume the receiver of Figure 15-3 has an oscillation on FM. Power supply filters $C1A$ and $C1B$ (Figure 15-6) and AGC filter capacitor $C50$ (Figure 15-3) are bridged with test units, but the oscillation remains. The circuit connections are inspected and resoldered as deemed necessary. No neutralizing capacitors are present.

We now attempt to locate the oscillating stage by using a 0.01-μF capacitor as a *shunting capacitor*. The capacitor is connected from the base of third FM IF amplifier $Q6$ to ground, and the oscillation disappears. The capacitor is moved to the base of the first IF amplifier $Q4$, and the oscillation disappears. The oscillation also disappears when the capacitor is connected to the RF amplifier $Q1$ emitter or collector. This indicates that the oscillation most likely begins in the RF amplifier, so the stage is examined in minute detail. A questionable solder joint is found at the ground side of $C15$, which holds the base of RF amplifier $Q1$ at signal ground. Normal operation is restored when the solder joint is repaired.

15-6 DISTORTED AUDIO ON FM; AM NORMAL

A receiver that functions normally in the AM mode but which has distorted audio in the FM mode will usually be found to have a problem in the FM detector. The problem can be caused by misalignment of the discriminator (or ratio detector) secondary or by a faulty component in the FM detector circuit. The troubleshooting procedure is very simple.

The adjustment for the discriminator transformer secondary is "tweaked" to see if the distortion can be removed. If the distortion remains, the components in the circuit are checked because there are only a few present. Referring to Figure 15-4, the diodes $X4$ and $X5$ are prime suspects and should be checked first. Distorted audio will result if one diode opens.

15-7 INSUFFICIENT AM AND FM SENSITIVITY

A receiver that only functions on strong, local stations for both AM and FM has a loss of AM and FM sensitivity. The problem can be caused by a defect in the power supply or in the common IF amplifier stages. The power supply is checked first, and then signal injection or signal tracing is used to determine which of the IF amplifier stages has insufficient gain. V-R analysis is then used to pinpoint the problem.

EXAMPLE 15-7: Assuming that the receiver of Figure 15-3 has normal B+ voltages, the receiver is placed in the AM mode and tuned to a strong, local station. A scope with a low capacitance probe is used for signal tracing. At the collector of a second IF amplifier $Q5$ (Figure 15-4), a 0.5-V p-p signal is observed. At the base of $Q5$, a 0.25-V p-p signal is found. This indicates that $Q5$ has a gain of about 2, which is much below normal. The DC levels around $Q5$ are measured and found to be:

$Q5$	Measured	Normal
C	8.8	9
B	2	2
E	1.8	1.4

The forward bias at the B-E junction is reduced to 0.2 V, while the emitter voltage is higher than normal, indicating increased conduction. The slightly reduced collector voltage is further evidence of the increased conduction. $Q5$ is removed, collector-emitter leakage is confirmed, and installation of a new unit restores normal sensitivity.

15-8 NOISY FM; AM NORMAL

When excessive noise is present in the FM mode but not in AM, the problem is most likely in one of the FM-only stages: the FM RF amplifier, FM converter, FM IF amplifier, or FM detector. The troubleshooting procedure is basically the same as with a noisy AM problem. Suspected tubes of tube-type receivers are replaced first, and then freeze spray is used to cool resistors and capacitors in suspected stages. For solid-state sets, freeze spray is used to cool suspected transistors, followed by cooling the components around the stages. If no noticeable reduction in noise is achieved, a scope or signal tracer can be used to determine the source of noise as before.

EXAMPLE 15-8: Noisy FM in the receiver of Figure 15-3 causes us to suspect the RF amplifier $Q1$, the FM converter $Q2$, the third FM IF amplifier $Q6$, and the diodes $X4$ and $X5$ of the ratio detector. In this case the noise is found to decrease drastically when $Q1$ is cooled with freeze spray, and the noise gradually returns as $Q1$ warms back up. Installing a new $Q1$ restores normal operation.

Incidentally, a voltmeter connected across the speaker provides a visual indication of the reduction of noise in a freeze-spray procedure. Obviously, the receiver is not tuned to a station while looking for a noise problem, and the external antenna should be disconnected to eliminate noise from afar. It might also be helpful to short the antenna terminals with a clip lead.

15-9 FM ALIGNMENT—AN INTRODUCTION

There are two basic methods used to align an FM receiver. One method, which we call the *AM-VTVM method*, uses an AM generator and a VTVM. This procedure is suitable for inexpensive, unsophisticated FM receivers, but it lacks the precision necessary for ensuring optimum performance of more sophisticated receivers. In general, we recommend that this procedure be used only when equipment for the sweep/marker method is not available.

The sweep/marker method uses a special generator and a scope to display the actual IF amplifier response curve. This method shows the technician exactly what is going on and naturally gives rise to more precise adjustments and, very likely, better performance. We recommend this procedure for all FM receivers, but we realize that the economics of every service organization may not justify the necessary generator and scope.

In the following section we present general procedures for both methods of FM IF alignment of receivers that employ either Foster-Seeley discriminators or ratio detectors. A general procedure is then given for FM front-end alignment. In all cases, the manufacturer's or Howard W. Sams' alignment instructions should be followed whenever an alignment is attempted. Receiver alignment is like playing the sweet potato—it is not to be taken lightly, and practice makes perfect.

15-10 FM IF ALIGNMENT; AM-VTVM METHOD

Discriminator

The block diagram of Figure 15-8 illustrates the adjustments, test points, and connections for a discriminator receiver when using the AM-VTVM alignment method. An AM generator set to produce a 10.7-MHz unmodulated signal is connected, as shown, to the mixer input. A voltmeter, set to measure DC volts, is connected to the limiter grid or AGC line (point *A*) to serve as an indicator.

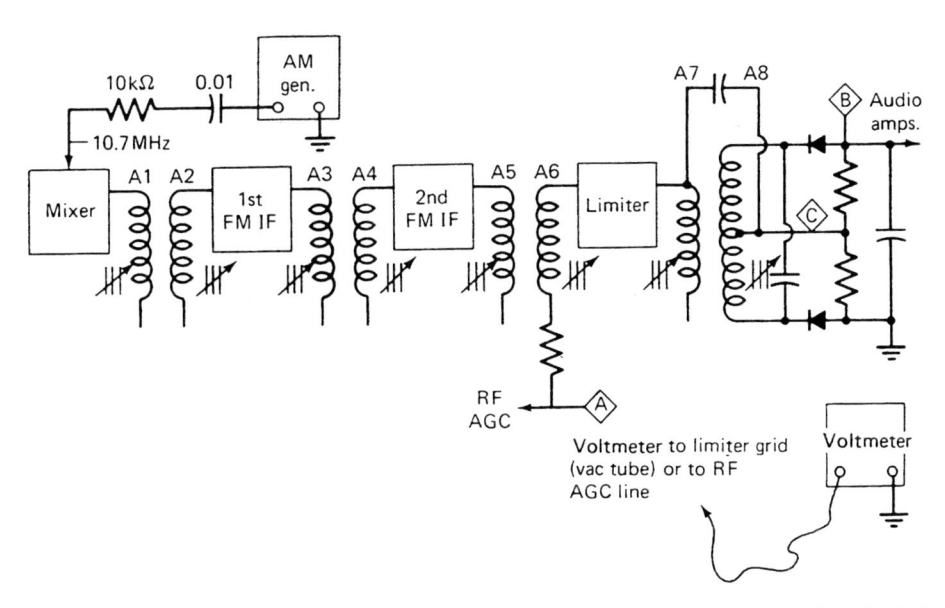

FIGURE 15-8 Block diagram illustrating the procedure for the AM generator/VTVM method of FM receiver alignment. An unmodulated 10.7-MHz signal is applied to the mixer input. The voltmeter may be connected either to the limiter grid (vacuum tube) or to the AGC line. Adjustments are made to give maximum indication of the voltmeter.

The voltmeter will indicate the signal strength at the limiter grid by measuring the negative grid bias developed by the signal.

Adjustments $A1$ through $A6$ are then made to produce maximum deflection on the voltmeter. The voltmeter is then moved to point C, where it measures the voltage across *one* of the discriminator load resistors. $A7$ is then adjusted for maximum deflection.

FM IF alignment using AM signal generator and VTVM: selector in FM position

	Signal-generator coupling	Signal-generator frequency	Radio dial setting	Indicator	Adjust	Remarks
4.	High side to ungrounded tube shield over V1, low side to chassis	10.7 MC (Unmod.)	(FM) Point of noninterference	DC probe of VTVM to point $\langle B \rangle$, common to chassis	A7, A8, A9, A10, A11, A12	Adjust for maximum.
5.	High side to ungrounded tube shield over V1, low side to chassis	10.7 MC (Unmod.)	(FM) Point of noninterference	DC probe to point $\langle C \rangle$, through 1 MΩ, common to chassis	A13	Adjust for zero reading. A positive and negative reading will be obtained on either side of the correct setting

FIGURE 15-9 Alignment instructions for vacuum-tube receiver of Figure 14-15.

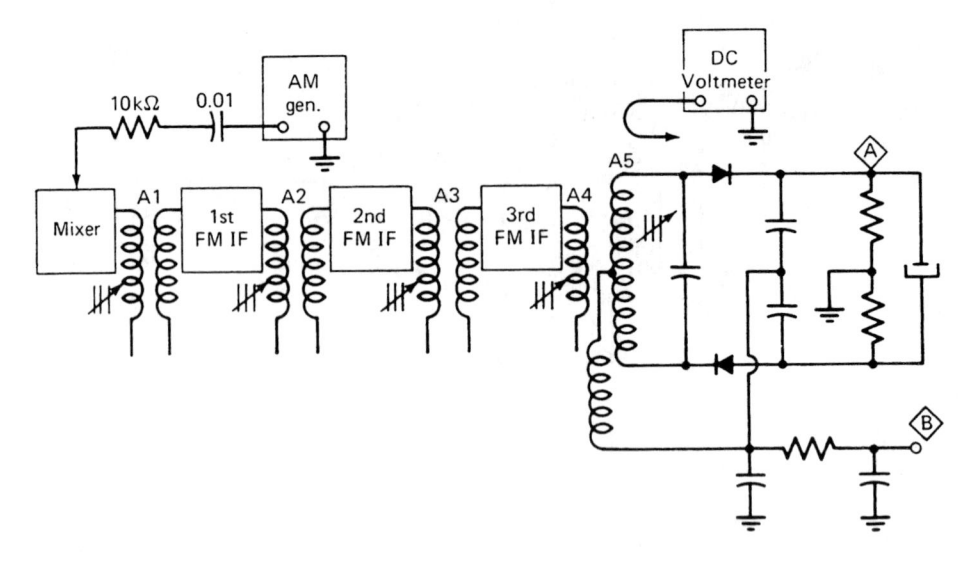

FIGURE 15-10 Block diagram illustrating the AM-VTVM method of aligning an FM receiver having a ratio detector. The AM generator is set to 10.7 MHz unmodulated, and the voltmeter is connected first to point A and then to point B, as described.

It is possible to make the $A1$–$A7$ adjustments with the meter at point C, but it is imperative that the generator output be kept as low as possible. If a large input signal is used, the limiting threshold may be exceeded so that the limiter action obscures the peaks of the adjustments $A1$–$A6$.

At this point the IF circuits are "peaked" at 10.7 MHz. It is now necessary to adjust the discriminator transformer secondary to give zero DC at the audio takeoff point. This is accomplished by placing the meter at point B and adjusting $A8$ for 0 V. It is possible to obtain either a positive or negative reading at this point by adjusting $A8$, but the adjustment is made to give 0 V.

FM if alignment using AM signal generator: selector in FM position

High side of generator through 0.001 µF to point Ⓒ, low side to ground.					
Generator frequency	Dial setting	Indicator	Adjust	Remarks
5. 10.7 MHz (unmod.)	Point of non-interference	DC probe of VTVM to point Ⓐ, common to ground	A7, A8, A9, A10	Adjust for maximum.
6. 10.7 MHz (unmod.)	Point of non-interference	DC probe to point Ⓑ, common to ground	A11	Adjust for zero reading. A positive or negative reading will be obtained on either side of the correct setting.

FIGURE 15-11 Alignment instructions for the receiver of Figure 14-3.

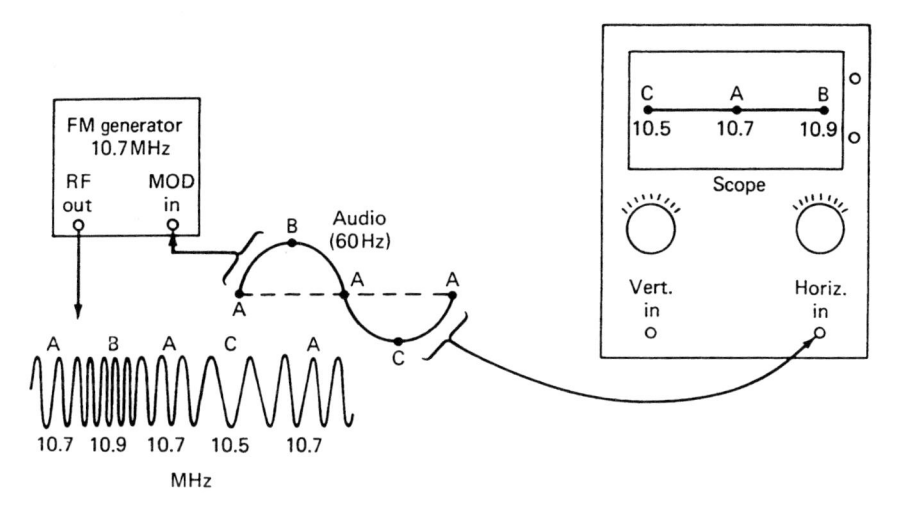

FIGURE 15-12 By using the same audio signal to modulate the generator and drive the horizontal sweep of the scope, a correspondence between horizontal position and frequency is established on the screen of the scope.

A set of alignment instructions utilizing this procedure is given in Figure 15-9 for the tube-type AM-FM receiver of Figure 14-15.

Ratio Detector

The block diagram of Figure 15-10 illustrates the test points, adjustments, and connections for performing an AM-VTVM alignment of the IF section of a ratio detector receiver. A 10.7-MHz unmodulated signal is injected at the mixer input as before. A voltmeter (+DC) is connected across the stabilizing capacitor, between point *A* and ground. Recall that the voltage developed across this capacitor is proportional to the signal strength applied to the ratio detector. Hence, adjustments *A*1–*A*4 are made to produce maximum deflection on the voltmeter.

The ratio detector secondary is tuned to 10.7 MHz by connecting the voltmeter to the audio takeoff (point *B*) and adjusting for 0 V with *A*5. This ensures that when the FM carrier is at the center frequency, no audio output will be produced. This completes the FM IF alignment. Figure 15-11 contains a set of AM-VTVM alignment instructions for the receiver of Figure 14-3.

15-11 SWEEP/MARKER GENERATOR OPERATION

A sweep/marker generator is actually an FM generator whose modulating frequency is 60 Hz, derived from the AC line. If a similar, in-phase signal is used to provide the horizontal deflection of a scope, there will be a direct correspondence between the generator output frequency and horizontal scale of the scope. This is illustrated in Figure 15-12.

When the 60-Hz modulating signal is instantaneously at 0 V, the generator output frequency will be the center frequency at 10.7 MHz. As the modulating signal goes positive, the generator frequency increases and the scope trace is deflected to the right. Subsequently, the modulating voltage will go negative, and the output frequency will fall below 10.7 MHz as the scope trace is deflected, proportionally, to the left. A typical setup may be such that the extreme left side of the scope screen corresponds to 10.5 MHz, the center to 10.7 MHz, and the right extreme to 10.9 MHz.

The FM IF frequency-response curve may be displayed on the scope by connecting the sweep/marker output to the FM mixer and also connecting the scope vertical input across one of the load resistors in a discriminator (for example). This is illustrated in Figure 15-13. The vertical deflection of the trace corresponds to the gain of the amplifier. Since the generator frequency sweeps completely across the FM IF bandwidth, and since the horizontal scale of the scope corresponds directly with frequency, the resulting trace is a plot of gain versus frequency—it is a frequency-response curve of the IF amplifier.

Generally speaking, the IF amplifier responds to frequencies between 10.6 and 10.8 MHz. When the generator output is 10.5 MHz, the scope trace will be at the left of the screen near the base line, since little voltage will be developed at the

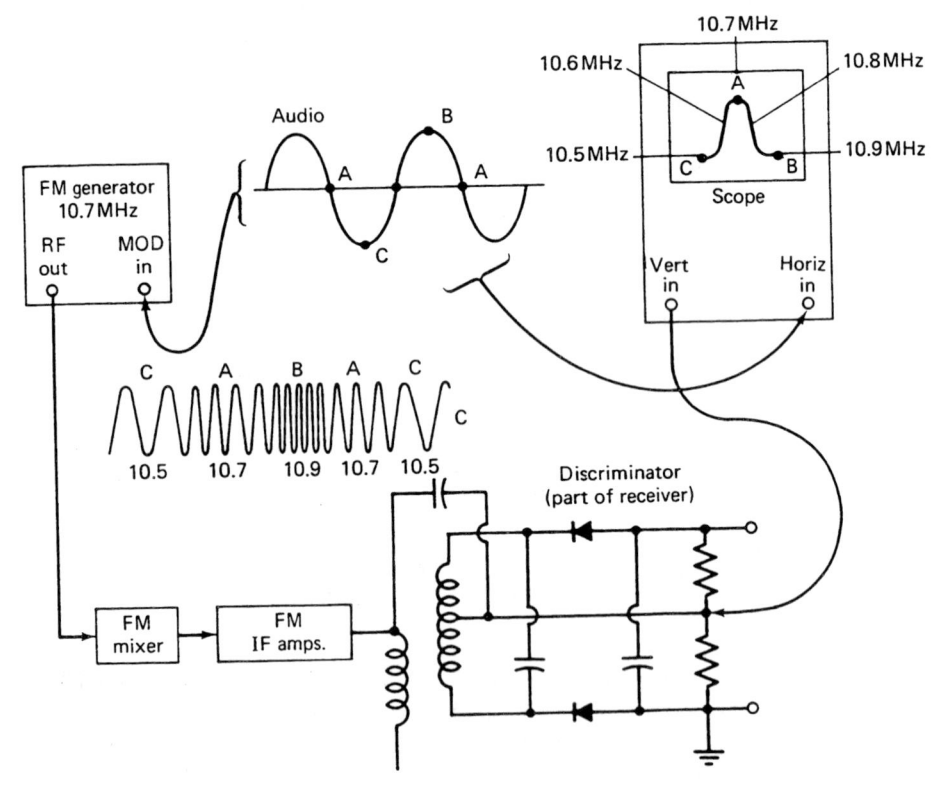

FIGURE 15-13 Displaying the FM IF amplifier frequency-response curve on a scope.

discriminator. As the generator frequency increases to 10.6 MHz, the IF amplifier gain increases and the scope trace is deflected upward accordingly. The IF amplifiers exhibit considerable gain until the generator frequency reaches the vicinity of 10.8 MHz. From this point the gain decreases gradually as the frequency approaches the upper limit of 10.9 MHz. The scope trace falls to the base line at or near the high-frequency limit.

Marker Generators

A marker generator is almost always used in conjunction with the sweep generator described above to accurately "mark" the point on the response curve that corresponds to the frequency of the marker. The marker generator is a crystal-controlled oscillator whose output is coupled to the output of the sweep generator. As the sweep frequency passes through the marker frequency, a beat note is produced that gives rise to a sharp variation of the amplitude of the sweep generator output. This produces a *blip* on the scope trace that identifies the sweep frequency at that point. The blip is called a *marker*, for obvious reasons.

Most modern generators incorporate the sweep and marker functions into one unit to form a sweep/marker generator. Several switchable markers are usually provided. Typical marker frequencies for FM alignment are 10.6, 10.7, and 10.8 MHz.

In a typical sweep/marker generator panel jacks are provided for the horizontal and vertical inputs of the scope. The "horizontal" jack provides the synchronized sweep voltage for the scope. The "vertical" jack provides the signal from the receiver, together with the crystal-controlled markers to locate the important frequencies. Controls are available for RF output level, marker amplitude, marker width, sweep direction, and polarity of the response curve.

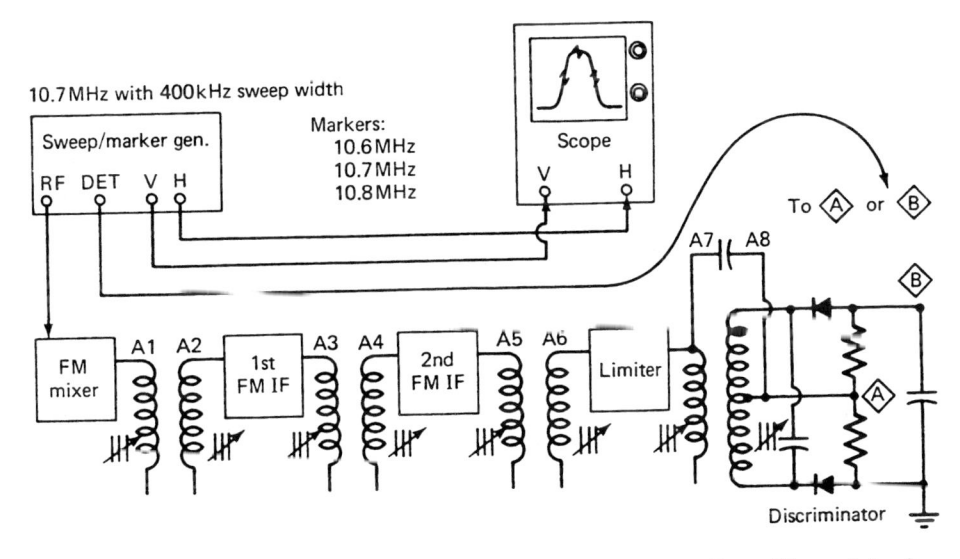

FIGURE 15-14 Block diagram of a sweep/marker alignment of an FM IF amplifier and discriminator.

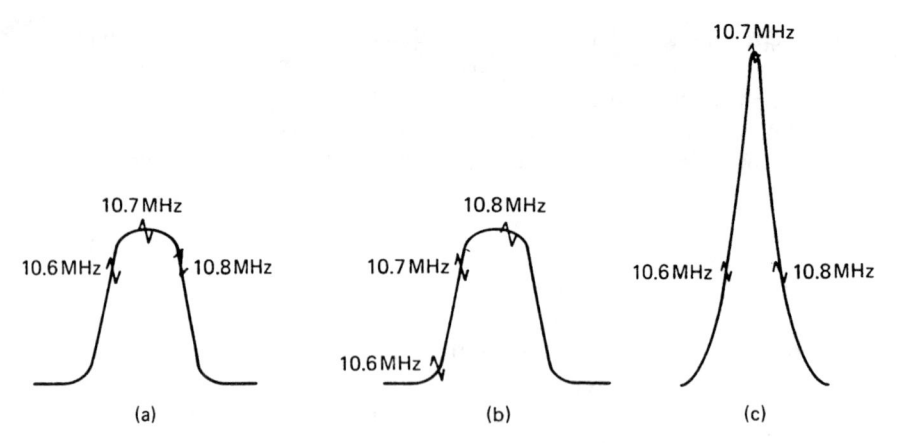

FIGURE 15-15 Frequency-response curves of an FM IF amplifier: (a) correct response; (b) and (c) incorrect responses.

15-12 SWEEP/MARKER FM IF ALIGNMENT

Discriminator

Figure 15-14 illustrates the connections necessary to perform a sweep alignment on a receiver that utilizes a discriminator. The scope horizontal control is set for external drive. The generator is set to produce a 10.7-MHz FM signal with about 400 kHz of sweep width. The detector input of the generator is connected to point A, and a response curve should be observed.

Adjustments $A1-A7$ are made to produce a symmetrical response curve of maximum amplitude, with the markers evenly spaced as shown in Figure 15-15(a). The incorrect response curves are shown in parts (b) and (c) of the figure.

To adjust the discriminator secondary, the detector input of the generator is moved to the audio takeoff point B, where an S-curve is observed. $A8$ is adjusted to place the 10.7-MHz marker at zero, and $A7$ is retouched to give maximum linearity to the curve. Figure 15-16 shows correct and incorrect S-curves. This completes the FM IF alignment. Figure 15-17 shows a set of alignment instructions for the receiver of Figure 14-15.

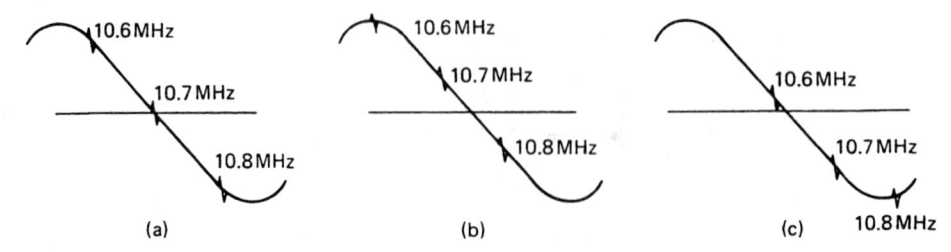

FIGURE 15-16 S-curve obtained at the audio takeoff point: (a) correct response; (b) and (c) incorrect responses.

456   Troubleshooting AM-FM Receivers   Chap. 15

If a sweep/marker generator is used that does not have the detector and vertical panel jack functions, the vertical input to the scope is connected to points A and B in the same manner as the detector input described above.

Ratio Detector

The procedure for aligning a receiver which uses a ratio detector is essentially the same as that for a discriminator. However, test point locations are different, and it is necessary to disconnect the stabilizing capacitor of the ratio detector, for the following reason.

The amplitude of the signal applied to the ratio detector varies as the sweep frequency traverses the bandwidth of the IF amplifier. Thus, during alignment, the input to the ratio detector contains a strong 60-Hz amplitude modulation component. This component is desirable since it is responsible for tracing the response curve on the scope. Recalling that the stabilizing capacitor is responsible for the amplitude-limiting properties of the ratio detector, we see that the capacitor must be disconnected to allow the ratio detector to follow the amplitude variations.

If the capacitor is not disconnected, it will charge to the peak level of the IF signal and will retain this charge. In such a case the scope will display only a straight

FM if alignment using FM signal generator and oscilloscope: selector in FM position

Signal generator coupling	Signal generator frequency	Radio dial setting	Indicator	Adjust	Remarks
High side to ungrounded tube shield over V1, low side to chasis	10.7 MHz (450 kHz Swp.)	(FM) Point of noninterference	Vertical amp. of scope to point Ⓑ, low side to chassis	A7, A8, A9, A10, A11, A12	Adjust for maximum gain and symmetry of response similar to (a) with marker as shown.
High side to ungrounded tube shield over V1, low side to chassis	10.7 MHz (450 kHz Swp.)		Vertical amp. to point Ⓒ through 47 kΩ low side to chassis	A13	Adjust to place marker at center of crossover lines similar to (b). Slightly retouch A7 for maximum amplitude and straightness of crossover lines.

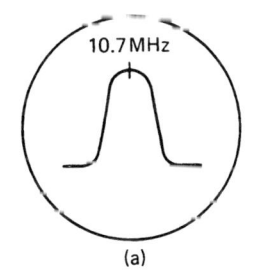

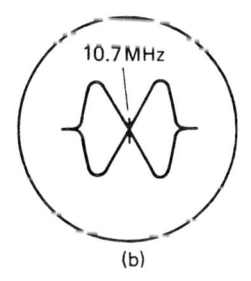

FIGURE 15-17 Alignment instructions for the receiver of Figure 14-15.

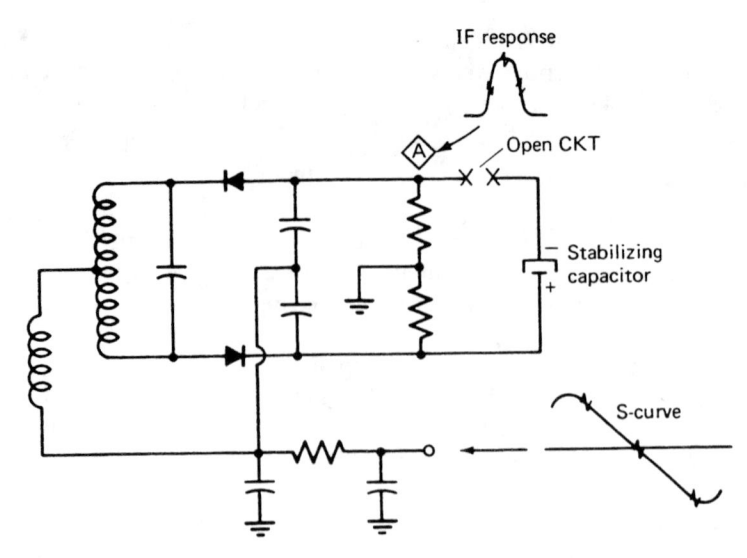

FIGURE 15-18 Stabilizing capacitor of a ratio detector must be disconnected to obtain a response curve in a sweep/marker alignment.

FM if alignment using FM signal generator: selector in FM position

High side of generator through 0.001 mF to point Ⓒ, low side to ground. Use only enough marker signal for indication. Use 60 Hz frequency modulated signal with 450 kHz sweep. Use 60 Hz sawtooth voltage in scope for horizontal deflection.

	Generator frequency	Dial setting	Indicator	Adjust	Remarks
5.	10.7 MHZ (450 kHz Swp.)	Point of noninterference	Vertical amp. of scope to point Ⓐ, low side to ground	A7, A8, A9, A10	Disconnect stabilizing capacitor C_3. Adjust for maximum gain and symmetry of response similar to (a) with marker as shown. Reconnect C_3.
6.	10.7 MHz (450 kHz Swp.)	Point of noninterference	Vertical amp. to point Ⓑ, low side to ground	A11	Adjust A_{11} (secondary) to place marker at center of "S" curve similar to (b). Adjust A_7 (primary) for maximum amplitude and straightness of line.

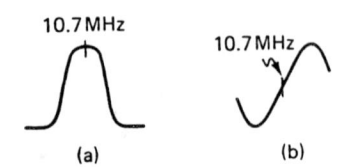

FIGURE 15-19 Alignment instructions for the receiver of Figure 14-1.

Alignment instructions

Maintain line voltage at 117 V. Use only enough generator output to obtain a suitable indication. Allow a 15-minute warmup for receiver and equipment.
Caution: Use isolation transformer, if available. If not, observe polarity when connecting test equipment.
Suggested alignment tools:
A_1-A_5, A_9, A_{12}, A_{13} .... General cement #8868, 8987, 9089..... Walsco #2531-X, 2541, 2587
A_6, A_8, A_{10} .... General cement #9440............. Walsco #2501
A_7, A_{11} .... General cement #9296, 9297, 9300..... Walsco #2510, 2546, 2547

AM alignment-selector in AM position

	Generator frequency (kHz)	Dial setting	Indicator	Adjust	Remarks
1.	455 kHz (400 Hz Mod.)	Tuning gang fully open.	Output meter across voice coil.	A_1, A_2, A_3	Adjust for maximum. Repeat until no further improvement can be made.
2.	1640 kHz	Tuning gang fully open.	Output meter across voice coil.	A_4	Adjust for maximum.
3.	1400 kHz	Tune to signal.	Output meter across voice coil.	A_5	Adjust for maximum.
4.	600 kHz	Tune to signal.	Output meter across voice coil.	A_6	Rock tuning gang and adjust for maximum. Repeat steps 2-4 until no further improvement can be made.

FM IF alignment using AM signal generator: selector in FM position

	Generator frequency (MHz)	Dial setting	Indicator	Adjust	Remarks
5.	10.7 (Unmod.)	Point of noninterference	DC probe of VTVM to point Ⓐ, common to ground	A_7, A_8, A_9, A_{10}	Adjust for maximum. Fig. 1
6.	10.7 (Unmod.)	Point of noninterference	DC probe to point Ⓑ, common to ground	A_{11}	Adjust for zero reading. A positive or negative reading will be obtained on either side of the correct setting.

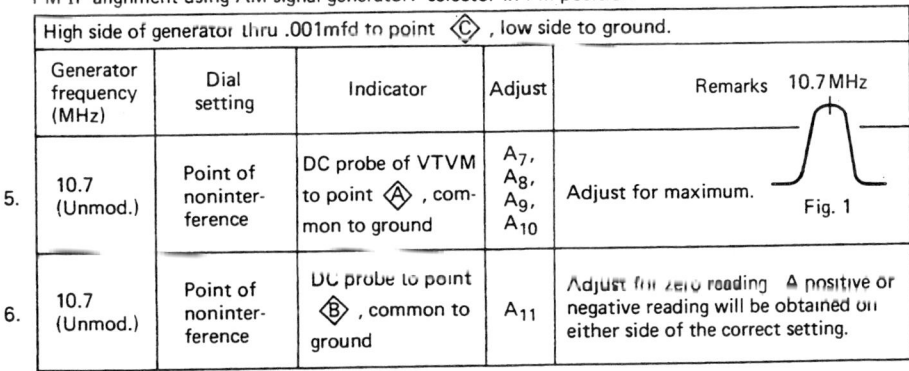

FIGURE 15-20 Complete set of alignment instructions for the receiver of Figure 15-3. (Courtesy of Howard W. Sams & Co., Inc.)

Sec. 15-12 Sweep/Marker FM IF Alignment

line. Disconnecting the capacitor, as in Figure 15-18, allows the varying output amplitude to trace the IF response curve on the scope.

Figure 15-19 shows a set of sweep/marker alignment instructions for the receiver of Figure 14-1. Incidentally, do not forget to reconnect the stabilizing capacitor.

15-13 FM FRONT-END ALIGNMENT

The procedure for setting the RF and oscillator frequencies is similar to that of AM, except that high frequencies are used and a VTVM connected at the FM detector is used as an indicator. An unmodulated AM generator signal is used. The VTVM is connected at the FM detector to indicate the strength of the signal arriving there. The VTVM is connected to measure the voltage across only one of the discriminator load resistors, or it is connected to measure the voltage across the stabilizing capacitor of a ratio detector.

Usually, the generator signal is injected by forming an injection loop of a few turns of wire as described in Sec. 12-14. The generator is then connected to the loop, and the signal is radiated to the receiver. If terminals are provided for an external antenna, the recommended procedure (from service data) will likely be to connect the generator through two 120-Ω resistors to the antenna terminals. A complete set of alignment instructions is given in Figure 15-20 for the receiver of Figure 15-3.

Required adjustments may include both capacitances and inductances of resonant circuits. Briefly, the procedure is to set both generator and receiver to 108 MHz and then adjust RF and oscillator capacitive adjustments to obtain maximum indication on the VTVM. The generator and receiver are then tuned to 90 MHz, where inductive adjustments are made to achieve maximum indication while rocking the tuning control of the receiver about 90 MHz. These steps are repeated several times until the best overall performance is obtained.

SUMMARY

1. Since the AM and FM modes share common stages of an AM-FM receiver, the function switch can be used to isolate or identify defective sections.

2. Common stages typically consist of the power supply, audio section, and one or more stages of IF amplification.

3. Typical signal tracers do not respond to the FM IF signal because signal tracers respond only to AM signals. Signal tracing in the IF section of an FM receiver must be done with a scope using a low-capacitance probe. Generally, signal injection procedures are preferred over signal-tracing procedures for the FM IF.

4. Signal injection must be performed carefully with FM receivers to avoid loading down the circuit with the stray capacitances associated with the signal generator leads. A resistor is frequently connected in series with the hot lead of the generator to minimize the loading effect.

5. Signal-injection procedures can often be performed successfully in an FM receiver by utilizing an AM signal generator. AM generators that do not go as high in frequency as the FM band (88 to 108 MHz) frequently produce harmonics of the output signals that are usable in the FM band.

6. Most troubleshooting procedures for specific symptoms for FM are quite similar to the procedures used for AM. Of course, the frequencies involved in the FM RF and IF sections are different.

7. FM IF transformer adjustments should not be "tweaked" during the preliminary phases of a troubleshooting procedure, because the adjustments are not always adjusted to give a maximum volume at the speaker, as in an AM receiver.

8. The front end (RF amplifier, mixer, local oscillator) of FM receivers is usually constructed within a metal enclosure to provide shielding at high frequencies. Considerable care is necessary when working within this enclosure so as not to deform any coils or inductors or to physically move components.

9. The two basic methods of aligning an FM receiver is the AM-VTVM method and the sweep/marker method. The sweep/marker method is the more accurate and is therefore preferred.

10. The AM-VTVM method utilizes an AM signal generator to inject a 10.7-MHz unmodulated test signal at the input to the mixer. A voltmeter at the FM detector is used to indicate the signal strength as the various adjustments are made to produce maximum signal strength at the FM detector.

11. The sweep/marker method utilizes a sweep/marker generator in conjunction with a scope to display the frequency response of the IF amplifier on the scope. Adjustments are then made to give the proper frequency response.

12. A voltmeter can be used to indicate the signal strength applied to a discriminator by connecting the voltmeter across only one of the discriminator load resistors.

13. Signal strength at the input of a ratio detector can be determined by connecting a voltmeter across the stabilizing capacitor.

14. The stabilizing capacitor of a ratio detector must be disconnected while performing a sweep/marker alignment.

15. The familiar S-curve is obtained at the audio output point of an FM detector.

16. FM front end alignment is generally straightforward, following the lines of an AM receiver. In all cases, specific alignment instructions for the particular receiver should be followed whenever possible.

QUESTIONS AND PROBLEMS

15-1. Why is a resistor sometimes connected in series with the hot lead of a signal generator used for FM IF signal injection?

15-2. Is it always necessary to make a physical connection of a generator to the circuit of an AM receiver in order to inject a signal into the circuit? What are some disadvantages of "radiating" a signal into the set?

15-3. What difficulty is encountered in using a typical service signal tracer in the FM IF section of a receiver?

15-4. Why is a low-capacitance probe required when a scope is used in the FM IF section of an FM receiver?

15-5. What precautions are in order when working inside the metal enclosure of an FM receiver front end?

15-6. What is the significance of the "level of hiss" present at the speaker of an FM receiver? How can the amount of hiss give a clue as to which, if any, FM IF amplifier may be defective?

15-7. Suppose that the local oscillator of an FM receiver is not working.

(a) What symptoms would the receiver most likely exhibit?

(b) What procedure would you recommend for isolating the defect?

15-8. Explain why an ordinary signal tracer might be able to isolate a source of noise in an FM IF amplifier.

15-9. A receiver lacks sensitivity in both the AM and FM modes.

(a) What defects in what stages could cause the loss of sensitivity?

(b) What procedure would you recommend for locating the defect?

15-10. The AM audio of an AM-FM receiver is normal, but the audio on FM is distorted. Which stage is most likely to be defective, and what is the defect likely to be?

15-11. Suppose that the stabilizing capacitor of a ratio detector somehow becomes open. What symptoms would this produce in the receiver?

15-12. Draw diagrams of various S-curves to show how a misaligned discriminator or ratio detector can cause distorted audio.

15-13. Give three defects that can cause an oscillation in an FM IF amplifier.

15-14. Describe the symptoms that may be caused by a defective function switch in an AM-FM receiver. Remember that the function switch may include power supply as well as signal connections.

15-15. An AM-FM receiver functions normally on FM but will not receive on AM. Loud erratic noises are produced when the tuning control is rotated. What is likely to be wrong with the receiver?

15-16. Describe a test for determining if the FM local oscillator is working.

15-17. In troubleshooting an FM front end, describe the complications that may arise if, in using an injection loop, the signal is radiated directly to the mixer stage rather than to the input to the RF amplifier.

15-18. In a discriminator or ratio detector, what is likely to be the symptom produced if one diode:

(a) Opens?

(b) Shorts?

15-19. What is a sweep generator?

15-20. What is a marker generator?

15-21. What is a sweep/marker generator?

15-22. Describe the general procedure for performing an FM IF alignment by the AM-VTVM method.

15-23. Describe the sweep/marker method of aligning an FM receiver.

15-24. Why is the voltmeter connected across only one of the discriminator load resistors while aligning the FM IF amplifier?

15-25. Describe the effect of the limiter stage on an alignment (improperly) performed by adjusting for maximum volume at the speaker.

15-26. What voltage should appear at the audio output of the discriminator (or other FM detector) if the input IF signal is at 10.7 MHz and the FM detector is properly aligned?

15-27. Must a scope used in conjunction with a sweep/marker generator for FM IF alignment have a bandwidth of at least 10.7 MHz?

15-28. Is it possible to use a VOM in place of a VTVM at the FM detector while performing an alignment of the FM IF amplifier?

15-29. Should the adjustments in the FM IF amplifier be made from detector to the mixer, or from the mixer to the detector? Does it make any real difference?

15-30. Refer to Figure 15-14. Suppose that a student mistakenly connects the generator to the output of the FM mixer rather than to the input. What effect would this have?

(a) On the tuning of the secondary, adjustment $A2$?

(b) On the tuning of the primary, adjustment $A1$?

15-31. Explain in detail why the stabilizing capacitor must be disconnected while aligning a ratio detector by the sweep/marker method.

15-32. What symptom would be exhibited by a receiver if the stabilizing capacitor of the ratio detector is not reconnected after performing an alignment procedure?

15-33. Must anything be disconnected about a discriminator during an alignment?

15-34. What are the marker frequencies typically used when aligning an FM IF amplifier?

15-35. Suppose that the sweep frequency of a sweep generator is 60 Hz, and suppose that the horizontal frequency of the scope is set to 120 Hz. What would be the effect on the S-curve? (See Figure 15-17.)

15-36. The probe of an EVM or scope is often connected to the FM detector through a resistor. What is the most likely reason for this?

Questions and Problems

16
PRINCIPLES OF FM STEREO

In this chapter we describe the multiplex system used to transmit two audio channels over a single FM carrier in the FM broadcast band. The system is ingenious and is rather complex when attention is given to detail. A step-by-step development is given which includes a brief description of a balanced modulator. The composite signal is described in detail, with detailed waveforms used as illustrations.

The operating principles of matrix and time-division decoders are treated in depth with detailed waveforms used to explain the rather subtle points of time-division decoder operation. Detailed circuit descriptions of the decoder circuits are given in Chapter 17.

16-1 COMPATIBILITY REQUIREMENTS

A primary concern of the FCC, when studying various proposals for FM stereo systems, was that the FM stereo system should be compatible with existing monophonic FM transmissions. The FM stereo system was not to affect monophonic reception to any perceptible degree. It was required that monophonic receivers were to be able to receive the stereo transmissions (as monaural) without noticeable change of the program content, while a stereophonic receiver was to be able to receive the stereo program with perfect separation (in theory at least) between the left (L) and right (R) channels. These were rather stringent requirements, but, as is well known, such a system was developed and is now widely enjoyed.

The FM "audio" spectrum as it was when FM stereo was proposed is shown in Figure 16-1. The main audio channel extended from 50 to 15,000 Hz (as presently), and a subcarrier was situated at higher frequencies, most often at 67 kHz.

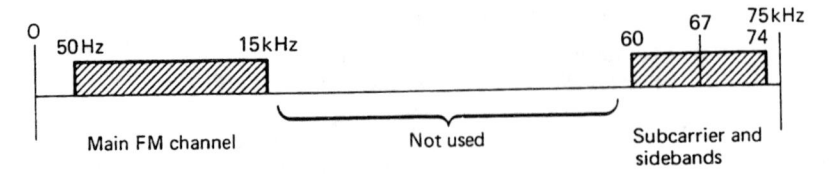

FIGURE 16-1 FM "audio" spectrum as it was when FM stereo was first proposed. The subcarrier located at 67 kHz is used for SCA transmissions.

The subcarrier and its sidebands occupied the spectrum between about 60 and 74 kHz. Thus, it was apparent that spectrum space was available from about 15 to 60 kHz which could possibly be used to transmit a second channel of audio information.

At this point we remind the reader that the spectrum of which we are speaking is the spectrum of the modulating signal. This signal is fed to the FM modulator to produce a maximum deviation of the FM carrier of 75 kHz. The practical limit of the modulating spectrum is 75 kHz, as determined by main carrier bandwidth considerations, and is only indirectly related to the maximum deviation of the main carrier of 75 kHz. Therefore, do not confuse the spectrum of the modulating signal with the deviation of the modulated FM carrier. They are entirely different.

Subcarrier Transmissions

A subcarrier is a part of the signal used to modulate the main carrier. In FM transmissions it is far above the audio spectrum, so it produces no sound at the receiver. For example, if a 67-kHz signal is added to the audio at the transmitter, the same 67-kHz signal will appear at the receiver at the output of the FM detector. It cannot be heard, of course, because of its high frequency. Generally, the subcarrier signal forms only a small portion of the total amplitude of the modulating signal.

It is possible to modulate the subcarrier (AM or FM), and when this is done, sidebands are formed on both sides of the subcarrier just as for a main carrier. Thus, the subcarrier is capable of carrying a separate channel of audio information. All that is required is that circuits be added at the output of the FM detector to pick off the higher-frequency subcarrier and associated sidebands for application to the appropriate subcarrier detector. Thus, an FM station may transmit two programs at the same time, one on the main carrier and another on the subcarrier, which, itself, rides on the main carrier. Such transmission is called *multiplex transmission*.

SCA

In 1955, the FCC approved subcarrier transmission by FM stations under the Subsidiary Communications Authorization (SCA) to broadcast music to private subscribers (stores, restaurants, etc.) provided with special SCA receivers. Initially, subcarrier frequencies ranged from 25 to 75 kHz, but now the SCA frequency is standardized at 67 kHz.

The SCA subcarrier is frequency-modulated with a maximum deviation of about ± 7 kHz. Compared to the main carrier deviation of ± 75 kHz, the SCA modulation is "narrow-band" FM. Consequently, the audio-frequency bandwidth of the SCA channel is much less than that of the main channel. Hence, the fidelity of SCA audio cannot match that of the main channel, but it is quite satisfactory for the purpose intended.

The amplitude of the SCA signal is maintained so that it is capable of modulating the main carrier a maximum of 10%. Thus, the signal of the main channel must not modulate the carrier by more than 90% to avoid a total modulation of more than 100%. When the SCA channel is being transmitted, the main channel modulation must be reduced by 10%.

Direct Approach to FM Stereo

The obvious approach to FM stereo is to use the main channel to transmit, say, the left channel and then place the right channel on a subcarrier above the audio spectrum of the modulating signal. Either AM or FM modulation may be used for the subcarrier.

Obviously, such an FM system is not compatible with existing nonaural receivers, because the monophonic listener would hear only the left channel. Further, the modulation percentage of the main channel would have to be reduced to provide equal representation for the subcarrier. The reduction of main channel modulation would seriously degrade monaural reception.

We see now that this approach is unacceptable. From the standpoint of compatibility, it is required that both L and R audio channels be represented in the main channel modulation of the FM carrier. Then, whatever is transmitted on the subcarrier must be capable of effecting a separation of the main channel components into pure L and pure R signals. We describe how this is done in the following sections.

16-2 L + R, L − R SCHEME OF MULTIPLEXING

A stereo program originates at two microphones (or comparable devices) which generate left (L) and right (R) audio signals representing the two stereo channels. If these two signals are combined in a linear network, an output is produced which is the sum of the two, $L + R$. We say they have been added. The $L + R$ output by itself is a monaural reproduction of the original program, acceptable in all respects.

Going further, if the R signal is passed through an inverting amplifier operating at unity gain, the resulting signal will be 180° out of phase with the original R. We denote the inverted signal as $-R$. If $-R$ is now added to the original L signal, the result is $L - R$. This signal is the difference between the left and right channels. By itself it does not have any entertainment value, but it is of paramount importance to FM stereo.

If the $L - R$ signal is passed through an inverting amplifier at unity gain, the result is $-(L - R)$ or $-L + R$. This signal is also important to FM stereo.

Now suppose that we have an $L + R$ and an $L - R$ signal. How can the original L and R signals be recovered from these two combinations of L and R? First, add the two combined signals:

$$(L + R) + (L - R) = L + R + L - R = 2L$$

We see that a pure left signal is obtained. The amplitude is twice the original, but this is of no consequence. Next, invert the $L - R$ signal and then add it to the $L + R$.

$$-(L - R) + (L + R) = -L + R + L + R = 2R$$

The result is a pure right signal of twice the original amplitude. Thus, it is possible to recover the original L and R signals from the combinations $L + R$ and $L - R$.

Circuitry for forming the $L + R$ and $L - R$ signals is shown in Figure 16-2. Signals are added by connecting respective sources through equal resistances to summing nodes. Note the inverting amplifiers. Also shown is a circuit for recovering the L and R signals from the $L + R$ and $L - R$ combinations.

This process is illustrated using simple waveforms in Figure 16-3. Study the waveforms carefully.

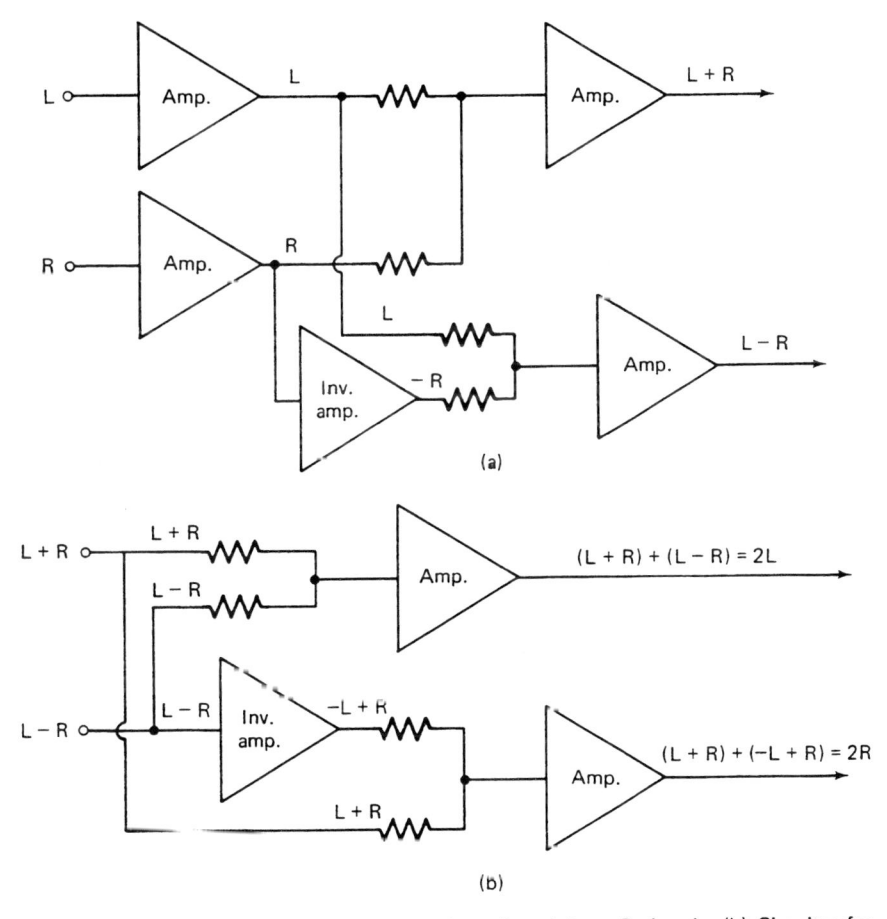

FIGURE 16-2 (a) Circuitry for forming the $L + R$ and $L - R$ signals. (b) Circuitry for recovering L and R from $L + R$ and $L - R$. Buffer amplifiers are shown in this figure that may not be needed in a given practical situation.

Special Property of (L + R) and (L − R)

If the $(L + R)$ and $(L - R)$ signals are formed for any L and R waveform, it turns out that when $(L + R)$ is at an instantaneous maximum, $(L - R)$ will be at a minimum. Conversely, when $(L - R)$ is at maximum, $(L + R)$ will be at a minimum. Further, the maximum amplitude of the combination $(L + R) + (L - R)$ will never exceed twice the maximum amplitude of either R or L alone,

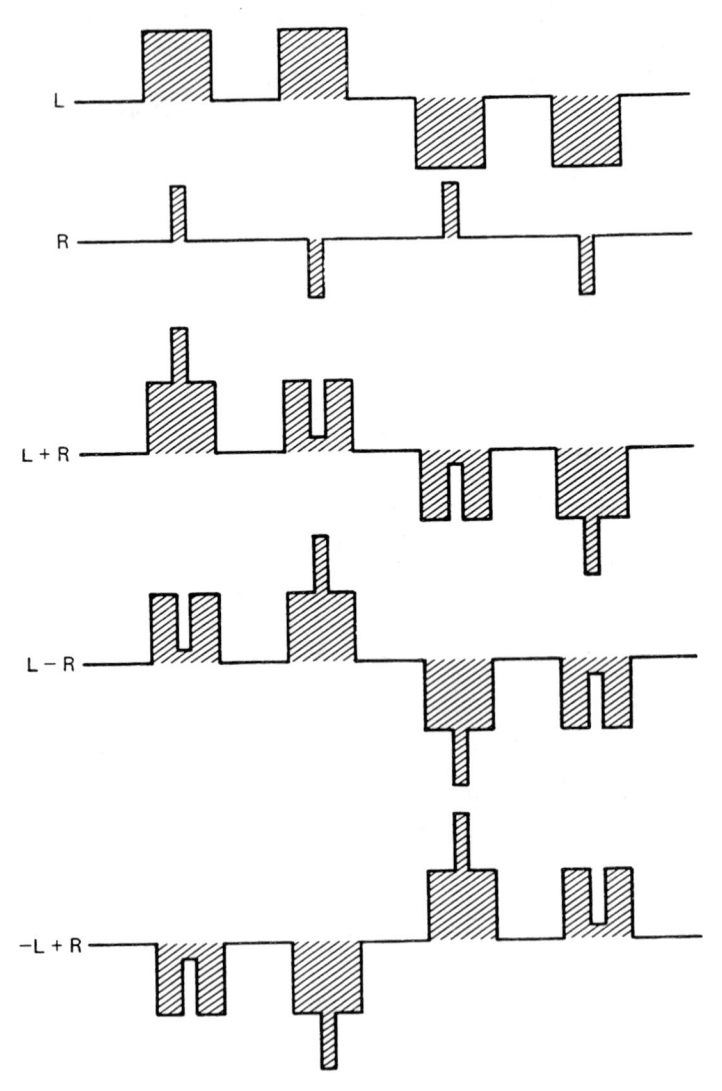

FIGURE 16-3 Resulting $L + R$, $L - R$, and $-L + R$ waveforms for simple square-wave L and R signals.

Principles of FM Stereo

L	R	L + R	L − R	(L + R) + (L − R)
5	5	10	0	10
5	−5	0	10	10
−5	5	0	−10	−10
−5	−5	−10	0	−10

When L + R is a maximum, L − R is a minimum.

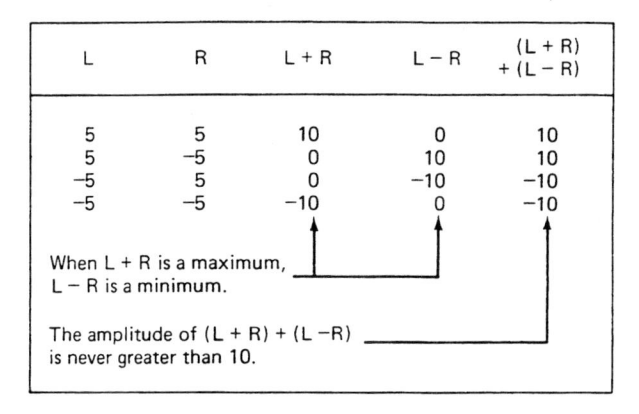

The amplitude of (L + R) + (L − R) is never greater than 10.

FIGURE 16-4 Extreme cases of the amplitude of the combined waveforms, assuming that the peak amplitude of the L and R signals is 5 units.

assuming that the maximum amplitudes of the L and R signals are the same. This is demonstrated in Figure 16-4, where it is assumed that the L and R signals may assume any value between +5 and −5 V. This property of $L + R$ and $L - R$ is important in stereo transmissions, as we shall see.

16-3 FM-STEREO SYSTEM

The system adopted for FM-stereo functions as follows. The combination $L + R$ signal modulates the main carrier so that monophonic listeners hear a combination of both L and R channels. A subcarrier located at 38 kHz carries the $L - R$ signal that, in stereo receivers, is used to effect channel separation. Monophonic receivers do not respond to the 38-kHz $L - R$ signal because of its higher-than-audio frequency.

A special type of AM modulation, called *double-sideband suppressed carrier* (DSSC), is used for the 38-kHz subcarrier. This is the same as ordinary amplitude modulation except that the carrier is suppressed. Only the sidebands are transmitted. Strictly speaking, a 38-kHz subcarrier is not transmitted; only the AM sidebands on both sides of 38 kHz are transmitted.

To detect or demodulate a DSSC signal, the carrier must be reinserted to restore the normal AM waveform. A 38-kHz signal may be generated and reinserted at the receiver, but, for high-fidelity demodulation, the frequency and phase of the locally generated carrier must exactly match that of the original. To meet this rather stringent requirement a *pilot carrier* at 19 kHz ($\frac{1}{2}$ of 38) is transmitted along with the other program material. This pilot carrier is used to "guide" or control the frequency and phase of the 38-kHz oscillator at the receiver. Or, the 19-kHz pilot may be amplified, doubled to 38 kHz, and reinserted. This is described in detail in a later section.

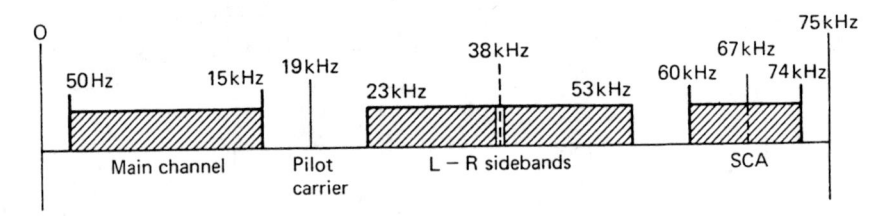

FIGURE 16-5 Frequency spectrum of the modulating signal applied to the main FM carrier for an FM-stereo station that also has SCA provisions.

Audio frequencies as high as 15 kHz are used to modulate the 38-kHz subcarrier. Therefore, the sidebands associated with the subcarrier range from 23 to 53 kHz. The frequency spectrum of the modulating signal applied to the main carrier is shown in Figure 16-5. Note the absence of the 38-kHz component and the presence of the 19-kHz pilot carrier.

Modulation Percentages

We now consider the main carrier modulation percentages produced by each component of the modulating signal. In this discussion we ignore the SCA, which, when present, produces a maximum modulation of 10%.

First, the 19-kHz pilot carrier modulates the main carrier a constant 10%, since it is a steady tone not varying in frequency or amplitude. Second, the $L + R$ component may modulate the main carrier to a maximum of 90%. Further, the $L - R$ sidebands, resulting from DSSC modulation of the 38-kHz subcarrier, are capable of modulating the main carrier to a maximum of 90%.

The preceding paragraph seems to imply that total modulation percentages exceeding 100% may be obtained. Recall, however, that the $L + R$ and $L - R$ signals never reach a maximum at the same time. Thus, when $L + R$ is modulating close to 90%, the $L - R$ sidebands will hardly be modulating at all. The converse is also true. Consequently, the total modulation will never exceed 100%.

We now see why DSSC modulation is preferred for the 38-kHz subcarrier. If the 38-kHz carrier component were not suppressed, the situation described above would not prevail. Recall that in AM, the carrier component contains twice the energy of the sidebands, but even so, it does not contain any information. If it were left in the FM modulating signal, a steady but useless modulation of the main carrier would occur, even when no $L - R$ signal was present. This would force a reduction of the main channel $L + R$ modulation which would degrade the reception on monophonic receivers. By suppressing the 38-kHz subcarrier, a great saving in modulation percentage is achieved. Further, when L equals R, as is the case when a monaural program is transmitted on a stereo station, the $L - R$ signal becomes zero, the $L - R$ sidebands disappear, and the transmitter automatically reverts back to monophonic transmission, except for the 19-kHz pilot carrier, which remains.

16-4 SIMPLE BALANCED MODULATOR

A double-sideband suppressed carrier (DSSC) signal may be produced by a balanced modulator such as that shown in Figure 16-6. We describe the operation of this circuit to provide insight into the DSSC waveform. A thorough understanding of the DSSC waveform is essential to the understanding of the FM stereo system. While balanced modulators assume many forms, the circuit of Figure 16-6 was selected because of its simplicity and applicability to this discussion.

An audio signal is delivered to point B via $T1$ and $R2$. This audio voltage is applied to diodes $D1$ and $D2$ as shown. An unmodulated RF carrier is also applied to the diodes via $T3$, as shown. The RF voltage appearing at points D and E is 180° out of phase since the input RF transformer $T3$ is center-tapped. RF voltages appearing at point B pass through $C2$ and $T2$ to the output. We now examine the operation of this circuit for the cases when point B is at zero, positive, and negative voltage due to the applied audio signal.

When point B is at zero voltage, both $D1$ and $D2$ conduct equally when E is made positive relative to D by the RF carrier. Consequently, no RF voltage appears at B, relative to ground, and no RF voltage is applied to the primary of $T2$. In this case no output occurs.

When an audio signal causes B to go positive, diode $D1$ becomes forward biased so that it conducts more than $D2$. Consequently, an RF voltage appears at point C that is in phase with the ϕ_+ end of $T3$. The amplitude of the RF voltage appearing at B is proportional to the instantaneous audio voltage applied.

When point B is driven negative by the audio signal, $D2$ conducts more than $D1$, and the RF voltage at C is in phase with the ϕ_- end of $T3$. Again, the amplitude of the RF voltage at B is proportional to the audio voltage.

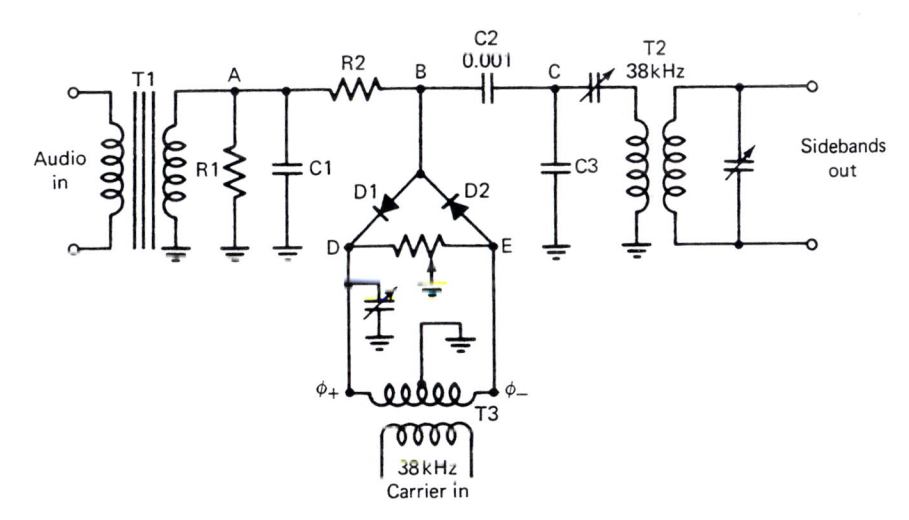

FIGURE 16-6 Balanced modulator. Only the sidebands appear at the RF output.

The RF voltage is coupled through *C2* to the RF transformer *T2*. An RF output occurs only when an audio voltage is present, and the amplitude of the RF output is proportional to the audio voltage. Further, the phase of the RF output voltage is determined by the polarity of the audio voltage.

The output waveforms for two audio signals are shown in Figure 16-7. Note very carefully that the phase of the RF changes by 180° when the audio voltage crosses the zero level.

The undulating waveform at the output of the balanced modulator is actually the combination of two sidetones symmetrically located about the carrier. (Refer to Sec. 8-4.) Here is an important point. Since the upper and lower sidetones are symmetrically positioned above and below the carrier frequency, the peaks of the DSSC waveform will occur either exactly in phase or exactly out of phase with the original carrier. This simple fact permits accurate waveforms to be graphically constructed, as we shall see. This makes it possible to understand in detail the complicated waveforms associated with FM multiplexing for stereo without having to resort to mathematical analysis. We hope this point is appreciated.

AM Waveform Restoration

The carrier component of a DSSC waveform must be reinserted before the waveform is demodulated. Otherwise, tremendous second harmonic distortion results. The effect of adding carrier components in gradually increasing amplitude is shown in Figure 16-8. Note that the amplitude of the reinserted carrier affects the percentage modulation of the resulting AM wave.

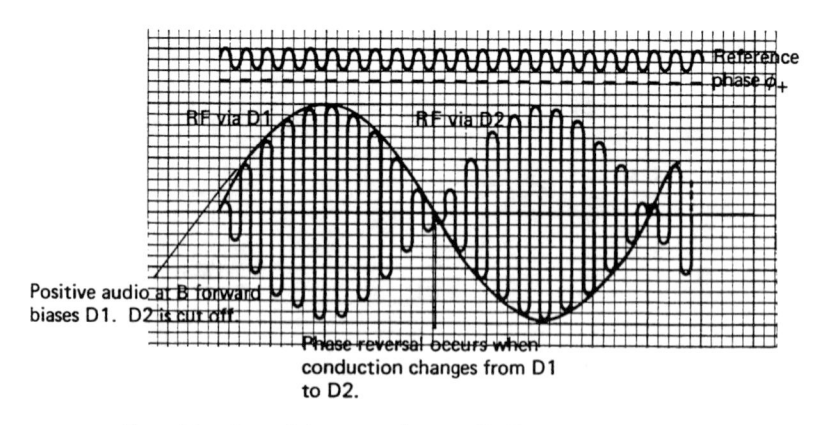

Note: The phase of the output is controlled by the polarity of the audio input.

FIGURE 16-7 Output waveform of the balanced modulator showing how the phase of the RF output signal depends upon the polarity of the audio input.

Audio signal
(a)

DSSC waveform
(b)

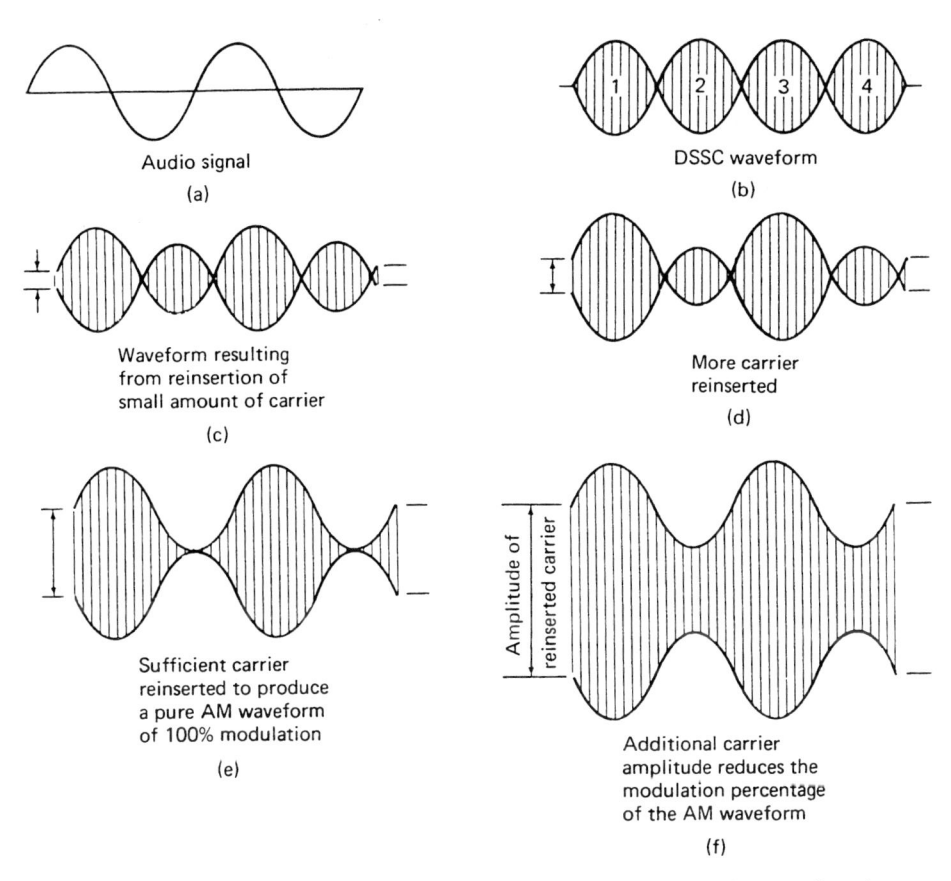

Waveform resulting from reinsertion of small amount of carrier
(c)

More carrier reinserted
(d)

Sufficient carrier reinserted to produce a pure AM waveform of 100% modulation
(e)

Additional carrier amplitude reduces the modulation percentage of the AM waveform
(f)

FIGURE 16-8 Waveforms showing how double-sideband suppressed carrier waveform is converted back to an AM waveform by reinserting the carrier. Lobes 2 and 4 decrease in amplitude, since in these lobes, the carrier is out of phase with the sideband sum. In lobes 1 and 3, the carrier is in phase with the sideband sum.

16-5 FREQUENCY DOUBLERS

A frequency doubler is an RF stage that will take a sine-wave RF input and, by utilizing the "flywheel effect" of resonant circuits, produce a sine wave output whose frequency is twice that of the input. Frequency doublers are used extensively in FM stereo receivers.

Two frequency doublers are shown in Figure 16-9. The circuit of part (a) utilizes a triode biased to near cutoff so that linear operation is not obtained. A fairly large 19 kHz (for example) sine-wave input drives the tube into heavy conduction only on the positive peaks of the input. Thus, at the plate, 19-kHz current pulses appear which are applied to the 38-kHz resonant circuit, causing the circuit to oscillate at

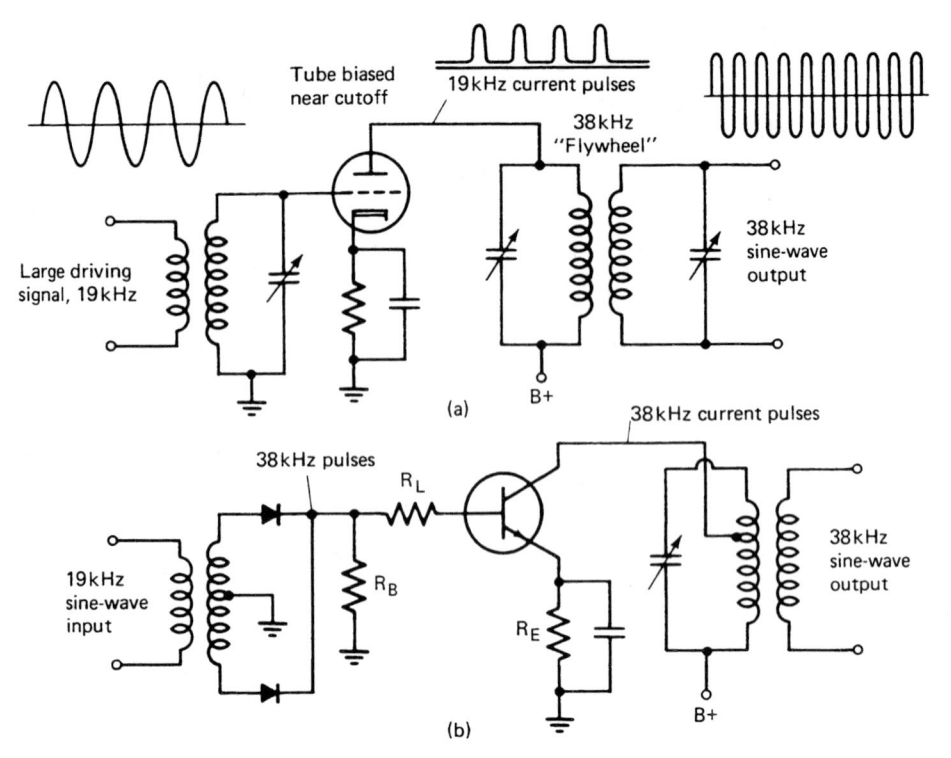

FIGURE 16-9 Frequency doublers: (a) vacuum-tube circuit that uses the flywheel effect of a resonant circuit to double the frequency; (b) solid-state circuit utilizing a full-wave rectifier at the input to provide pulses at twice the input frequency.

38 kHz. The 19-kHz pulses supply energy on alternate 38-kHz cycles to sustain the 38-kHz oscillation, as shown in the figure. The flywheel effect refers to the ability of a resonant circuit to take pulses (of the proper frequency) and convert the pulses to a sine-wave output.

Figure 16-9(b) is a frequently encountered circuit that involves a full-wave rectifier at the input to produce pulses at a rate twice that of the input frequency. No steady bias is applied to the transistor base, and therefore the transistor conducts only during the positive peaks of the input pulses. Thus, 38-kHz current pulses are applied to the resonant circuit at the output, and the flywheel effect produces a sine-wave output at 38 kHz. Resistor R_B is the load for the diodes, and R_L limits the base current that is drawn on the positive peaks.

16-6 COMPOSITE FM-STEREO SIGNAL

We now examine in detail the nature of the composite signal that is sent to the modulator of an FM stereo transmitter. This same composite signal will appear at the output of the FM detector at the receiver.

The composite signal consists of the following: the $L + R$ audio frequency component, the 19-kHz pilot carrier, and the $L - R$ sidebands centered about 38 kHz. If SCA is simultaneously transmitted, the SCA components will be present in the composite signal also. We prefer to ignore the SCA in this discussion. Further, the 19-kHz pilot carrier is a constant "add-on" to the composite signal, and we would like to ignore it also. Thus, we are left to investigate the $L + R$ signal and the $L - R$ sidebands.

Refer to Figure 16-10, in which square waves are selected for their simplicity as the L and R signals. Note carefully how the $L + R$ and $L - R$ are derived from the original L and R. The $L - R$ audio signal, as such, does not appear in the composite signal. The $L - R$ component is used to modulate the 38-kHz subcarrier to form the $L - R$ sidebands.

The DSSC waveform resulting from the $L - R$ audio is shown in part (e) of the figure. Observe the small sine wave and the series of dashes that represents the phase of the 38-kHz subcarrier. This is only a graphical aid and is not part of the composite signal. We shall call this the *phase reference*.

Note that the $L - R$ sideband envelope follows the amplitude of the $L - R$ signal. Here is the most important point of all: When the $L - R$ signal is *positive*, the sideband signal is in *phase* with the 38-kHz phase reference; when the $L - R$ is *negative*, the sideband signal is *out of phase* with the 38-kHz phase reference. This can be understood by referring back to the balanced modulator described earlier.

After the $L + R$ audio signal and the $L - R$ sideband signal is obtained, the two are added together to form the composite (ignoring the 19-kHz pilot and SCA). This is shown in part (f) of the figure.

Second Example

We now consider a slightly more complicated example, shown in Figure 16-11. Note that the R signal is now greater in amplitude than the L. We shall see that the effect of this is to cause a phase reversal in the composite.

The $L + R$ and $L - R$ audio waveforms are obtained as before, but the waveforms are slightly more complex. Note very carefully the phase reversal that occurs in part (e) when the $L - R$ signal goes negative.

The composite is also obtained as before, by adding the $L + R$ audio and the $L - R$ sideband waveform. Observe that the phase reversal of the $L - R$ sideband signal also appears in the composite.

Composite signals for L-only and R-only sine-wave audio signals are shown in Figure 16-12.

Transmitter Block Diagram

A block diagram of an FM stereo transmitter is shown in Figure 16-13. The audio sources may consist of studio microphones, tape and record units, and so forth. Either stereo or monaural sources may be utilized. The audio processing section

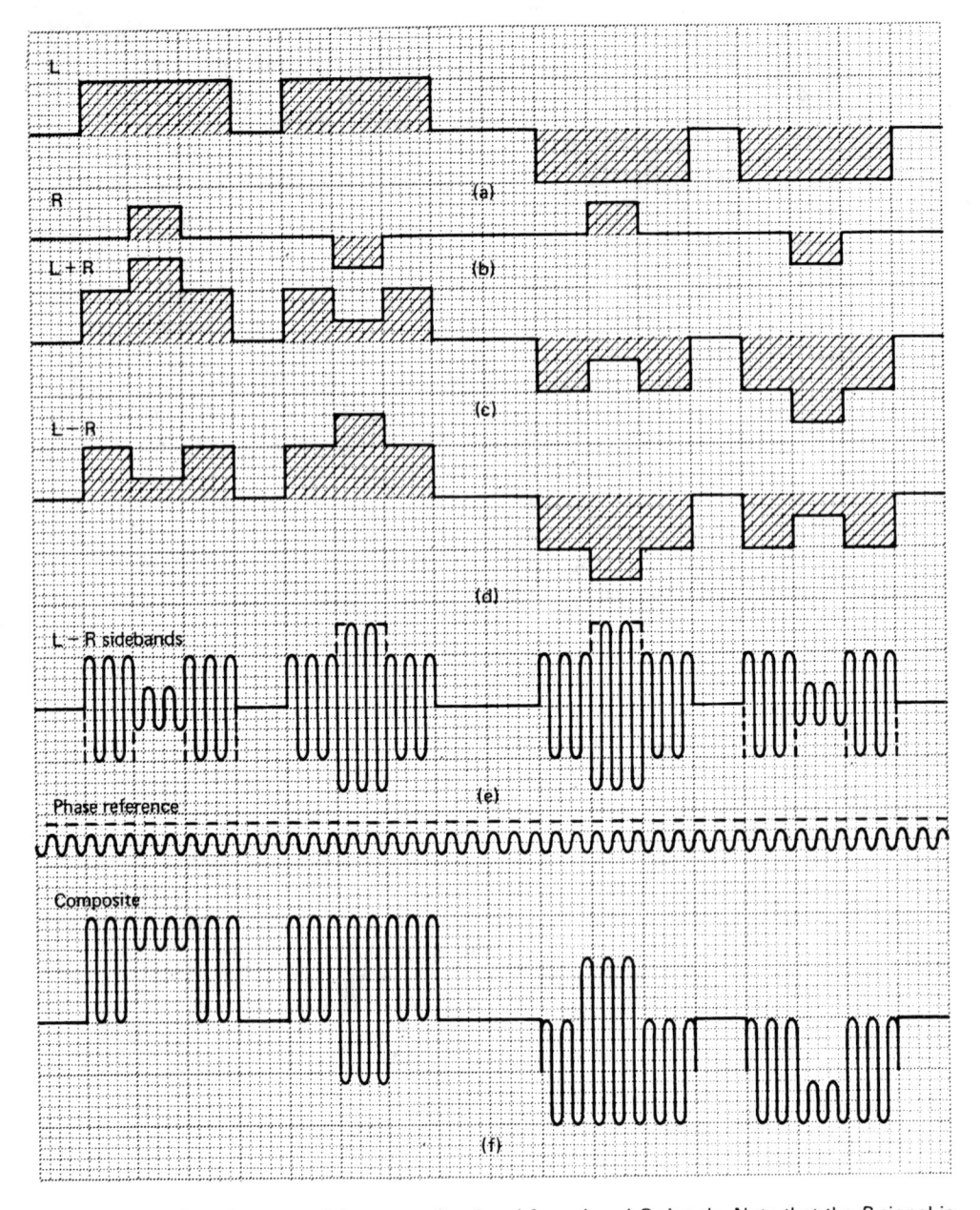

FIGURE 16-10 Development of the composite signal from L and R signals. Note that the R signal is smaller in amplitude than the L signal in this example.

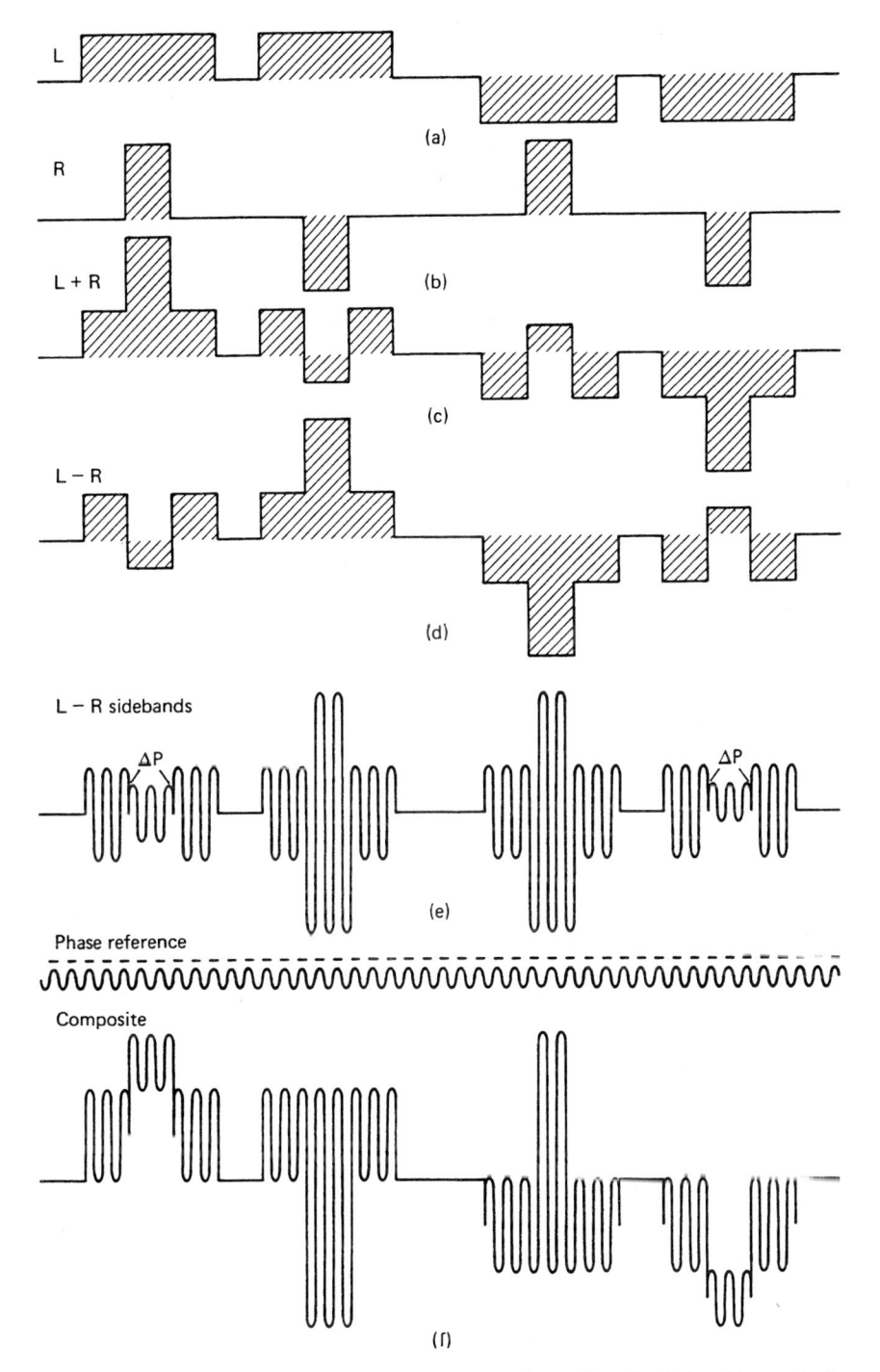

FIGURE 16-11 Development of the composite waveform when the R signal is greater in amplitude than the L signal. Note the phase shifts $\triangle P$ introduced into the $L - R$ sideband signal.

Sec. 16-6 *Composite FM-Stereo Signal* 477

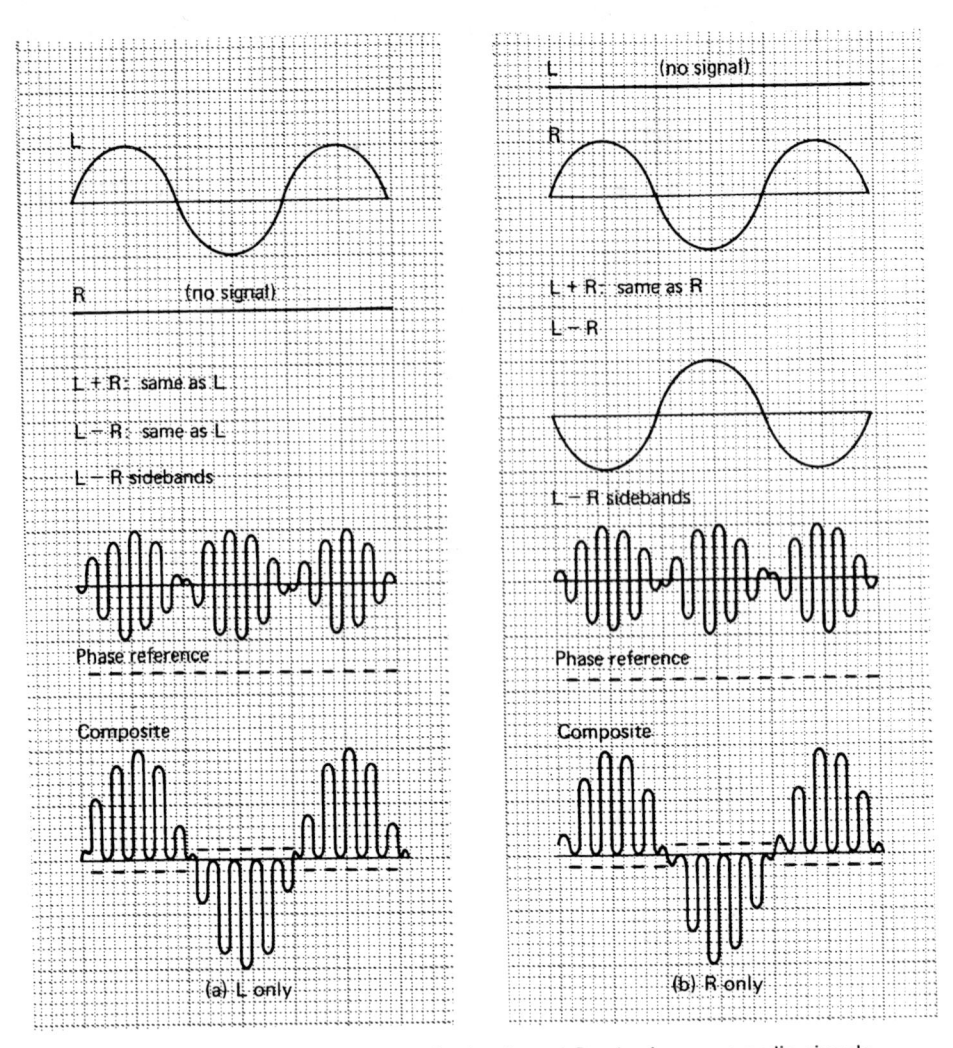

FIGURE 16-12 Composite signals for *L*-only and *R*-only sine-wave audio signals.

may be an elaborate array of audio mixers, tone control networks, preemphasis networks, and switching arrangements. For monaural broadcasts, the L and R channels are tied together. In such case, the $L - R$ output is zero.

The $L + R$ output of the audio processing block goes to the adder input of the main FM modulator. The amplitude controls are indicated symbolically as variable resistors in series with each line going to the adder input.

The FM stereo section begins with the 19-kHz oscillator, which provides the pilot carrier to the main carrier modulator input, and also applies the 19-kHz signal to a frequency doubler to obtain the 38-kHz subcarrier. The DSSC balanced modulator receives the $L - R$ audio and the 38-kHz subcarrier and forms the $L - R$ sidebands. The sidebands are then applied to the adder input of the main FM modulator.

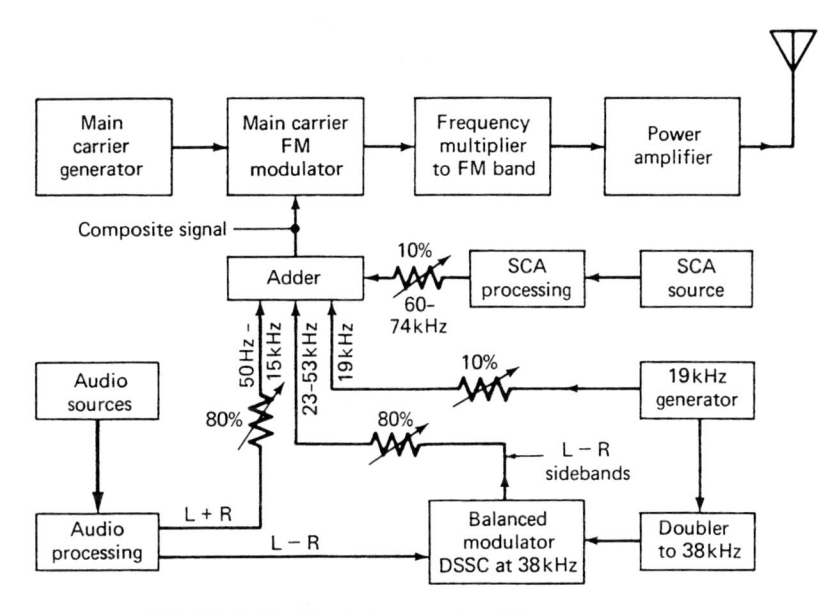

FIGURE 16-13 Block diagram of an FM-stereo transmitter.

The SCA channel includes an SCA source, which, incidentally, may be a receiver that receives the SCA program from an originating point far away. The SCA signal is processed and is applied to the main FM modulator as frequency-modulated 67-kHz subcarrier and sidebands.

The main FM carrier channel begins with a carrier generator that produces a carrier whose frequency usually is only a few MHz. This low-frequency carrier is then FM-modulated by the composite signal. A series of frequency multipliers then raises the frequency of the FM carrier to the proper frequency in the FM band. A power amplifier raises the power level to the proper level, and the antenna flings the multiplexed FM signal out into radioland.

16-7 MATRIX DECODER PRINCIPLES

Matrix decoders represent the direct approach to recovering the L and R signals from the composite waveform. They are found, however, only in early tube type FM stereo receivers because more elegant decoders have been developed. In this section we describe the block diagram and principle of operation of a typical matrix decoder, shown in Figure 16-14.

The most common FM detector found in FM stereo receivers is a ratio detector quite similar to those described in Chapter 13. The one significant difference is that the deemphasis network is removed from the ratio detector output and is placed at the output of each decoder output channel.

The ratio detector output is the composite signal, containing $L + R$, $L - R$, pilot carrier, and SCA components. The composite signal is applied to the multiplex

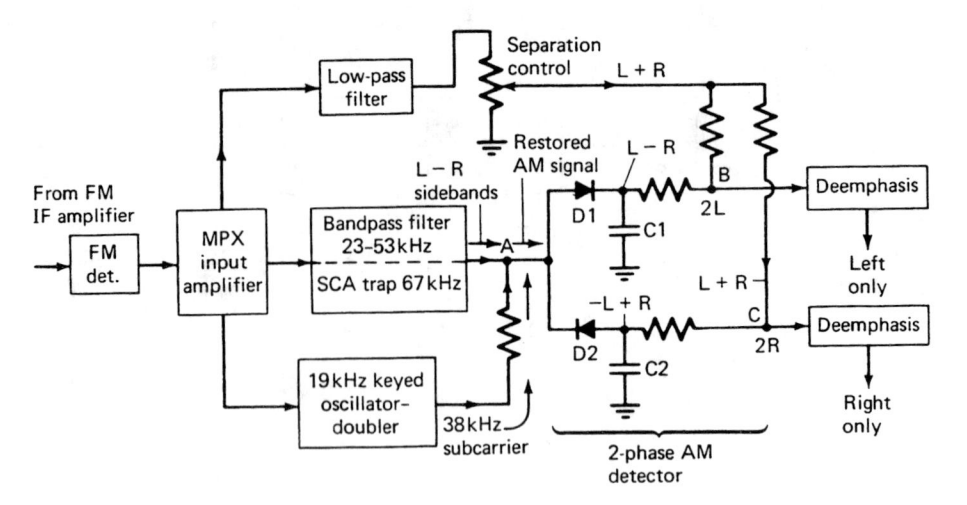

FIGURE 16-14 Block diagram of a matrix decoder.

(MPX) input amplifier, which is also known as the composite amplifier. At this amplifier the composite is separated into three components by appropriate filters and tuned circuits.

To recover the $L + R$ audio signal from the composite, it is only necessary to apply the composite to a low-pass filter to remove frequencies above the audio spectrum. The frequencies removed from the composite are the pilot carrier at 19 kHz; the $L - R$ sidebands, which extend from 23 to 53 kHz; and the SCA frequencies, which range from about 60 to 75 kHz. The $L + R$ branch is the top branch in Figure 16-14. The *separation control* simply controls the amplitude of the $L + R$ signal applied to the matrix network. Its function is described in more detail below.

A tuned circuit picks up the 19-kHz pilot carrier from the composite (MPX input) amplifier and applies it to a 19-kHz local oscillator. The pilot carrier causes the local oscillator to "lock in" in phase and frequency with the pilot carrier. The oscillator output circuit incorporates a resonant circuit tuned to 38 kHz so that the stage serves both as an oscillator and as a frequency doubler. Its output is a 38-kHz sine wave.

The remaining branch at the composite amplifier output is that taken by the $L - R$ sidebands. A combination bandpass filter and SCA trap causes the $L + R$, pilot carrier, and SCA signals to be rejected; only the $L - R$ sidebands (23 to 53 kHz) are allowed to pass through. Thus, the waveform at the output of the bandpass filter is the DSSC waveform containing the $L - R$ information.

At point A in Figure 16-14, the 38-kHz subcarrier is recombined with the $L - R$ sidebands to produce a conventional AM waveform whose modulation is the $L - R$ audio signal. This AM waveform is then applied to an ordinary AM detector to obtain the $L - R$ audio signal.

A two-phase AM detector is indicated in the figure. Actually, the circuit is that of two conventional AM detectors in parallel, with the diodes connected in opposite directions. The effect of this is to produce outputs 180° out of phase. One output is $L - R$ and the other is $L + R$, as indicated.

Points B and C are summing nodes where the $L + R$ and $L - R$ signals are recombined to form the pure L and pure R audio signals. Note that the $L + R$ signal is applied to each node. The $L - R$ signal is also applied to point B, and the $-L + R$ signal is also applied to point C. Adding the two signals gives, at point B: $(L + R) + (L - R) = 2L$; and, at point C: $(L + R) + (-L + R) = 2R$. Thus, the L and R signals are recovered. They are then applied to deemphasis networks to restore the high-frequency components of the audio to the proper relative amplitudes. Incidentally, the resistors connecting to points B and C form the "adder" networks, so that linear addition of the signals is ensured.

Separation Control

In order to achieve perfect separation of the L and R signals, the $L + R$ and $L - R$ audio signal amplitudes must be exactly equal. If they are unequal, the L channel will also contain a little R, and conversely. The separation control allows the $L + R$ level to be adjusted so that it exactly equals the $L - R$ component.

16-8 TIME-DIVISION (TD) DECODING

Time-division stereo decoders are now used almost universally. In comparison with matrix decoders, they are less expensive and are less critical to design and to align. In Chapter 17, we will see that integrated circuit technology has been adapted to TD decoders in at least two basic forms. In this section we describe in detail the operation of one form of TD decoder and thereby establish the operating principles of all TD decoders.

The block diagram of the TD decoder that we shall investigate is shown in Figure 16-15. The composite and possibly an SCA signal appear at the output of the FM detector. An SCA rejection filter removes the SCA component, and the composite signal then passes to the MPX input amplifier (or composite amplifier), where it is amplified. A tuned circuit at the amplifier output picks off the 19-kHz

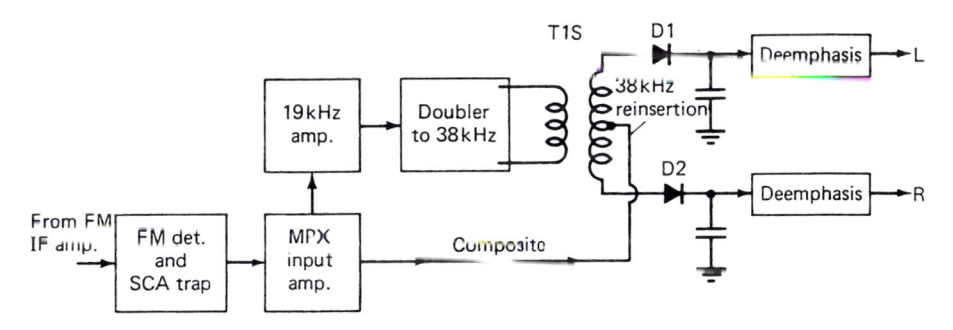

FIGURE 16-15 Block diagram of a time-division stereo decoder.

pilot carrier and applies it to an amplifier and doubler as indicated. The output of the doubler, a 38-kHz signal of proper amplitude and phase, is coupled into the secondary $T1S$ of transformer $T1$.

Back at the composite amplifier output, the composite signal is not separated into the various components as in the matrix decoder. It is left intact and is applied to the center tap of $T1S$. We shall see that the 38-kHz subcarrier and the composite signal combine in such a manner that only the left audio channel appears at the output of diode $D1$, and only the right channel audio appears at the output of $D2$.

Diode Voltages

The composite signal voltage is applied via $T1S$ equally to diodes $D1$ and $D2$ since the composite signal is applied to the center tap of $T1S$. The diodes connect to opposite ends of $T1S$, however, and receive opposite phases of the 38-kHz subcarrier that is magnetically induced into $T1S$. Thus, the diodes receive signal voltages from two sources, the composite and the subcarrier.

We emphasize that the phase of the 38-kHz subcarrier signal appearing at $T1$ is held very strictly in phase with the original 38-kHz subcarrier at the transmitter. But the center-tapped secondary of $T1$ applies 38-kHz voltages of opposite phase to the diodes $D1$ and $D2$. Therefore, one diode receives the 38-kHz subcarrier in phase with the original subcarrier, while the other diode receives the 38-kHz component exactly out of phase with the original. In this discussion we assume that $D1$ receives the subcarrier component that is in phase with the original. This is illustrated in Figure 16-16.

If we examine the composite waveform relative to the phase of the original subcarrier, we observe a rather striking and important feature. The peaks of the composite (whether positive or negative peaks) that occur in coincidence with the positive peaks of the original subcarrier trace out the waveform of the left audio signal, L. Further, the peaks of the composite that occur in coincidence with the negative peaks of the original subcarrier trace out the waveform of the right audio signal, R. This is illustrated in Figure 16-17.

SC, subcarrier
SC_{IN}, subcarrier component in-phase with the original
SC_{OUT}, subcarrier component out-of-phase with original
Com, composite signal
V_{D1}, SC_{IN} + Com
V_{D2}, SC_{OUT} + Com

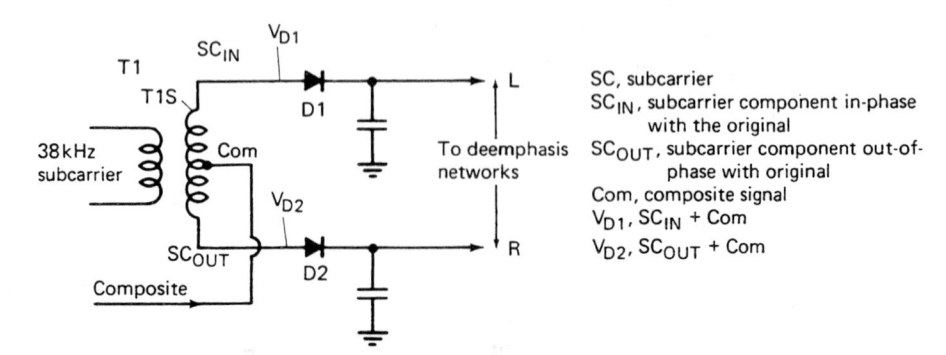

FIGURE 16-16 When the composite signal is added to the reconstructed subcarrier, either the left- or right-channel audio is produced, depending upon whether the subcarrier is in phase or out of phase with the original subcarrier at the transmitter.

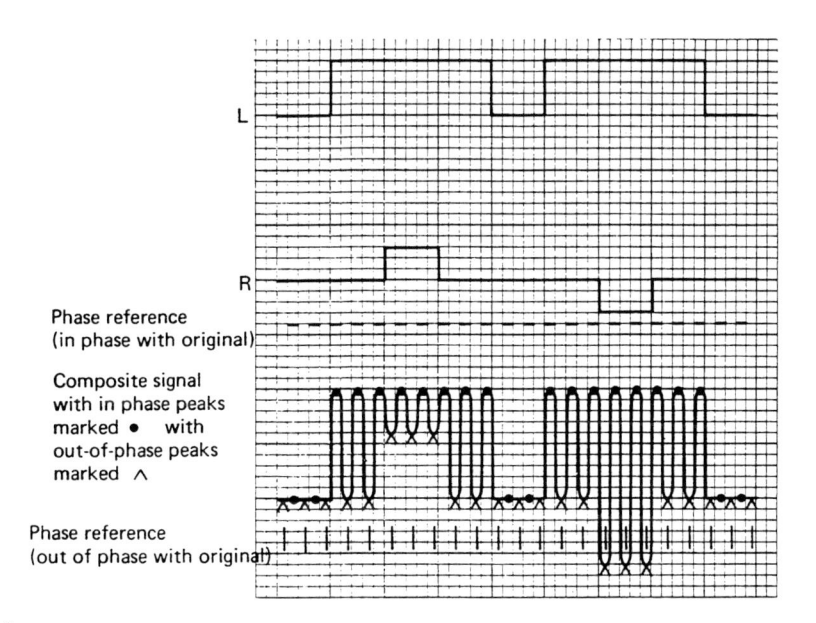

FIGURE 16-17 Peaks of the composite signal that occur in phase with the original subcarrier trace out the left audio signal. Peaks that occur out of phase with the original subcarrier trace out the right audio signal. (This example is extremely simple. Verify the relationship in Figures 16-11 and 16-12.)

Returning to the voltage at diodes $D1$ and $D2$, the voltage at $D1$ is obtained by adding to the composite a fairly large component of SC_{in} (see Figure 16-16). On the other hand, the voltage at $D2$ is obtained by adding to the composite a fairly large component of SC_{out}. The difference described here does not at first seem particularly significant. However, the effect is to cause *entirely different waveforms* to be formed at $D1$ and $D2$. The waveforms are not completely independent, but they are in no way identical.

This is what happens. When the composite is added to the positive phase of the 38-kHz subcarrier (SC_{in}), a waveform is produced whose positive peaks trace out the left channel audio signal, L. At the same time, the negative peaks of the same waveform trace out the right channel audio information, R.

On the other hand, when the composite is added to the negative phase of the 38-kHz subcarrier (SC_{out}), a waveform is produced whose positive peaks trace out the right channel audio signal while the negative peaks of the same waveform trace out the left channel audio signal. These two situations are illustrated in Figure 16-18. Compare the waveforms. Is one the inverse of the other? How are they similar?

Observe now that both diodes are connected so that they conduct on the positive peaks of the applied waveform, clipping off the negative portion of the waveform. The action of each diode is the same as for a conventional AM detector. The audio waveform produced by the positive peaks is recovered. Diode $D1$ produces the left stereo channel, and diode $D2$ produces the right stereo channel.

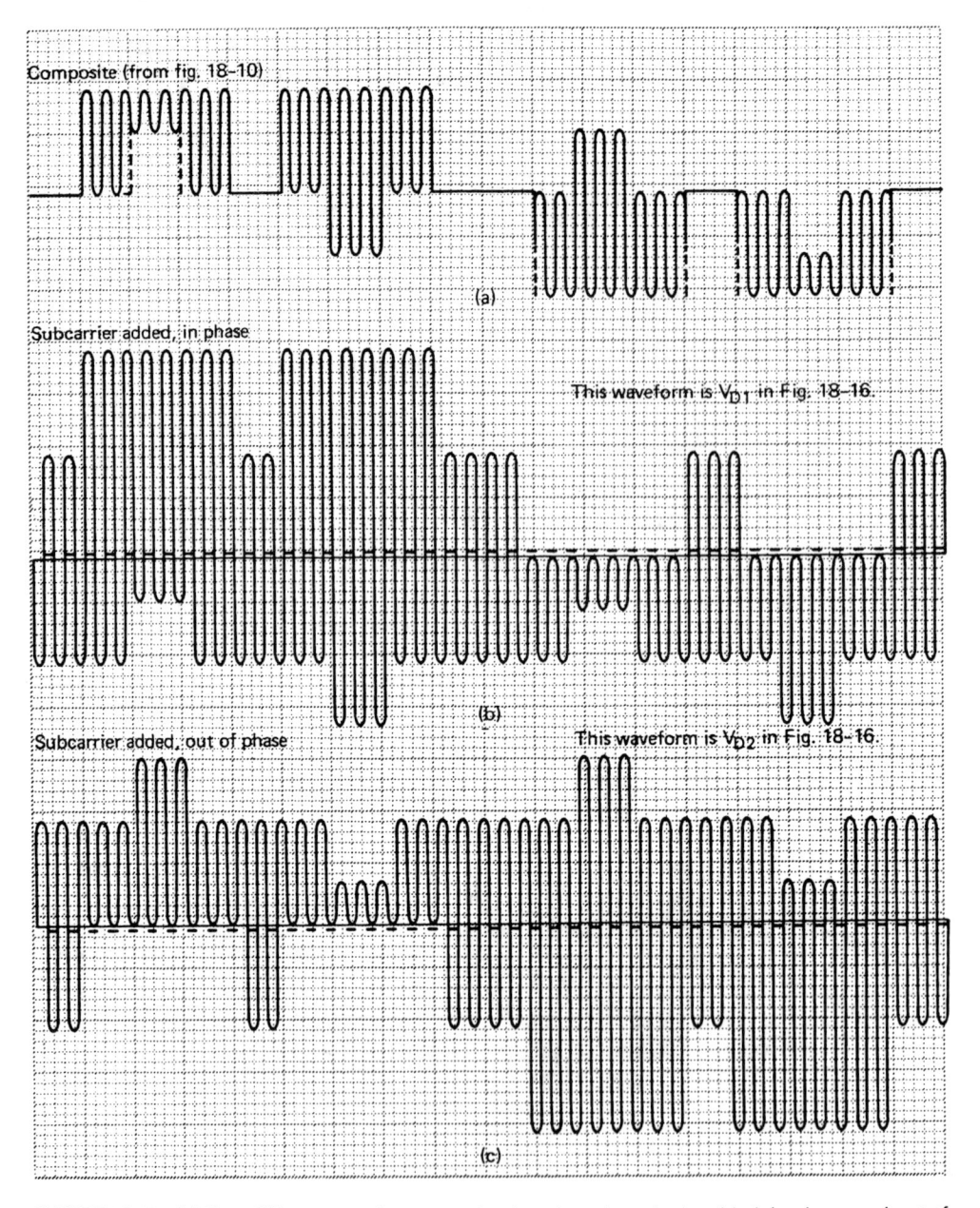

FIGURE 16-18 (a) Two different waveforms result when the subcarrier is added, in phase and out of phase, to the composite. (b) The left signal appears on top and the right is on the bottom. (c) The right appears on top and the left is on the bottom. (Refer to Figure 16-16.)

Illustration

We now present the waveforms that result from the transmission of two sine waves. The waveforms are shown in Figure 16-19. In Figure 16-19(a) the left and right waveforms are drawn, and note that the higher frequency signal is the R signal. The waveforms are superimposed simply to conserve space in the figure. Part (b) is the $L + R$ signal, obtained by adding the two signals point by point The $L - R$ signal is shown in part (c) and is obtained by subtracting the R amplitude from the L amplitude and plotting the result, point by point.

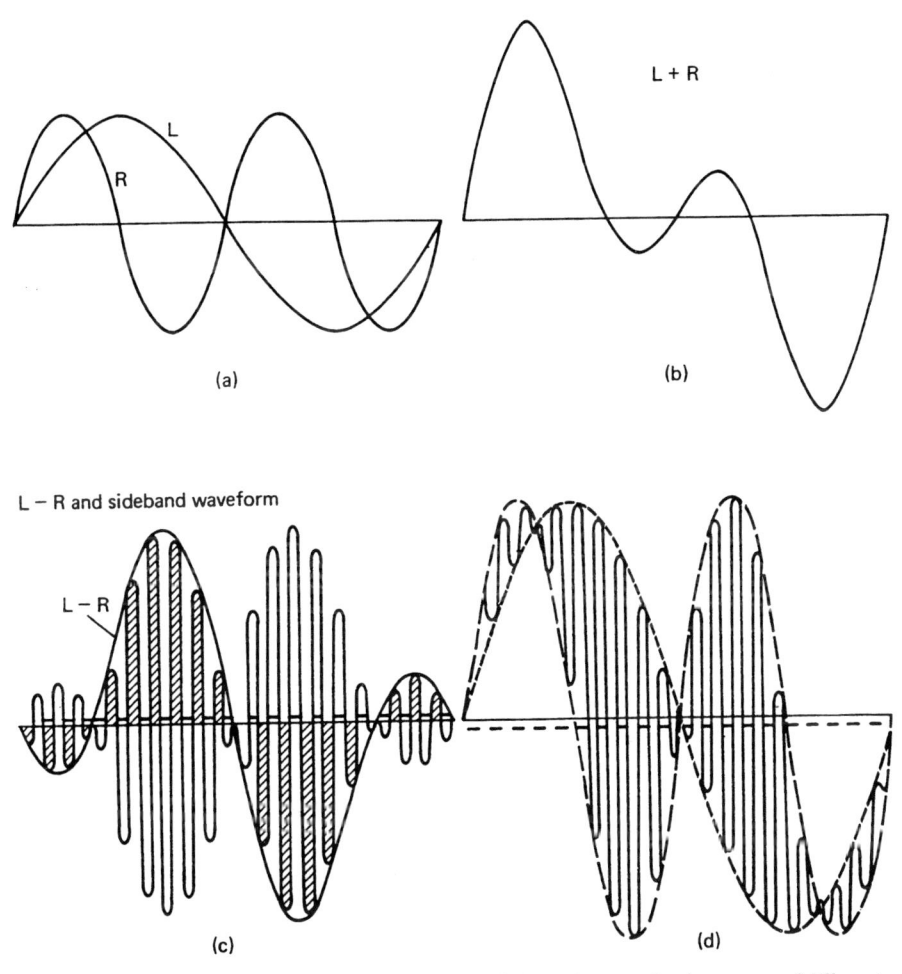

FIGURE 16-19 Waveforms for the FM stereo transmission of two audio sine waves of different frequency: (a) L and R signals, shown superimposed; (b) $L + R$ signal; (c) $L - R$ signal shown superimposed upon the $L - R$ sideband signal; (d) composite signal, minus the 19-kHz pilot carrier; (e) receiver waveform obtained by adding a 38-kHz subcarrier to composite with a subcarrier in phase with the original subcarrier; (f) receiver waveform obtained by adding a 38-kHz subcarrier to composite with a subcarrier out of phase with the original subcarrier.

Sec. 16-8   Time-Division (TD) Decoding

Subcarrier reinserted in phase with the original

Subcarrier reinserted out of phase with the original

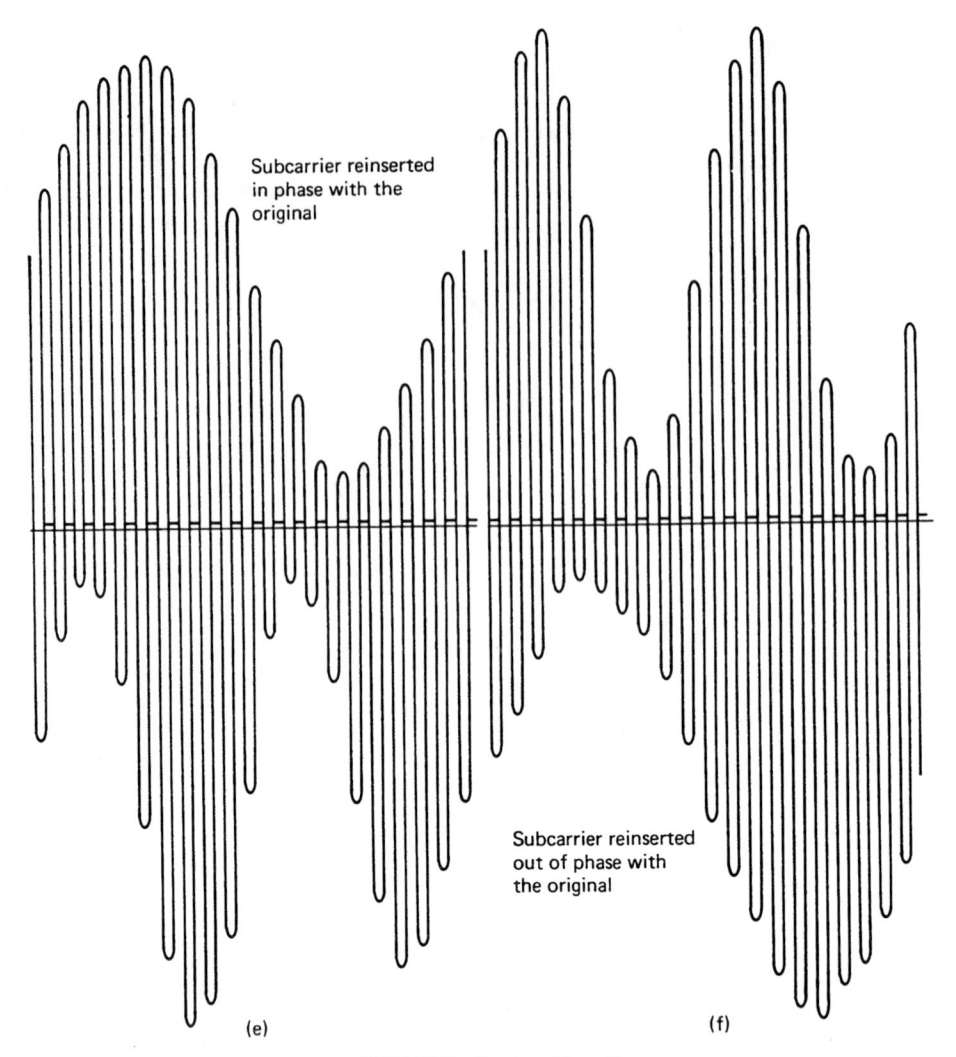

(e) (f)

FIGURE 16-19 (continued)

The $L - R$ sideband waveform is also shown in part (c). The phase reference must be available in order to construct this waveform, and it is shown in the figure. The amplitude of the envelope follows the amplitude of the $L - R$ signal. Note that when $L - R$ is positive, the sine-wave variation is in phase with the reference. When $L - R$ is negative, the sine-wave variation is out of phase with the reference. Note the phase reversals that occur when $L - R$ changes polarity.

The composite signal is obtained in part (d) by adding the $L + R$ and the $L - R$ sideband signal of part (c). This is accomplished by picking a particular peak of the sideband signal and noting its amplitude, paying attention to the algebraic sign. To this amplitude is added the amplitude of the $L + R$ signal at the same instant. The result of the addition is then plotted as one point of the composite waveform. The process is repeated for each peak, positive and negative, for the sideband waveform.

The presence of the L and R signals is indicated in part (d) by dotted and dashed lines drawn on the envelope. This is for illustrative purposes only.

The composite signal of part (d) is the signal transmitted by the FM receiver. As has been our practice, however, we have not included the 19-kHz pilot carrier in the waveform, in order to keep things as simple as possible. The composite signal is reproduced at the output of the FM detector at the receiver and is then combined with the positive and negative phases of the 38-kHz subcarrier as part of time-division decoding. The result of the combination is shown in the figure.

In part (e) the composite is combined with the positive phase of the subcarrier. Observe that the L signal appears on the top of the resulting waveform, and the R signal appears on the bottom. This waveform is applied to diode $D1$ of the time-division decoder of Figure 16-15.

In part (f) the composite is combined with the negative phase of the 38-kHz subcarrier. In this case the R signal appears on top, and the L signal appears on the bottom. This waveform is applied to diode $D2$ of Figure 16-15. We have already described how the diodes $D1$ and $D2$ recover the signal riding on the positive peaks of the applied voltages, so our discussion is now complete.

16-9 QUALITATIVE DESCRIPTION OF TD DECODING

Here is a simpler explanation that avoids the use of waveforms altogether. At the transmitter, let us imagine the subcarrier to call out "Left!" when it goes positive, and to call out "Right!" when it goes negative. When the composite signal hears "Left!" it immediately rushes to the value of the L signal. But soon thereafter it hears "Right!" and rushes to the value of the R signal. This process continues at a 38-kHz rate. The subcarrier calls out "Left! Right! Left! . . ." and the composite rushes back and forth, in step, between the L and R audio signal waveforms. In this manner the composite waveform is produced.

At the receiver, and in particular at the diode section of the TD decoder, the objective is to reproduce the L and R signals from the incoming composite waveform. The 38-kHz subcarrier is available to call the signals, but in this case, the diodes "open" and "close" in step with the signals called. When "Left!" is called, $D1$ conducts an amount determined by the level of the composite signal at that instant. This level, back at the transmitter, was determined by the level of the L audio signal when the "Left!" signal was called. Hence, the $D1$ conduction is controlled by the L signal at the transmitter.

When the subcarrier calls "Right!" at the TD decoder, diode $D2$ conducts an amount determined by the level of the composite. At this instant the level of the composite will correspond to the level of the R signal at the transmitter. Therefore, the $D2$ conduction is controlled by the R signal. Consequently, because of the instep switching action of the composite and of the diodes, the L signal is produced at $D1$ and the R signal is produced at $D2$.

Because the composite signal must "divide its time" between the L and R signals, this process is called *time-division multiplexing*, and the decoder is called a *time-division decoder*.

Credibility

To lend credibility to our simplified explanation, let us compare it with the explanation given earlier. Consider the excursions of the composite between the instantaneous L and R signal amplitudes. These occur at a frequency of 38 kHz, and the excursions will be greatest when the L and R signals differ in amplitude by the largest amount. In other words, the 38-kHz excursions of the composite are proportional to $L - R$. We recognize these excursions as being the $L - R$ sideband component of our original explanation.

When the L and R signals are of exactly the same amplitude, no excursion of the composite will be necessary. The composite will assume the common value of L and R, and no 38-kHz excursions will be present. This corresponds to the situation where $L - R$ is zero, for which the $L - R$ sidebands disappear.

The references above to the 38-kHz excursions is not to imply that the composite contains a 38-kHz signal component. Recall, however, that the sideband components are symmetrically located in frequency about 38 kHz, and the result of adding the component of a sideband pair is to produce a waveform whose peaks (+ or −) occur at a 38-kHz rate. These peaks represent the excursions referred to above, and they do indeed occur at 38 kHz.

Graphical Construction of the Composite Waveform

We now present a shortcut method of constructing the composite waveform. This method gives an exact waveform with the exception that the amplitude is only one-half the amplitude obtained with the longer procedure given earlier using the $L + R$ and $L - R$ sidebands. We ignore the 19-kHz pilot carrier to avoid unnecessary complication.

This procedure is based on the idea that the composite signal rushes back and forth between the L and R audio signals. Begin by constructing the L and R audio signals on the same axis as shown in Figure 16-20(a). Then construct the phase reference for the 38-kHz subcarrier as shown.

On the L signal, mark the points that coincide (in time) with the positive peaks of the phase reference. On the R signal, mark the points that coincide with the negative peaks of the phase reference. This is illustrated in Figure 16-20(b). Then, use these marks to draw in the composite waveform as shown.

When L is above R, the composite will be in phase with the reference, and conversely. Pay attention to the points where L and R become equal. Phase changes of the sideband component usually occur at these points.

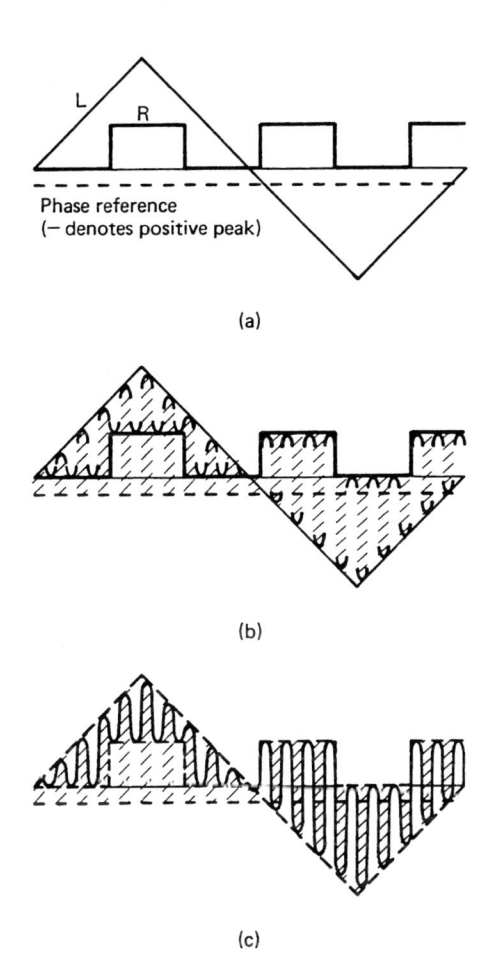

FIGURE 16-20 Shortcut method of constructing the composite waveform. (a) The L and R signals are drawn, superimposed. The phase reference is also constructed. (b) Upright or inverted cusps are used to mark the L signal at points corresponding to the positive peaks of the phase reference. Inverted cusps are used when L falls below R on the graph. A similar procedure is used to mark the points on the R signal corresponding to the negative peaks of the phase reference. (c) Finally, the cusps are connected to construct the composite waveform.

Sec. 16-9   *Qualitative Description of TD Decoding*

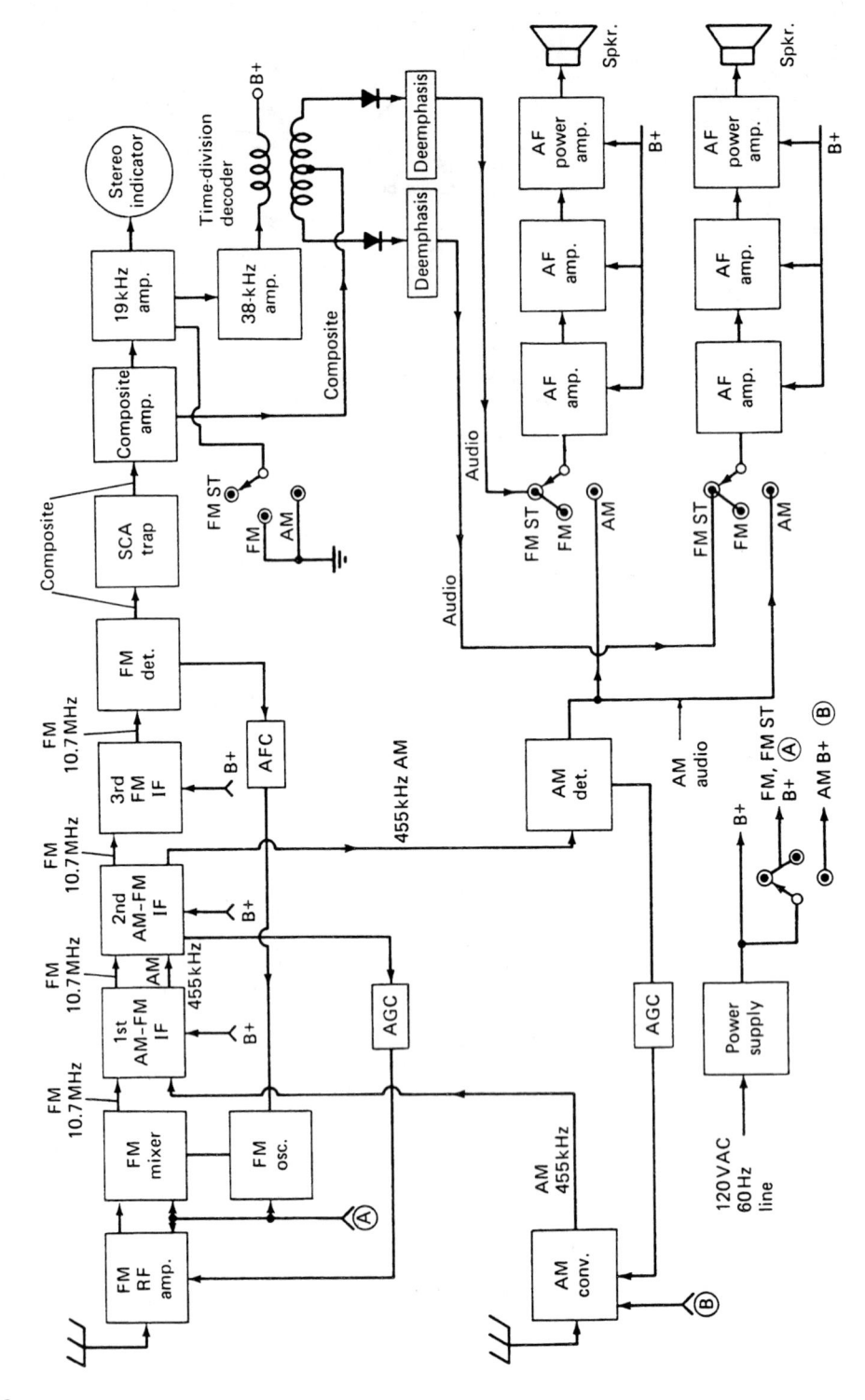

FIGURE 16-21 Block diagram of an AM, FM, FM-stereo receiver that utilizes a time-division decoder.

16-10 BLOCK DIAGRAM OF A TYPICAL FM-STEREO RECEIVER

Figure 16-21 is a block diagram of an FM-stereo receiver that utilizes a time-division stereo decoder. The first and second IF amplifiers are common to the AM and FM sections. A switched B+ power supply is utilized.

A separate FM oscillator is used with the FM frequency changer, and the frequency of the oscillator is controlled by the AFC voltage derived from the FM detector. An AGC voltage developed at the second FM IF amplifier is applied to the FM RF amplifier.

The output of the FM detector is applied through an SCA trap to the composite amplifier. The composite signal is applied to the center tap of the decoder transformer secondary. The 19-kHz pilot carrier is taken from the composite amplifier, and is amplified and doubled to obtain the 38-kHz switching voltage for the decoder.

The pilot carrier is shunted to ground by the function switch in the FM mode, and a stereo indicator circuit is tapped onto the 19-kHz signal path. Deemphasis networks follow the decoder diodes, and the audio is applied through the function switch to the stereo audio amplifier. In the FM monaural mode, the audio still passes through the stereo decoder.

The AM portion of the diagram is straightforward. AGC is provided to the AM converter, and the input to the stereo audio amplifiers are tied together in the AM mode.

SUMMARY

1. The FCC required the FM-stereo system to be compatible with the existing FM monaural system. Compatibility was achieved by using the $L + R$, $L - R$ scheme of multiplexing.

2. In the frequency spectrum of the modulating signal for the main FM carrier, space existed between 15 and 60 kHz which could be utilized to transmit a second channel of audio information by using a subcarrier.

3. The direct approach of placing the L signal on the main FM audio channel and placing the R signal on the subcarrier is unacceptable because monaural receivers would receive only half the stereo broadcasts.

4. By forming $L + R$ signal for modulating the main channel, monaural receivers reproduce audio information from both the L and R channels.

5. An $L - R$ signal is used to modulate a subcarrier located at 38 kHz in the spectrum of the modulating signal for the main FM carrier. FM-stereo receivers respond to this subcarrier and receive the $L - R$ signal. The $L + R$ and $L - R$ signals are combined in the stereo decoder section of an FM receiver to form L-only and R-only signals.

6. A double-sideband suppressed carrier scheme of modulation is used for the 38-kHz subcarrier so that the 38-kHz subcarrier component is not transmitted. Only the sidebands, which extend from 23 to 53 kHz, are transmitted.

7. When the L and R signals are identical, as in monaural transmissions, the $L - R$ signal is zero. Consequently, because of the DSSC modulation, no subcarrier sidebands are formed, and the transmission reverts automatically back to a monaural transmission. (The 19-kHz pilot may remain, however.)

8. A 19-kHz pilot carrier is transmitted at a level of 10% modulation of the main carrier so that the FM receiver can reconstruct the 38-kHz component of the subcarrier.

9. A property of the $L + R$ and $L - R$ signals is that both never reach a maximum at the same time. Thus, each signal is allowed to modulate the main FM carrier up to 90% without exceeding the total modulation limit of 100%.

10. A balanced modulator is used to produce a double-sideband suppressed carrier signal. The RF output amplitude of the modulator is controlled by the amplitude of the input audio signal, while the phase of the RF output is controlled by the polarity of the audio input. An AM waveform may be restored by reinserting the carrier component that was originally suppressed.

11. At the receiver, the 38-kHz subcarrier may be obtained either by using the 19-kHz pilot to synchronize a 38-kHz oscillator or by doubling and amplifying the 19-kHz pilot directly. The phase of the 38-kHz component must be accurately controlled.

12. The composite FM stereo signal consists of the main FM audio channel $(L + R)$, the pilot carrier at 19 kHz, the $L - R$ sidebands ranging from 23 to 53 kHz, and, possibly, a SCA subcarrier and sidebands which range from about 60 to 74 kHz.

13. A matrix decoder uses the direct approach to decoding the FM-stereo composite signal to form the L and R signals. Filter circuits separate the $L + R$ audio signal and $L - R$ sidebands. The 19-kHz pilot controls a 38-kHz subcarrier generator which reproduces the 38-kHz subcarrier. Reinsertion of the subcarrier generates an AM waveform from the $L - R$ sidebands, and this AM waveform, when detected, provides the $L - R$ signal. The $L + R$ and $L - R$ signals are then combined in a "matrix" to form the L-only and R-only signals.

14. In a time-division decoder, the composite is not separated into the $L + R$ and $L - R$ components, but rather the composite is added to the positive phase and to the negative phase of the regenerated 38-kHz subcarrier. The positive phase produces a waveform with the L-only signal represented by the positive peaks. The negative phase produces a waveform with the R-only signal represented by the positive peaks. When the waveforms are half-wave-rectified (AM-detected), the L and R audio signals are produced from the respective waveforms.

15. Time-division decoders are now in almost universal use in several forms. Matrix decoders are found primarily in older vacuum-tube receivers.

16. Since the peaks (positive or negative) of the DSSC waveform occur coincidentally with the peaks of the suppressed carrier component, it is possible to draw (with reasonable effort) the DSSC waveform produced by a given modulating signal. This makes it possible to graphically construct the composite signal, as was done in this chapter.

17. The composite signal resulting from any L and R waveforms may be graphically constructed by a procedure given in the latter part of this chapter.

QUESTIONS AND PROBLEMS

16-1. What is the allocated audio frequency range for the main audio channel?

16-2. (a) What is SCA and for what is it intended?

(b) What is the standard carrier frequency for the SCA?

(c) What type of modulation is used for SCA?

(d) What is the approximate bandwidth of the SCA component of the modulating signal of the main FM carrier?

16-3. Practice until you can draw from memory the circuits shown in Figure 16-2 for forming the $L + R$ and $L - R$ signals and for recovering the L and R signals from the $L + R$ and $L - R$.

16-4. In practice, how might an $L - R$ signal be converted to a $-L + R$ signal?

16-5. (a) What signal components make up the composite waveform?

(b) What is the frequency or range of frequencies represented by each component?

16-6. Follow the pattern of Figure 16-4 and calculate the appropriate items for the following combinations of L and R, given in the form $[L, R]$.

(a) $[4, 1]$.

(b) $[5, 3]$.

(c) $[2, -2]$.

(d) $[-2, 2]$.

16-7. What is the maximum percentage modulation that the following components may modulate the main FM carrier (in the absence of SCA)?

(a) $L + R$.

(b) $L - R$ sidebands.

(c) 19-kHz pilot.

16-8. (a) When the SCA is present, what percentage modulation of the main FM carrier is devoted to SCA?

(b) What must be done to the modulation percentages of the other components?

16-9. What accounts for the 180° change in phase of the RF signal when the modulating signal changes polarity as in Figure 16-7? (Refer to Figure 16-6.)

16-10. How is it possible to convert a DSSC waveform back to an AM waveform?

16-11. What is the "flywheel effect" of a resonant circuit, and how is it used in frequency doublers?

16-12. What is the purpose of the full-wave rectifier circuit of Figure 16-9(b)? Why are no filter capacitors used in conjunction with the rectifier?

Questions and Problems

16-13. In Figure 16-12, how does the composite for the *R*-only signal differ markedly from the composite for the *L*-only signal?

16-14. Practice until you can draw from memory the block diagram of an FM-stereo transmitter shown in Figure 16-13.

16-15. What is the purpose of the low-pass filter in the matrix decoder of Figure 16-14?

16-16. In a stereo receiver, why must the deemphasis networks come after the stereo decoder rather than immediately after the FM detector?

16-17. Describe a two-phase AM detector such as that used in Figure 16-14.

16-18. In the matrix decoder of Figure 16-14, give the function of the following:

(a) The separation control.

(b) The 23 to 53-kHz bandpass filter.

(c) The SCA trap.

(d) The 19-kHz keyed oscillator doubler.

(e) The deemphasis networks.

16-19. In a time-division decoder, is the composite signal separated into the $L + R$ and the $L - R$ sidebands?

16-20. The typical time-division decoder does not utilize a keyed oscillator. What is the approach taken to develop the 38-kHz subcarrier?

16-21. In Figures 16-15 and 16-16, what determines whether the *L* or the *R* signal will be developed at diode $D1$?

16-22. In a time-division decoder, why is it essential for the regenerated 38-kHz component to be exactly in phase with the original 38-kHz component at the transmitter?

16-23. Suppose, in a time-division decoder, that something mysterious happened so that the regenerated 38-kHz component was shifted in phase by 180°.

(a) What effect would this have upon channel separation?

(b) What would be the effect upon the left and right channels?

16-24. Describe the role of the phase reference in graphically constructing the $L - R$ sideband signal and the composite.

16-25. What happens to the phase of the high-frequency component (38 kHz) of the composite when the $L - R$ audio signal changes polarity?

16-26. Explain how the waveforms shown in Figure 16-19(e) and (f) are obtained. Is it possible to obtain the waveform of part (f) simply by inverting the waveform of part (e)? (Careful!)

16-27. On a sheet of graph paper, construct an *L* and an *R* waveform of your own choosing, superimposed as in Figure 16-20. Then, use the shortcut method of constructing the composite waveform for the *L* and *R* signals.

16-28. Which blocks of the block diagram of Figure 16-21 do not appear in a nonstereo AM-FM receiver?

16-29. Practice until you can draw from memory the block diagram of Figure 16-21.

17
FM-STEREO DECODER CIRCUIT ANALYSIS

The addition of the FM-stereo decoder section to form an FM-stereo receiver requires essentially no alteration of the receiver sections up to the FM detector. At the FM detector, a bandwidth of 50 Hz to 53 kHz must be maintained for full stereo capabilities, and therefore the deemphasis network must be placed at the decoder output rather than at the output of the FM detector. Naturally, appropriate function switching and stereo audio amplifiers must be provided.

Since the front end, IF amplifiers, audio amplifiers, and power supply sections remain essentially the same as for AM-FM monophonic receivers, this chapter is devoted to the two basic types of multiplex decoders, circuit variations of each type, and new circuitry that may be found in FM-stereo receivers. The new circuitry includes FM muting, IC IF amplifiers, FM quadrature detectors, and phase-locked-loop time-division decoders.

17-1 TYPICAL MATRIX DECODER

A typical matrix decoder section is shown in Figure 17-1. Three triode stages provide the necessary amplification and oscillator functions. Three tuned signal paths are used to divide the signal information before it is decoded by the two-phase AM detector and resistor matrix. The decoded L and R signals are deemphasized and then passed on to the stereo audio amplifier.

Input Amplifier and Cathode Follower

The multiplex (MPX) input amplifier, $V8A$, receives the composite signal from a discriminator through $C52$. All components of the composite signal ($L + R$, $L - R$ sidebands, 19-kHz pilot) are amplified and coupled through $C53$ to the cathode follower 19-kHz amplifier, $V8B$.

Frequency separation is achieved at the two output points of $V8B$. In the plate circuit, $L18$ is tuned to 19 kHz and couples only the pilot signal to the oscillator coil, the $L18$ secondary. Since the 19-kHz signal is taken from the plate of $V8B$, the stage acts as an amplifier for the 19-kHz component. However, since the other components are removed from the cathode circuit, no amplification of these components occurs. In this capacity, $V8B$ acts as a cathode follower.

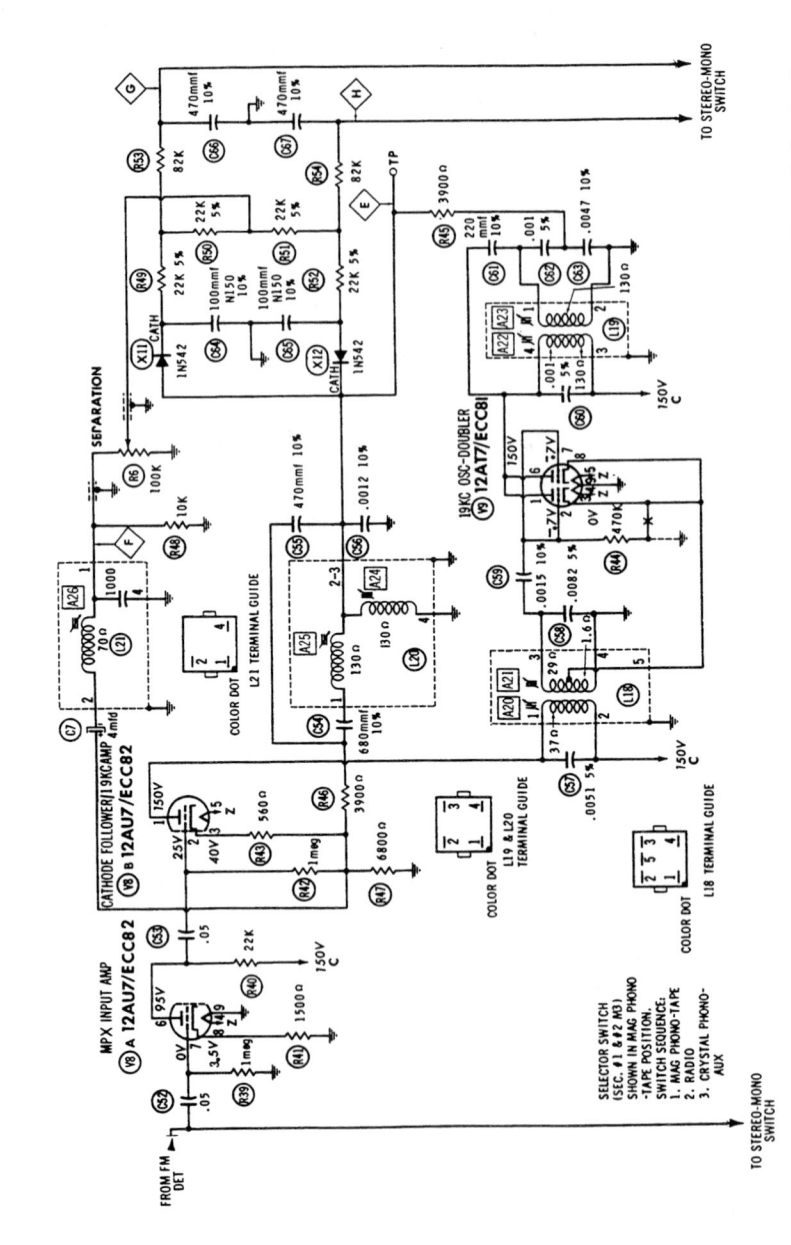

FIGURE 17-1 Matrix-type FM-stereo decoder (a Photofact schematic, courtesy of Howard W. Sams & Co., Inc.).

Resistor $R47$ serves as the load for the cathode follower, and the output is developed across the resistor. The $L - R$ double-sideband suppressed carrier signal (23 to 53 kHz) is removed by a bandpass filter composed of $C54$, $C55$, $L20$, and $C56$. This filter rejects frequencies below 23 kHz and above 53 kHz. In particular, it very sharply attenuates the 67-kHz SCA component that may be present as a result of SCA transmissions.

The $L + R$ signal is removed from the cathode circuit by a low-pass filter, $L21$. This filter rejects frequencies above 15 kHz, so that only the $L + R$ component is applied to separation control $R6$. The separation control simply varies the amount of $L + R$ signal applies to the decoding matrix.

Since a large load resistance (6800 Ω) is used in the cathode circuit, grid resistor $R42$ is returned to the high side of the load resistor $R47$ to avoid having $R47$ affect the bias of the tube. As connected, only $R43$ determines the self-bias developed at the tube cathode. Resistor $R46$ separates the filter networks ($L20$ and $L21$) to avoid interaction of the filter responses.

19-kHz Oscillator-Doubler

The two triode sections of $V9$ are connected in parallel as a modifed Hartley oscillator. Feedback is achieved by the auto-transformer action of the $L18$ secondary, which also determines the oscillator frequency of 19 kHz. Grid leak bias is provided by $C59$ and $R44$.

The oscillator is forced to run at exactly the correct frequency by the 19-kHz pilot signal applied to the $L18$ primary. Since the natural frequency of the oscillator is set during alignment as close as possible to 19 kHz, the oscillator will "lock in" and synchronize with the pilot carrier when it is present. This represents a useful application of oscillator pulling, described in Sec. 7-7.

The output of $V9$ is tuned to twice the frequency of the oscillator, and it therefore acts as a frequency doubler (Sec. 16-5). Consequently, the output of the circuit is a signal exactly twice the frequency of the 19-kHz pilot; it is the regenerated 38-kHz subcarrier.

Two-Phase Detector and Matrix

The 38-kHz signal is combined with the $L - R$ double-sideband signal at the input to detector diodes $X1$ and $X12$. The result is that a conventional AM waveform is produced whose modulation is the $L - R$ audio component. Since the diodes are oppositely connected, $X11$ will produce the positive phase of $L - R$, while $X12$ develops the negative phase, $-L + R$. Capacitors $C64$ and $C65$ serve as conventional AM detection capacitors and remove most of the 38-kHz component from the output of the diodes.

Resistors $R49$, $R50$, $R51$, and $R52$ constitute the "matrix" and serve actually as summing resistors for the $L - R$ and $L + R$ components. The $L + R$ component is applied through $R50$ and $R51$ to each diode output. $L - R$ is applied through $R49$ to the summing node at the $R49$-$R50$ junction. $-L + R$ is applied through

R52 to the summing node at the R51-R52 junction. The left and right audio signals are developed at these nodes as described in Sec. 16-2. Deemphasis networks composed of R53, C66, R54, and C67 restore the proper amplitude to the high audio-frequency components, and the L and R signals then pass from <G> and <H> to the function switch and to the stereo audio amplifiers.

FM/FM Stereo Switching

In the FM-stereo mode, the signal developed at test point <G> is fed to the left audio amplifier, and the signal developed at test point <H> is fed to the right audio amplifier. In the FM monaural mode, however, <G> and <H> are disconnected from the audio inputs, and the composite signal developed at the FM detector output is coupled to both audio amplifiers by the function switch. The 19-kHz pilot and the 38-kHz sidebands are above the audio range and therefore will not be heard, even when stereophonic transmissions are received in the FM monaural mode. Only the L + R information will be heard. Actually, the high-frequency components at 19 and 38 kHz will never reach the speakers, because of the frequency response of the audio amplifiers.

17-2 MATRIX-DECODER VARIATION

A common variation of a matrix decoder is shown in Figure 17-2. This is a plug-in unit with all inputs, outputs, and power supply voltages coming through the MPX plug. This unit utilizes a different circuit configuration to achieve the same objectives as that of the previous decoder.

Signal Separation

Signal separation occurs in the plate circuit of V8A. The 19-kHz signal is applied through C54 to L9, the oscillator coil. The L − R sideband signal sees a low-impedance path through the bandpass filter between R46 and M13. The composite signal is applied to R42, where the high-frequency components are removed by C50, leaving only the L + R audio signal.

19-kHz Oscillator-Doubler

A modified Hartley oscillator (V8B) operates at the exact frequency of the pilot carrier, since the pilot is coupled to the oscillator tuned circuit by C54. The plate circuit, tuned to twice the 19-kHz value, produces the 38-kHz component necessary for restoring the AM waveform from the L − R sidebands.

498 FM-Stereo Decoder Circuit Analysis Chap. 17

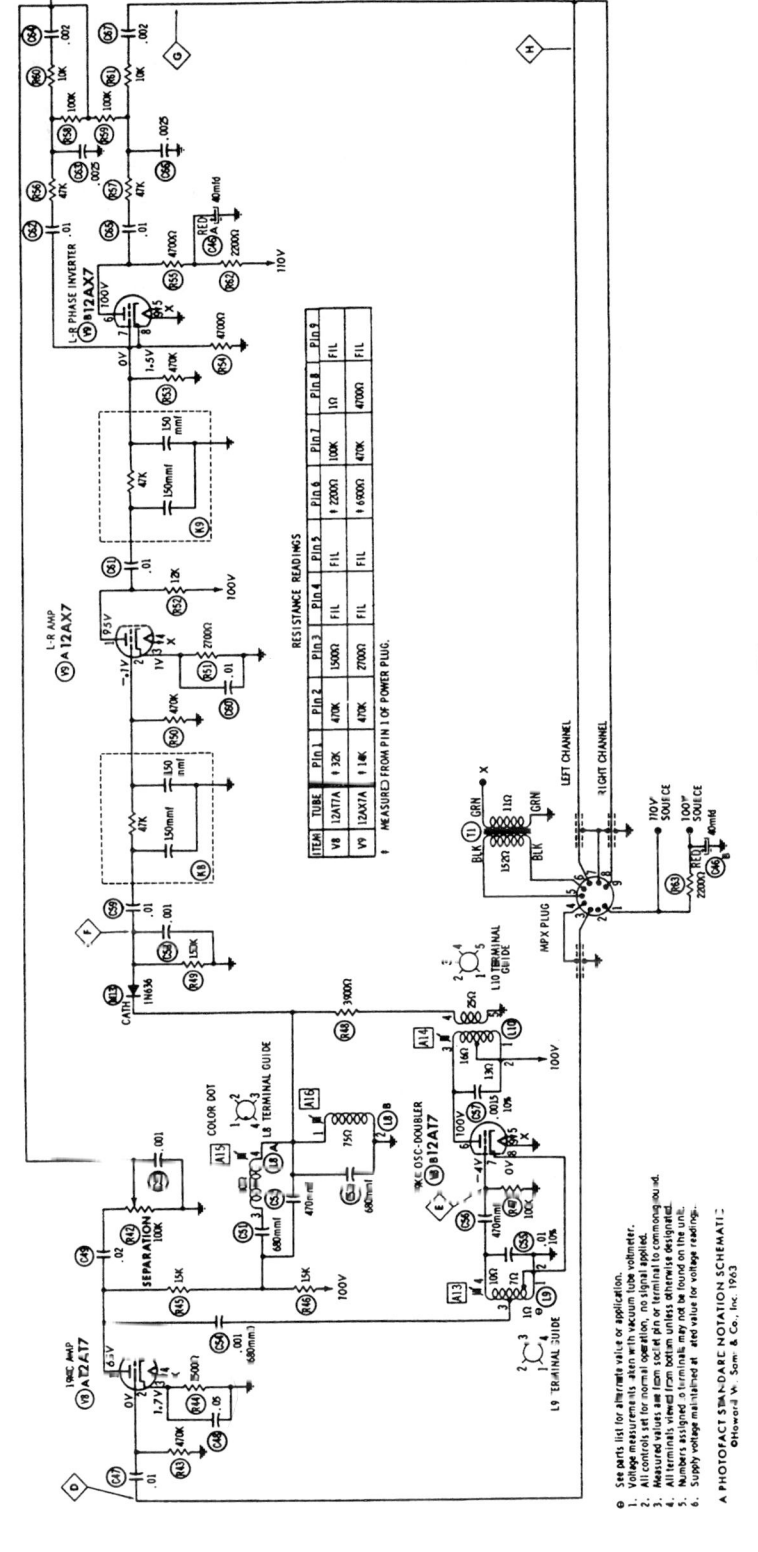

FIGURE 17-2 Variation of matrix-type FM-stereo decoder.

Single-Phase Detector

The restored AM waveform is demodulated by $M13$ to obtain the $L - R$ audio signal. More specifically, since the diode passes the negative portion of the signal, a $-L + R$ signal is developed across $R49$. Capacitor $C58$ and couplate $K8$ remove the 38-kHz component appearing at $<F>$.

Triode $V9A$ amplifies the $-L + R$ signal and inverts its phase so that an $L - R$ signal appears at the output. $V9B$ is a phase splitter (see Sec. 5-6) that produces an $L - R$ signal at the cathode while producing a $-L + R$ signal at the plate. These outputs are applied to summing resistors $R56$ and $R57$, which tie in with the $L + R$ component applied to the summing nodes through $R58$ and $R59$. The L and R audio signals appear at test points $<H>$ and $<G>$, respectively.

17-3 STANDARD TIME-DIVISION MULTIPLEX DECODER

A standard time-division multiplex decoder is shown in Figure 17-3. The MPX input amplifier provides separation of the 19-kHz pilot carrier from the composite signal. The composite signal (less 19-kHz pilot) is then applied to the dual-diode decoder circuit. The 38-kHz carrier is developed from the 19-kHz pilot after it is amplified and then applied to a frequency doubler. A stereo indicator amplifier senses the presence of the 19-kHz pilot and drives a stereo indicator lamp accordingly.

MPX Input Amplifier and Signal Separation

The composite signal developed at the ratio detector is applied to the input to the MPX amplifier through a 67-kHz trap that rejects the SCA component. The MPX stage is universal-biased, and the collector load consists of a 19-kHz resonant circuit and a collector load resistor $R58$. The 19-kHz pilot is taken off by $L15$ and applied to the base of $Q8$. The $L + R$ audio and the $L - R$ sidebands are developed across $R58$ and are coupled to the decoder via CTP 63.

19-kHz Amplifier and Doubler

The 19-kHz amplifier $Q8$ is a tuned amplifier whose output is applied to diodes $X10$ and $X11$, which are connected in the same configuration as a full-wave rectifier. In this application, however, no filter capacitor is provided for the "rectifier," so that positive-going pulses rather than a DC voltage appear at the output (test point $<D>$). $R64$ is the load resistor for the full-wave rectifier. The pulse rate is twice the input frequency of 19 kHz, and the pulses are applied through $R65$ to the base of $Q9$. The output circuit of $Q9$ is tuned to 38 kHz and develops a sine wave from the pulses.

Observe that neither the 19-kHz amplifier nor the 38-kHz amplifier is provided with a fixed DC forward bias at the B-E junction. In fact, the emitter of $Q8$ is held at a very small positive potential by the voltage divider, composed of $R6$, $R61$, and

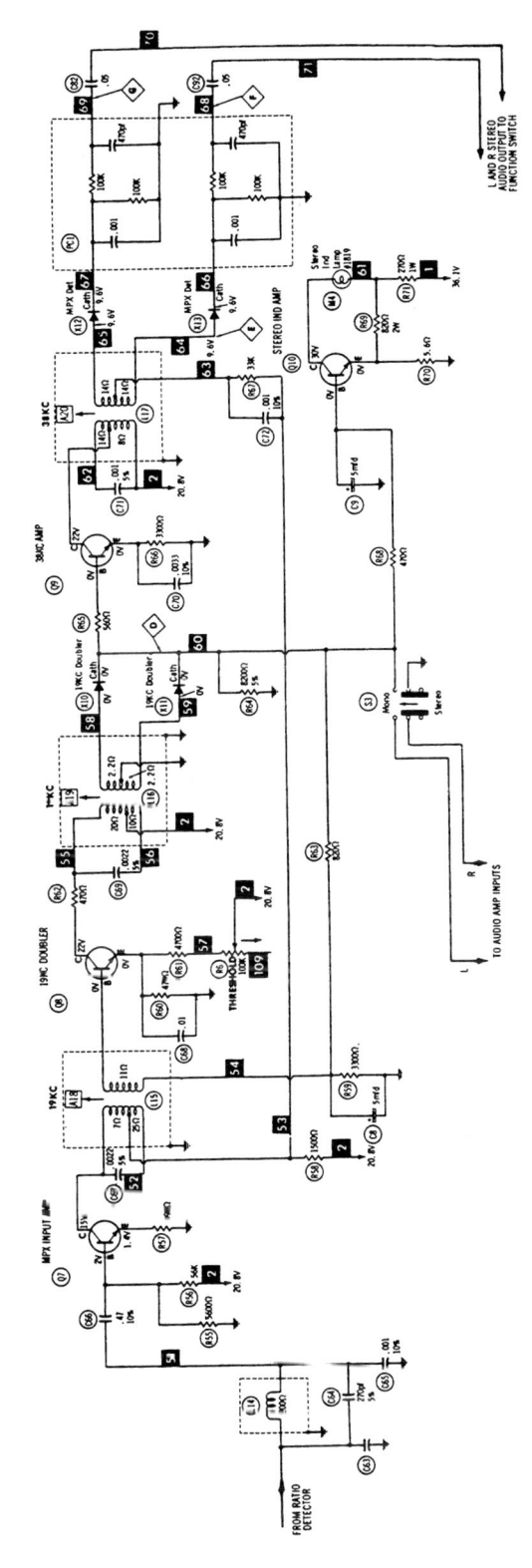

FIGURE 17-3 Time-division FM stereo decoder (a Photofact schematic, courtesy of Howard W. Sams & Co., Inc.).

$R60$, so the B-E junction is slightly reverse-biased. Consequently, $Q8$ will not conduct until a 19-kHz signal is applied that is large enough in amplitude to drive the transistor into conduction on the positive peaks of the input signal.

When a 19-kHz signal larger than the threshold amplitude is applied to $Q8$, the positive pulses will appear at the output of the doubler rectifier, test point $<D>$. These pulses are sampled by $R63$ and applied to filter capacitor $C8$, which smooths the pulses to a positive DC voltage. This voltage is applied to the base of $Q8$ and tends to increase the conduction through the transistor.

The operation of doubler transistor $Q9$ depends upon pulsed operation, and therefore no significant DC component is provided to the $Q9$ base. At this point we note that voltages given on the schematic (Figure 17-3) are obtained under no-signal conditions.

Decoder Circuit

The operation of this type of decoder was described in detail in Chap. 16. The composite signal ($L + R$ audio and $L - R$ sidebands) is applied to the center tap of the L secondary, and the 38-kHz subcarrier component is induced across the secondary. At CTP 65, the composite signal is added to one phase of the 38-kHz signal to produce a waveform with the left audio information appearing on the positive envelope, as shown in Figure 16-18. At CTP 64, the composite is added to the opposite phase of the 38-kHz signal to produce a waveform with the right audio information appearing on the positive envelope.

MPX detector diodes $X12$ and $X13$ conduct only during the positive portion of the applied waveforms, and therefore produce the left and right channel information as described in Chap. 16. The filter networks in PC1 remove the 38-kHz component and provide the proper deemphasis.

Stereo Indicator

The positive-going, 38-kHz pulses at the output of the 19-kHz doubler (test point $<D>$) are coupled through $R68$ to the stereo indicator amplifier $Q10$. Since no fixed DC bias is applied to $Q10$, conduction will occur only when the pulses are present. Capacitor $C9$ filters the pulses to produce a constant voltage at the base so that a steady current flows through the stereo indicator lamp $M4$ in the collector circuit. When the input pulses disappear, the voltage on $C9$ falls to zero, conduction through the transistor ceases, and the stereo lamp goes out.

$R69$ and $R70$ form a voltage divider that raises the voltage at the emitter so that the B-E junction of $Q10$ is reverse-biased when no input pulses are applied. When pulses are applied and the transistor begins to conduct, the voltage at CTP 61 decreases so that the reverse bias is relieved. This produces a sharp turn-on and turn-off characteristic for the indicator.

Mono-Stereo Switch

When switch S3 is placed in the mono position, the 38-kHz pulses at the doubler input are shorted to ground and the inputs to the left and right channels of the stereo audio amplifier are tied together. This time-division decoding function is killed since no 38-kHz signal is present at the MPX diodes, but the L + R information still passes through the decoder section to the audio amplifiers. The L + R signal passes through the MPX diodes without being rectified, because the diodes are biased into the forward conduction region by voltage applied through R58 and R67 near CTP 63. In the FM stereo mode, the diodes are switched by the 38-kHz signal to reverse-bias the diodes so that rectification occurs.

17-4 BRIDGE-TYPE TIME-DIVISION DECODER

Figure 17-4 shows a time-division decoder that utilizes a bridge-type switched detector. The circuitry (except the bridge) is similar to the previous decoder except that a 19-kHz amplifier is not used before the doubler, and a fixed bias is provided for the 38-kHz amplifier.

Composite Amplifier

The composite signal from the ratio detector is coupled by the SCA trap to the base of Q7, the composite amplifier (also known as the MPX input amplifier). Three outputs are taken from the composite amplifier. The amplified 19-kHz pilot is developed across T8 in the collector circuit and is applied to the doubler diodes, D9 and D10. The composite signal is developed across R203 in the emitter circuit and is coupled to the center-tapped secondary of 38-kHz transformer T9.

Another composite signal, 180° out of phase, is developed across the separation control, VR216, and is applied to the bridge decoder circuit. The purpose of these out-of-phase composite signals is described along with the bridge detector in a following section.

Doubler, 38-kHz Amplifier, and Indicator Control

The 19-kHz pilot signal appearing at the center-tapped secondary of T8 is applied to the full-wave frequency-doubler diodes D9 and D10. The resulting 38-kHz positive pulses are coupled to Q8, where they are amplified and applied to T9 to produce the 38-kHz sinusoidal switching voltage.

The 38-kHz signal at the collector of Q8 is coupled through C215 to the indicator control rectifier D15. The negative-going portion of the signal is removed by D15, and the remaining positive-going pulses drive Q9 into conduction. This causes the stereo lamp to illuminate. The voltages shown at the base and collector of Q9 have

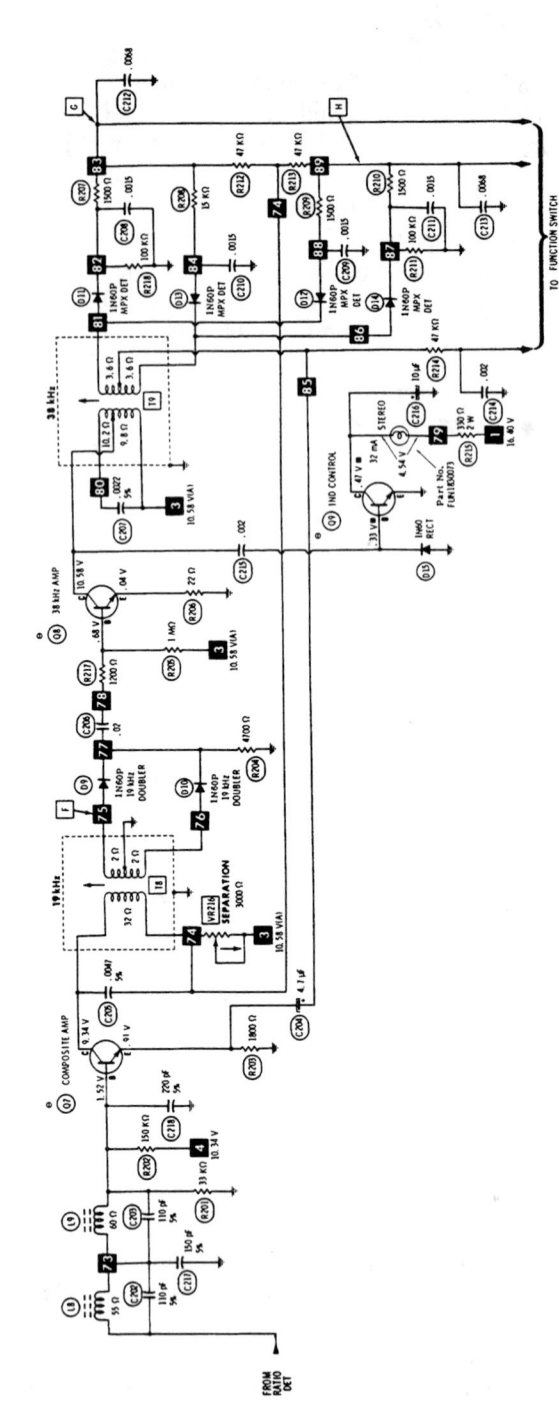

FIGURE 17-4 Bridge-type time-division decoder (a Photofact schematic, courtesy of Howard W. Sams & Co., Inc.).

a small, solid square beside them to indicate that they appear only when a stereo station is being received.

The 0.33 V at the base appears to be insufficient to turn on transistor $Q9$, but recall that the meter used to measure the voltage indicates the average value of the 38-kHz pulses. The peaks go beyond the 0.6 V necessary to turn on $Q9$. Electrolytic capacitor $C216$, connected to the collector, smooths the 38-kHz pulses primarily to keep the pulses out of the power supply. It is a decoupling capacitor.

Bridge MPX Detector

The bridge switched detector circuit of Figure 17-4 is redrawn in Figure 17-5 to emphasize the bridge arrangement and to show the signal input points. The bridge provides excellent cancellation of the 38-kHz switching signal, which simplifies the filtering necessary at the left and right audio output points.

The left channel audio is developed by conduction through $D11$ and $D13$ and a resulting current flow through $R218$. The right-channel audio is developed by conduction through $D12$ and $D14$ and a resulting current flow through $R211$. Current flow through the diodes is determined primarily by voltages applied to CTPs 81 and 86.

Voltages at these points result from the 38-kHz subcarrier and the composite applied to the center tap of the $T9$ secondary. The composite signal applied to CTP 74 at the center of the bridge performs only a corrective function and contributes only an $L + R$ component to the left and the right output points.

Consider the diode conduction when only the 38-kHz subcarrier voltage is applied, that is, when the composite is zero. Equal and out-of-phase voltages will

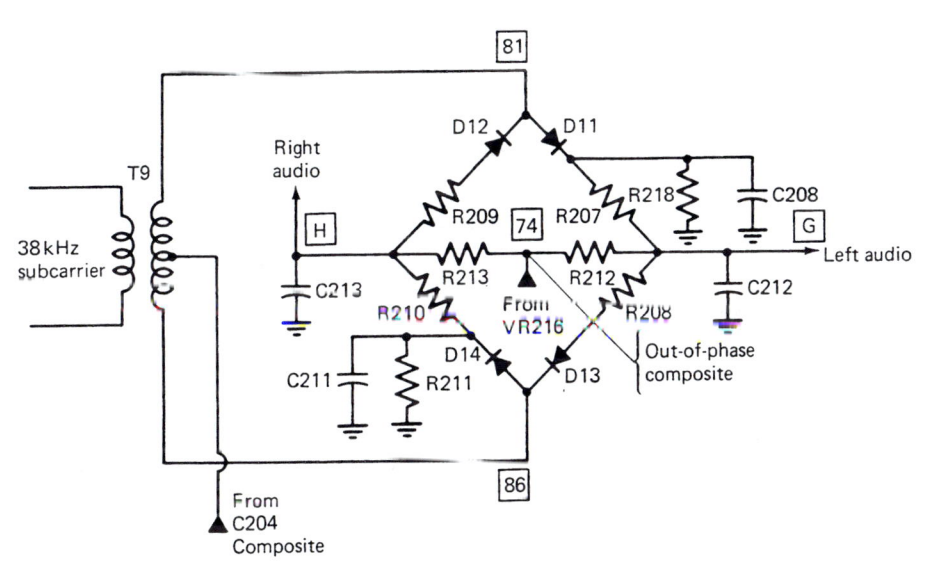

FIGURE 17-5 Bridge decoder redrawn from Figure 17-4.

be applied to CTPs 81 and 86. Diodes $D11$ and $D13$ will conduct equally when CTP 81 is positive relative to CP 86. No current will flow through $R218$, and no voltage will be developed at $<G>$. When the polarity reverses, neither $D11$ nor $D13$ will conduct, and the voltage at point $<G>$ will remain at zero.

Since $D12$ and $D14$, on the other side of the bridge, are connected with opposite polarity to $D11$ and $D13$, conduction through these diodes will occur 180° out of phase with conduction through $D11$ and $D13$. Also, no voltage will occur at point $<H>$ when the composite is zero.

Thus, we conclude that no output voltage will be developed at either $<G>$ or $<H>$ when the composite is zero, because equal conduction occurs in the diodes. An unbalance in conduction must occur before an output voltage is developed. It is only in this case that a current flow will occur through $R218$ or $R211$.

Recall from Sec. 16-8 that different waveforms are formed when the composite is added to the positive and negative phase of the 38-kHz subcarrier. Consequently, different waveforms are formed at CTP 81 and CTP 86 in Figure 17-5. With the L and R waveforms of Figure 16-17, the waveforms shown in Figure 17-6 will be produced.

Figure 17-6(a) is the waveform at CTP 81. Diode $D11$ conducts on the positive

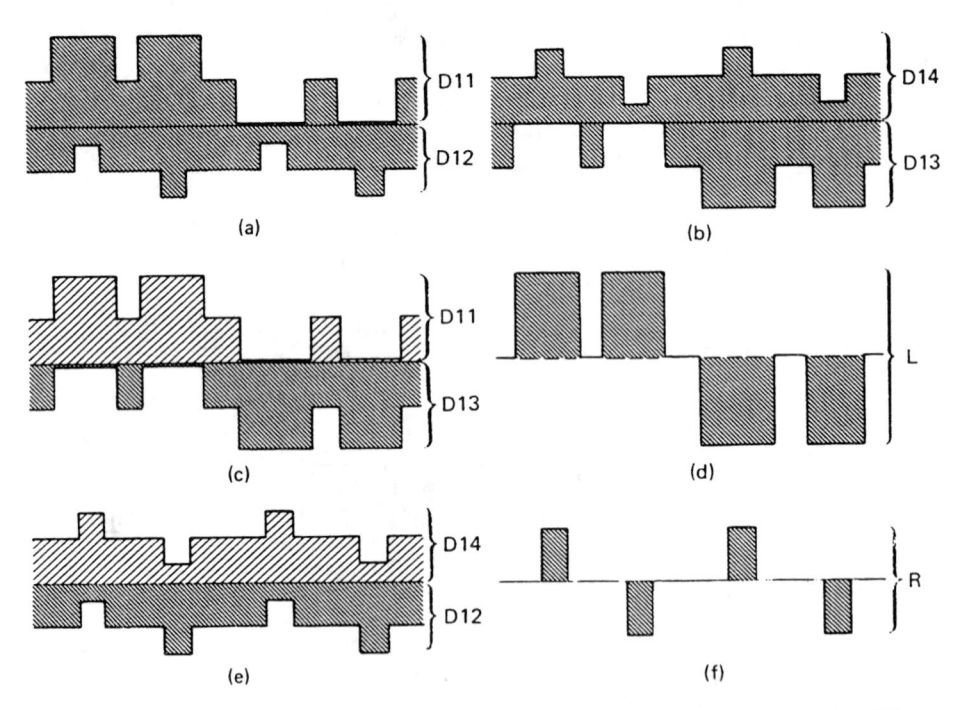

FIGURE 17-6 (a) and (b) Typical waveforms applied to the diodes of the bridge decoder of Figure 17-5. (c) Conduction of $D11$ and $D13$. (d) Resulting L signal. (e) Conduction of $D14$ and $D12$. (f) Resulting R signal.

portion, and $D12$ conducts on the negative portion. Part (b) is the waveform at CTP 86. Diodes $D14$ and $D13$ conduct on the positive and negative portions, respectively.

The L output developed at point $<G>$ results from the conduction of diodes $D11$ and $D13$. The output-voltage contribution of each diode is shown in part (c), and the result of the combined $D11$ and $D13$ conduction is shown in part (d). This is obviously the L signal, shown previously in Figure 16-17.

A similar combination of voltages from $D12$ and $D14$ reduces the R signal, as illustrated in Figure 17-6(e) and (f). Thus, the bridge decoder recovers the L and R signals.

Correction

Let us now consider why the composite signal is added at CTP 74 at the center of the bridge in Figure 17-5. In practice, the $L - R$ sidebands of the composite signal are usually several decibels smaller in amplitude than the $L + R$ component. Owing to the higher frequencies, the $L - R$ sidebands are more prone to attenuation by the circuit capacitances. Thus, in the bridge circuit, the $L - R$ component appearing at the stereo outputs is smaller in amplitude than the $L + R$. If left uncorrected, this inequality would result in decreased separation between the left and right channels.

If the proper amount of out-of-phase $L + R$ signal $(-L - R)$ is added to the outputs, a portion of the $L + R$ amplitude will be canceled, and the $L + R$ and $L - R$ components become equal. This produces improved stereo separation. Hence, the control that varies the amplitude of the out-of-phase $L + R$ component is called the *separation control*. This is VR216 in Figure 17-4, and in Figure 17-5, the correction composite signal is added equally to $<G>$ and $<H>$ via $R212$ and $R213$.

Recall, however, that a composite signal is more than just $L + R$; it also includes the $L - R$ sidebands, so that the sidebands also appear at CTP 74. However, capacitors $C212$ and $C213$ attenuate the sideband components of the correction composite so that only the $L + R$ component is added to the outputs.

FM/FM-Stereo Function Switch

In the FM-stereo mode, the outputs from test points $<G>$ and $<H>$ are connected through the function switch to their respective audio amplifiers. When a monaural FM signal is received, the audio information has to pass through $C204$, $T9$, and the diode circuitry to reach the audio amplifiers. When the function switch is placed in the FM monaural position test points $<G>$ and $<H>$ are disconnected from the audio stages, and the $L + R$ audio information is coupled through $R214$ and the function switch to both audio amplifiers. Capacitor $C215$ removes the $L - R$ sidebands if a stereo broadcast is being received.

17-5 BIPLEX DETECTOR

In this section we describe a stereo decoder that is notable both for the simplicity of the circuit and for subtlety of operation. Desirable features of the biplex detector include an improvement of demodulation efficiency by a factor of about two, improved channel separation at the higher audio frequencies, and the ability of the detector to receive both stereo and monophonic transmissions at approximately the same volume level.

The general configuration of a biplex detector is shown in Figure 17-7, and the schematic of a complete stereo decoder that utilizes a biplex detector is shown in Figure 17-8. A conventional biasing arrangement is *not* used. In the no-signal condition, both the B-E and B-C junctions are reverse biased.

A transistor having *bilateral* characteristics is utilized. A bilateral transistor will function with the collector used as the emitter, and with the emitter used as the collector. All bipolar transistors are bilateral to a certain degree, but special symmetrical transistors have nearly the same characteristics in both directions.

Collector-emitter voltage is provided by the 38-kHz regenerated subcarrier switching voltage. Since this voltage changes polarity at the 38-kHz rate, the collector and emitter functions of the transistor are "switched" at the same rate. The transistor will conduct in both directions when the applied voltages are sufficient to overcome the reverse bias of the base-emitter junction.

Note that the composite signal is applied to the base of the transistor, and recall that the composite consists of the $L + R$ audio component and the $L - R$ sidebands. The composite signal passes through R_{C2} to the center tap of the transformer secondary. At this point, however, the $L - R$ sideband component of the composite is shunted to ground by C_{SF} in Figure 17-7. Thus, only the $L + R$ component remains at the center tap, and this component is applied equally to the emitter and

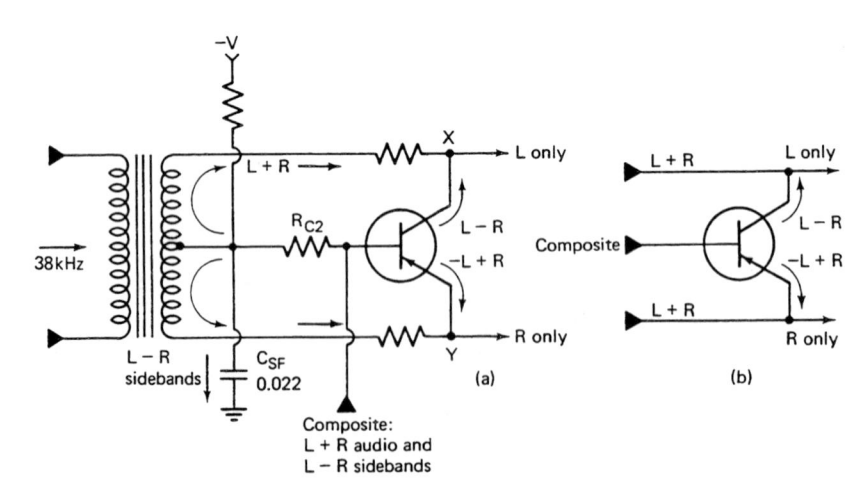

FIGURE 17-7 (a) General configuration of a biplex detector. (b) Illustration of the manner in which the signals combine to produce the *L* and *R* signals.

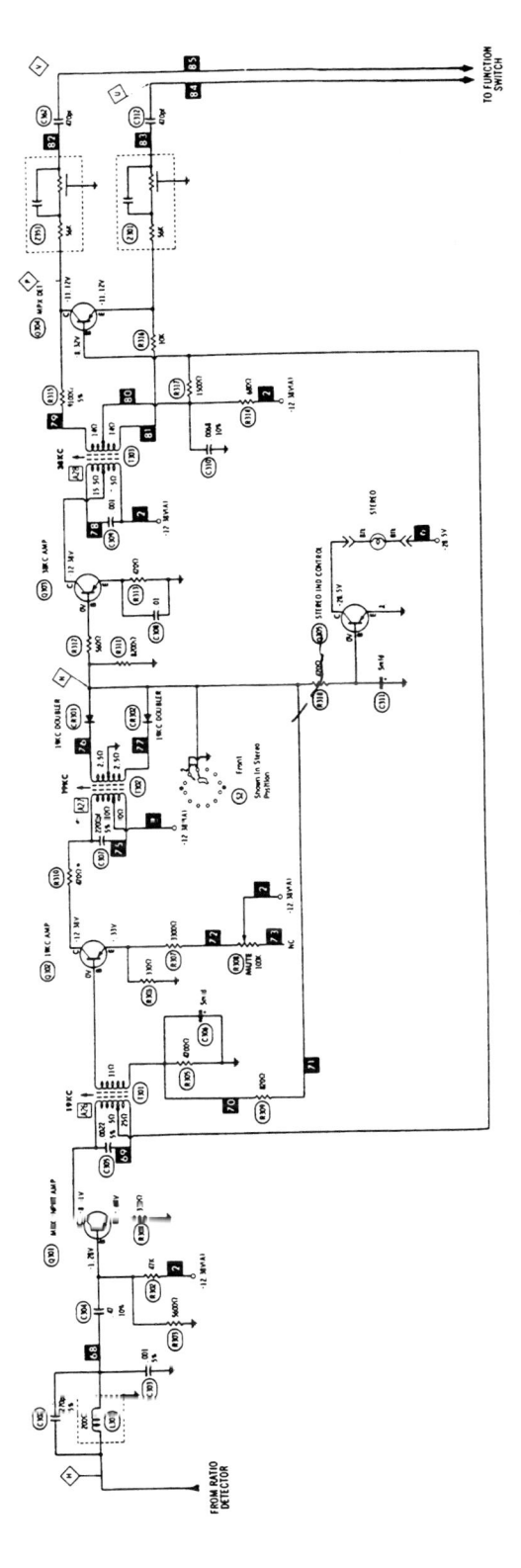

FIGURE 7-8 Stereo-decoder section that uses a biplex detector (courtesy of Zenith Radio Corporation).

collector of the transistor via R_{L1} and R_{L2}. Therefore, the $L + R$ component appears in equal amplitude and phase at all three terminals of the transistor.

We see now that only the $L - R$ sideband components will appear across the base-emitter junction of the transistor, and consequently the conduction of the transistor will be dependent upon the $L - R$ sideband component. Further, a phase of the 38-kHz switching voltage will be added to the $L - R$ sideband component to overcome the reverse bias of the B-E junction. Hence, the switching voltage brings the transistor into the conduction region, but the amount of conduction is modified by the level of the $L - R$ sideband component of the composite signal.

The 38-kHz voltage that appears at the emitter and collector of the transistor is modified by the conduction of the transistor, which, as stated above, depends upon the sideband component of the composite. The effect of this is to simultaneously recover the $L - R$ audio signal from the 38-kHz sidebands and to perform matrix addition of the $L + R$ and $L - R$ components to form the original L and R signals.

Summarizing, the circuit is such that $L + R$ component appears at each terminal of the transistor and therefore has no effect upon the conduction of the transistor. Transistor conduction depends upon the $L - R$ sideband component of the composite, and the effect of this conduction is to contribute an $L - R$ to the collector and a $-L + R$ audio component to the emitter. These components add to the $L + R$ component to form left-only and right-only audio signals at the collector and emitter of the transistor.

FM Monaural Signal Path

In the FM monaural mode, the 19-kHz pilot signal path is interrupted so that no switching voltage is applied to the biplex detector. In such a case the bilateral transistor does not conduct, and only the $L + R$ audio component appears at the collector and emitter. Hence, the $L + R$ audio is applied to both left and right stereo amplifiers, resulting in monophonic operation.

When a monophonic (nonstereo) FM signal is received in the FM stereo mode, no $L - R$ sidebands will be present, and no conduction will occur in the biplex transistor. Therefore, only $L + R$ information will be applied to the audio amplifiers.

Typical Circuit

A stereo decoder section that uses a biplex detector is shown in Figure 17-8. The composite signal is applied to the base of the MPX input amplifier through an SCA trap. The output circuit of the MPX amplifier ($Q301$) consists of a 19-kHz resonant circuit and collector load resistors $R317$ and $R314$ in the biplex detector circuit. The 19-kHz pilot signal is applied to an amplifier and doubler circuit that is quite similar to circuits already described. The 38-kHz output of the doubler is applied to the biplex detector circuit at the $T303$ secondary.

The biplex detector transistor is $Q304$, and the 38-kHz switching voltage is applied to the collector and emitter through load resistors $R315$ and $R316$. These resistors are slightly different in ohmic value to compensate for the imperfect bilateral characteristic of the transistor. The composite signal is developed at the base of $Q304$ by the collector current of the MPX input amplifier $Q301$. Capacitor $C310$ removes the sideband components from the signal applied to the center tap of the $T303$ secondary so that an $L + R$ signal is applied to the collector and emitter of the biplex detector transistor.

The $L - R$ audio signal is added to the collector of the biplex detector, and the $-L + R$ audio signal is added to the emitter. These components combine with the $L + R$ component to form the L and R signals as previously described. The L and R outputs are then applied to filter circuits $Z351$ and $Z301$ to remove the 38-kHz component and to provide deemphasis. The outputs of the filters are applied to the function switch and to the stereo amplifiers.

The stereo indicator lamp is driven by a circuit that has already been described. In the FM monaural mode, switch $S2$ grounds the output of the full-wave doubler so that monophonic operation results.

17-6 INTEGRATED-CIRCUIT TIME-DIVISION DECODER

Figure 17-9 is the schematic of a time-division multiplex decoder in which all the active stages are contained in an integrated circuit. The composite signal from the ratio detector is applied to the input (pin 3) of the IC through an SCA filter at CTP 56. The left- and right-channel audio output points are at pins 11 and 12, respectively. Note that power supply voltage must be applied to these pins through "pull-up" resistors $R306$ and $R307$.

Resonant circuits for the 19-kHz pilot carrier and the 38-kHz regenerate subcarrier are, of course, external to the chip. While it is not apparent from the schematic, the two 19-kHz resonant circuits are electrically located at the input and output of a 19-kHz amplifier. The 38-kHz tuned circuit works in conjunction with a frequency doubler, as expected.

The integrated circuit provides for a stereo indicator lamp and an interstation muting function. The mute adjustment is $R302$. The FM monaural function is selected by grounding CTP 59 by the function switch, killing the 19-kHz subcarrier. In either the FM monaural or the FM stereo modes, the information from the ratio detector must pass through the integrated circuit. In the monaural side, both outputs receive the same information.

Incidentally, voltages given on the Howard W. Sams schematics are normally taken with no signal applied. The voltages on the schematic of Figure 17-9 that are designated with a small, solid square were taken while a stereo station signal was being received.

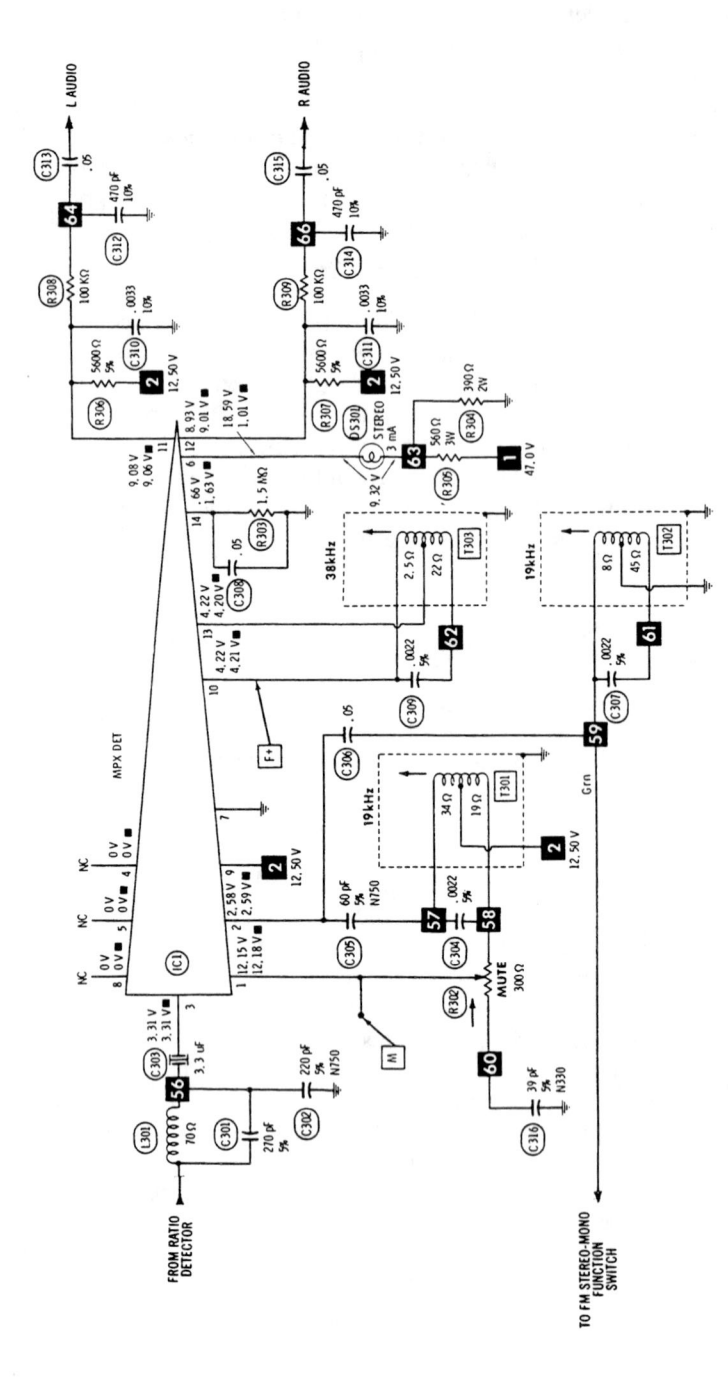

FIGURE 17-9 Time-division stereo decoder in which all active stages are contained in an integrated circuit (a Photofact schematic, courtesy of Howard W. Sams & Co., Inc.).

17-7 PHASE-LOCKED-LOOP DECODER PRINCIPLES

In Sec. 13-9 we described the fundamentals of a phase-locked loop (PLL). We saw that a voltage-controlled oscillator (VCO) could be made to oscillate at the frequency of an input reference by applying the proper control voltage derived from a phase detector.

Figure 17-10(a) is a block diagram showing how a multiple of the reference frequency can be generated. The VCO is adjusted so that its natural frequency is in the vicinity of twice the reference. Then, a frequency divider divides the VCO output by 2 and applies the result to the phase detector.

This principle can be extended as shown in Figure 17-10(b) by placing frequency dividers in series. In short, any multiple of the reference frequency can be obtained by using sophisticated divider circuits.

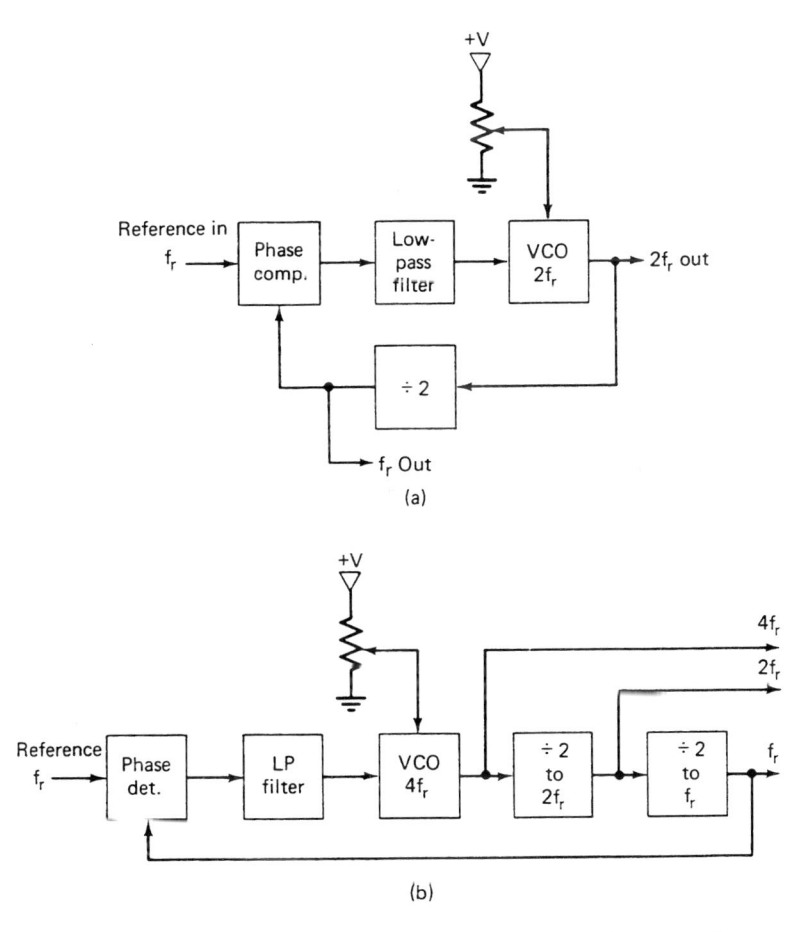

FIGURE 17-10 Block diagrams illustrating how multiples of the input reference frequency may be obtained from a phase-locked loop.

PLL Decoder

The 38-kHz signal necessary for decoding the multiplexed information of an FM stereo signal can be generated by a PLL circuit that uses the 19-kHz pilot carrier as a reference. Use of the PLL eliminates the 19- and 38-kHz tuned circuits, which greatly simplifies the alignment procedure. Better stereo separation is also generally achieved with PLL decoders, because of the better phase control of the 38-kHz subcarrier.

Figure 17-11 is block diagram of the internal circuitry of a typical multiplex PLL decoder. This decoder is used in the receiver described in the next section.

The composite signal from the FM detector is applied at the input, pin 2. An audio amplifier amplifies the composite and sends it (less 19-kHz pilot) to the demodulator. Capacitor $C308$, connected externally, couples the 19-kHz pilot to the phase detector of the PLL circuit. A low-pass filter and a DC amplifier couple the phase-detector output to the VCO, which operates at 76 kHz. $R317$ connected to pin 14 is the adjustment that is used to set the VCO "natural frequency" during alignment. The VCO output is applied to two frequency dividers that divide the 76 kHz by 4 to obtain a 19-kHz signal that is fed back to the phase detector for use in generating the correction voltage applied to the VCO.

The 38-kHz output of the first divider following the VCO is applied to a stereo switch that controls the application of the 38-kHz signal to the demodulator. When the switch is closed, the 38-kHz switching voltage is applied to the demodulator, and the L and R outputs are produced. When the stereo switch is open, no 38 kHz is applied to the demodulator, and, in such case, the same information is present at both the L and R outputs of the demodulator.

Directing our attention to the control circuitry for the stereo switch, we note

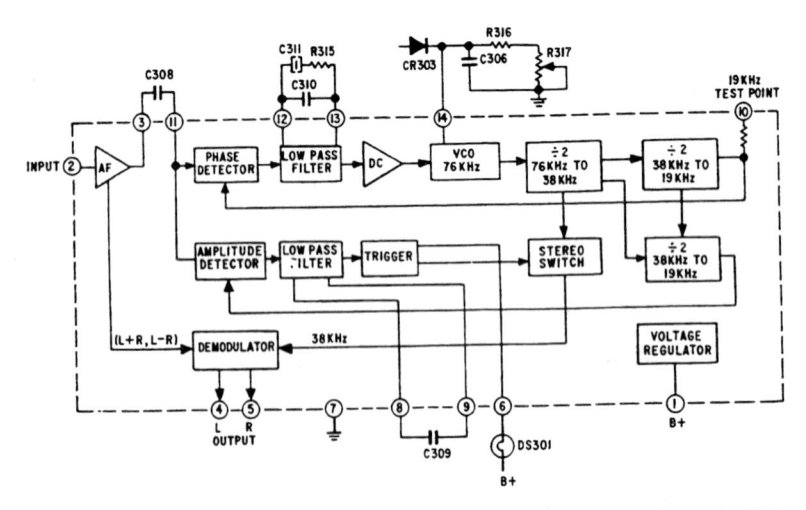

FIGURE 17-11 Block diagram illustrating the internal circuitry of a typical PLL FM stereo decoder (courtesy of Zenith Radio Corporation).

that the 19-kHz pilot at pin 11 is applied also to an amplitude detector. This detector determines if the 19-kHz signal is present in sufficient amplitude for stereo operation to be realized, and it also checks to see that the 19-kHz signal derived from the PLL circuit is also present in the proper phase. If these conditions are met, the amplitude detector provides an output to the trigger circuit via a low-pass filter. The trigger circuit then activates the stereo indicator lamp and closes the stereo switch to allow the 38-kHz switching voltages to pass to the demodulator.

If a monophonic broadcast is being received, no 19-kHz pilot will appear at the pin 11 input to the amplitude detector, but the 76-kHz VCO will continue to run at its natural frequency. Hence, the amplitude detector will detect the absence of 19-kHz pilot at the input, and will open the stereo switch so that the free-running 38-kHz switching voltage is not applied to the demodulator.

17-8 AM, FM, FM-STEREO RECEIVER

Figure 17-12 is the schematic of the RF, IF, and stereo decoder circuitry of a receiver manufactured by Zenith Radio Corporation. In this section we describe the most prominent features of this circuit. We consider the FM section first.

A dual-gate MOSFET $Q1$ is used as a common source, tuned RF amplifier. Source resistor $R4$ provides a component of self-bias while $R3$ and $R6$ provide a fixed bias for $G1$. Gate $G2$ receives a bias voltage from the FM AGC circuit. A complex coupling network couples the RF output to the input of the FM converter $Q2$. A 10.7-MHz IF trap is formed by $L3$ and $C9$.

A universal-biased, common-base stage is used as an FM converter. The 10.7-MHz FM IF signal is coupled by $T201$ to the IF amplifier section. The oscillator coil $L4$ is in series with the collector lead and responds to the oscillator component of the collector current. The oscillator signal is coupled to the emitter by $C10$. Varactor diode CR1 serves as the AFC element and is driven by the DC output of the FM detector, IC202. Resistors $R9$ and $R10$ bias the anode of CR1 to a level compatible with the output of IC202, the FM detector.

The output of the FM converter is passed through 10.7-MHz ceramic filter Y201 to the input (pin 1) of FM IF amplifier IC201. The ceramic filters are primarily responsible for setting the FM bandwidth and selectivity. The output of IC201 is applied to Y202, a second ceramic filter, and then to FM limiter/detector, IC202.

The AGC amplifier, $Q201$, receives a sample of the FM IF signal from pin 7 of IF amplifier IC201. This sample is amplified by $Q201$ and is then coupled to a voltage doubler consisting of CR202 and CR203. The output of the doubler depends upon the amplitude of the amplified IF sample and is applied through $R203$ and $R2$ to the gate $G2$ of the FM RF amplifier $Q1$.

A voltage divider consisting of $R15$, $R201$, and $R202$ sets a fixed voltage at the cathode of CR202 so that the initial voltage at $G2$ of $Q1$ is +5.6 V. When a signal voltage appears at the collector of $Q201$, the AGC amplifier, the voltage doubler develops a negative output which lowers the voltage at $G2$ of $Q1$. This reduces the gain of $Q1$ as desired.

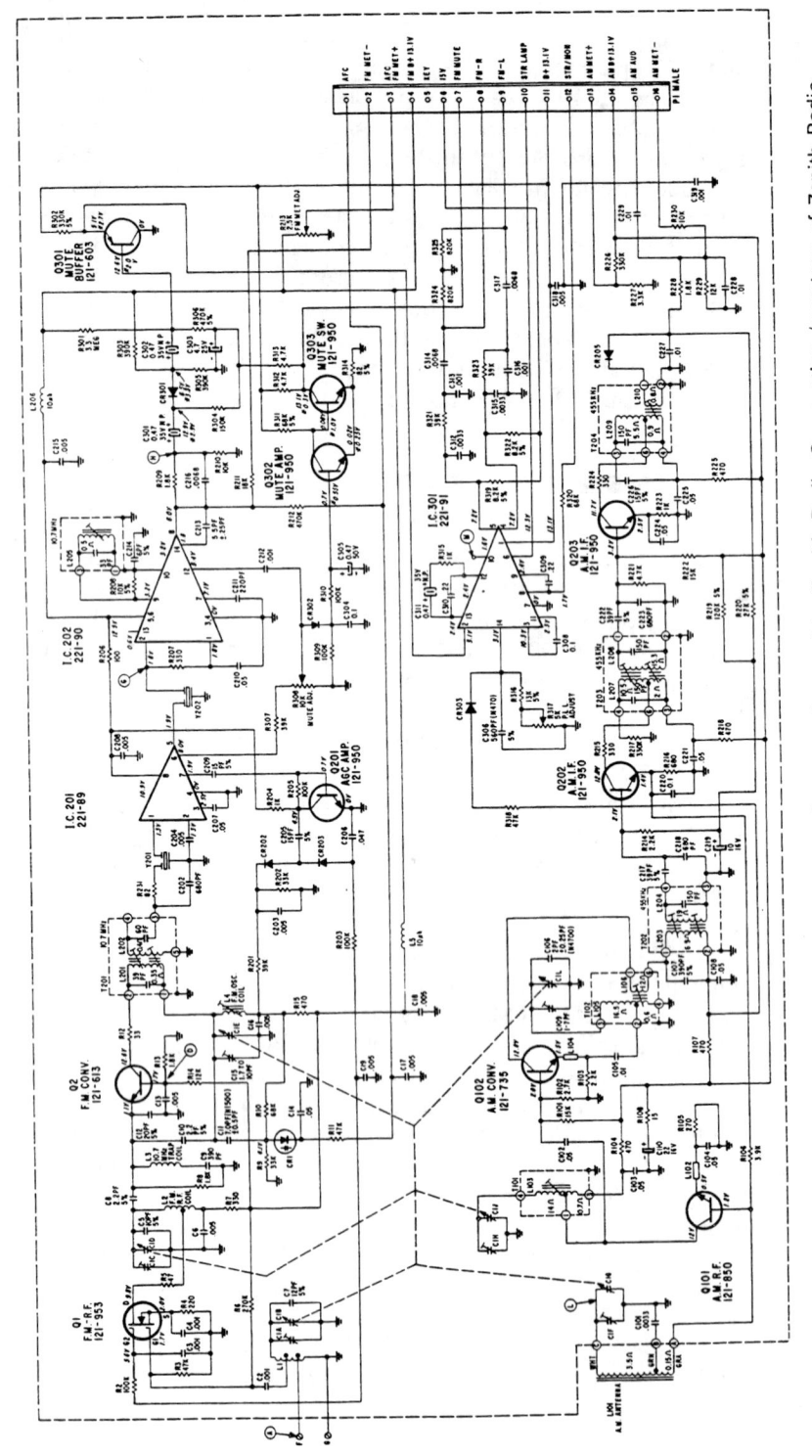

FIGURE 17-12 RF, IF, and stereo-decoder circuitry of a receiver manufactured by Zenith Radio Corporation (courtesy of Zenith Radio Corporation).

Incidentally, because of the high input impedance of $G2$ of $Q1$, the voltage atop $R202$ will be quite readily conveyed to $G2$ of $Q1$ by the small but sufficient leakage currents through CR202 and CR203. It does not matter that the diodes are reverse-biased.

FM Limiter/Detector, Mute Source

IC202 serves primarily as a limiter and detector, but it also contains a noise amplifier for use in conjunction with interstation muting. On the schematic, the single 10.7-MHz tuned circuit identifies the circuit as being a quadrature detector (see Sec. 13-8). The detector output appears at pin 8.

IC202 is also actively involved in the interstation muting function that opens the FM signal path when the interstation noise is detected at the FM detector output.

Capacitor $C213$ picks off the high-frequency noise components at the FM detector output (pin 8) and applies the noise voltage to a noise amplifier within IC202. The output of this amplifier is pin 12. The amplified noise signal is rectified by CR302 to produce a positive-going voltage ultimately applied to mute amplifier $Q302$.

Pin 6 of IC201 is the output of a voltage regulator that provides a regulated voltage for the mute adjust $R308$, which sets the noise threshold for the muting action. As CR302 is brought closer to the conduction region by $R308$, a smaller noise voltage is required to mute the output.

Mute Amplifier

Transistors $Q302$ and $Q303$ form a direct-coupled DC amplifier whose output is used to reverse-bias diode CR301 when the FM detector output is to be muted. When the noise voltage exceeds the threshold set by $R308$, CR302 conducts, producing a positive voltage at the base of $Q302$. Henceforth, $Q302$ conducts; its collector voltage drops and turns $Q303$ off. The lack of conduction through $Q303$ causes its collector voltage to rise. CR301 is then reverse-biased so that the FM detector output cannot pass through to the base of $Q301$. Further, the forward bias is decreased at the base of $Q301$, which also reduces the passage of the FM detector output.

The FM muting feature may be defeated by closing the front-panel mute override switch. This switch grounds the $Q303$ collector.

Mute Buffer

$Q301$ is an emitter follower used to match impedances between CR301 and the input to the stereo decoder IC301. The output of $Q301$ is the FM-stereo composite signal and is applied to pin 2 of IC301, the input to the stereo decoder.

PLL Stereo Decoder

IC301 is the PLL decoder described in block diagram form in the preceding section. Only a few comments are made here. First, the components connected to pins 8, 9, 12, and 13 set the frequency response of the low-pass filters. $C308$, between pins 3 and 11, couples the 19-kHz pilot to the PLL circuit inside the IC.

The network connected to pin 14 sets the natural frequency of the VCO of the PLL. When B+ is applied to the AM section, CR303 conducts and produces a positive voltage on pin 14. This turns off the 76-kHz VCO to avoid possible interference between harmonics of the VCO output and the various AM signals.

Attached to pin 8 of IC301 is $R320$, which leads out to the front panel stereo-monaural mode switch. This manually controls the application of the 38-kHz switching voltage in the decoder IC, and therefore determines whether a given stereo-broadcast will be rendered in stereo or monaural.

AM Section

Even though the appearance of the AM section in Figure 17-12 is unconventional, analysis reveals that the AM section is typical. An RF amplifier $Q101$ is complex-coupled to the AM converter $Q102$. The RF amplifier is a common-emitter stage (not common-base). Two stages of IF amplification precede the AM detector CR205.

An AGC voltage is derived from the detector by $R220$ and is filtered by $C219$. The AGC voltage is applied to the base circuits of the IF stages. The driving voltage for the base of RF amplifier $Q101$ is obtained via $R106$ from the emitter of $Q202$. AGC action is applied also to $Q101$, since the $Q202$ emitter voltage will vary with the AGC voltage.

The load for the AM detector is $R228$ and $R229$. The audio is taken off via $C229$. A portion of the DC component of the detector output is obtained for the AM tuning meter through $R230$.

The connection via $R318$ of the AM section to CR303 at the input to the stereo decoder simply applies a small positive voltage to the decoder in order to turn off the VCO of the PLL when B+ is applied to the AM section.

SUMMARY

1. In FM-stereo receivers, the output of the FM detector, the composite signal, is applied to the input stage of the stereo decoder. The input stage is called the multiplex (MPX) input amplifier.

2. Matrix decoders are commonly found only in older tube-type receivers. In these decoders, the second stage is typically a frequency separation stage, involving a cathode follower, in which the $L + R$, 19-kHz pilot, and $L - R$ sidebands are separated.

3. An SCA trap tuned to 67 kHz is employed to reject the SCA signal if it is present.

4. In matrix decoders, the $L - R$ sideband signal is converted to an AM signal, which is then detected by an ordinary AM detector to recover the $L - R$ signal. The AM signal is restored by reinserting the 38-kHz component obtained from a 19-kHz oscillator doubler keyed by the 19-kHz pilot carrier. The L and R signals are obtained by combining the $L + R$ and $L - R$ signals in a resistor matrix.

5. Some matrix decoders use a two-phase AM detector to obtain the $L - R$ and $-L + R$ signals. Others use an inverter to convert the $L - R$ to a $-L + R$ signal.

6. In time-division decoders the composite signal is not separated into the $L + R$ and $L - R$ components, but the 19-kHz pilot is isolated and applied to a frequency doubler to obtain the 38-kHz subcarrier. The composite and subcarrier are then combined via center-tapped secondary of a 38-kHz transformer. Two waveforms are thus obtained, one of which carries the L signal on the positive peaks while the other carries the R signal on the positive peaks.

7. The bridge-type time-division decoder is a more elaborate decoder that provides excellent cancellation of the 38-kHz switching signal.

8. The biplex detector is a time-division decoder that utilizes the bilateral characteristics of a BJT to achieve separation of the L and R signals. The voltage across the transistor (from E to C, or vice versa) is provided by the 38-kHz subcarrier, which controls the direction of current conduction through the device. The amount of conduction is controlled by the $L - R$ sideband component of the composite. The end result is that the L-only signal will appear at the collector (typical), while the R-only signal appears at the emitter.

9. Almost all the circuitry except the tuned circuit components of a time division decoder may be fabricated in an integrated circuit. While the details of the IC are not at hand (ordinarily), the circuit may be recognized by the tuned circuits associated with the IC.

10. A PLL stereo decoder utilizes a PLL to develop the 38-kHz subcarrier, generally without the encumbrance of tuned circuits. Such decoders are simpler from the servicing standpoint than the preceding decoders, and therefore give rise to more nearly trouble free operation.

QUESTIONS AND PROBLEMS

17-1. The circuit of Figure 17-1 contains filter circuits $L18$, $L20$, and $L21$. To what frequency or what range of frequencies does each filter respond?

17-2. Identify by number and ohmic value the resistors composing the adder circuits

(a) Of the matix decoder of Figure 17-1.

(b) Of Figure 17-2.

17-3. What components form the deemphasis networks

(a) In Figure 17-1?

(b) In Figure 17-2?

17-4. Describe the waveforms appearing at the following points in Figure 17-1:

    (a) $V8$, pin 6.

    (b) Test point $<F>$.

    (c) Test point $<E>$.

    (d) Test point $<G>$.

17-5. Explain the operation of the separation control of Figure 17-2.

17-6. How is the $L - R$ signal and its inverted $-L + R$ signal obtained in Figure 17-2?

17-7. In Figure 17-2, $M13$ is a typical AM detector. How is the AM waveform obtained since no AM waveform as such is present in the composite waveform?

17-8. Describe the waveforms appearing at the following points in Figure 17-3:

    (a) At the base of $Q9$.

    (b) At CTP 63.

    (c) At test point $<D>$.

    (d) At test point $<E>$.

17-9. Describe the operation of the stereo indicator circuit of Figure 17-3.

17-10. Explain the operation of the separation control of Figure 17-4.

17-11. Explain the operation of the stereo indicator circuit of Figure 17-4.

17-12. in Figure 17-5, under what circumstances will a current flow through $R211$ or $R218$?

17-13. Why is it that different waveforms appear at CTP 81 and CTP 86 in the bridge-type decoder of Figure 17-5?

17-14. Practice until you can draw from memory the basic biplex detector circuit of Figure 17-7.

17-15. Trace the path of the composite signal in the biplex detector decoder of Figure 17-8.

17-16. Explain why a biplex detector is a time-division type decoder.

17-17. Compare the integrated circuit decoder of Figure 17-9 with the discrete component decoder of Figure 17-8. In particular, note the similarities of $L301$, $T301$, $T302$, and $T303$.

17-18. In Figure 17-9, determine which pins of the integrated circuit IC1 have voltages supplied from an external source, and determine which pins have voltages supplied from inside the IC.

17-19. Explain how a PLL can be used to generate multiples of a given input frequency [see Figure 17-10(b)].

17-20. In the PLL stereo decoder of Figure 17-11, what function does the PLL perform?

17-21. How is the 38-kHz subcarrier regenerated in Figure 17-11?

17-22. In Figure 17-11:
- **(a)** Trace the path of the composite signal.
- **(b)** Trace the path of the 19-kHz pilot carrier.

17-23. What is the function of the components, in Figure 17-11, connected to
- **(a)** Pin 14?
- **(b)** Pins 12 and 13?
- **(c)** Pins 8 and 9?
- **(d)** Pin 6?

17-24. In Figure 17-12:
- **(a)** What type of FM RF amplifier is used?
- **(b)** What are the functions of IC201, IC202, and IC203?
- **(c)** What type of stereo decoder is used?

17-25. Describe the operation of the voltage doubler associated with the output circuit of $Q201$, the AGC amplifier.

17-26. What is the function of pin 12 of IC202 in the FM detector circuit of Figure 17-12?

17-27. Describe the operation of the muting circuit of Figure 17-12. What role does CR301 play?

17-28. How is the VCO of the PLL decoder turned off when B+ is applied to the AM section of the receiver of Figure 17-12?

18
TROUBLESHOOTING FM-STEREO RECEIVERS

This chapter deals primarily with problems arising from the stereo-decoder portion of an FM-stereo receiver. Procedures are given for determining when a decoder malfunction exists and for isolating the defective component. Our emphasis upon the stereo-decoder section is not to imply that this section is especially troublesome. It is because other sections of the receiver may be handled with procedures established in earlier chapters.

Several examples are related to stages whose active device is an integrated circuit (IC). This is done to present the techniques applicable to troubleshooting ICs. Stereo-decoder alignment is covered in the last sections of this chapter.

18-1 INTRODUCTION

While the addition of the FM multiplex circuitry adds to the complexity of the receiver as a whole, the troubleshooting procedures for an FM stereo receiver are generally no more difficult than for monaural FM. The additional section provides more opportunities for defects to occur, but the same procedure of signal injection and signal tracing are applicable.

Several new symptoms may be encountered as a result of defects in the multiplex circuitry. These are shown with their most likely cause in Figure 18-1.

The AM, FM, audio, and power supply sections receivers maintain the same form as in AM-FM receivers, but frequently the sections are more elaborate. Integrated circuits, ceramic filters, and FM muting are commonly encountered.

We now proceed to a description of commonly encountered symptoms, the likely causes of each, and procedures for isolating the defect. Again, as in Chapter 15, we urge the reader to try to anticipate the likely causes and applicable procedures for each symptom.

18-2 ONE CHANNEL INOPERATIVE ON ALL FUNCTIONS

When one channel is inoperative on all functions, the problem is almost certain to be in the audio amplifier section for that channel. Since both audio channels use the same B+ sources, the power supply can be assumed to be operating normally since the other channel functions normally.

Symptom	Most likely cause
1. One channel inoperative on all functions.	Audio amplifier section of the defective channel; foreign object in headphone jack.
2. AM normal; FM and FM stereo inoperative.	Any FM-only stages: composite amplifier; FM muting; FM detector; FM front end.
3. One channel inoperative in FM-stereo mode.	Stereo-decoder circuitry: matrixing resistors; multiplex detector; PLL chip.
4. Stereo separation completely lacking.	Stereo-decoder circuit: 19 kHz or 38 kHz amplifier; bandpass filter; VCO frequency control voltage.
5. Nonfunctioning stereo indicator.	19 kHz source; stereo indicator driver stage; stereo indicator output of IC decoder.

FIGURE 18-1 Common symptoms and their most likely causes for the stereo decoder section of an FM-stereo receiver.

Signal tracing can be used to determine the defective stage, and a V-R analysis is then used to determine the defective component. The signal tracing should begin at the speaker and proceed toward the volume control. This is the logical approach, since the audio output stage handles a large amount of power and is therefore more likely to fail.

Headphone jacks are usually found between the audio output stage and the speakers. On several occasions, foreign objects, such as crayons or pencils, have been discovered in the headphone jack. These objects may render either one or both channels inoperative. This defect is quickly discovered with signal tracing, even when the foreign object is not readily visible.

Troubleshooting IC Stages

In IC stages the circuitry external to the IC should be checked very carefully before a suspected IC is removed from the circuit. ICs are usually soldered directly onto the circuit board and are notoriously tedious to remove. Further, removal of an IC always involves some risk to the device. Defective resistors and capacitors can cause an IC stage to be inoperative, and it is far better to discover a defective resistor or capacitor before the IC is removed.

The voltages on all the pins of a suspected IC should be compared with normal values given in the service data. Ascertain that power supply voltages are being applied to the IC. Voltages appearing at certain pins may originate inside the IC. When such a voltage is low or missing, the external circuitry connected to the pin must be carefully checked for shorts to ground.

When it is deemed necessary to remove an IC, a desoldering device must be used to remove the solder from the pins to avoid damage to the PC board. This operation must be performed with great care. After the replacement is installed, check carefully for solder bridges between pins before applying power. The pins

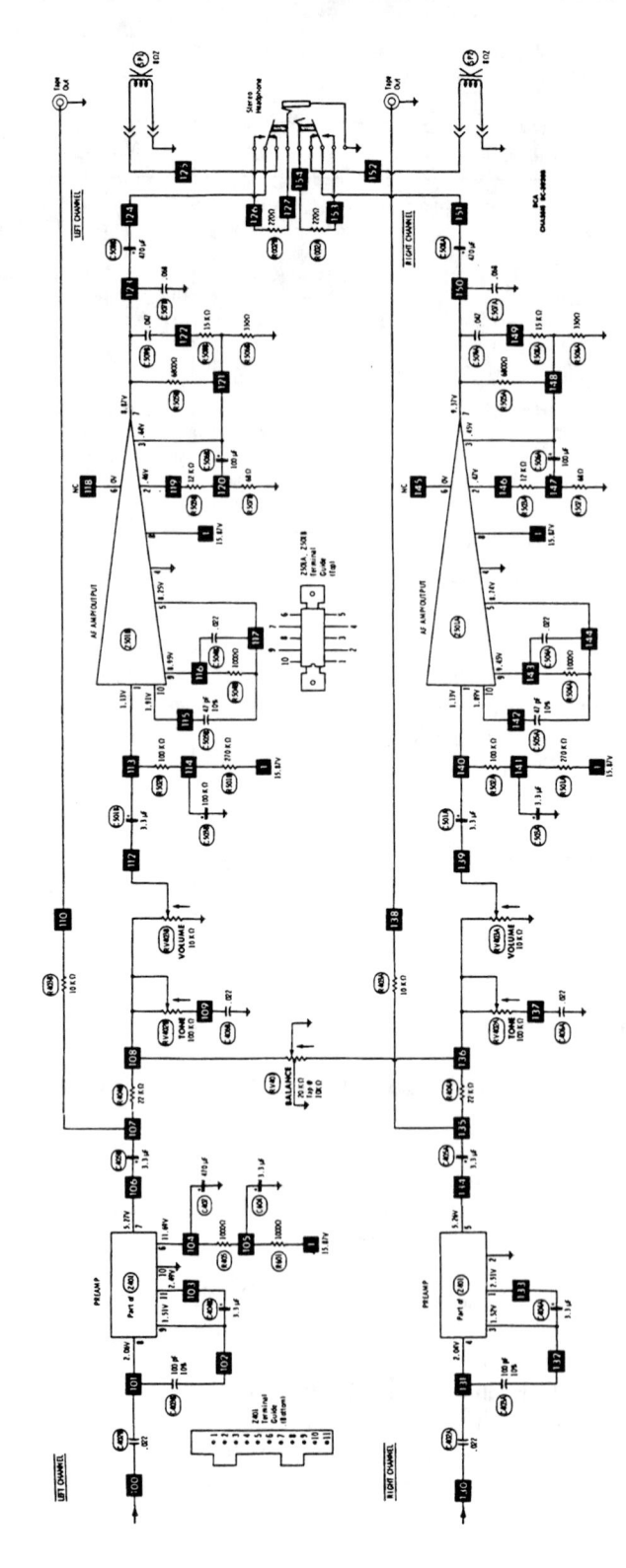

FIGURE 18-2 Audio section of an FM-stereo receiver that utilizes IC audio preamplifiers and also IC audio power amplifiers (a Photofact schematic, courtesy of Howard W. Sams & Co., Inc.).

are very close together, and it is easy for solder to short two pins together. We note here, in passing, that ICs installed with the pins *reversed* seldom give fully satisfactory operation.

EXAMPLE 18-1: We assume the receiver whose audio section (partial) is shown in Figure 18-2 produces no sound from the left channel. The receiver is disassembled, and test speakers mounted on the service bench are connected for use while servicing the unit. This eliminates the speaker of the left channel as a possible cause of the problem.

To eliminate the headphone jack, we connect a scope to the input to the jack. No signal is present. Also, no signal is present at the output of the audio power IC, pin 7 of Z501B. Next, we move the scope to the input to the IC, pin 1, and find an audio signal of about 1 V p-p. We conclude that the IC is not amplifying.

The DC voltage at the B+ pin, pin 8, is checked and found to be normal. At the output, pin 7, however, no voltage is found where there should be 8.87 V. Since this voltage originates within the IC, we measure the resistance from pin 7 to ground. It is very nearly 7 kΩ, which is normal. Suspicion now falls heavily upon the IC. It is carefully replaced with a new unit, and normal operation of the receiver is obtained.

18-3 AM NORMAL; FM AND FM STEREO INOPERATIVE

Any of the FM stages, including the composite amplifier (sometimes called the *multiplex input amplifier*) and the FM muting section, can cause the FM and FM stereo functions to be inoperative. The level of hiss from the speaker can be useful in locating the problem. A loud hiss indicates a defect in the FM front end, and signal injection can be started at the input to the IF amplifier as described in Chapter 15. If the level of hiss is very low, or if no hiss is present, signal tracing is started at the FM detector output. The signal-injection or signal-tracing procedure is continued until the defective stage is located.

It should be noted that in some receivers, a defect in the composite amplifier will *not* cause the FM to be inoperative. This depends upon the method employed to switch from FM monaural to FM stereo. In most cases, however, failure of the composite amplifier will cause both the monaural and stereo FM functions to be inoperative.

EXAMPLE 18-2: Refer to the receiver of Figure 18-3, which we assume to have inoperative FM and FM stereo. Note the use of ICs and the FM muting feature. If the muting circuit is working properly, no hiss or other sound will be heard from the speaker until a signal is tuned in. Thus, we must turn the muting circuits off with the mute switch before rendering an opinion based upon the hiss level.

When the mute switch is turned off, a loud hiss appears at the speaker. We suspect the defect to be in the front end. An FM generator is used to inject a 10.7-MHz FM signal at the collector of $Q2$, the FM converter, and a tone is heard at the speaker.

Both generator and receiver are then tuned to 90 MHz. The generator is connected to the emitter of $Q2$, and the signal is heard. A signal is also heard when injected at the drain of FM RF amplifier $Q1$. But when the signal is applied to the gate $G1$ of $Q1$, no signal is heard. The FET seems not to be working.

Normal DC voltages are found at the gates, zero voltage is found at the source, and the

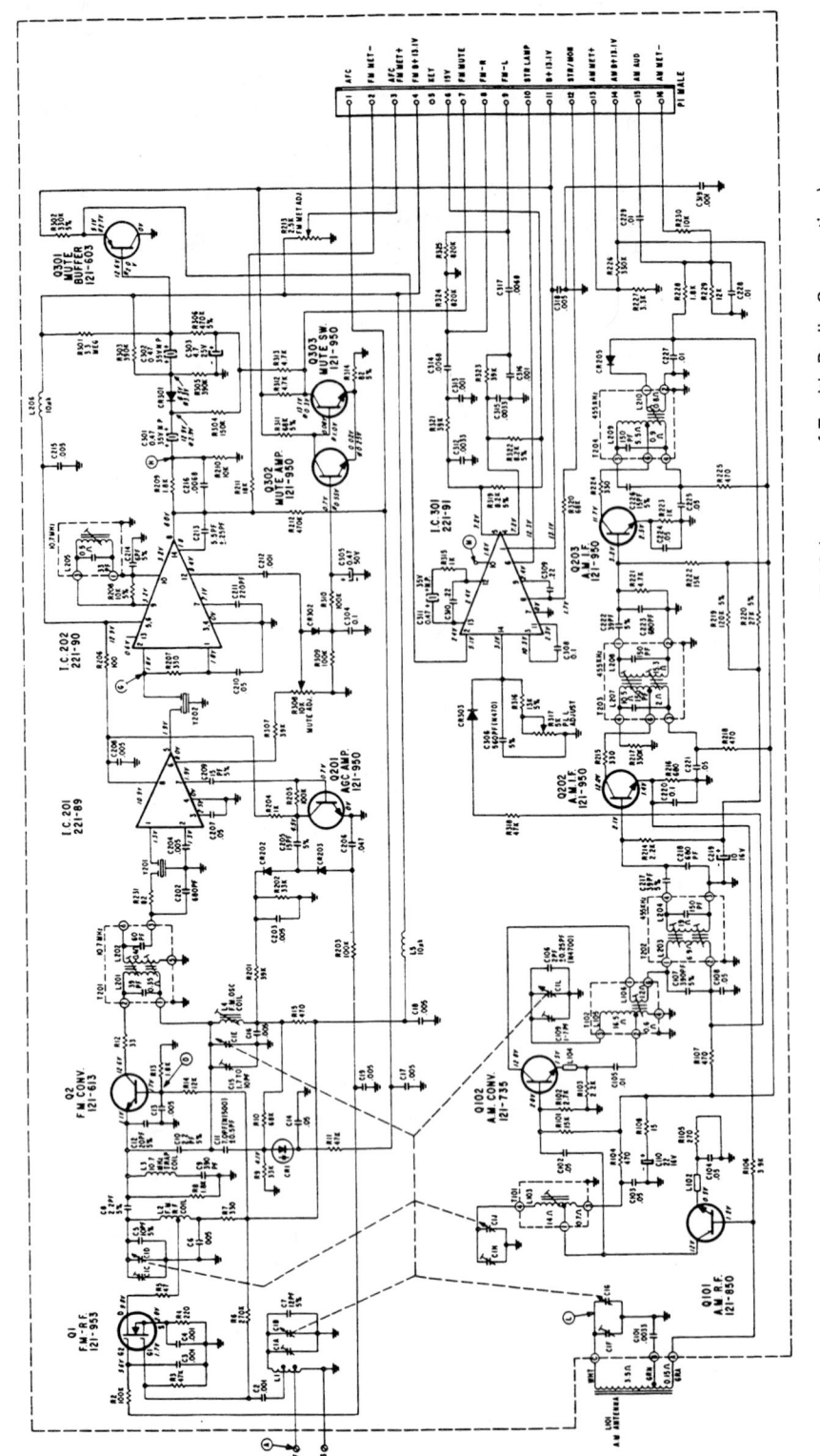

FIGURE 18-3 RF, IF, and stereo-decoder circuitry of the receiver of Figure 17-12 (courtesy of Zenith Radio Corporation).

drain voltage is higher than normal. This indicates a lack of conduction from source to drain. The device must be replaced.

Since the device is a dual-gate MOSFET, care must be taken not to damage the replacement by discharges of static electricity. The tip of the soldering iron should be connected to a good earth ground, and the technician should ground his body to the same earth ground, possibly by connecting a clip lead to the band of a metal wristwatch. Also, a ground lead should be connected to the chassis of the receiver. The replacement part should be removed from the package and placed directly onto the PC board.

After the soldering operation is completed, the ground leads are removed, and normal operation of the receiver is restored. Normal sensitivity is obtained, so realignment is deemed unnecessary.

Precaution

A technician who connects an earth ground to his body as suggested above must be extremely careful not to contact any *hot* chassis of either the receiver or a service instrument that is improperly grounded. One should ground himself only when necessary, and then remove the ground immediately when the critical job is completed.

EXAMPLE 18-3: Referring to the same receiver (Figure 18-3), let us assume that the hiss level was very low when the muting circuit was disabled. We check the output of the FM detector, pin 8 of IC202, and a large noise signal is present. Tuning the receiver to a known station produces an audio signal on the scope.

The scope is moved to the cathode of CR301, and the audio signal remains. No signal is found, however, at the anode side of CR301. Since the FM muting switch removes the mute amplifier and mute switch transistor $Q303$ from the circuit (by placing a fixed bias on CR301), these components can be eliminated as possible causes. CR301 is found to be open, and installation of a new unit restores normal operation.

EXAMPLE 18-4: We now suppose that CR301 in Example 18-3 was functioning normally. The signal-tracing procedure is continued, and an audio signal is found at the base and collector of $Q301$ and at pin 2 of IC301, the FM stereo decoder. No signal is found at pin 4 or pin 5, the decoder outputs for the L and R channels. Evidently, the decoder is not functioning normally. Voltages around the "chip" are measured. Supply voltage is present, but the voltage at pin 3 is missing. Since this voltage originates inside the chip, and since no shorts to ground are in evidence, we conclude that the IC is bad.

The unit is replaced, and near-normal operation is obtained. Adjusting the PLL ADJUST $R317$ according to the alignment instructions restores normal operation. This adjustment is described in a later section of this chapter.

18-4 ONE CHANNEL INOPERATIVE IN FM-STEREO MODE

When one channel fails to operate in the stereo mode but functions normally on all other functions, we suspect a defect in the decoder circuitry. In time-division decoders the FM multiplex detector is probably at fault. In matrix decoders we look for an open circuit after the matrixing resistors. In PLL decoders we suspect a defective PLL decoder chip.

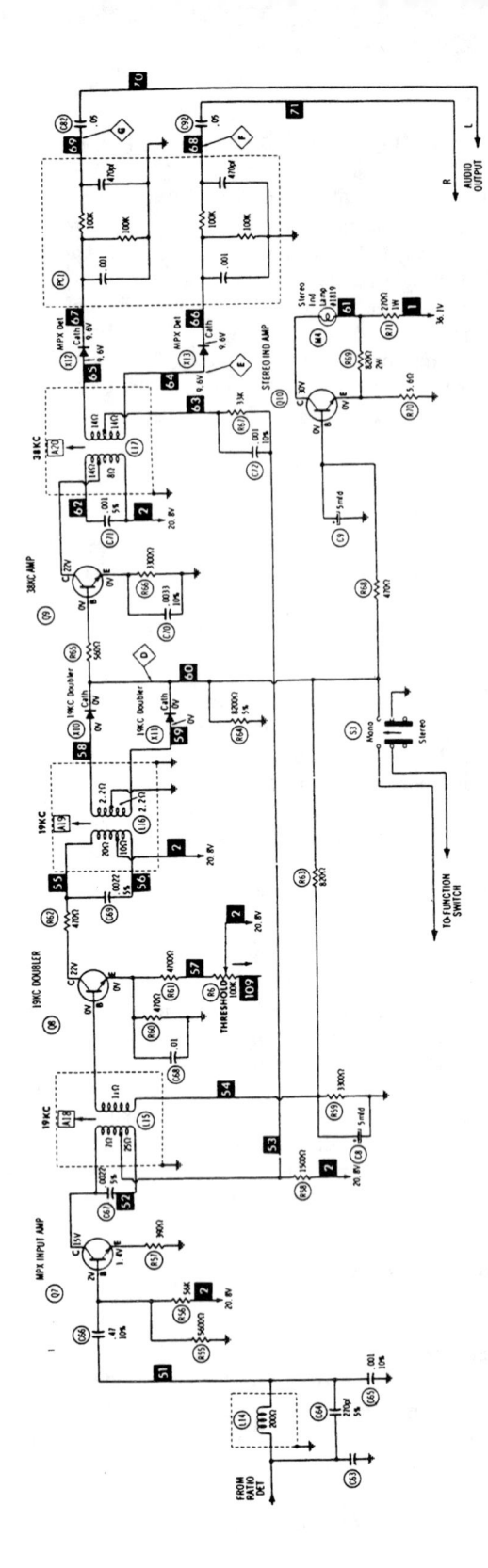

FIGURE 18-4 Time-division stereo decoder of Figure 17-3 (a Photofact schematic, courtesy of Howard W. Sams & Co., Inc.).

The same channel may also be inoperative in the FM mode, depending upon the switching arrangement used to change from FM to FM stereo. If switching is accomplished by disabling the 19-kHz stage only, and not by tying the audio inputs together or by taking the FM audio signal directly from the FM detector, the same channel will be inoperative also in the FM mode.

A simple signal-tracing procedure is applicable. Signal tracing is begun at the multiplex diodes of a time-division decoder, at the AM detector diodes in a matrix decoder, and at the audio output of a PLL decoder. The signal tracing proceeds toward the speaker until the signal is lost. V-R analysis then is used to locate the defective component.

EXAMPLE 18-5: The receiver of Figure 18-4 is used to illustrate this procedure. The receiver functions normally in all modes except FM stereo, for which the left channel is inoperative. The FM mode operates normally because the audio inputs are tied together by the function switch in the FM mode.

A scope is connected to MPX detector $X12$ at CTP 65, and a multiplexed signal is found. At the cathode of $X12$, however, no signal whatsoever appears. We reason that $X12$ is probably open, and an ohmmeter check confirms this. Normal operation is restored by replacing the defective diode.

18-5 MONAURAL FM STEREO

When stereo separation is completely lacking so that a monaural response is obtained from the speakers when the receiver is tuned to an FM stereo station with the function switch in the FM-stereo mode, the multiplex decoder section is probably defective. In a time-division decoder the defect will probably be in the 19-kHz or 38-kHz amplifier. In a matrix decoder the bandpass filter, as well as the 19-kHz oscillator and doubler, could be responsible. In a PLL decoder a defect in the PLL IC or associated circuitry is indicated.

The troubleshooting procedure varies, depending upon the type of decoder used. For a time-division decoder, signal tracing is used to follow the 19-kHz signal through the 19-kHz amplifier, doubler, and 38-kHz amplifier stages to determine the defective stage. V-R analysis is then used to locate the defective component.

In a matrix decoder the scope is used to observe the waveform at the input to the AM MPX detector diodes. If a 38-kHz signal is observed, the band-pass filter is checked. If a double-sideband suppressed carrier signal is observed, the 19-kHz oscillator and doubler are checked.

For PLL decoders the voltage-controlled oscillator (VCO) frequency and the VCO control voltage are checked. If they are normal, the decoder IC is replaced.

In many cases, in a time-division circuit, the stereo indicator light can be used to help isolate the defective stage. Usually, the driver for the stereo indicator receives a signal from the doubler stage. If the light is not on, the 19-kHz amplifier or doubler will be at fault. If the light is on, the 38-kHz amplifier is at fault.

Another important point is that a certain minimum signal strength is necessary for proper stereo operation. A lack of sensitivity at the receiver front end can cause the FM stereo section to be inoperative for lack of signal amplitude.

The bias for the 19-kHz and 38-kHz amplifier stages in a time-division circuit depends upon the level of 19-kHz pilot signal recovered. If insufficient signal is produced at the FM detector output, these stages will not be turned on, and there will be no FM stereo.

Many people listen to the same station most of the time. Hence, a problem in the front end or IF amplifier that reduces FM sensitivity will bring forth the complaint of a lack of stereo separation as opposed to the complaint of insufficient sensitivity. The following illustrates this point.

Some years ago, a technician, who chooses to remain anonymous, encountered an FM multiplex receiver for which the complaint was a lack of stereo separation. A local station was tuned in, and sure enough, the stereo light did not come on, and the same audio content was present at both speakers. The noise level was noticeable, but this point was overshadowed by the complete lack of stereo separation. Several hours of signal tracing and V-R analysis uncovered nothing, and frustration was rapidly setting in. Luckily, and for no real purpose, an attempt was made to tune another, more distant station, and the lack of sensitivity became immediately apparent. Shortly thereafter, a defective FM RF amplifier was located and replaced, and normal operation was restored.

EXAMPLE 18-6: The procedure for treating a lack of stereo separation is illustrated for a time-division decoder by referring to the schematic of Figure 18-4. The stereo light is on, so the problem is more likely in the 38-kHz amplifier. A scope connected to the base of $Q9$ (38-kHz amp) reveals 38-kHz pulses from the diode doubler, but no 38-kHz signal appears at the collector.

A V-R analysis turns up a missing voltage at the collector of $Q9$, and this causes us to suspect the $L17$ primary winding. A resistance check indicates the winding is open, as expected. The unit is removed from the circuit, the shield is removed, and the unit is inspected carefully for broken wires. Nothing is found, so a new part is ordered. Installation of the new component and a stereo alignment restores normal operation.

EXAMPLE 18-7: The receiver of Figure 18-5 is used to illustrate the procedure for treating a matrix decoder for a lack of stereo separation. A scope placed at the cathode of MPX detector $X12$ shows a 38-kHz sine wave and nothing more. (We assume that the receiver is tuned to a stereo station.) This indicates that the $L - R$ sideband information is not getting through the band-pass filter to combine with the 38-kHz signal to form the $L - R$, reconstituted AM signal. The scope is moved to pin 1 of band-pass filter $L20$, and no signal is found. A composite waveform is found at the junction of $R46$ and $R43$. Subsequent V-R analysis determines that $R46$ is open. Replacement restores normal operation.

EXAMPLE 18-8: The receiver of Figure 18-3 is assumed to have a lack of stereo separation. The receiver uses a PLL decoder. A scope is placed at pin 10 of IC301, which is a 19-kHz test point provided on the chip, and no 19-kHz signal is found. This indicates a defect within the chip in either a frequency divider or the VCO. Normal operation is restored when the IC is replaced.

18-6 NONFUNCTIONING STEREO INDICATOR

Receivers that have stereo indicator lamps usually employ a separate transistor stage that uses the 19-kHz pilot carrier to activate the lamp. Troubleshooting is limited to this one stage, since, by definition of this symptom, the audio signal

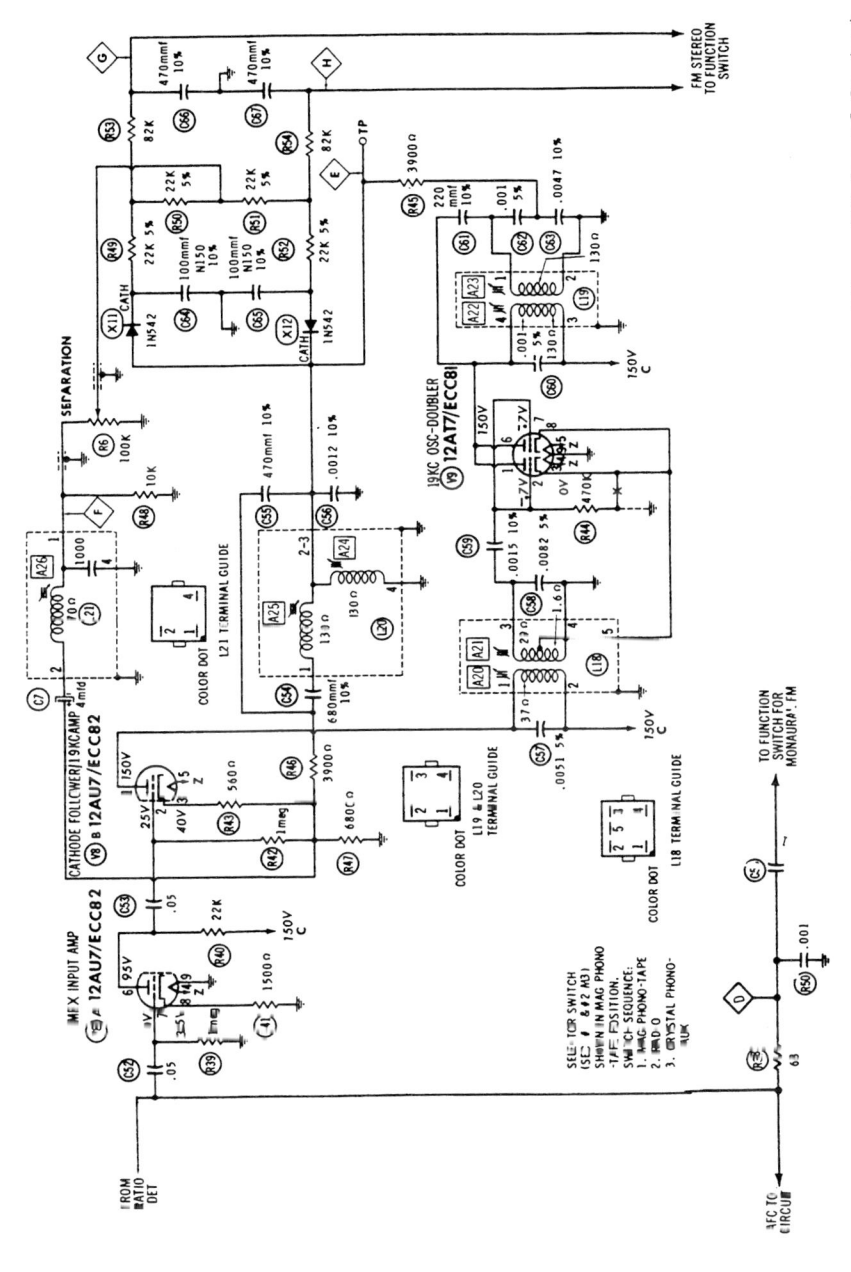

FIGURE 18-5 Matrix-type FM-stereo decoder of Figure 17-1 (a Photofact schematic, courtesy of Howard W. Sams & Co., Inc.).

appears in stereo at the speakers, thereby indicating that the decoding section is working normally. The DC voltages are checked first, and if they are found to be normal, the 19-kHz source is checked.

In receivers that employ ICs for decoders, the lamp circuitry is checked first. If it is found to be normal, the IC becomes suspect. Chances are that it will have to be replaced.

EXAMPLE 18-9: In checking the stereo indicator lamp circuitry of Figure 18-4, no voltage is found at the $Q10$ collector, but 27.5 V is found at the B+ side of the indicator lamp $M4$. An ohmmeter reveals that $M4$ is open, and replacement of $M4$ restores normal operation.

It should be noted that collector-emitter leakage in $Q10$ could cause the lamp to be lighted even when no stereo signal is present.

18-7 STEREO-DECODER ALIGNMENT—AN INTRODUCTION

Stereo-decoder alignment involves setting the tuned circuits of the decoder section to the proper frequencies and making any other adjustments required for optimum performance. This serves to achieve the best possible stereo separation between the left and right channels, to prevent SCA interference, and to establish the minimum signal strength for stereo operation. Decoder alignment can be performed using several methods. A stereo generator and scope will provide the best results. If a stereo generator is unavailable, a signal generator, a signal from a stereo station, and a scope can be used.

A stereo generator produces a signal that contains a 19-kHz pilot signal, a $L + R$ audio signal, a $L - R$ double-sideband signal, and a 67-kHz signal that can be injected at the input to the stereo-decoder section. Further, the generator can be set to produce a left-channel-only or a right-channel-only signal so that the stereo separation can be checked and optimized.

General alignment procedures are given in the following for each of the three basic stereo decoders. First, a general procedure is given that requires a stereo generator. Then, a general procedure is given for use when no stereo generator is available. As for all alignment operations, the manufacturer's or Howard W. Sams' alignment instructions should be followed for the particular unit at hand.

18-8 MATRIX-DECODER ALIGNMENT

Figure 18-6 shows a block diagram representation for aligning a matrix decoder with a stereo generator and scope. The 19-kHz oscillator is disabled to avoid interference between the oscillator and injected signals. Since matrix decoders are found only in older tube-type receivers, the oscillator can, in most cases, be disabled by removing the oscillator tube.

The generator is set to produce a 19-kHz signal, and the scope is connected to test point (TP) $<A>$ in Figure 18-6. $A1$ and $A2$ are adjusted for maximum 19-kHz response. This sets the frequency of the oscillator to 19 kHz.

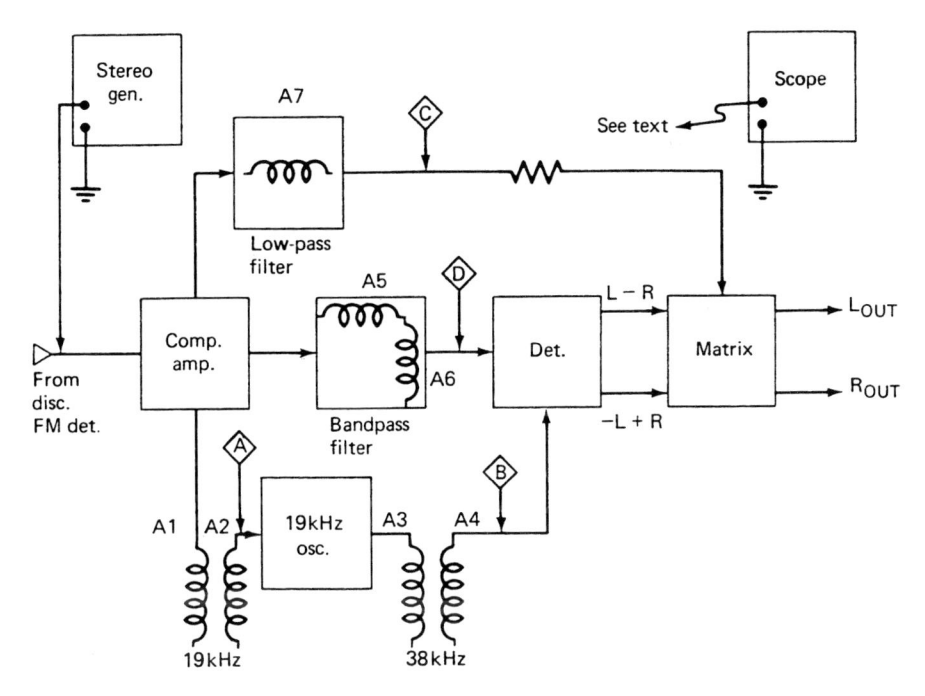

FIGURE 18-6 Block diagram representation for aligning a matrix decoder using a stereo generator and scope.

The scope is then moved to TP $<C>$, and $A7$ is adjusted for minimum 19-kHz response. This establishes the high cutoff frequency for the low-pass filter that separates the $L + R$ signal from higher-frequency components of the composite waveform.

Next, the scope is placed at TP $<D>$, and $A6$ is adjusted for minimum response to 19 kHz. This sets the low-frequency cutoff for the band-pass amplifier that separates the sidebands centered about 38 kHz from the composite waveform. To set the high-frequency cutoff for this filter, the generator is set to 67 kHz, the SCA frequency, and $A5$ is adjusted to give minimum response, with the scope still at TP $<D>$.

The oscillator tube is reinserted and the scope is placed at TP $<B>$. $A3$ and $A4$ are adjusted to give maximum 38-kHz response on the scope. These adjustments tune the output of the frequency doubler.

To optimize the stereo separation, the generator is set to produce a left-only signal, and the scope is placed on the right-channel output from the matrix. Separation control $A8$ is adjusted for minimum signal.

The generator is then set for a right-only signal, and the scope is connected to the left-channel output of the matrix. $A8$ is readjusted for minimum output, and, if necessary, a compromise is made between this and the previous adjustment to obtain the best overall stereo separation. This concludes the matrix-decoder alignment.

Multiplex alignment using multiplex generator

	Signal generator	Scope	Adjust	Remarks
9.	Inject 19-kHz signal ±0.01% into pin 7 of V_8, low side to chassis	Connect vertical input of scope to pin 2 of V_9, low side to chassis	A_{20}, A_{21}	Remove V_9. Adjust A_{20}, A_{21} for max maximum deflection. Adjust scope to lock in 2-cycle waveform. Replace V_9
10.	Inject 19-kHz signal ±0.01% into pin 7 of V_8, low side to chassis	Connect vertical input of scope to point Ⓔ, low side of chassis	A_{22}, A_{23}	Adjust A_{22}, A_{23} for maximum deflection of 4-cycle waveform.
11.	Inject 67-kHz signal from audio oscillator into pin 7 of V_8, low side to chassis	Connect vertical input of scope to point Ⓔ, low side of chassis	A_{24}, A_{25}	Remove V_9 and adjust A_{24}, A_{25} for minimum deflection.
12.	Inject 19-kHz signal from audio oscillator into high side of L_{21}, low side to chassis	Connect vertical input of scope to point Ⓕ, low side to chassis	A_{26}	Adjust for maximum deflection.
13.	Inject 19-kHz signal ±0.01% into pin 7 of V_8, low side to chassis	Connect vertical input of scope to point Ⓖ, Horiz. input of scope to point Ⓗ, low sides to chassis	Separation control R_6	Adjust maximum clockwise and counterclockwise to produce diagonal responses shown in Fig. 3. Set control to produce round pattern shown in Fig. 3.

Multiplex alignment using air signal

Make sure the FM section of receiver is properly aligned.			
	Signal generator	Scope	Adjust	Remarks
14.	Tune in a strong FM stereo multiplex signal.	Connect vertical input of scope to pin 2 of V_9, low side to chassis	A_{20}, A_{21}	Remove V_9 and adjust A_{20}, A_{21} for maximum deflection. Adjust scope to lock in 2-cycle waveform. Replace V_9
15.	Tune in a strong FM stereo multiplex signal	Connect vertical input of scope to point Ⓔ, low side to chassis	A_{22}, A_{23}	Adjust A_{22}, A_{23} for maximum deflection of 4-cycle waveform.
16.	Tune in a strong FM stereo multiplex signal	Connect vertical input of scope to point Ⓖ, Horizontal input of scope to point Ⓗ, low side to chassis.	Separation control R_6	Adjust maximum clockwise and counterclockwise to produce diagonal responses shown in Fig. 3. Set control to produce round pattern shown in Fig. 3.
17.	Tune in a strong FM stereo multiplex signal	None	A_{25}	Adjust A_{25} to eliminate whistle or interference.

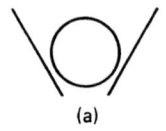

(a)

FM stereo multiplex alignment using FM-stereo signal generator (±0.0001%) accuracy)

High side to point ⟨D⟩, low side to ground
Suggested alignment tools:

	Generator frequency	Indicator	Adjust	Remarks
6.	19 kHz	Vertical amp. to point ⟨E⟩, low side to ground	A_{13}	Adjust for maximum. Set scope to lock in 2 cycles of 19-kHz waveform
7.	19 kHz	Vertical amp. to point ⟨F⟩, low side to ground	A_{14}	Adjust for maximum 4-cycle waveform
8.	67 kHz	Vertical amp. to point ⟨F⟩, low side to ground	A_{15}	Use audio oscillator if necessary. Adjust for minimum.
9.	23 kHz	Vertical amp. to point ⟨F⟩, low side to ground	A_{16}	Adjust for maximum
10.	Modulated left channel	Vertical amp. to point ⟨G⟩, low side to ground	R_{42}	Adjust for minimum.
11.	Modulated right channel	Vertical amp. to point ⟨H⟩, low side to ground	R_{42}	Check for minimum. Make compromise adjustment if necessary.

To align multiplex section using an air signal, first make sure FM section is properly aligned. Tune in a strong FM stereo signal. Follow steps 6-8, except in step 8 adjust to eliminate whistle or interference.

(b)

FIGURE 18-7 (a) Multiplex alignment instructions for the receiver of Figure 17-1. (b) Alignment instructions for the stereo decoder of Figure 17-2 using a stereo generator (courtesy of Howard W. Sams & Co., Inc.).

Matrix-Decoder Alignment without Stereo Generator

When a stereo generator is unavailable, the following process can be used for matrix-decoder alignment using the signal from a FM stereo station and an audio signal generator capable of producing the 19-kHz and 67-kHz signals. The block diagram of Figure 18-6 is referenced for adjustments and test points.

A stereo station is tuned in, and the 19-kHz oscillator is disabled. The scope is connected at TP <A>, and adjustments $A1$ and $A2$ are set for maximum 19-kHz response.

The receiver is then tuned off-station, and a signal generator is used to inject a 19-kHz signal at the input to the matrix decoder. The scope is connected to TP <C>, and $A7$ is adjusted for a minimum to set the high cutoff frequency for the low-pass filter.

The scope is moved to TP <D>, and $A6$ is adjusted for minimum 19-kHz response. This sets the low frequency cutoff for the bandpass filter. To set the high-frequency cutoff, the generator is set to 67 kHz, and $A5$ is adjusted for minimum 67-kHz response.

To tune the doubler, the scope is moved to TP <B>, and the oscillator is restored to normal operation. $A3$ and $A4$ are then adjusted for maximum 38-kHz response.

To obtain the best stereo separation, a stereo station is tuned in and separation control $A8$ is adjusted to produce the best possible stereo separation.

Figure 18-7 is a set of alignment instructions for the matrix decoders of Figures 17-1 and 17-2.

Sec. 18-8   Matrix-Decoder Alignment   535

18-9 TIME-DIVISION-DECODER ALIGNMENT

Figure 18-8 illustrates the setup for a time-division decoder alignment using a stereo generator. The procedure begins by adjusting the SCA filter.

The stereo generator is connected to the decoder input and set to produce a 67-kHz signal. The scope is connected to TP <A>, and A1 is set to produce minimum 67-kHz response.

To align the 19-kHz amplifier, connect the scope to TP <B> and adjust A2 and A3 for maximum 19-kHz response.

The doubler is aligned by connecting the scope to TP <C> and adjusting A4 for maximum 38-kHz response.

To obtain the best possible stereo separation, the generator is set to produce a left-only signal, and with the scope connected at the right channel output of the MPX detector, adjustments A2, A3, and A4 are readjusted slightly to produce minimum output. We emphasize here that only slight adjustments of A2, A3, and A4 are made at this point. These adjustments shift the phase of the 38-kHz carrier.

Continuing, the generator is set to produce a right-only signal, and the scope is connected to the left-channel output of the MPX detector. A2, A3, and A4 are again readjusted slightly to produce minimum output. A compromise setting of the adjustments between this and the previous paragraph may be required to obtain the best overall stereo separation.

The generator and scope are now disconnected as preparation is made to set the stereo threshold adjustment A5. It is set so that the stereo light comes on for weak stereo stations but does not come on because of random noise as the receiver is tuned across the FM band. This completes the alignment.

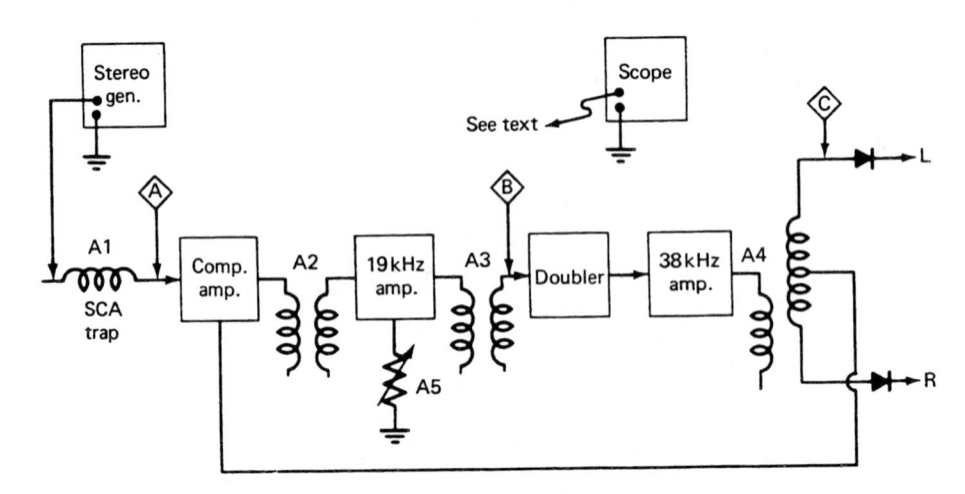

FIGURE 18-8 Block diagram representation of a time-division decoder alignment using a stereo generator.

FM-stereo multiplex alignment using FM-stereo signal generator (±0.0001% accuracy)

		High side of generator through 47 kΩ to point ⟨B⟩, low side to ground.		
	Generator frequency	Indicator	Adjust	Remarks
8.	19 kHz	Vertical amp. through 47 kΩ to point ⟨D⟩, low side to ground	A_{18}, A_{19}	Adjust for maximum.
9.	19 kHz	Vertical amp. through 47 kΩ to point ⟨E⟩, low side to ground	A_{20}	Adjust for maximum 38 kHz response.
10.	Modulated left channel	Vertical amp. to point ⟨F⟩ low side to ground	A_{18}, A_{19}, A_{20}	Adjust for minimum. This step should require only slight adjustment.
11.	Modulated right channel	Vertical amp. to point ⟨G⟩, low side to ground		Check for minimum. If necessary make compromise adjustments of A_{18}, A_{19}, and A_{20}

FM-stereo threshold adjustment (R_6)

Tune across dial (with AFC off) noting operation of stereo indicator lamp. Adjust R_6 until lamp glows on weak stereo stations but does not glow on noise pulses.

(a)

FM-stereo multiplex alignment using FM-stereo signal generator (±0.0001% accuracy)

	High side of generator through 47 kΩ to point ⟨D⟩, low side to ground.		
Generator frequency	Indicator	Adjust	Remarks
19 kHz	Vertical input of scope through 47 kΩ to point ⟨F⟩, low side to ground	T_8	Adjust for maximum.
19 kHz	Vertical input of scope through 47 kΩ to point ⟨G⟩, low side to ground	T_9	Adjust for maximum 38-kHz response.
Modulated left channel	Vertical input of scope to point ⟨H⟩, low side to ground	VR_{216}	Adjust for minimum. This step should require only slight adjustment.
Modulated right channel	Vertical input of scope to point ⟨G⟩, low side to ground		Check for minimum. If necessary, make compromise adjustment of VR_{216}

(b)

FIGURE 18-9 Alignment instructions for the stereo-decoder sections of (a) Figure 17-3 and (b) Figure 17-4.

Time-Division-Decoder Alignment without a Stereo Generator

When a stereo generator is unavailable, a signal generator and a station signal can be used to align the decoder section. Refer again to Figure 18-8 for the following description.

As before, SCA rejection is optimized by injecting a 67-kHz signal to the decoder input and adjusting $A1$ for minimum response as indicated by a scope connected at TP $<A>$. The generator is then disconnected.

A stereo station is tuned in. The scope is placed at TP $<B>$ and adjustments $A2$ and $A3$ are set to produce maximum 19-kHz response. The scope is then moved to TP $<C>$, and $A4$ is adjusted to produce maximum amplitude of the composite waveform. $A4$, $A3$, and $A2$ are then readjusted slightly to produce the best stereo separation as perceived by ear. The threshold adjustment $A5$ is made in the same manner as before.

Figure 18-9 contains a set of alignment instructions for the time-division circuits in Figures 17-3 and 17-4.

18-10 PLL-DECODER ALIGNMENT

Alignment of a PLL decoder requires only one adjustment, and this adjustment can be made with a station signal or by using a frequency counter. With a stereo station tuned in, the PLL control ($R317$ near IC301 of Figure 18-3) is turned from one extreme to the other, noting the settings where the stereo light comes on and then goes off. The control is then set halfway between these two points.

If a frequency counter is available, the receiver is tuned off-station, and the frequency counter is connected to the 19-kHz test point. The PLL control is then adjusted to produce 19 kHz on the frequency counter.

SUMMARY

1. When one channel of a stereo is inoperative on all functions, the audio amplifier is almost certain to be defective. Frequently, voltage and signal-level comparisons between the good and bad channels are helpful in locating the defect.

2. The following general procedure is applicable to integrated circuit stages. First, try signal injection to determine if a signal applied to the input is delivered in its proper form at the output. If it is not, check for the proper voltage at all pins of the IC. If a voltage is found off-value, check the components associated with the IC for shorts, opens, or for solder bridges between PC conductors. As a last resort, suspect the IC to be defective and replace it with a new part.

3. The method employed to switch from FM monaural to FM stereo will determine whether the FM monaural will be functional when the composite amplifier is defective.

4. The muting feature (if present) must be considered when forming conclusions based upon hiss at the speaker. The presence of significant hiss indicates a problem in the FM front end.

5. When an insulated gate MOS device is to be removed from a circuit, the device being removed should be given the same care to avoid static discharges as the new device to be installed. The original device may not be defective.

6. A technician must guard against shock hazards when grounded while working with MOS devices.

7. A defect in the stereo decoder may cause one channel to be inoperative in the FM-stereo mode.

8. A loss of stereo separation may be caused by a defect in the 19-kHz/38-kHz amplifier/doubler circuitry. Also, misadjustment of the separation control or a loss of sensitivity in the FM front end may be responsible for lack of channel separation.

9. A stereo generator is required for optimum alignment of stereo decoders even though a less precise alignment can be achieved without a stereo generator. A scope is almost essential if no stereo generator is available.

10. Alignment of a matrix decoder involves adjustment of the filter circuits that separate the $L + R$, $L - R$, and 19-kHz components. The doubler and the separation control also require adjustment. All decoders require that the SCA trap be properly tuned.

11. Time-division decoders require alignment of the 19-kHz/38-kHz signal path. Stereo separation is frequently optimized by fine-tuning the coupling transformers so as to vary, slightly, the phase of the regenerated subcarrier.

12. PLL decoders are the simplest to align, requiring only an adjustment of the free-running frequency of the VCO. A frequency counter may be helpful in this regard.

QUESTIONS AND PROBLEMS

18-1. Suppose that one channel of a stereo amplifier is out while the other functions normally. Explain how the good channel can provide valuable information about voltages and signal levels that can be used in locating the defect in the faulty channel.

18-2. What problems are sometimes encountered with headphone jacks?

18-3. Give a general procedure for troubleshooting IC stages such as the audio amplifiers of Figure 18-2. At what point in the procedure should the IC be removed?

18-4. Under what circumstances will a defect in the composite amplifier not cause the FM monaural function to be inoperative?

18-5. (a) What is the purpose of a mute circuit?

(b) How can the mute circuit cause a technician to form a wrong conclusion based on the level of hiss in the speaker?

18-6. It is often suggested that a technician should connect an earth ground to his or her body while replacing insulated gate MOS devices to avoid damaging the device by static electricity. Explain the hazard involved in connecting oneself to a ground. What safety tips might be recommended?

18-7. Suppose MPX detector diode $X12$ opens in Figure 18-4. What symptom would likely result?

18-8. What symptom would likely result if diode $X11$ in Figure 18-4 opened so that the 19-kHz doubler input rectifier operates as a half-wave rather than a full-wave rectifier? What would be the effect upon the pulse rate at the base of $Q9$?

18-9. What may be the effect of a "weak" RF amplifier upon stereo separation?

18-10. What defects can cause a lack of stereo separation?

18-11. What adjustments must typically be made in aligning the following decoders?

(a) Matrix.

(b) Discrete component, time division.

(c) IC time division.

(d) PLL time division.

(e) Biplex detector decoder.

18-12. Describe the test instruments required to align a typical time-division stereo decoder.

18-13. Most stereo alignment generators provide an SCA signal at 67 kHz. For what is this test signal used in aligning an FM stereo receiver that has no provisions for receiving SCA transmissions?

18-14. Describe the procedures typically employed for adjusting the separation control.

18-15. Describe a typical matrix-decoder alignment procedure that does not require a stereo generator.

18-16. How does the alignment procedure for a time-division decoder differ from the procedure for a matrix decoder?

18-17. What is the typical procedure for aligning a PLL decoder?

18-18. For what purpose is a frequency counter used in PLL-decoder alignment?

SUGGESTED BOOKS FOR FURTHER READING

For more information on transistors and simple mathematical analysis of basic transistor circuits:

MALVINO, A. P.: *Transistor Circuit Approximations*, 4th ed. McGraw-Hill Book Company, New York, 1985.

The fundamentals of operational amplifiers and the applications of linear integrated circuits (no emphasis on communications circuits) are given in:

JACOB, J. MICHAEL: *Applications and Design with Analog Integrated Circuits*, Reston Publishing Company, Inc., Reston, Va., 1982.

Two texts that cover a broad range of topics in electronic communication:

MILLER, GARY M.: *Modern Electronic Communication*, 2nd ed. Prentice-Hall, Inc., Englewood Cliffs, N.J., 1983.

SHRADER, ROBERT L.: *Electronic Communication*, 5th ed. McGraw-Hill Book Company, New York, 1985.

An older text based on vacuum-tube technology but which provides excellent descriptions of receiver circuitry and troubleshooting procedures:

MARCUS, W., and A. LEVY: *Practical Radio Servicing*, 2nd ed. McGraw-Hill Book Company, New York, 1963.

An excellent book on the theoretical aspects of FM receivers, but which is not overly mathematical, is:

COOK, A. B., and A. A. LIFF: *Frequency Modulation Receivers*. Prentice-Hall, Inc., Englewood Cliffs, N.J., 1968.

An easy-to-read text on the FM system is:

KIVER, M. S.: *FM Simplified*, 3rd ed. D. Van Nostrand Company, Princeton, N.J., 1960.

And, finally, for the person who wishes to round-out his or her technical education by studying physics, an easy-to-read book by a well-known author:

GREEN, CLARENCE R., *Technical Physics*. Prentice-Hall, Inc., Englewood Cliffs, N.J., 1984.

SOLUTIONS TO SELECTED QUESTIONS AND PROBLEMS

Chapter 2

2-3. A hole represents a deficiency of one negative charge, and in the environment of the crystal, appears positive   **2-5.** Electrons and holes, respectively   **2-7.** It becomes thinner   **2-8.** 0.7 V   **2-9.** Very sharp; voltage regulators   **2-10.** Positive; collector   **2-13.** The base region is very thin and lightly doped   **2-14.** Base current; the number by which the base current is multiplied   **2-15.** Nearly equal; base current is small compared to the collector current   **2-16.** Base and emitter   **2-18.** If forward biased, channel to gate conduction might destroy the device   **2-20.** Negative   **2-21.** Negative   **2-23.** (a) Yes; (b) yes; (c) no   **2-24.** To prevent conduction from the channel to the substrate   **2-25.** Positive   **2-27.** The difference in voltage applied to the $(-)$ and $(+)$ inputs is amplified   **2-28.** Negative feedback   **2-29.** So that the output may be zero volts when the input is zero volts   **2-30.** Slew rate, frequency response, and noise production   **2-31.** The plate is not heated to thermionic emission temperatures   **2-32.** Decrease   **2-33.** Negative

Chapter 3

3-1. (a) 20 V; (b) none; (c) 2 A; (d) 0.333 A   **3-2.** (a) 25 V; (b) 5 A; (c) 125 V-A; (d) 125 V-A; (e) 1.25 A   **3-3.** (b) 10:1; (c) 2 A, 3 A; (d) 24 V-A, 36 V-A; (e) 60 V-A; (f) 0.5 A   **3-4.** 30 A   **3-5.** 1 A   **3-6.** (a) 1:1   **3-7.** 15.27 V   **3-8.** (a) 60; (b) 120   **3-9.** (a) Charges the filter capacitors; (b) use a surge-limiting resistor   **3-10.** (a) 12 V; (b) 6 V; (c) 12 V; (d) +6 V, −6 V   **3-11.** (a) 60 Hz; (b) 120 Hz; (c) 120 Hz; (d) 120 Hz   **3-12.** Bridge would operate as a half-wave rectifier with reduced output voltage and increased ripple   **3-13.** Transformer would overheat; one other diode would be destroyed   **3-14.** Very sharp reverse breakdown characteristic   **3-15.** Reverse   **3-16.** 0.288 W   **3-17.** 100%; the power dissipation is also doubled   **3-18.** 0.5 W   **3-19.** The input voltage to the zener drops below the zener voltage   **3-20.** Current loading of the zener circuit is reduced   **3-22.** Yes   **3-23.** To avoid regulator oscillations   **3-24.** The internal circuitry of the regulator   **3-27.** Open would produce half-wave operation; short would apply AC to load   **3-28.** High B+ would be applied to the other tube filaments, probably causing subsequent filament-to-cathode shorts   **3-29.** Low B+ or B++ with possible overheating of the rectifier tube; open filter resistor causes B+ to become zero   **3-30.** Fuse would blow in the external circuit   **3-31.** Both cases: capacitor would be destroyed

Chapter 4

4-1. (a) Positive; (b) negative; (c) positive; (d) negative; (e) positive; (f) negative   **4-2.** About 0.7 V   **4-3.** (a) 625 µA; (b) 630 µA; (c) 0.992   **4-4.** 25 mA   **4-5.** From 480 µA to 880 µA   **4-6.** Power supply polarity is reversed   **4-7.** All silicon PN junctions have very similar conduction characteristics   **4-8.** See Sec. 4-3; yes, yes, yes   **4-10.** Decrease; increase   **4-11.** (a) Decreases; (b) increases; (c) increases   **4-13.** In any arrangement that utilizes collector feedback   **4-14.** Yes; see text for formulas for calculating V_c **4-15.** (a) No effect; (b) decreases; (c) increases; (d) no effect   **4-16.** To allow for the swamping of the AC junction resistance   **4-17.** I_c = 1.012 mA; V_c = 5.66 V; r'_e = 24.7 Ω; gain = 133; Z_{in} = 2.463 kΩ   **4-18.** I_c = 1.771 mA; V_c = 3.16 V; r'_e = 14.116 Ω; gain = 233; Z_{in} = 2.463 kΩ (change in r'_e resulting from increased collector current nullifies changes in Z_{in} due to changes in beta)   **4-19.** I_c = 0.902 mA; V_c = 6.023 V; gain = 3.211; Z_{in} = 91.27 kΩ   **4-20.** I_c = 1.46 mA; V_c = 4.183 V; gain = 3.245; Z_{in} = 146 kΩ; parameters most affected are collector current, collector voltage, and input impedance   **4-21.** I_c = 1.60 mA; V_c = 6.72 V; gain = 211; r'_e = 15.625 Ω; Z_{in} = 1.3 kΩ; the input impedance increases primarily due to an increase in r'_e caused by a decrease in collector current   **4-22.** I_c = 1.01 mA; V_c = 9.34 V; r'_e = 24.75 Ω: gain = 2.77; Z_{in} = 136 kΩ   **4-23.** V_b = 1.09 V; V_e = 0.39 V; I_c = 1 mA; I_b = 8 µA; V_c = 6.4 V; gain = 13.49; Z_{in} = 1.926 kΩ   **4-24.** I_c = 2.54 mA; V_c = 3.07 V; V_e = 1.27 V; V_b = 1.97 V; r'_e = 9.84 Ω; gain = 9.22; Z_{in} = 38.92 kΩ   **4-25.** I_c = 2.85 mA; V_c = 1.3 V; V_e = −8.23 V; V_b = −7.53 V; r'_e = 8.772 Ω; gain = 9.685; Z_{in} = 31.52 kΩ   **4-26.** (a) Drop to zero; (b) rise to that of the power supply; (c) drop to zero; (d) remain essentially unchanged   **4-27.** Cause it to decrease   **4-28.** (a) Remain essentially unchanged; (b) rise to that of the power supply; (c) drop to zero; (d) increase due to the increased collector voltage   **4-29.** Less than 2.0 V   **4-30.** Reverse-biased gate-channel junction   **4-31.** Zero   **4-32.** Decrease; drain current will increase   **4-33.** Slope becomes less; becomes smaller   **4-34.** Toward the power supply point   **4-35.** No effect   **4-36.** Very small   **4-37.** 3 mS   **4-38.** Larger   **4-39.** 8.8   **4-40.** Slightly greater   **4-41.** Greater   **4-42.** Less positive   **4-43.** More negative   **4-44.** 0 V   **4-45.** Yes; due to the plate current flowing through the cathode resistor   **4-46.** Electrons are trapped on the grid   **4-47.** The attraction of electrons to the grid depends upon the amplitude of the input signal

Chapter 5

5-1. Gain is reduced. Total gain is the product of the individual gains   **5-2.** 30   **5-3.** Any departure of the Q-point of stage 1 from the desired value is amplified by subsequent stages   **5-4.** Use complementary transistors   **5-6.** 10,000   **5-7.** No; must have an emitter and collector resistor   **5-8.** No   **5-9.** No; must have an emitter resistor   **5-10.** (a) 1; (b) no   **5-11.** Replace R_c with an output transformer   **5-12.** Beam power pentode   **5-13.** 1548.8 Ω   **5-14.** Plate current goes to zero; cathode voltage goes to zero; grid voltage is not affected   **5-15.** The production of an identical signal 180° out of phase   **5-16.** Transformer; phase inverter stage   **5-17.** DC current flow through the transformer   **5-18.** Crossover distortion; use crossover diode, etc. (see the text)   **5-19.** Output transistors may overheat   **5-20.** A temperature-dependent resistor. Stabilize DC parameters against variations in temperature   **5-21.** They are low-impedance devices (i.e., they operate at low voltages and high currents)   **5-22.** The initial charging of the output coupling capacitor to the midpoint voltage   **5-23.** Because of the complementary (opposite polarity) characteristics of *NPN* and *PNP* transistors **5-24.** Drivers are complementary; output transistors are the same type

Chapter 6

6-2. Response broadens as double-hump characteristic develops **6-3.** Tends to oscillate. Neutralize triode or use pentode **6-4.** An effective amplification of capacitance between the plate and grid of a tube **6-11.** Reduces the sensitivity of the receiver **6-12.** No; see Sec. 6-6 **6-14.** Base region of a transistor acts as a shield between collector and emitter **6-18.** No; carrier is still present **6-19.** Sec. 6-7; Forward: increases collector current to reduce gain. Reverse: decreases collector current to reduce gain **6-20.** Base becomes more positive to reduce gain **6-21.** AVC **6-22.** (a) 50.3 kHz; (b) 503 kHz; (c) 5.03 MHz; (d) 99.5 MHz **6-23.** (a) 316 Ω; (b) 316 Ω; (c) 316 Ω; (d) 1 kΩ **6-24.** (a) 45.5 kHz; (b) 18.2 kHz; (c) 9.1 kHz **6-25.** 45.5 **6-26.** 10 kHz **6-27.** F_r = 11,254 Hz; Q = 70.7; Q_p = 17.55; BW = 641 Hz **6-28.** F_r = 11.254 Hz; Q = 70.7; BW = 159 Hz **6-29.** 45.5 **6-30.** Increases; decreases **6-31.** 29.1 kHz **6-32.** 500 **6-36.** 2 **6-37.** (a) 1.67; (b) 2.23 **6-38.** 2 **6-39.** (a) 2.33; (b) 4.29

Chapter 7

7-1. Converter is both mixer and oscillator in one stage **7-3.** Transfer characteristic of a mixer is nonlinear **7-4.** 3 kHz, 4 kHz, 7 kHz, 1 kHz **7-5.** Yes **7-8.** Oscillators use positive feedback, amplifiers negative **7-9.** Hartley uses a tapped coil, whereas Armstrong uses two separate coils; both achieve feedback via magnetic coupling **7-10.** Split capacitors **7-11.** Trimmer capacitors used in alignment **7-12.** Universal bias, base bias with emitter feedback **7-14.** Oscillator signal **7-15.** Serves as the plate of the oscillator **7-16.** A few volts DC, negative **7-17.** Uses two independent windings to achieve feedback via magnetic coupling **7-18.** Emitter **7-20.** Lower **7-24.** (a) Plate current will be in phase with the grid voltage. Therefore, the plate current leads the plate voltage by 90° **7-25.** A simpler means is available for obtaining a controlled reactance, namely a varactor diode **7-27.** If forward-biased, the varactor would conduct and appear as a shorted capacitor **7-28.** Decreases **7-29.** (a) Yes; (b) a varying collector voltage (with base voltage nearly constant) will alter the thickness of the depletion layer causing the collector-base capacitance to vary; (c) the varying capacitance can alter the resonant frequency

Chapter 8

8-4. Ultimately, the B+ voltage **8-5.** The strength of the waves passing the receiving antenna **8-6.** See Figure 8-9 **8-9.** Atmospheric-induced variations of signal strength at the receiver would be interpreted as temperature variations at the north pole **8-10.** (a) 306 m; 1004 ft; (b) 1.02 μs; (c) 306 m; (d) 134 μs; (e) 131 **8-11.** 30% **8-12.** 6 W **8-13.** (a) 5 W; (b) 5 W; (c) 20 W; (d) 20 W **8-14.** 120 W **8-16.** 136, 138, and 140 kHz **8-17.** 1725 **8-18.** (a) 1695 kHz; (b) 3390 kHz; (c) 2935 kHz; 3845 kHz; 4630 kHz; 5540 kHz; (d) 2150 kHz **8-19.** 262.5 kHz **8-20.** 499.4 s (8 min, 19 s); same

Chapter 9

9-1. Fuse; surge limiting; voltage drop **9-2.** (a) *NPN;* (b) silicon **9-3.** High-base and low-base resistors for universal-biased *Q*3 **9-4.** A decoupling network to keep 455-kHz signals out of the power supply **9-5.** Line filter to remove RF on the AC line **9-6.** Prevents AC feedback **9-7.** Varistor; limits maximum voltage applied to the *Q*6 collector **9-8.** Thermistor; provides thermal stability for the output transistors **9-9.** All four bridge diodes are in one package **9-10.** Negative voltage sources are used to supply the *PNP* transistors **9-11.** *C*19 is part of the tone control circuit **9-12.** Emitter bypass capacitor **9-13.** Clock motor winding is used as a step-down transformer **9-15.** Half-wave **9-16.** By moving the ground point to a midpoint on the voltage divider (see Figure 9-14) **9-17.** Decoupling **9-18.** To reduce crossover distortion **9-19.** Darlington pair **9-20.** Makes CTP 4 a signal ground **9-21.** In Figure 9-1, AVC is negative-going; in Figure 9-7, AVC is positive going **9-22.** More positive (actually less negative) **9-23.** High-base and low-base resistors for *Q*1 universal bias **9-24.** Emitter bypass capacitor **9-25.** Positive **9-26.** AVC filter capacitor **9-27.** Negative **9-28.** Remove 455-kHz carrier from output of detector **9-29.** Ohm's law applied to *R*28 yields 1.2 mA; to *R*29 yields 0.4 mA. Normal variations of measured values confuse the issue **9-30.** 14.4 mA **9-31.** *C*11 blocks DC bias voltage in Figure 9-3, allowing the low end of *L*1 to be grounded **9-32.** 2.8 mA **9-33.** Simple base bias **9-34.** Signal level is large compared to ripple voltage **9-35.** Oscillator uses grid leak biasing; signal grid connects to the AVC line **9-36.** Check for a negative voltage of several volts on the oscillator grid **9-37.** No **9-38.** Remove any 455-kHz carrier signal that may remain **9-39.** 0.047 μF, 0.05 μF, 3.3 MΩ, 2.2 MΩ **9-40.** Divide cathode voltage by cathode resistance (this is only approximate for tetrodes and pentodes due to the screen current) **9-41.** (1) acts as a fuse; (2) limits surge current **9-42.** Tone compensation

Chapter 10

10-3. A metal screwdriver may change the resonant frequency of a tuned circuit **10-7.** For VTVM is 11 MΩ vs. 100 kΩ for VOM **10-9.** Would probably cause the oscillator to stop oscillating **10-14.** Audio; no **10-21.** AC parameters **10-22.** (a) Use a DC VOM or VTVM to measure output voltage of ohmmeter. The ohmmeter polarity can be deduced from an up-scale deflection of the voltmeter; (b) *NPN* and *PNP* determinations would be reversed **10-23.** Maintain a larger stock of replacement transistors **10-24.** Cool the transistor and note any change in the noise level. Most noisy transistors are temperature sensitive **10-25.** Heat-sink compound to enhance heat transfer **10-26.** Freeze spray is ineffective on tubes and is not intended for use on tubes **10-31.** 99 MΩ **10-33.** The frequency response of the vertical section **10-36.** Shorted junction

Chapter 11

11-5. Signal injection; signal tracing; V-R analysis **11-10.** See Figure 11-2; audio test signal **11-12.** 455 kHz, 30-50% modulated RF generator **11-13.** Nothing **11-14.** The high plate voltage may damage the generator unless a coupling capacitor is used **11-15.** If a signal whose frequency is in the broadcast band (to which the receiver is tuned) will pass through the converter and IF amplifiers, the local oscillator must be working because the signal has been converted to the IF frequency **11-18.** No. Station

Solutions to Selected Questions and Problems

must be tuned in before signal tracing can be done **11-19.** Demodulator **11-20.** Most scopes will pick up the oscillator signal if the probe is simply held near the oscillator coil **11-21.** Signal levels are very low **11-22.** No. No audible tone will be produced since the oscillator signal is not modulated **11-23.** Voltages at A and B would increase slightly; voltage at C would decrease **11-24.** Same **11-25.** The voltages at B and C would decrease greatly; the voltage at A would decrease slightly **11-27.** Check for shorts on the B+ line, paying particular attention to the filter capacitors **11-28.** Disconnect $C1$ from the grid and measure the voltage on the free side with the EVM as explained in Sec. 10-2 **11-38.** (a) Decreases; (b) increases **11-39.** (a) Increases; (b) decreases; (c) may increase somewhat **11-40.** High and low emitter and collector voltages, respectively, with possible overheating of the device **11-42.** An excessive voltage difference between the base and emitter may indicate an open B-E junction **11-43.** Measure voltage between the base and the emitter

Chapter 12

12-3. B-E voltage is normal for a silicon transistor. B-E junction is probably good and, unless the B-C junction is leaky, entire transistor is probably good. This test does not check AC parameters, however **12-4.** Not necessarily. DC bias is normal, but AC characteristics may be degraded **12-5.** A B-E voltage as high as 4 V usually indicates an open B-E junction; device will be inoperative at all frequencies **12-6.** C-B junction is leaky **12-7.** (a) Would equal supply voltage; (b) lower than normal; (c) lower than normal **12-18.** Yes. By excessive reduction of the gain of the RF amplifier, mixer, or IF amplifiers **12-21.** Loss of sensitivity and/or loss of volume **12-27.** Electrolytic capacitors are polarized and will indicate power supply polarity **12-29.** (a) Not advisable as a first step; try a replacement transistor from another line of semiconductors if possible; (b) increase $R29$ and $R32$; (c) increase $R26$ **12-30.** $R26$ and $X2$ should be checked. If necessary, $R26$ can be increased in value. **12-47.** Setting each tuned circuit to a slightly different frequency lowers the total effective Q of the amplifier. The selectivity will be reduced because of the increased bandwidth, and the sensitivity will be reduced because of the lower Q **12-50.** Touch the center terminal with a metal screwdriver **12-51.** AGC

Chapter 13

13-6. (a) 40%; (b) 30; (c) 60 kHz; (d) 80%, 60, 60 kHz **13-10.** 101.501 MHz; 101.502 MHz; 101.503 MHz; no **13-11.** Increases **13-12.** No **13-13.** No. An AM carrier component is constant in amplitude **13-14.** Figure 13-4(b) **13-17.** To control the frequency of the local oscillator for FM **13-18.** Limiter action regulates signal amplitude **13-20.** See Figures 13-6 and 13-7

Chapter 14

14-1. (a) Universal bias modified by FM RF AGC **14-2.** (a) Universal; (c) negative; (d) $A12$ is oscillator trimmer; $A13$ is trimmer for RF amplifier collector-load tuning capacitor; $A10$ tunes the first IF input transformer; $A3$ is the counterpart of $A10$ for the AM section **14-3.** (a) Universal; (b) prevents direct current flow through $L6S$; (c) a loopstick (ferrite rod) with two windings; (d) $A4$, $A5$, $A6$ **14-4.** (b) Ground return for the

Chapter 9

9-1. Fuse; surge limiting; voltage drop **9-2.** (a) *NPN;* (b) silicon **9-3.** High-base and low-base resistors for universal-biased $Q3$ **9-4.** A decoupling network to keep 455-kHz signals out of the power supply **9-5.** Line filter to remove RF on the AC line **9-6.** Prevents AC feedback **9-7.** Varistor; limits maximum voltage applied to the $Q6$ collector **9-8.** Thermistor; provides thermal stability for the output transistors **9-9.** All four bridge diodes are in one package **9-10.** Negative voltage sources are used to supply the *PNP* transistors **9-11.** $C19$ is part of the tone control circuit **9-12.** Emitter bypass capacitor **9-13.** Clock motor winding is used as a step-down transformer **9-15.** Half-wave **9-16.** By moving the ground point to a midpoint on the voltage divider (see Figure 9-14) **9-17.** Decoupling **9-18.** To reduce crossover distortion **9-19.** Darlington pair **9-20.** Makes CTP 4 a signal ground **9-21.** In Figure 9-1, AVC is negative-going; in Figure 9-7, AVC is positive going **9-22.** More positive (actually less negative) **9-23.** High-base and low-base resistors for $Q1$ universal bias **9-24.** Emitter bypass capacitor **9-25.** Positive **9-26.** AVC filter capacitor **9-27.** Negative **9-28.** Remove 455-kHz carrier from output of detector **9-29.** Ohm's law applied to $R28$ yields 1.2 mA; to $R29$ yields 0.4 mA. Normal variations of measured values confuse the issue **9-30.** 14.4 mA **9-31.** $C11$ blocks DC bias voltage in Figure 9-3, allowing the low end of $L1$ to be grounded **9-32.** 2.8 mA **9-33.** Simple base bias **9-34.** Signal level is large compared to ripple voltage **9-35.** Oscillator uses grid leak biasing; signal grid connects to the AVC line **9-36.** Check for a negative voltage of several volts on the oscillator grid **9-37.** No **9-38.** Remove any 455-kHz carrier signal that may remain **9-39.** 0.047 µF, 0.05 µF, 3.3 MΩ, 2.2 MΩ **9-40.** Divide cathode voltage by cathode resistance (this is only approximate for tetrodes and pentodes due to the screen current) **9-41.** (1) acts as a fuse; (2) limits surge current **9-42.** Tone compensation

Chapter 10

10-3. A metal screwdriver may change the resonant frequency of a tuned circuit **10-7.** For VTVM is 11 MΩ vs. 100 kΩ for VOM **10-9.** Would probably cause the oscillator to stop oscillating **10-14.** Audio; no **10-21.** AC parameters **10-22.** (a) Use a DC VOM or VTVM to measure output voltage of ohmmeter. The ohmmeter polarity can be deduced from an up-scale deflection of the voltmeter; (b) *NPN* and *PNP* determinations would be reversed **10-23.** Maintain a larger stock of replacement transistors **10-24.** Cool the transistor and note any change in the noise level. Most noisy transistors are temperature sensitive **10-25.** Heat-sink compound to enhance heat transfer **10-26.** Freeze spray is ineffective on tubes and is not intended for use on tubes **10-31.** 99 MΩ **10-33.** The frequency response of the vertical section **10-36.** Shorted junction

Chapter 11

11-5. Signal injection; signal tracing; V-R analysis **11-10.** See Figure 11-2; audio test signal **11-12.** 455 kHz, 30–50% modulated RF generator **11-13.** Nothing **11-14.** The high plate voltage may damage the generator unless a coupling capacitor is used **11-15.** If a signal whose frequency is in the broadcast band (to which the receiver is tuned) will pass through the converter and IF amplifiers, the local oscillator must be working because the signal has been converted to the IF frequency **11-18.** No. Station

Solutions to Selected Questions and Problems

must be tuned in before signal tracing can be done **11-19.** Demodulator **11-20.** Most scopes will pick up the oscillator signal if the probe is simply held near the oscillator coil **11-21.** Signal levels are very low **11-22.** No. No audible tone will be produced since the oscillator signal is not modulated **11-23.** Voltages at A and B would increase slightly; voltage at C would decrease **11-24.** Same **11-25.** The voltages at B and C would decrease greatly; the voltage at A would decrease slightly **11-27.** Check for shorts on the B+ line, paying particular attention to the filter capacitors **11-28.** Disconnect $C1$ from the grid and measure the voltage on the free side with the EVM as explained in Sec. 10-2 **11-38.** (a) Decreases; (b) increases **11-39.** (a) Increases; (b) decreases; (c) may increase somewhat **11-40.** High and low emitter and collector voltages, respectively, with possible overheating of the device **11-42.** An excessive voltage difference between the base and emitter may indicate an open B-E junction **11-43.** Measure voltage between the base and the emitter

Chapter 12

12-3. B-E voltage is normal for a silicon transistor. B-E junction is probably good and, unless the B-C junction is leaky, entire transistor is probably good. This test does not check AC parameters, however **12-4.** Not necessarily. DC bias is normal, but AC characteristics may be degraded **12-5.** A B-E voltage as high as 4 V usually indicates an open B-E junction; device will be inoperative at all frequencies **12-6.** C-B junction is leaky **12-7.** (a) Would equal supply voltage; (b) lower than normal; (c) lower than normal **12-18.** Yes. By excessive reduction of the gain of the RF amplifier, mixer, or IF amplifiers **12-21.** Loss of sensitivity and/or loss of volume **12-27.** Electrolytic capacitors are polarized and will indicate power supply polarity **12-29.** (a) Not advisable as a first step; try a replacement transistor from another line of semiconductors if possible; (b) increase $R29$ and $R32$; (c) increase $R26$ **12-30.** $R26$ and $X2$ should be checked. If necessary, $R26$ can be increased in value. **12-47.** Setting each tuned circuit to a slightly different frequency lowers the total effective Q of the amplifier. The selectivity will be reduced because of the increased bandwidth, and the sensitivity will be reduced because of the lower Q **12-50.** Touch the center terminal with a metal screwdriver **12-51.** AGC

Chapter 13

13-6. (a) 40%; (b) 30; (c) 60 kHz; (d) 80%, 60, 60 kHz **13-10.** 101.501 MHz; 101.502 MHz; 101.503 MHz; no **13-11.** Increases **13-12.** No **13-13.** No. An AM carrier component is constant in amplitude **13-14.** Figure 13-4(b) **13-17.** To control the frequency of the local oscillator for FM **13-18.** Limiter action regulates signal amplitude **13-20.** See Figures 13-6 and 13-7

Chapter 14

14-1. (a) Universal bias modified by FM RF AGC **14-2.** (a) Universal; (c) negative; (d) $A12$ is oscillator trimmer; $A13$ is trimmer for RF amplifier collector-load tuning capacitor; $A10$ tunes the first IF input transformer; $A3$ is the counterpart of $A10$ for the AM section **14-3.** (a) Universal; (b) prevents direct current flow through $L6S$; (c) a loopstick (ferrite rod) with two windings; (d) $A4$, $A5$, $A6$ **14-4.** (b) Ground return for the

FM and AM signals, respectively; (c) 70 Ω, 3 Ω; (d) negative; (e) compensation resistors (see Sec. 13-7); (f) completes circuit for tertiary winding   **14-7.** (a) It is not used; (b) signal voltage for neutralization; (c) none; (d) coupling capacitors; (e) review Sec. 3-8   **14-8.** (a) It is untuned; (b) broaden bandwidth   **14-10.** $X8$ and $X9$ are silicon diodes used to obtain bias voltage for $Q8$ **14-13.** These components constitute the FM antenna   **14-15.** (d) So limiter tube $V5$ will saturate readily; (e) applying Ohm's law to $R14$ yields 4.17 mA; ... to $R15$ yields 3.0 mA   **14-18.** No; Figure 14-11 is a grounded grid triode, whereas Figure 14-14 is a pentode, neither of which requires neutralization

Chapter 15

15-2. Input signal amplitude and exact point of signal injection are not precisely determined   **15-3.** Typical signal tracer with demodulator probe will not respond to the FM IF signal   **15-7.** (a) Lots of hiss at speaker but no signals received; (b) signal injection   **15-8.** Noise is primarily manifested as amplitude variations, and therefore may be detected by an ordinary signal tracer   **15-11.** Receiver will be noisy because the stabilizing capacitor is responsible for the AM rejection properties of a ratio detector. AM sensitivity results in excessive noise   **15-14.** No AM and/or no FM   **15-15.** AM tuning capacitor plates are shorted together   **15-17.** One might erroneously assume the RF amplifier to be working when actually it is not   **15-25.** The action of the limiter is to mask the effect of signal amplitude variations for signals that exceed the limiting threshold   **15-26.** Sec. 15-10; zero volts   **15-27.** Sec. 15-11; no   **15-28.** No; a high-impedance meter is required   **15-29.** For the most part, this is a matter of personal preference   **15-30.** (a) Would tune very broadly; (b) would fail to show significant tuning action   **15-33.** No   **15-34.** 10.6 MHz, 10.7 MHz, 10.8 MHz   **15-36.** To prevent circuit loading

Chapter 16

16-1. 50 Hz to 15 kHz   **16-2.** (b) 67 kHz; (c) FM; (d) 14 kHz   **16-4.** Use an inverting amplifier of unity gain   **16-5.** Refer to Figure 16-5   **16-6.** (a) 5,3; (b) 8,2; (c) 0,−4   **16-7.** (a) 90%; (b) 90%; (c) 10%   **16-8.** (a) 10%; (b) must be reduced proportionately   **16-9.** DSSC phase is determined by the polarity of the audio signal (Sec. 16-4)   **16-13.** For the L-only signal, the positive peaks are in phase with the phase reference. For the R-only signal, the negative peaks are in phase with the phase reference   **16-16.** To avoid attenuation of the 19-kHz pilot and L-R sidebands   **16-19.** Sec. 16-8; no   **16-20.** Via frequency doubling the 19-kHz pilot   **16-21.** Sec. 16-8; the phase of the reinserted carrier   **16-23.** (a) None; (b) L and R would be reversed   **16-25.** Changes by 180 degrees   **16-26.** Sec. 16-8; no   **16-28.** SCA trap; composite amplifier; 19-kHz amplifier; 38-kHz amplifier; stereo indicator; one audio channel

Chapter 17

17-1. $L18$, 19 kHz; $L20$, 50 Hz to 15 kHz; $L21$, 23 kHz to 53 kHz   **17-2.** (a) $R49$, $R50$, $R51$, and $R52$ all 22 kΩ; (b) $R56$, $R57$, 47 kΩ; $R58$, $R59$, 100 kΩ   **17-3.** (a) $R53$-$C66$ and $R54$-$C67$; (b) $R56$-$C63$ and $R56$-$C66$ **17-4.** (a) Composite; (b) audio ($L + R$); (c) $L - R$ AM signal; (d) L audio   **17-5.** Varies the $L + R$ signal amplitude to match $L - R$ signal   **17-8.** (a) 38-kHz positive-going pulses; (b) $L + R$ and the $L - R$ DSSC waveform; (c) same as part (a); (d) composite plus one phase of the 38-kHz carrier (see Figure 16-19) **16-13.** Composite is recombined with opposite phases of the 38-kHz carrier **16-18.** (E = external; I = internal) 1, E; 2, I; 3, I; 4, none; 5, none; 6, E; 7, none; 8, none; 9, E; 10, I; 11, I; 12, I; 13, I; 14, I **17-20.** Generates the 38-kHz carrier in the proper phase **17-21.** By PLL circuit **17-23.** (a) VCO frequency adjustment; (b) determine the response of the low-pass filter; (c) determines response of filter for stereo switch circuit; (d) stereo indicator lamp **17-26.** Sec. 17-8; output of the noise amplifier

Chapter 18

18-1. Comparisons of AC and DC levels can be made at corresponding points in each channel **18-4.** If the monaural signal is passed directly to the audio amplifiers from the FM detector **18-6.** See precaution in Sec. 18-3 **18-7.** There would be no response from the left channel in the FM stereo mode **18-8.** Pulses would be applied to 38 kHz resonant circuit $L17$ at 19 kHz rather than 38 kHz, but $L17$ would continue to resonate at 38 kHz, but at reduced amplitude. Each symptom produced is hard to predict, but a likely effect is loss of stereo separation with stereo illuminator failing to illuminate **18-13.** To adjust the SCA trap to the correct frequency **18-16.** It is simpler. Only 19-kHz and 38-kHz adjustments are necessary (possibly with an SCA trap). There are no 23- to 53-kHz or low-pass filter adjustments.

INDEX

A

AC junction resistance, 80
   demonstration circuit, 83
   gain, dependence upon, 89
AC line filter, 69, 236
AC load line, 105, 116
Active ripple filter, 68, 418
AFC, (see Automatic frequency control)
AGC, (see Automatic volume control)
Air check, 302
Alignment:
   AM receivers, 359
   FM receivers, 450
   stereo decoder, 532
Alpha, 25
Alpha-cutoff frequency, 166
AM antennas, 215
AM broadcast band, 4
AM demodulation, 220
AM receiver:
   block diagram, 5, 226
   simplest, 219
   TRF, 5, 225
AM transmitter, 220
Amplifier:
   cascaded, 123
   complementary, 126
   complementary-symmetry, 146
   controlled-gain, 109
   direct-coupled, 125
   grounded grid, 165
   IC power, 323
   IF amplifier, 6, 200, 227
   operational amplifier (op amp), 36
   OTL, 145, 247
   push-pull, 136, 262
   quasi-complementary, 149, 252
   RF amplifier, 5, 158
   single-ended, 132
Amplifier configurations, 129
Amplitude limiting, 373
Amplitude modulation, 3, 217
   characteristics, 221
   power distribution, 225
   sidebands, 222
   two-phase detector, 497
Anode, 40
Antennas:
   AM, 215
   FM, 374
Armstrong oscillator, 190, 425
Audio output stage, 132
   troubleshooting, 334
Autodyne converter, 421
Automatic frequency control (AFC), 202, 205, 406, 425
Automatic volume control (AVC), 159, 171, 227, 243, 261
   filter, 172
   forward and reverse, 173

B

Balanced power supply, 68
Basic troubleshooting procedure, 300

Beam power pentode, 132
Beta, 24
Biasing arrangements, 76, 102
   base bias, 79
   base bias with collector and emitter feedback, 90
   base bias with collector feedback, 88
   base bias with emitter feedback, 86
   base bias with separate supply, 91
   emitter bias with two supplies, 97
   JFET arrangements, 102
   modified universal bias, 94
   MOSFET arrangements, 112
   NPN arrangements, 77
   PNP biasing, 97
   universal bias, 92, 320
   universal bias with collector feedback, 94, 318
   vacuum tube biasing, 115
   voltage-divider bias, JFET, 110
Biplex detector, 508
BJT (bipolar junction transistor), 21
   alpha, 25
   beta, 24
   characteristic curves, 25
   current gain, 24
Block diagram:
   AM, FM, FM-stereo receiver, 490
   AM-FM receiver, 379
   AM IF alignment, 360
   FM demodulator and audio section, 8
   FM IF alignment, 451, 455
   matrix decoder, 480
   matrix decoder alignment, 533
   phase-locked loop, 397, 513
   PLL stereo decoder, 514
   superheterodyne receiver, 6, 226, 378
   time-division decoder, 481
   time-division decoder alignment, 536
Breakdown region, 18
Bridge neutralization, 175
Bridge rectifier, 245

Broadcast band:
   AM, 4
   FM, 4, 373

C

Capacitor leakage check, 283
Capacitor replacement, 291
Capacitors, 291
   temperature coefficient, 292
Carrier, 3
Cascaded stages, 123
   noise in, 180
Cathode, 40
   virtual, 196
Cathode follower, 130, 495
Ceramic filter, 413, 415, 526
Channel (FET), 26
Choke, 56
Collector current stability, 86
Collector feedback, 88
Collector identification, 289
Collector stabilizing resistor, 169
Collector voltage, stacking, 126
Colpitts oscillator, 193, 205
Commercial amplifiers, 150, 524
Common base, 129, 167
Common collector, 131
Complementary transistors, 146
Component checker, 282
Composite (FM stereo), 474
   graphical construction, 488
Contact bias, 115
Converters, 189, 244
   for AM, 195, 249
   BJT, 198
   for FM, 406, 421
   pentagrid, 195
Core saturation, 138
Crossover distortion, 141
Crystal filter, 413

D

Damped sine wave, 190
Darlington pair, 128, 252, 255

DC load line, 104
Decoupling, 235
Deemphasis, 398
Demodulator, 5, 219
Depletion layer, 16
Depletion mode, 33, 113
Detection capacitor, 219, 220
Detector, 5, 219, 227, 243, 260
Deviation, FM, 465
Difference amplifier, 37
Diffusion current, 23
Diode:
   light emitting, 20
   overload, IF and RF, 410
   peak inverse voltage, 19
   sensitivity to light, 359
   vacuum, 40
   varactor, 206
   zener, 19, 60, 182
Diode equation, 19
Dipole antenna, 210
Direct-coupled amplifiers, 125, 146, 149, 153, 238, 240, 344
   DC stabilized, 127, 152
   troubleshooting, 344
Discriminator, 383, 430
   alignment of, 450, 455
Distortion:
   crossover, 141
   harmonic, 139
   symptom, 344, 448
Doping, 14
Double-humped response, 163
Doubler:
   frequency, 473, 497, 500
   voltage, 57
Drain characteristics, 29
Driver transistor, 141
DSSC modulation, 469
Dual-gate MOSFET, 114, 175
Dynamic range, 77, 110, 116
Dynamic transfer curve, 107, 172

E

Electromagnetic waves, 211

Electron-hole pairs, 12
Emitter bypass capacitor, 100
Emitter feedback, 86
Emitter follower, 131
Enhancement mode, 33, 113
Equipment, recommended, 278
EVM, 277, 282

F

Failure mechanisms:
   transistors, 287, 319
   tubes, 285
Field-effect transistor, 26
Filament string, 69, 331
Filter capacitor, 51, 53
Flicker noise, 178
FM sidebands, 375
Forward bias, 16
Foster-Seeley discriminator, 383, 430
   alignment of, 450, 456
Freeze spray, 358
Frequency and wavelength, 212
Frequency doublers, 473
Front end, 5, 194

G

Gain-bandwidth product, 40
Generator coupling, 324
Grid, 41
   control, 132, 195
   screen, 132, 169, 196
   signal, 195
   suppressor, 132, 169, 196
Grid-leak bias, 115, 380
Ground, relocation of, 99

H

Harmonic distortion, 139
Hartley oscillator, 191, 216
Heat and semiconductors, 287
Heater (vacuum tube), 40
Heat-sink compound, 294

Heat sinks, 274, 294
Heterodyning, 187
Hole, 12
Hole conduction, 13

I

IC audio amplifier, 323
IC stage troubleshooting, 523
IF amplifier, 153, 167, 230, 243
    for FM, 408, 422, 428
IF frequency, 187
    selection, 230
Image frequency, 229
Impedance matching, 133
Inductor, 56
    testing, 282
Injection loop, 361
Input impedance, 84
Interelectrode capacitance, 163, 167, 200
    Miller effect, 165
Interlock, 69
Intermittent problems, 357
Intrinsic semiconductors, 11
Inverting amplifier, 39
Ionosphere, 214

J

JFET (junction field-effect transistor), 26
    amplifier design, 103
    biasing arrangements, 102
    characteristic curves, 30
Johnson noise, 177

L

LED indicator circuit, 21
Limiters, 378
Line filter capacitor, 236

Load line:
    AC, 105, 116
    DC, 104
Local oscillator, 188
    radiation, 228
    tracking, 228
Loopstick, 215
Loudspeaker, 3

M

Majority carriers, 15
Marker generator, 455
Matrix decoder, 479, 495
    alignment, 532
Microphonics (tube), 286
Miller effect, 165
Minority carriers, 15
Mixer, 189
    self-oscillating, 421
Modulation index, 375
Modulator, 3
    AM, 220
    balanced, 471
    FM, 372
MOSFETs, 32
    biasing requirements, 34
    dual-gate, 35, 175
    handling, 36
Motorboating, 354
MPX detector, bridge, 505
Multimeters, 277
Multiplexing, time-division, 488
Multiplex transmission, 465
Mute amplifier, 517
Mute switch, 525

N

Negative feedback, 37, 87, 241, 257
Neutralization, 164
    bridge, 175
Noise, 177
    effects, 181

factor, 180
limiting in FM, 373
troubleshooting, 356, 449
Noisy reception, 356
Noninverting amplifier, 38
NPO temperature coefficient, 292
N-type materials, 14

O

Offset-null, 37
Ohmmeter probe polarity, 280
Ohmmeter testing:
   BJTs, 288
   B+ power supply, 312
   capacitors, 282
   filament string, 334
   transformer, 339
Operational amplifiers, 36
   inverting amplifier, 39
   noninverting amplifier, 38
   voltage follower, 37
Oscillator, 188
   drift, 200
   pulling, 201
   testing (tube), 342
Oscilloscope, 284
Output transformer, 133
   core saturation, 138
   reflected impedance, 134
Overload diode, 410

P

Partition noise, 178
Pentode, 132
   beam power, 132
   RF amplifiers, 169
Pentode amplifier, 169, 316
Phase detector, 397
Phase inversion, 136, 266
Phase-locked loop, 397
   stereo decoder, 513, 518
   stereo decoder alignment, 538

Phase reference, 475
Phasors, 381
Pilot carrier, 469
Pinch-off voltage, 28
Plate (tube), 40
Plate characteristics, 42
Plop, turn on, 334
PN junction, 15
   breakdown region, 18
   conduction characteristics, 18
   forward biased, 15
   reverse biased, 17
Polarization, 211
Potentiometers, 291
Power amplifiers, 123
   IC audio, 323
Power supplies:
   balanced, 68
   bridge, 57, 418
   complicated, 249
   full-wave, 55, 418
   half-wave, 50, 53, 70, 311
   op amp regulated, 63
   regulated, 60
   vacuum tube, 69, 258
Power supply V-R analysis, 311
Preemphasis, 398
P-type materials, 14
Pulsating DC, 55
Push-pull amplifier:
   solid state, 140, 245, 247
   vacuum tube, 136, 262

Q

Q (resonant circuits), 160
Q-point, 77, 107, 116
Quadrature detector, 395
Quasi-complementary amplifier, 149, 252, 256

R

Radiation:
   electromagnetic, 1

Radiation (*cont.*)
  infrared, 1
  ultraviolet, 1
Radio frequency, 5
Radio waves, 1, 210
  field strength, 214
  reception of, 213
Ratio detector, 390, 409, 424
  alignment, 453, 457
  limiting action, 393
  response, 392
R-C coupling, 115
Reactance tube, 203, 205, 372, 425
Rectifier, 16
  bridge, 57
  full-wave, 55
  half-wave, 50, 236
Recurrent sweep, 284
Reflected impedance, 134
Regulated power supplies, 60
  negative, 67
  op-amp, 63, 64
  pass-transistor, 62
  three-terminal, 66
  two-transistor, 65
Resistors, replacement, 290
  tolerance, 290
Resonant circuits, 159
  bandwidth, 160
  loosely-coupled, 384
  Q (quality factor), 160
  as signal sources, 189
Reverse bias, 17
RF amplifier, 158
  common base, 167
  for FM, 403, 420
  gain of, 170
  JFET, 173
  MOSFET, 174
  pentode, 169
  triode, 165
RF probe, 281
Ripple, 52
  active filter, 68, 418
  as a symptom, 342

S

Saturation current, 18
SCA, 465
Screen grid, 169
S-curve, 425, 456
Secondary emission, 169
Self bias, JFET, 103
Self-oscillating mixer, 421
Semiconductors:
  intrinsic, 11
  N-type, 14
  P-type, 14
Sensitivity, loss of, 346, 352, 446, 449, 529
Separation control, 480
Service data, 272
Shot noise, 178
Siemens (unit), 108
Signal (definition), 76
  amplification, 80
Signal generator, 281
  accuracy, testing of, 361
  AM used for FM, 437
Signal injection, 303
  charts, 304, 306, 307
  example, 340, 350
  injection loop, 361
  modulation percentage, 347
  standard output, 347
Signal/noise ratio, 179
Signal tracing, 308
  charts, 309
  example, 344
Slew rate, 40
Soldering devices, 274
Space charge, 40
Spark gap, 405
Speaker, 3
  testing, 334
Speed of light, 2
Speed of sound, 2
Spurious response, 229
Square-law response, 30
Squeals, howls, 354, 448

Stereo amplifier, 524
Stereo decoder, 7
Stereo indicator, 502, 530
Stereo separation, lack of, 529
Subcarrier, 465
Substrate, 32, 34
Superheterodyne receivers, 6, 226, 378
Suppressor grid, 169
Surge current, 54
Surge-limiting resistor, 54
Swamping resistor, 89, 100
Sweep/marker generator, 453
Symptoms (charts):
   AM-FM receivers, 438, 439
   FM stereo decoders, 523
   solid-state AM receivers, 333
   tube-type AM receivers, 332

T

Taper, 291
Temperature coefficient, 292
Tetrode, 169
Thermal noise, 177
Thermionic emission, 40
Thermistor, 143
Time-division decoding, 481, 500
   alignment, 536
   bridge decoder, 503
   IC decoder, 511
   qualitative description, 487
Tolerance, resistor, 282, 290
Tone compensation, 239, 263
Tone control, 238, 247
Tools, recommended, 273
Total harmonic distortion, 140
Transconductance, 108
Transducer, 3
Transformers, 47, 292
   isolation, 49, 276
   output, 133
   power relationships in, 48
   tuned, 162

Transmitter systems:
   AM, 216
   FM, 475, 479
   simple AM, 216
TRF receiver, 5, 225
Triode, 41
   audio amplifier, 314
   characteristic curves, 42
   grounded grid, 165
   Miller effect in, 165
   RF amplifier, 163
Tuned circuits, 159
   loosely-coupled, 384
   overcoupled, 163
Tuning meter, 413

U

Universal bias, 92, 247

V

Valence band, 12
Varactor diode, 206, 373
VCO, 514
Virtual ground, 39
Voice transmission system, 217
Voice waves, 2
Voltage-controlled oscillator, 397, 514
Voltage-dependent resistor, 241
Voltage-divider bias, 110
Voltage doubler, 57
Voltage follower, 37
Voltage regulators:
   negative, 67
   op-amp, 63
   pass-transistor, 62
   three-terminal, 66
   two-transistor, 65
Voltage-resistance analysis, 310
Voltage-variable capacitor, 206, 373
Volt-amp, 48
VOM, 277
VTVM, 49

W

Waveforms:
  AM, 218
  of balanced modulator, 472
  FM, 372
  FM stereo, 474
Wavelength, 212

Z

Zener diode, 19, 60
  conduction characteristic, 20, 60
  noise in, 182
  reverse breakdown, 60
  voltage regulators, 60

[Handwritten notes:]
RF μ chapter... 6
Converter/mixer/LO... 7
IF μ 6
Demodulator...... 8, 13
audio amp....... 4, 5
Power supply... 3
Stereo........ 16, 17